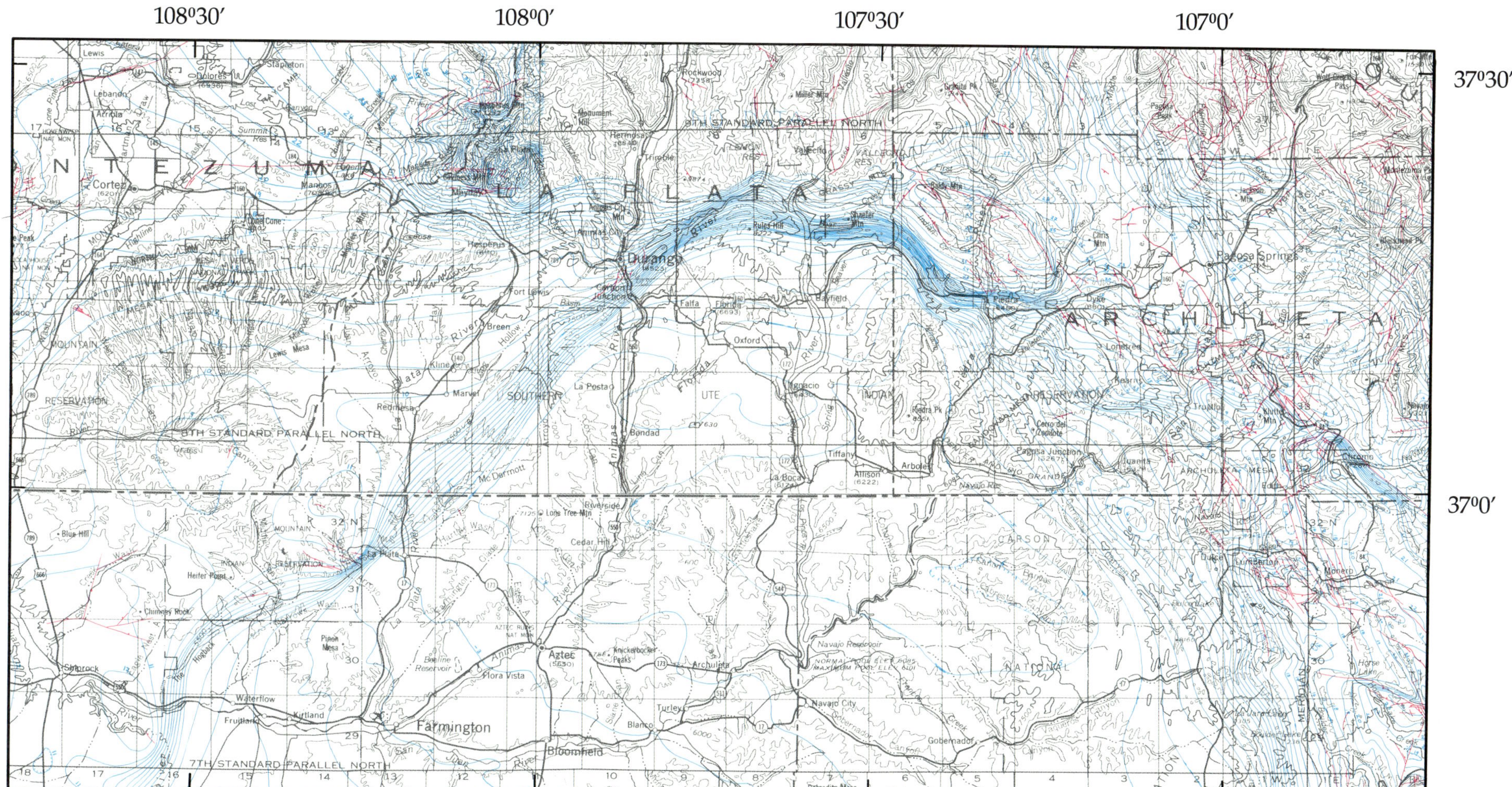

Northern San Juan Basin
Generalized structure contour map
(modified from Thaden and Zech, 1984) Structure contours in blue, faults in red. Contour interval [100 m.]

Library of Congress Catalogue Card Number 88-060746 ISBN 0-933979-10-X

Satellite photo and structure map: The U. S. Geological Survey's Branch of Geophysics remote sensing laboratory in Denver, Colorado, prepared the color-composite Landsat image on the left-front endsheet. The image is a digital mosaic of two Landsat multispectral scanner scenes (2584-17060 28 August 1976 and 2637-16584 20 October 1976) formed by compositing image mosaics of bands 7, 5, and 4 in red, green, and blue, respectively. The image mosaic was registered to a Universal Transverse Mercator grid and merged with an image map showing latitude and longitude coordinates and other accessory annotations. Color saturations in this standard color-infrared composite image were digitally increased to enhance subtle color differences and a slight edge enhancement was applied to emphasize the high-frequency component of the digital data. Vegetation appears as various shades of red and water is dark blue to black on the image. Strongly limonitic rocks and soils are yellow, yellowish-green, or white, while non-limonitic rocks and soils are mostly light to medium blues.

The structure contour map on the right-front endsheet was modified from the Preliminary Structure Contour Map of the San Juan Basin by R. E. Thaden and R. S. Zech, U. S. Geological Survey Miscellaneous Investigations Map MF-1673. The contours in this area were derived from published sources and elevations were converted to metric by interpolation.

About the cover: The circle encompassing the pump jack, production incline curve, and bear tracks represents the Circle of Life. The Ute people consider the Circle of Life to be the balance required in nature to continue life. The bear tracks represent power and guidance provided by the bear for Ute people. Eagle feathers from the great Golden Eagle also contain great power. The eagle is the only animal that can reach the Creator to deliver messages from the Utes. The Indian in headdress is looking over the land as landlord to see that all within the Circle of Life is prayed for. The tepees represent the homes of the Ute people, always set up in circles and constructed of natural materials found within the Circle of Life.

Artist is Orian L. Box; Council Member, Southern Ute Indian Tribe

GEOLOGY AND COAL-BED METHANE RESOURCES OF THE NORTHERN SAN JUAN BASIN, COLORADO AND NEW MEXICO

EDITED BY:

JAMES E. FASSETT

PUBLISHED BY:

ROCKY MOUNTAIN ASSOCIATION OF GEOLOGISTS
DENVER, COLORADO

1988

SOUTHERN UTE INDIAN TRIBAL COUNCIL

Leonard C. Burch Chairman

Vida B. Peabody Vice Chairman
Lilian I. Seibel Council Member
Richard L. Jefferson Council Member
Guy Pinnecoose Council Member
Orian L. Box Council Member
Gerald D. Howe Council Member

TRIBAL CHAIRMAN'S WELCOME

Welcome to the first Fruitland Formation coal-bed methane Field Conference and Symposium to be held in the San Juan Basin. As your host, the Southern Ute Indian Tribe extends best wishes for your full enjoyment and comprehension of both the technical sessions and the beauty of the lands on which this meeting is being held.

The spirit and intent of the symposium was to open avenues for thought and discussion regarding the recently identified and economically exciting coal-bed methane resource here in the San Juan Basin. We recognize the field conference and technical sessions as a forum for focusing on a technically suitable and prudent approach to the development, production, and conservation of what appears to be a vast and valuable resource.

The Southern Ute Indian Tribe, as one of the larger land-and mineral-resource owners in the northern San Juan Basin, is keenly interested in the economic outcome of the coal-bed methane development. We are also concerned with the impact of the overall operation on both the natural and human resources of the area. It is our hope that these meetings will provide a basis for establishing mutually acceptable methods for the development, production, and conservation of all resources. Good hunting.

Leonard C. Burch, Chairman
Southern Ute Indian Tribe

ROCKY MOUNTAIN ASSOCIATION OF GEOLOGISTS

ROCKY MOUNTAIN ASSOCIATION OF GEOLOGISTS PRESIDENT'S MESSAGE

This guidebook, and the accompanying field trip and symposium, have been the products of a cooperative effort involving the Southern Ute Indian Tribe, the U. S. Geological Survey, the Energy and Minerals Division of the Bureau of Indian Affairs, the Rocky Mountain Association of Geologists, and the Gas Research Institute. It has been with pleasure and satisfaction that I have witnessed the organization and development of these endeavors, especially the production of this guidebook. Guidebooks such as this make important contributions through the dissemination of scientific and technical knowledge. This guidebook is especially important in that it is the first of its kind on the subject of coal-bed methane in a Western United States basin. Gas from coal beds is a resource that is just beginning to be developed and this book will help lay the foundation for advancing the understanding of the origin, occurrence, and utilization of this resource for the needs and benefit of society. I am proud and pleased that the Rocky Mountain Association of Geologists was a partner in its publication.

Robert T. Sellars, Jr., President
Rocky Mountain Association of Geologists

COMMITTEES

Organizing Committee

James E. Fassett, *General Chairman* .... U.S. Geological Survey
Bradford C. Boyce .... Southern Ute Indian Tribe
Clarence L. Harr .... Consulting Petroleum Geologist
Stephen A. Manydeeds .... Energy and Minerals Division, B.I.A.
Robert S. Zech .... U.S. Geological Survey

Guidebook Committee

James E. Fassett, *Editor* .... U.S. Geological Survey
Clarence L. Harr, *Assistant Editor* .... Consulting Petroleum Geologist
Robert S. Zech, *Assistant Editor* .... U.S. Geological Survey
Brent W. Hale, *Assistant Editor* .... Northwest Pipeline Corporation
Annette L. Primrose, *Computer Operations Editor* .... ORI Systems and Applied Sciences Corporation
Raymond G. Marvin, *Publications Manager* .... Consulting Geologist

Program Committee

Bradford C. Boyce, *Chairman* .... Southern Ute Indian Tribe
Lenora Tracy .... Scout Land Company
John A. Campbell .... Fort Lewis College
Stephen A. Manydeeds .... Energy and Minerals Division, B.I.A.
Darlene Vigil .... Southern Ute Indian Tribe
Terry L. Logan .... Resource Enterprises, Inc.
Marvin E. Cook .... Southern Ute Indian Tribe

Road Logging Committee

James E. Fassett, *Chairman* .... U.S. Geological Survey
Bradford C. Boyce, *Art Editor* .... Southern Ute Indian Tribe
Clarence Harr .... Consulting Petroleum Geologist
Robert S. Zech .... U.S. Geological Survey

Logistical Support Group

Dee Tyler .... Rocky Mountain Association of Geologists
Dianne Fox .... U.S. Geological Survey
Florence Vallejos .... Southern Ute Indian Tribe

Supporting Institutions

Southern Ute Indian Tribe
Gas Research Institute
Bureau of Indian Affairs, Energy and Minerals Division
U.S. Geological Survey
Scout Land Company
Fort Lewis College
Rocky Mountain Association of Geologists

ACKNOWLEDGEMENTS

Planning, organizing, and participating in the 1988 symposium and field trip on the geology and coal-bed methane resources of the northern San Juan Basin, Colorado and New Mexico, will be a long-remembered experience for all of us involved in this project. The work involved in putting this meeting together often lasted well into the evening hours and many deadlines were met at the last minute. However, the long hours were enjoyable because of the many hard-working people involved.

It is my pleasure as Program Chairman (logistics nightmare coordinator) to recognize those individuals and organizations who contributed their time, effort, and funds to bring this event from conception to completion. Many of these individuals and organizations had never been involved in organizing such an event and were on a steep learning curve. This truly added to the excitement. The people involved, though, believed they could accomplish the required tasks, and under that philosophy, the meeting was organized. Therefore, many thanks to all who helped. I wish to express my gratitude to Jim Fassett for his willingness to accept the monumental task of serving as General Chairman of the Symposium, Editor of the Guidebook, and Chairman of the Road Logging Committee. Jim's professionalism, integrity, and technical experience has produced another great Guidebook and Field Conference. Additionally, I extend thanks on behalf of all of the Symposium committee members to the San Juan Basin oil and gas operators who contributed to this event. The input and guidance you provided is appreciated. A special thanks to the Southern Ute Indian Tribe which provided funding, facilities, and manpower and to the Gas Research Institute which helped underwrite the publishing costs for the guidebook and helped to advertise the symposium and field conference. I would also like to thank my patient wife, Ineke, for her input and support throughout the months of effort involved in putting this meeting together.

I have enjoyed the part I have played in this venture and I hope that the efforts of the Program Committee members will help to make your stay in the northern San Juan Basin memorable. Should you encounter any of the folks listed who contributed to this event, do not hesitate to give them a well-deserved pat on the back (or buy them a beer!). They have worked hard. My thanks to all.

Bradford C. Boyce
Program Chairman

EDITOR'S MESSAGE

This guidebook is the first of its kind: A focused look at coal-bed methane in a large Rocky Mountain Laramide basin. The papers in this volume cover every aspect of coal-bed methane in the San Juan Basin, including: The geology, environments of deposition, and geometry of the coal beds that contain the resource; the origin and migration history of the gas; basin-wide resource estimates; the engineering aspects of getting the gas out of the ground; the marketing and economics of producing coal-bed methane in the San Juan Basin; the legal ownership of the gas; state regulations governing well spacing and field rules; disposal of produced water; and land and mineral ownership patterns in the northern part of the basin. Also included are detailed papers on all of the major coal-bed methane fields in the basin, and in a paper on the history of Fruitland gas production, a discussion of most of the not-so-major fields. A small section of the book deals with geophysical methods, as yet still experimental, for surface detection of underground hydrocarbon resources.

Having read all of these papers very carefully (two, or three, or more times each!) I can report with confidence that we are all at the bottom of the learning curve when it comes to understanding coal-bed methane in the San Juan Basin. The good news, however, is that after you have read this book, you will know more overall about this exciting new resource than any of the individual operators in the basin has known up until now. As editor of this book, I have experienced a great deal of satisfaction in watching this collection of papers come together into a primer on coal-bed methane in the San Juan Basin, and as usual, it is the authors of all of these papers who deserve the credit. Without their energy, intelligence, and ability and willingness to share their knowledge with the rest of us, this book would not exist. Without the publication of ideas there can be no progress.

This project, including the coal-bed methane symposium, the field trip, and the guidebook, was first conceived of by Brad Boyce and Clancy Harr as they worked with the Southern Ute Indian Tribe to understand and promote the development of coal-bed methane on the Reservation in the northern San Juan Basin. They realized very early on that this resource had an enormous economic potential, not only on the Reservation, but basin-wide and their initial thoughts were to hold a symposium on this topic and get as much information out as possible about this resource as quickly as possible. Their philosophy was that the more everyone knew about this resource, the better off everyone in the industry would be, from mineral owner, to investor, to consumer, to developer. A full and detailed discussion of the evolution of this project is fascinating, but beyond the space I have here. Suffice it to say, this project went from its conception to final flower through the cooperative efforts of the Southern Utes, the Energy and Minerals Division of the Bureau of Indian Affairs, the U.S. Geological Survey, the Rocky Mountain Association of Geologists, and the Gas Research Institute. Innovative methods were used to produce this book involving having all authors submit their manuscripts on diskettes and utilizing computerized typesetting. We were all on a steep learning curve in this process, but working together, we did learn and in the end the process worked beautifully, thanks especially to Annette Primrose, ORI Systems and Applied Sciences Corporation, and Ray Marvin of the RMAG.

In a technical sense, there is little to say about the book. If follows the format of most other RMAG Guidebooks and style guidelines are generally those used by the U.S. Geological Survey, Geological Society of America, and American Association of Petroleum Geologists. I have worked hard to eliminate mysterious abbreviations and the ones that are present are only the very standard ones in general use in normal technical or scientific writing. All of us involved in producing this book hope you learn something from it and then go out and build on that new knowledge.

James E. Fassett
Editor

CONTENTS

HISTORY

GEOLOGY AND GEOCHEMISTRY

WELL LOG INTERPRETATION

PRODUCTION ENGINEERING

COAL-BED METHANE FIELD STUDIES

ECONOMICS, LAW, AND REGULATIONS

GEOPHYSICS AND REMOTE SENSING

ROAD LOGS

Chief Ignacio of the Ute Indians. The town of Ignacio, Colorado was named in his honor. Photo courtesy of the Durango Public Library.

History of Gas Produced from Coal Seams in the San Juan Basin

Thomas A. Dugan[1]
and
Barbara L. Williams[1]

ABSTRACT

The San Juan Basin has been producing commercial quantities of oil and gas for over 65 years. Abandoned gas wells drilled as early as 1896 have been remedially plugged by the State of New Mexico. The basin has a rich and varied history concerning all aspects of the oil and gas industry. This paper presents a historical look at the recovery of coal-bed gas from the San Juan Basin. The two geologic formations of interest are the Menefee Formation of the Mesaverde Group and the Fruitland Formation, both of which contain sandstone, shale, and coal beds. The sandstones were the prime target in the early completions, but over the years, the coals have become a significant contributor to the gas production in the basin. The early open-hole completions of Mesaverde Group wells were probably the first source of coal gas in the San Juan Basin. Until the mid-1970's, Fruitland coal beds were sporadically developed if sandstones were not found or if the coals had minimal water production associated with the gas recovery. Producers began researching and designing large scale coal gas projects over thirteen years ago. In the 1980's, the first projects designed to dispose of the large quantities of water produced from the coals were put into operation. Tax breaks for coal-gas production have been an added incentive for development of the coals but the current low spot-market prices for natural gas have affected many of the projects.

INTRODUCTION

Many different operators in the San Juan Basin have tried their hand at gas well discoveries. On October 21, 1921, the first gas well to commercially produce in New Mexico was completed in the Farmington Sandstone Member of the Kirtland Shale south of Aztec, New Mexico. Gas, overall, was not considered commercially significant on a large scale until the construction of Southern Union's pipeline in 1931 and El Paso Natural Gas Company's (EPNG) pipeline to California in 1951. With these markets, active drilling for gas began. One early main target was the Mesaverde Group. Gas was also producible from the Pictured Cliffs Sandstone and Dakota Sandstone, but at that time the Fruitland Formation was undeveloped and viewed as nonproductive. The Blanco Mesaverde field was created in 1949 and extensive reserve studies were done to justify the construction of the interstate pipelines. Though gas-kicks were almost always noted from the Fruitland coals and sandstones, it wasn't until early 1952 that the Fruitland Formation was formally completed and it wasn't until 1954 that the first Fruitland pools were created. Early Fruitland wells were completed in the sandstones and the coal seams were either ignored or seen only as a last chance target to make a well. Ironically, the Fruitland coals, which were once avoided by everyone in the industry if a Fruitland sandstone bed was found, are now the zones of expanding development and technical experimentation. The much sought-after sandstones that originally caused the discovery of so many Fruitland pools have now been brushed aside.

OPEN-HOLE MESAVERDE COMPLETIONS

Most of the early Mesaverde completions (from 1948 through 1955, particularly those in the Blanco field) were open-hole completions through the Cliff House Sandstone, the coal-bearing Menefee Formation, and the Point Lookout Sandstone; the Mesaverde Group. By the end of 1955, there were 922 wells in the Blanco Mesaverde pool, which had produced 329,726,152 MCF of gas and 767,763 barrels of distillate. It is impossible to speculate or estimate how much gas was produced from Menefee coals or how much is still being produced from Menefee coals in other open-hole completions

[1] Dugan Production Corporation
P.O. Box 208
Farmington, New Mexico 87499

Some of the information in this paper came from files of the New Mexico Oil Conservation Division and the Colorado Oil and Gas Conservation Commission and from the Annual Report of the New Mexico Oil and Gas Engineering Committee: years 1950 through 1986. Other information came form the following: The Southern Ute Tribe, Ignacio, Colorado; Emery Arnold, Aztec, New Mexico; Charles F. "Chief" Brown, Farmington, New Mexico; William B. Hanson, Amoco, Denver, Colorado; Jim Hill, ARCO, Farmington, New Mexico; Jim Jacobs, Dugan Production Corp., Farmington, New Mexico; Delasso Loos, Durango, Colorado; Bernard Mahony, Horizontal Recovery Specialists, Inc., Midland,TX; Phil McGrath, Farmington, New Mexico; Val Reese, Albuquerque, New Mexico; Bob Sagle, Tenneco, Denver, Colorado; Leo Schuller, Carbon River Energy Partnership, Seattle, Washington; Richard Ullrich, Farmington, New Mexico; Hurchel Williams, Farmington, New Mexico

Figure 1. Spudder rig and Halliburton cement pump truck on the Blackwood and Nichols Northeast Blanco Unit No. 7 well in the Mesaverde Group, sec. 21, T. 30 N., R. 7 W. (Photo courtesy of Delasso Loos)

in the Mesaverde Group. These Mesaverde wells may have been the first coal-gas wells in the basin with, of course, no knowledge at that time that any of the gas was coming from coal beds.

The method of completion and stimulation for many early Mesaverde Group wells was to set 7-inch casing in a mud-drilled hole, to the top of the Cliff House Sandstone. This helped to control the gas encountered in the Fruitland Formation and Pictured Cliffs Sandstone. As one operator mentioned, the Fruitland in the northern part of the basin was hard to ignore because it usually kicked you in the face. On more than one occasion, the pressure in the Fruitland was enough to collapse the 7-inch casing. Increasing the mud weight to prevent blow-outs in the Fruitland could cause more lost circulation in the Mesaverde, which was already a problem in some areas. One early operator would let the mud blow out of the hole when the Fruitland Formation was penetrated, mud up again and drill on ahead until the depth for the 7-inch casing was reached. The hole was then cable-tool drilled or, later on, gas drilled and left open for stimulation throughout the Mesaverde to the bottom of the Point Lookout Sandstone. This completion practice helped to alleviate the severe lost circulation problems encountered in the Mesaverde Group in some areas. Figure 1 shows a spudder rig that was used in the basin about this time period.

Up until the mid-1950's, the use of solidified nitroglycerine (SNG) as a stimulation method was wide-spread. Shooting a hole with SNG was done by filling the hole with a column of explosives. Common practice in the Blanco Mesaverde field was to place four quarts of SNG per foot through the Cliff House and Point Lookout Sandstone, but only one and a half quarts of SNG per foot through the Menefee Formation, which was enough to maintain the chain reaction of explosions between the two more desirable sandstones. On the top of the column of SNG was 20-30 lbs of a more highly sensitive explosive with a time clock and blasting cap detonator. The time clock was normally set to go off about 12 hours later, or during daylight hours. The detonator was set that far ahead to allow for problems. The shot cans or tins were loaded into the hole 15-20 feet at a time. If problems developed, the cans could be removed but once the detonator was in place, one didn't want to have to pull it back out of the hole for obvious reasons. The idea was to maintain a continuous explosion that would detonate all of the explosive in the hole, with the upper explosion setting off the SNG packed in a tin below which would in turn set off the tin under it. In those days, the blow-out preventers on the rigs were manually operated and they would build a 7-inch blowline to divert the explosion away from the derrick. When they got the shot in the hole, they took off the rotating head and hooked up the blowline. Following the explosion, gas, dust, sand, rocks, pebbles, chunks of coal and anything else in the hole would come flying out of the blowline in the derrick and rain rubble all over the location (see fig. 2). The shower would last from minutes to hours depending on how much gas was coming into the hole from the formation. The trick for the workers was to stay out of the way of this shower. More than one time, the wind shifted and the workers were bombarded. It

Figure 2. Shooting a Mesaverde Group well with solidified nitroglycerine; Blackwood and Nichols' Northeast Blanco Unit No. 8 well, sec. 10, T. 30 N., R. 7 W. The well was drilled in September 1953. (Photo courtesy of Delasso Loos)

was not uncommon to find rocks or lumps of coal that didn't seem to fit back into the hole they had come out of. This method of stimulation would use up to 2,500-3,000 quarts of explosive and they would not know if all of the explosive was detonated. When SNG was used in other fields throughout the country they would tamp the shot by using an inverted "umbrella" that held 50-60 feet of gravel above the detonator with a column of oil topping it. This would force more of the explosion into the formations. Because of the lost circulation problems in the Mesaverde Group, SNG shots in this formation were not tamped and filled with fluid.

After shooting the well with SNG, the chore of cleaning out the open hole began with cable tools or, in later wells, circulating with natural gas. It sometimes took months to cleanup the rubble and, possibly, undetonated SNG in the hole. Crews were understandably uneasy when bits or chunks of SNG began appearing with the drill cuttings. Jetting out a hole with gas made cleaning out quicker and easier and EPNG began using gas extensively. This type of cleanup was slightly easier in the Blanco, New Mexico area because the Mesaverde Group does not produce as much water there.

There are numerous coal beds in the Menefee Formation that were fractured with SNG even though the shot was lighter through the interval. These coals were enlarged by the shot. Many drillers noticed natural gas increases while drilling with gas through the Menefee coal seams; the stream of flared gas would increase, indicating gas was present in the coals of the Menefee. When drilling through these coals with gas, the coal drill cuttings ignited as tiny sparks in the flare. These displays must have been very spectacular at night. Though the sands of the Mesaverde were the target, the coals in the Menefee Formation have been an important (but now unquantifiable) contributor to the gas production in the San Juan Basin.

EARLY COMPLETIONS

Beginning in the fall of 1950, Phillips Petroleum Co. cored and ran drill-stem tests on test wells in T. 30 N., R. 6 W. and T. 30 W., R. 5 W. in northern New Mexico, for a project called the Gobernador Strat Test. Information from the New Mexico Oil Conservation Division (NMOCD) in Aztec, New Mexico indicates that the purpose of the test was to evaluate the coals and adjacent sandstone beds down to the Pictured Cliffs Sandstone. The coals blew out on them in every hole. On one hole, the coal blew out and the roughneck in the derrick caught a chunk of coal in his leg. The story goes that after several years, he still had small pieces of coal embedded in his leg. Phillips reported the results of the coring and drill-stem tests to the NMOCD. The records state that in several coal intervals the coals were bleeding gas and also produced brackish water. The coal thicknesses reported were up to approximately 30 feet. The shut-in bottom hole pressures were from 880 psi to 1,780 psi. A good deal of seismic work and stratigraphy work was done by Phillips, EPNG, and Stanolind during these early years.

EPNG began large-scale drilling for gas in the basin because they needed gas to fill their pipeline to California, which was completed in 1951. They extensively drilled and completed hundreds of Mesaverde and Pictured Cliffs wells. Many Mesaverde wells were completed open hole as previously mentioned. The EPNG Pictured Cliffs wells are historically interesting because EPNG would drill approximately 2 feet into the Pictured Cliffs Sandstone, set casing, cable tool to the desired depth into the Pictured Cliffs and then shoot the open hole with SNG. They would place the detonator and two time clocks about 10 feet below the casing and tamp the shot with 10 feet of gravel and a column of water. In areas where the Fruitland coal is immediately above the Pictured Cliffs, there is no doubt that the coals were also affected by the shot and coal gas must have contributed to the overall gas production. Water production was not necessarily an indicator of coal gas production in these wells since some Pictured Cliffs wells produced water and some didn't. EPNG's goal was to drill wells that produced 300 MCFPD so any added coal gas was a bonus.

The first true attempts to complete wells in coal beds in the San Juan Basin were probably Stanolind Oil and Gas Co.'s (today's Amoco Production Company) Ignacio Blanco Fruitland-Pictured Cliffs wells that were drilled in the early 1950's. The wells were in the area of the Ignacio anticline in southern Colorado, around T. 32 N. or T. 33 N. and R. 6 W. or R. 7 W. Most were open-hole completions with definite coal-gas contributions and had high pressures. The discovery well in 1951 for the Ignacio Fruitland-Pictured Cliffs field was Stanolind's Ute Indians No. D-1 which was drilled and shot with SNG in the Pictured Cliffs and the casing was perforated in the Fruitland. The initial potential for both zones as stated on the completion report was 1,900 MCFPD (absolute open flow) and the Southern Ute Indian Tribe records show an initial potential of 2,200 MCFPD from the Fruitland zone with a shut-in pressure of 1,240 psi. Other wells drilled by Stanolind were strictly Fruitland wells. For example, the Garcia No.1,

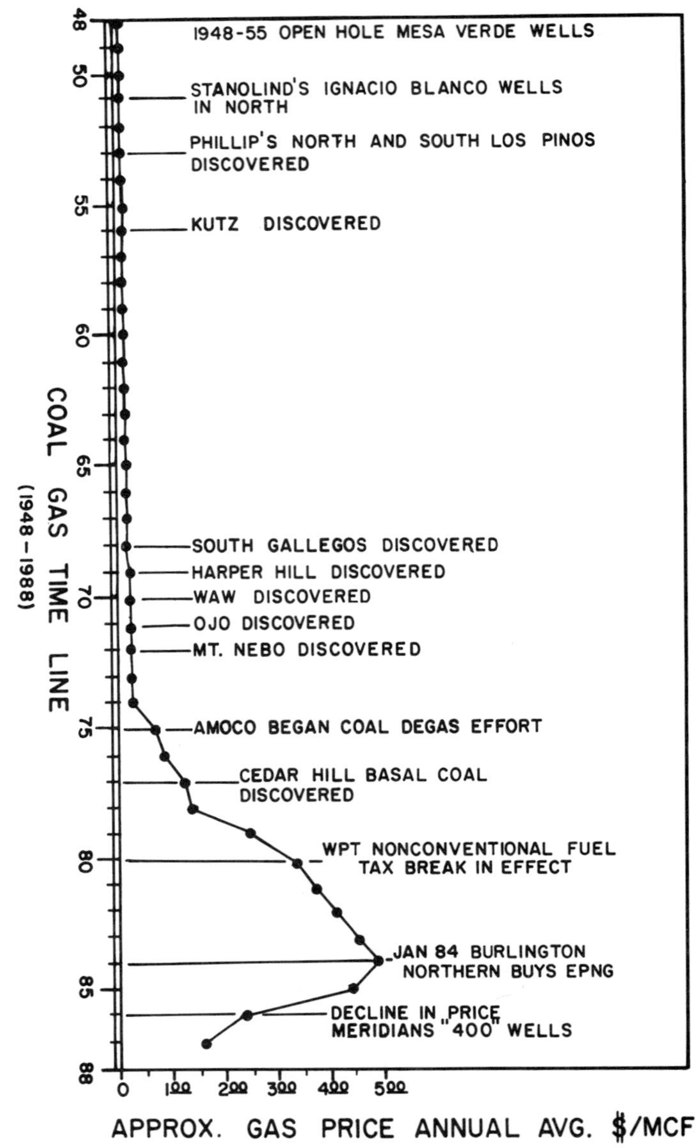

Figure 3. Coal-gas time line with annual average gas prices.

drilled in 1953 and located in sec. 7, T. 33 N., R. 7 W., was an Ignacio Fruitland test. Core information shows shale and coal with some bleeding gas. The well was completed open hole for 241 feet through the Fruitland. The initial potential for the well was 4,500 MCFPD. It was plugged in 1969. The wells originally drilled in the area produced little water but as the bottom hole pressure decreased, water production increased. When these first Fruitland wells began producing more water than was economical to handle, Stanolind plugged them. At the time, no one felt the gas production was great enough to justify rigging up the equipment to handle the volumes of water, for at the time, gas was selling for approximately $0.11 per MCF (see fig. 3). Also, typical conventional-sandstone-reservoir experience tended to curtail coal-gas development because when large amounts of water were encountered in these reservoirs it usually meant that oil or gas recovery would decrease and water production would increase. In 1960, approximately 10 million MCF had been produced from what the Oil and Gas Conservation Commission of the State of Colorado called abandoned leases in the

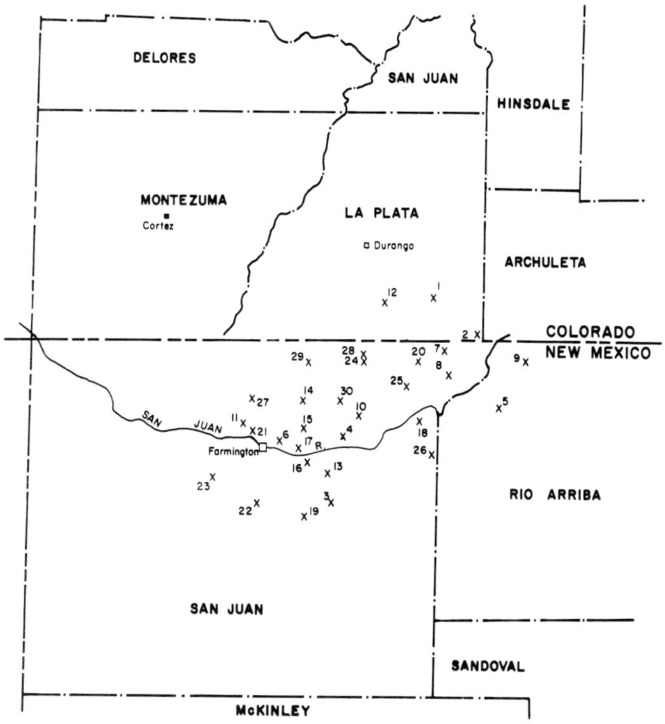

Figure 4. Index map of the San Juan Basin, Colorado and New Mexico (Modified from Marvin L. Matheny and Del A. Talley, 1978, 1983), showing locations of discovery wells for Fruitland pools. Numbers correspond to numbered list in table 1.

Ignacio Blanco Fruitland-Pictured Cliffs field in the northern San Juan Basin.

The first Fruitland Formation wells in New Mexico were completed in what was known as the Fruitland "sand". The Gallegos, Aztec, La Jara, and West Kutz Fruitland pools were all discovered by wells completed in the "sand". The discovery wells were drilled in 1952 but the pools were not established until 1954 for all but the Gallegos Fruitland which was established in 1955 (see fig. 4 and table 1). These wells were completed much like the Pictured Cliffs Sandstone wells of the day with either open-hole completions through the Fruitland which exposed the coals or perforations in the sandstones. Stimulation through perforations was not started until later; as operators used to say, "if you have to do something to a well to get it to produce, you don't have a well".

In June of 1953, Phillips Petroleum Co. completed drilling the San Juan Unit 32-7 No. 3 well in sec. 18, T. 32 N., R. 7 W. This well started out as a Mesaverde well but they lost the hole. Casing (7-inch) had been set through the Fruitland, so on the strength of Stanolind's work in the Ignacio Blanco Fruitland-Pictured Cliffs in southern Colorado; their own assessments of the coal production possibilities in T. 30 N., R. 5 W. and T. 30 N., R. 6 W., and the gas kick regularly experienced going through the coals, Phillips decided to complete the well in the coal since the sandstones did not appear to be worthwhile. They were desperate to make a well. The well started off poorly when it was first completed. There was not a lot of water or gas, but production steadily increased (fig. 5). Completion of the San Juan Unit 32-7 No. 3 estab-

Table 1. Fruitland Formation pools and discovery well names, completion dates, and locations. Locations of discovery wells are shown on figure 4.

1. Ignacio Blanco Fruitland-Pictured Cliffs
   Stanolind Oil and Gas Co. Ute D No.1
   September 4, 1951,
   NE1/4NW1/4 sec. 7, T. 33 N., R. 7 W., Colorado
2. Allison Fruitland-Pictured Cliffs
   (combined into Blanco Fields)
   Name and location of discovery well unavailable
   1952
   T. 32 N., R. 6 W., Colorado
3. Gallegos Fruitland
   Western Development Co. (British American) Douthit No.2
   March 16, 1952
   NW1/4NW1/4 sec. 27, T. 27 N., R. 11 W., New Mexico
4. Aztec Fruitland
   F. L. Harvey Hare No.1
   June 19, 1952
   SE1/4SE1/4 sec. 14, T. 29 N., R. 11 W., New Mexico
5. La Jara Fruitland
   El Paso Natural Gas Co. Abraham No. 1
   (recompleted as Mesaverde)
   June 30, 1952
   NW1/4NE1/4 sec.13, T. 30 N., R. 6 W., New Mexico
6. West Kutz Fruitland
   Locke Taylor Tycksen No. 1
   October 22, 1952
   NE1/4NE1/4 sec. 23, T. 29 N., R. 13 W., New Mexico
7. North Los Pinos Fruitland
   Phillips Petroleum Co. San Juan Unit 32-7 No. 3
   July 31, 1953
   NE1/4NE1/4 sec. 18, T. 32 N., R. 7 W., New Mexico
8. South Los Pinos Fruitland
   Phillips Petroleum Co. San Juan Unit 32-7 No. 6
   August 24, 1953
   NE1/4NE1/4 sec. 17, T. 31 N., R. 7 W., New Mexico
9. Cottonwood Fruitland
   Stanolind Oil and Gas Co. San Juan Unit 32-5 No. 2
   November 21, 1953
   SW1/4SW1/4 sec. 35, T. 32 N., R. 5 W., New Mexico
10. Aztec Fruitland, North
    El Paso Natural Gas Co. Gage No. 1
    April 24, 1954
    NW1/4NW1/4 sec. 20, T. 30 N., R. 10 W., New Mexico
11. Twin Mounds Fruitland-Pictured Cliffs
    Pubco Petroleum Corp. Russell Federal No. 2
    July 28, 1954
    NE1/4NW1/4 sec. 33, T. 30 N., R. 14 W., New Mexico
12. Bondad Fruitland (Combined w/Blanco fields)
    U. S. Smelting Refining and Mining Co. Brown No. 1-X
    November 12, 1954
    SW1/4SE1/4 sec 15, T. 33 N., R. 9 W., Colorado

13. Kutz Fruitland
    R. G. Drilling Schlosser No. 25
    October 30, 1956
    NW1/4SE1/4 sec. 27, T. 28 N., R. 11 W., New Mexico

14. Flora Vista Fruitland
    Northwest Production Corp. Blanco 30-12 No. 3-10
    December 28, 1956
    NE1/4SW1/4 sec. 10, T. 30 N., R. 12 W., New Mexico

15. Crouch Mesa Fruitland
    Devonian Oil and Gas Co. Federal No. 1
    June 26, 1959
    SE1/4NW1/4 sec. 4, T. 29 N., R. 12 W., New Mexico

16. Pinon Fruitland
    Pan American Petroleum Corp. Gallegos Canyon No. 220
    June 14, 1966
    NE1/4SE1/4 sec. 13, T. 28 N., R. 12 W., New Mexico

17. Pinon Fruitland, North
    Pan American Petroleum Corp. Gallegos Canyon Unit No. 82
    August 1966
    SW1/4SW1/4 sec. 28, T. 29 N., R. 12 W., New Mexico

18. Blanco Fruitland
    Tenneco Oil Co. Florance No. 101
    March 1, 1968
    SE1/4SW1/4 sec. 29, T. 30 N., R. 8 W., New Mexico

19. Gallegos Fruitland, South
    Skelly Oil Co. Navajo No. 1-G
    May 27, 1968
    NE1/4SW1/4 sec. 12, T. 26 N., R. 12 W., New Mexico

20. Pump Mesa Fruitland
    Aztec Oil & Gas Rattlesnake Canyon No. 1
    August 25, 1969
    SE1/4SW1/4 sec. 32, T. 32 N., R. 8 W., New Mexico

21. Harper Hill Fruitland-Pictured Cliffs
    Dugan Production Corp. Federal "I" No. 4
    October 3, 1969
    NE1/4NW1/4 sec. 1, T. 29 N., R. 14 W., New Mexico

22. WAW Fruitland-Pictured Cliffs
    Dugan Production Corp. WAW No. 1
    June 30, 1970
    NW1/4SW1/4 sec. 32, T. 27 N., R. 13 W., New Mexico

23. Ojo Fruitland-Pictured Cliffs
    Dugan Production Corp. Pet Inc. No. 2
    April 20, 1971
    NE1/4NE1/4 sec. 36, T. 28 N., R. 15 W., New Mexico

24. Mount Nebo Fruitland
    Amoco Production Co. Holmberg Com "A" No. 1
    December 15, 1972
    NW1/4NE1/4 sec. 28, T. 32 N., R. 10 W., New Mexico

25. Sedro Canyon Fruitland
    Southern Union Production Co. Seymour No. 9
    July 26, 1973
    SE1/4NW1/4 sec. 23, T. 31 N., R. 9 W., New Mexico

26. Jasis Canyon Fruitland
    Mesa Petroleum Co. State Comm No. 39
    June 6, 1976
    NE1/4NW1/4 sec. 36, T. 29 N., R. 8 W., New Mexico

27. Conner Fruitland
    Odessa Natural Gas Little Federal No. 1
    December 28, 1976
    NW1/4SW1/4 sec. 1, T. 30 N., R. 14 W., New Mexico

28. Cedar Hill Fruitland Basal Coal
    Amoco Production Co. Cahn Gas Com No. 1
    May 10, 1977
    NE1/4NW1/4 sec. 33, T. 32 N., R. 10 W., New Mexico

29. Glades Fruitland
    Mesa Petroleum Co. State Com-S No. 15-A
    May 25, 1978
    NW1/4NW1/4 sec. 36, T. 32 N., R. 12 W., New Mexico

30. Farmer Fruitland
    Manana Gas, Inc. Bobbie Herrera No. 1
    February 14, 1979
    NE1/4SW1/4 sec. 4, T. 30 N., R. 11 W., New Mexico

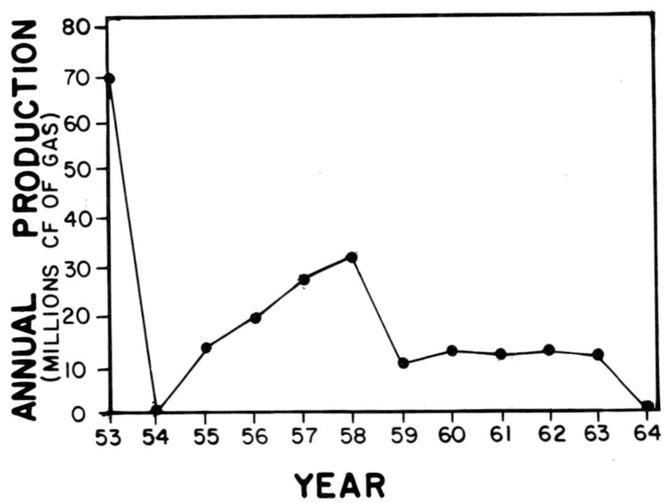

Figure 5. Annual production curve for the San Juan Unit 32-7 No. 3 well, sec. 3, T. 32 N., R. 7 W., North Los Pinos Fruitland field.

lished a new pool, the North Los Pinos, named after the Pine River. The pool was formally established May 25, 1954. The cumulative production for the well is 430,201 MCF of gas. The cumulative production for the pool on January 1, 1987 was 1,742,035 MCF from six wells. In August of 1953, Phillips completed drilling another Fruitland well, the San Juan Unit 32-7 No. 6 (now called San Juan Unit No.6-17) well, sec. 17, T. 31 N., R. 7 W. It was the discovery well for the South Los Pinos Fruitland field. This well with its legendary incline curve has been highly publicized as a classic coal-bed methane well; this curve is shown on the northern San Juan Basin coal-bed methane symposium poster and brochure and on the cover of this guidebook. The well has produced over 1,300,000 MCF of gas. The South Los Pinos field has a cumulative production as of January 1, 1987, of 10,036,394 MCF from 39 wells.

The early 1950's were a busy time for gas-well drillers. Several new Fruitland pools were discovered but exploration activity from 1955 to 1965 was limited with drilling confined to development of existing fields. Exploration proved that Fruitland sandstone beds were spotty in nature and development was confined to known trends. Gas exploration picked up in the late 1960's to mid-1970's with several new Fruitland pools being discovered. In general, the Fruitland gas wells that were being drilled were targeting sandstones and coal completions were either last resort measures or done in conjunction with sandstone completions. The Fruitland was basically viewed as a nuisance requiring mudding up to prevent blowouts as holes were drilled through it.

SOUTHWESTERN SAN JUAN BASIN

Several areas in the southwest part of the basin were beginning to be developed in the late 1960's to mid-1970's. Many small operators were drilling for gas in fields that were shallow and comparatively inexpensive to develop. The most attractive targets were the Pictured Cliffs or Fruitland sandstones. The South Gallegos, Harper Hill, WAW, and Ojo fields all began with wells completed in the sandstones. Beginning in 1971, Dugan Production completed several Pictured Cliffs wells in T. 28 N., R. 15 W., which later became the Ojo Fruitland-Pictured Cliffs pool. These wells were drilled with air using a slim-hole technique and using 2 7/8-inch pipe as the production string. The wells were a maximum of 750 feet deep with an open-hole packer set above the Pictured Cliffs, but below the Fruitland coal in the shale between the two. It was thought that the Pictured Cliffs was wet further down, so only the first few feet of the Pictured Cliffs were drilled. When the Pet. Inc. No. 5 well in section 36 was completed, more gas came up the backside or Fruitland coal side than up the tubing or the Pictured Cliffs Sandstone side. The packer was moved above the coal and afterward all other wells on the lease were completed in the same manner. No water production problems have been encountered on any of these wells. Because the field is low pressure, it has been producible only with a two-stage compressor installed on the lease. The compressor allows low pressure to be maintained on the wells and still produce the wells into a high-pressure gas line. The pool, which was established by the Oil and Gas Commission of New Mexico in January of 1978, has produced over one billion cubic feet (BCF) of gas. It is interesting to note that the later wells that were drilled on the lease were not as good as the earlier wells. One explanation may be that the field was in the latter stages of gas recovery and the reserves have almost been depleted. When the field was first drilled, Tom Dugan gave the field a tentative life of ten years and thought he was being optimistic.

Another operator involved in drilling shallow gas wells was Jerome P. McHugh who completed the Clay No. 1 located in sec. 12, T. 26 N., R. 12 W. which is in the South Gallegos Fruitland field. It was to be a Pictured Cliffs test. Pipe was set and the Pictured Cliffs was perforated with no success and the operator ended up having to perforate the Fruitland coal to get a well. In the beginning the well wasn't very encouraging. It screened off during the frac treatment and after cleanout the well was frac'd again and made very little gas. Tubing was run and the well was completed. During the test, a finger placed over the choke was enough to shut it off, which demonstrated the weakness of the flow. Over the next few months, production gradually increased. As can be seen from figure 6, the well exhibits a typical coal-gas incline curve. Production stabilized at approximately 10,000 MCF per month for four years before it began declining. There were no detectable problems with coal fines in the production equipment but when the well was blown to the atmosphere, chunks of coal would blow out. Another South Gallegos Fruitland well drilled by McHugh, the Nassau No. 6 located in sec. 36, T. 27 N., R. 12 W., is completed in coal beds and has produced approximately 1,340,000 MCF of gas since 1976. After the experiences with wells like the Clay No. 1, the Nassau No. 6, and the Pet. Inc. wells, the Fruitland coal-bed gas in the southwestern portion of the basin area was often produced commingled with gas from the Pictured Cliffs with the hope that additional Fruitland coal gas might augment gas production.

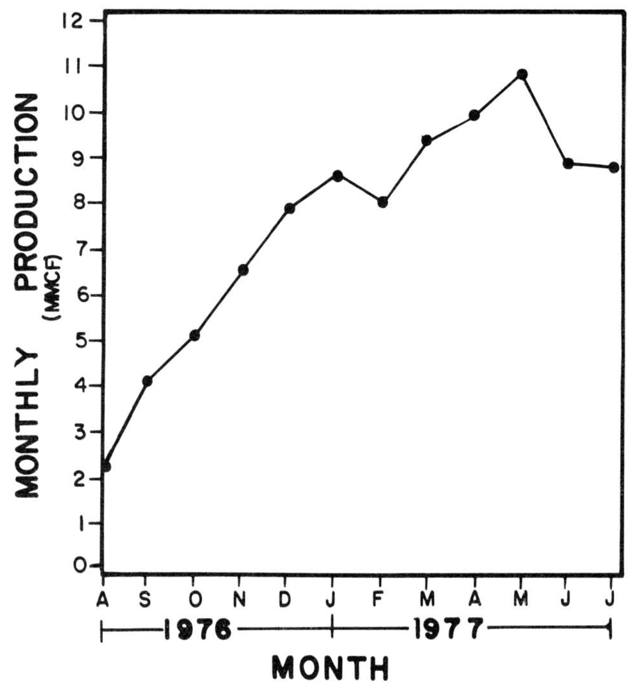

Figure 6. Monthly production curve for the Clay No. 1 well, sec. 12, T. 26 N., R. 12 W., South Gallegos Fruitland field.

The coal-gas wells drilled in the southwestern portion of the basin were shallow and comparatively inexpensive to complete. However, the life of these wells seems to be unpredictable. Many are still producing but for others, having produced regularly, production fell dramatically for no apparent reason, and they became plugging candidates. The production equipment is very rarely fouled with coal fines in the southwest; a problem which seems to affect the wells in the northern San Juan Basin. And the water that is so evident in the deeper coals of the basin seems to be absent from the shallower southwestern coals. The shallow wells have not produced extremely large quantities of gas like the northern wells but they have been easy to maintain and comparatively inexpensive to operate.

NORTHERN SAN JUAN BASIN

Amoco Production Company has done extensive work with coal-bed methane recovery. Amoco began their coal-bed methane efforts in 1975 with the coordination of a task force that prepared research reports outlining the potential. These reports summarized coal provinces and resulted in drilling in four basins, including the San Juan Basin. Amoco also tested a Cedar Hill well where bituminous rank coal was found to have a very high gas content. The first Cedar Hill well was completed in Fruitland coal during 1977. The Cahn Gas Com No. 1 in sec. 33, T. 32 N., R. 10 W., had an initial production rate of less than 100 MCFPD with 100 BWPD. In 1978, Amoco cored a southern Colorado well in T. 33 N., R. 5 W. and tested the coals which were found to be of high rank with a high gas content. That year the Cahn Gas Com No. 1 well produced up to 500 MCFPD but because of production problems, the well could only be tested for three months. About the same time, K. C. Bowman, a Northwest Pipeline Corp. energy consultant, recognized that a gas well, the San Juan Unit No. 6-17 was producing coal-bed methane. This much publicized well, the previously mentioned discovery well for the South Los Pinos Fruitland-Pictured Cliffs Field, has produced over one BCF of gas. (This well was originally drilled by Phillips in 1953.) Mr. Bowman's research prompted many other companies besides Amoco to look again at the Fruitland coal possibilities in the northern area of the San Juan Basin.

In 1979, Amoco's Cahn Gas Com No. 1 well had numerous pump changes and workovers due to coal fines plugging the equipment. The well, however, did produce 200 to 700 MCFPD that year. Amoco began testing low rank New Mexico coals from wells in T. 25 N., R. 14 W. and the gas-desorption tests showed a very low gas content. Amoco began an extensive desorption-testing program in the southern San Juan Basin in 1980. The results showed that the coals were low rank with low to moderate gas content. Work was done which resulted in extensive coal characterization information. A well was tested in the Blanco, New Mexico area, in T. 29 N., R. 9 W., with the results being that the coal was low rank and had low gas content. This well produced a trace of water and negligible gas.

On January 1, 1980, producers were subject to the Crude Oil Windfall Profits Tax Act of 1980 imposed by the Federal Government. The Tax Act contained a tax credit for gas produced from a non-conventional fuel source, which in 1986 was $0.7526 per million Btu. This tax credit applies to coal-seam gas and made the prospect of drilling a Fruitland coal gas well even more economical. **[Editor's note: For more on the coal-gas tax credit, see the paper by Soot in this volume.]** This tax credit may have been a prime reason for many operators in the San Juan Basin to consider drilling coal-gas wells and they felt that any water problems could be handled. Tenneco had also been looking at coal-gas possibilities and, like the rest of the industry, watched the Amoco coal-bed methane wells develop in the Cedar Hill area. Tenneco made plans in 1982 to begin a Fruitland coal project on their Florance lease. The plan was to dually complete existing in-fill Blanco Mesaverde wells in the Fruitland coals. From 1983 to 1985, they recompleted 20 wells on their Florance lease and wells previously operated by EPNG in the Cedar Hill area. The wells produce from 100 to 600 MCFPD with water production in the less than five BWPD range.

The Amoco Cahn Gas Com No. 1 well continued to improve its production rate and by 1983, it was making 1,000 MCFPD with less than 100 BWPD. Four other Amoco wells at Cedar Hill were producing at commercial rates. Amoco also began testing on the Ignacio anticline in southern Colorado. The early results were problematical. Amoco's Denver Regional Office continued the coal-bed methane exploration and production research program in several basins with coal geology and petrology work. They also did drilling, cementing, completion, stimulation, and production research for coal-bed methane wells.

William Perlman, of Houston, Texas, began development in 1982 of the Fruitland coal in three, basically separate areas in southwestern Colorado: the Arboles area; the area west of Ignacio, Colorado; and the area south of the town of Bayfield, Colorado. Perlman drilled over 50 wells in these three areas in 1983 and 1984. Many of the wells were completed in 1985. Mr. Perlman developed and patented a completion procedure which he applied to the Fruitland-Pictured Cliffs. The fracture procedure used 20-, 40- and 100-mesh sand in 2 to 10 lb/gallon slurries. The slurries were alternated with hydrochloric acid pads. The patented procedure used approximately 500,000 pounds of sand and 2,000 barrels of water.

Problems developed with the quantities of water being produced from the coals in the Perlman wells. Stock tanks were installed at the compressor sites to store the water. Applications were filed by Perlman with the La Plata County Commission for permission to construct and maintain an evaporation pit to handle the produced water. The pit was designed and the location obtained but Perlman could not get the Commission's permission to go ahead with the project. Public sentiment may have contributed to his failure to secure permission for the evaporation pit. Water problems continued to plague the wells and many were shut in. Problems also developed in some of the equipment due to frac sand coming back from the formation. Perforations were often covered by sand and cleanup was necessary in many of the wells.

In 1985 Amoco purchased the Perlman properties in the northern San Juan Basin. Amoco acquired more than 60 wells plus an undisclosed amount of undeveloped leases. Amoco also acquired leases from other sources than Perlman, making them probably the major lease holder in the area. That year, they also developed the Ignacio anticline by drilling 15 wells. They constructed a water disposal facility and a water gathering system and installed compressors.

After Burlington Northern bought out EPNG, their subsidiary, Meridian, Inc., began extensively evaluating coal-bed methane possibilities in the basin. Meridian began development of properties in an area that had always blown out dramatically when EPNG had done their Mesaverde drilling, specifically in the San Juan 30-6 Unit. It seems to have been a good choice. Meridian's "400" wells may well be some of the largest coal-bed methane wells in the world. In November of 1987, seven "400" wells produced approximately 407,000 MCF of gas. One well, the San Juan 30-6 Unit No. 413, produced 171,367 MCF in 30 days of production for an average of 5,712 MCFPD. The drawback has been the water production. November's water production for the seven wells was 73,188 BW or 2,440 BWPD. Meridian originally transported the water by truck to injection sites but now has an elaborate system of pipelines to transport the water to a central tank battery where it is either used for drilling new wells or injected into a deeper well into the Entrada Sandstone and Morrison Formation. More wells are planned in the area and it will be interesting to follow the results of Meridian's massive dewatering program in future years.

ARCO began drilling four wells to be completed in the Fruitland coals but stopped any progress in 1986. Amoco, ARCO, and all gas producers in the San Juan Basin experienced the same cold hard fact: the price received for gas fell below $2.00 per MCF. With this turn of events, Amoco has continued their testing programs but they have mostly concentrated their efforts in the southern Colorado portion of the northern San Juan Basin. Many operators feel that even with the tax credit which is even more attractive with low spot market prices, the expense to produce gas with the quantities of water possible is too great compared to the returns.

Many smaller operators are completing wells in Fruitland Formation coal beds and almost every operator has at least one well which is producing coal-gas. One such company is Carbon River Energy Partnership of Seattle, Washington; Carbon River has acreage in southern Colorado near Bondad Hill (T. 32 N., R. 10 W.). Their well was hydraulically fractured with straight cross-link gel. Only the sand volume was increased during the frac treatment rather than the water. The well is shut-in until an injection well becomes available to dispose of the water. Carbon River tends to think that the more water, the better the well will be.

A new technique being tried in the coals of the San Juan Basin is horizontally drilling through the coal seams in order to contact and degas as much coal as possible. This method is used in the thickest Fruitland coal seam or the lowest Fruitland coal bed just above the Pictured Cliffs Sandstone. One major problem with drilling through coal seams is the severe well-bore caving that is so characteristic of coals. Meridian has only recently completed their horizontal drilling of a well in sec. 23, T. 30 N., R. 7 W. Details of the results of this drilling have not been made public. Horizontal Recovery Specialists, of Midland, Texas is planning to drill two horizontal Fruitland tests in sec. 28, T. 32 N., R. 5 W. They plan to run a 4.5-inch slotted liner horizontally through the basal Fruitland coal seam. To try to combat the cave-in problems, they will displace the mud in the hole with a gas while maintaining the bottom-hole pressure. Since this type of experimental drilling and completing is a new technique in the basin, it will be an exciting addition to the history of coal-bed methane recovery. **[Editor's note: See the paper by Logan elsewhere in this volume for more on horizontal drilling in coal beds.]**

CONCLUSION

Exploration for gas in the San Juan Basin was made profitable by the construction, in 1951, of pipelines to transport the gas to the large California markets. The gas wells were originally completed in the sandstone formations in the basin. The recovery of coal gas in the basin began in the early 1950's with open-hole Mesaverde Group completions or with Fruitland wells that were either completed in the coal because the sandstone potential was not apparent or in areas that produced gas from the coals without significant water-production problems. The odd thing about the coal gas development in the basin is that even though virtually every well drilled in the northern portion of the basin blew out in the Fruitland coals (and some *really* blew out!), no one saw the coal as a long-term, long-range prospect because the wells blew so fiercely and produced so much water, that it was felt that they were watered out and that the life of such wells was unknown. We in the industry have now reached a time to look back on the early wells that are still producing coal gas and evaluate the technical problems that have arisen related to production of the gas. The southwest part of the basin has produced coal gas without significant water production but only recently has any company rigged up to handle the large quantities of water produced with the equally large quantities of gas in the northern portion of the basin. Development in these northern areas has been made more attractive by the large quantities of gas and the tax break for unconventional fuel sources. The recent chapters in the history of coal-gas recovery show that we are just beginning to fully realize the potential of the vast resources of coal-bed methane contained in the San Juan Basin coals. Though many projects, some conventional and some experimental, have been postponed until the price of natural gas increases to an amount which will sustain large scale water-disposal projects or justify expensive experimentation, we now know that the reserves are great enough to maintain production and are waiting for the advanced technologies and the capital required to remove them. **[Editor's note: For an analysis of coal-bed methane resources in the San Juan Basin, see the paper by Kelso and Wicks in this volume.]**

REFERENCES CITED

Kelso, B. S., and Wicks, D. E., 1988, A geologic analysis of the Fruitland coal and coal-bed methane resources of the San Juan Basin, southwestern Colorado and northwestern New Mexico, *in* Fassett, J. E., ed., Geology and coal-bed methane resources of the northern San Juan Basin, Colorado and New Mexico: Rocky Mountain Association of Geologists Guidebook

Logan, T. L., 1988, Horizontal drain-hole drilling techniques used in Rocky Mountain coal seams *in* Fassett, J. E., ed., Geology and coal-bed methane resources of the northern San Juan Basin, Colorado and New Mexico: Rocky Mountain Association of Geologists Guidebook

Matheny, M. L., and Talley, D. A., 1978, Oil and gas field discovery lists and index maps *in* Fassett, J. E., ed., Oil and gas fields of the Four Corners Area: Four Corners Geological Society, Durango, Colorado, Volume I, p. 25-39.

Matheny, M. L., and Talley, D. A., 1983, Oil and gas field discovery lists and index maps *in* Fassett, J. E., ed., Oil and gas fields of the Four Corners Area: Four Corners Geological Society, Durango, Colorado, Volume III, p. 811-829.

Soot, P. M., 1988, Non-conventional fuel tax credit, *in* Fassett, J. E., ed., Geology and coal-bed methane resources of the northern San Juan Basin, Colorado and New Mexico: Rocky Mountain Association of Geologists Guidebook

If the ballad "Mother was a Lady" which begins "Two drummers sat at dinner in a grand hotel one day," had had Durango for its setting, it would have been the Strater. "Strictly First Class," it had been a Durango landmark since 1888. The horsedrawn street-car dates the photograph to 1891. It was advertisements such as those tacked onto the poles that aroused the council's ire, and the ill-repaired wooden sidewalks that antagonized their constituents. (Photo and caption courtesy of the Center of Southwest Studies.)

The Southern Utes: An Economically and Socially Successful Indian Nation Building upon its History and Challenging the Future

Bradford C. Boyce[1]
Leonard C. Burch[2]

ABSTRACT

Today, the Southern Ute Indian Tribe persists in the goal of making the lands productive and the people economically and socially successful. Following the leadership of their early chiefs, the present Tribal Council is demonstrating vision and realism in planning and expediting a modern and sophisticated program of economic and social improvement. The Tribe's assets; its natural resources and its people, are committed to a full, responsible, and effective participation in the affairs of the local tri-ethnic community, the State of Colorado, and the Nation. Tribal management has demonstrated farsighted program development including educational policies, economic improvements, and energy-mineral management. The Southern Ute Indian Tribe is proud of its accomplishments, determined in its purpose, and deeply aware of its responsibilities to its members, to its neighbors, to its business partners, and to the whole Nation. This program constitutes the pledge and commitment of the modern Southern Ute Indian Tribe.

INTRODUCTION

Historically, Indian mineral properties have been leased to industry by the Federal Government. Following the leasing of Indian mineral property, the Federal Government was responsible for the management of the property, including tracking activity obligations and overseeing royalty payments. With the passing of the Indian Mineral Development Act in 1982, some Tribes responded by actively participating in the leasing, exploration, and development of their mineral estates. The Federal Government also responded by creating the Energy and Minerals Division of the Bureau of Indian Affairs (BIA) Lakewood, Colorado, to assist tribes in evaluating, exploring, and developing their mineral resources. Data and evaluations of a tribe's mineral resources can be provided to the tribe by the Energy and Minerals Division, BIA.

Tribes vary substantially in their participation in evaluating, promoting, exploring, and developing tribal mineral resources. The Southern Ute Indian Tribe has decided to pursue aggressively the evaluation-through-development phases of their mineral resources programs. Utilizing a permanent professional staff and working with the Energy and Minerals Division of the BIA, the Southern Ute Indian Tribe has made substantial progress in their mineral-development program.

This paper is intended to provide the reader with a summary of the history of the Ute Indians as well as specific information on the Southern Ute Indian Tribe, its goals and objectives, governmental structure, Federal interaction, business relationships, and specific Tribal divisions and departments which may be of interest.

[1]Bradford C. Boyce Manager, Exploration and Production
Southern Ute Indian Tribe Tribal Affairs Building
P.O. Box 737
Ignacio, Colorado 81137

[2]Leonard C. Burch Chairman,
Southern Ute Tribal Council
Tribal Affairs Building
P.O. Box 737
Ignacio, Colorado 81137

HISTORY OF THE UTE INDIANS[3]

Prior to the Spanish

The Four Corners area is rich in the history of the American Indian, including some of its earliest inhabitants, the Anasazi. The Ute Indians are recognized as the oldest continuous residents of the State of Colorado. The original Ute lands included the mountainous areas of Colorado and Utah, and the northern sections of New Mexico and Arizona. The Ute Indians were nomadic and subsisted by hunting big game and gathering grasses, berries, and fruit in the mountains during summer.

The language of the Utes is Shoshonean, considered a branch or dialect of the Uto-Aztecan language. It is believed that the people who speak Shoshonean separated from other Uto-Aztecan speaking groups about the time of the birth of Christ. Other Indian groups in the U.S. who speak Shoshonean are the Paiutes, Goshutes, Shoshones, Bannocks, and a few tribes in California. Though much has been researched and theorized about American Indians, much remains unknown regarding their cultural and genetic histories. Lack of evidence will likely continue to shroud the clarity of the Southern Ute's history as well.

The early organization of the Utes consisted of small family units, associated in a loose confederation of seven bands, as follows: 1. The MOUACHE band resided in Southern Colora-

[3] The historical section of this paper is modified from Jefferson and others, (1973) and from Delaney (1974). The use of this material was approved by the Southern Ute Indian Tribe, 1988.

do and in Northern New Mexico, occasionally traveling as far south as Santa Fe; 2. The CAPOTE band inhabited the San Luis Valley of Colorado near the headwaters of the Rio Grande, and parts of New Mexico, especially the region where the towns of Chama and Tierra Amarilla are now located; 3. The WEEMINUCHE occupied the valley of the San Juan River and its northern tributaries in Colorado and northwestern New Mexico; 4. The TABEHUACHE (also called UNCOMPAHGRE) lived in the valleys of the Gunnison and Uncompahgre Rivers in Colorado; 5. The Grand River Utes (also called PARIANUC) lived along that River in Colorado and Utah; 6. The YAMPA band inhabited the Yampa River Valley and adjacent land; 7. The UINTAH UTES inhabited the Uintah Basin, especially the western portion.

Of the bands mentioned above, the first two (Mouache and Capote) make up the present day Southern Ute Indian Tribe with headquarters at Ignacio, Colorado. The Weeminuche are now called the Ute Mountain Ute Indian Tribe with headquarters at Towaoc, Colorado. The last four bands mentioned (Tabehuache, Grand, Yampa, and Uintah) now comprise the Northern Utes on the Uintah-Ouray Reservation with headquarters at the town of Fort Duchesne, Utah.

Spanish Influence

The arrival of the Spanish in the Southwest in the 1500s brought the horse and a change of lifestyle for the Ute people. They became buffalo hunters, dependent upon the herds that roamed the high plains. They also became more warlike and developed as skilled warriors because of the need to obtain more horses from the Spanish and from the plains tribes, especially the Arapahoes, Kiowas, and Cheyennes. Furthermore, horses allowed the Utes to increase their trade with other Indian groups, especially the Shoshonian nations to the northwest and the Pueblos to the south. The availability of horses also made possible food gathering expeditions, thereby enabling the Utes to band together for longer periods of time. Although the family unit continued to be the basic unit of Ute society, the longer time periods of grouping resulted in a restructured Ute society in which a single powerful leader emerged. The leader of the band or of the larger group directed the movement of the camp, the hunt, and all the raids and war parties.

The first encounter with the Anglo-American society occurred when Lt. Zebulon Pike entered the San Luis Valley in 1806. Following this encounter, many trappers and traders traveled onto Ute land. The Utes welcomed them because they understood they were only visitors to the Ute domain.

The early relationships between the Mexican government and the Utes were peaceful, resulting in the opening of the Old Spanish Trail as a regular trading route in the 1820s. However,this peace was disrupted in the 1830s when Mexican farmers (Spanish settlers of the southwest controlled by Mexico) attempted to settle on Indian territory. The Utes and Navajos joined forces in conducting raids on Mexican settlements in northern New Mexico during the 1830s and 1840s, in an attempt to remove the farmers.

Mexican control of the area ended in 1848 with the Mexican-American War. The U.S. government, concerned about maintaining open lines of supply and communication, initiated peace talks between Ute leaders and military authorities. The first treaty between the Utes and the United States was signed December 30, 1849 at Abiquiu, New Mexico. In this treaty the Utes recognized the sovereignty of the United States and agreed not to depart from their accustomed territory without permission, however, boundaries defining the Ute Territory were not established at this time.

U.S. Control

Friction developed between the Ute people and the increasing numbers of permanent settlers of the San Luis Valley during the mid 1800s. The U.S. Government decided to solve the "Ute problem" by removing them from the San Luis Valley, an area the Utes had considered their homeland for centuries. To accomplish this, a treaty was negotiated and signed on March 2, 1868, at Washington, D.C. By this treaty, a single reservation was provided for all the Ute Indian bands, an area comprised roughly of the western one-third of Colorado (fig. 1). A Federal agency to provide food, clothing, education, and direction, for the three bands of the Northern Utes was established on the White River near the present town of Meeker, Colorado. Another agency was established on the Los Pinos River for the Tabeguache, Mouache, Capote, and Weeminuche bands, with a third agency stationed in Denver, from 1871 to 1876, for the Utes of that area. Under the 1868 treaty, education, clothing, and rations were to be provided by the United States until the Utes were capable of supporting themselves. The Utes were assured that this reservation would forever be theirs and they would be protected from all trespassers.

After the signing of the 1868 treaty, it was discovered that the San Juan Mountains of southwestern Colorado held huge deposits of rich minerals. In 1873, Chief Ouray was convinced to sign the Brunot Agreement, relinquishing the Ute's claim to about 6,000 square miles in the San Juan Mountains, almost one quarter of the entire reservation of 1868 (fig. 1).

In September of 1879, a small band of Utes, led by Chief Douglas of the White River Utes, attacked the White River Ute Indian Agency and killed all the male inhabitants, including agent Nathan Meeker. This was the Ute's last attempt at united fighting in an effort to retain their lands.

Final Settlements

In response to demands by white settlers, the U.S. government again negotiated with the Utes, specifically for their removal from Colorado. In 1880 an agreement was signed by leaders of the Utes and approved by Congress. The Utes agreed not only to relocate but also to relinquish all claims to their reservation. The Northern Ute bands were relocated onto the Uintah Reservation in Utah. The Southern Utes were to be allotted lands along the La Plata River. The designated Southern Ute lands were found to be unsuitable for farming and incapable of supporting the Utes. In a series of proposed agreements, the fate of the Southern Ute bands was negotiated. Final disposition of the Utes was decided by the Hunter Act of 1895 in which the Utes were to be located in their old reservation in southwestern Colorado. Individual allotments were to be distributed to Ute families, and any lands not taken by the Utes were to be opened to white settlement. The Weeminuche band, under the leadership of Ignacio, was against the planned allotments and established themselves in a camp on the western end of the old Southern Ute Reservation. Therefore,

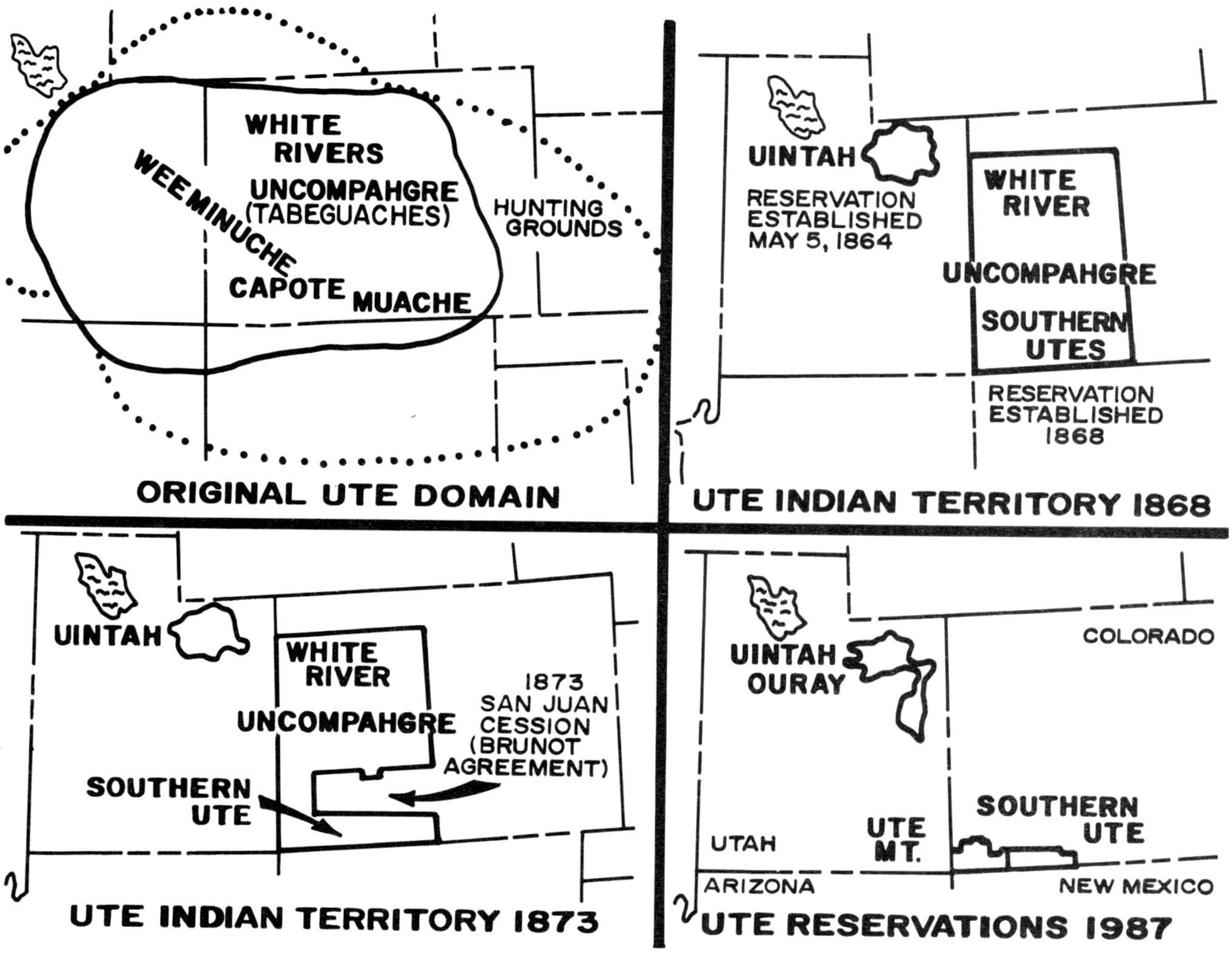

Figure 1. Diminishment of Ute Lands through time, (Modified from Jefferson, Delaney, Thompson, 1973).

the western end of the reservation was retained as land-in-common for Ignacio and his band.

This separation initiated the regrouping of the three bands of the Southern Ute Indians into two tribes. The Mouache and Capote bands became known as the Southern Utes and the Weeminuche band was referred to as the Ute Mountain Ute Tribe, in reference to the Sleeping Ute Mountain area of their settlement. The Southern Utes were located on the eastern portion of the former reservation on individually allotted land, while the Ute Mountain Utes were located on the western end of the reservation on land held in common. After the turn of the century, the sections became known as the Southern Ute Reservation and the Ute Mountain Ute Reservation.

The Southern Ute Indian Reservation became a strip of land, 15 miles wide and approximately 73 miles long. The reservation is not all Indian-owned lands. Homesteading by non-Indians from 1899 to 1938 resulted in the majority of the prime agricultural lands, mostly along the river bottoms, becoming the property of homesteaders. Thus, the Southern Ute Indian Reservation is now checker-boarded with Indian and non-Indian land ownership. Today, more non-Indians than Indians live within the boundaries of the reservation (fig. 2).

In 1906, the United States government ceded approximately 50,000 acres of the Ute Mountain Ute Reservation for the establishment of Mesa Verde National Park. These lands are world-renowned for their extensive cliff-dwelling Anasazi ruins. In exchange for the park lands, the Ute Mountain Utes received lands near the Utah border, including most of the area in and around Sleeping Ute Mountain.

RESERVATION LIFE OF THE 20TH CENTURY

Government

The first years of the 20th century were not kind to the Southern Utes. Their traditional lifestyle was gone and they found it exceedingly difficult to make a living by farming. They did only what they had to do to keep meager crops alive, working only as long as great need existed. Gradually however, members of the Southern Utes adapted to the new way of life. By 1915 they had made great advances in farming their lands and had a total of 2,000 acres planted in alfalfa, wheat, oats, beans, and potatoes. Some families supplemented their agricultural income by raising sheep and cattle and by wage work pursuits.

On November 4, 1936, the Southern Ute Indian Tribe organized and approved its constitution and by-laws. The adjacent Ute Mountain Ute Tribe approved its constitution and by-laws on June 6, 1940. The Southern Ute Tribal constitution provides for a governing body of elected Tribal Council

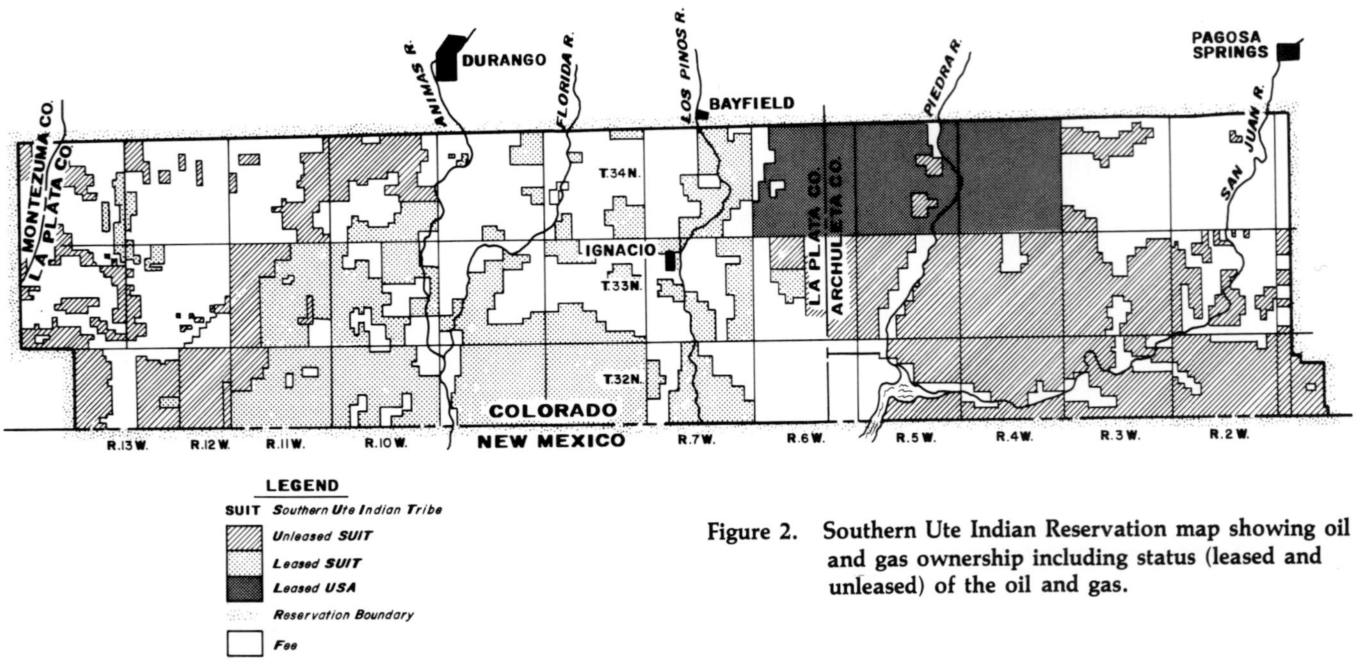

Figure 2. Southern Ute Indian Reservation map showing oil and gas ownership including status (leased and unleased) of the oil and gas.

officials consisting of seven Tribal members, each serving a three year term. Tribal Council elections occur each year with two council members elected yearly during two years and three council members in the third. When elected, the Chairman chooses from the council members a Vice-Chairman to serve concurrently with the Chairman. This election system has provided a sound and consistent government for the Southern Ute Indian Tribe.

Tribal Goals

Perhaps the modern social and economic system for the Southern Ute people today was established on July 13, 1950 when the United States Court of Claims judged that lands had been taken illegally from the Utes from 1892 to 1938 and, therefore, the bands of Utes were entitled to monetary settlement. Public Law 120 directed the Secretary of the Interior to divide settlement funds among the Northern, Ute Mountain, and Southern Utes. The share for the Southern Utes amounted to approximately six million dollars.

The Southern Utes took the entire responsibility for the formulation and development of rehabilitation plans which would determine the best use of the settlement funds. They received from the Department of Interior, Bureau of Indian Affairs, (B.I.A.) only that assistance requested. Fifteen major goals were established with the overall objective of the Southern Ute people to function successfully within the new Anglo-American community. In order to attain these goals, it was necessary to create organizations within the Tribal structure. Tribal divisions and departments were structured to orient social and economic programs in a sound direction.

In the thirty-seven years since the beginning of the rehabilitation plan of the Southern Ute Indian Tribe, the Tribe has made continual progress. Many problems remain to be solved, yet the human and natural resources of the Tribe provide a sound basis for confidence in the future.

The Southern Ute Indian Tribe, determined in its purpose and deeply aware of its responsibilities, has established a strong, consistent governmental organization. Departments, working to improve the quality of life for the Southern Ute Indian Tribe, include the Energy Resources Division, Economic Development and Planning Division, Education Division, Enterprises Division, Natural Resources (Surface and Wildlife) Division, Justice Division, and others (fig. 3). These departments accomplish tasks independently while working together to resolve many Tribal issues. Tribal departments in many cases must also integrate with Federal programs. For example, the Land and Exploration and Production Departments of the Energy Resources Division work closely, not only with the Tribal Natural Resources Division, but with the B.I.A. and Bureau of Land Management (B.L.M.) in processing oil and (or) gas well permits. Such integration requires clear communication and timely action from each party.

OIL AND GAS

Operations

Application processing for oil and (or) gas wells on Southern Ute Tribal leased properties follow these steps:

1. Permission to survey and stake filed with B.I.A. Realty Office. Applicable administration and archaeological fees are paid at this time at the B.I.A.
2. The on-site well location inspection is scheduled by the operator with the Southern Ute Energy Resources Land Department, B.I.A. Realty Office, B.L.M., and Archaeologists. Operator supplies survey plats of well location(s) and access roads to the four offices at least 48 hours (excluding weekends and holidays) in advance of the on-site inspection. Archaeological surveys cannot be conducted when snow covers more than 30 percent of the location to be surveyed.
3. Application for Permit to Drill (A.P.D.) submitted to B.L.M. by operator. B.L.M. submits the A.P.D. to B.I.A. for approval. B.I.A. attaches Southern Ute Energy Department surface use stipulations, B.I.A. Threatened-and-Endangered-Species clearance letter, and B.I.A. Area

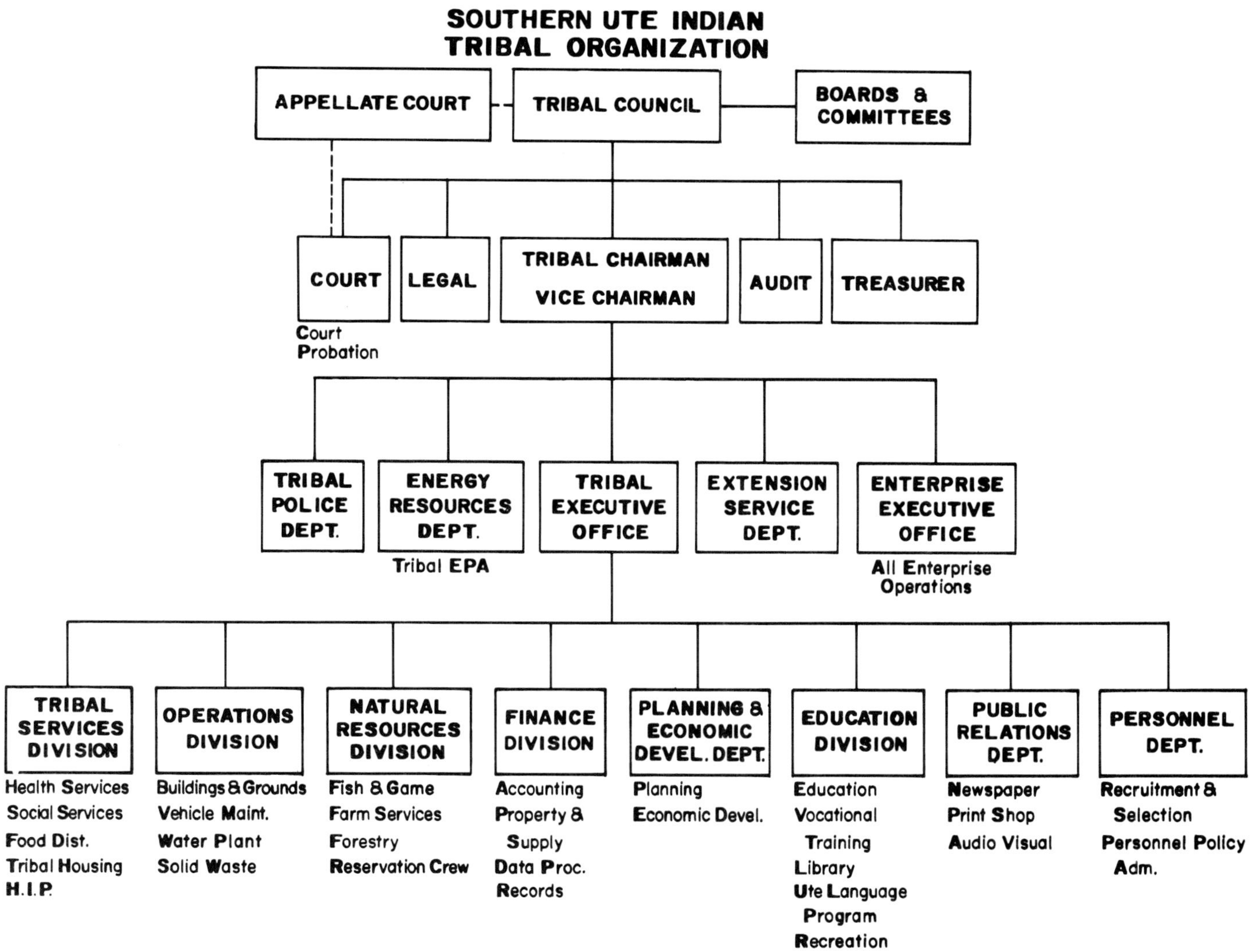

Figure 3. Southern Ute Indian Tribe organization chart.

Archaeologist clearance letter. A.P.D. is returned to the B.L.M.

4. Following the on-site well location inspection, Tribal Energy Resources Land Department assesses the operator for surface damages.
5. B.L.M. approves or disapproves A.P.D.
6. If A.P.D. is approved, construction may begin after 48 hours notice (excluding weekends and holidays) to Tribal Energy Resources Division Land Department.

Production pipeline rights-of-way (oil, gas, or water) are similar to well site application processing. Permission to survey, filed with the B.I.A. Realty Office and on-site inspection scheduling with survey plats supplied at least 48 hours in advance, follow the same procedures as for well sites. Following the on-site inspection, the operator files a Right-of-Way (R.O.W.) application with the B.I.A. and a Sundry Notice and Reports on Wells (B.L.M. form 3160-5) with the B.L.M., if applicable. The operator is then assessed for appropriate surface damages and R.O.W. fees. The Energy Resources Division Land Department then requests final R.O.W. approval from the Tribal Council and if approved, a Council resolution is provided to B.I.A., which then issues the Grant of Easement for R.O.W. Construction may begin following 48 hours (excluding weekends and holidays) notice to the Tribal Energy Resources Division Land Department.

The previously-described procedures require a competently staffed Division to quickly and efficiently complete the processing of oil-and-gas-related paperwork. The Southern Ute Indian Tribe has created the Energy Resources Division to fulfill this need and as a result all applications are now processed in a timely manner.

Southern Ute Energy Resources Division

Natural gas was discovered within the Southern Ute Reservation at Ignacio Dome in 1950 (See Harr, this volume). Gas and oil exploration, development, and production activities are important factors in the Southern Ute Indian Tribe's program of economic and social improvement. Not only does oil and gas activity provide direct monetary benefit through landowner royalties, but numerous indirect benefits result, including job opportunities for Tribal members. The Southern Ute Indian Tribe included a drilling rig within the Tribal Seal (fig. 4), in recognition of the value of the Tribe's oil and gas resources and associated industries.

The Southern Ute Tribal Council created the Energy Resources Division in 1979 to oversee established oil and gas

Figure 4. Seal of the Southern Ute Indian Tribe.

exploration, development, and production activities on Tribal mineral properties. Originally, the division consisted of a staff of three. In 1984, following an independent review and recommendation, the Southern Ute Indian Tribe restructured the Energy Resources Division, creating and staffing Oil and Gas Accounting, Exploration and Production, and Land departments.

Specific responsibilities have been delegated to each department. The Oil and Gas Accounting department is responsible for the collection, accounting, and auditing of Tribal mineral revenue (fig. 5). The Land department is not only responsible for controlling oil and gas surface use, but additionally is responsible for land ownership, lease, and agreement records. The Exploration and Production department reviews all drilling activity, monitors production, promotes activity both on leased and unleased properties, and formulates, evaluates, and implements short and long-range exploration and development programs. All Energy Resource Division departments interact closely with each other and with the Tribal Council, with all final decisions resulting from joint efforts.

Congress implemented the Indian Mineral Development Act in 1982, providing tribes the right to negotiate for mineral development on tribal lands. The Southern Ute Energy Resources Division was delegated the responsibility to negotiate with interested energy resource companies for exploration and development of Tribal mineral properties, with Tribal Council approval or disapproval during stages of the negotiations. Working together, the Southern Ute Tribal Council and Energy Resources Division staff have negotiated and consummated oil and gas exploration agreements on certain tribal properties.

Potential for Further Development

The Southern Utes own 139,000 acres of oil and gas properties which are leased and producing (fig. 2). Unleased tribal properties consist of 192,000 acres, including acreage on the Four Corners Platform, a broad, structurally-high area separating the San Juan Basin from the Paradox Basin. A substantial area of non-developed property is also present in the eastern portion of the Reservation. Producing Tribal mineral properties are predominantly located in the Colorado section of the San Juan Basin within the established Ignacio Blanco gas field. As discussed previously, surface ownership and oil and gas ownership is checker-boarded throughout the reservation. However, coal ownership is not as fragmented because the U.S. Government reserved title to the coal during the homesteading years and transferred those coal rights to the Tribe in 1938. Tribal coal ownership consists of approximately 520,000 acres (fig. 6).

The Four Corners Platform, within and immediately adjacent to the Reservation, contains gas fields producing from Pennsylvanian rocks as at Barker Dome which has produced 205 billion cubic feet (BCF) of gas (cumulative), Ute Dome which has produced 50 BCF (cumulative), and Alkali Gulch, which has produced 25 BCF (cumulative) (Harr, this volume). Additionally, shallow Cretaceous oil and gas production, though relatively small, exists in the platform area in the western part of the Reservation. Further exploration efforts, particularly with seismic data, have a good chance of identifying new gas reserves on the Four Corners Platform (fig. 7).

The eastern half of the Southern Ute Reservation remains relatively unexplored. The Archuleta anticlinorium (Wood and others, 1948), a 10 to 25 mile-wide system of folds and faults (fig. 7), defines the northeastern boundary of the San Juan Basin and merges eastward and northward into the San Juan Mountains uplift. The reservation boundaries extend from the San Juan Basin proper towards the Archuleta anticlinorium where folds, faults, and outcropping dikes are present. Source rocks within the eastern section of the reservation include, but are not limited to, the Todilto Limestone, Mancos Shale, Lewis Shale, and Fruitland Formation; especially the coals in the Fruitland, which contain substantial methane gas reserves (Choate and others, 1984). Reservoir rocks exist throughout the strata of the area (Gries, 1985). The Gramps oil field has produced six million barrels of oil, cumulative, from the Dakota Sandstone at less than 2,000 feet and is only about fifteen miles east of the reservation boundary. The field lies along the faulted Gramps anticline, a structural style that may exist elsewhere in the area (Ryder, 1985). Potential oil and gas discovery in the eastern section of the reservation is considered good to excellent.

Coal-Bed Methane

Coal-bed methane exploration and development on Tribal mineral properties is increasing rapidly. Industry tax credits, a major economic incentive, expire on coal-bed wells drilled after January 1, 1990 (Windfall Profits Tax). For industry to utilize this tax credit, coal-bed methane properties must be developed prior to the tax credit deadline. **[Editors note: For more on the fuel tax credit see the paper by Soot, elsewhere in this volume.]**

The Southern Ute Tribal Council recognizes the need for a well-coordinated interaction among parties involved in exploration and development of Tribal coal-bed methane properties to take advantage of the tax incentive, a credit unavailable to the Southern Ute Indian Tribe. Tribal Energy Resources Division personnel work diligently to process all applicable permits in a timely manner. Weather always provides some difficulties in permitting and drilling; however,

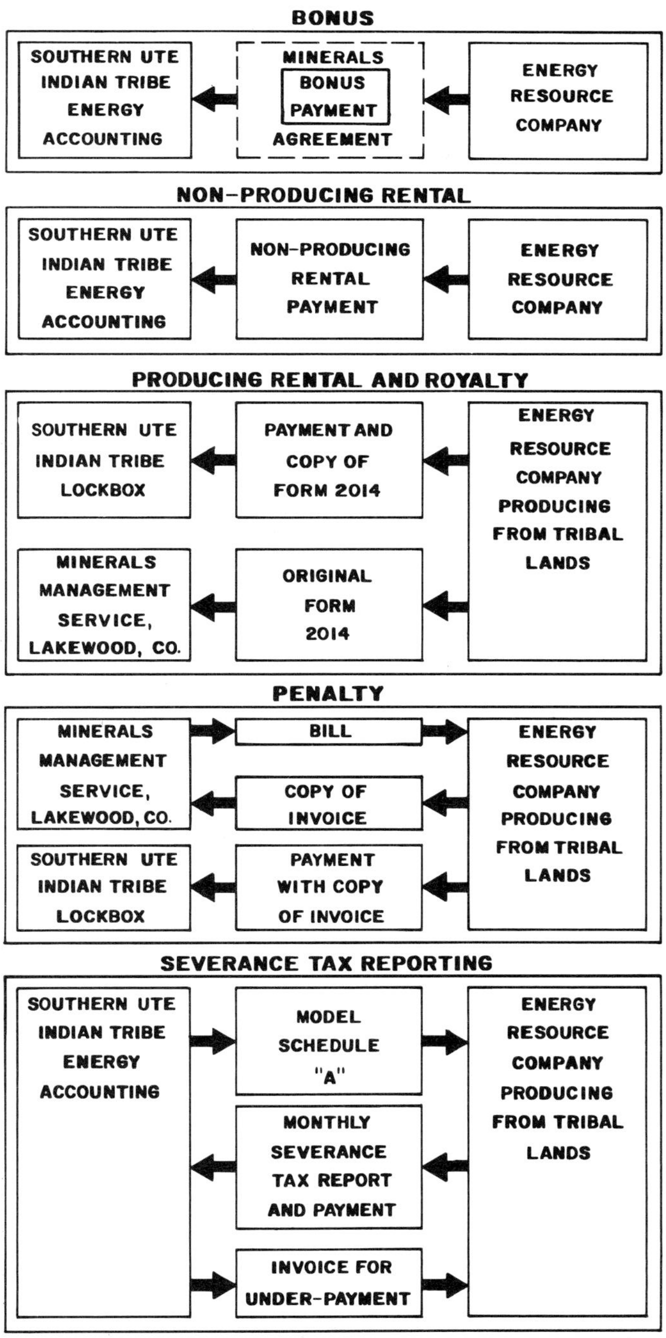

BONUS, RENTAL, ROYALTY & PENALTY

Bonus payments and non-producing rentals payable to the Tribe should be submitted to the Tribe's general mailing address (P.O. Box 737, Ignacio, CO 81137, ATTN: Energy Accounting). Payments of Tribal royalty, producing rentals, and penalties on late payments of royalty should be mailed to the Tribe's Lockbox (P.O. Box 696, Ignacio, CO 81137). Royalties and producing rentals must be reported to the Minerals Management Service in the manner described in the MMS payor handbook. A courtesy copy of each royalty/rental report should accompany the payment mailed to the lockbox.

TRIBAL SEVERANCE TAX

On February 23, 1982, the Council of the Southern Ute Indian Tribe adopted Ordinance No. 33 by Resolution No. 82-19 which provides for the imposition of a tax on the severance of gas and oil from Reservation lands. The tax established by the Ordinance is based on the value of the minerals removed and is payable by the well operator.

The Ordinance, as adopted in 1982, is currently under review for possible revision and any questions regarding severance tax reporting should be directed to the Energy Accounting Department.

AUDITS

Desk reviews of severance tax payments are performed by the Energy Accounting Department routinely on a monthly basis. When discrepancies are identified, the Energy Accounting Department will issue a bill to the responsible taxpayer.

Royalty audits are conducted either jointly by the Tribe and the Minerals Management Service or independently by MMS, under MMS audit standards and procedures.

CORRESPONDENCE

Correspondence to the Energy Accounting Department should be mailed to P.O. Box 737, Ignacio, CO 81137. Department staff may be contacted at (303) 563-4525, extension 444, 445, and 380.

Figure 5. **Energy Accounting flow chart and description of payment and audit procedures of the Southern Ute Tribe. Energy Accounting is a department of the Energy Resources Division which is responsible for the collection, accounting, and auditing of Tribal mineral revenue.**

most stumbling blocks causing delays result from poor initial planning. Developers who meet and discuss exploration and development projects on Tribal properties with the Tribal Energy Resources Division staff can identify, and generally resolve potential difficulties early, thus eliminating or reducing costly delays. For operators concerned about project confidentiality, the Tribal Energy Resources Division adheres to strict confidentiality policies.

Tribal Exploration and Development Processes

The Energy Resources Division follows a program that provides energy-resource data to the Tribal Council and identifies the present and future potential of the leased and unleased tribal mineral properties. Current evaluation and assessment studies of Tribal mineral properties include in-house studies combined with studies at the United States Geological Survey and the B.I.A. Energy and Minerals Division. The Tribal Council integrates such information into the goals and objectives of the Tribe with action plans developed and implemented for the Tribe's energy mineral properties.

The Tribe, as of January 1988, has initiated study projects in which data on the unleased mineral acreage is being developed, reviewed, and processed by Energy Resources Division staff. Data and recommendations are then provided to the

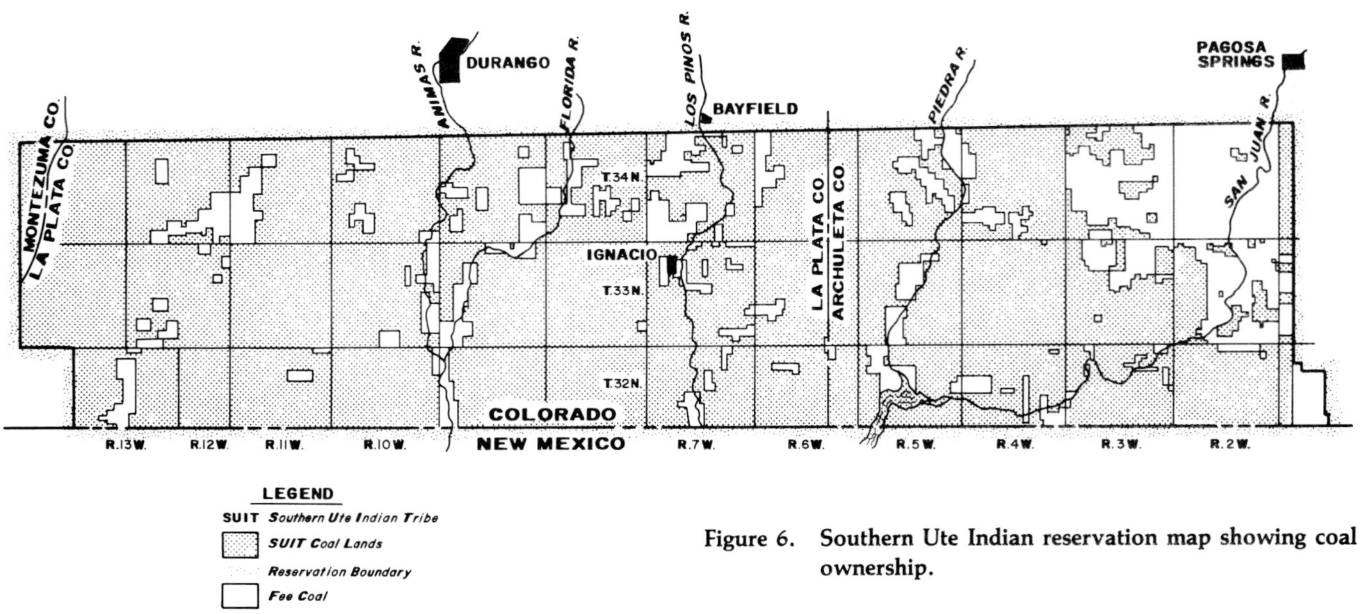

Figure 6. Southern Ute Indian reservation map showing coal ownership.

Tribal Council. Following discussions of the data and Energy Resource Division recommendation, the Tribal Council formulates and implements exploration and development plans for the unexplored tribal acreage. Many factors, including cultural, social, and environmental aspects contribute to the final plans, which range from no exploration and development to full-scale exploration and development.

The unleased eastern acreage block requires an extensive investigation prior to final determinations on exploration and development of its mineral resources. Preliminary plans may be established by January, 1989. Action plans for specific Tribal properties have been initiated on the Four Corners Platform. Further evaluation is continuing for the remaining properties, with exploration and development plans expected to be formulated by January, 1989.

CONCLUSION

The Southern Utes continue to pursue their objectives of providing opportunities for their people to become economically and socially successful, and building a desirable environment in which the Utes may live and prosper. Throughout the Tribe's history, the Utes have faced many challenges, and expect to face many more in the future. With sound planning practices in place, the Tribal Government is prepared to meet future challenges and attain Tribal goals.

Planned land and mineral development provides a sound base from which the Southern Utes may reach their objectives. Today, oil and gas exploration and production provides a substantial portion of the Tribe's economic and social environment. Future mineral development of oil, gas, coal, and potentially other mineral resources will provide an important economic-resource base from which the Tribe will prosper. Proper planning and management of the Tribe's resources today will ensure prosperity into the future.

REFERENCES CITED

Choate, R., Lent, J., and Rightmire, C.T., 1984, Upper Cretaceous Geology, Coal, and the Potential for Methane Recovery from Coalbeds in San Juan Basin—Colorado and New Mexico: American Association of Petroleum Geologists Studies in Geology Series, No. 17, p 185-223.

Delaney, R.W., 1974, Southern Ute people: Indian Tribal Series, Phoenix, Arizona.

Gries, R.R., 1985, San Juan Sag: Cretaceous rocks in a volcanic-covered basin, south central Colorado: Rocky Mountain Association of Geologists, p 167-179.

Harr, C.L., 1988, The Ignacio Blanco gas field, northern San Juan Basin, Colorado, *in* Fassett, J. E., ed., Geology and coal-bed methane resources of the northern San Juan Basin, Colorado and New Mexico: Rocky Mountain Association of Geologists Guidebook

Jefferson, J.E., Delaney, R.W., and Thompson, G.C., 1973, (edited by Floy A. O'Neil), The Southern Utes: A Tribal History, Southern Ute Indian Tribe, Ignacio, Colorado: published by the Southern Ute Tribe, Ignacio, Colorado.

Ryder, Robert T., 1985, Oil and Gas Potential of the Chama-Southern San Juan Mountains Wilderness Study Area, Colorado, U.S. Geological Survey Bulletin 1524, p 81-118.

Soot P.M., 1988, The non-conventional fuel tax credit, *in* Fassett, J. E., ed., Geology and coal-bed methane resources of the northern San Juan Basin, Colorado and New Mexico: Rocky Mountain Association of Geologists Guidebook

Wood, G.H., Kelley, V.C., and MacAlpin, A.J., 1948, Geology of the Southern Part of Archuleta County, Colorado, U.S. Geological Survey Oil and Gas Investigations Preliminary Map 81.

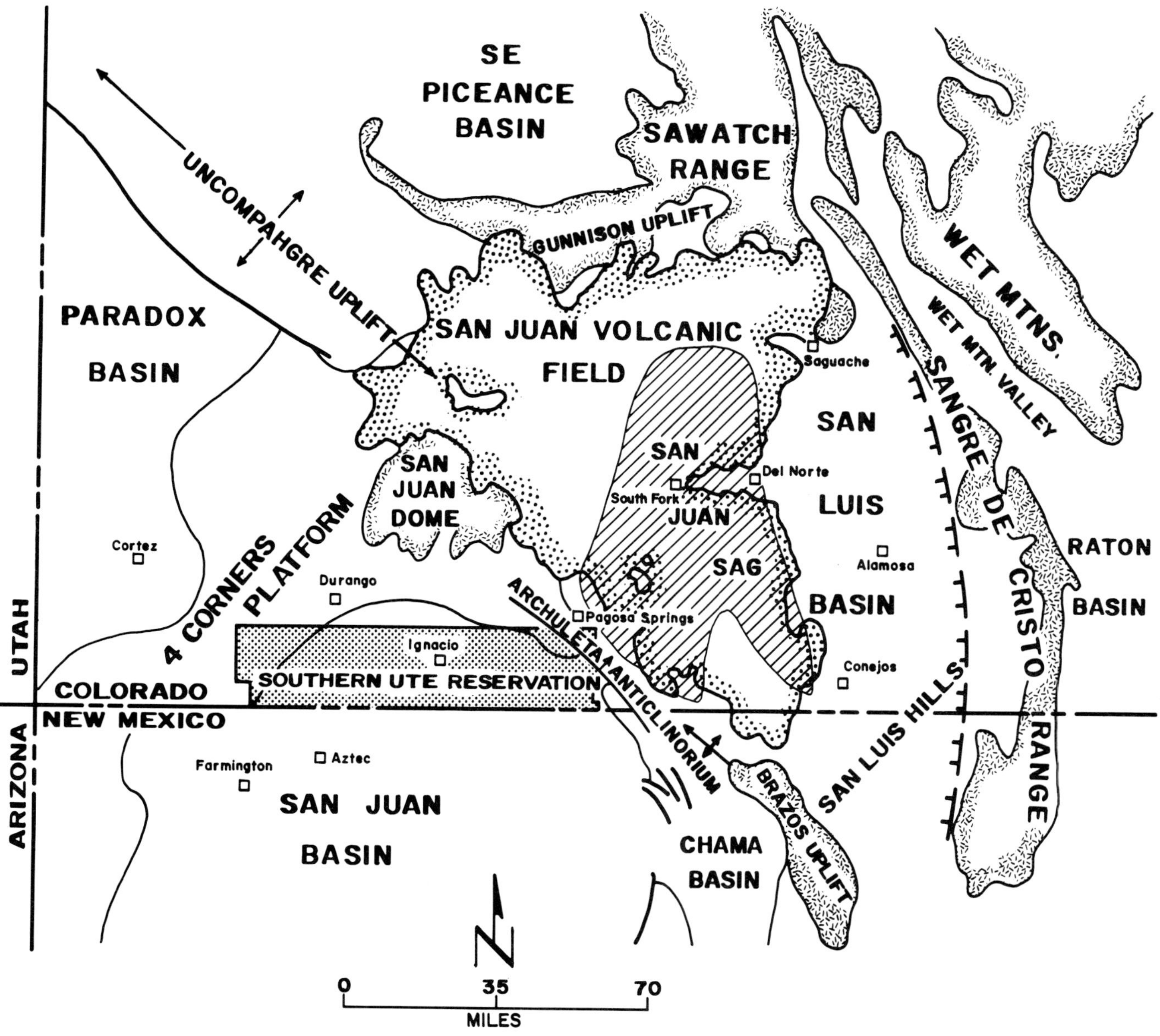

Figure 7. Regional map showing the location of the Southern Ute Indian Reservation in relationship to the structural configuration of southwest Colorado (Modified from Gries, 1985).

The ladies and gentlemen of the Durango Wheel Club enjoy an outing at Baker's Bridge, June 16, 1895. Near here, a generation before, stood Animas City (No. 1). Its founder, Charles Baker, also left his name at Baker's Park, site of Silverton. In June, 1895, two local stores advertised bicycles at prices ranging from $40 to $100. (Photo and caption courtesy of the Durango Library.)

Stratigraphic Cross Section Showing Upper Cretaceous Rocks Across the San Juan Basin, New Mexico and Colorado

C. M. Molenaar[1]

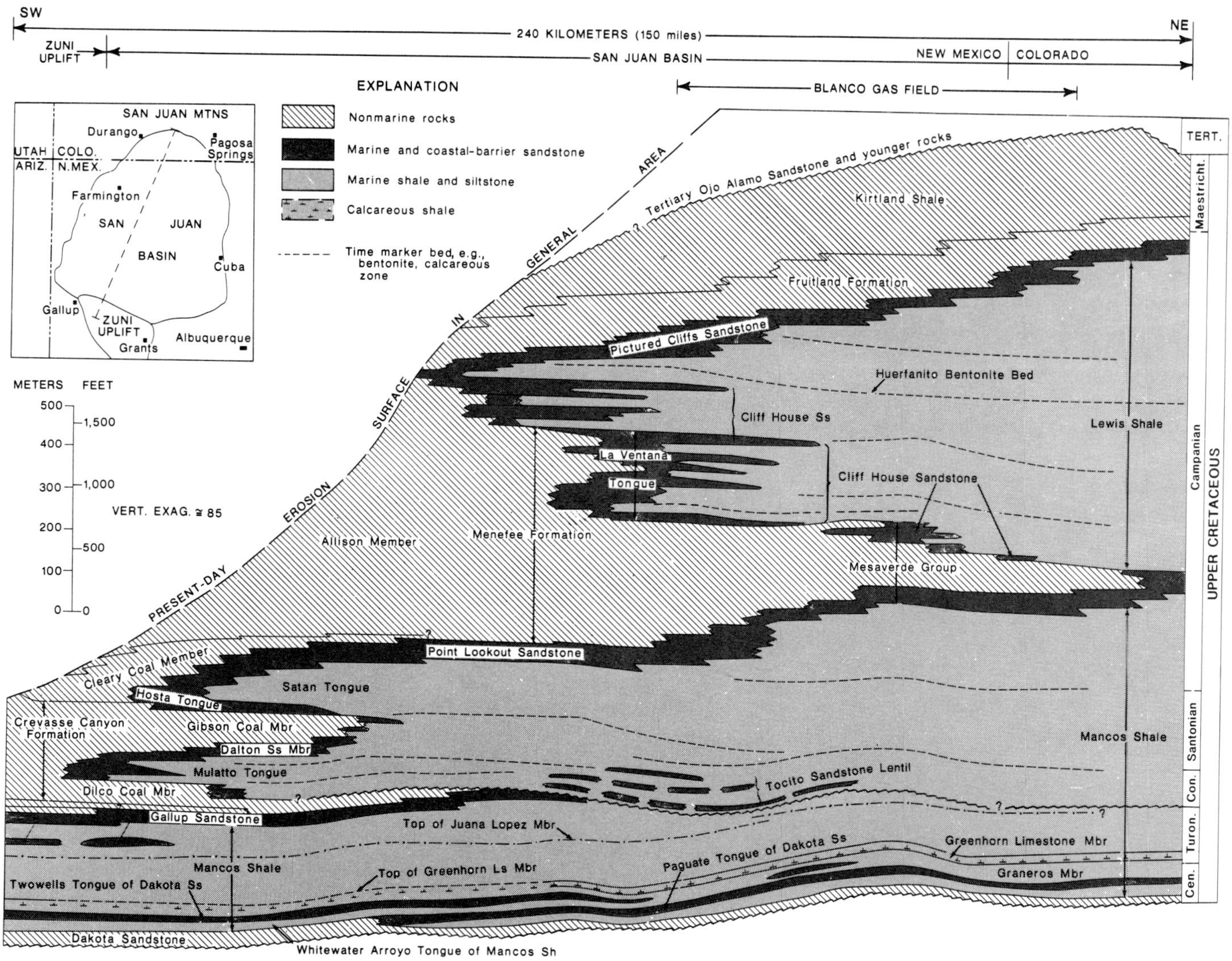

Figure 1. Northeast-trending stratigraphic cross section showing Upper Cretaceous rocks of the San Juan Basin of New Mexico and Colorado. Location of line of section is shown on inset map in upper-left corner. Control for section is based on detailed measured outcrop sections and well-log interpretation. Cen. = Cenomanian, Turon. = Turonian, Con. = Coniancian, Maestricht. = Maestrichtian

[1] U. S. Geological Survey
Box 25046, MS 940, DFC
Lakewood, Colorado 80225

"In the good old summer time" men wore straw hats and rode No. 5 as it passed the Newman Block, built by Charles Newman, using part of his Rico mining profits. In winter, sides replaced the curtains to keep passengers somewhat warmer. The banners indicate the Knights of Pythias were having a celebration. (Photo and caption courtesy of the Center of Southwest Studies.)

Geometry and Depositional Environment of Fruitland Formation Coal Beds, San Juan Basin, New Mexico and Colorado: Anatomy of a Giant Coal-Bed Methane Deposit

James E. Fassett[1]

ABSTRACT

Coal-bed methane resources in Fruitland Formation coal beds, San Juan Basin, New Mexico and Colorado, have recently been estimated to exceed 50 trillion cubic feet of gas. This enormous resource is only beginning to be tapped in the northern part of the basin, principally on the Southern Ute Indian Reservation in southwest Colorado. In order to develop exploration plans for the production of this resource in the most efficient and orderly manner, the detailed geometry of individual Fruitland coal beds must be carefully mapped. Previously published studies relate environments of deposition of the Fruitland and associated rocks to Fruitland coal-bed geometry throughout the San Juan Basin and provide guidelines that should make such coal-bed mapping easier and more accurate. The principal guidelines are: 1) Coal beds can only be correlated by using the Huerfanito Bentonite Bed of the Lewis Shale as a datum. 2) Depositional strike of Fruitland coals trended northwest. 3) Fruitland coals occur in an en echelon arrangement and rise 1,200 feet stratigraphically from southwest to northeast across the basin. 4) Coals generally thicken and terminate abruptly northeastward. 5) Coals generally thin and split southwestward. 6) Along depositional strike, coals may terminate abruptly against or be cut out by stream-channel deposits.

INTRODUCTION

The San Juan Basin of northwest New Mexico and southwest Colorado (fig. 1) contains large tonnages of coal in several Upper Cretaceous rock units. One of these units, the Fruitland Formation, has been described in detail in a paper by Fassett and Hinds (1971). That study showed that the distribution of the thickest total coal in the Fruitland seemed to coincide with the locations of large vertical buildups of the underlying regressive-marine Pictured Cliffs Sandstone. Because of this relationship, it was possible to develop a model describing how the distribution of coal in the Fruitland Formation was closely tied to the depositional history of the Pictured Cliffs Sandstone.

In this report I present a summary of the Fruitland coal depositional model of Fassett and Hinds (1971) and discuss the basinwide distribution and geometry of the Fruitland coal beds to illustrate some of the problems in coal-bed correlation and present apparent physical constraints on the movement of coal-bed methane through Fruitland coal beds and associated rocks throughout geologic time. Most of the data, all of the illustrations, and some of the text in this report are from Fassett and Hinds, 1971 and more recently, Fassett, 1986. (This report is a slightly revised version of a paper by Fassett published in 1987.)

[1] U.S. Geological Survey
Box 25046, MS 939, DFC
Lakewood, Colorado 80225

GENERAL GEOLOGY

The central part of the structural San Juan Basin is outlined in figure 1 by the outcrop of the Fruitland Formation. The basin outline is roughly circular in shape, except for the linear north-trending east side. The basin structure is asymmetric (fig. 2) with its northwest-trending axis located near, and parallel to, the northeast rim of the basin. Dips around the northwest, north, and east parts of the basin are 20-40 degrees, whereas dips on the west and south sides are quite gentle and average less than a degree. The diameter of the basin is about 100 mi and its surface area is about 7,500 mi^2.

Upper Cretaceous rocks have produced most of the oil and gas in the San Juan Basin; they range from 0 to 6,000 feet thick and consist of cyclically interbedded marine and nonmarine rocks. These rocks represent a remarkably complete record of a series of transgressions and regressions of the western shoreline of the Western Interior Upper Cretaceous seaway into and out of the San Juan Basin area (the location of the basin relative to this seaway is shown on fig. 3). The Fruitland Formation and its coal beds are intimately associated with the last regression of the Western Interior seaway from the San Juan Basin area.

GAS AND OIL RESOURCES

The San Juan Basin is a vast storehouse of gas, oil, and coal. In the conterminous United States the San Juan Basin gas field is second in size only to the Hugoton field of Oklahoma, Texas, and Kansas. Total cumulative gas production in the basin (through 1982 for the New Mexico part and through

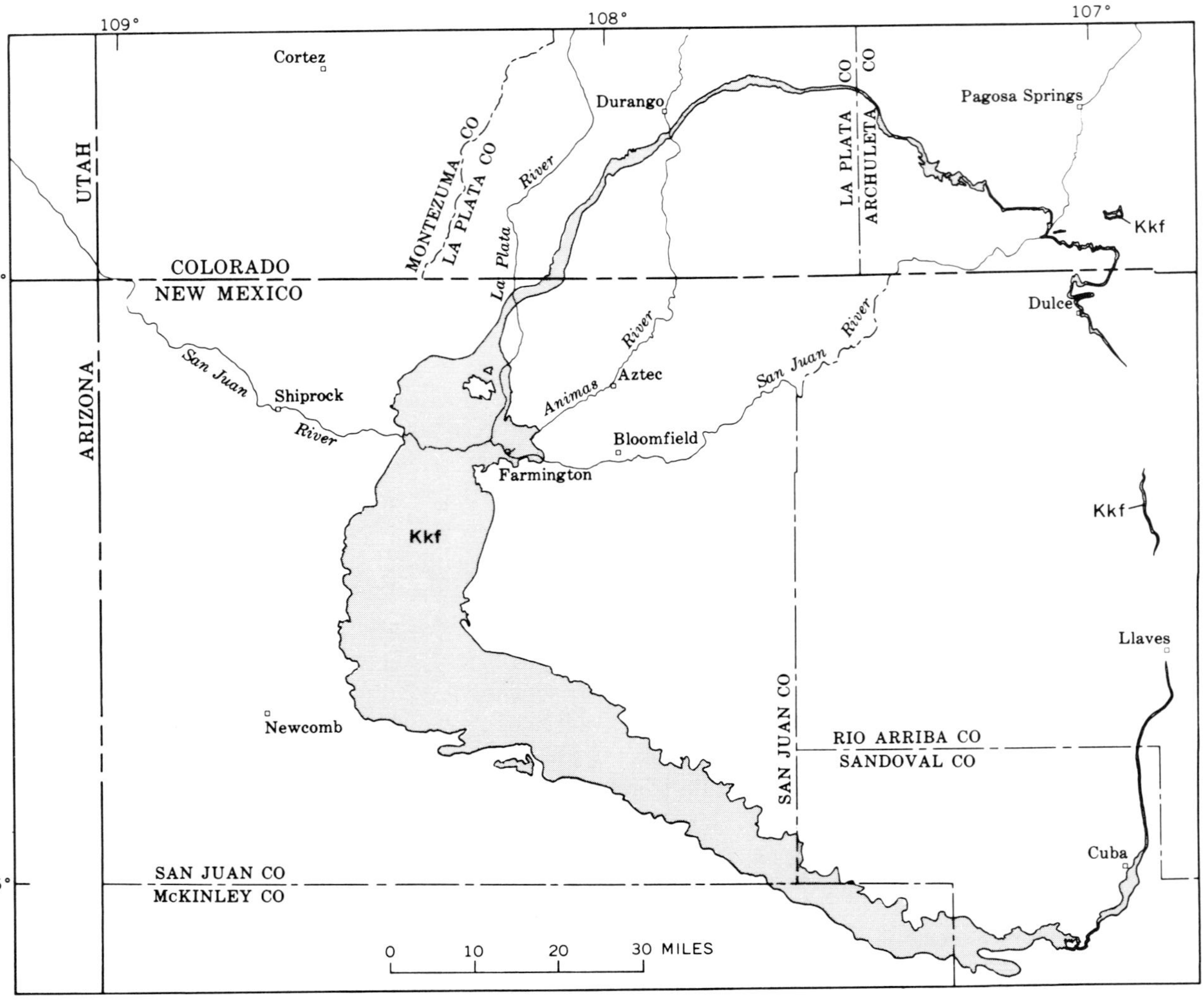

Figure 1. Index map showing the location of the San Juan Basin. Shaded area (Kkf) outlining the basin is the Fruitland Formation outcrop.

June 1983 for the Colorado part) was more than 14 trillion cubic feet of gas and 250 million barrels of oil. Nearly all of this production has come from conventional sandstone reservoirs in the New Mexico part of the basin. Of the 14 trillion cubic feet of gas produced, only about 90 billion cubic feet (about .6 percent) has come from the Fruitland Formation. About half of the Fruitland production has come from one field in Colorado, the Ignacio Blanco field. A large percentage of this Fruitland Formation gas has doubtlessly been coal-bed methane. (For detailed summaries of oil and gas production from the San Juan Basin, see reports by Fassett (1983) and Irwin (1983). Indeed, recent papers by Rice (1983) and Meissner (1984) suggest that most of the total gas produced so far in the San Juan Basin is actually coal-bed methane that has migrated into sandstone reservoirs.

Two recent papers by Meissner (1984) and Choate and others (1984) estimate that the Fruitland Formation contains 26 and 31 trillion cubic feet, respectively, of coal-bed methane in the San Juan Basin. More recent studies by Kelso and others (1987) and Kelso and Wicks (this volume), indicate that the Fruitland Formation contains 50 trillion cubic feet of coal-bed methane. These estimates of the coal-bed methane production potential for the San Juan Basin are enormous and if only half of this gas is recovered, the total ultimate gas production for the San Juan Basin will be more than double old production estimates, which did not include coal-bed methane. In view of this potential, it is essential that the geometry of Fruitland coal beds is clearly understood so that future production of methane from Fruitland coal beds can be carried out in the most orderly and efficient manner possible.

FRUITLAND FORMATION

Physical Characteristics

The stratigraphic relations of the Fruitland Formation to associated rocks are shown in figure 4. Throughout most of the basin, the Fruitland is underlain by the Pictured Cliffs Sandstone; throughout the western two-thirds of the basin it is overlain conformably by the Kirtland Shale. In the eastern part of the basin the Fruitland Formation is unconformably overlain by early, but not earliest, Tertiary Ojo Alamo Sandstone (Fassett and others, 1987). Figure 1 shows that the Fruitland is missing in two areas on the eastern edge of the basin where the formation has been removed by the pre-Ojo Alamo erosion cycle.

Where overlain by the Kirtland Shale, the Fruitland is 300 to 500 feet thick, and generally thickens to the northeast. The

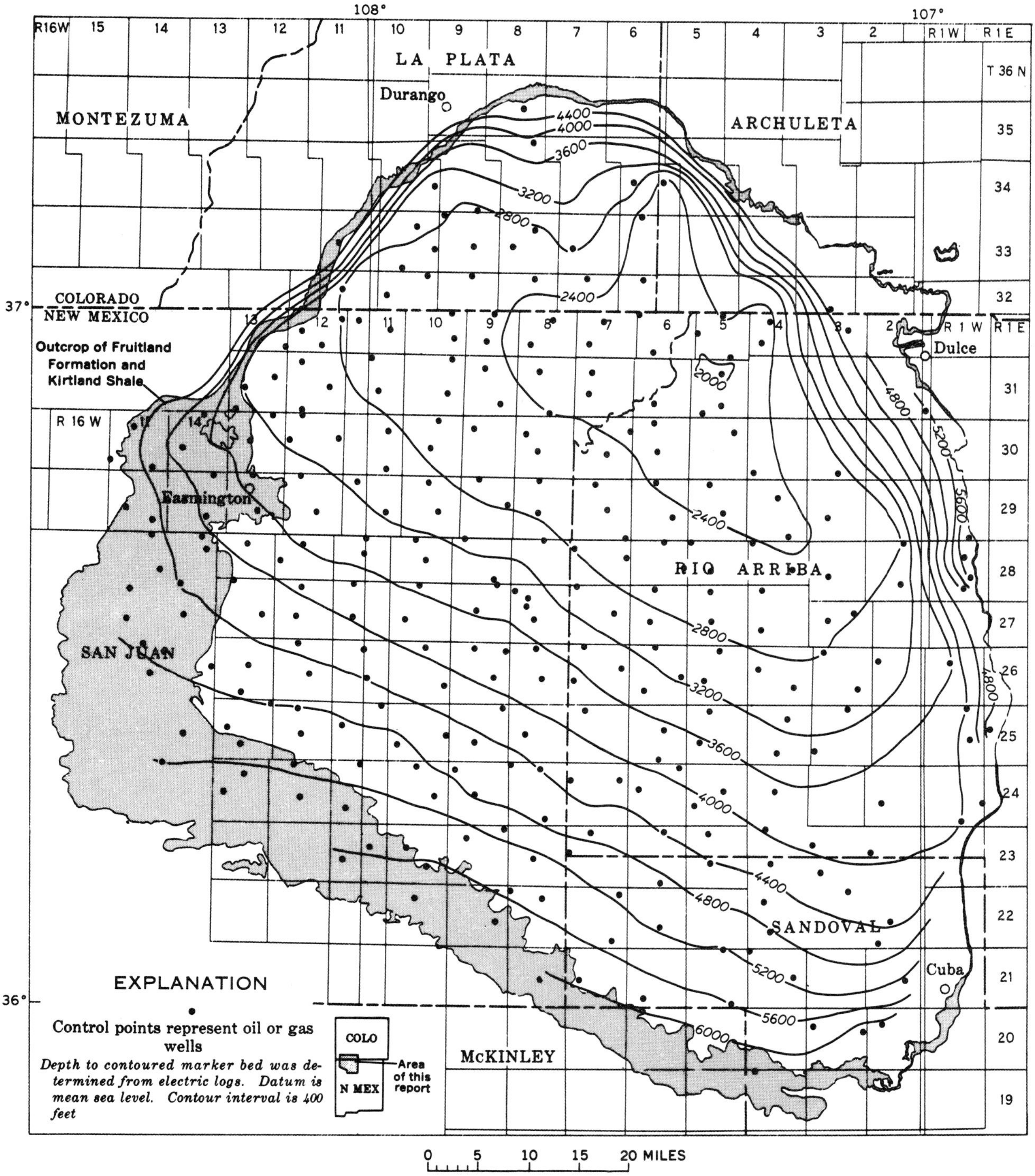

Figure 2. Contour map of the Huerfanito Bentonite Bed of the Lewis Shale.

contact between the Fruitland and Kirtland is gradational and often difficult to locate, both on the outcrop and on geophysical logs in the subsurface. As no apparent continuous lithologic break separates the two units basin wide, the contact is generally placed at the top of the highest coal or carbonaceous shale bed in the section (Fassett and Hinds, 1971). Even this criterion is fallible because thin carbonaceous shales or coaly beds are locally present in the lower part of the Kirtland.

Where overlain by the Ojo Alamo Sandstone in the eastern part of the basin, the upper contact of the Fruitland is an unconformity. The Fruitland thins eastward beneath the Ojo Alamo and is truncated along or near the east side of the basin. The relationship between the Fruitland Formation and the underlying Pictured Cliffs Sandstone must be understood to relate Fruitland coal-bed geometry to environments of deposition and to try and create a predictive model that will allow for better coal-bed correlation; that relationship is discussed at length in a later section of this report.

Coal Beds

Fruitland Formation coal beds range from thin stringers to beds more than 40 feet thick. Coal zones may contain several

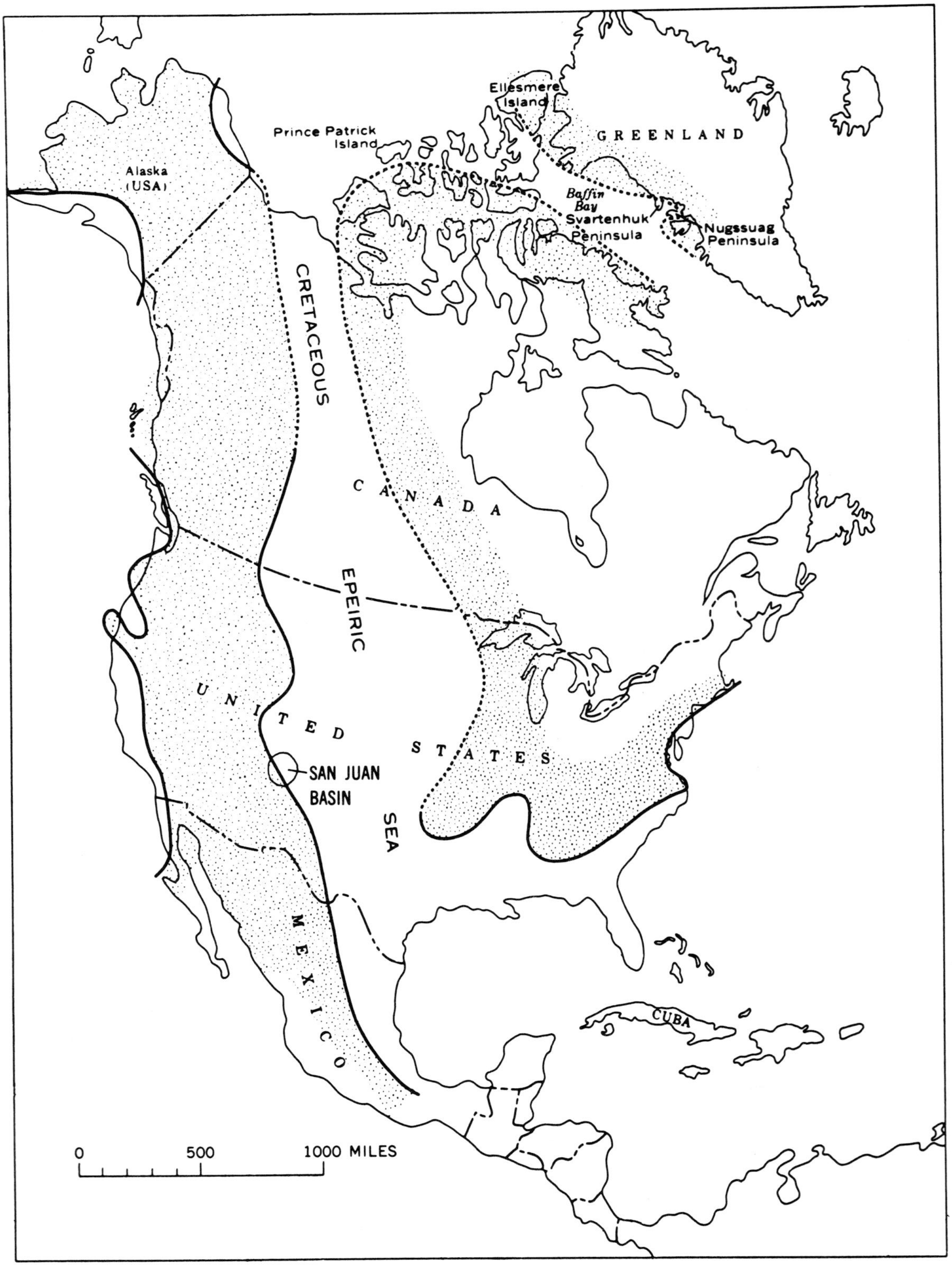

Figure 3. Probable configuration of the North American epeiric seaway at the time that the Upper Cretaceous rocks of the San Juan Basin were being deposited.

coal beds that total as much as 80 feet of coal. Heating values of the coal generally range from about 9,000 Btu to more than 13,000 Btu, on an as-received basis (Fassett and Hinds, 1971, and Hinds, 1965). These values are for pure coal exclusive of non-coal partings. Channel samples or core samples through an entire coal bed generally yield lower values, between 9,000 and 10,000 Btu, because of the ash content contributed by non-coal partings in most coal beds. Vitrinite reflectance studies (Rice, 1983) indicate that Fruitland coal rank ranges from subbituminous to medium- to low-volatile bituminous.

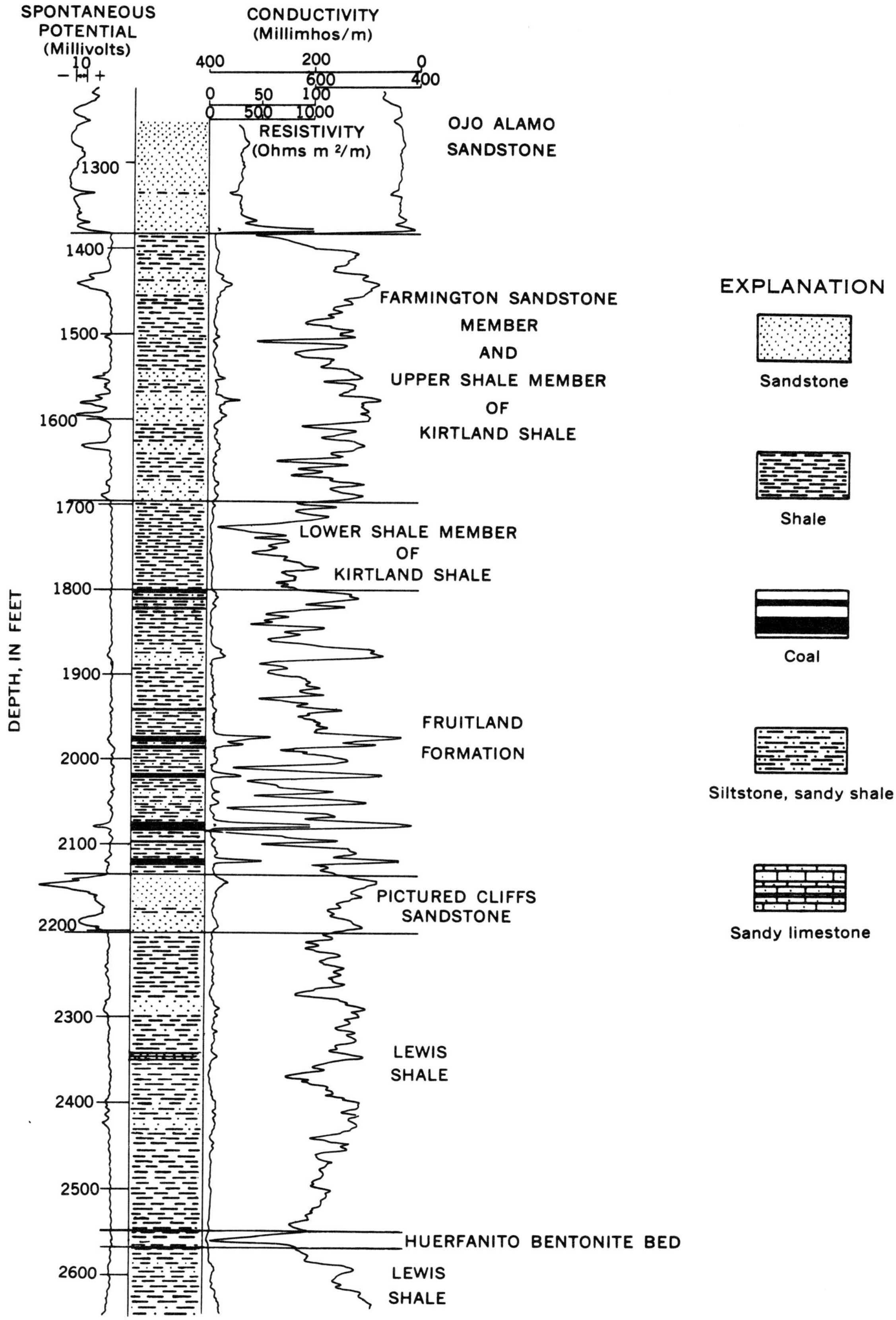

Figure 4. Induction-electric log and lithologic column of the type well of the Huerfanito Bentonite Bed of the Lewis Shale showing the interval from below the Huerfanito through the lower part of the Ojo Alamo Sandstone. Lithologies are based on an interpretation of the three geophysical logs shown.

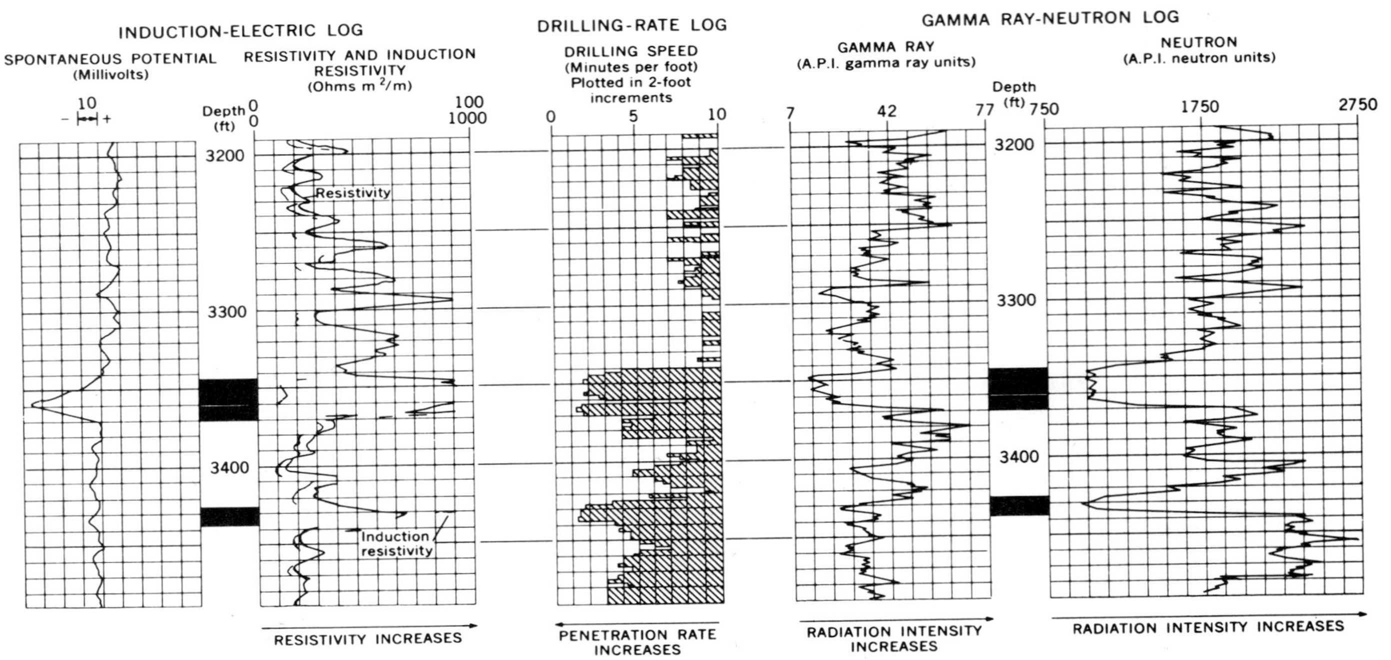

Figure 5. Induction-electric, gamma ray-neutron, and drilling-rate logs through the lower part of the Fruitland Formation, showing characteristic responses to coal.

Fruitland coal is non-coking except for the Chimney Rock mine area in the northeast part of the basin. See the discussion at the Chimney Rock mine stop (mile 49.4) in the second day field trip road log elsewhere in this guidebook for more information on the Fruitland coal at the Chimney Rock mine.

The ash content of Fruitland coals is generally high, even in samples free of macroscopic non-coal partings. Ash content for such samples generally ranges from slightly less than 8 percent to as high as 30 percent. There seems to be a general increase in ash content eastward across the basin. Moisture content of the coal averages from 2 to 5 percent. Sulfur content is low, averaging less than 1 percent, with a range of 0.5 to 2.5 percent. Amounts of fixed carbon and volatile matter vary across the basin, with a general increase in fixed carbon from around 50 percent in the southwestern half to more than 70 percent in the area of the present structural axis of the basin (Fassett and Hinds, 1971). This trend indicates that fixed carbon in Fruitland coals is generally related to the present depth of burial of the coal beds, although Rice (1983) points out that coal with the highest known vitrinite reflectance in the basin (1.45 percent) comes from a point slightly north of the present structural axis of the basin.

The distribution of Fruitland coal throughout the San Juan Basin was determined by Fassett and Hinds (1971) to a large extent through the use of geophysical logs from oil or gas drill holes. Figure 5 shows how coal beds appear on induction-electric, gamma ray-neutron, and drilling rate logs from a well in the northeast part of the San Juan Basin. These kinds of logs were run in large numbers in wells throughout the basin and still constitute the bulk of the San Juan Basin log file. Since 1971 large numbers of density logs have been added to the files, which make the identification of coal beds in the subsurface easier and more certain but coal-bed thickness measurements are probably slightly less precise on oil and gas drill hole density logs than on resistivity logs.

In conducting their subsurface coal evaluation study, Fassett and Hinds (1971) were able to confirm directly the presence and thickness of coal in the subsurface at 60 sites by being present on oil and gas drill rigs as holes were drilled through Fruitland Formation coal beds. Coal thicknesses were determined on the basis of noting the depths where coal cuttings were discharged from the drill hole and, more precisely, on the basis of the drilling-rate record obtained at the drill rig (see fig. 5).

Later, when the geophysical logs (usually electric) for these wells became available, it was possible to directly compare the sample log and drilling-rate record with the electric log so as to make a comparison of the top and bottom of a coal bed on these independent records (fig. 5). Having this suite of 60 carefully evaluated logs scattered across the basin made it possible to measure coal in intervening areas on the basis of geophysical logs alone with a high degree of confidence.

In hole GB-1 (Gasbuggy 1), in the northeast part of the San Juan Basin, the Fruitland Formation was cored. A direct comparison of the coal depths and thicknesses measured from that core with a full suite of geophysical logs from the same hole further confirmed the accuracy of coal bed identification and measurement based on geophysical logs. (See reports by Fassett, (1968a, 1968b) and Fassett and Hinds (1971) for descriptions of the Gasbuggy core and a depiction of some of the geophysical logs from hole GB-1.)

Figure 6 is an isopach map of Fruitland Formation total coal thicknesses throughout the San Juan Basin. Thicknesses shown at each control point represent the total of all coal beds 2 feet or more in thickness at that point. Each thickness value may represent anywhere from one to as many as a dozen or more coal beds. A total of 324 control points (representing both subsurface and outcrop measurements) were used to construct the isopach map.

The coal distribution shown on figure 6 reveals an interesting pattern consisting of a thick pod of coal in the southwest part of the basin; a swath of thin coal trending northwestward

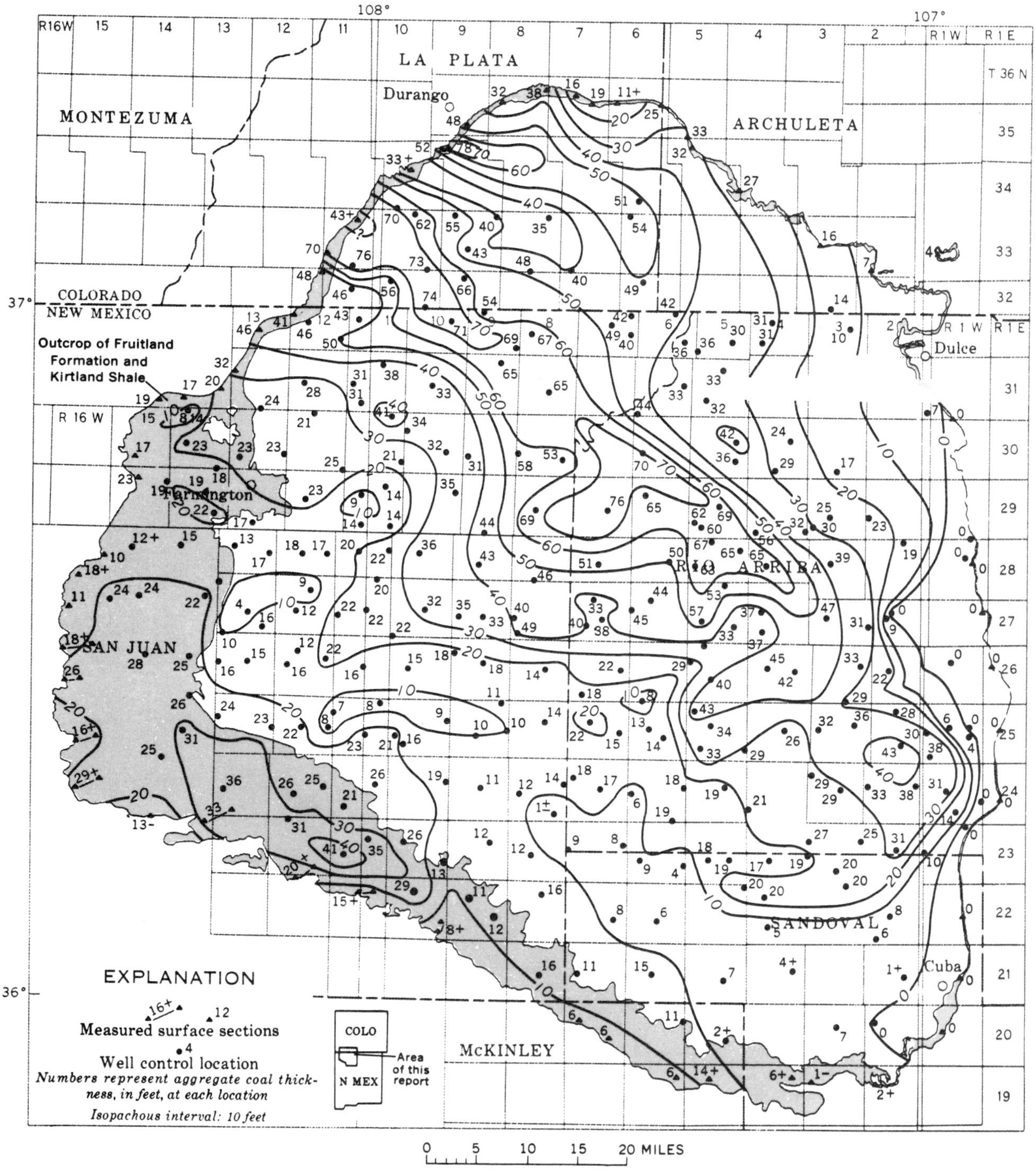

Figure 6. Isopach map of total thickness of coal in the Fruitland Formation.

across the southwestern part of the basin; a band of thicker coal across the central basin area; and an area of thin coal in the northeasternmost part of the basin. What do these coal thickness patterns mean and what might they be telling us about the environments of coal deposition during Fruitland time? To respond to this question, a brief discussion of the underlying Pictured Cliffs Sandstone is required.

PICTURED CLIFFS SANDSTONE

The Pictured Cliffs Sandstone underlies the Fruitland Formation throughout most of the San Juan Basin. The Pictured Cliffs is a very fine to fine-grained quartzose sandstone that, at most exposures, is made up of an upper part consisting of one or two ledges of massive sandstone and a lower part consisting of interbedded sandstones and shales. Sandstone beds in the lower part become thinner and finer grained downward in the section. The Pictured Cliffs Sandstone was deposited during the final regression of the Late Cretaceous epeiric seaway from the San Juan Basin area; it was deposited as a strandline sandstone as the sea retreated northeastward out of the area. Figure 3 shows the paleogeography of North America during deposition of the Pictured Cliffs Sandstone.

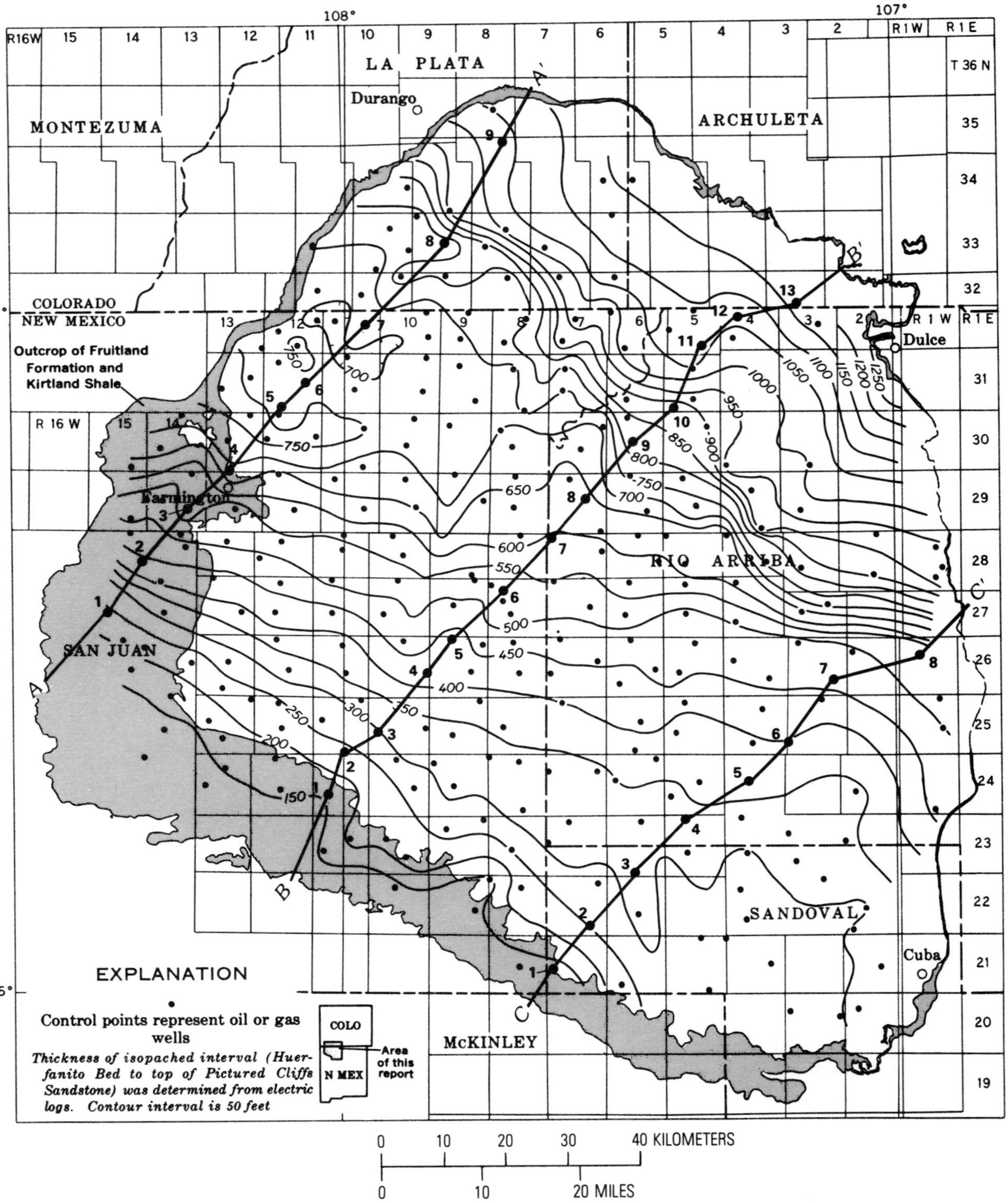

Figure 7. Isopach map of the interval between the Huerfanito Bentonite Bed of the Lewis Shale and the top of the Pictured Cliffs Sandstone. Lines of cross sections A-A′, B-B′, and C-C′ of figure 8 are shown.

The model for deposition of the Upper Cretaceous rocks of the San Juan Basin associated with transgressing and regressing shorelines over a period of some 25 million years was first described by Sears and others in 1941 in a classic, landmark paper. Their model, simply put, supposed a continuously subsiding trough (the seaway of fig. 3) receiving sediment at a varying rate. A high rate of sediment influx resulted in outbuilding of the shoreline (regression), and a low rate of sediment influx resulted in landward advance of the shoreline (transgression). Thus, the Pictured Cliffs Sandstone strandline regressed from the southwest part to the northeast part of the San Juan Basin area in response to an increase in sediment being delivered to the shoreline by northeast-flowing streams. Figure 7 is an isopach map of the interval between the Huerfanito Bentonite Bed of the Lewis Shale and the top of the Pictured Cliffs Sandstone. The isopach lines are probably a fairly

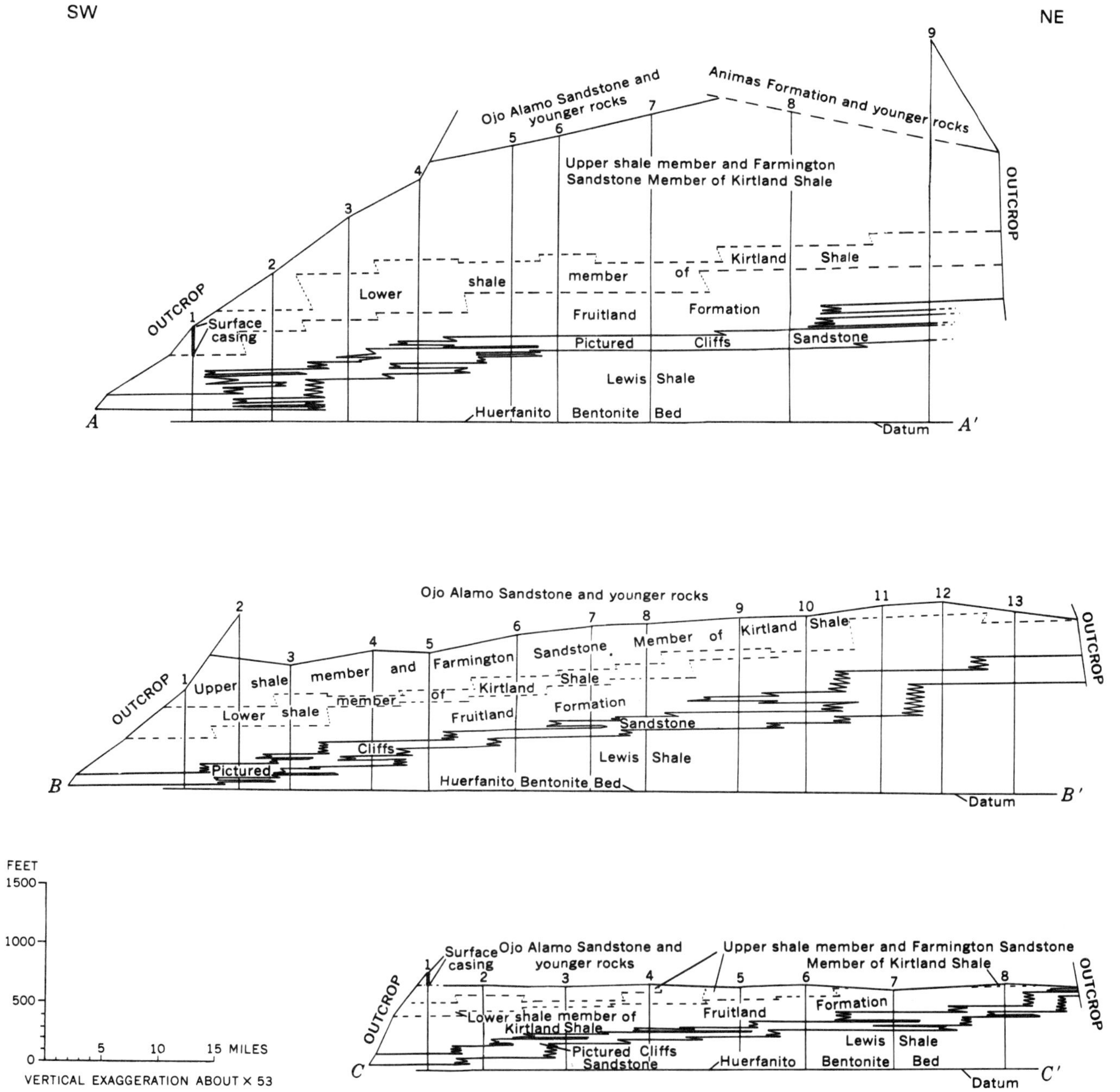

Figure 8. Northeast-trending stratigraphic cross sections showing the northeastward stratigraphic rise of the Pictured Cliffs Sandstone and associated rocks. (Lines of cross sections are shown on figure 7.)

accurate portrayal of the positions of the Pictured Cliffs Sandstone shorelines throughout the time the sea was retreating across the basin area.

Figure 8 shows three stratigraphic cross sections oriented northeastward across the basin showing the relations of the Lewis Shale and its Huerfanito Bentonite Bed Member, Pictured Cliffs Sandstone, Fruitland Formation, and overlying rocks. These sections show the time-transgressive nature of the Pictured Cliffs Sandstone, which becomes younger northeastward across the basin. These sections also show that the sea did not retreat steadily and evenly across the basin. For example, between wells 8 and 9 on cross section A-A′ and 10 and 11 on B-B′, there is a stratigraphic rise in the top of the Pictured Cliffs of about 200 ft. These rises indicate that, for a time, the shoreline in these areas ceased to regress northeastward and stayed in the same general geographic area, building the strandline sands into a thick vertical stack. Figure 7, the isopach map of the interval from the Huerfanito Bed to the top of the Pictured Cliffs Sandstone also shows this large stratigraphic rise of the top of the Pictured Cliffs Sandstone along a northwest trend across the north-central part of the basin.

FRUITLAND COAL DEPOSITION MODEL

In 1941, Sears and others suggested that thick coal deposits should form landward of upbuilding beach-sand deposits. That suggestion was only theoretical at the time it was made

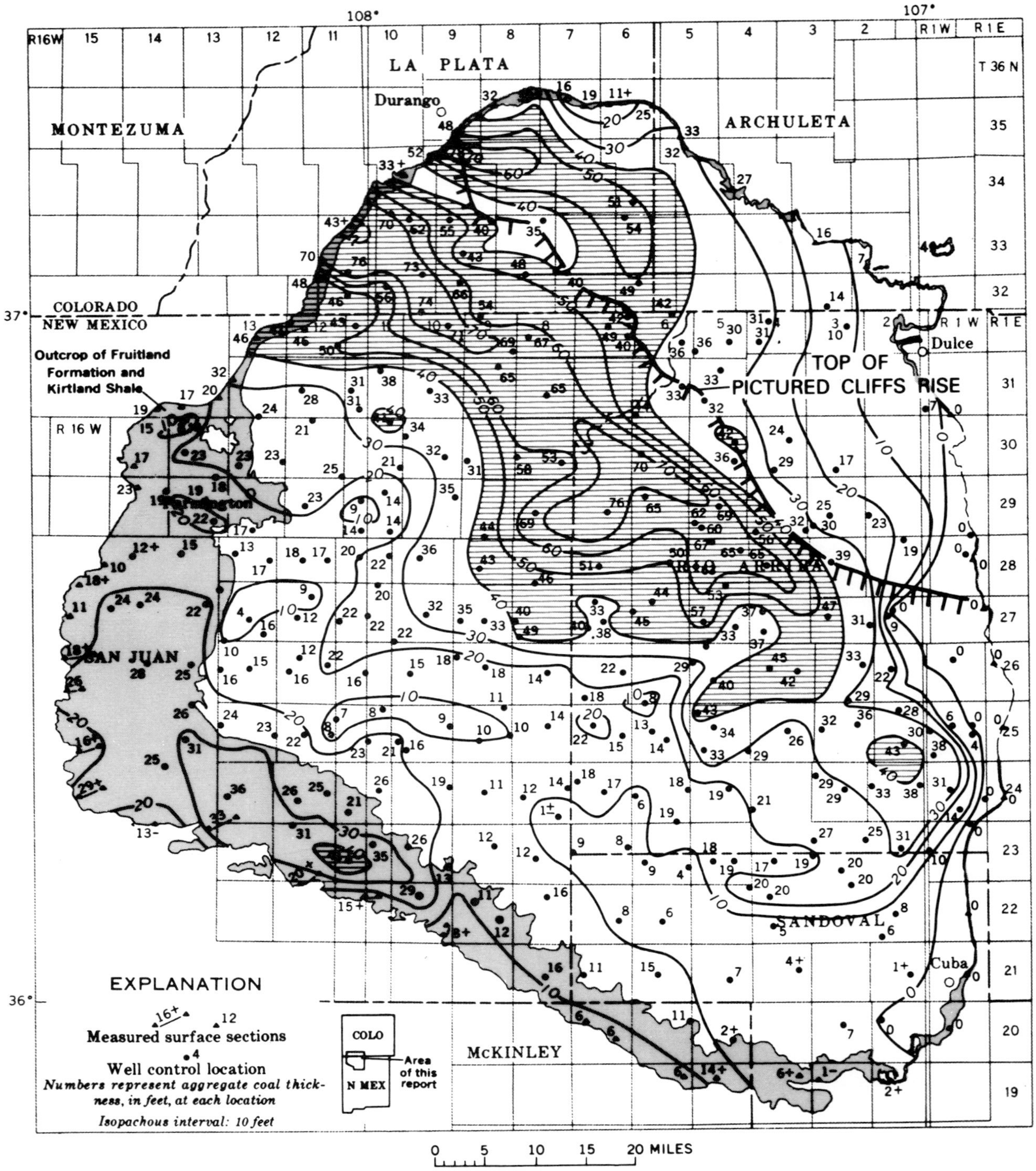

Figure 9. Composite map showing the total-coal-thickness isopachs for Fruitland Formation coal (from figure 6) and the top of the largest stratigraphic rise of the Pictured Cliffs Sandstone in the north-central part of the basin (from figure 7). This composite map shows that the area of thickest Fruitland coal is located southwest of the greatest stratigraphic rise of the Pictured Cliffs. Areas where total-coal thickness is greater than 40 feet are patterned.

because such an occurrence had not actually been documented. Recently, however, with the availability of thousands of geophysical logs through the Fruitland Formation throughout the San Juan Basin, Fassett and Hinds (1971) were able to demonstrate that the theoretical prediction of Sears and others (1941) is valid for the Pictured Cliffs-Fruitland rocks and that thick Fruitland coals did actually develop shoreward of upbuilding, shoreface Pictured Cliffs sands. (It is important to note, however, that this model is not universal, not even in the San Juan Basin, as demonstrated by Fassett in 1986.)

Figure 9 is a composite map of the Pictured Cliffs-Huerfanito isopach map (fig. 7) and the total-Fruitland-coal isopach map (fig. 6) showing that the thickest zone of Fruitland coal is located southwest of the largest vertical buildup of the Pictured Cliffs Sandstone. The correspondence is not perfect, but considering the variability in natural systems, it is

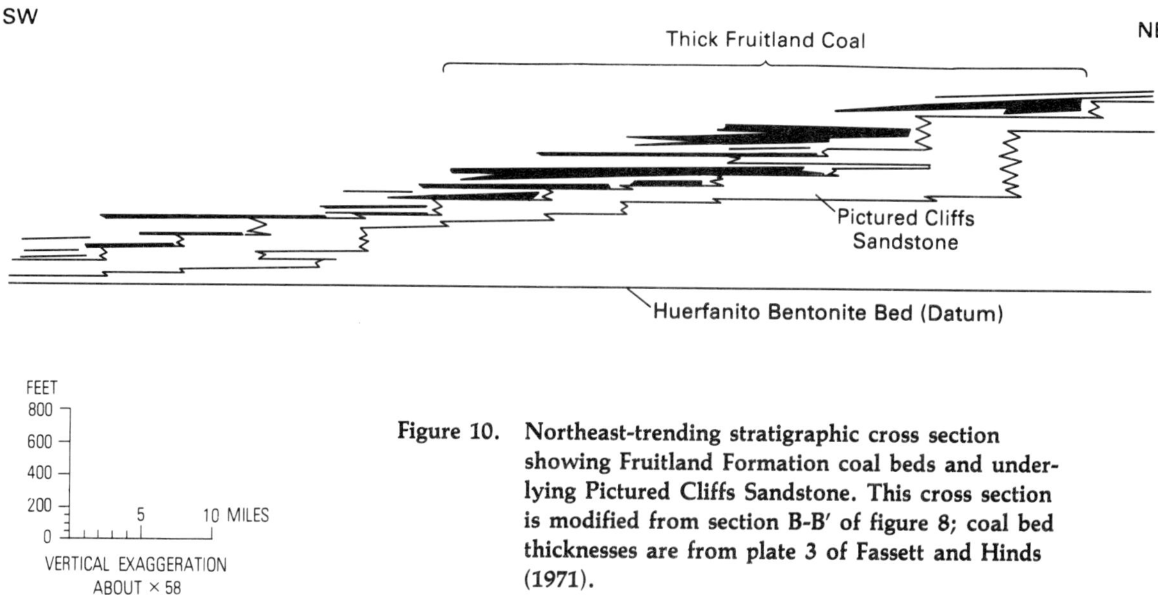

Figure 10. Northeast-trending stratigraphic cross section showing Fruitland Formation coal beds and underlying Pictured Cliffs Sandstone. This cross section is modified from section B-B′ of figure 8; coal bed thicknesses are from plate 3 of Fassett and Hinds (1971).

impressive. One of the problems with this depiction is that the total-coal isopach map represents several coal beds in the Fruitland that occur through a stratigraphic interval as much as 300 feet thick. A series of isopach maps of individual coal beds in the Fruitland would probably demonstrate more clearly the relationship of thick Fruitland coal beds to this large stratigraphic rise in the Pictured Cliffs Sandstone. Such a series of maps, however, is beyond the scope of this report.

Another way to portray the relations of thick Fruitland coals to stratigraphic rises in the Pictured Cliffs is in cross section. Figure 10 is a northeast-trending cross section showing the Fruitland Formation coal beds and the Pictured Cliffs Sandstone. The thickest Fruitland coals are clearly related to the largest stratigraphic rises of the Pictured Cliffs. Figure 11 is a diagrammatic portrayal of how thick coal beds were developed in the Fruitland Formation in association with vertical build-ups of strandline sands in the Pictured Cliffs Sandstone.

GEOMETRY OF FRUITLAND COAL BEDS

Correlation Problems

The realization that Fruitland Formation coal beds were difficult to correlate dates back to the earliest surface studies of the formation in the early 1920s. Fassett and Hinds (1971) provided a chronicle of the problems early coal geologists experienced in trying to correlate Fruitland coal beds on the outcrop throughout the San Juan Basin. The early workers generally concluded that Fruitland coals were so discontinuous and so variable in thickness over short distances as to make reliable widespread correlation of Fruitland coal beds difficult, if not impossible. The detailed subsurface studies of Fassett and Hinds (1971) tended to confirm the fact that Fruitland coals were not easy to correlate. A recent publication by Sandberg (1986), however, demonstrates that Fruitland Formation coal-bed correlation is possible in the subsurface if a large number of closely-spaced logs are used. The Sandberg study covered a relatively small area (about 400 square miles) in the southwest part of the basin. A more recent paper by Sandberg (this volume), correlates Fruitland coal beds on the Southern Ute Indian Reservation in the northern San Juan Basin of Colorado.

Even though Fruitland coal bed correlation is difficult, our vastly improved knowledge of the details of the physical distribution of these coals, both geographic and spatial—especially as related to the general model explaining the environment of deposition of these coals (Fassett and Hinds, 1971, Fassett, 1986)—provides powerful new tools for predicting the extent, continuity, and thickness of these coal beds. The following material adapted from Fassett and Hinds (1971) summarizes our knowledge of Fruitland coal beds throughout the San Juan Basin.

The Fassett and Hinds (1971) study of Fruitland coal beds was done mainly through the use of logs from drill holes spaced from 2 to more than 6 miles apart throughout the San Juan Basin. The same coal-bed correlation problems found on the outcrop by previous workers were also found to exist throughout the basin in the subsurface, in spite of the fact that the subsurface data was so much more plentiful. Correlation of coal beds at angles to the generally-northwest-trending Pictured Cliffs paleoshorelines is complicated because coal beds wedge out abruptly northeastward against intertonguing sandstones of the Pictured Cliffs Sandstone; correlation to the southwest is also difficult because the beds are first split by, and then grade into, nonmarine sandstones and shales, which were deposited landward and along the trailing edges of Fruitland swamps. These facies transitions generally take place within a distance of only a few miles. Correlation of coal beds in a northwest direction is often more reliable because this is the trend of the long axis of most coal bodies and most coal beds retain their identity farther in this direction. Between widely spaced control points, however, correlation along the northwest trend has proved only slightly more feasible than correlation at angles to it.

Some of the causes of lack of continuity of Fruitland coal beds are: (1) Channel sandstones and associated flood plain deposits were laid down by streams that flowed between coal

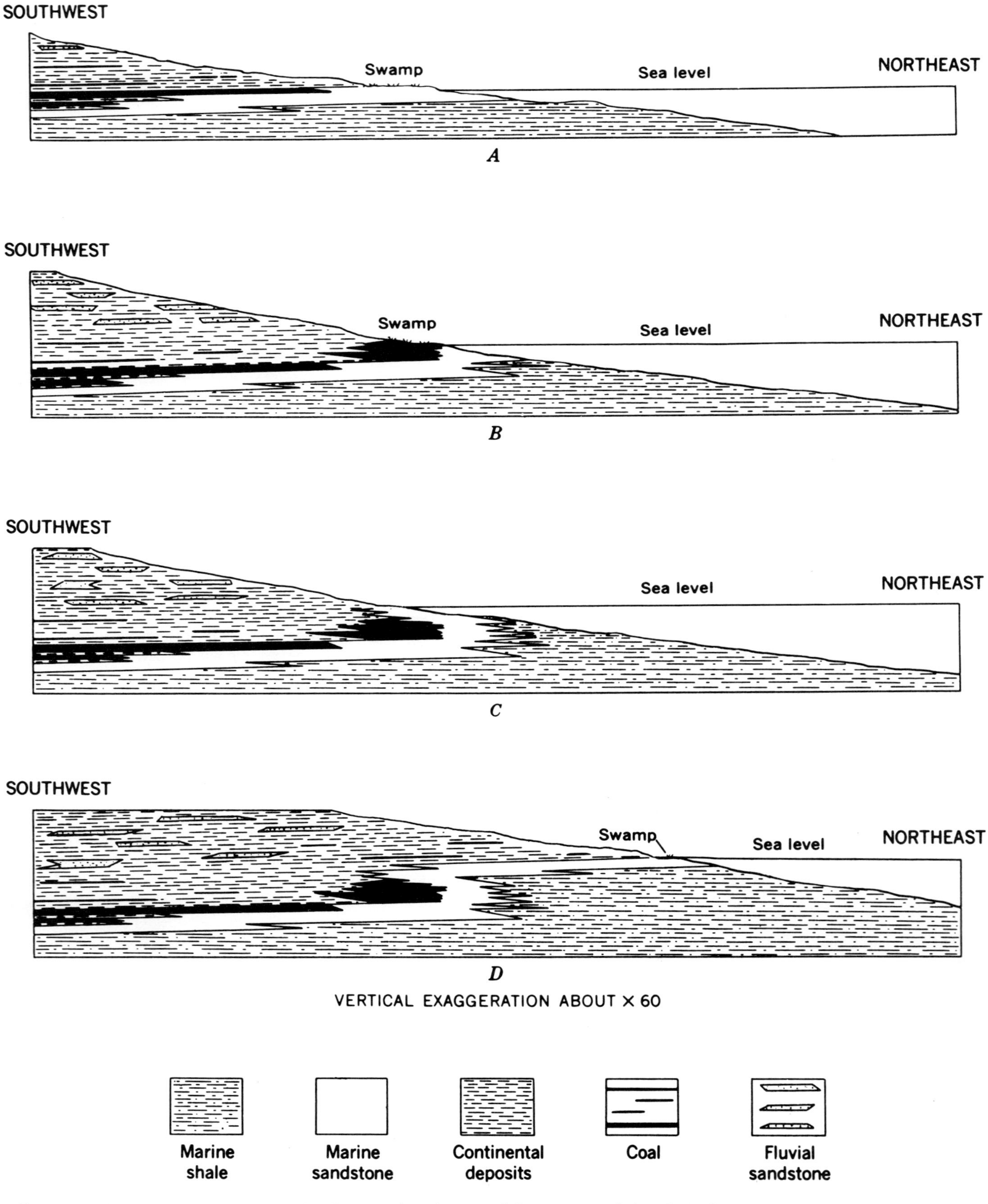

Figure 11. **Diagrammatic cross sections showing the relations of the continental, beach, and marine deposits of Pictured Cliffs time after (A) shoreline regression, (B) shoreline stability, (C) shoreline transgression, and (D) shoreline regression.**

swamps to the coast. These deposits were contemporaneous with adjacent coal-swamp deposits and thus the coal beds thin or pinch out toward the channels by lateral facies change. The horizontal separation of time-equivalent coals adjacent to such channel deposits may be several thousand feet. Other smaller streams eroded swamp deposits and the deposits within such channels now represent breaks in coal-bed continuity of no more than a few tens to a few hundred feet. (2) Scattered areas among the swamps that had a few feet of topographic relief were apparently above the swamp water level and were too well drained to allow preservation of organic material. These areas may have been localities of beach dunes or bars or the preserved levees of abandoned stream channels. The coal beds thin or split over these areas or terminate on their flanks. (3) At other places among the coastal swamps, fine clastics filled lakes and lagoons. Such areas are reflected in the present deposits by coal beds that terminate laterally into a sequence of carbonaceous mudstone and shale. (4) The boundaries between flood plain and coal swamps were irregular due to breakouts of streams from their channels. The resultant crevasse-splay deposits extended into the swamplands, resulting in gaps in the continuity or abrupt changes in the thickness of some coal beds, particularly those in the upper part of the Fruitland Formation. (5) Local irregularities of the seaward edges of swamps in the form of small deltas, bays, and estuaries interrupted the continuity of coal-swamp deposits. (6) Differential compaction of sediments during diagenesis has probably caused some contemporaneously-deposited coal beds to now occupy different stratigraphic positions and beds of different ages to occupy the same stratigraphic plane.

To summarize, the band of backshore coal swamps in Fruitland time was generally continuous landward of the Late Cretaceous Pictured Cliffs shoreline. Individual swamps were generally limited in area and separated from one another by the kinds of paleogeographic features described above. These features shifted laterally through time, causing the coal swamps to migrate over all parts of the coastal area. The result was an averaging of the total amount of coal deposited in belts or zones paralleling the shore so that the total coal thickness in these belts is due more to the stability of the position of the shoreline rather than to the more transitory paleoenvironmental changes that occurred within the backshore-coal-swamp belt. Fruitland coals as a whole form a decipherable general pattern, but few individual coal beds are traceable at the surface or in the subsurface for long distances. Individual beds can be traced for from a few to at most about 25 miles.

There is no doubt that the most serious difficulty in correlating Fruitland Formation coal beds in the San Juan Basin, prior to the Fassett and Hinds study (1971), was the lack of a recognizable, basin-wide time plane or datum to which the geometry of Fruitland coals could be related. Such a datum, the Huerfanito Bentonite Bed of the Lewis Shale, was discovered, described, named, and traced throughout the basin in the Fassett and Hinds study of 1971. Using this datum it was possible for the first time to understand the three-dimensional distribution of Fruitland coals throughout the basin and to make confident, valid correlations of individual coal beds where that was possible, and equally importantly, to identify those areas where Fruitland coals did not correlate.

Spatial Distribution of Coal

As has been discussed above, the geographic distribution of Fruitland coal beds in the San Juan Basin, as portrayed on coal isopach maps, is a series of northwest-trending bands of thick and thin coal (Fassett and Hinds, 1971, Fassett, 1986) (see figures 6 and 9). When viewed in a stratigraphic cross section, using the Huerfanito Bentonite Bed as a datum (fig. 10), the third dimension or spatial distribution of the Fruitland coal beds becomes evident. Figure 10 shows that the Fruitland coal beds of the northeast part of the San Juan Basin are more than 1,000 feet stratigraphically higher than the Fruitland coal beds in the southwest part of the basin. (Figure 7 shows that the maximum stratigraphic difference in the base of the Fruitland, from southwest to northeast across the basin is 1,200 feet.) In an earlier report (Fassett, 1987) I showed that this stratigraphic rise in the level of the base of the Fruitland (top of the Pictured Cliffs Sandstone) represents approximately 1.2 m.y. In other words, the Fruitland coal beds in the northeast part of the basin are 1.2 m.y younger than the Fruitland coal beds in the southwest part of the basin.

Figure 10 also shows that the Fruitland coals form an en echelon pattern with progressively younger (stratigraphically higher) coal beds progressively overlapping older (stratigraphically lower) coal beds from southwest to northeast across the basin. Individual coal beds parallel the underlying time plane of the Huerfanito Bed and many of the coals terminate abruptly northeastward into the risers of the stratigraphic rises of the top of the Pictured Cliffs. In a landward (southwest) direction, several of the Fruitland coals split and thin.

Fassett and Hinds (1971) stated that in general, Fruitland coals could be correlated more reliably in a northwest direction (parallel to the Pictured Cliffs paleoshorelines) than in a southwest direction (at an angle to the Pictured Cliffs paleoshorelines). That statement is true only in terms of the correlation of the thicker total-coal zones, as shown on figure 6. When it comes to correlation of individual coal beds, especially the thicker beds, continuity is frequently greater at right angles to the Pictured Cliffs paleoshoreline. Figure 10 shows that the four major coal beds depicted on this cross section extend for more than 15 miles (one bed goes for 25 miles) at right angles to the Pictured Cliffs shoreline. Interestingly, each of these beds is on top of, or very close to the top of, the Pictured Cliffs to the northeast and from 200-250 feet above the top of the Pictured Cliffs to the southwest; in other words, these beds maintain their continuity across major stratigraphic rises of the underlying Pictured Cliffs Sandstone. It is no wonder that early surface mappers had problems correlating Fruitland coal beds using the top of the Pictured Cliffs as a reference plane—there was no way they could know then that the time-transgressive Pictured Cliffs is rising at such a great rate to the northeast, whereas individual coal beds are essentially time-equivalent throughout their extent (and the very discontinuous surface exposures of Fruitland coal beds certainly didn't help matters).

Some studies of Fruitland coal beds have used terms such as "the lower Fruitland coal zone" or "the upper Fruitland coal zone". Figure 10 clearly shows that in a basin-wide sense, there really are no such zones because of the stratigraphic rise in the position of the Fruitland coal beds from southwest to northeast across the basin. As was pointed out in the preceding paragraph, a coal bed that is at the base of the Fruitland at its northeast end may be 250 feet above the base of the Fruitland at its southwest end; theoretically the same bed could be characterized as both a "lower" and an "upper" coal zone bed in different places. In a limited area along a northwest trend, the use of "upper and lower coal zone" terminology might be correct along that trend, but the danger in thinking of Fruitland coals in that sense is that one tends to then think of the coal beds in reference to the top of the Pictured Cliffs, which could result in serious miscorrelation of coals.

The only way to correlate Fruitland coal beds accurately is to set up stratigraphic cross sections using the Huerfanito Bentonite Bed as a datum, as illustrated on figure 10. Some earlier basin-wide studies of Fruitland coals have used the top of the Pictured Cliffs Sandstone as a datum. Such a methodology should yield total resource values comparable to those published by Fassett and Hinds (1971), however, the spatial configuration of the coal beds and the ability to accurately correlate individual beds is totally lost using such a procedure.

Migration of Coal Bed Methane

Publications by Rice (1983) and Meissner (1984) have discussed the thermal history of the Fruitland and its coal beds and estimated the volume of coal-bed methane that could have been generated in the Fruitland coals since the time of their deposition. Both authors suggested that much of the natural gas that has been produced from the Pictured Cliffs Sandstone has been coal-bed methane produced from Fruitland coals. A detailed discussion of these two excellent and detailed papers is beyond the scope of this report; however, it is difficult to understand how large volumes of Fruitland coal-bed methane could have migrated downward from Fruitland Coal beds into the Pictured Cliffs, given the geometry of these units. (A more recent report by Rice in this volume indicates that there has been little to no movement of Fruitland coal-bed methane out of the coal beds in which the gas was generated.)

Figure 12 shows two present-day structural cross sections; one north-trending and the other east-trending. Because of the small scale of these sections, individual Fruitland coal beds are not shown. Figure 2, a structure contour map of the San Juan Basin, shows that the axis of the basin has a northwest trend and is located in the northern part of the basin. The band of thickest Fruitland coal is just southwest of the basin axis, although there is abundant thick Fruitland coal northeast of the basin's structural axis, as well (fig. 9). The Pictured Cliffs gas field also has a northwest trend, but is located about 15 miles southwest of the present basin axis (see fig.6 of Brown, 1973).

Meissner (1984) has suggested that large volumes of coal-bed methane were generated in Fruitland coal beds in the deeper, and more thermally mature, north-central part of the San Juan Basin and that this gas ultimately migrated southwestward through Fruitland coal beds and charged the underlying Pictured Cliffs Sandstone. Given the fact that when Fruitland coals are correlated and traced to the southwest they become higher and higher above the Pictured Cliffs, it becomes difficult to envision a mechanism for the migration of Fruitland coal gas downward into the Pictured Cliffs. Another difficulty with the Meissner scenario of southwesterly migration of Fruitland gas through Fruitland coals is that individual Fruitland coal beds not only become higher above the Pictured Cliffs to the southwest, they also thin and pinch out in that direction into finer-grained, impervious mudstones and siltstones. And finally, Rice (this volume) suggests that the chemistry of Fruitland coal-bed methane is different from the chemistry of Pictured Cliffs Sandstone gas and that the Pictured Cliffs gas chemistry suggests that the underlying marine Lewis Shale is the source rock for Pictured Cliffs gas.

It is ironic to note that the migration of Fruitland coal gas into the Pictured Cliffs on the northern flank of the basin would seem to have been a much more easily explainable event, with the Fruitland coal beds getting closer to and actually abutting against the Pictured Cliffs sandstones in that direction. It turns out that there is very little Pictured Cliffs gas production on the northern flank of the San Juan Basin. The San Juan Basin gas field continues to be an enigma and much more research will be required to understand its mysteries.

CONCLUSION

The San Juan Basin has produced more than 14 trillion cubic feet of natural gas, principally from conventional (or unconventional?) sandstone reservoirs. Recent studies (Kelso and others, 1987 and Kelso and Wicks, this volume) have shown that coal beds within the Fruitland Formation contain as much as 50 trillion cubic feet of coal-bed methane in the San Juan Basin. This enormous resource is only beginning to be produced, principally in the northern part of the basin on the Southern Ute Indian Reservation. A careful detailed study and correlation of individual coal beds using the Huerfanito Bentonite Bed as a datum throughout the San Juan Basin will insure the most efficient and orderly production of this important resource.

REFERENCES CITED

Brown, C.F., 1973, A history of the development of the Pictured Cliffs Sandstone in the San Juan Basin of northwestern New Mexico, *in* Fassett, J.E., ed., Cretaceous and Tertiary Rocks of the Southern Colorado Plateau: Four Corners Geological Society Memoir, p. 178-184.

Choate, R., Lent, J, and Rightmire, C.T., 1984, Upper Cretaceous geology, coal, and the potential for methane recovery from coalbeds in San Juan Basin—Colorado and New Mexico, *in* Rightmire, C.T., Eddy, G.E., and Kirr, J.N., eds., Coalbed Methane Resources of the United States: American Association of Petroleum Geologists Studies in Geology Series no. 17, p. 185-222.

Fassett, J.E., 1968a, Core description from GB-1 (Gasbuggy 1) in the northeastern part of the San Juan Basin, Rio Arriba County, New Mexico: U.S. Geological Survey Open-File Report, 37 p.

__________, 1968b, Summary of geologic data obtained from borehole GB-1, Project Gasbuggy, *in* San Juan-San Miguel-La Plata Region: New Mexico Geological Society 19th Field Conference Guidebook, p. 24-27.

__________, 1983, Stratigraphy and oil and gas production of northwest New Mexico updated through 1983, *in* Fassett, J.E., ed., Oil and Gas Fields of the Four Corners Area: Four Corners Geological Society, Durango, CO, p. 849-863.

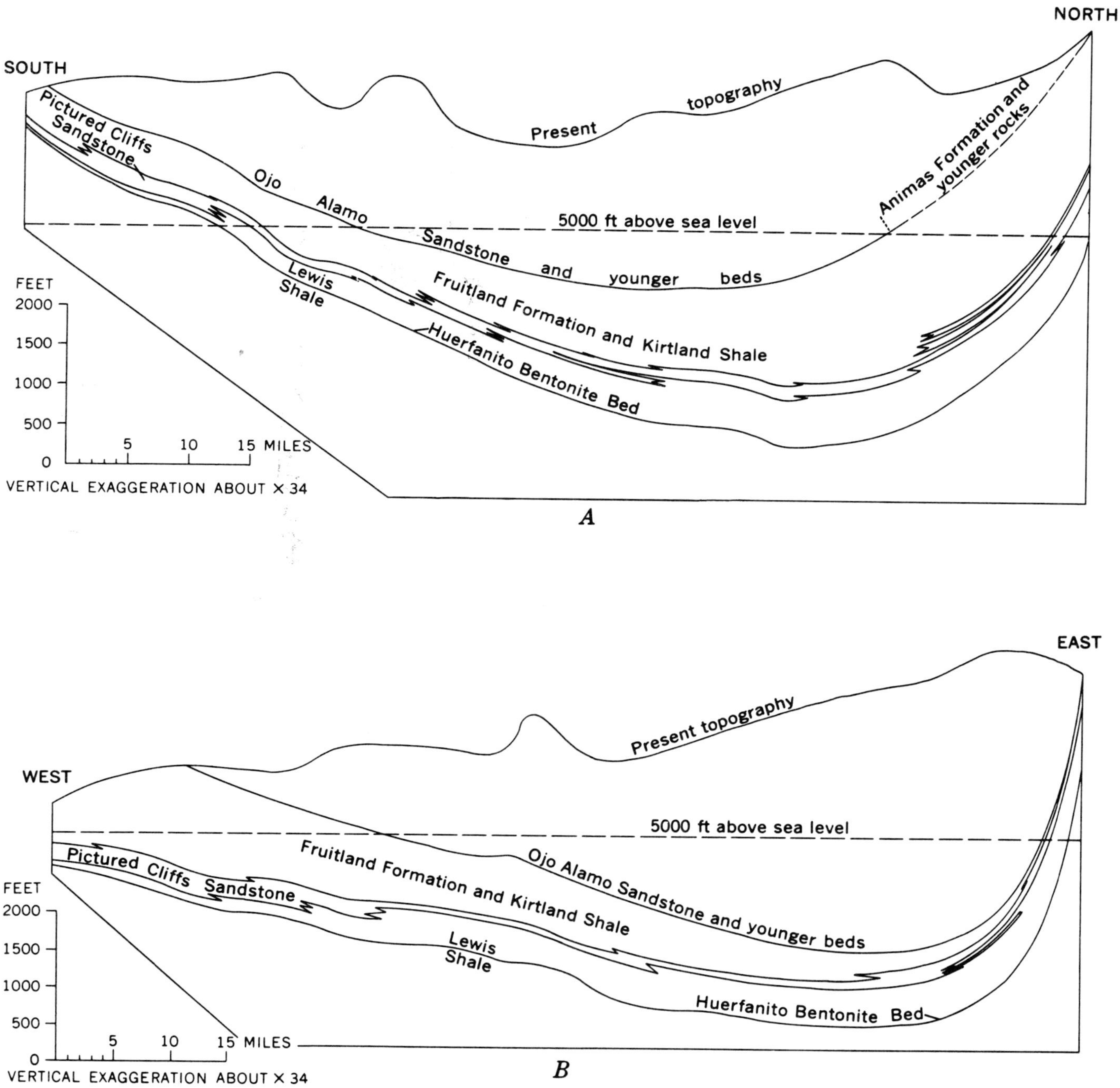

Figure 12. North-trending (A) and east-trending (B) structural cross sections across the San Juan Basin showing the present basin structure. (From Fassett, 1985).

__________, 1985, Early Tertiary paleogeography and paleotectonics of the San Juan Basin area, New Mexico and Colorado, *in* Flores, R.M., and Kaplan, S.S., eds., Cenozoic Paleogeography of the West-Central United States: Rocky Mountain Section of the Society of Exploration Paleontologists and Mineralogists, p. 317-334.

__________, 1986, The non-transferability of a Cretaceous coal model in the San Juan Basin of New Mexico and Colorado, *in* Lyons, P.C and Rice, C.I., eds., Paleoenvironmental and Tectonic Controls in Coal-Forming Basins in the United States: Geological Society of America Special Paper 210, p. 155-171.

__________, 1987, The ages of the continental, Upper Cretaceous Fruitland Formation and Kirtland Shale based on a projection of ammonite zones from the Lewis Shale, San Juan Basin, New Mexico and Colorado, *in* Fassett, J.E., and Rigby, J.K., Jr., eds., The Cretaceous-Tertiary Boundary in the San Juan and Raton Basins, New Mexico and Colorado: Geological Society of America Special Paper 209, p. 5-16.

__________, 1987, Geometry and depositional environment of Fruitland Formation coal beds, San Juan Basin, New Mexico and Colorado: Anatomy of a giant coal-bed methane deposit, *in* Proceedings—The 1987 Coalbed Methane Symposium, November 16-19, 1987, The University of Alabama, Tuscaloosa, Alabama: Gas Research Institute, The University of Alabama, and U.S. Department of Labor, sponsors, p. 19-35.

Fassett, J.E., Lucas, S.G., and O'Neill, F.M., 1987, Dinosaurs, pollen and spores, and the age of the Ojo Alamo Sandstone, San Juan Basin, New Mexico, *in* Fassett, J.E., and Rigby, J.K. Jr., eds., The Cretaceous-Tertiary Boundary in the San Juan and Raton Basins, New Mexico and Colorado: Geological Society of America Special Paper 209, p. 17-34.

Fassett, J.E., and Hinds, J.S., 1971, Geology and fuel resources of the Fruitland Formation and Kirtland Shale of the San Juan Basin, New Mexico and Colorado: U.S. Geological Survey Professional Paper 676, 76 p.

Hinds, J.S., 1965, Btu values of Fruitland Formation coal deposits in Colorado and New Mexico, as determined from rotary-drill cuttings, *in* Geological Survey Research 1964: U.S. Geological Survey Professional Paper 501-D, p. D90-D94.

Irwin, C.D., 1983, Stratigraphy and oil and gas production of southwest Colorado, *in* Fassett, J.E., ed., Oil and Gas Fields of the Four Corners Area: Four Corners Geological Society, Durango, CO, p. 844-845.

Kelso, B. S., Decker, A.D., Wicks, D.E., and Horner, D.M., 1987, Geologic and economic appraisal of coalbed methane in the San Juan Basin, *in* Proceedings—The 1987 Coalbed Methane Symposium, November 16-19, 1987, The University of Alabama, Tuscaloosa, Alabama: Gas Research Institute, The University of Alabama, and U.S. Department of Labor, sponsors, p. 119-130.

Kelso, B.S., and Wicks, D.E., 1988, A geologic analysis of the Fruitland Formation coal and coal-bed methane resources of the San Juan Basin, southwestern Colorado and northwestern New Mexico, *in* Fassett, J.E., ed., Geology and coal-bed methane resources of the northern San Juan Basin, Colorado and New Mexico: Rocky Mountain Association of Geologists Guidebook

Meissner, F.J., 1984, Cretaceous and lower Tertiary coals as sources for gas accumulations in the Rocky Mountain area, *in* Woodward, J., Meissner, F.J., and Clayton, J.L., eds., Hydrocarbon Source Rocks of the Greater Rocky Mountain Region: Rocky Mountain Association of Geologists Guidebook, p. 401-431.

Rice, D.D., 1983, Relation of natural gas composition to thermal maturity and source rock type in San Juan Basin, northwestern New Mexico and southwestern Colorado: American Association of Petroleum Geologists Bulletin V. 67, No. 8, p. 1199-1218.

__________, 1988, Identification and significance of coal bed gas, San Juan Basin, northwestern New Mexico and southwestern Colorado, *in* Fassett, J.E., ed., Geology and coal-bed methane resources of the northern San Juan Basin, Colorado and New Mexico: Rocky Mountain Association of Geologists Guidebook

Sandberg, D.T., 1986, Correlation of coal beds in the Fruitland Formation as interpreted from geophysical logs, east-central San Juan County, New Mexico: U.S. Geological Survey Miscellaneous Field Studies Map MF-1848.

__________, 1986, Coal resources and coal-bed geometry, Fruitland Formation, Southern Ute Indian Reservation, Archuleta and La Plata Counties, Colorado and New Mexico, *in* Fassett, J.E., ed., Geology and coal-bed methane resources of the northern San Juan Basin, Colorado and New Mexico: Rocky Mountain Association of Geologists Guidebook

Sears, J.D., Hunt, C.B., and Hendricks, T.A., 1941, Transgressive and regressive Cretaceous deposits in southern San Juan Basin, New Mexico: U.S. Geological Survey Professional Paper 193-F, p. 101-121.

Coal Resources and Coal-Bed Geometry, Fruitland Formation, Southern Ute Indian Reservation, Archuleta and La Plata Counties, Colorado

Dorothy T. Sandberg[1]

ABSTRACT

Coal resources of the Fruitland Formation (Upper Cretaceous) on the Southern Ute Indian Reservation are estimated to total more than 16 billion short tons of bituminous coal in beds at least 2 feet thick. These resources underlie about 700 square miles of the Reservation. The total coal-resource estimate includes 15 billion short tons of identified resources, based on data points less than 3 miles apart, and about 1 billion short tons of undiscovered or hypothetical resources, based on data points more than 3 miles apart. Most of the estimated resources are at depths of at least 2,000 feet. The apparent rank of Fruitland coal on the Southern Ute Reservation is bituminous, but the coal is characterized by high ash content.

In this report, the coal-bearing interval of the Fruitland Formation is divided into three coal zones that overlap; from oldest in the southwestern part of the Reservation, to youngest in the central and eastern part of the Reservation and eastward. Distribution of these coal zones is closely related to, and was influenced by, the deposition of underlying regressive shoreline deposits of the Pictured Cliffs Sandstone. Structural control of the Pictured Cliffs and the coal zones is indicated in the area of the Ignacio anticline.

Coal beds of the lower zone are mostly in the southwestern part of the Reservation and include two of the thickest and most extensive coal beds on the Reservation. Aggregate thickness of coal in the lower zone is locally more than 50 feet. The middle coal zone, which overlies and overlaps the lower zone in places, is thickest north of the area where it overlies the lower zone. The middle zone is made up generally of thinner coal beds, but maximum aggregate thickness is more than 20 feet. The upper coal zone overlies and overlaps the middle zone mostly near the northeastern boundary of the middle zone. The thickest aggregate coal of the upper zone is in a broad northwest-trending band across the central part of the Reservation area. Within this area, aggregate thickness of coal is generally more than 20 feet. The upper zone overlies about three-fourths of the Fruitland area on the Reservation.

INTRODUCTION

The coal-bearing Fruitland Formation (Upper Cretaceous) underlies about 700 square miles of the Southern Ute Indian Reservation in southwestern Colorado (fig. 1). The Reservation is in the northern part of the San Juan Basin, which is a structural basin that underlies northwestern New Mexico and southwestern Colorado.

Sedimentary rocks associated with the coal-bearing Fruitland Formation crop out in a semicircular band around the northern edge of the basin. Stratigraphically higher Upper Cretaceous and Tertiary formations are exposed toward the more central part of the basin. In general the rocks dip basinward toward the south central part of the Reservation. Most thick coal resources in the Fruitland Formation underlie the area at depths of more than 2,000 feet. In the western part of the Reservation, where the Fruitland crops out in the Hogback monocline, the dip of coal beds is as much as 26° SE but flattens out within about 2 miles of the outcrop.

The purpose of this paper on coal beds of the Fruitland Formation is to provide a general estimate of coal resources, to discuss the geometry of the coal beds, and to show that the distribution of the coal beds was controlled by the depositional setting of the swamps where plant material accumulated near the Late Cretaceous shoreline. This report is based mostly on the study and interpretation of about 500 geophysical logs. Page-size maps in this report (figs. 3, 5, 6, 7, and 9) will be available in 1988 at a scale of 1:100,000 from the U.S. Geological Survey as Open-File Report 88-230 by D.T. Sandberg.

Formations considered in this study of Fruitland coal beds include, in ascending stratigraphic order, the Lewis Shale and its contained Huerfanito Bentonite Bed, Pictured Cliffs Sandstone, Fruitland Formation, and Kirtland Shale, all of Late Cretaceous age (fig. 2). These formations were deposited during the final regressive cycle of a sea that covered most of the region during Late Cretaceous time. The Lewis Shale is an offshore marine unit that was deposited largely as muds; the Pictured Cliffs Sandstone is a shoreline deposit that was laid down as the sea regressed to the northeast. The nonmarine

[1] U. S. Geological Survey
P.O. Box 25046, M.S. 939, DFC
Lakewood, Colorado 80225

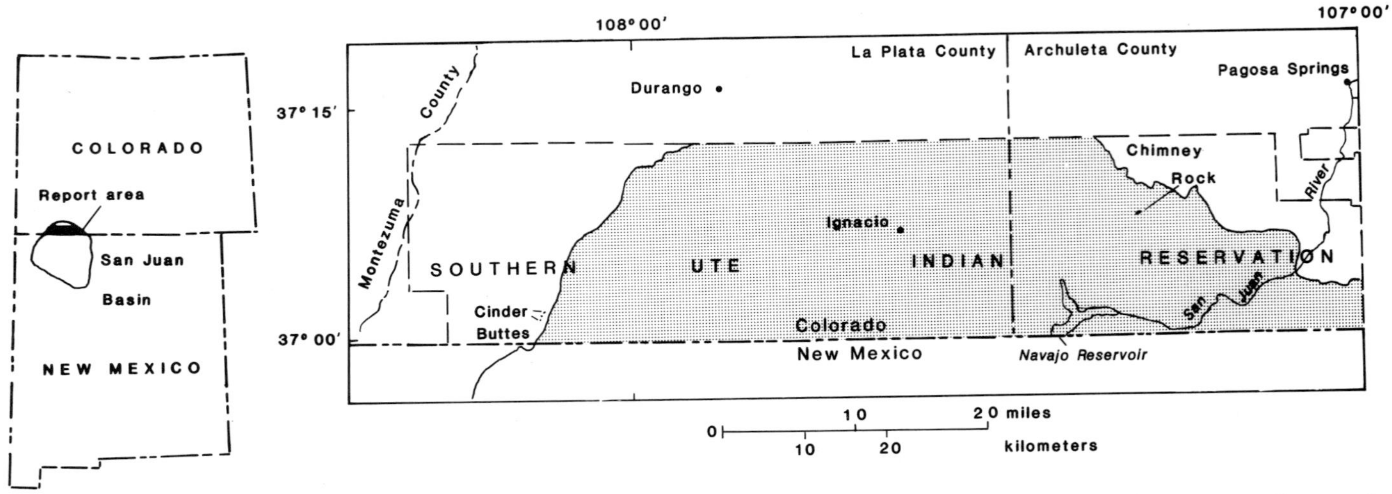

Figure 1. Index map showing location of Southern Ute Indian Reservation and area underlain by Fruitland Formation. Patterned area is underlain by the Fruitland Formation.

Fruitland Formation originated in an environment of coastal swamps and distributary river systems landward of the Pictured Cliffs shoreline (Fassett and Hinds, 1971, p. 19). In some of the swamps, conditions were favorable for the accumulation of great thicknesses of plant remains that gradually were compressed and altered into thick and extensive coal deposits. The Kirtland Shale, also nonmarine, was deposited even farther landward than the Fruitland Formation, probably in relatively higher areas of better drainage (Fassett and Hinds, 1971, p. 25).

The Huerfanito Bentonite Bed of the Lewis Shale was identified in the subsurface and named as a marker bed by Fassett and Hinds (1971, p. 6). This unit is characterized by a distinctive geophysical log response and has been identified in the subsurface throughout the San Juan Basin. The Huerfanito is believed to have been deposited everywhere at the same time and is, therefore, useful as a datum to show depositional patterns of the regressive Pictured Cliffs Sandstone and locally to correlate coal beds of the Fruitland Formation. Because the Pictured Cliffs is time-transgressive and the Huerfanito is isochronous, the Pictured Cliffs rises stratigraphically, northeastward, relative to the Huerfanito. Fassett and Hinds (1971) first illustrated this relationship between the Pictured Cliffs and the Huerfanito with an isopach map and cross sections of the San Juan Basin.

Figure 3 is an isopach map of the same interval as that map constructed by Fassett and Hinds but limited to the Southern Ute Indian Reservation. This map was prepared to show the close relationship between the distribution of Fruitland coal beds to the configuration of the top of the Pictured Cliffs Sandstone and thus, to attempt detailed correlation of coal beds, at least locally. In this area the thickness of the interval between the Pictured Cliffs and the Huerfanito ranges from about 680 feet in the southwestern part of the Reservation to more than 1,140 feet in the northeastern part. An area (screened in fig. 3) of large-scale intertonguing in the upper part of the Pictured Cliffs is interpreted from logs. A stratigraphic rise of as much as 150 feet is recognized over a distance of about 2-4 miles. The lower part of the stratigraphic rise extends west of the screened area across the northern part of T. 33 N., Rs. 9 and 10 W., beginning approximately with

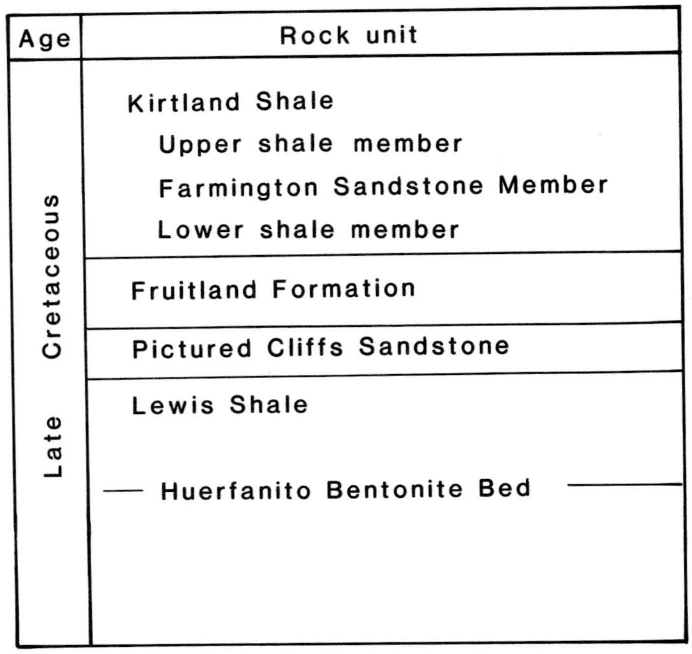

Figure 2. Stratigraphic section of geologic rock units adjacent to the coal-bearing Fruitland Formation on the Southern Ute Indian Reservation.

the 800-foot isopach. The suggested trend of the Pictured Cliffs paleoshoreline is about N. 20°-30° W. South and west of the 800-foot isopach, however, northeast- and east-trending sandstone thickness patterns are suggested by the isopachs. A pause in the general northeastward retreat of the shoreline and vertical stacking of the Pictured Cliffs, suggested in the stratigraphic rise, is one of the conditions necessary for the accumulation of thick deposits of peat in swamps in the southwestern part of the Reservation. The stacking of Pictured Cliffs sands and concurrent accumulation landward of thick deposits of vegetal matter while the shoreline appeared to remain in essentially one position, were identified and illustrated in the San Juan Basin by Fassett and Hinds (1971, p. 11.)

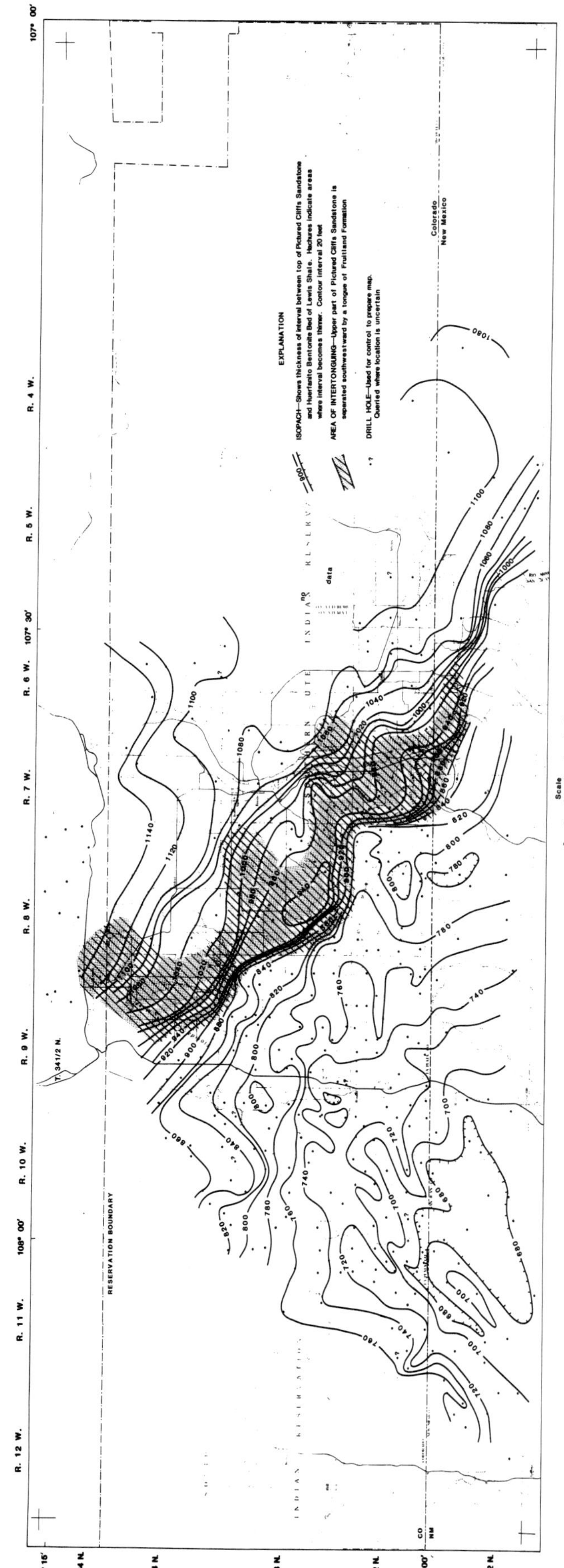

Figure 3. Isopach map showing thicknesses of the interval between the top of the Pictured Cliffs Sandstone and the Huerfanito Bentonite Bed of the Lewis Shale, Southern Ute Indian Reservation and vicinity, Archuleta and La Plata Counties, Colorado, and Rio Arriba and San Juan Counties, New Mexico. Contour interval is 20 feet.

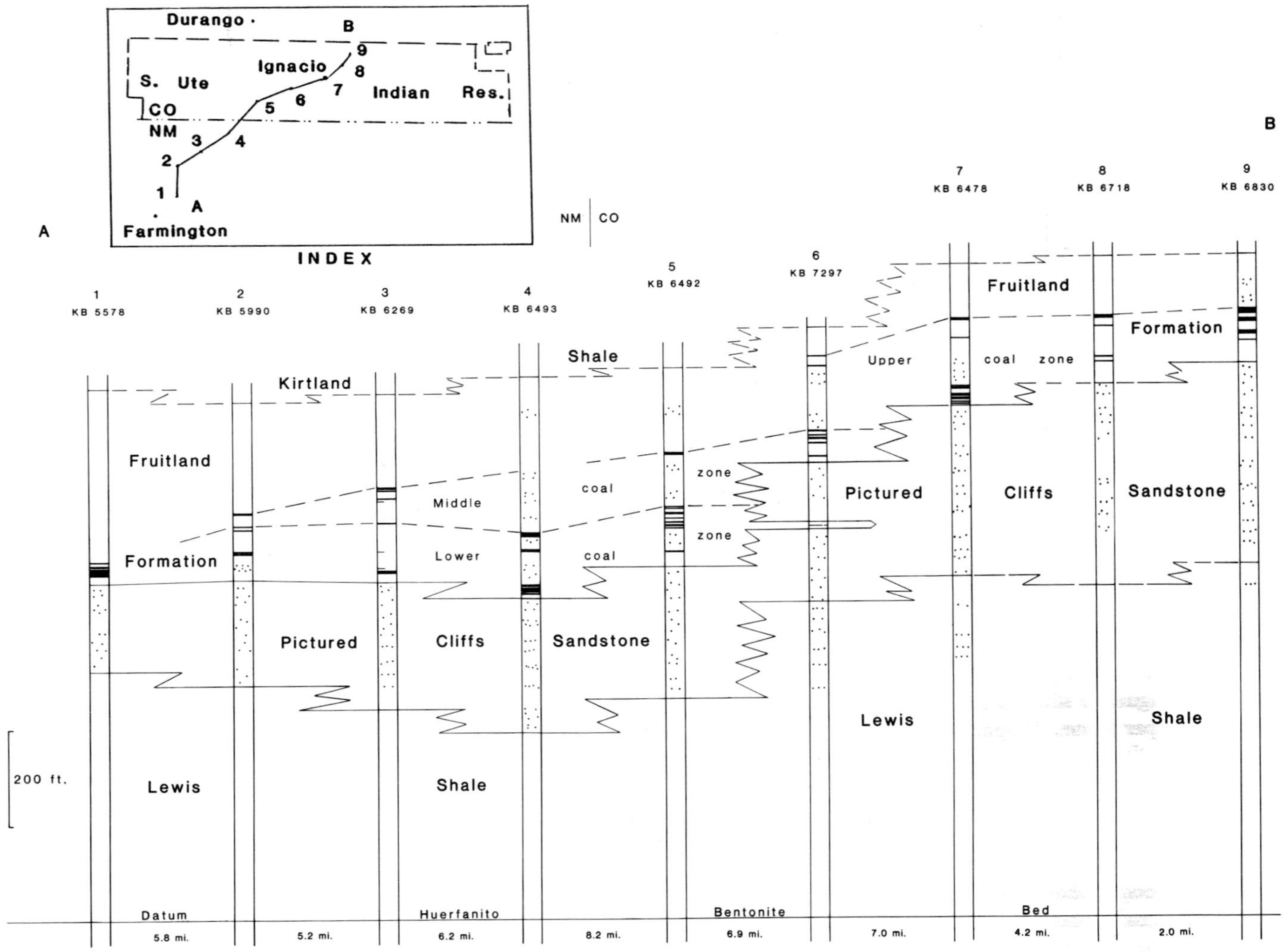

Figure 4. Cross section A-B which extends from northern New Mexico northeastward across the Southern Ute Indian Reservation in Colorado (see inset). The names of the wells used to construct this section are listed in table 1.

TABLE 1. Drill holes in line of section A-B (fig. 4) from northern New Mexico northeastward across Southern Ute Reservation, Colorado. Well numbers are keyed to figure 4.

Number	Location in section	Township and range	Company, well name, and number	Year drilled
			New Mexico	
1.	NE SE 30	30 N., 12 W.	Amoco Production Co. Federal E No. 1-E	1980
2.	NE SE 30	31 N., 12 W.	John Hill and others Taliferro No. 5M	1981
3.	SW SW 12	31 N., 12 W.	Southland Royalty Co. Davis No. 10E	1980
4.	NE NW 26	32 N., 11 W.	El Paso Natural Gas Co. Barnes No. 4A	1978
			Colorado	
5.	SE NE 6	32 N., 9 W.	St. Louis Car Co.-Menlo Corp. No. 5-6 Block 6	1960
6.	SW SW 20	33 N., 8 W.	ARCO Oil and Gas Co. Southern Ute Herrera No. 2	1980
7.	NE SE 8	33 N., 7 W.	Amoco Production Co. Southern Ute 1-8 No. 1A	1983
8.	NW SE 26	34 N., 7 W.	Fuelco Southern Ute No. 1	1977
9.	NW SE 13	34 N., 7 W.	Natomas North America, Inc. Harper No. 1-13	1981

The geometry and stratigraphic rise of the coal beds and of the Pictured Cliffs Sandstone are shown in section A-B, figure 4, using as a datum the Huerfanito Bentonite Bed of the Lewis Shale. This figure shows that the stratigraphically lowest (and oldest) coal beds (lower coal zone) are in the southwest; progressively higher (and younger) coal beds (middle and upper coal zones) appear northeastward. This line of section extends from northern New Mexico northeastward across the Southern Ute Reservation. Only coal beds at least 2 feet thick are shown. Drill holes used in figure 4 are listed in table 1.

COAL DEPOSITS

Thick coal beds generally are in the lower half of the Fruitland Formation but thinner coal beds are common throughout the formation. The thickest beds are generally in the lowest part of the formation, in places lying directly on top of the underlying Pictured Cliffs Sandstone.

Coal beds of the Fruitland Formation are divided for convenience in this paper into lower, middle, and upper zones (fig. 4). These zones overlap from the oldest, or stratigraphically lowest, in the southwestern part of the Reservation, northeastward to the youngest, or stratigraphically highest, in the central part of the Reservation. Distribution, aggregate coal thickness, and overburden thickness of these zones are shown in figures 5, 6, and 7; figure 5 shows the lower zone, figure 6 shows the middle zone, and figure 7 shows the upper zone. Isopach lines used for thickness categories are those needed to estimate resources of bituminous coal. The northwest limit, or approximate line of outcrop of the zones, shown by dashed lines on the isopach maps, was determined from points of intersecting structure contours on the top of the zone and topographic contours on 7.5-minute topographic maps.

The lower coal zone (fig. 5) is located in the southwestern part of the Reservation and contains the two thickest and stratigraphically lowest coal beds; aggregate coal thickness is more than 50 feet in places. The middle coal zone (fig. 6), which overlies and overlaps the lower zone in places, is generally thickest north of the area underlain by the lower coal zone. The middle zone is made up of thinner coal beds that reach a maximum aggregate thickness of more than 20 feet. The upper coal zone (fig. 7) partially overlies and overlaps the middle zone, and is generally made up of thinner beds than the lower zone, and covers the largest area of the Reservation. The greatest aggregate coal thickness in the upper coal zone is in the central part of the study area underlain by the Fruitland Formation; this area is east of the area underlain by the lower and middle zones. Aggregate thickness of coal in the upper zone is generally more than 20 feet, but it is considerably thicker near the northern boundary of the Reservation.

Aggregate thickness of coal shown on the isopach maps is in places somewhat thinner than those shown in reports by others, such as Fassett and Hinds (1971), Shomaker and Holt (1973), and Kelso and others (1980). This is in part the result of not including intervals of extremely impure coal (33 percent ash or more on the dry basis), in coal exploratory holes near the western edge of Fruitland outcrop (Shomaker and Feldman, 1978). In this area proximate analyses of coal samples were determined for each 2 feet of coal core in at least 24 coal exploratory holes drilled by Sunoco Energy Development Company (SUNEDCO), near the western outcrop of the Fruitland Formation. The suite of geophysical logs used with the coal exploratory drill holes included gamma ray, resistivity, and density logs, which reflect the changes in quality shown in the proximate analyses of coal samples. Eastward and deeper into the subsurface, geophysical logs of many oil and gas test wells also suggest that coal quality varies considerably within coal beds and that noncoal partings and impure coal are common. Where the quality of logs was judged to be adequate, parts of coal beds in these holes were determined to be high in ash content and thus not included in the aggregate coal thicknesses. Figure 8 is an example of part of a geophysical log from a drill hole in the southwestern part of the Reservation that shows coal thickness used in this study of coal resources in the Fruitland Formation of the Southern Ute Indian Reservation.

[Editor's note: The purpose of this paper was to determine the coal resources according to the coal resource classification system of the U.S. Geological Survey (Wood and others, 1983), thus, coal with more than 33 percent ash was not included. For evaluating coal-bed methane potential, however, these high-ash coals should be considered, with perhaps some discount factor applied. See the papers cited above for coal resource numbers that include high-ash coal. See Kelso and Wicks (this volume) for another estimate of Fruitland coal resources on the Southern Ute Indian Reservation.]

Lower Coal Zone

The lower coal zone, generally in the southwestern part of the Fruitland area, contains thick coal in essentially two fairly extensive beds. This zone directly overlies, or is in general within 50 feet above, the top of the Pictured Cliffs Sandstone. The area in which the aggregate thickness of coal in the lower zone is 20 feet thick or more, is patterned in figure 5; locally the aggregate thickness is more than 50 feet. The configuration of the top of the Pictured Cliffs Sandstone (fig. 3) and some of the coal beds suggest a northeast trend or alignment. Toward the east and north the coal beds split, become thinner, and terminate into the Pictured Cliffs Sandstone (fig. 4).

The stratigraphic interval between the two thick beds ranges from about 100 to 0 feet, where apparently they merge. Most of the area of thickest aggregate coal is more than 2,000 feet beneath the surface. Near the Fruitland outcrop along the western part of the study area where the coal is not as deep, the dips range from about 8° to 26° SE.

East of the area of thickest aggregate coal in figure 5 (near the stratigraphic rise of the Pictured Cliffs Sandstone in T. 32 N., R. 8 W.), sandstone replaces coal beds in the lower part of the Fruitland Formation. There, the lower coal zone essentially ends and is replaced by as much as 100 feet of sandstone. The curves on the geophysical logs are not characteristic of the Pictured Cliffs in this area and suggest that possibly a Fruitland estuary system drained into the Pictured Cliffs sea in this area.

Middle Coal Zone

The middle coal zone overlies the lower coal zone throughout most of the study area but extends farther eastward than the lower zone (fig. 6). The lowest coal bed of the middle zone ranges from directly overlying, to as much as 100 feet above the top of the lower coal zone. This middle zone contains two fairly persistent beds and several others that are more local.

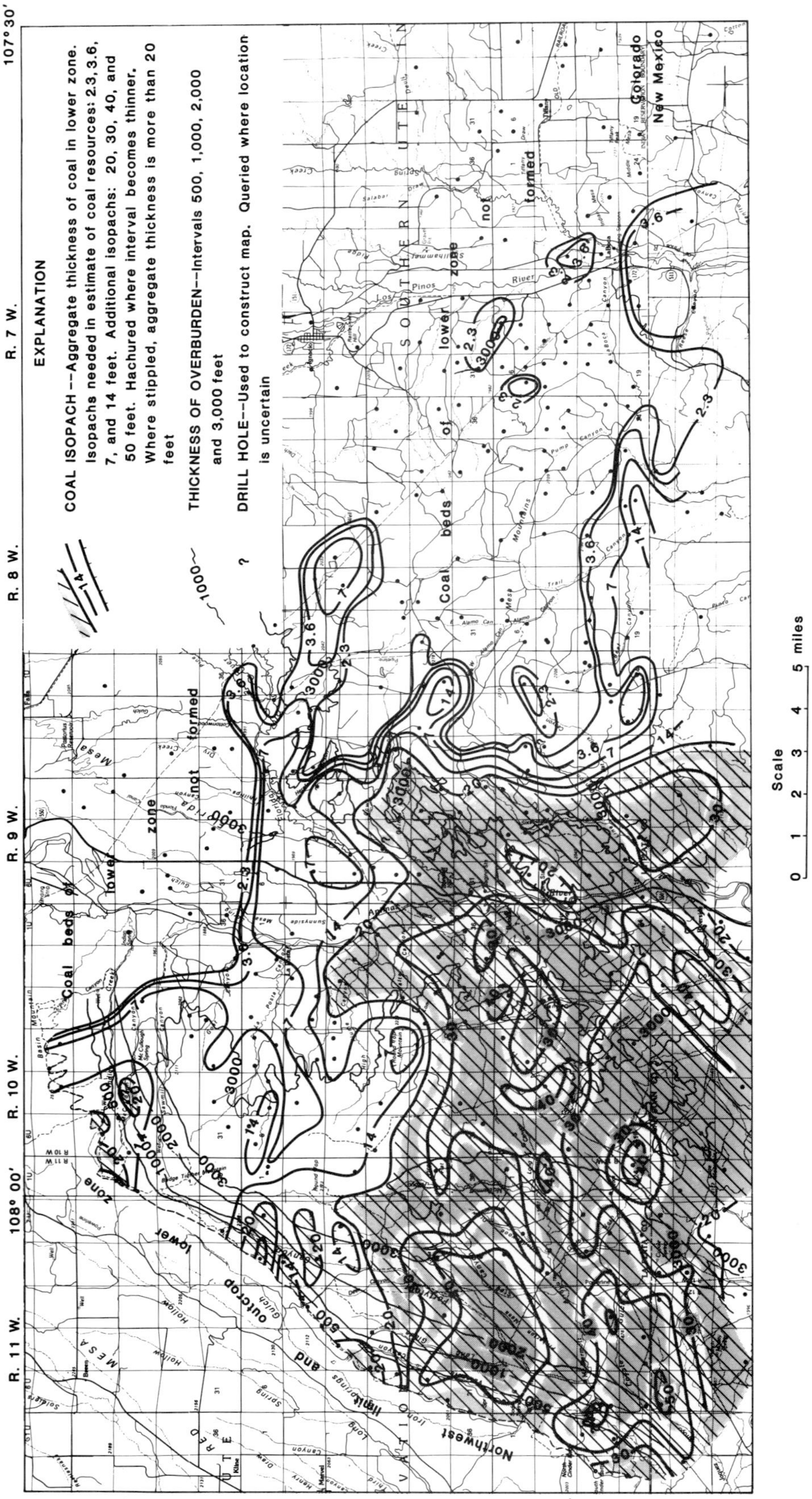

Figure 5. Isopach map showing thickness of overburden and aggregate thickness of coal beds in the lower coal zone of the Fruitland Formation on the Southern Ute Indian Reservation. Areas where aggregate coal thickness is greater than 20 feet are patterned.

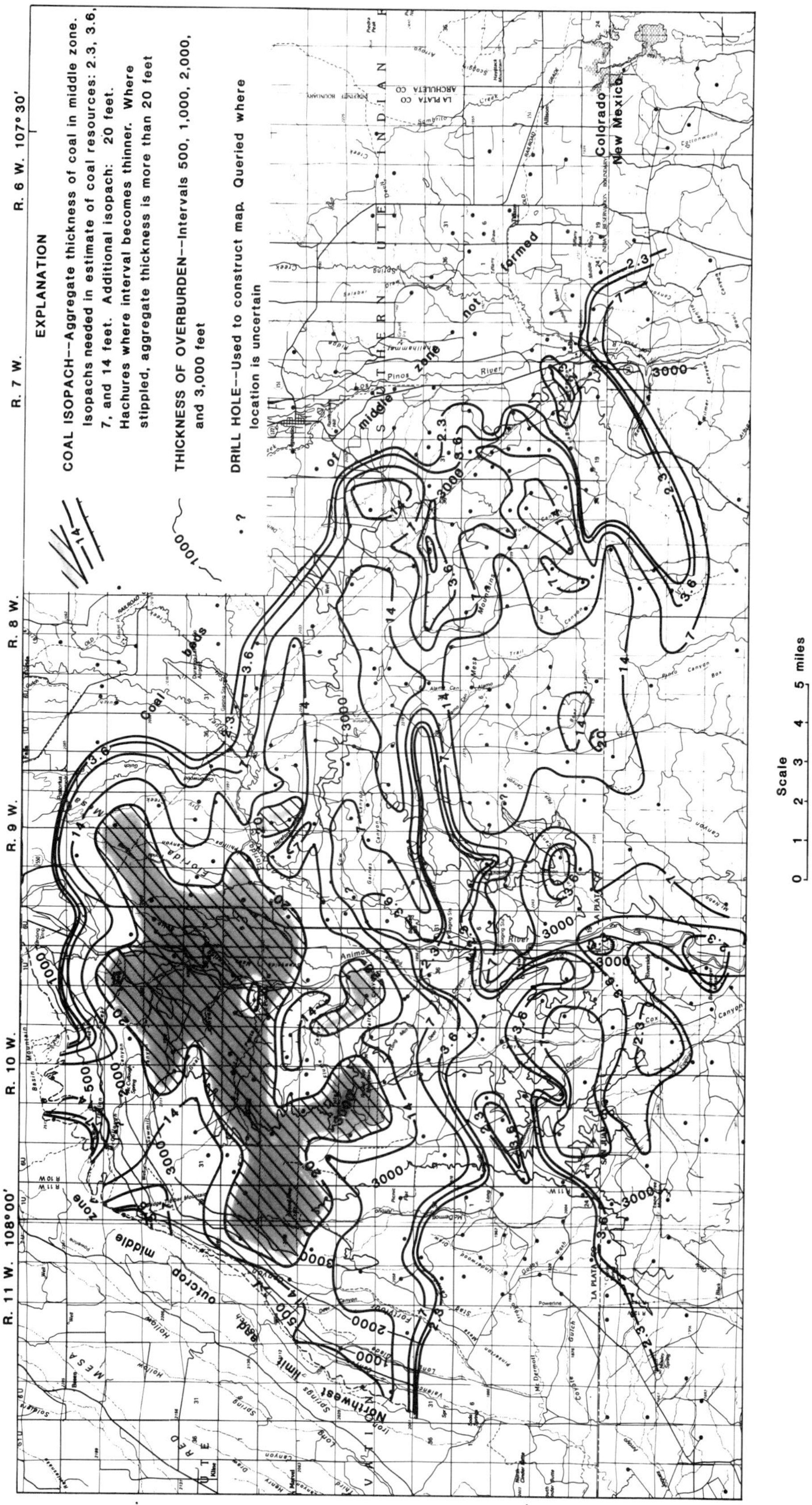

Figure 6. Isopach map showing thickness of overburden and aggregate thickness of coal beds in the middle coal zone of the Fruitland Formation on the Southern Ute Indian Reservation. Areas where aggregate coal thickness is greater than 20 feet are patterned.

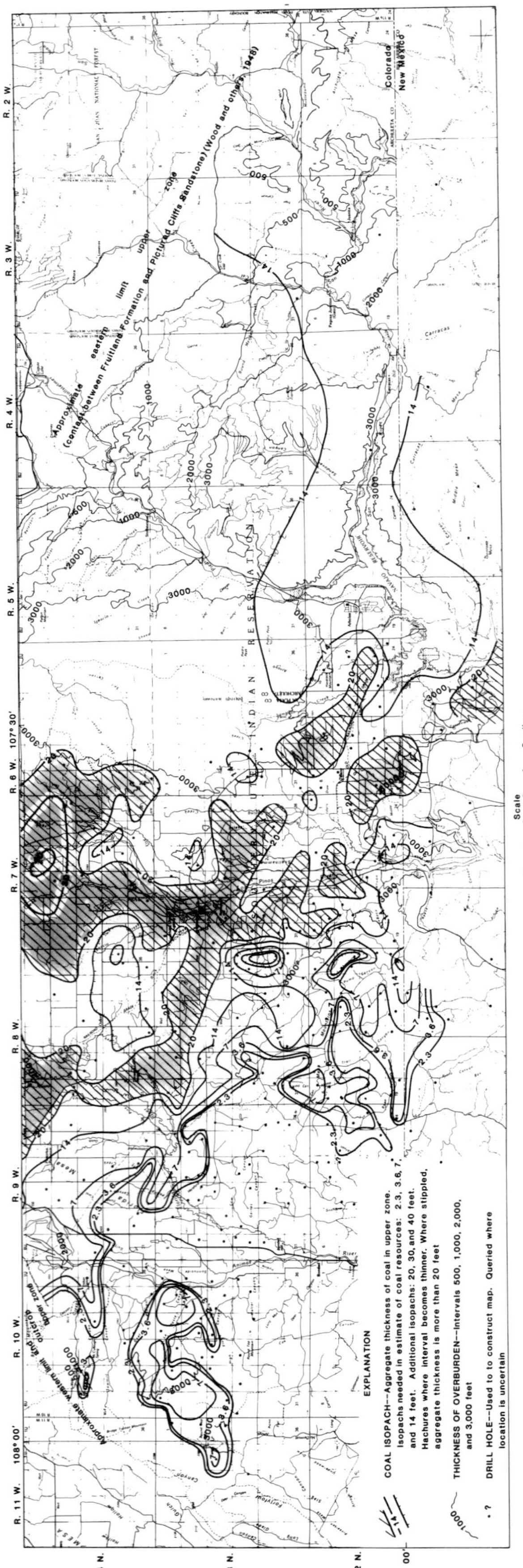

Figure 7. Isopach map showing thickness of overburden and aggregate thickness of coal beds in the upper coal zone of the Fruitland Formation on the Southern Ute Indian Reservation. Areas where aggregate coal thickness is greater than 20 feet are patterned.

The area of maximum aggregate thickness of more than 20 feet of coal covers an area mostly in Tps. 33 and 34 N. (patterned in figure 6). Areas of thickest aggregate coal of the middle zone coincide with areas of thin aggregate coal in the lower zone. In general the middle zone is more than 2,000 feet beneath the surface. Other areas of thick coal of the middle zone, where the thickness is more than 14 feet, extend in a northwest trending band, roughly southwest of the large stratigraphic rise of the Pictured Cliffs Sandstone. This northwest trending band of thick coal is intersected by a northeast trending area, in which the coal is less than 2 feet thick or missing (fig. 6). The area of thin or missing coal in the middle zone coincides with the area of thickest coal in the underlying lower zone. The northeast trend also coincides with the direction of trends of sandstone in the Pictured Cliffs (fig. 3) and trends of coal in the lower zone. The distribution pattern suggests that possibly a northeast-trending drainage system crossed the backshore swamps here; this drainage pattern has been traced to the south into New Mexico.

The middle coal zone extends into the lower part of the large stratigraphic rise of the Pictured Cliffs, gradually thins, and ends against the Pictured Cliffs (figs. 3, 6).

Upper Coal Zone

The upper coal zone occupies the largest area of the Fruitland Formation on the Southern Ute Reservation (fig. 7). The location of the upper zone in the western part of the Reservation coincides in part with the underlying middle zone, but only local and comparatively thin beds of the upper zone are present. The lowest, upper-zone coal bed ranges from 35 feet to more than 100 feet above the top of the middle coal zone. The area of thickest upper-zone coal beds in aggregate is in a roughly northwest-trending band across the central part of the Reservation. The upper-zone coal attains a maximum aggregate thickness of about 40 feet near the northern boundary of the Reservation in T. 34 N., R. 6 W., but this area of thick coal lies more than 2,000 feet beneath the surface. East of this area of thickest coal (patterned in figure 7), data are generally sparse but suggest 10-20 feet of aggregate thickness of coal beds.

Where the upper zone is thickest, it generally includes several beds within a stratigraphic interval typically between 100 and 200 feet thick. Most of the thick coal beds appear at or near the top of the large stratigraphic rise of the Pictured Cliffs Sandstone. In places some individual coal beds probably can be correlated for short distances.

COAL QUALITY

The apparent rank of coal in the Fruitland Formation in the northern part of the San Juan Basin is generally considered to range from high-volatile A bituminous to medium-volatile bituminous (Fassett and Hinds, 1971, Goolsby and others, 1979, and Khalsa, 1981). This designation is in accordance with the definitions of the American Society for Testing and Materials (ASTM, 1984). Table 2 lists selected chemical analyses of coal samples from the Fruitland Formation in, and north of, the Southern Ute Reservation.

Vitrinite reflectance of coal samples is another method used to help determine coal rank (McCartney and Teichmuller, 1972). Results from this method, which is particularly useful for various ranks of bituminous coal, show that Fruitland coal

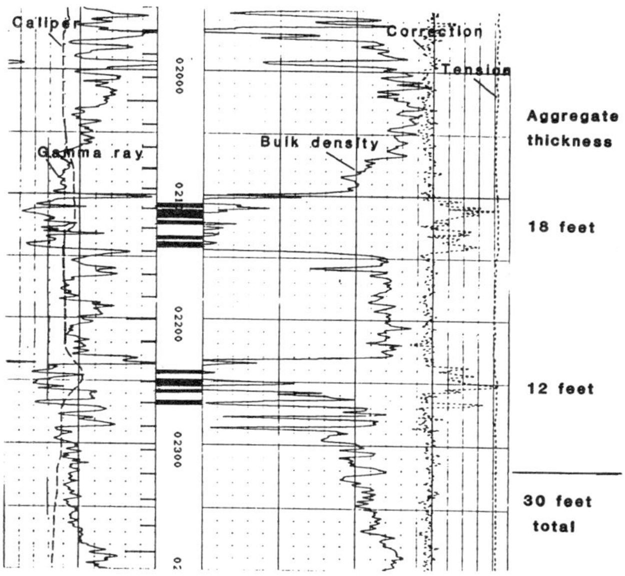

Figure 8. Geophysical logs through the lower coal zone of the Fruitland Formation from the ARCO Oil and Gas Co., Southern Ute No. 17-3 well in NW¼ NW¼ sec. 17, T. 32 N., R. 11 W. This figure shows the way coal thicknesses were measured from well logs for this study.

in the Southern Ute Reservation is high- to medium-volatile bituminous (Kelso and Rushworth, 1982).

Fruitland coal in most parts of the Southern Ute Reservation is reported by some authors to be coking coal. In discussing coal in the Fruitland of the San Juan region, Goolsby and others (1979, p. 40) stated: "In Colorado, the coal resources in this formation are predominantly marginal to latent grade high-volatile A bituminous coking coal." They further stated that the high ash content generally prevents coal in the Fruitland from being premium grade coking coal. To determine which coal constitutes coking coal in their report, these authors combined a number of criteria, but used especially ash and sulfur percentages and ASTM determination of rank.

COAL RESOURCES

Estimated original coal resources in the Fruitland Formation of the Southern Ute Reservation total 16 billion short tons. This total includes 15 billion short tons of identified resources and about 1 billion short tons of undiscovered resources. Identified resources are based on data that are spaced three miles apart or less; undiscovered resources are based on more widely spaced data and knowledge of the coal zone in the region. The undiscovered resources are limited to the eastern half of the Reservation, where there is little subsurface information. This coal resource estimate is based on aggregate coal thicknesses (minimum 2 feet thick) and is mostly based on an interpretation of geophysical logs of drill holes shown on the isopach maps of the respective zones in figures 5, 6, and 7. Impure coal (judged to contain more than one-third ash) and partings between coal layers were excluded from measurements of coal thickness.

The method used to estimate the coal resources is in accordance with that prescribed in Wood and others (1983). To estimate the resources, thickness of coal, depth, and elevation were plotted on 7.5-minute topographic quadrangle maps. Structure contours and overburden isopachs were drawn on the topographic maps; coal isopachs and lines delineating resources according to reliability of information (measured, indicated, and inferred) were constructed on separate overlays of the respective quadrangles. Coal isopachs used in this report for thickness categories of bituminous coal are 2.3, 3.6, 7, and 14 feet. The thinnest category, 1.2 to 2.3 feet, generally used for bituminous coal resource estimates, could not be assessed with confidence on most of the geophysical logs of oil and gas test wells and, therefore, these thin coal beds were not considered in this report. An average weight of 1,800 short tons per acre-foot, which is generally assumed for bituminous coal, was used in the calculations. Each category delineated on the respective quadrangles was planimetered to determine areal extent, and then multiplied by the weighted-average thickness in feet and by 1,800 short tons per acre-foot. Where the dip of coal beds is less than 10°, the coal beds are treated as flat lying. Where the dip is more than 10°, as in places in the western part of the Reservation, estimated resources were divided by the cosine of the dip to obtain final figures.

The resources were estimated using thickness categories suggested (Wood and others, 1983) for bituminous coal, with the exception of the thinnest category, 1.2-2.3 feet. A summary of coal resources by aggregate thickness of coal is presented in table 3.

Distribution of lower, middle, and upper coal zones is shown in figures 5, 6, and 7, respectively. The upper coal zone contains nearly half of the identified resources but extends over a much larger area than the lower and middle zones. The lower zone includes much thicker beds within a smaller area. Estimated resources of each zone are summarized in table 4.

Classification of estimated coal resources by overburden category shows that by far most of the coal is more than 2,000 feet beneath the surface (table 5).

POSSIBLE STRUCTURAL INFLUENCE

A major structure within the area of subsurface control on the Reservation is a feature called the Ignacio anticline (Steven and others, 1974). This structure is illustrated in figure 9, which is a structure contour map on the top of the Pictured Cliffs Sandstone. Barnes in 1953 noted the presence of a "roughly oval dome southwest of the town of Ignacio" that trends northwest. Subsurface information available later has confirmed that this structure is part of an anticline extending across several townships within the Reservation. The anticline forms a broad arc, extending in a westerly direction from T. 32 N., R. 5 W., turning northwest to about T. 33 N., R. 7 W. as far as the town of Ignacio, where the axis reaches its maximum elevation (fig. 9). In the vicinity of Ignacio the axis turns west and continues as far as T. 33 N., R. 9 W., where it forms the west end of the arc by turning to the southwest, but is less well defined in this western area.

Evidence suggests that the Ignacio anticline may have existed in some form at least as early as Pictured Cliffs time, and that it possibly influenced the distribution of the Pictured Cliffs Sandstone and the three Fruitland coal zones. The main part of the stratigraphic rise of the Pictured Cliffs Sandstone is southwest of and parallel to the Ignacio anticline where it

TABLE 2. Chemical analyses of coal samples from the Fruitland Formation in the Colorado portion of the San Juan Basin.

[A, as-received; B, moisture free; C, moisture and ash free; dash (—), not applicable]

Location and sample source	Lab. no.	Condition	Proximate analyses (percent)				Ultimate analysis (percent)	
			Moisture	Volatile matter	Fixed carbon	Ash	Sulfur	Heat value Btu/lb
1.								
Sec. 30,	K88634	A	4.9	27.7	54.2	13.2	0.8	11,740
T. 34 N., R. 4 W.	D205233	B	—	29.1	57.0	13.9	0.8	12,340
channel		C	—	33.8	66.2	—	1.0	14,330
	K88635	A	1.5	31.0	56.7	10.8	.9	13,530
	D205234	B	—	31.5	57.6	11.0	.9	13,740
		C	—	35.3	64.7	—	1.0	15,430
	K88640	A	2.3	25.5	45.1	27.1	.9	10,710
	D205239	B	—	26.1	46.2	27.7	.9	10,960
		C	—	36.1	63.9	—	1.3	15,170
2.								
Sec. 5,	H55350	A	.8	21.5	56.4	21.3	.7	12,140
T. 32 N., R. 7 W.		B	—	21.6	56.9	21.5	.7	12,240
well cuttings		C	—	27.6	72.4	—	.8	15,600
2.								
Sec. 15,	H46452	A	2.3	23.6	54.6	19.5	.7	12,070
T. 32 N., R. 10 W.		B	—	24.2	55.9	19.9	.7	12,360
well cuttings		C	—	30.2	69.8	—	.9	15,440
2.								
Sec. 36,	H38041	A	.9	20.8	51.7	26.6	.8	11,230
T. 34 N., R. 10 W.		B	—	21.0	52.2	26.8	.8	11,330
well cuttings		C	—	28.7	71.3	—	1.1	15,480
3.								
Sec. 27,	3551	A	3.1	32.7	47.4	16.8	1.3	11,900
T. 35 N., R. 9 W.		B	—	33.7	49.0	17.3	1.3	12,270
mine		C	—	40.8	59.2	—	1.6	14,840
4.								
Sec. 25,	204	A	1.1	16.4	56.1	26.4	.4	11,280
T. 34 N., R. 8 W.		B	—	16.6	56.7	26.7	.5	11,410
well core		C	—	22.6	77.4	—	.6	15,560
	209	A	.9	15.1	59.1	24.9	.6	11,580
		B	—	15.3	59.6	25.1	.6	11,680
		C	—	20.4	79.6	—	.6	15,590

Sources of sample-analysis data:

1 Khalsa, 1981. Tests of samples show coal to be agglomerating.
2 Fassett and Hinds, 1971
3 U.S. Bureau of Mines, 1937
4 Kelso and Rushworth, 1982

TABLE 3. Resources estimated by aggregate coal thickness.

	Aggregate coal thickness (feet)	Millions of short tons	Percent of total
Identified coal resources	2.3- 3.6	135	1
	3.6- 7.0	520	3
	7.0-14.0	2,890	19
	14.0	11,770	77
Undiscovered coal resources	7.0-14.0	70	6
	>14.0	1,080	94

TABLE 4. Estimated coal resources according to coal zone.

Coal zone	Estimated coal resources (millions of short tons	Percent of total
Identified coal		
Upper	7,170	47
Middle	3,540	23
Lower	4,600	30
Undiscovered coal		
Upper	1,140	100

TABLE 5. Coal resources by overburden category in each zone.

Coal zone and overburden category (feet)	Millions of short tons	Percent of zone total
Identified resources		
Upper zone		
Less than 500	158	2
500-1,000	290	4
1,000-2,000	523	7
2,000-3,000	4,627	65
>3,000	1,575	22
Total	7,173	100
Middle zone		
Less than 500	69	2
500-1,000	46	1
1,000-2,000	106	3
2,000-3,000	2,002	57
>3,000	1,312	37
Total	3,535	100
Lower zone		
Less than 500	273	6
500-1,000	175	4
1,000-2,000	293	6
2,000-3,000	2,193	48
>3,000	1,664	36
Total	4,598	100
Undiscovered resources		
Upper zone		
1,000-2,000	110	10
2,000-3,000	330	29
>3,000	700	61
Total	1,140	100
Grand total	16,446	

trends northwest. The area of intertonguing of the Pictured Cliffs and Fruitland ends southwest of the structurally highest part of the anticline. The lower coal zone does not extend as far east as the northwest trending section of the anticline. Between the main part of the lower zone and the axis of the anticline are some scattered small lenses of coal and an unusual thickness of sandstone beds that apparently are not part of the Pictured Cliffs Sandstone. The middle coal zone does not extend as far east as the highest part of the anticline near Ignacio, but where the axis of the anticline extends west and southwest, the area roughly coincides with some of the thin and missing areas of the middle zone. The upper coal zone overlies the highest part of the anticline where it trends northwest, but thinning locally of the upper coal zone is indicated along the axis of the anticline. These observations lead to an interpretation that the Ignacio anticline may have existed in some form that could have influenced and delayed the Pictured Cliffs sea withdrawal while Fruitland backshore swamps were accumulating plant material. The subtle growth of the anticline may have upset the delicate balance needed for peat accumulation and could be one factor that resulted in deposition locally of muds, silts, and sands, which are now represented by coal intervals of high ash content.

CONCLUSION

Estimated coal resources in the Fruitland Formation total more than 16 billion short tons, most of which are more than 2,000 feet beneath the surface. Fruitland coal beds are in overlapping zones, the oldest of which is in the southwestern part of the Reservation and the youngest of which underlies the eastern three-fourths of the Fruitland area. The lower zone contains the thickest coal, generally in two beds. The upper zone contains the largest percentage of estimated resources but includes about three-fourths of the Fruitland area. Areal extent of coal beds was controlled by depositional patterns of the Pictured Cliffs Sandstone and possibly by the growth of an incipient Ignacio anticline in Late Cretaceous time.

REFERENCES CITED

American Society for Testing and Materials, 1984, Standard classification of coals by rank (ASTM Designation D 388-84), in Gaseous fuels; coal and coke; atmospheric analysis: Annual book of ASTM standards, part 26, p. 190-195.

Barnes, H., 1953, Geology of the Ignacio area, Ignacio and Pagosa Springs quadrangles, La Plata and Archuleta Counties, Colorado: U.S. Geological Survey Oil and Gas Investigations Map OM-138, scale 1:63,360.

Fassett, J.E., and Hinds, J.S., 1971, Geology and fuel resources of the Fruitland Formation and Kirtland Shale of the San Juan Basin, New Mexico and Colorado: U.S. Geological Survey Professional Paper 676, 76 p.

Goolsby, S.M., Reade, N.S., and Murray, D.K., 1979, Evaluation of coking coals in Colorado: Colorado Geological Survey Resource Series 7, 72 p.

Kelso, B.S., Goolsby, S.M., and Tremain, C.M., 1980, Deep coal bed methane potential of the San Juan River coal region, southwestern Colorado: Colorado Geological Survey Open-File Report 80-2, 56 p.

Kelso, B.S., and Rushworth, Peter, 1982, Southern Ute/Department of Energy coalbed methane test wells: Colorado Geological Survey Open-File Report 82-4, 21 p.

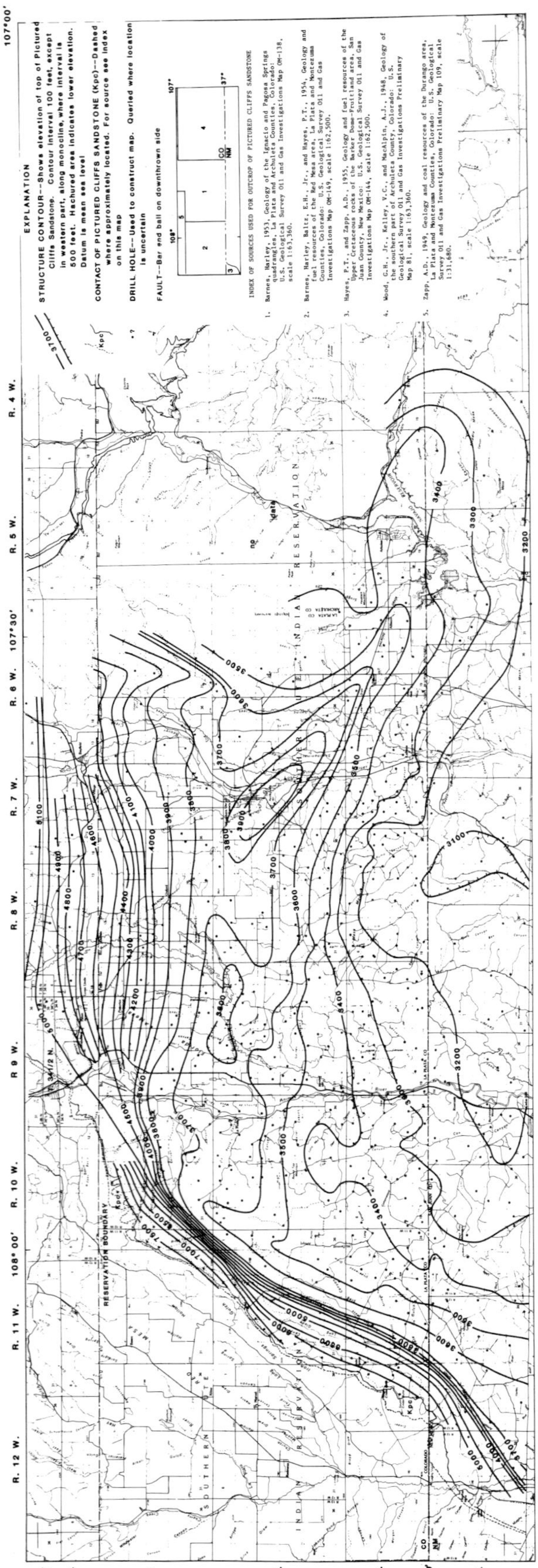

Figure 9. Structure contour map on the top of the Pictured Cliffs Sandstone, Southern Ute Indian Reservation and vicinity, Archuleta and La Plata Counties, Colorado, and Rio Arriba and San Juan Counties, New Mexico. Contour interval is 100 feet.

Kelso, B.S. and Wicks, D. E., 1988, A geologic analysis of the Fruitland Formation coal and coal-bed methane resources of the San Juan Basin, southwestern Colorado and northwestern New Mexico, *in* Fassett, J. E. ed., Geology and coal-bed methane resources of the northern San Juan Basin, Colorado and New Mexico: Rocky Mountain Association on Geologists Guidebook

Khalsa, N.S., 1981, Chemical analyses of coal samples from San Juan River Region, in Colorado coal analyses 1976-1979: Colorado Geological Survey Information Series 10, p. 214-239.

McCartney, J.T., and Teichmuller, M., 1972, Classification of coals according to degree of coalification by reflectance of vitrinite component: FUEL, v. 51-1, p. 64-68.

Shomaker, J.W., and Feldman, S.C., 1978, 1978 re-evaluation of original strippable coal reserves, Picnic Flats area, Southern Ute project: unpublished report prepared for the Southern Ute Tribe, 21 p.

Shomaker, J.W., and Holt, R.D., 1973, Coal resources of Southern Ute and Ute Mountain Ute Indian Reservations, Colorado and New Mexico: New Mexico State Bureau of Mines and Mineral Resources Circular 134, 22 p.

Steven, T.A., Lipman, P.W., Hail, W.J., Jr., Barker, Fred, and Luedke, R.G., 1974, Geologic map of the Durango quadrangle, southwestern Colorado: U.S. Geological Survey Miscellaneous Investigations Series Map I-764, scale 1:250,000.

U.S. Bureau of Mines, 1937, Analyses of Colorado coals: U.S. Bureau of Mines Technical Paper 574, 327 p.

Wood, G.H., Jr., Kehn, T.M., Carter, M.D., and Culbertson, W.C., 1983, Coal resource classification of the U.S. Geological Survey: U.S. Geological Survey Circular 891, 65 p.

Identification and Significance of Coal-Bed Gas, San Juan Basin, Northwestern New Mexico and Southwestern Colorado

Dudley D. Rice, Charles N. Threlkeld, April K. Vuletich, and Mark J. Pawlewicz[1]

ABSTRACT

Coal beds are considered to be a major source of nonassociated gas in the Rocky Mountain basins. In the San Juan Basin, located in northwestern New Mexico and southwestern Colorado, significant quantities of natural gas are being produced from coal beds of the Upper Cretaceous Fruitland Formation and from adjacent sandstone reservoirs. Analysis of gas samples from the various producing intervals provided a means of determining their origin and of evaluating coal beds as source rocks.

The rank of coal beds in the Fruitland Formation in the central basin, where major production occurs, increases to the northeast and ranges from high-volatile bituminous A to medium-volatile bituminous (R_0 values range from 0.7 to 1.4 percent). On the basis of chemical and isotopic data and coal rank, the gases are interpreted to be thermogenic. Gases from the coal beds show little isotopic variation ($\delta^{13}C_1$ values range −43.6 to −40.5 ppt), are chemically dry ($C_1/C_{1\text{-}5}$ values are >0.99), and contain significant amounts of CO_2 (as much as 6 percent). These gases are interpreted to have resulted from devolatilization of the humic-type coal composed mainly of vitrinite. The primary products of this process are methane, CO_2, and water.

The coal-generated gas is usually contained in the coal beds, and has not been expelled and has not migrated into the adjacent sandstone reservoirs. In addition, the coal bed reservoirs produce a distinctive bicarbonate-type connate water and have higher reservoir pressures than adjacent sandstones. The combination of these factors indicates that coal beds are a closed reservoir system created by the gas, water, and associated pressures being trapped in the micropore structure.

In contrast, gases produced from overlying sandstones in the Fruitland Formation and the underlying Pictured Cliffs Sandstone have a wider range of isotopic values ($\delta^{13}C_1$ values range from −43.5 to −38.5 ppt), are chemically wetter ($C_1/C_{1\text{-}5}$ values range from 0.85 to 0.95), and contain less CO_2 (<2 percent). These gases are interpreted to have been derived from type III kerogen dispersed in marine shales of the underlying Lewis Shale and nonmarine shales of the Fruitland Formation.

In the underlying Upper Cretaceous Dakota Sandstone and Tocito Sandstone Lentil of the Mancos Shale, another gas type is produced. This gas is isotopically lighter and chemically wetter at the intermediate stage of thermal maturity than gases derived from dispersed type III kerogen and coal; it is interpreted to have been generated from type II kerogen.

Occasionally, gas produced from Fruitland coal beds or sandstone reservoirs in the Pictured Cliffs or Fruitland is typical of that produced from the adjacent reservoir. These gases probably either migrated from or were produced from the adjacent reservoir because of recent well drilling and completion practices.

INTRODUCTION

The San Juan Basin is a major hydrocarbon-producing province in the Rocky Mountains. Cumulative production for both the New Mexico and Colorado parts of the basin, as of January 1, 1983, was: 190 million bbls of oil, 14 trillion cubic feet (TCF) of nonassociated gas, 425 billion cubic feet (BCF) of associated gas, and 60 million bbls of condensate (Fassett, 1983; Irwin, 1983). Ultimate recovery of nonassociated gas was estimated to be 23 TCF (Pritchard, 1972). On the Four Corners platform and Chaco slope (fig. 1), oil and associated gas occur in structural and stratigraphic traps. In the generally structureless central basin (fig. 1), nonassociated gas and con-

[1] U. S. Geological Survey
Box 25046, MS 971, DFC
Lakewood, Colorado 80225

The study was partially funded by the Bureau of Indian Affairs. We are thankful to several companies, especially Amoco and Tenneco, for helping us obtain gas and core samples.

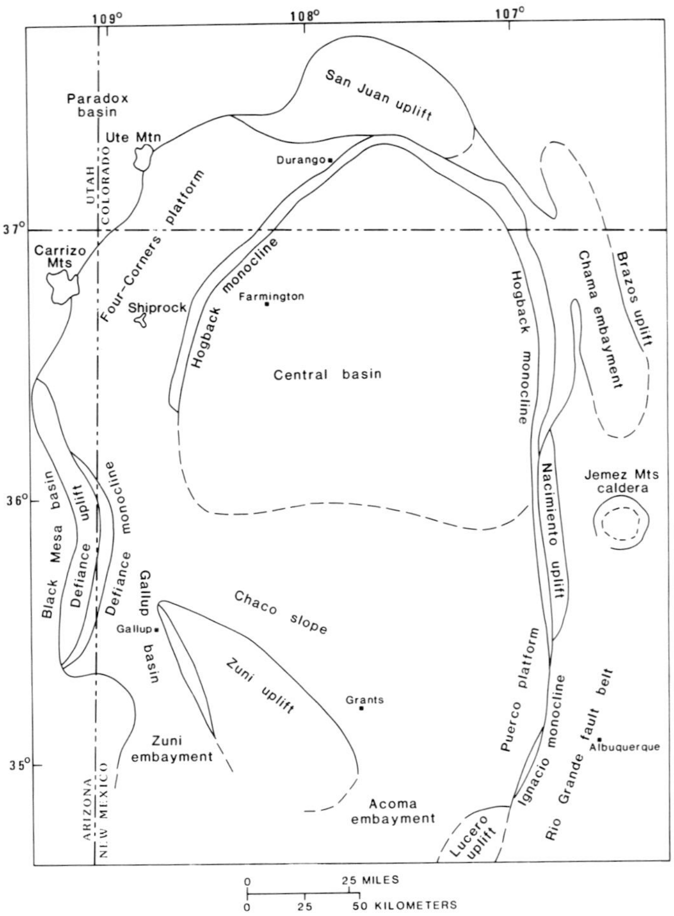

Figure 1. Index map of the San Juan basin showing major structural elements. After Fassett and Hinds (1971).

densate occur in underpressured sandstone reservoirs. Hydrodynamics (Berry, 1959) and low permeability are responsible for trapping the gas in this unusual type of accumulation. Coal beds have been considered as the source rocks for much of this gas.

Recently, there has been active exploration for gas from coal beds of the Upper Cretaceous Fruitland Formation, with the coal beds serving as both the reservoir and source rocks. While coal is present in other units, the major resources are in the Fruitland Formation. Very large gas resources are estimated to be contained in these unconventional, yet promising reservoirs. Meissner (1984) calculated that the Fruitland coal beds have generated about 26 TCF of gas in the thermally mature, north part of the basin. Choate and others (1984) estimated that the Fruitland coal beds have in-place gas resources of 31 TCF, while more recently, Kelso and others (1987) and Kelso and Wicks (this volume) calculated the in-place resources to be 56 and 50 TCF, respectively.

The objectives of this study are to: (1) characterize the chemical and isotopic composition of the natural gases produced from coal beds of the Fruitland Formation and the adjacent sandstone reservoirs, (2) determine the origin of these natural gases, and (3) characterize the coal petrography. Particular emphasis will be placed on identifying gas generated from coal beds and determining its contribution in charging adjacent reservoirs.

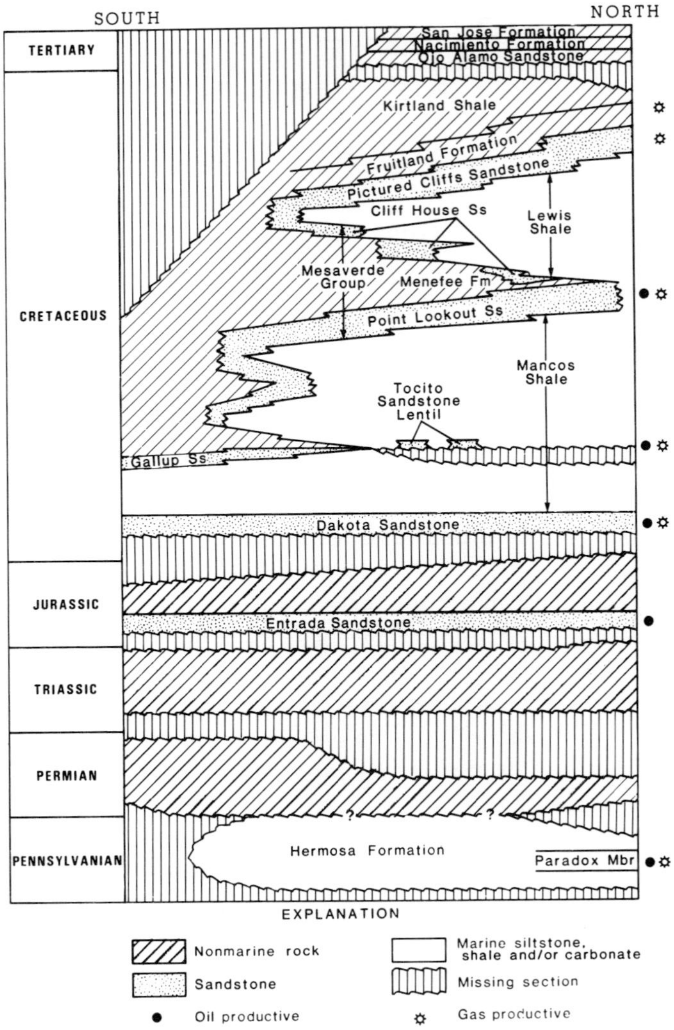

Figure 2. Correlation chart of Pennsylvanian and younger rocks in the San Juan basin showing productive intervals. After Molenaar (1983).

STRATIGRAPHY AS RELATED TO HYDROCARBON OCCURRENCE

The Cretaceous section is as much as 6,000 feet thick in the San Juan basin and contains the source and reservoir rocks for most of the hydrocarbon resources in the basin, especially the central basin. A correlation chart of Pennsylvanian and younger rocks is presented in figure 2. The Cretaceous forms a sequence of intertonguing marine and nonmarine deposits laid down along a fluctuating northwest-trending shoreline. The main source area was highlands to the southwest. Widespread coastal sandstones have been the primary reservoirs for natural gas. Fractured marine shales and fluvial sandstones are less important reservoirs. However, coal beds are becoming an increasingly important exploration target for gas accumulations. Marine and nonmarine shales and coals are the main source rocks. There are many excellent articles on the Cretaceous stratigraphy as related to hydrocarbon occurrences in the basin. They are summarized by Fassett and others (1978) and Fassett (1983) in papers on northwestern New Mexico, which are generally applicable to the Colorado part of the basin. Following is a discussion of producing intervals in the coal-bearing Fruitland Formation and adjacent units.

Pictured Cliffs Sandstone

The Pictured Cliffs Sandstone gradationally overlies the thick Lewis Shale and is the youngest marine unit in the basin. It was deposited at the time of final retreat of the epeiric sea from the basin area. In cross section, the Pictured Cliffs displays a significant stratigraphic rise from southwest to northeast across the basin (Fassett, 1987, and this volume). Hydrocarbon production is mainly nonassociated gas with very little condensate from sandstone reservoirs along a northwest-trending belt located mainly within the New Mexico part of the central basin. Low permeability caused by authigenic clay (Cumella,1981) and hydrodynamics are the main trapping mechanisms.

Fruitland Formation

The Fruitland Formation is composed of lower flood-plain and swamp deposits of interbedded sandstone, siltstone, shale, carbonaceous shale, and coal that intertongue with and overlie the Pictured Cliffs Sandstone. The Fruitland contains the largest reserves of coal in the basin (200 billion tons) and individual coal beds are thickest and most continuous in the lower part of the formation where they intertongue with shoreline sandstones of the Pictured Cliffs Sandstone (Fassett and Hinds, 1971). Because of the stratigraphic rise of the Pictured Cliffs, the basal coals become younger to the northeast (Fassett, 1987, and this volume). These thick, basal Fruitland coals are the primary target for gas exploration.

The Fruitland Formation also produces small amounts of nonassociated gas from lenticular fluvial channel sandstones in stratigraphic traps. These sandstones are above the basal coal zone and intertongue with shale, carbonaceous in part, and coal. In the central basin of Colorado, commingled production from the Pictured Cliffs Sandstone and Fruitland Formation is mainly structurally controlled (Craney, 1978).

NATURAL GAS GENERATION AND COMPOSITION

Natural gas is generated by two distinct processes. At shallow depths of burial and low temperatures, methane-rich gas is generated during decomposition of organic matter by microorganisms. This gas is formed in environments free of dissolved oxygen and sulfate, and is referred to as biogenic gas. Natural gas is also generated by thermal degradation and cracking processes with increasing temperature and advancing geologic time. This type of gas is referred to as thermogenic. Most gas produced from commercial oil and gas fields is probably of thermogenic origin.

Figure 3 illustrates the different gases that are generated from sapropelic and humic organic matter at various temperature ranges, assuming a normal geothermal gradient of about 1.7×F/100 ft. In general, sapropelic organic matter is comparable to types I and II kerogen, and humic organic matter is composed of type III kerogen. Both types of organic matter generate biogenic methane at low temperatures. With a sapropelic source, thermogenic methane generation is accompanied by heavier hydrocarbons, including oil, during the intermediate stages of thermal history (catagenesis). With increasing temperatures, lighter hydrocarbons are formed more rapidly because of thermal cracking processes affecting residual kerogen and heavier hydrocarbons previously formed. During the late stages of thermal history (metamorphism), methane is the main hydrocarbon product again resulting from cracking

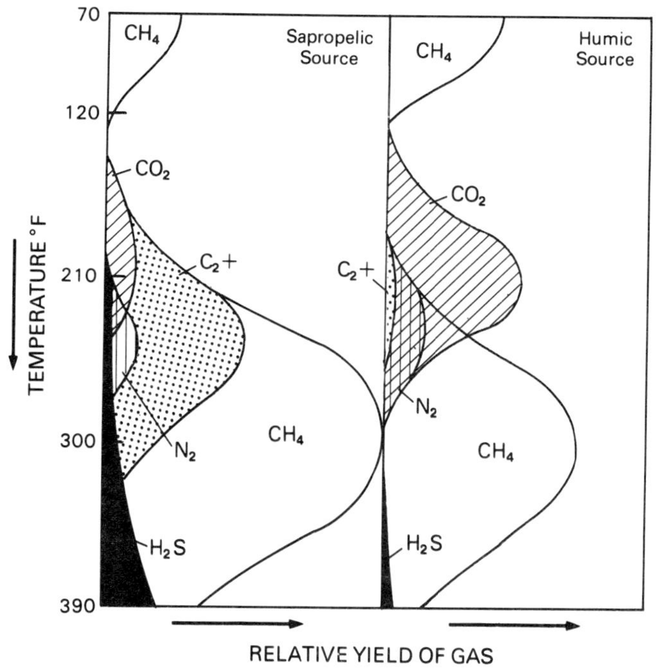

Figure 3. Diagram showing generation of gases from sapropelic and humic organic matter with increasing temperatures. Modified from Hunt (1979).

of existing hydrocarbons.

Coal, which is generally made up of humic organic matter, goes through progressive steps of devolatilization with increasing rank. Only small amounts of volatile matter are lost at the lower ranks. However, at ranks of high-volatile bituminous and higher, large amounts of volatile matter are lost yielding methane, carbon dioxide (CO_2), nitrogen, and water as the primary products. As shown in figure 3, the major generation stage of CO_2, which is also accompanied by large amounts of water, occurs prior to the main event of methane generation by thermogenic processes. Details of coal-bed gas generation are reviewed by Meissner (1984) and Rightmire (1984).

Natural gases can be distinguished by their chemical composition, as just discussed, and by their isotopic composition (Schoell, 1983). Biogenic gas consists predominantly of methane that is isotopically light ($\delta^{13}C_1$ values lighter than -55 ppt) because of biological ^{12}C-enrichment (Rice and Claypool, 1981). During intermediate stages of thermal history, thermogenic methane, which is often accompanied by heavier hydrocarbons, is isotopically heavier than biogenic methane ($\delta^{13}C_1$ values of about -55 to -35 ppt) because of smaller kinetic isotopic effects associated with thermal cracking processes. During late stages of thermal history and at high temperatures, natural gas becomes devoid of all heavier hydrocarbons and the isotopic composition of the methane approaches that of the original matter ($\delta^{13}C_1$ values are heavier than -35 ppt).

THERMAL MATURITY AND COAL RANK PATTERNS

Across the central basin, vitrinite reflectance values (R_0) of Fruitland coal samples increase to the northeast from 0.46 percent to 1.45 percent near the northern end in Colorado (Rice, 1983). Thermal maturity patterns for older rock units show a

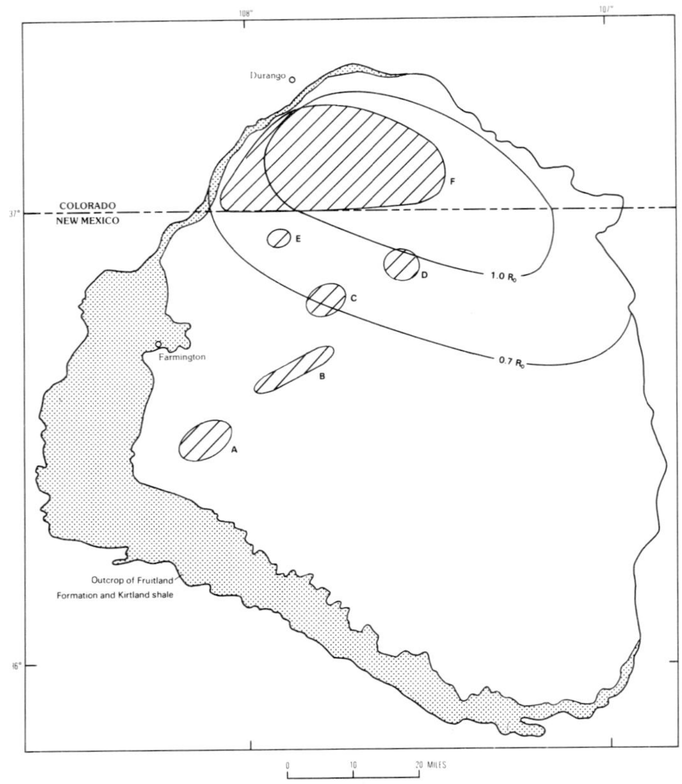

Figure 4. Map of central basin showing isoreflectance (Ro) lines on coals of Fruitland Formation. Study areas are labeled with letters and hachured.

similar trend to that of the Fruitland coals, except that they are higher because of greater depth of burial. On the basis of vitrinite reflectance values, the coal beds increase in rank from subbituminous in the southern part to medium-volatile bituminous at the extreme northern end. For sapropelic organic matter, an R_o value of 0.6 percent is considered as the beginning of major oil generation, while a value of 1.3 percent is thought to be the end of the oil generation window. For coals, significant thermal methane generation probably begins at an R_o value of about 0.7 percent or a rank of high-volatile A bituminous (Meissner, 1984). On figure 4, isoreflectance lines of 0.7 and 1.0 percent are shown. Individual data points can be obtained in Rice (1983) and Rice and Threlkeld (in press).

The northeast trend of increasing thermal maturity within the central basin can generally be explained by the structure of the basin and the overburden. That is, subsidence and the resulting thickness of the deposits overlying the Fruitland gradually increase to the northeast in the central basin (Fassett and Hinds, 1971). However, maximum depth of burial combined with present-day geothermal gradients cannot totally account for the very high coal ranks at the extreme north end. Both Rice (1983) and Bond (1984) concluded that the maturation process was enhanced by an increased geothermal gradient associated with emplacement of the nearby San Juan batholith complex during mid-Tertiary time (Tweto, 1975). Clarkson and Reiter (1987, and this volume) postulated that heat advection by groundwater movement may have also influenced the thermal maturation levels. They proposed that both magmatic activity associated with the emplacement of the San Juan batholith and elevated geothermal gradients associated with the steady-state thermal source at depth may have heated the groundwater.

There is a sudden reversal in the northeast pattern of increasing thermal maturity on the north rim of the basin where the level rapidly drops off (fig.4). This results in a pattern where the isoreflectance lines probably parallel the structure of the steeply dipping west, north, and east flanks. This rapid decrease on the north rim is attributed to the greatly reduced amount of overburden that was deposited in this area during Late Cretaceous through mid-Tertiary time when the structure of the basin was evolving.

GAS GEOCHEMISTRY AND ORIGIN

Gas samples were collected and analyzed from 40 producing wells from 6 separate areas within the central basin (fig. 4) where production has been established in the Upper Cretaceous Fruitland coal beds. In these areas, the vitrinite reflectance values of the coals increase to the northeast from a minimum of about 0.5 percent in area A to a maximum of 1.4 percent in area F. Gas samples were taken and analyzed from both the coal-bed and adjacent sandstone reservoirs. The following numbers of samples were analyzed from each producing interval: 13 from the Pictured Cliffs Sandstone, 22 from the basal Fruitland coals, 3 from the Fruitland sandstones above the basal coals, and 2 from commingled Fruitland and Pictured Cliffs.

Gas analyses are summarized in table 1. The volume percentage of selected components is reported, together with the proportion of methane in the hydrocarbon fraction ($C_1/C_{1\text{-}5}$) and the stable carbon isotopic ratio ($\delta^{13}C_1$) of the methane component. In addition, gas samples were also collected and analyzed from the Upper Cretaceous Dakota Sandstone, Tocito Sandstone Lentil of the Mancos Shale, Mesaverde Group, and Tertiary Nacimiento Formation (fig. 2). Detailed chemical and isotopic analyses of these samples are given in Rice (1983) and Rice and Threlkeld (in press). The emphasis of this paper is on comparing gases produced from the Fruitland coal beds and adjacent sandstone reservoirs in the Pictured Cliffs and Fruitland.

In general, natural gases from all producing intervals in the basin become isotopically heavier ($\delta^{13}C_1$ values range from -48.7 to -31.4 ppt) and chemically drier ($C_1/C_{1\text{-}5}$ values range from 0.75 to 0.99) with increasing depth of burial (1,100 to 8,600 feet) (table 1; Rice, 1983; Rice and Threlkeld, in press). Also, the gases from the individual producing intervals display a similar trend of becoming isotopically heavier and chemically drier in a northeast direction. These gas data, in addition to associated vitrinite reflectance values, indicate that the gases are thermogenic in origin and were generated at intermediate and late stages of thermal history. The scatter in the chemical and isotopic composition of the gases with increasing depth of the gas pool is probably due to gases of distinct composition being generated from different types of source rocks (fig. 5). These gas types will become obvious when the individual producing intervals are discussed next.

The Pictured Cliffs Sandstone produces nonassociated gas and minor amounts of condensate (0.006 gal/MCF) in the central basin. This liquid to gas ratio is much lower than ratios of nonassociated gas production from the Dakota Sandstone and Mesaverde Group. The gases from the Pictured Cliffs in the 6 areas shown on figure 4 were generated at

TABLE 1. **Chemical and isotopic composition of natural gases in the six study areas shown on figure 4. Wells in each area are arranged by producing interval, township, range, and section. tr = trace.**

Well name	Location	Depth to Gas Pool (ft)	Producing Interval	N_2 and (or) air	C_1	CO_2	C_2	C_3	$_iC_4$	$_nC_4$	$_iC_5$	$_nC_5$	C_1/C_{1-5}	$^{13}C_1$ (ppt)
Area A														
Hard Deal No. 3	sec. 18, T. 26 N., R. 12 W.	1,124	Pictured Cliffs	5.64	85.81	0.95	4.74	1.91	0.44	0.28	0.18	0.03	0.919	-42.8
Mike No. 1J	sec. 20, T. 26 N., R. 12 W.	1,090	------do----------	17.42	75.16	0.88	4.68	1.4	0.31	0.14	0.12	0	0.92	-42.3
Western Federal No. 7	sec. 7, T. 26 N., R. 11 W.	1,210	Fruitland coal	4.89	94.38	0.13	0.49	0.08	0.02	0	0	0	0.994	-46.5
Chaco Plant No. 28	sec. 13, T. 26 N., R. 12 W.	1,098	------do-----------	5.72	93.35	0.12	0.64	0.15	0.02	0	0	0	0.991	-46.6
Nassau No. 5R	sec. 36, T. 27 N., R. 12 W.	1,434	Fruitland/ Pictured Cliffs	8.98	88.04	0.22	1.85	0.64	0.13	0.1	0.03	0.01	0.97	-43.7
Area B														
Florence No. 86	sec. 26, T. 29 N., R. 9 W.	2,334	Pictured Cliffs	3.71	85.3	0.33	6.2	2.78	0.58	0.72	0.05	0.32	0.889	-42.1
Knauff No. 1	sec. 31, T. 28 N., R. 10 W.	1,515	Fruitland coal	11.24	82.33	0.18	4.24	1.35	0.33	0.19	0.06	0.08	0.929	-42.4
Florence No. 27A	sec. 26, T. 29 N., R. 9 W.	2,042	------do-----------	1.79	88.42	0.19	5.63	2.52	0.46	0.57	0.25	0.17	0.902	-42.1
Area C														
Elliot Gas Com. "W" No. 1	sec. 9, T. 30 N., R. 9 W.	2,972	Pictured Cliffs	6.77	85.13	7.45	0.34	0.29	0.01	0.01	0	0	0.993	-42.9
Florence No. 81	sec. 14, T. 30 N., R. 9 W.	2,672	------do-----------	1.62	88.5	1.19	4.84	2.63	0.53	0.63	0.24	0.16	0.907	-42.5
Mansfield No. 2	sec. 19, T. 30 N., R. 9 W.	2,886	------do-----------	2.75	82.43	1.23	7.67	3.65	0.64	0.97	0.38	0.21	0.859	-41.9
Florence No. 59	sec. 23, T. 30 N., R. 9 W.	2,641	------do-----------	2.15	84.35	1.01	7.28	3.31	0.6	0.79	0.3	0.2	0.871	-41.6
Florence No. 20A	sec. 24, T. 30 N., R. 9 W.	2,544	------do-----------	6.9	82.05	0.87	6.22	2.62	0.47	0.57	0.2	0.1	0.89	-41.4
Pritchard No. 9	sec. 1, T. 30 N., R. 9 W.	2,470	Fruitland coal	3.82	88.52	6.13	0.96	0.41	0.06	0.06	0.02	0.01	0.983	-43.1
Florence No. 4A	sec. 10, T. 30 N., R. 9 W.	2,715	------do-----------	4.44	89.36	5.66	2.66	0.18	0.05	0.03	0.01	0.01	0.994	-43.6
Florence No. 8A	sec. 14, T. 30 N., R. 9 W.	2,518	------do-----------	4.15	87.66	0.33	4.01	2.57	0.56	0.51	0.02	0.19	0.918	-43.9
Mansfield No. 1A	sec. 19, T. 30 N., R. 9 W.	2,667	------do-----------	4.2	84.03	0.46	6.13	3.24	0.68	0.75	0.32	0.19	0.881	-42.4
Florence No. 6A	sec. 23, T. 30 N., R. 9 W.	2,641	------do-----------	2.15	84.35	1.01	7.28	3.31	0.6	0.79	0.3	0.2	0.871	-41.6
Elliot "C" No. 1	sec. 9, T. 30 N., R. 9 W.	2,608	Fruitland sandstone	8.52	85.46	1.12	2.83	1.33	0.32	0.27	0.11	0.04	0.946	-42.6
Area D														
N.E. Blanco Unit No. 215	sec. 26, T. 31 N., R. 7 W.	3,274	Pictured Cliffs	5.74	79.82	0.57	7.84	4.13	0.7	0.87	0.05	0.27	0.852	-43.5
N.E. Blanco Unit No. 211	sec. 27, T. 31 N., R. 7 W.	3,300	------do-----------	5.29	85.18	0.93	5.1	2.32	0.38	0.49	0.17	0.13	0.908	-42.9
N.E. Blanco Unit No. 212	sec. 10, T. 31 N., R. 7 W.	3,174	Fruitland coal	5.26	88.78	5.23	0.41	0.16	0.01	0.02	0.01	0.12	0.992	-43.5
San Juan 32-7 Unit No. 6	sec. 17, T. 31 N., R. 7 W.	3,054	------do-----------	11.5	84.4	3.67	0.29	0.11	0.01	0.01	0	0	0.995	-42.4
N.E. Blanco Unit No. 213	sec. 12, T. 31 N., R. 7 W.	3,200	Fruitland/ Pictured Cliffs	6.05	78.01	0.61	9.38	4.1	0.68	0.78	0.26	0.13	0.836	-42.6
Area E														
Gardner Gas Com. No. 1	sec. 33, T. 32 N., R. 10 W.	2,709	Pictured Cliffs	6.62	84.84	1.67	4.56	1.55	0.28	0.33	0.1	0.05	0.925	-39.1
Schneider Gas Com. "B" No. 15	sec. 29, T. 32 N., R. 10 W.	2,826	Fruitland coal	8.41	87.5	3.89	0.18	tr	0	0	0	0	0.998	-42.8
State Gas Com. "BW" No. 1	sec. 32, T. 32 N., R. 10 W.	2,680	------do-----------	7.21	87.36	5.18	0.8	0.05	0	0	0	0	0.997	-43
Cahn Gas Com. No. 1	sec. 33, T. 32 N., R. 10 W.	2,795	------do-----------	6.99	86.8	5.94	0.19	0.04	0	0	0	0	0.997	-43
Keys Gas Com. "E" No. 1	sec. 27, T. 32 N., R. 10 W.	2,402	Fruitland sandstone	8.89	83.91	1.93	3.42	1.26	0.2	0.24	0.09	0.04	0.941	-38.5
Area F														
Trail Canyon No. 3-10	sec. 10, T. 32 N., R. 8 W.	4,066	Pictured Cliffs	0.96	87.76	0.74	7.0	2.27	0.45	0.48	0.21	0.14	0.893	-41.9
Ute No. 11	sec. 36, T. 33 N., R. 9 W.	3,919	------do-----------	5.16	84.15	0.73	6.61	2.1	0.45	0.44	0.2	0.13	0.894	-41.9
Glover No. 1	sec. 2, T. 32 N., R. 6 W.	3,048	Fruitland coal	0.93	95.63	2.87	0.39	0.17	0	0	0	0	0.994	-40.5
So. Ute Tribal No. 5-1	sec. 3, T. 33 N., R. 6 W.	3,908	------do-----------	5.6	87.07	0.58	5.38	1.07	0.16	0.13	0	0	0.928	-41.6
So. Ute Tribal No. D-1	sec. 17, T. 33 N., R. 7 W.	2,547	------do-----------	7.36	90.03	2.48	0.09	0.05	0	0	0	0	0.999	-42.0
So. Ute Gas Com. No. E-3	sec. 21, T. 33 N., R. 7 W.	2,344	------do-----------	9.18	88.58	2.15	0	0	0	0	0	0	0.999	-42.2
Mayfield No. IOU-2	sec. 10, T. 34 N., R. 8 W.	2,282	------do-----------	6.09	93.53	0.19	0.18	0	0	0	0	0	0.998	-42.6
Rea No. 18-1	sec. 18, T. 34 N., R. 8 W.	2,534	------do-----------	42.3	57.5	0.1	0	0	0	0	0	0	1	-42
Lemon No. 18-1	sec. 18, T. 34 N., R. 8 W.	2,403	------do-----------	2.45	97.03	0.52	0	0	0	0	0	0	1	-42.8
Ute No. 1	sec. 12, T. 34 N., R. 9 W.	2,396	------do-----------	7.89	91.63	0.37	0	0	0	0	0	0	1	-43.6
Ute No. 15	sec. 21, T. 32 N., R. 11 W.	2,624	Fruitland sandstone	9.69	76.8	1.55	6.36	3.56	0.79	0.73	0.3	0.19	0.865	-39.0

moderate levels of thermal maturity (R_0 values range from 0.5 to 1.1 percent), are of intermediate isotopic composition ($\delta^{13}C_1$ values range from -43.5 to -39.5 ppt), and are chemically wet ($C_1/C_{1\text{-}5}$ values range from 0.85 to 0.93). This composition suggests generation during intermediate stages of thermal history, when oil should have been generated if type II kerogen was present in sufficient amounts. These nonassociated gases are interpreted to have been generated from dispersed type III kerogen in the underlying marine Lewis Shale. This type III kerogen resulted from either in situ oxidation of marine type II kerogen and (or) from transportation and subsequent deposition of terrestrial kerogen into the offshore setting of Lewis Shale deposition. The other possibility is that the gases were generated in and migrated from the adjacent coals in the intertonguing Fruitland Formation. However, that gas produced in the coal beds is distinctly different in composition, which will be discussed later.

The Pictured Cliffs gases also contain as much as 2 percent CO_2, which is interpreted to have been generated by the thermal decomposition of the oxygen-containing functional groups in the kerogen (Hunt, 1979). Type III kerogen is considered to be the major source of this organically derived CO_2.

The Fruitland Formation produces natural gas from the basal coal beds in all 6 areas (fig. 4), and occasionally from stratigraphic traps in overlying channel sandstones. Gas is produced and was sampled from coal beds ranging in rank from high-volatile C bituminous (R_0 value of about 0.5 percent) to medium volatile bituminous (R_0 value of 1.4 percent),

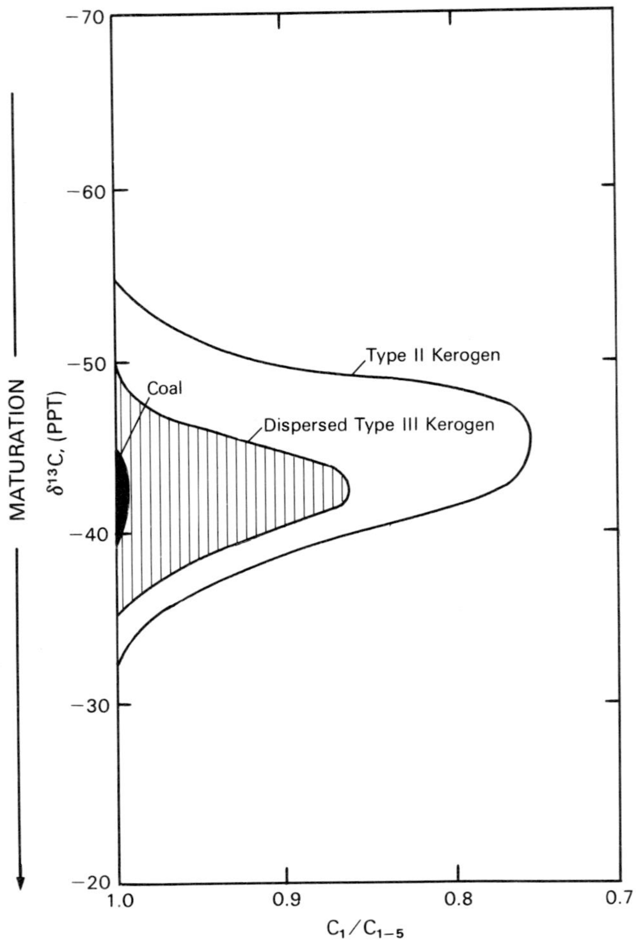

Figure 5. Diagram showing changes in chemical and isotopic composition of natural gas with different source rock types.

which is equivalent to intermediate and high levels of thermal maturity. In the northeast part of area C and all of areas D-F and, where reflectance values are 0.7 percent or greater, gases from the Fruitland coals are distinct from those previously described. They are characterized by: (1) high amounts of methane ($C_1/C_{1\text{-}5}$ values are generally greater than 0.99), (2) significant amounts of CO_2 (as much as 6 percent), except the extreme north end of area 6 where they are less than 1 percent, and (3) a narrow range of carbon isotopic values for the methane fraction ($\delta^{13}C_1$ values range from -40.5 to -43.6 ppt). In addition, there is a decrease or loss of C_{2+} gas components to the northeast where levels of thermal maturity are higher. Gas desorbed from the Fruitland coal core in area F is similar to nearby produced gas with the following exceptions: (1) the volume percentage of CO_2 is higher and (2) the carbon isotopic composition of the methane is heavier (Rice and Threlkeld, in press).

These gases produced from the coal beds are interpreted to be the product of coalification and coincident devolatilization, which yields methane, CO_2, nitrogen, and water. Although most coals are generally composed of type III kerogen, the gas generated from the Fruitland coals is different in composition from that generated from dispersed type III kerogen in the Lewis Shale and that produced from the Pictured Cliffs Sandstone (fig.5). This distinct type of gas generally occurs only in the coal beds and has not migrated into adjacent reservoirs.

All of the coal-bed gas wells in areas C-F produce large quantities of methane, CO_2, and water. With increasing production time, the wells generally show an increase in gas production and a decrease in water production. Although the amount of CO_2 decreases with increasing rank, the overall volume percentage in produced coal-bed gas is much less than that generated, because it is highly soluble in water and is very mobile. Nitrogen migrates more easily than the other gases with which it is generated because of its small molecular size, and is only a minor constituent of the Fruitland coal-bed gas.

This coal-bed gas is, at a glance, quite similar in chemical composition to nonassociated gases produced from the Dakota Sandstone and Mesaverde Group at the north end of the central basin in Colorado as described by Rice and Threlkeld (in press). The hydrocarbon fraction consists mainly of methane, and CO_2 is the chief nonhyrocarbon component (as much as 6.6 percent). However, the gases from the Dakota and Mesaverde are not interpreted to have been derived from coal beds, and they were generated at much higher levels of thermal maturity. As a result, where these gases are mostly methane ($C_1/C_{1\text{-}5}$ values greater than 0.99), the $\delta^{13}C_1$ values are heavier than -35 ppt. The higher amounts of CO_2 in gases of these older units, which increase with depth, are attributed to a volcanic or magmatic origin.

In area B, the southwest part of area C, and infrequent wells in the other northern areas, gas produced from the Fruitland coal beds is similar in chemical and isotopic composition to gas produced from the adjacent Pictured Cliffs reservoirs. This suggests that, because of methods of drilling and completing wells, (1) gas of a different composition and origin has migrated into the coal beds from the adjacent Pictured Cliffs Sandstone reservoir, or (2) gas is actually draining from the Pictured Cliffs reservoir. In area A, the gas produced from Fruitland coal beds, at R_0 value of about 0.5 percent, is distinct from that typical of the underlying Pictured Cliffs Sandstone or that of the coal beds at R_0 values of 0.7 percent or greater. Here, the gas from the coal beds is isotopically light ($\delta^{13}C_1$ values of -46.6 to -46.5 ppt), chemically dry ($C_1/C_{1\text{-}5}$ values of 0.99), and has minute amounts of CO_2 (0.1 percent). The origin of this gas is interpreted to be either early thermogenic or biogenic with the heavy $\delta^{13}C_1$ values resulting from generation by acetate dissimilation, instead of CO_2 reduction (Jenden and Kaplan, 1986; Whiticar and others, 1986).

In contrast to the typical coal-bed gas, gases analyzed from channel sandstones in the Fruitland Formation above the basal coal zone in areas C, E, and F are chemically wetter ($C_1/C_{1\text{-}5}$ values are 0.87 to 0.95), and generally isotopically heavier ($\delta^{13}C_1$ values range from -42.6 to -38.5 ppt). The gas is similar in composition to that produced from the Pictured Cliffs Sandstone, which was interpreted to have been generated from dispersed type III kerogen in the underlying Lewis Shale. The sandstone reservoirs in the upper part of the Fruitland are separated from the Pictured Cliffs reservoirs by productive coal beds with a different type of gas. As a result, the upper Fruitland gases are interpreted to have been generated from dispersed type III kerogen in the enclosing nonmarine shales, which are carbonaceous in part.

In the Ignacio Blanco field, gas production has been commingled from the Pictured Cliffs and Fruitland. Because most of the wells are depleted, no gas samples were obtained for our analysis. Based on published (Craney, 1978) and unpublished chemical analyses from commercial laboratories, the gases are interpreted to have been generated from the coal beds and not from dispersed type III kerogen in the Lewis Shale or Fruitland Formation. This assumption is confirmed by the fact that Pictured Cliffs production does not extend into Colorado because of low permeability and because production from stratigraphic traps in the Fruitland sandstones is localized.

The two gas types generated from coal and dispersed type III kerogen, as described previously, are dissimilar from those produced from the underlying Dakota Sandstone and Tocito Sandstone Lentil of the Mancos Shale (Rice, 1983). This latter type of gas is isotopically lighter and chemically wetter at equivalent levels of thermal maturity, especially during intermediate stages, than gases derived from coal and dispersed type III kerogen (fig.5). This gas is interpreted as having been generated from type II kerogen.

The possibility exists that the gases produced from sandstone reservoirs in the Pictured Cliffs and Fruitland were actually generated in and migrated from the Fruitland coal beds. Recent work by Fuex (1980) suggests that isotopic fractionation of gases by migration is minor and only ^{13}C is depleted slightly. Bernard and others (1977) and Coleman and others (1977) have noted a significant change in the chemical composition of gases with migration, that is, gases become chemically drier or more enriched in methane during migration.

Although the gases produced from the coal-bed and sandstone reservoirs are similar in isotopic composition, they are quite different in chemical composition. The coal gas is mostly methane with as much as 6 percent CO_2; the gas in the sandstone reservoirs contains as much as 15 percent C_{2+} components and less than 2 percent CO_2. Based on the chemical composition alone, it does not seem possible to have derived by migration the gas produced from the sandstone reservoirs from gas generated in the coal beds.

COAL PETROGRAPHY

Core samples of Fruitland coal were examined from 2 wells located in areas E and F (fig.5). Macroscopically, the coal samples are made up of clarain, containing alternating bright and dull bands. The bright or highly reflective bands indicate a high percentage of vitrinite, while the relatively dull bands suggest some material from the exinite and inertinite maceral groups.

The coal samples were prepared by crushing to a minus 60 fraction, riffling several times, and quartering. The remaining coal was cast with epoxy in one inch molds and centrifuged at about 2,800 RPM for 2 minutes to settle the particles. The resulting pellets were allowed to set up over night. The pellets were then ground under a stream of water by 240 grit paper followed by 600 grit paper. Coarse polishing was done by hand with 0.3 micron slurry on a nylon cloth-covered wheel turning at about 60 RPM. Final polishing was with a 0.03 micron polish on a felt cloth at the same speed. The pellets were cleaned ultrasonically and dried before examination with reflected light under oil immersion.

Point counting indicated an abundance of vitrinite and much lower amounts of the exinite and inertinite maceral groups. In all of the samples, the vitrinite component exceeded 80 percent. The inertinite maceral varied from less than 5 percent to more than 16 percent and occurred as small shards scattered throughout the vitrinite and with some larger pieces of fusinite distinguished by its characteristic cellular structure. Exinite was always a minor component ($<$2 percent).

Based on maceral composition, the coals are classified as humic. Geochemically, humic coals are composed of type III kerogen which generates mostly methane-rich gas. The Fruitland coals, then, are considered to be their own hydrocarbon source and the gas was probably generated in situ with little migration.

The most striking petrographic aspect of the samples was the holes or voids in the vitrinite. Some occur as ovoids and long, thin stringers that parallel bedding indicative of exinite macerals, others in sawtoothed shapes indicative of cutinite. The filling of the holes or voids was first thought to be clay, mineral matter, or possibly micrinite, which may be a degradation product of exinite. A positive identification could not be made under white light.

As much as 70 percent of the individual void fillings of one sample fluoresced with a dull orange-brown color when examined under ultraviolet (UV) light. All of the voids in the samples contain some hydrogen-rich (aliphatic) material that fluoresced. In some cases, the material was definitely exinite. The remainder of the fluorescing material is probably bitumenite.

If the void fillings had been oil, it would have been destroyed during sample preparation. Oil generated in situ and type I organic matter would have dissolved when immersion oil, used for petrography, or organic solvent, used to clean the pellets, was applied. Considering the occurrence of the material in the coal's pores and the low permeability of the coal, it is unlikely that the material would have entered by migration. The fact that the fluorescent material is prevalent, and that the fluorescence is uniform, suggests that it was generated in situ. The presence of the fluorescent material, both within large voids in the coal and within the cells of the fusinite, also implies in situ generation. The large size of the voids, about 2 to 8 microns, suggests generation early in the coal's diagenetic history.

SUMMARY AND CONCLUSIONS

The San Juan basin of northwestern New Mexico and southwestern Colorado contains major reserves of hydrocarbons, particularly nonassociated gas. The gas occurs mainly in the central basin where a combination of low permeability and hydrodynamics are the main trapping mechanisms.

Cretaceous rocks are the main reservoir and source rocks in the central basin. Coastal marine sandstones are the most important reservoir rocks, and enclosing marine and nonmarine shales and coals are considered as the major source rocks. Coal beds in the Fruitland Formation are estimated to contain major resources of nonassociated gas and have been the target for recent exploration.

Within areas of current production from coal beds of the Fruitland Formation, vitrinite reflectance values increase to the northeast from subbituminous to medium-volatile bituminous. The high level of thermal maturity in the north probably resulted from a combination of maximum depth of burial and a higher geothermal gradient during mid-Tertiary time due to emplacement of the San Juan batholith.

Gases analyzed from Cretaceous reservoirs throughout the central basin show a trend of becoming isotopically heavier and chemically drier with increasing depth of burial and thermal maturity. The gases are interpreted to be thermogenic in origin and were generated during the mature and post-mature stages of hydrocarbon generation. Scatter in the general trend is attributed to differences in the type of source rock.

Gases produced from the Pictured Cliffs Sandstone are nonassociated, and were generated during the mature stage of hydrocarbon generation. The gas was probably generated from type III kerogen dispersed in the underlying, marine Lewis Shale. Minor amounts of CO_2 resulted from thermal decomposition of oxygen-containing functional groups in the kerogen.

Small quantities of nonassociated gas produced from sandstone reservoirs in the upper part of the Fruitland Formation are similar in composition and origin to gases of the Pictured Cliffs Sandstone. This gas was probably derived from dispersed type III kerogen in the enclosing, nonmarine shales.

Nonassociated gas produced from the Fruitland coal beds is different from that produced from adjacent reservoirs. The coal gas is composed mainly of methane with decreasing amounts of CO_2 accompanying increasing levels of thermal maturity. The gases are the product of devolatilization and significant methane generation probably began at a rank of high-volatile bituminous A. This distinct gas is generally confined to the coal beds and has not migrated to adjacent reservoirs, either above or below.

In addition to containing a distinct type of gas, the Fruitland coal-bed reservoirs are also characterized by a different produced water composition and higher reservoir pressures than adjacent sandstone reservoirs. These factors indicate that the coal beds form a closed reservoir system. The coal beds produce a bicarbonate-type connate water, which probably resulted from geochemical processes associated with the early stages of coalification (Decker and others, 1987, and this volume). Hydrocarbon generation is regarded as the principal mechanism for overpressuring in the Rocky Mountain region (Spencer, 1987). In the San Juan Basin, higher reservoir pressures in the coal-bed reservoirs are considered to be residual overpressuring caused by gas generation in the low permeability coal beds. The adjacent sandstone reservoirs, such as the Pictured Cliffs and Fruitland, have lower pressures because of uplift, erosion, and cooling since active hydrocarbon generation in mid-Tertiary time. However, the coal beds have retained at least part of the higher pressures because the gas is retained in the coal's micropore structure. This storage phenomenon has prevented the gas from being expelled and migrating into adjacent reservoirs.

Occasionally, the coal beds and associated sandstones produce gas that is characteristic of the adjacent reservoir. As a result of drilling and completing wells, these gases are interpreted to have either migrated from or have drained from an adjacent reservoir.

The Fruitland coals are composed mainly of vitrinite and are capable of generating mostly methane-rich gas. Therefore, they can be considered as their own source. The coals have numerous voids that are filled with a hydrogen-rich material that fluoresces and was probably generated in situ.

REFERENCES CITED

Bernard, B., Brooks, J.M., and Sackett, W.M., 1977, A geochemical model for characterization of hydrocarbon gas sources in marine sediments: 9th Annual Offshore Technology Conference Proceedings, v. 3, p. 435-438.

Berry, F.A., 1959, Hydrodynamics and geochemistry of the Jurassic and Cretaceous Systems in the San Juan Basin, northwestern New Mexico and southwestern Colorado: Ph.D. dissertation, Stanford University, 192 p.

Bond, W.A., 1984, Application of Lopatin's method to determine burial history, evolution of the geothermal gradient, and timing of hydrocarbon generation in Cretaceous source rocks in the San Juan basin,northwestern New Mexico and southwestern Colorado, *in* Woodward, J., Meissner, F.F., and Clayton, J.L., eds., Rocky Mountain Association of Geologists Guidebook, p. 433-448.

Choate, R., Lent, J., and Rightmire, C.T., 1984, Upper Cretaceous geology, and the potential for methane recovery from coalbeds in San Juan basin-Colorado and New Mexico, *in* Rightmire, C.T., Eddy, G.E., and Kirr, J.N., eds., Coalbed methane resources of the United States, American Association of Petroleum Geologists Studies in Geology Series No.17, p. 185-222.

Clarkson, G., and Reiter, M., 1987, The thermal regime of the San Juan basin since Late Cretaceous times and its relationship to San Juan Mountains thermal sources: Journal of Volcanology and Geothermal Research, v. 31, p. 217-237.

______, 1988, An overview of geothermal studies in the San Juan Basin, New Mexico and Colorado, *in* Fassett, J. E., ed., Geology and coal-bed methane resources of the northern San Juan Basin, New Mexico and Colorado: Rocky Mountain Association of Geologists Guidebook

Coleman, D.D., Meents, W.F., Liu, C.L., and Keogh, R.A., 1977, Isotopic identification of leakage gas from underground storage reservoirs-a progress report: Illinois State Geological Survey, Illinois Petroleum III, 10p.

Craney, D.L.,1978, Ignacio Blanco Fruitland-Pictured Cliffs, *in* Fassett, J.E., Thomaidis, N.D., and Matheny, M.L., eds., Oil and gas fields of the Four Corners area: Four Corners Geological Society, v.1, p.134-136.

Cumella, S.P., 1981, Sedimentary history and diagenesis of the Pictured Cliffs Sandstone, San Juan basin, New Mexico and Colorado: Master's thesis, University of Texas, Austin, 219p.

Decker, D., Klusman, R., and Horner, D.M., 1987, Geochemical techniques applied to the identification and disposal of connate coal water: Proceedings of the 1987 Coalbed Methane Symposium, p.229-242.

Decker, D., Cooper, J. D., and Wicks, D. E., 1988, Geology, geochemistry , reservoir engineering, and completion methods at the Cedar Hill field, San Juan County, New Mexico: A field study of classic coal-degasification behavior, *in* Fassett, J. E., ed, Geology and coal-bed methane resources of the northern San Juan Basin, New Mexico and Colorado: Rocky Mountain Association of Geologists Guidebook

Fassett, J.E., 1983, Stratigraphy and oil and gas production of northwest New Mexico updated through 1983, *in* Fassett, J.E., ed., Oil and gas fields of the Four Corners area: Four Corners Geological Society, v.1, p.849-863.

______, 1987, Geometry and depositional environments of Fruitland Formation coal beds, San Juan basin, New Mexico and Colorado: Anatomy of a giant coal-bed methane deposit: Proceedings of the 1987 Coalbed Methane Symposium, p.19-35.

______, 1988, Geometry and depositional environments of Fruitland Formation coalbeds, San Juan basin, New Mexico and Colorado: Anatomy of a giant coal-bed methane deposit, *in* Fassett, J. E., ed, Geology and coal-bed methane resources of the northern San Juan Basin, New Mexico and Colorado: Rocky Mountain Association of Geologists Guidebook

Fassett, J.E., and Hinds, J.S., 1971, Geology and fuel resources of the Fruitland Formation and Kirtland Shale of the San Juan Basin, New Mexico and Colorado: U.S. Geological Survey Professional Paper 676, 76p.

Fassett, J.E.,Arnold, E.C., Hill, J.M., Hatton, K.S., Martinez, L.B., and Donaldson, D.A., 1978, Stratigraphy and oil and gas production of northwest New Mexico, *in* Fassett, J.E., Thomaidis, N.D., and Matheny, M.L., eds., Oil and gas fields of the Four Corners area: Four Corners Geological Society, v.1, p.46-61.

Fuex, A.N., 1980, Experimental evidence against an appreciable isotopic fractionation of methane during migration, *in* Advances in organic geochemistry, 1979: New York, Pergamon Press, p.725-732.

Hunt, J.M., 1979, Petroleum geochemistry and geology: W.H. Freeman and Co., San Francisco, 617p.

Irwin, C.D., 1983, Stratigraphy and oil and gas production of southwest Colorado, *in* Fassett, J.E., ed., Oil and gas fields of the Four Corners area: Four Corners Geological Society, p.844-845.

Jenden,P.D., and Kaplan, I. R., 1986, Comparison of microbial gases from the Middle America Trench and Scripps Submarine Canyon: implications for the origin of natural gas: Applied Geochemistry, v.1, p.631-646.

Kelso, B.S., Decker, A.D., Wicks, D.E., and Horner, D.M., 1987, GRI geologic and economic appraisal of coalbed methane in the San Juan basin: Proceedings of the 1987 Coalbed Methane Symposium, Tuscaloosa, Alabama, p.119-130.

Kelso, B.S., and Wicks,D.E., 1988, A geologic analysis of the Fruitland Formation coal and coal-bed methane resources of the San Juan Basin, southwestern Colorado and northwestern New Mexico, *in* Fassett, J. E., ed, Geology and coal-bed methane resources of the northern San Juan Basin, New Mexico and Colorado: Rocky Mountain Association of Geologists Guidebook

Meissner, F.F., 1984, Cretaceous and lower Tertiary coals as sources for gas accumulations in the Rocky Mountains area, *in* Woodward, J.J., Meissner, F.F., and Clayton, J.L., eds. Rocky Mountain Association of Geologists 1984 Guidebook, p.401-431.

Molenaar, C.M., 1983, Time-stratigraphic nomenclature chart for the San Juan basin, *in* Fassett, J.E., ed., Oil and gas fields of the Four Corners area: Four Corners Geological Society, p.830.

Pritchard, R.L., 1972, The San Juan basin, *in* Mallory, W.W., ed., Geologic atlas of the Rocky Mountain region: Rocky Mountain Association of Geologists, p.284-286.

Rice, D.D., 1983, Relation of natural gas composition to thermal maturity and source rock type in San Juan basin, northwestern New Mexico and southwestern Colorado: American Association of Petroleum Geologists Bulletin, v.67, p.1199-1218.

Rice, D.D., and Claypool, G.E., 1981, Generation, accumulation, and resource potential of biogenic gas: American Association of Petroleum Geologists, v.65, p.1199-1218.

Rice, D.D., and Threlkeld, C.N., in press, Composition and origin of natural gases, Southern Ute Indian Reservation and vicinity, southwestern Colorado: U. S. Geological Survey Bulletin.

Rightmire, C.T., 1984, Coalbed methane resource, *in* Rightmire, C.T., Greg, G.E., and Kirr, J.N., eds., Coalbed methane resources of the United States, American Association of Petroleum Geologists Studies in Geology Series No.17, p.1-14.

Schoell, M., 1983, Genetic characterization of natural gases: American Association of Petroleum Geologists Bulletin, v.67, p. 2225-2238.

Spencer, C.W., 1987, Hydrocarbon generation as a mechanism for overpressuring in Rocky Mountain region: American Association of Petroleum Geologists Bulletin, v.67, p. 2225-2238.

Tweto, O., 1975, Laramide (Late Cretaceous-early Tertiary) orogeny in the southern Rocky Mountains: Geological Society of America Memoir 144, p.1-44.

Whiticar, M.J., Faber, E., and Schoell, M., 1986, Biogenic methane formation in marine and freshwater environments: CO_2 reduction vs. acetate fermentation—isotopic evidence: Geochimica et Cosomochimica Acta, v.50, p.693-709.

By 1923, the housewife found a greater selection when she went to the market, and the friendly clerks and owners Thomas Mason (second from right) and Walt Ambold (left) of the Mason & Ambold grocery were ready for her. According to their advertisement, they were exclusive agents for Chase & Sandborn coffee and teas, besides offering staple and fancy groceries, meats, and bakery products. (Photo and caption courtesy of the Center of Southwest Studies.)

Joint Patterns on the Northwest Side of the San Juan Basin (Southern Ute Indian Reservation), Southwest Colorado

Steven M. Condon[1]

ABSTRACT

Joints in sandstone and cleats in coal were measured at 37 stations on and near the western part of the Southern Ute Indian Reservation. The study area is in the northwest part of the San Juan Basin in southwestern Colorado. The main structural elements within the study area are the Four Corners platform on the west, the Hogback monocline, and the San Juan Basin on the east. Geologic units that were studied were the Cretaceous Cliff House Sandstone of the Mesaverde Group, Pictured Cliffs Sandstone, Fruitland Formation, Farmington Sandstone Member of the Kirtland Shale, and McDermott Member of the Animas Formation; Tertiary units studied in less detail were the Nacimiento Formation and the San Jose Formation. Joints of all the rock units are similar, and can be discussed together. Four discrete sets of joints present in this area were designated as J1, J2, J3, and J4. The J1 set is the oldest and best-defined set; its average strike is N. 15° E. Joints of the J1 set, as well as those of the other sets are subvertical. Joints of the J2 set meet J1 joints at approximately right angles and terminate against J1 joints; the J2 joints strike N. 78° W. on average. The J3 set was formed later than both the J1 and J2 sets because J3 joints terminate against J1 and J2 joints. J3 joints strike N. 53° E. Joints of the J4 set are the least well formed and were formed last. They terminate against the joints of the other three sets and strike N. 43° W. Cleats in coal of the Fruitland Formation are subvertical and have orientations similar to the J3 and J4 joint sets; however, the relative age of the face and butt cleats suggests that they did not form at the same time as the J3 and J4 joint sets.

INTRODUCTION

This study of joint patterns on the Southern Ute Indian Reservation (fig. 1) was initiated in 1987 to determine whether hydrocarbon production on the Reservation was controlled by natural fractures. Because of the importance of coal-bed methane resources in this area, the Pictured Cliffs Sandstone, Fruitland Formation coals, and the Farmington Sandstone Member of the Kirtland Shale were the units of principal interest. Joint data were also collected from the Cliff House Sandstone of the Mesaverde Group, the McDermott Member of the Animas Formation, and the Nacimiento and San Jose Formations (fig 2). The study area is in the northwest part of the San Juan Basin and the eastern part of the Four Corners platform, southwest Colorado. These structural elements are separated by the Hogback monocline (fig. 1).

The purpose of this study is twofold: First, knowledge of joint characteristics, and particularly joint orientations, in different-aged rocks and in different areas may aid in the siting of new wells for optimum hydrocarbon recovery. (See other papers in this volume for a discussion of oil and gas production in this area.) Second, a determination of the sequence of joint development will provide data to help reconstruct the tectonic history of this area.

Although the orientations of joint sets differ slightly in each of the rock units studied, the joints of all units will be summarized together here. Within each set, there appears to be as

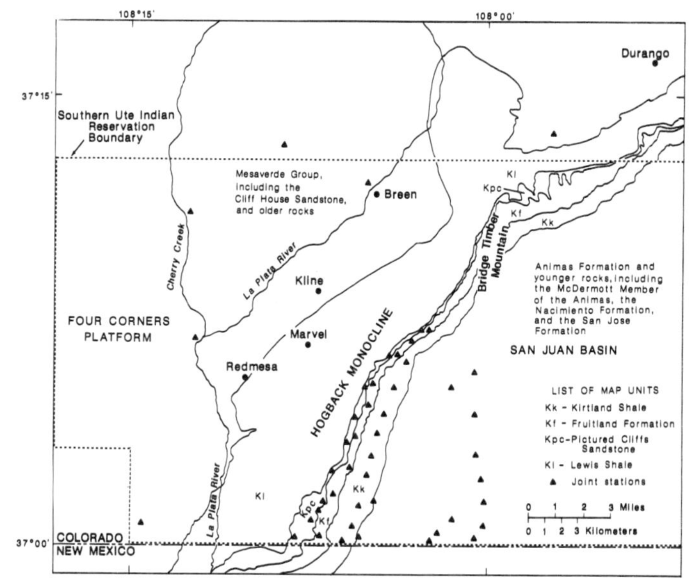

Figure 1. Generalized geologic map of the study area showing joint station locations and major structural elements.

[1] U.S. Geological Survey
P. O. Box 25046, MS 939 DFC
Lakewood, Colorado 80225

Thanks are extended to C. C. Barton and R. L. Wheeler for their helpful discussions of joints and methods of studying joints. E. R. Verbeek, M. A. Grout, and J. E. Fassett critically reviewed this manuscript and provided many instructive comments. Particular thanks are expressed to M. A. Grout for her interest in this project and for her time spent tutoring in both the office and in the field.

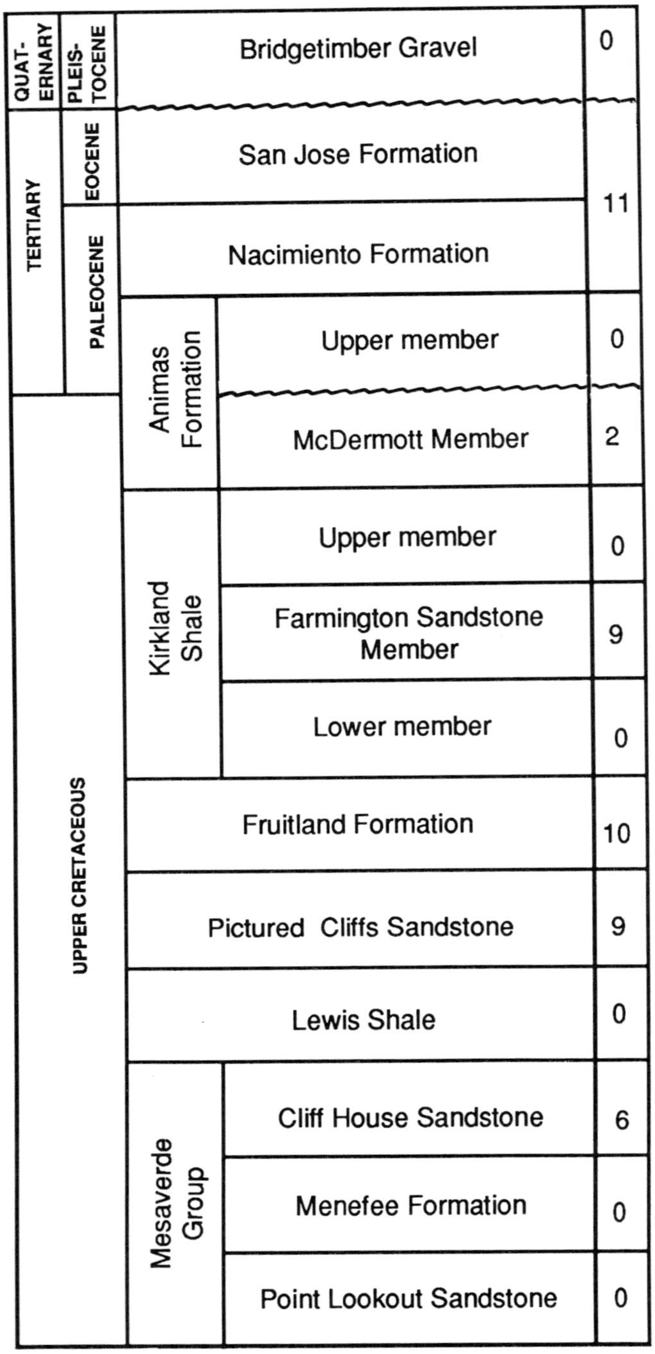

Figure 2. Chart showing stratigraphic units exposed at the surface in the study area. The number of joint stations in each unit is indicated in the right column.

much variability in strike between joints in different geographic areas as there is between joints in different-aged rocks. Because this study is not yet complete no attempt is made in this report to interpret the tectonic history of the northern San Juan Basin. This study will continue on the eastern side of the Reservation where the monoclinal rim of the San Juan Basin trends to the southeast, and in the south- central part of the Reservation, the area of greatest gas production.

PREVIOUS WORK

This is the first detailed study of joints concentrated on the Southern Ute Reservation; however, earlier studies of larger areas included the present study area. Kelley and Clinton (1960) mapped linear features from aerial photographs over most of the Colorado Plateau, which includes the Southern Ute Reservation. These linear features included traces of fractures seen directly on exposed rock surfaces, vegetation patterns, drainage patterns, and other features such as veins and dikes. All of these features were assumed to represent fractures or to be controlled by fractures, though little field work was performed to confirm that assumption. The fracture network along the Hogback monocline was characterized as being heterogeneous by Kelley and Clinton (1960, p. 20).

Badgley (1962, 1965) recorded lineament and fracture trends from aerial photographs and made field observations in the Four Corners area of southwest Colorado, northwest New Mexico, southeast Utah, and northeast Arizona. On the basis of his own studies and those of Kelley and Clinton (1960), Badgley defined four main lineament trends that had strikes of N. 5° E., N. 75° W., N. 50° E., and N. 40° W. Badgley also defined seven joint sets that had strikes of N. 5° E., N. 70° W., N. 50° E., N. 45° W., N. 15° W., N. 75° E., and N. 55° W. Badgley (1965, p. 134) noted that the joint pattern was dominated by longitudinal joints that parallel the Hogback monocline. He considered other joints to be diagonal shear fractures; however, this interpretation was based on the geometric relationship of the diagonal joints to longitudinal joints and was not confirmed by other features of the joints such as terminations of one set against another or slickenside striations.

Hodgson (1961) studied joints in an area of Utah and Arizona just to the west of the Southern Ute Reservation. His study was a field-based examination of joint-surface features, spacings, and terminations of joints that serves as the basis for many present-day joint studies. The orientations of joints mapped by Hodgson (1961, pl. I) were presented in neither tabular form nor plotted at individual stations on maps, but the strike of joints on his interpretive maps are similar to patterns determined by the present study of the Southern Ute Reservation.

SETTING

This study was conducted in the area along and adjoining the Hogback monocline (fig. 1). Most of the readings were taken in the Redmesa and Pinkerton Mesa 7½ minute quadrangles in T. 32 and 33 N., R. 11 and 12 W. (Colorado). Other scattered readings extended as far north as T. 35 N., and from R. 10 W. to R. 13 W. Rocks both east and west of the Hogback monocline dip from 1 to 5 degrees to the southeast. Along the monocline, dips commonly range from 6 to 15 degrees to the southeast, and increase to a maximum of about 30 degrees locally. Strikes along the monocline commonly range from about N. 20° E. to N. 50° E., although there are numerous segments that strike nearly due north. There are few faults in the study area; those found are located on the Hogback monocline transverse to the strike of the beds. These normal faults have offsets ranging from a few feet to a few tens of feet.

Rocks from which readings were taken are Late Cretaceous and early Tertiary in age (fig. 2). The oldest unit studied is the Upper Cretaceous Cliff House Sandstone of the Mesaverde Group. The Cliff House is pale to medium brown, very fine to fine grained, well-sorted sandstone. The unit changes from a massive sandstone, 150-250 feet thick, in the southwest part of the study area to a sequence of thin, interbedded sandstones and shales, 50 feet thick or less, in the northeast part of the area. Sandstone beds are moderately cemented with calcite; quartz overgrowths are also a constituent of the cement. Bedding of the Cliff House is very thin to very thick; the geometry of individual sandstone beds ranges from tabular to lenticular. The Cliff House is locally disconformable on the underlying Menefee Formation of the Mesaverde Group and is conformably overlain by the Lewis Shale. The Cliff House was deposited as near-shore sands during a southwestward transgression of the Late Cretaceous sea. Readings of joints in the Cliff House Sandstone were taken at six stations arrayed on the Four Corners platform to the west, northwest, and north of the Hogback monocline (fig. 1).

The Pictured Cliffs Sandstone is pale gray to pale brown and is divisible into two parts, a lower part that consists of interbedded sandstone and shale, and an upper part that is massive sandstone. The lower part is 25-50 feet thick and contains tabular sandstone beds ranging from a few inches to 2-3 feet thick. Sandstone beds in the lower part are firmly cemented with calcite. The upper part of the Pictured Cliffs is 75 feet or more thick, and is commonly less firmly cemented than the lower part, although there are scattered areas where the unit is very calcareous. The Pictured Cliffs conformably overlies the Lewis Shale and was deposited as regressive-marine sands. Readings were taken at nine stations in the Pictured Cliffs Sandstone along the Hogback monocline (fig. 1).

The Fruitland Formation is a heterogeneous continental deposit of lenticular sandstone beds, shale and mudstone, and coal. For this study the joints in coal, which are called cleats, were the features that were measured. The coal beds of the Fruitland Formation occur mainly at or near the base of the unit, although some beds occur more towards the middle. Because the Fruitland weathers readily, complete sections of coal beds are rarely exposed; thicknesses of as much as 10-15 feet were recorded for this study. The coal is highly jointed and commonly contains thin beds of shale and carbonaceous shale. The Fruitland Formation conformably overlies the Pictured Cliffs Sandstone and is conformably overlain by the Kirtland Shale. Readings of coal cleat orientations in the Fruitland Formation were taken at ten stations along the Hogback monocline (fig. 1).

The Farmington Sandstone Member of the Kirtland Shale is a sandy, fluvial interval that lies in the middle of the Kirtland Shale (fig. 2). The Farmington Member consists of a series of lenticular channel sandstone beds separated by mudstone. For this study, at any given locality the lowest channel sandstone in the sequence was examined because of its proximity to the potential source of gas in the underlying Fruitland coals.

Channel deposits of the Farmington Member consist of greenish-gray to grayish-brown, fine- to medium-grained, moderately sorted sandstone. Each channel sandstone is composed of smaller sandstone lenses separated by clay drapes. Individual lenses are crossbedded and commonly have concentrations of clay clasts at their bases. The small lenses are 1-4 feet thick and the composite channels are up to 20 feet thick; they pinch out laterally within a few hundred feet. The Farmington is weakly to moderately cemented, with a corresponding moderate to high porosity. The upper foot or two of the channel sandstones is locally well-cemented with calcite. Readings of joints in the Farmington Sandstone Member of the Kirtland Shale were taken at nine stations along the Hogback monocline (fig. 1).

The McDermott Member of the Animas Formation is a unique and interesting geologic unit and is the youngest Upper Cretaceous rock unit of the San Juan Basin. The McDermott consists of greenish-gray, grayish-yellow, and pale purple, very poorly sorted conglomerate, conglomeratic sandstone, sandstone, and mudstone. Pebbles and cobbles in the McDermott, which measure up to about 8 inches in diameter in the study area, are composed of quartzite and andesite, whose source was possibly in the area of the present-day La Plata Mountains to the north (Fassett, 1985, p. 319). The McDermott was deposited in a high energy fluvial environment. Cementation of the McDermott is moderate to weak. In the course of taking readings from one outcrop, it was discovered that parts of the McDermott are highly magnetic. Only two stations were established in the McDermott (fig. 1), and only the northern one displayed aberrant magnetic properties, so the extent of this magnetic anomaly is not known. The McDermott Member conformably overlies the Kirtland Shale and is unconformably overlain by the Tertiary upper member of the Animas Formation (Fassett, 1985, p. 323). The stations in the McDermott are in the south-central part of the study area (fig. 1), just to the east of the upper contact of the Kirtland Shale.

The Tertiary Nacimiento and San Jose Formations are the youngest formations studied for this project. The formations were deposited as fluvial channel sands and overbank muds. These units are similar lithologically and are difficult to distinguish in the field; the contact between them is indefinite in many places (Haynes and others, 1972; Steven and others, 1974). In the southern part of the study area the Nacimiento and San Jose Formations consist largely of massive channel sandstones. These sandstones are yellowish gray to pale brown, fine- to medium-grained, and well sorted. Channel deposits are as thick as 25-30 feet in places, but are generally not laterally continuous. Channel-filling sandstones of the Nacimiento and San Jose are moderately to weakly cemented with calcite, resulting in high porosity.

Sandstone decreases to the north in the Nacimiento and San Jose where both units consist mainly of mudstone with minor thin sandstone ledges. The Nacimiento conformably overlies and grades laterally into the upper part of the Animas Formation. The San Jose unconformably overlies the Nacimiento in the northern part of the San Juan Basin, although there is no distinctive angular relationship between the formations where they were studied. To the northeast of the study area, at Bridge Timber Mountain (fig. 1), an angular relationship between the San Jose and older formations is apparent. Readings in the Nacimiento and San Jose were taken at eleven localities in the study area (fig. 1). All the eastern joint stations shown on figure 1 are in the upper part of the Nacimiento Formation or in the lower part of the San Jose Formation.

JOINTS IN SANDSTONE

Methods of study

In his studies of Jurassic rocks in southeast Utah and northeast Arizona, Hodgson (1961) pioneered the modern study of fractures. Other more recent examples of joint studies that introduced new techniques include Wheeler and Stubbs (1979), Kulander, Barton, and Dean (1979), Grout and Verbeek (1983, 1985), and Verbeek and Grout (1983, 1984a, 1984b). Of particular benefit to the present study was the paper of Grout and Verbeek (1983) which detailed the techniques used to study Cretaceous and Tertiary rocks farther to the north in Colorado.

Following the precedent set in earlier studies of joints, several features of joints and their host strata on the Southern Ute Indian Reservation were recorded. These included the orientation of bedding, tortuosity or shape of joints (planar versus sinuous), dimensions of exposed joint faces, spacing between joints, surface markings, aperture, terminations, mineral fillings or coatings, number of beds cut by individual joints, sense and magnitude of any offset, mode of failure, effect of rock fabric, and relative age of joint sets. For a detailed discussion of the measurable features of joints that are listed above, the reader is referred to Grout and Verbeek (1983). Because the present study was an attempt to relate joint patterns to possible hydrocarbon production, it focused on the orientations of joints, abutting relationships of joints, and relative ages of joint sets.

Readings were taken of about 1,600 joints and coal cleats at 37 stations and at additional localities in the study area. At each station different joint sets were first identified by observing orientations and abutting relationships and then, where possible, the orientation of about 12-15 joints of each set were recorded. (The exception to this is in the Nacimiento and San Jose Formations. In these formations the readings were more of a reconnaissance nature because of poor exposures and irregular development of joints in friable sandstones. More work will be done on these units in subsequent studies.) The style of each joint set was then characterized by recording the features listed above, such as shape, spacing, and terminations of the measured joints. At stations where the dip of the rocks was greater than 6 degrees the dip was mathematically rotated back to horizontal. The recorded joint orientations were also adjusted an equal amount at these stations. Additionally, the entire outcrop of the Pictured Cliffs Sandstone was walked from the Colorado-New Mexico State line northeastward to the northernmost station in that unit, and readings of joints were taken at approximately ¼ mile intervals at 50 supplementary locations. All of the readings taken were from sandstone beds; mudstone beds were generally too poorly exposed to get an adequate number of readings.

The joint data presented in this paper were compiled using a microcomputer program developed by R.A. Thompson of the U.S. Geological Survey. After readings of individual joints were combined and an average orientation for each joint station was calculated, the structural dip was rotated out where appropriate. These individual station resultants were then combined to get regional averages for the joint sets. The computer program also was used to generate stereonet plots. These plots are equal-area, lower-hemisphere projections of the station averages, not projections of all the individual joint readings.

Joint set designations

On the basis of orientations and abutting relationships among coexisting joints, four discrete sets, termed J1, J2, J3, and J4, were identified in this study. The J1 set is the most well-developed and oldest set of the four; the J4 set is the least well-developed and the youngest set. Joints of all four sets are vertical to sub-vertical in the sub-horizontally bedded rocks of this area. Where the rocks dip more steeply, the majority of the joints are also inclined, but remain perpendicular to bedding surfaces. Rock fabric, such as crossbedding or current lineations, did not have any discernible effect on the joint sets.

J1 joints

The average regional orientation of joints of the J1 set for all geologic units studied is N. 15° E., with a dip of 86° to the northwest (table 1). Figure 3 shows the average orientation of J1 joints at the individual station locations throughout the area; figure 4a shows a stereonet plot of poles to J1 joints. Because the dip of J1 joints and joints of the other sets are vertical to subvertical, only the strikes of the joints are shown on figure 3 and subsequent figures.

Joints of the J1 set are long, planar, and cut multiple beds. The true length of J1 joints is rarely exposed because of cover; however, exposed lengths of 20-30 feet are common, and lengths of over 100 feet were recorded at several stations. The averaged orientations of J1 joints show relatively little scatter from one station to another (fig. 4a). Spacings depend on the thickness of the beds and their degree of cementation and range from as close as 6 inches to as wide as 20 feet. Spacings of 1 to 3 feet are most common. Thinner, more firmly cemented beds have closer-spaced joints than do thicker beds that are only moderately cemented. Most of the joint surfaces examined were quite weathered, which destroyed surface features; however, feather-like plumose structures, arrest lines, and twist hackle were noted on some joint surfaces, showing that they formed as extension joints. (See Kulander, Barton, and Dean, 1979, for a discussion of surface features of joints.)

The aperture of joints of the J1 set ranges from no measurable aperture to ⅛ inch. As used here, aperture refers to open space between opposite joint walls, regardless of whether or not there are mineral coatings. Successful measurement of

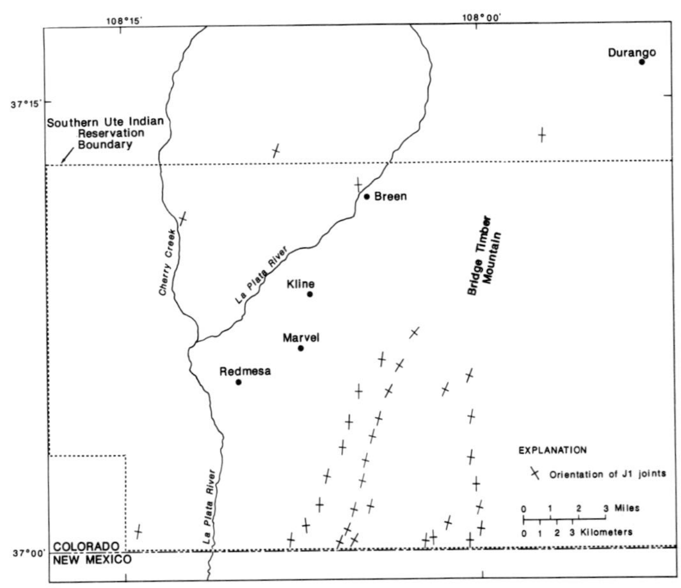

Figure 3. Map showing orientations of J1 joints. Each plotted orientation represents an average of 12 to 15 readings, where possible. See text for exceptions.

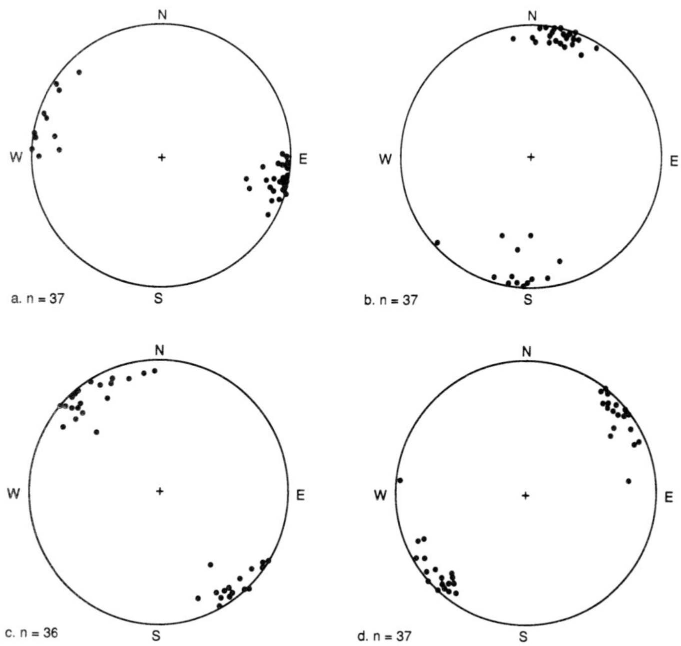

Figure 4. Stereonet plots of poles to joints. Plots are equal-area, lower-hemisphere projections of averages at individual stations. a. Joints of the J1 set. b. Joints of the J2 set. c. Joints of the J3 set. d. Joints of the J4 set.

aperture of joints at the surface depends on the character of the outcrop; apertures of joints seen only in plan view are more difficult to measure than of joints in cross-section view because dirt and other debris tends to accumulate on horizontal rock surfaces, which obscures the joints. The recorded apertures of joints in surface outcrops are probably greater than those which would be found at depth. Weathering of joint surfaces and small movements of individual rock blocks undoubtedly contribute to increasing apertures of exposed joints with time. Where calcite fillings are present, J1 joints are commonly cemented closed and have no open space. J1 joints are present in all the outcrops examined and are extensive both laterally and vertically. J1 joints commonly die out as hairline cracks within the rock, or the ends of J1 joints hook to meet adjacent, parallel J1 joints. J1 joints are not crossed by joints of the other sets; the J1 joints appear to have formed the earliest, and joints of other sets terminate against J1 joints. There were no slickenside striations noted on any joints of the J1 set, suggesting that there was no lateral movement of the joint surfaces.

J2 joints

Joints of the J2 set strike N. 78° W. and dip 87° SW on average (table 1). Figure 5 shows J2 joint orientations at the stations. Like J1 joints, a plot of poles to J2 joints is fairly tightly clustered (fig. 4b). J2 joints are also planar and cut multiple beds; their lengths are limited by the spacings of joints of the J1 set, which range from 6 inches to 20 feet. Spacings of the J2 joints also range from 6 inches to 20 feet or more; spacings of 2-4 feet are common, so that J2 joints in most places are about as abundant as J1 joints. Surfaces of the J2 joints are also commonly weathered; however, some

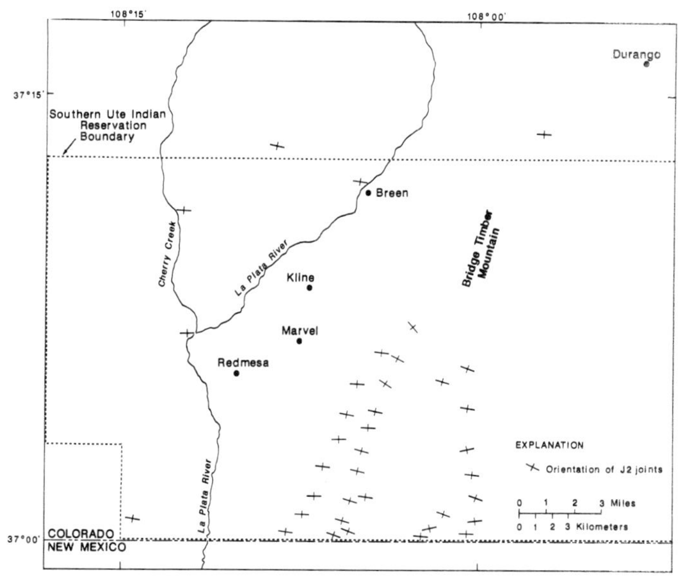

Figure 5. Map showing orientations of J2 joints. Each plotted orientation represents an average of 12 to 15 readings, where possible. See text for exceptions.

plumose structures and arrest lines were noted. Apertures of J2 joints are similar to those of the J1 joints, and calcite fillings are present locally. J2 joints terminate only against J1 joints. There also were no slickenside striations noted on any of the J2 joints.

J3 joints

Joints of the J3 set are oriented an average of N. 53° E. and dip 87° SE (table 1, fig. 6). In most areas joints of the J3 set are well developed; however, they are least well expressed and are locally nearly absent in the Farmington Sandstone Member of the Kirtland Shale. Poles to station averages of J3 joints are somewhat more scattered than those of J1 or J2 joint sets (fig. 4c). J3 joints are relatively long and planar and commonly have exposed lengths of 3-10 feet; true lengths are rarely seen. Where present, the J3 joints cut multiple beds. Spacings between J3 joints range from 6 inches to 10 feet or more but they are spaced from 2-5 feet apart in many places. Surface features are rarely seen because of the weathered nature of the joint surfaces, although J3 joints in one outcrop showed excellent twist hackle and others displayed plumose structures. Apertures are variable but are as wide as ⅛ inch; some J3 joints are completely cemented closed by calcite fillings. J3 joints terminate against both J1 and J2 joints where all three sets are present; however, it is most common for the pairs of J1-J2 and J3-J4 joints to occur together with only moderate overlap of the pairs of sets. In the areas where only the J3 and J4 sets are present, the J3 set dominates the fracture network. No evidence for lateral movement on J3 joints was seen. On the basis of the observed abutting relationships, the J3 set appears to be younger than the J1 and J2 sets, and older than the J4 set.

J4 joints

The overall orientation of joints of the J4 set in the area is N. 43° W. and a dip of 89° SW (table 1). Figure 7 shows the readings taken at individual stations. As with the J3 set, J4

TABLE 1. Average joint and cleat orientations and dips in Upper Cretaceous and lower Tertiary rocks on the western side of the Southern Ute Indian Reservation.

	Joint set			
Geologic unit	**J1**	**J2**	**J3**	**J4**
Cliff House Sandstone	N9E/88NW	N83W/85SW	N44E/81SE	N49W/87SW
Pictured Cliffs Sandstone	N4E/87NW	N73W/88NE	N41E/88NW	N42W/88NE and N4E/90 (secondary)
Farmington Sandstone Member of Kirtland Shale	N18E/81NW	N72W/84SW	N64E/88NW	N45W/86NE
McDermott Member of Animas Formation	N23E/81NW	N73W/88SW	N58E/89NW	N24W/83SW
Nacimiento and San Jose Formations	N12E/89NW	N82W/70NE	N56E/89SE	N39W/88SW
Regional averages	N15E/86NW	N78W/87SW	N53E/87SE	N43W/89SW
	Coal cleats			
	Face		**Butt**	
Fruitland Formation	N21W/90		N48E/89NW	

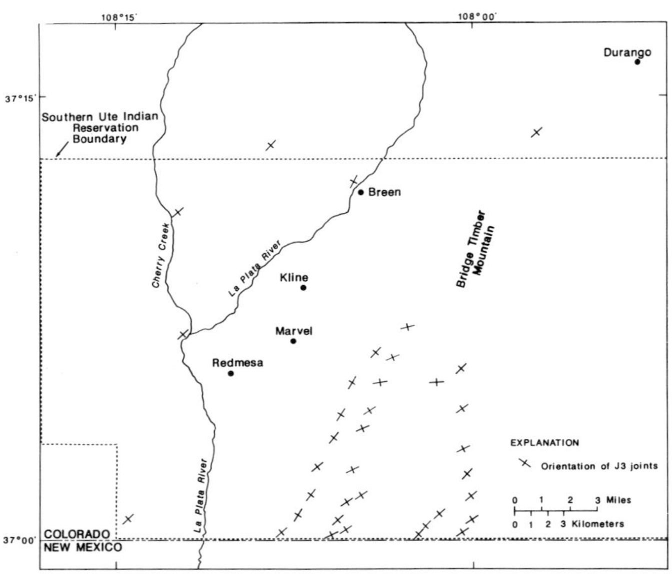

Figure 6. Map showing orientations of J3 joints. Each plotted orientation represents an average of 12 to 15 readings, where possible. See text for exceptions.

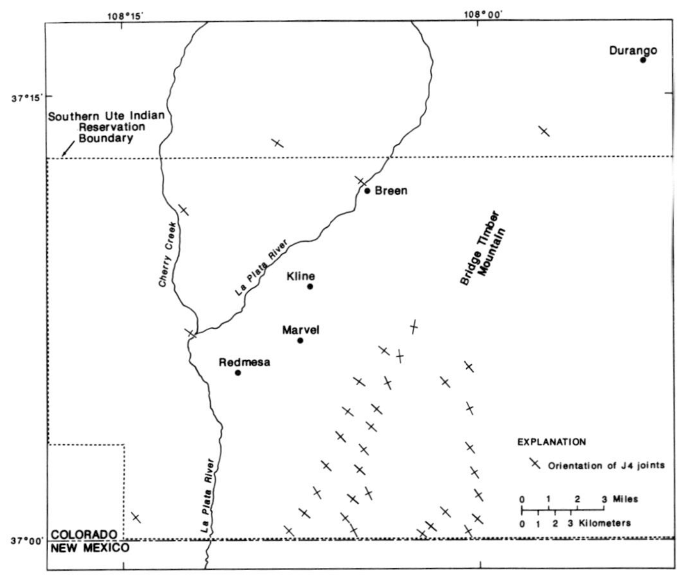

Figure 7. Map showing orientations of J4 joints. Each plotted orientation represents an average of 12 to 15 readings, where possible. See text for exceptions.

joints are present in most areas and in most geologic units; they are scarce and are difficult to locate, however, in much of the Farmington Sandstone Member. Poles to station averages of J4 joints are somewhat scattered (fig. 4d). The J4 set is the least well formed joint set. J4 joints have exposed lengths of commonly less than 5 feet, are curved in many places, and cut only one bed in interbedded sequences of sandstone and shale. Spacings are not consistent, and in some outcrops joints of the J4 set are widely scattered in contrast to the well-formed and abundant J1 and J2 joints. As with the other joint sets, the surfaces of the J4 set are weathered and generally do not retain original surface features; possible plumose structures were noted on some J4 joints. Apertures are inconsistent, and are commonly affected by the type of exposure and the degree of weathering; calcite coatings on joint walls are not common, but were seen in some J4 joints. The J4 joints terminate against the joints of all the other sets studied in the area. As with joints of the other sets, no slickenside striations were noted on any of the J4 joints.

Other joints

Other joints are present in the study area, but they could not be grouped into sets using the criteria of orientation or abutting relationships. Some of these fractures occurred adjacent to cliff faces and could be attributed to release of stress. Others could be due to freeze-thaw cycles or other types of weathering phenomena that affect exposed rocks. These types of joints are characteristically only a few feet long, are curved

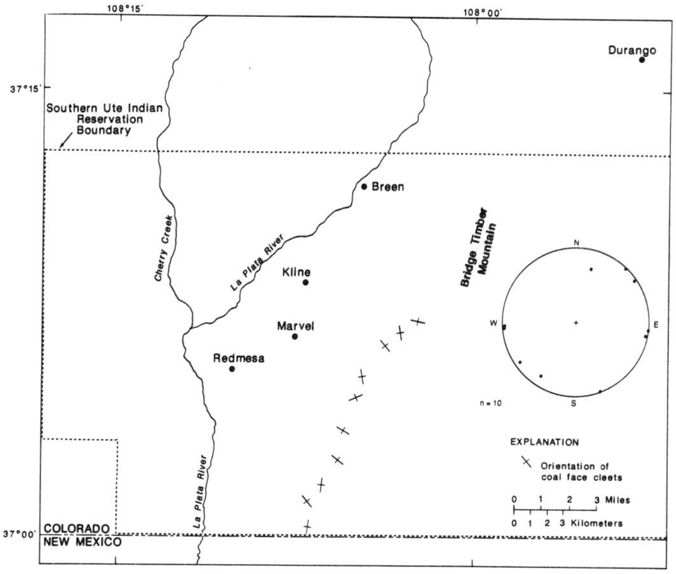

Figure 8. Map showing orientations of face cleats of coal in the Fruitland Formation, and an equal-area, lower-hemisphere projection of averages at individual stations. Each plotted orientation represents an average of 12 to 15 readings.

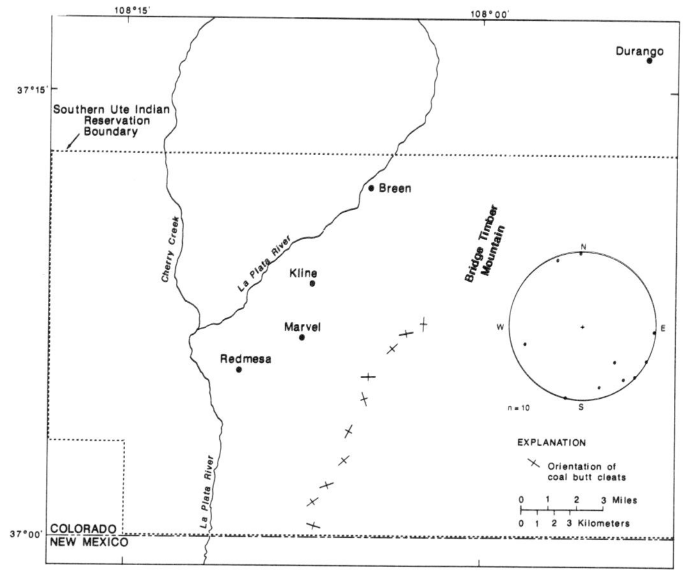

Figure 9. Map showing orientations of butt cleats of coal in the Fruitland Formation, and an equal-area, lower-hemisphere projection of averages at individual stations. Each plotted orientation represents an average of 12 to 15 readings.

or sinuous as opposed to being planar, and cut only single beds. They are interpreted to be of surficial, non-tectonic origin, and consequently are of limited use in understanding the fracture network of buried source or reservoir rocks of this area.

COAL CLEATS

Figures 8 and 9 show the orientations of face and butt cleats in coal beds of the Fruitland Formation. The overall resultant of all face-cleat readings is N. 21° W., dipping 90°, and that of butt cleats is N. 48° E. dipping 89° NW, but there is significant variability between stations. The dip of both face and butt cleats is subvertical, or is perpendicular to bedding surfaces where the coal is dipping. Other poorly expressed cleats also are present in minor numbers at some stations.

Face cleats are planar and cut multiple stacked coal beds. True lengths of face cleats are never exposed in the poorly exposed outcrops of this area, but exposed lengths at one locality are as much as 15 feet. Exposed lengths are only 1 foot or less in many places. The face cleats are usually closely spaced, commonly ¼-½ inch. Some coal outcrops had more widely spaced face cleats; however, the widest spacing noted was only 7 inches to 1 foot. Face cleats are the major joints of coal beds; butt cleats terminate against face cleats.

Butt cleats in coal beds of the Fruitland are less than 1 foot long, have an irregular, hackly surface, and commonly terminate against the face cleats; vertically butt cleats terminate against upper or lower bedding surfaces or shale breaks. Spacing of the butt cleats is commonly slightly less than that of the face cleats at any given outcrop. Butt cleats are commonly, but not always, perpendicular to face cleats.

The data collected in this study thus far indicate that the joints and the cleats did not form at the same time, even though the orientation of the face cleats generally coincides with that of the J4 joints in sandstones (fig. 7), and the butt cleats have orientations similar to those of the J3 joints (fig. 6). Age relations between joints in sandstones in this area show that the joints of the J3 set formed before those of the J4 set, a relationship that is opposite to that shown by the coal cleats. This suggests that either the cleat system in the coals of the Fruitland Formation was formed prior to the J3 and J4 joint sets in sandstones in this area, or alternatively, that the butt cleats are even younger than J4 joints and are only coincidentally parallel to the J3 joint set. Further study on both the west and east sides of the Reservation will be undertaken to better interpret the relationship, if any, between joint sets in sandstones and coal cleats.

SUMMARY

Four discrete sets of sub-vertical joints have been identified in sandstones on and adjacent to the Hogback monocline on the west side of the Southern Ute Indian Reservation. These sets were defined on the basis of joint orientations and abutting relationships. Previous workers (Kelley and Clinton, 1960; Badgley, 1962, 1965; and Hodgson, 1961) mapped somewhat similar joint trends in nearby areas. Joint sets commonly occur in the pairs J1-J2 and J3-J4. J2 joints are nearly perpendicular to J1 joints; J4 joints are nearly perpendicular to J3 joints. Most outcrops that were examined had numerous occurrences of all four joint sets, and there was enough overlap of the two pairs of joints to establish age relations.

In all areas and in all geologic units examined, J1 joints were the first formed and the most well developed. J1 joints generally parallel the strike of the outcrop along the Hogback monocline; they have a pronounced northerly trend in areas away from the monocline. J2 joints terminate against J1 joints and in many respects, other than orientation, are similar to J1 joints. Another exception is in their length, which is always shorter than that of J1 joints, and is controlled by the spacing between J1 joints. J2 joints always post-date J1 joints.

J3 joints also are similar to J1 joints in that they are the dominant set in the J3-J4 pair. In parts of some outcrops the J3 set is the main joint set present. There is strong evidence that J3 joints postdate both J1 and J2 joints because J3 joints were observed to terminate against J1 and J2 joints. J4 joints form the youngest and most poorly developed set in the study area. They always terminate against other joints, and commonly are spaced widely across outcrops. The limited expression of J4 joints is due to the presence of older joints, which constrained their development.

No features, such as slickenside striations, were noted on any of the joint surfaces that would indicate lateral shear movement. The only features of joint surfaces observed in this study are plumose structures, arrest lines, and twist hackle; features that are diagnostic of formation by extension. Bedding thickness and degree of cementation are the dominant controls on the number, size, and spacing of joints in sandstone, regardless of location or lithologic unit. Thin-bedded, firmly cemented sandstone beds invariably have more abundant joints than do thick-bedded, weakly cemented strata. Because members of all the joint sets defined in this study are nearly perpendicular to the bedding surfaces of the host rocks, even in rocks dipping as much as 30 degrees, it is felt that the joints predate folding along the Hogback monocline.

Cleats in coal in the study area are problematic, and relations between coal cleats and joints in sandstone are still unclear. Many butt cleats have orientations parallel to J3 joints, and many face cleats have orientations similar to J4 joints. The age relationships of the cleats, however, are opposite to those indicated by the joints in sandstone.

The lack of correspondence between coal cleats and joints, and the lack of studies that demonstrate that joint networks at the surface correspond to joint networks at depth is of significance to those exploring for coal-bed methane in this area. The wide range in strike of both the face and butt cleats of Fruitland coal strongly suggests that 1) cleats did not form contemporaneously with any one joint set, 2) cleats of one area may not correspond at all with those of other areas, and 3) cleats do not have an apparent relationship to the tectonic events that formed the San Juan Basin, at least in the area discussed in this report.

It seems advisable then, that exploration stratagems for oil and gas resources in this area should not depend solely on surface studies of joints to predict the fracture network at depth. Detailed local studies, ideally using oriented core, is the recommended approach to correctly identify and map local variations in the regional joint and cleat systems.

REFERENCES CITED

Badgley, P.C., 1962, Analysis of structural patterns in bedrock: Society of Mining Engineers of AIME, Transactions of the American Institute of Mining, Metallurgical, and Petroleum Engineers (incorporated), v. 223, p. 381-389.

Badgley, P.C., 1965, Structural and Tectonic Principles: Harper and Row, New York, 521 p.

Fassett, J.E., 1985, Early Tertiary paleogeography and paleotectonics of the San Juan Basin area, New Mexico and Colorado, *in* Flores, R.M., and Kaplan, S.S., eds., Cenozoic paleogeography of the west-central United States: Rocky Mountain Section of the Society of Economic Paleontologists and Mineralogists, Rocky Mountain Paleogeography, Symposium 3, p. 317-334.

Grout, M.A., and Verbeek, E.R., 1983, Field studies of joints—insufficiencies and solutions, with examples from the Piceance Creek basin, Colorado, *in* Gary, J.H., ed., Proceedings, 16th Oil Shale Symposium, Golden, Colorado: Colorado School of Mines Press, Golden, Colorado, p. 68-80.

________1985, Fracture history of the Plateau Creek and adjacent Colorado River valleys, southern Piceance Basin: Implications for predicting joint patterns at depth: U.S. Geological Survey Open-File Report 85-744, 17 p.

Haynes, D.D., Vogel, J.D., and Wyant, D.G., 1972, Geology, structure, and uranium deposits of the Cortez quadrangle, Colorado and Utah: U.S. Geological Survey Miscellaneous Investigations Series, Map I-629, scale 1:250,000.

Hodgson, R.A., 1961, Regional study of jointing in Comb Ridge-Navajo Mountain area, Arizona and Utah: American Association of Petroleum Geologists Bulletin, v. 45, no. 1, p. 1-38.

Kelley, V.C., and Clinton, N.J., 1960, Fracture systems and tectonic elements of the Colorado Plateau: University of New Mexico Publications in Geology, Number 6, 104 p.

Kulander, B.R., Barton, C.C., and Dean, S.L., 1979, The application of fractography to core and outcrop fracture investigations: U.S. Department of Energy, METC/SP-79/3, 174 p.

Steven, T.A., Lipman, P.W., Hail, W.J., Jr., Barker, Fred, and Luedke, R.G., 1974, Geologic map of the Durango quadrangle, southwest Colorado: U.S. Geological Survey Miscellaneous Investigations Series, Map I-764, scale 1:250,000.

Verbeek, E.R., and Grout, M.A., 1983, Fracture history of the northern Piceance Creek basin, northwestern Colorado, *in* Gary, J.H., ed., Proceedings, 16th Oil Shale Symposium, Golden, Colorado: Colorado School of Mines Press, Golden, Colorado, p. 26-44.

________1984a, Fracture studies in and around the Piceance Basin, Colorado: Preliminary results and their bearing on a fracture-controlled natural gas reservoir at the MWX site: U.S. Geological Survey Open-File Report 84-156, 30 p.

________1984b, Prediction of subsurface fracture patterns from surface studies of joints—an example from the Piceance Creek Basin, Colorado, *in* Spencer, C.W., and Keighin, C.W., eds., Geologic studies in support of the U.S. Department of Energy's Multiwell Experiment, Garfield County, Colorado: U.S. Geological Survey Open-File Report 84-757, p. 75-86.

Wheeler, R.L., and Stubbs, J.L., Jr., 1979, Style elements of systematic joints: statistical analysis of size, spacing, and other characteristics, *in* Podwysocki, M.H., and Earle, J.L., eds., Basement tectonics committee, International Conference on Basement Tectonics, 1976 Proceedings: Denver, Colo., p. 491-499.

A Geologic Analysis of the Fruitland Formation Coal and Coal-bed Methane Resources of the San Juan Basin, Southwestern Colorado and Northwestern New Mexico

Bruce S. Kelso[1]
and
Donald E. Wicks[2]

ABSTRACT

The Upper Cretaceous Fruitland Formation contains large coal-bed methane resources in the San Juan Basin of New Mexico and Colorado. These resources were calculated by interpretation of extensive subsurface geologic data. Empirical gas-content equations were derived correlating desorbed-gas-content values, coal depths, and coal ranks specific to the Fruitland Formation coals. The combination of coal-resource and gas-content data resulted in a Fruitland Formation coal-bed methane in-place estimate of nearly 50 trillion cubic feet of gas contained in 219 billion tons of coal.

In addition to the gas-in-place estimate, an analysis of some geologic controls affecting methane extraction potential was conducted. The results of this analysis, combined with the gas-in-place study, identify the north-central portion of the basin as a promising area for development of this vast coal-bed methane resource. Much of the north-central portion of the basin is in the Southern Ute Indian Reservation; coal and gas-in-place resource estimates for the Tribal acreage are presented for comparison with other studies.

INTRODUCTION

Located in southwestern Colorado and northwestern New Mexico (fig. 1), the San Juan Basin has the highest coal-bed methane development level in the western United States. The Phillips Petroleum Co., No. 6-17, SJ 32-7 Unit well has produced methane from Fruitland Formation coals since 1953. More recent activity dates back to the mid-1970s and records indicate that more than 200 wells have been drilled to develop the Fruitland Formation coals which underlie an area of approximately 7,500 square miles. Two regional cross sections and several geologic maps were constructed to evaluate the vertical and lateral distribution of the coals. New work, specific to Fruitland Formation coals, correlating gas content with depth and coal rank was also performed. The combination of the geologic investigation and gas-content relationships resulted in a regional gas-in-place estimate for the Cretaceous Fruitland Formation coals. In addition to identifying a gas-in-place resource, regional analysis of the geologic parameters which affect recovery, specifically permeability and pressure, was completed. This analysis for the Fruitland Formation coals defines areas of the basin which may geologically favor development of the resource.

[1] B. S. Kelso
ICF-Lewin Energy Division
ICF-Incorporated
165 South Union Blvd.
Suite 802
Lakewood, CO 80228

[2] D. E. Wicks
ICF-Lewin Energy Division
ICF-Incorporated
930 Lee Highway
Fairfax, VA 22031

GEOLOGIC SETTING OF THE SAN JUAN BASIN

Regional Setting

The San Juan Basin is located in northwestern New Mexico and southwestern Colorado; the area studied for this report is within the outcrop of the Pictured Cliffs Sandstone. The basin is approximately 90 miles wide, west to east, 100 miles long, north to south, and covers 7,500 square miles.

Stratigraphy and Depositional Environments

The coals of the San Juan Basin are Upper Cretaceous in age and are located in the Dakota Sandstone, Mesaverde Group, and Fruitland Formation (fig. 2). The Fruitland, the youngest of these formations, contains the largest coal resource. Deposition of the Fruitland coals occurred predominantly in lagoons, landward of the Pictured Cliffs Sandstone barrier strandline. The thickest and most continuous coal seams are located in the lowermost 70 feet of the Fruitland and often are associated with stratigraphic rises in the Pictured Cliffs (Fassett and Hinds, 1971).

The Fruitland Formation is a coastal plain deposit of paludal carbonaceous shales, siltstones, sandstones, and coals deposited shoreward of the regressing Pictured Cliffs strandline. The formation thickness ranges from less than 100 to greater than 600 feet and contains evidence of fresh and brackish water environments. Deposited in lagoons, marshes, swamps, and abandoned channels, the coals are overlain by fluvial shales and sandstones. The coals are often the most correlative units of the fluvial Fruitland sediments.

The underlying Pictured Cliffs is a regressive, coastal-barrier sandstone. The thickness of the Pictured Cliffs varies from 125

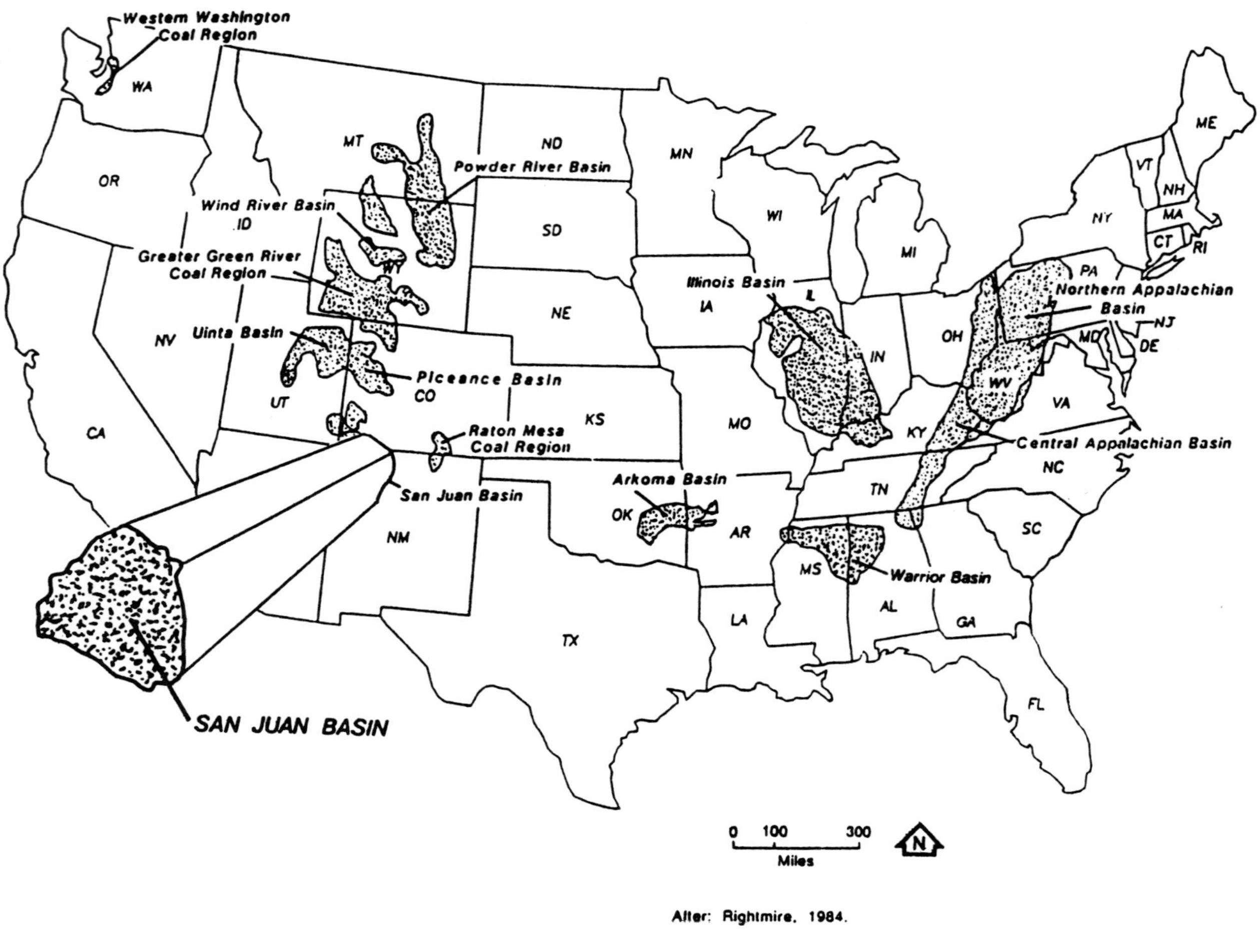

Figure 1. Major coal basins of the United States.

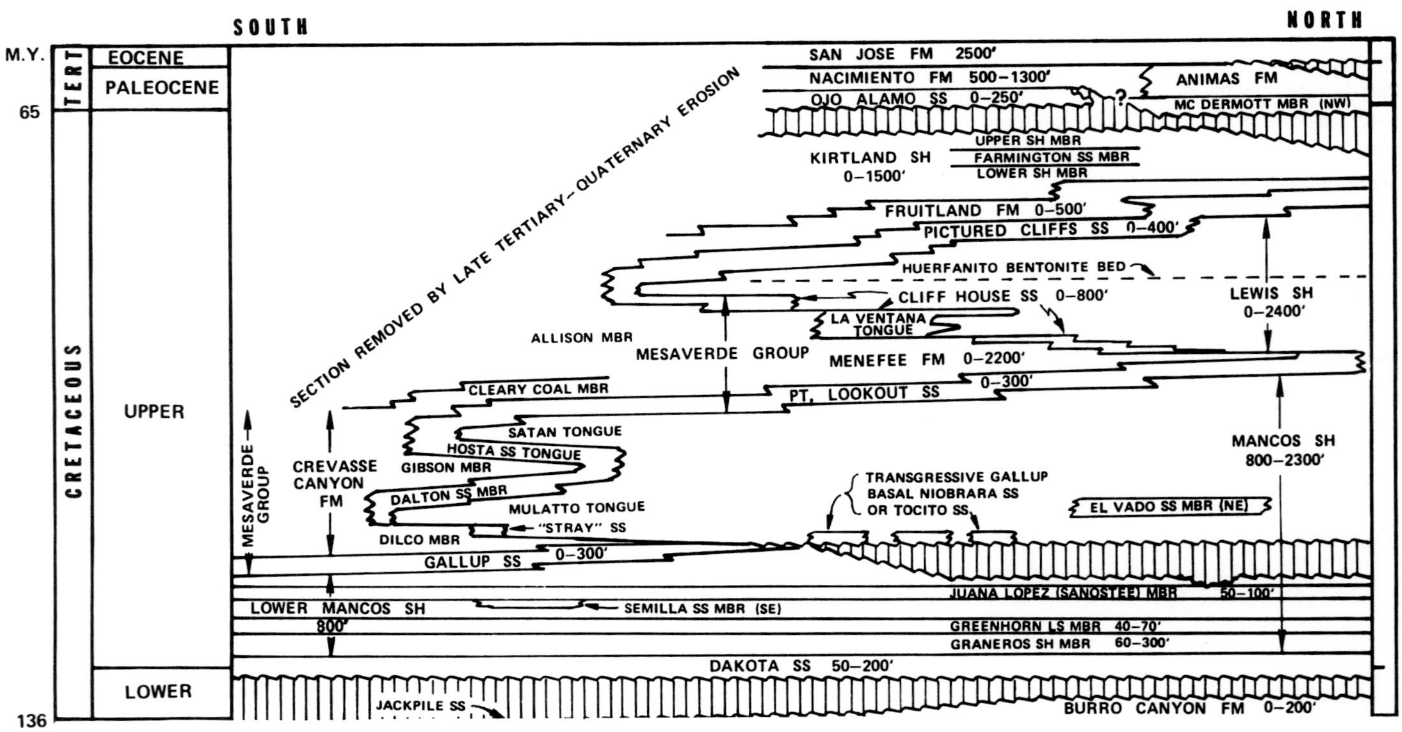

Figure 2. San Juan Basin time-stratigraphic nomenclature chart (modified from C. M. Molenaar, 1977).

to 400 feet; due to minor transgressive episodes, the Pictured Cliffs locally intertongues with the Fruitland Formation. The lower portion of the Pictured Cliffs is primarily interbedded sandstones and shales, with the upper unit being a quartzitic, fine-to-medium grained sandstone.

Structure

The San Juan Basin's arcuate structural axis lies just south of the Colorado-New Mexico State line. The U-shaped Hogback Monocline forms the western and northern rims of the basin. To the east, the Nacimiento uplift and Archuleta arch bound the basin. The south and southwestern boundaries of the basin are not structurally defined and sediments gently dip northward from the Chaco Slope (fig. 3).

The few major structural elements within the San Juan Basin are of Laramide age. Located in the north-central portion of the basin, the Ignacio anticline is the largest and best documented structure. Minor northwest-trending en echelon folds and northeast-trending, high angle, low displacement faults are found on the eastern and southeastern edges of the basin. Radial folds, which plunge toward the basin's center, can be found around the perimeter. Minor structures resulting from Tertiary intrusive activity are located in the basin. Kelly, 1951, and Woodward and Calender, 1977, provided complete structural and tectonic descriptions of the San Juan Basin.

STUDY METHODOLOGY

This San Juan Basin geologic study used two regional cross sections and subsurface data developed from 549 well logs, which penetrated the Fruitland Formation (fig. 4), to determine the lateral and vertical distribution of the Fruitland coals. Data pertaining to structure, overburden, coal thickness, and coal rank were obtained from regional mapping of each parameter.

The final step in the geologic analysis was the development of correlations relating gas-content measurements to coal rank and depth. Twenty-eight data points which are specific to Fruitland Formation coals were used in the analysis. The correlations, in conjunction with the geologic data, were used to establish gas in place on a township level throughout the basin and these results were summed to determine the regional gas-in-place estimate.

Because numerous geologic and reservoir factors are important in coal-bed methane development and extraction, this study also assessed permeability and pressure variation to help delineate favorable areas within the basin for development of the resource.

An area of interest for this symposium is the Southern Ute Indian Reservation in southwestern Colorado, located in the north-central portion of the basin. The coal and coal-bed methane resources for the reservation are presented in this paper, based on the regional study of the Fruitland Formation coals.

RESULTS OF THE GEOLOGIC ANALYSIS

Stratigraphic Cross Sections

Two regional, basin-wide, geologic cross sections (not included in this report) were constructed using density-porosity or gamma-density geophysical logs (fig. 4). Both sections were segmented due to their lengths and the number of logs used. Section A-A'', trending northwest to southeast, is parallel to the depositional strike of the Pictured Cliffs strandline. Section B-B'' is constructed perpendicular to depositional strike, or southwest to northeast. Orientation of the cross sections was designed to support the concept that coals parallel to depositional strike can be followed for greater distances than those perpendicular to strike. Interpretations of the cross sections were used to understand the lateral continuity of the Fruitland coals and to determine if the Fruitland Formation could be divided into zones, based on vertical coal distribution. No grouping or zone determination was established due to the lack of correlative marker units within the formation.

Vertical and Lateral Extent of Fruitland Formation Coals

A series of maps was constructed for the Fruitland Formation to establish regional variations in net-coal thickness, overburden, and coal rank. The following sections discuss the results of these maps:

Net-Coal Thickness

A Fruitland Formation net-coal thickness map was constructed using data from gamma-density, density-porosity, and a limited number of gamma ray-neutron geophysical logs (fig. 5). Net-coal values exclude coals thinner than 2 feet, and shaly partings were subtracted from individual coals. The zero net-coal line lies along the eastern margin of the basin, the result of non-deposition or erosion of the Fruitland. The northwest region of the basin has the maximum net-coal values which exceed 100 feet. A northwest trend in the north-central portion of the basin reflects net-coal thickness values between 60 and 80 feet. The trend parallels the basin's structural axis and is partially associated with a stratigraphic rise of the Pictured Cliffs. In general, the southern portion of the basin contains net-coal values between 10 and 30 feet.

Overburden

An overburden map was constructed on top of the Pictured Cliffs Sandstone (not included in this report) and was used to establish gas containment boundaries for the gas-in-place evaluation. Depths of the basal Fruitland Formation coals range from 0 feet along the basin's perimeter to more than 4,200 feet along the basin's structural axis. Depths change rapidly along the northern, northwestern, and northeastern margins of the basin due to structural monoclines. Coal depths in the southern area of the basin are extensively shallower and gradually increase to the northeast.

Coal Rank

Rank determination in this investigation was based on the thermal maturity of the coals, measured by vitrinite reflectance values. In the San Juan Basin, coal rank generally increases from south to north, with the higher rank coals located in the north-central portion of the basin. The higher ranks in the north are associated with the coal's proximity to the San Juan Mountain intrusive complex (Choate and Rightmire, 1982). A Fruitland Formation vitrinite reflectance map of the San Juan Basin (Rice, 1983) was modified to include additional data and established coal ranks ranging from high volatile C bituminous to low volatile bituminous, with 0.46 to 1.51 reflectance values, respectively (not included in this report).

GAS CONTENTS OF FRUITLAND COALS

The gas contents of Fruitland Formation coals are affected by numerous geologic and depositional factors. Three of the

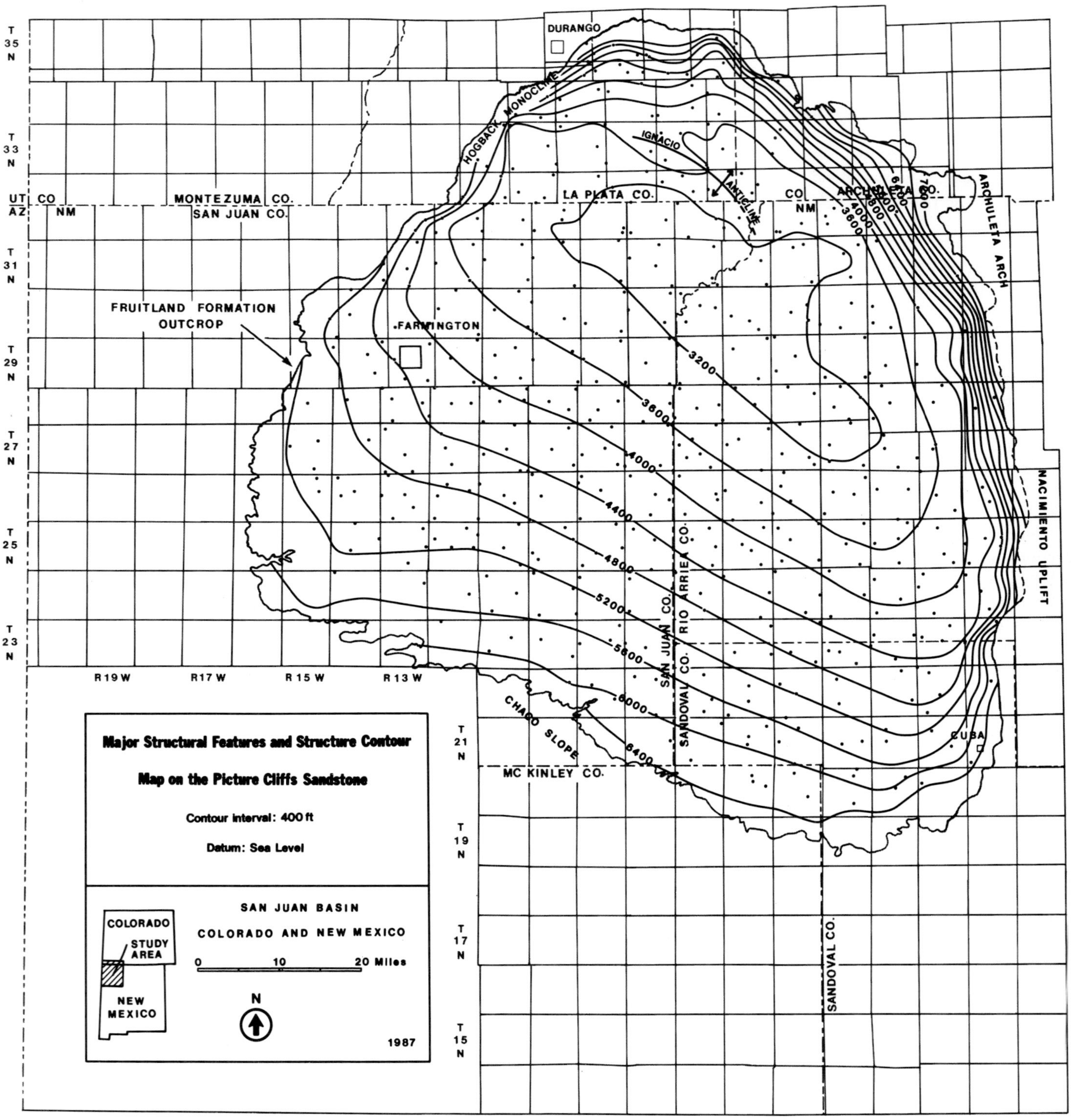

Figure 3. Structure contour map on the Pictured Cliffs Sandstone and major structural features.

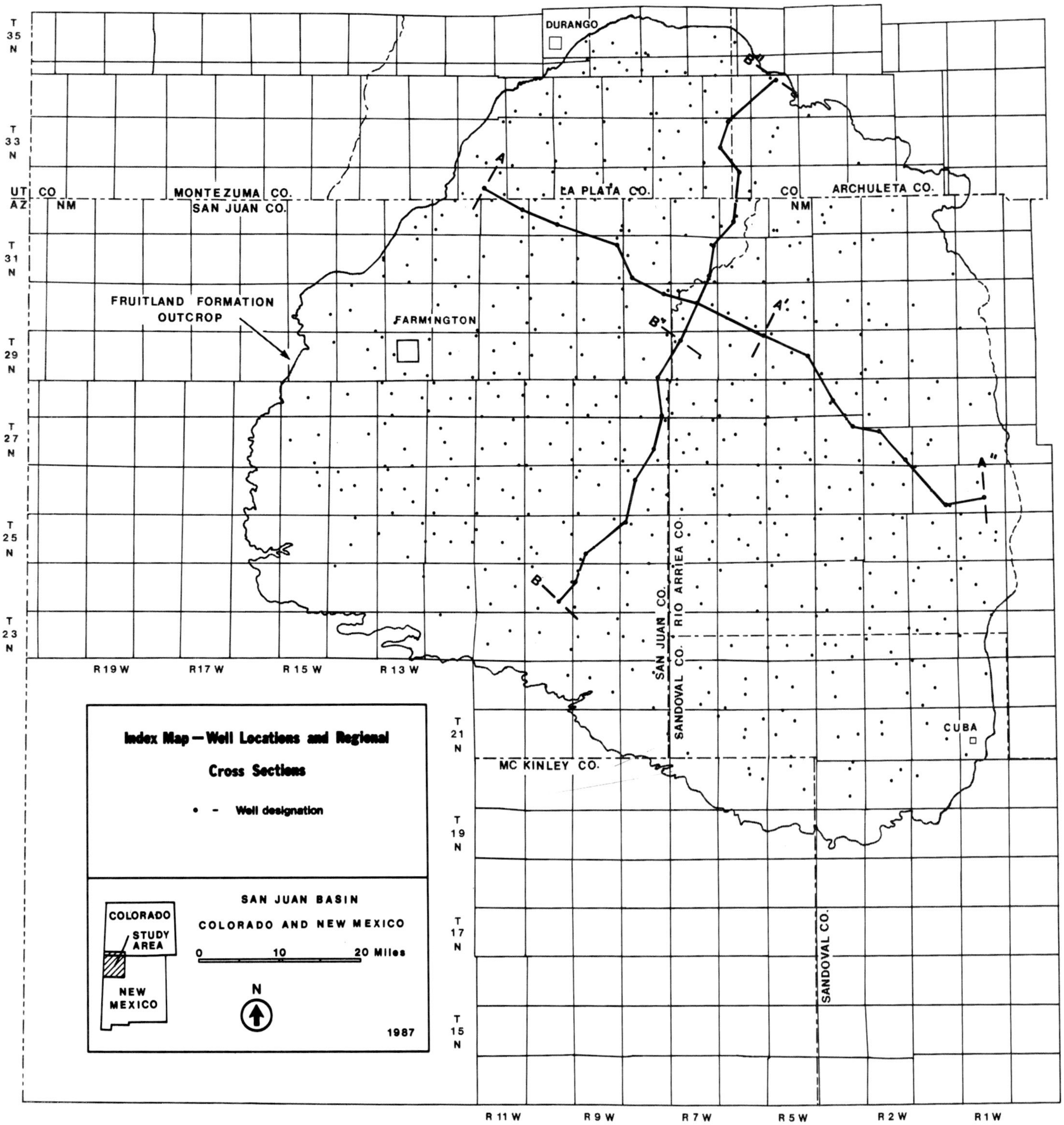

Figure 4. Map showing locations of wells studied in this report and lines of cross sections.

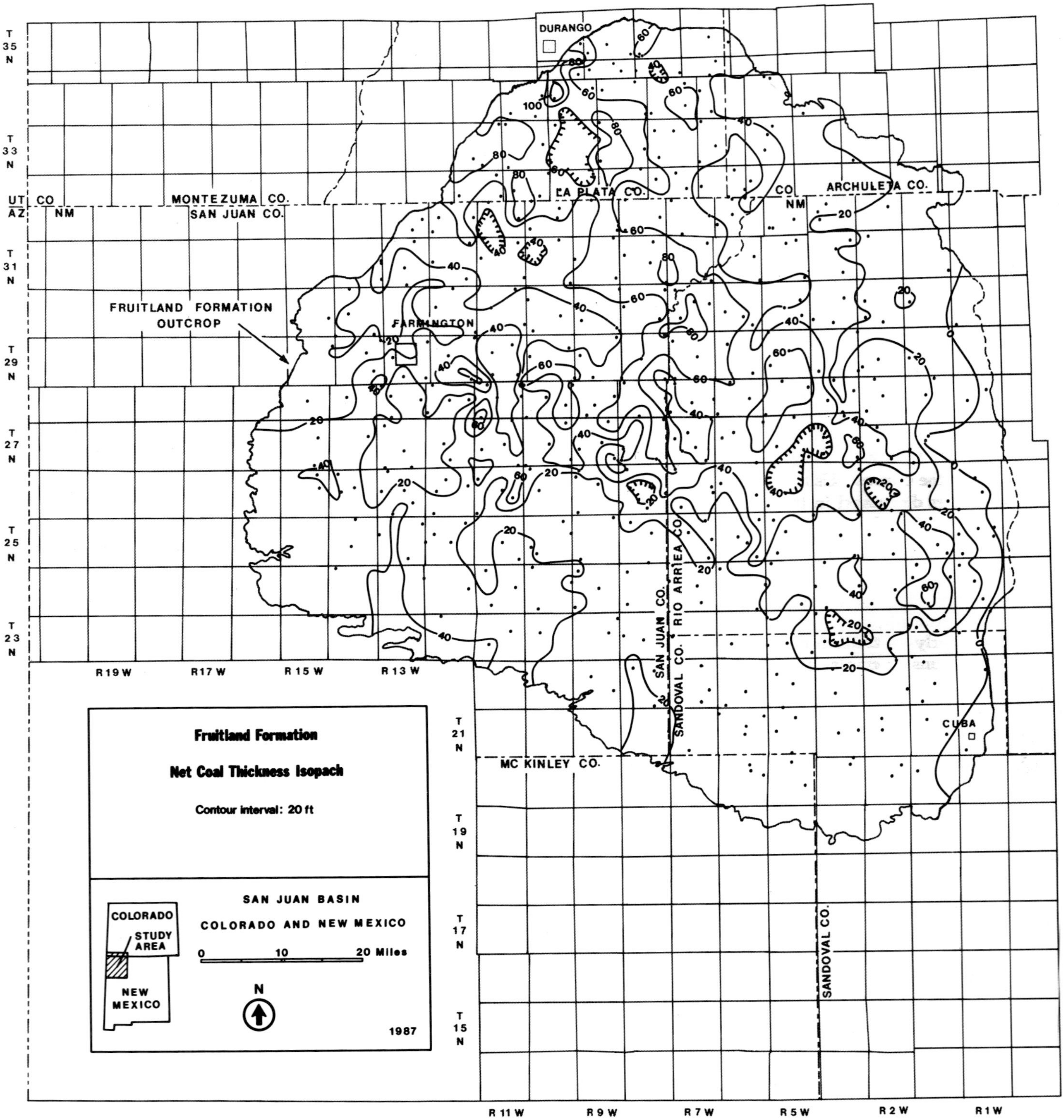

Figure 5. Fruitland Formation net-coal-thickness isopach map.

more important factors related to methane generation and accumulation are degree of coalification (coal rank), coal quality (ash percentage), and extent of containment (overburden). A relationship was established in this study between measured-gas contents, depth, and coal rank, to determine gas-in-place estimates on a site-specific level throughout the entire basin.

A data base of 28 desorption samples was compiled from the U. S. Bureau of Mines, the Colorado Geological Survey, and private industry. The data was standardized for ash content, temperature, and pressure and was sorted by rank and sample depth to develop equations correlating gas contents with depth and rank. Equations were established for high volatile A bituminous coals and combined high volatile B and C bituminous coals. Since little or no data was available for medium and low volatile bituminous coals, equations for these were estimated using the results from a similar study in the Piceance Basin, Colorado (McFall and others, 1986).

A significant factor affecting estimated gas content values is the percentage of ash contained within the coal. Coals of the Fruitland Formation have relatively high ash contents, ranging from less than 10 percent to more than 40 percent, with a regional average between 15 to 20 percent. The ash-percentage values of Fruitland coals were accounted for in the gas content, depth, and rank evaluation, resulting in a downward adjustment of the curves. Using the adjusted curves and geologic parameters discussed in the previous section, average gas contents were assigned to 210 townships within the basin.

GAS IN PLACE IN THE FRUITLAND COALS

The regional geologic evaluation of the Fruitland Formation coals, San Juan Basin, concluded with a gas-in-place resource estimate of nearly 50 trillion cubic feet of gas (TCF) contained in 219 billion tons of coals. This is only a resource estimate, not a reserve estimate, and no recovery factor has been applied. Figure 6 is a gas-in-place contour map which represents the data generated for each of the 210 townships in the basin.

The variables affecting the gas-in-place estimate are gas contents, coal thickness, drillable area, and coal density. The source of the gas content values has been discussed in the previous section. Average net-coal thickness values for each township were obtained from the net coal thickness map developed in this study. Drillable area excludes acreage already containing coal-gas wells and areas of the basin with less than 400 feet of overburden. Coal density varies with rank and values for high volatile A, B, and C; medium volatile; and low volatile are 1,800, 1,850, and 1,900 tons/acre-foot, respectively (Averitt, 1975).

GEOLOGIC CONTROLS AFFECTING METHANE EXTRACTION

Numerous geologic and reservoir parameters have significant effects on the potential for methane extraction from coal-bed reservoirs. Two of the more important factors are permeability and reservoir pressures. The occurrence and significance of these in Fruitland Formation coals of the San Juan Basin are discussed below.

Areas of Enhanced Permeability

Ideally, permeability measurements of the Fruitland Formation coals are needed to identify favorable areas for development of the coal-bed methane resource. Because this data is not available, alternate methods were used to regionally quantify the coal's permeability, based upon structural enhancement. Structural geology has been used in other basins to explain and delineate areas of enhanced permeability (McFall and others, 1986). The basis of this work is founded in the theory that coal-cleat and rock-fracture systems are better developed and more frequent in areas containing structure, providing enhanced flowpaths for gas and water.

The structural geology of the San Juan Basin, when mapped on the Pictured Cliffs Sandstone (fig.3), reveals few major structural features. Only the Ignacio anticline and monoclines in the northern portion of the basin are evident on this map. Therefore, a Pictured Cliffs structural analysis in the San Juan Basin provides little help in identifying areas with potentially higher permeability.

Structures associated with deeper, underlying formations and even the geologic basement may have some effect on the permeability of the Fruitland Formation coals. This may be the case at two areas in the northern San Juan Basin where enhanced permeability is suspected. The Cedar Hill field, T. 32 N. R. 10 W., New Mexico and an undesignated Meridian Oil Company field, T. 30 N. R. 7 W., New Mexico exhibit tremendous gas and water production rates (fig. 7). Very little structural relief is mappable on the Pictured Cliffs Sandstone at either area. A synclinal structure is apparent on the Dakota Sandstone at the Meridian field, while little structure is evident on the Dakota Sandstone at the Cedar Hill field. The Cedar Hill field may be enhanced due to basement related structures but sufficient data is not available to confirm this. When trying to determine areas in the San Juan Basin that might contain more permeable coals, a structural evaluation on several horizons is recommended.

Reservoir Pressures

Coal-gas reservoir pressures are important when developing a coal-bed methane field. The reservoir pressures of the Fruitland Formation coals range from normal to overpressured. Overpressured Fruitland coals have been documented by operators in certain areas of the San Juan Basin as significant gas kicks and even blow-outs. To delineate the overpressured reservoir areas of Fruitland coals, drilling fluid weights were correlated with pressure data from wells in the Fruitland and Pictured Cliffs. This analysis of measured Fruitland reservoir pressures indicated that drilling fluid weights in excess of 11.0 pounds per gallon usually mark overpressured coal reservoirs (fig. 8).

Numerous mechanisms explaining the generation of methane and overpressured reservoirs have been discussed by Meissner (1984) and Spencer (1987). Two of these mechanisms seemingly related to coal reservoirs are active gas generation and artesian conditions associated with topographical relief. In order for active gas generation to cause overpressuring, present thermal conditions of the reservoir must exceed 200° F (Spencer, 1987). The formation temperatures of the Fruitland coals rarely exceed 120° F, which is significantly lower than the temperature needed to explain overpressuring by gas generation. The Fruitland coals, which have been identified as overpressured, are possibly explained by hydrodynamic overpressuring related to topographic relief. Regional uplift and down-cutting in the

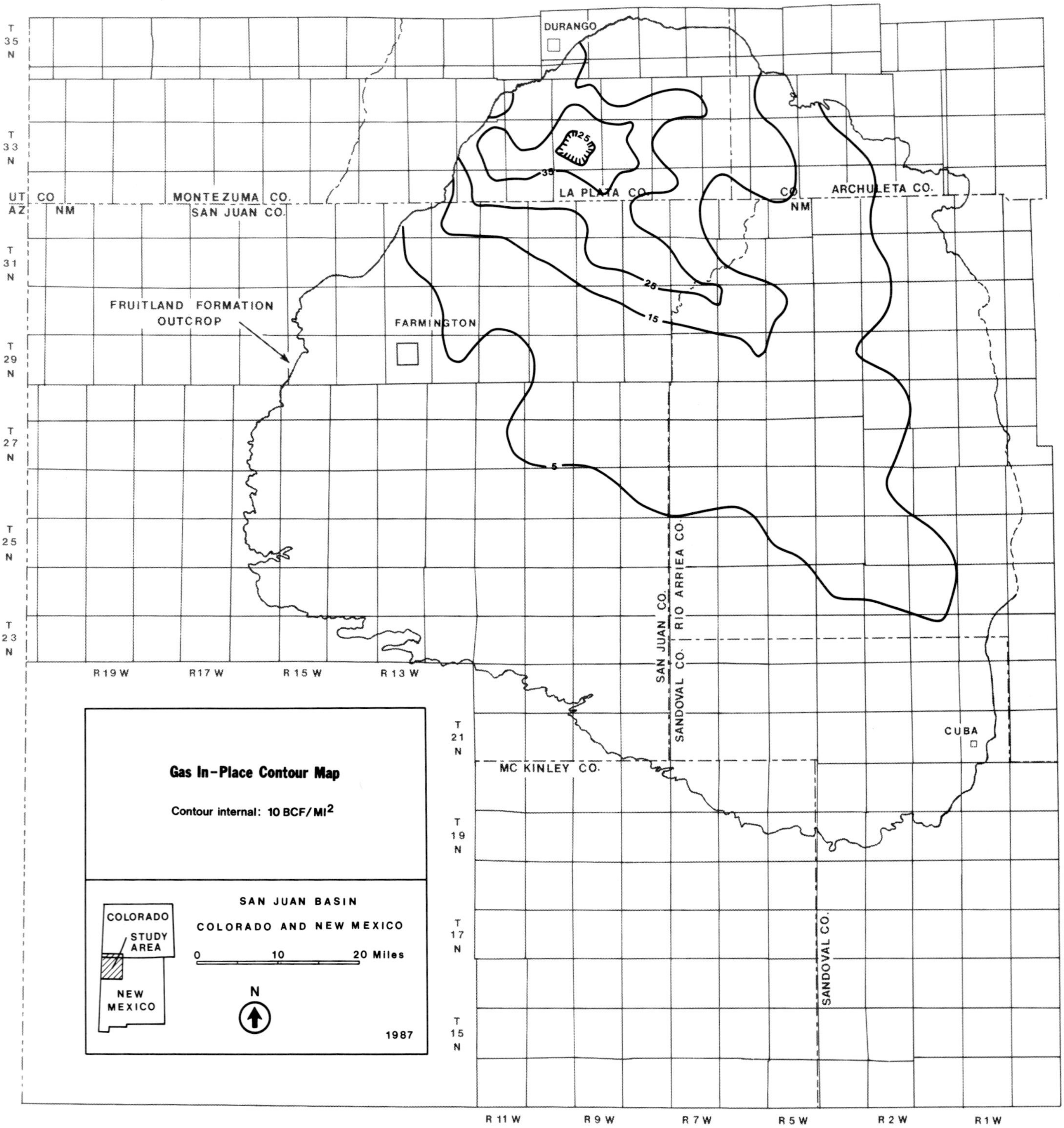

Figure 6. Gas-in-place contour map.

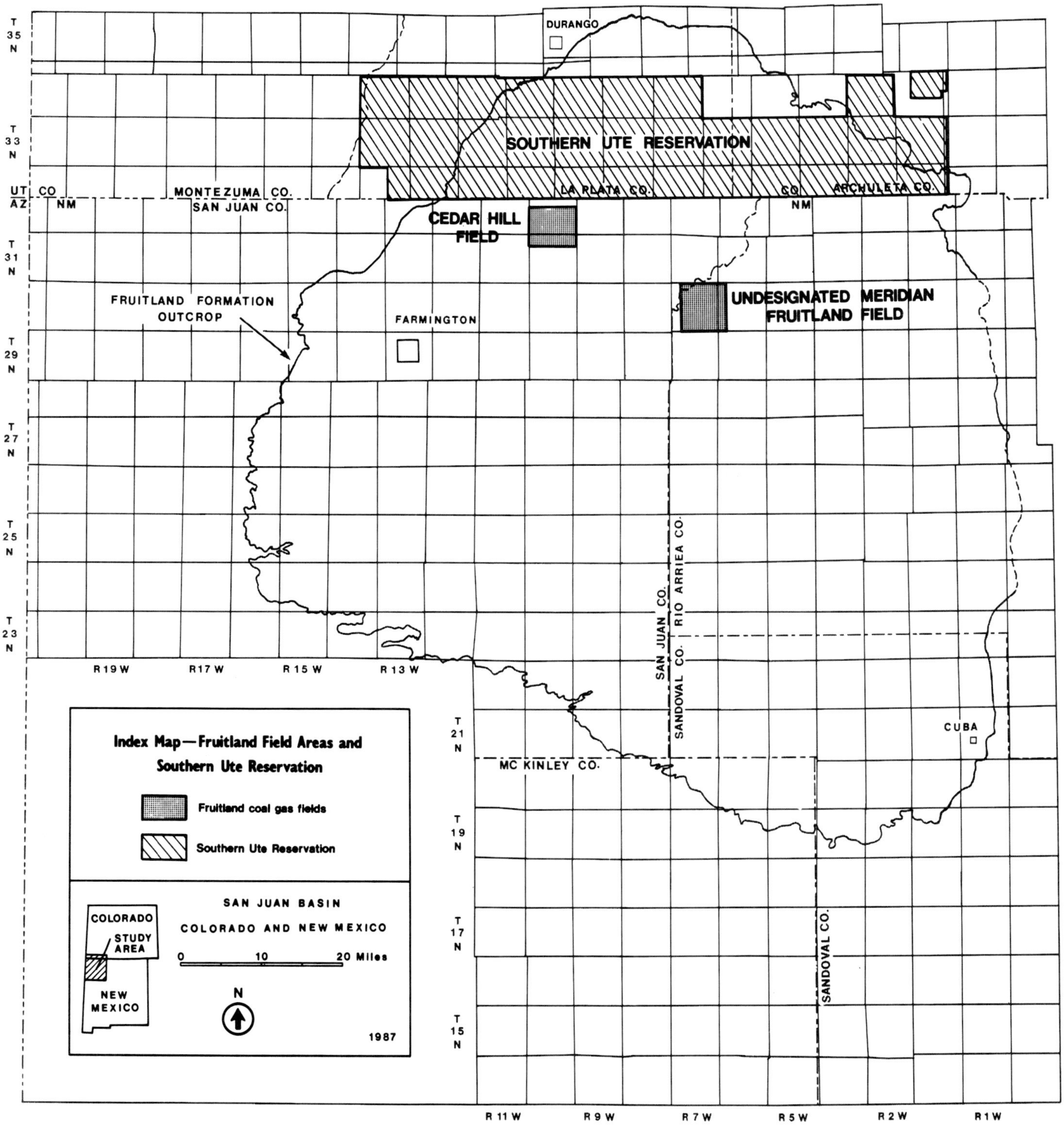

Figure 7. Map showing 2 Fruitland gas-field areas and the Southern Ute Indian Reservation.

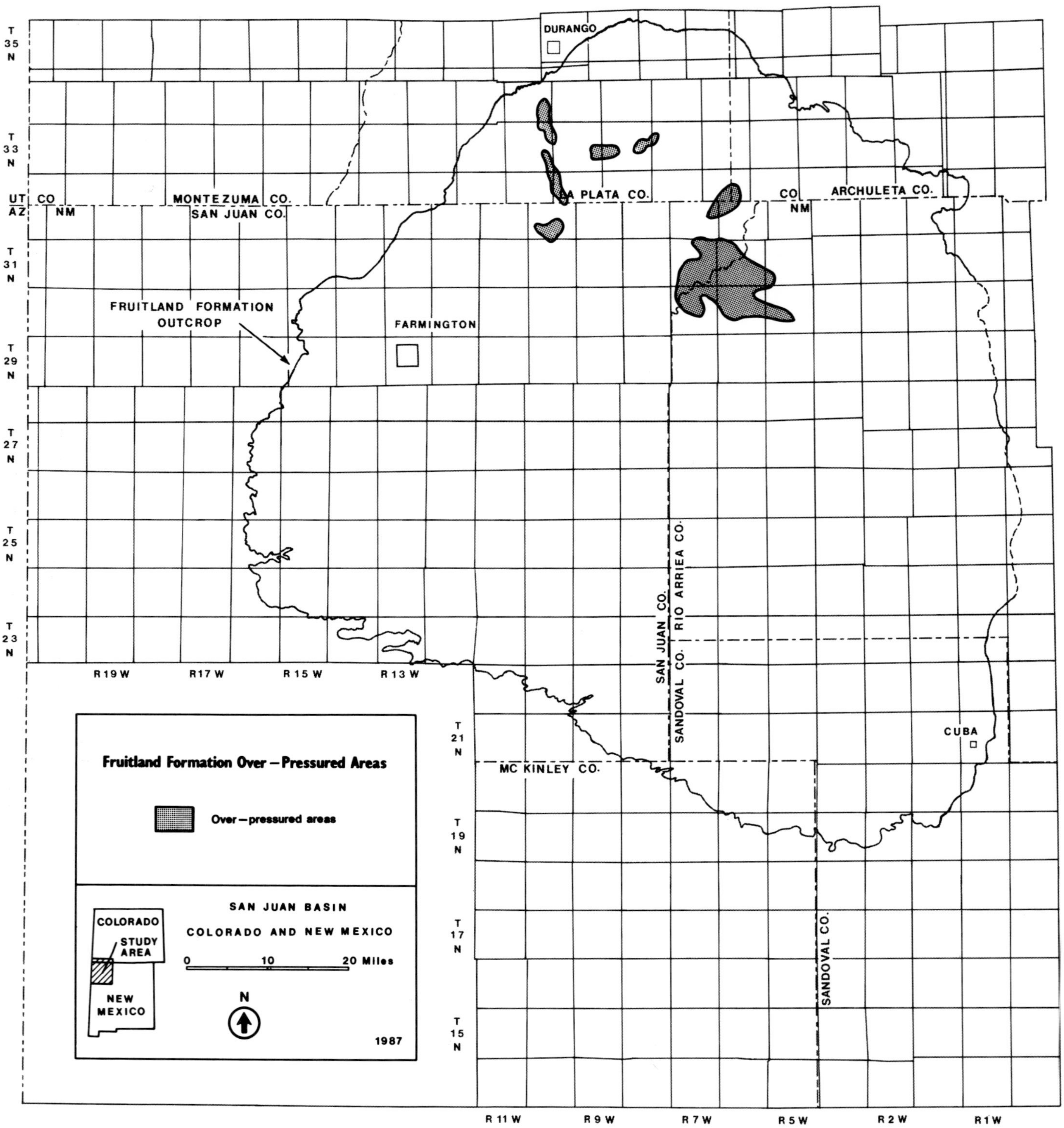

Figure 8. Map of Fruitland Formation overpressured areas.

major river valleys of the San Juan Basin coincide with many of the areas identified in figure 8 as overpressured.

The geologic setting which most favors coal-bed methane extraction is where the coal seams have enhanced permeability and the reservoir is overpressured. Additional site-specific geological and geophysical research and analysis of well-drilling and test data are required to establish accurate boundaries for these favorable areas for Fruitland Formation coals in the San Juan Basin.

NORTHERN SAN JUAN BASIN AREA OF INTEREST

The scope of this symposium and guidebook is coal-bed methane in the northern San Juan Basin. Located in the Colorado portion of the basin is the Southern Ute Indian Reservation (fig. 7). Many of the favorable geological conditions discussed in the previous section, as well as areas of significant net-coal thickness, are found within the Reservation boundaries. A discussion of the coal and coal-bed methane resources of the tribal lands is presented in this section.

Coal Resources

Several reports present Fruitland Formation coal resource estimates for the Southern Ute Indian Reservation (Shoemaker and Holt, 1973 and Sandberg, this volume). While various authors have used a variety of methodologies to calculate the Fruitland coal resources, the purpose here is not to discuss the merits and pitfalls of each. In summary, the methodology in this study, simply stated, was to include coals greater than two feet in thickness (minus partings) from individual coal beds identified on geophysical logs. A reduction of coal thickness to correct for ash content was not performed. A Fruitland Formation coal resource of 39.1 billion tons is estimated for 612 square miles of Tribal lands.

Coal-bed Methane Resource

Contained within the 39.1 billion tons of Fruitland coal on the Southern Ute Reservation is an estimated 14.5 TCF of gas. This represents nearly 30 percent of the San Juan Basin's total resource estimate and was calculated by separating the Tribal acreage and gas-in-place estimates from the larger regional evaluation. Significant development of this resource is already underway and future development will certainly continue to benefit the Southern Ute Tribe.

CONCLUSIONS

The geologically diverse and complex reservoirs of the Fruitland Formation coals in the San Juan Basin contain an estimated 50 trillion cubic feet of methane gas in 219 billion tons of bituminous-rank coal. Although the regional subsurface analysis was unable to establish and correlate Fruitland Formation coal groups, some individual seams appear to extend over tens of square miles. The southern portion of the basin is where the thinner, shallower, lower rank coals of the Fruitland are located. The north-central portion of the basin contains the thicker, deeper, and higher rank coals and their higher associated gas contents, which results in more concentrated accumulations of the resource in this area.

An evaluation of producing coal-bed methane wells and fields to determine Fruitland Formation geologic controls that affect coal-bed methane extraction have identified reservoir pressure and permeability as major factors. Overpressured Fruitland coals with structurally enhanced permeability appear to favor production of coal-bed methane. Additional research is needed to delineate high potential areas for coal-bed methane development in the San Juan Basin.

REFERENCES CITED

Averitt, P., 1975, Coal resources of the United States, January 1, 1974: U.S. Geological Survey Bulletin 1412, 131 p.

Choate, R. and Rightmire, C.T., 1982, Influence of the San Juan Mountain geothermal anomaly and other Tertiary igneous events on the coalbed methane potential in the Piceance, San Juan, and Raton Basins, Colorado and New Mexico, in Proceedings of the Unconventional Gas Recovery Symposium, May 16-18, 1982, Pittsburgh, PA: Society of Petroleum Engineers/U. S. Department of Energy, 10805, p. 151-164.

Fassett, J.E. and Hinds, J.S., 1971, Geology and fuel resources of the Fruitland Formation and Kirtland Shale of the San Juan Basin, New Mexico and Colorado: U.S. Geological Survey Professional Paper 676, 76 p.

Kelly, V.C., 1951, Tectonics of the San Juan Basin, *in* Guidebook of the south and west sides of the San Juan Basin, New Mexico and Arizona: New Mexico Geological Society 2nd Field Conference, 1951, p. 124-131.

McFall, K.S., Wicks, D.E., Kuuskraa, V.A., and Sedwick, K.B., 1986, A geologic assessment of natural gas from coal seams in the Piceance Basin, Colorado: Topical Report (September 1985-1986), prepared for Gas Research Institute under contract number 5084-214-1066, 76 p.

Meissner, F., 1984, Cretaceous and Lower Tertiary coals as sources for gas accumulations in the Rocky Mountain area, in Woodward J., Meissner, F., and Clayton, J., eds., Hydrocarbon source rocks of the Greater Rocky Mountain Region: Rocky Mountain Association of Geologists, 1984, p. 401-431.

Molenaar, C.M., 1977, Stratigraphy and depositional history of Upper Cretaceous rocks of the San Juan Basin area, New Mexico and Colorado, with a note on economic resources, *in* San Juan Basin III, New Mexico Geological Society 28th Field Conference Guidebook, 1977, p. 159-166.

Rice, D.D., 1983, Relation of natural gas composition to thermal maturity and source rock type in San Juan Basin, northwestern New Mexico and southwestern Colorado: American Association of Petroleum Geologists Bulletin, V. 67, no. 8, p.1199-1218.

Sandberg, D. T., 1988, Coal Resources and coal-bed geometry, Fruitland Formation, Southern Ute Indian Reservation, Archuleta and La Plata Counties, Colorado, *in* Fassett, J. E., ed., Geology and coal-bed methane resources of the northern San Juan Basin, Colorado and New Mexico: Rocky Mountain Association of Geologists Guidebook

Shomaker, J.W. and Holt, R.D., 1973, Coal resources of Southern Ute and Ute Mountain Ute Indian Reservations, Colorado and New Mexico: New Mexico State Bureau of Mines and Mineral Resources Circular no. 134, 22 p.

Spencer, C.W., 1987, Hydrocarbon generation as a mechanism for overpressuring in Rocky Mountain Region: American Association of Petroleum Geologists Bulletin, vol. 71, no. 4, p. 368-388.

Woodward L. and Calender, J., 1977, Tectonic framework of the San Juan Basin, in San Juan Basin III: New Mexico Geological Society 28th Field Conference Guidebook, 1977, p. 209-212.

Aerial view toward the west-northwest with the San Juan Mountains west of the Animas River and north of Purgatory Ski Area on the skyline. Foreground south dipping ridges, the northern boundry of the San Juan Basin, are from left to right; Fruitland and Pictured Cliffs Formation sands, Mesaverde Point Lookout Formation sands, and the shallower dipping Dakota Formation. Photo by Bruce Conrad.

Microlithotype Analysis of Three Coals from the Upper Cretaceous Menefee Formation near Durango, Colorado

Mark J. Pawlewicz[1]

ABSTRACT

Three column samples of coal from the Upper Cretaceous Menefee Formation near Durango, Colorado, were examined petrographically by using microlithotype analysis. This kind of analysis is performed on polished coal blocks and yields information on the growing conditions of the flora in the paleoswamp from interpretations of the associations of the three coal maceral groups; liptinite or exinite, vitrinite, and inertinite. The physical condition of the macerals is also useful, giving information on degradation by subaerial oxidation and (or) transportation, each indicative of specific environmental conditions.

The changes in the coals are believed to be due to ground water fluctuations within a mainly terrestrial forest containing woody plants and trees. Occasionally, the environment was subjected to deeper water which supported primarily a herbaceous flora. Infrequent dry periods or periods of reduced water levels exposed the peat to oxidation. The absence of pyrite, which can occur as a result of marine water influence, suggests these coals were developed in an upper delta plain environment.

INTRODUCTION

The accumulation of thick peat deposits is not a simple matter of plant succession over a period of time. Numerous geological, biological, and chemical conditions must exist and be maintained for the system to continue. Mackowsky (1968) noted the importance of two factors; climate and subsidence (or the rise in the ground water table): Climate to promote heavy growth and the ground water table to aid growth but also to protect the fallen vegetation from oxidation for long periods of time immediately after deposition. Subsidence provides the space necessary for continued accumulation of peat. Spackman and others (1969) and Ting and Spackman (1971) pointed out the need for balance between the geological and biological conditions. Each plant community requires, and contributes to, its own geochemistry. Environmental areas change in size and shape as the result of the phenomenon of plant succession and migration.

Geological factors affecting coal swamps include rate of subsidence. Taylor (1981) noted that with adequate vegetation, an area that subsides rapidly will accumulate more peat than a stable area. Stach (1982) reported an annual peat accumulation of 0.5 to 1.0 mm per year in temperate climates, whereas Fisk (1960) reported 1.3 and 1.0 mm per year accumulation in subtropical swamps of Florida and the Mississippi River delta, respectively. According to Anderson (1964) peat growth may reach 3 to 4 mm per year in the forest swamps of northwestern Borneo. Ultimately, the thickness of a coal seam is determined by the degree of subsidence during the accumulation of peat in the basin of deposition (Smith, 1968). Differential compaction can provide the conditions of poor drainage and subsidence necessary for the development of thick peats. Probably the most favorable areas for accumulation are interdistributary troughs or bays and along levees of active and abandoned river channels.

Wanless and others (1969) stated that the study of coals provides convincing evidence that some horizons of coal deposition were very widespread. In general, peat accumulation takes place at the same time throughout most of the areal extent of the coal bed, although peat accumulation probably does not begin and end simultaneously throughout a coal basin. Within a peat swamp, conditions may differ enough to cause the coal to vary widely in thickness or petrographic character.

The purpose of this paper is to report on the petrographic character of the microlithotypes of coal from column samples of three Menefee Formation coals, collected near Durango, Colorado (figs. 1 and 2), and to relate the petrographic characteristics to the subenvironments in which the peats were formed. (A column sample of coal is collected by cutting ordered blocks through the entire seam.) Also, the petrography is evaluated as to the methane-generation potential of the coals.

Methods

One use of coal petrography is to identify microscopic entities, or macerals, that make up coal. It is performed with transmitted or reflected light, the latter being more common. This study was conducted with polished blocks of coal, sampled from the outcrop so as to preserve the stratigraphic integrity (Pawlewicz, 1984). The petrographic composition of coal is dependent on several factors: 1) the ecological conditions of the swamp in which the peat was deposited, such as the level of ground water, the oxygen supply, the pH of the water, and any marine influence; 2) the climate; 3) the taxonomic composition of the peat-forming communities; and 4) the degree of coalification or rank (Stach, 1982). Changing one factor will cause the plant community to readjust itself to

[1] U. S. Geological Survey
P.O. Box 25046, MS 940, DFC
Lakewood, Colorado 80225

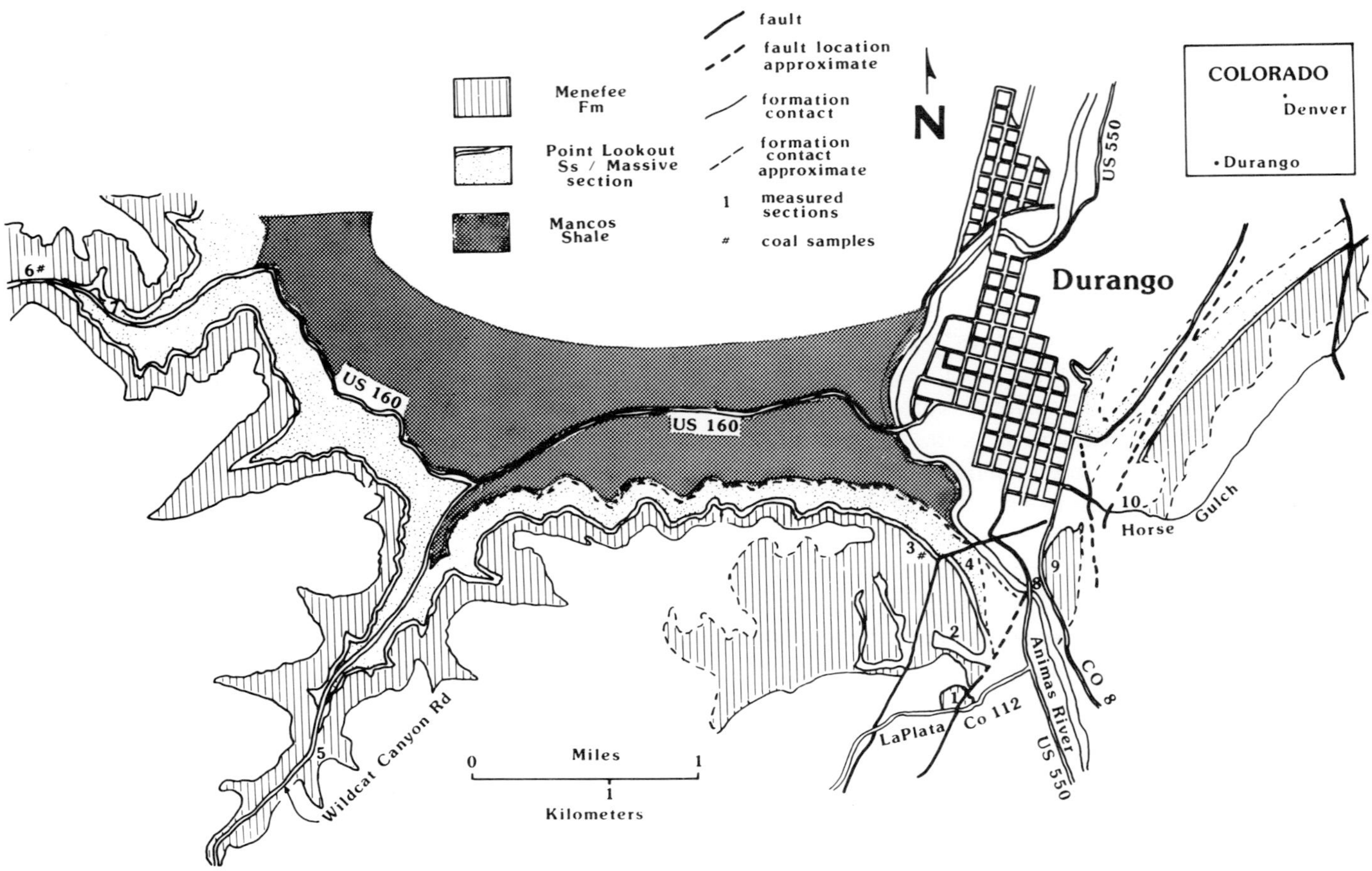

Figure 1. Geologic map and location of study area with sample locations. From Zapp (1949).

the new prevailing conditions and a change in the types of peat will follow. Severe changes will result in the death of the peat-forming swamp (Ting and Spackman, 1971). The result of any change then, is that the individual zones are superimposed on top of others (fig. 3) so that the vertical profile of the coal seam reflects the original, horizontally alternating facies (Overbeck 1950).

The identification of the macerals in subbituminous and high volatile bituminous coals is performed primarily on the basis of morphology and color. The inertinite, perhaps the most easy to identify, stands out because of its bright, white color, as opposed to the shades of gray of the vitrinite and exinite. The inertinite can occur as small shards, "T" or "I" shaped. The most characteristic occurrence is as large pieces showing remnant plant cell structure.

The vitrinite is usually the most common constituent in the coal. It is intermediate in color between the brighter inertinite and the darker exinite. The other macerals are often included or contained within the vitrinite. The vitrinite can be homogeneous in color or show mottling.

The exinite or liptinite group displays the darkest color of the three maceral groups; the shape of the maceral is characteristic and helps distinguish the individual macerals in this group. Sporinite usually appears as flattened ovals with slits, resinite as ovoid blobs, and cutinite as elongate pieces with saw-toothed edges. Fluorescence can be used to distinguish the exinite when identification is in question.

Coal Lithotypes and Macerals

Stopes (1919) was the first to consider coal from a petrological standpoint and named, for descriptive purposes, the four visibly different ingredients in banded bituminous coal: vitrain, clarain, durain, and fusain. In 1935, Stopes introduced the maceral concept to designate the microscopic constituents of coal (table 1). Theissen (1919) also formulated nomenclature for coals related to his own work with coal thin-sections (thus providing a fundamental difference between his and Stopes' terminology) which became widely used in America. The Stopes system gained acceptance throughout Europe and because of the relative simplicity, later became more and more widely used in America. The term "lithotype" was first proposed by Seyler in 1954 and was accepted by the International Committee for Coal Petrology (ICCP) to designate the "four macroscopically visible ingredients in banded bituminous coal" of Stopes. These bands are recognized by the difference in their luster which is imparted by the percent composition of the various macerals (table 1). The term microlithotype was also introduced by Seyler in 1954 to designate the typical associations of macerals, the minimum band-width has been fixed at 50u; microlithotypes have the suffix "ite" (table 1). The ICCP handbook is the authority on coal nomenclature and covers macerals, microlithotypes, and lithotypes in their entirety.

Coal macerals were first named by Stopes (1935) for the

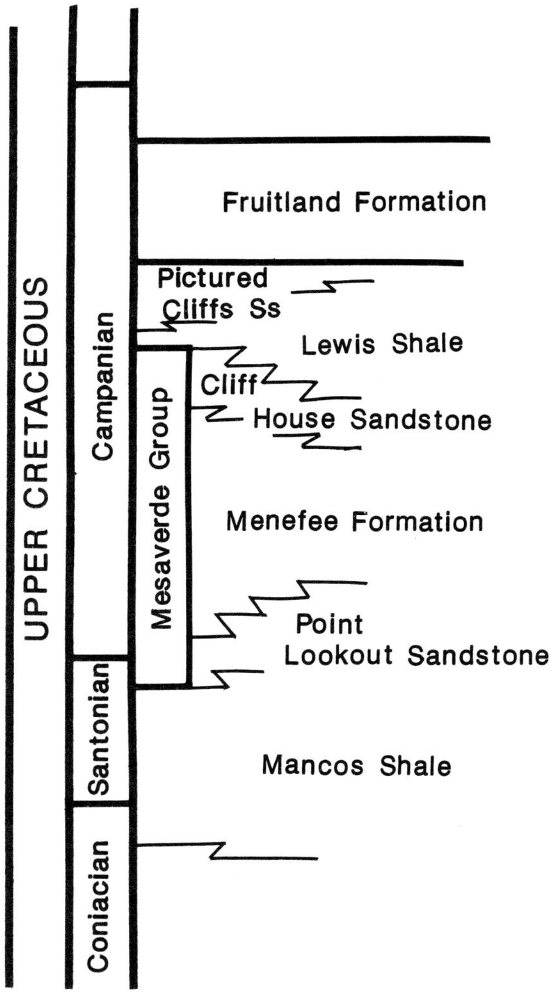

Figure 2. Stratigraphic section showing the Menefee Formation in relation to Fruitland Formation for northwestern San Juan Basin. Modified from Campbell (1971).

Table 1. Determination of the various microlithotypes, both composition and percent. V is vitrite; E is liptite; I is inertite; F is fusite.

MACERAL COMPOSITION	MICRO-LITHOTYPE	MACERAL GROUP COMPOSITION	MICRO-LITHOTYPE GROUP
MONOMACERAL			
V > 95%	Vitrite	V > 95%	Vitrite
F > 95%	Fusite	F > 95%	Fusite
BIMACERAL			
V + F > 95%		V + F > 95%	Vitrinertite
V + E > 95%		V + E > 95%	Clarite
I + E > 95%		I + E > 95%	Durite
TRIMACERAL			
V + I + E > 95%	Duroclarite	V > I or E	Trimacerite
	Vitrinertoliptite	E > I or V	
	Clarodurite	I > V or E	

microscopically recognizable constituents of coal that originate from the various plant parts; they are analogous to the minerals in rocks. There are three groups of macerals: vitrinite, exinite or liptinite, and inertinite (table 1). They are grouped according to their petrographic character.

The vitrinite maceral group is the largest of the three. It is composed of two common macerals and several less common ones. The most common are collinite and telinite (Crelling and Dutcher, 1980). The two macerals are not often distinguished from each other, and are then collectively referred to as vitrinite. Their origin is the cell wall material of the woody tissue of plants. Vitrinite makes up from 50 to 90 percent of most North American coals. Chemically, vitrinite is intermediate in hydrogen and carbon content with respect to exinite and inertinite.

The exinite or liptinite maceral group is derived from the resinous and waxy parts of plants such as leaf cuticles, spores, and resins. Exinite generally makes up about 5 to 15 percent

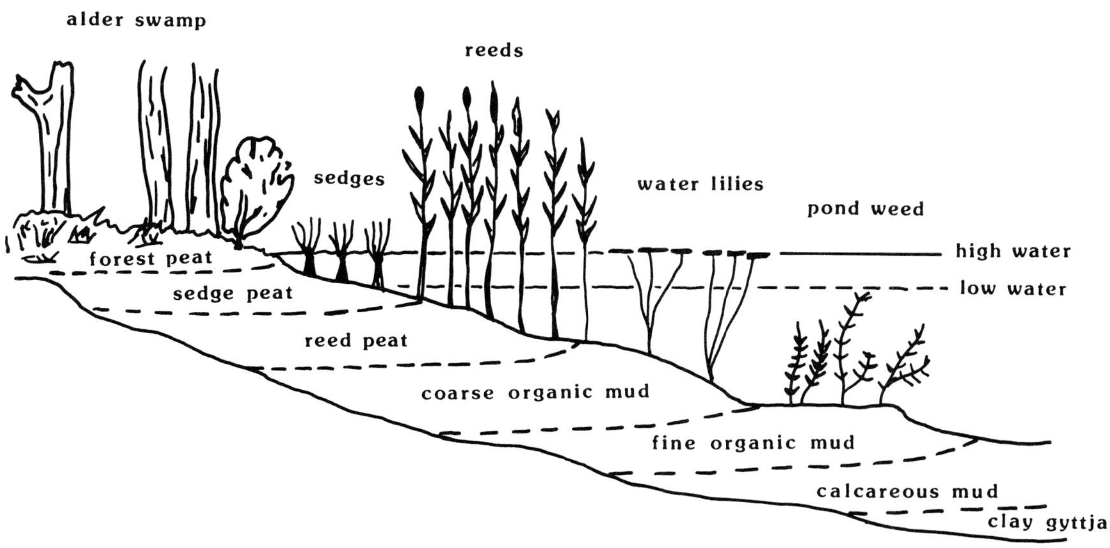

Figure 3. Illustration of fresh water environment profile showing the vertical stacking of peat types. Modified from Overbeck (1950).

of most North American coals. The exinite group is high in hydrogen compared with the other two groups, and the high hydrogen content causes them to fluoresce under ultraviolet light.

The inertinite group has the highest carbon content of the three groups. It comes from plant material that has been strongly altered and degraded in the early stages of peat formation, either by fire or subaerial oxidation. The inertinite content in North American coals ranges from 5 to 40 percent.

Petrography and Environments

Coal beds are commonly subdivided into benches during mining on the basis of fusain, mineral partings, and other natural breaks in the beds. Implicitly these subdivisions indicate a change or break in the environment of sedimentation (Ting and Spackman, 1971). Early publications dealing with coal composition as related to the environments were by Teichmuller (1950) and Teichmuller and Thompson (1958). These two studies have formed the basis for all subsequent petrography-environment studies. Hacquebard and others (1965), Hacquebard and others (1967), Hacquebard and Donaldson (1969), Cameron and Birmingham (1971), and Broughton (1972) have performed petrography studies on Canadian coals and Marchioni (1980) has done so with Australian coal.

Interpretation of the petrography is made according to the relationship of swamp environment and the associated vegetation. These relationships are based on the works of Teichmuller (1950) and Teichmuller and Thompson (1958). The facies Teichmuller and Thompson established are:

Forest-moor: contains a high proportion of vitrite, clarite, and vitrinertite. These constituents approach 50 percent or more of any seam, the remainder comprising the microlithotypes duroclarite, and clarite. These coals have a high concentration of fusinite.

Reed-moor: the proportion of non-vitrinitic material increases to greater than 80 percent. The highest concentrations of exinite and particulate vitrinite are in this environment.

Open-moor: also called open-water. This environment tends to be high in vitrinite and sporinite, but is characterized by high ash yields. The lithotype fusite is scarce.

Besides the maceral composition and the percent volume of each maceral, consideration also was given to the distribution of the macerals within the bands, that is, whether the macerals were randomly distributed or in microbands or lenses. Also, the physical condition of the organic material was given strong consideration, that is, whether the macerals showed signs of weathering or transportation.

MENEFEE FORMATION COAL SEAMS

Column samples were taken from three beds (fig. 1); one from an outcrop on the north face of Smelter Mountain south of Durango and two from the Arness-McGriffin surface mine on the west end of the study area (Z-28 and Z-28 AP and BP). The Z-28 designation for the samples from the mine area refer to the location studied by Zapp (1949). The locations were dictated primarily by exposure and accessibility.

Smelter Mountain Sample

The first sample, taken from near the top of Smelter Mountain, is an unnamed 1 foot (25 cm) thick coal bed which is 70 ft (21 m) above the Point Lookout Sandstone. The bed occurs in the middle of a repetitious sequence of siltstone, medium to dark gray and carbonaceous shale, and other thin coal beds. The bed appears to be limited in lateral extent, possibly less than 300 ft (92 m), although soil cover and the steepness of the slope prevented tracing the seam.

The repetitious vertical assemblage of shale, siltstone, and coal and the lack of burrows and other biological indicators suggest that deposition was in a fresh-water environment. The absence of pyrite in the coal blocks, as observed petrographically, also suggests fresh water.

The number of thin coal beds, (about 10 within 100 feet or 30 m), and the cyclical nature of the sedimentation indicates a fairly rapid but discontinuous subsidence of the depositional site. This, in turn, caused fluctuating water levels in the peat bogs, which may be reflected in the petrographic composition of the coal (Hacquebard and Donaldson, 1969).

Macroscopic Seam Characteristics

Figure 5 illustrates the macroscopic composition of the Smelter Mountain coal bed. The bed is composed primarily of the lithotype clarain with two durain and two thin vitrain bands. Clarain and durain are primarily composed of varying proportions of two macerals; vitrinite plus exinite and exinite plus inertinite, respectively. The presence of shiny vitrinite lenses in the clarain, and the inherent high vitrinite content of the clarain itself, make the distinction between clarain and durain clear. Clarain, because of its vitrinite content, is commonly called "bright" coal, and durain is called "dull" because of its lack of vitrinite-rich layers. The lithotype bands are medium to thick, ranging from 1 to 10 cm, except the vitrain bands, which are only 3 to 6 mm thick.

Petrography of Smelter Mountain Seam and Inferred Environment

The results of the macroscopic and microscopic analysis are illustrated in figure 5. The microlithotypes are expressed in percentages, which are cumulative along the horizontal axis. The abundance of the lithotype clarain, in sections I, IV, VI, and VIII, indicates that the environmental conditions within the swamp were fairly consistent, being interrupted, perhaps by dry spells, only twice for short periods of time, as indicated by the durain bands, sections II and VII.

The constituents of the clarain are distributed equally throughout; the proportion of the vitrite and vitrinertite vary somewhat with the clarodurite and the durite, the only two microlithotypes not common to all four bands. The macerals fusinite and sporinite are common in the clarain bands, but not as common as the petrography may suggest. The macerals, apparently randomly, are scattered throughout the lithotypes, that is, they are not in distinct microbands. If they were in bands, the vitrite percentage would be much higher.

The durain bands, sections II and VII, are significantly reduced in vitrinite. The fusite microlithotype is more common, and mineral matter, probably clay, is present in the durites. The inertinite group maceral, fusinite, is disseminated throughout, but in section II and IV it is in large pieces, with characteristic cellular structure.

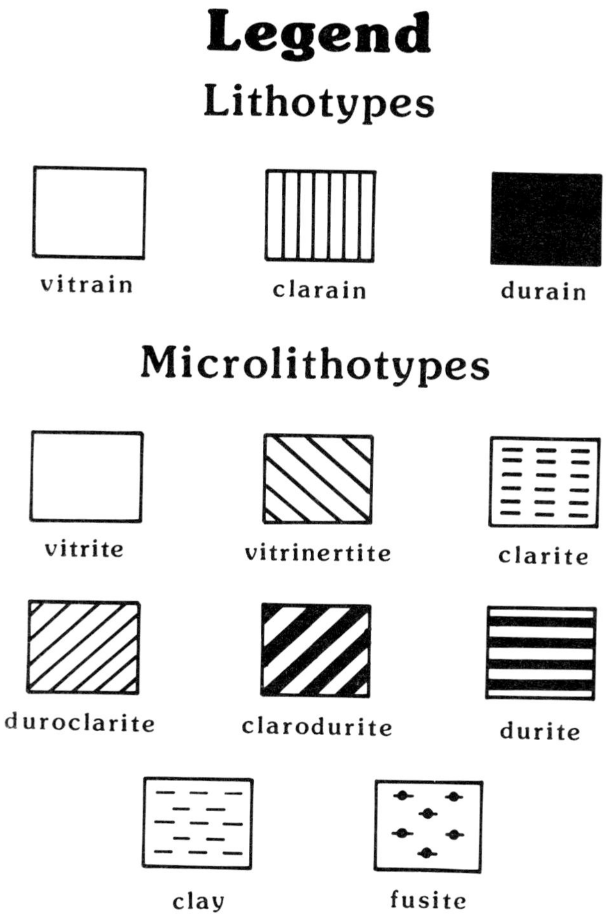

Figure 4. **Explanation of lithotype and microlithotype patterns for figures 5, 6, 7, and 8.**

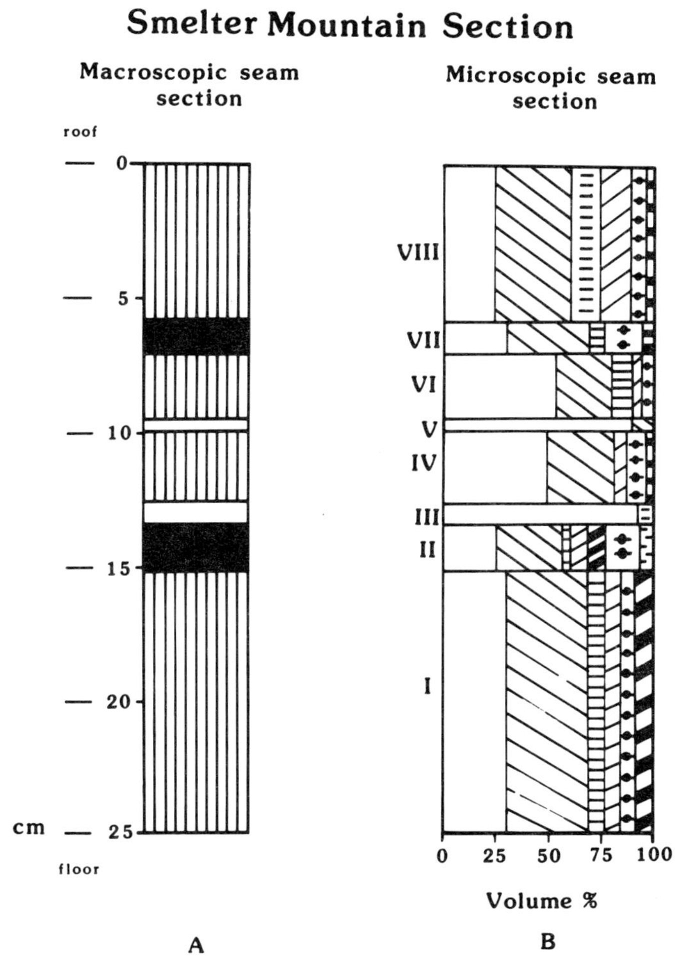

Figure 5. **Macroscopic and microscopic section for Smelter Mountain sample. Roman numerals refer to individual sections discussed in the text. See figure 4 for explanation of symbols.**

The high cumulative percentage of vitrite, vitrinertite, and clarite suggests a terrestrial environment with a high water table, probably just above the soil surface. This situation describes a forest-moor facies. The predominance of the maceral resinite over sporinite supports this conclusion. The durain bands, with relatively high percentages of exinite and fusinite and less vitrinite, suggest a forest-moor environment but with a reduced water table. The low water would allow fallen vegetation to be exposed to subaerial oxidation.

The change from wet to dry conditions is the difference between the poorly drained and the well-drained swamp. This change probably reflects a drop or rise in the water level possibly resulting from changes in the drainage patterns.

Seam-Sample Z-28

Sample Z-28 comes from the Arness-McGriffin surface mine (formerly the Sunshine mine) four miles west of Durango, Colorado on US 160. The bed is approximately 54 feet (17 m) above the Point Lookout Sandstone (Zapp, 1949) and 24 feet (7 m) above the main bed; the seam is 2.2 feet (66 cm) thick at the sampling site. A siltstone overlies the coal, and the contact with the coal is sharp. The lower contact is with a claystone which, in turn, overlies a fine-grained sandstone or siltstone with root structures. All indications suggest the enclosing sediments are fresh-water in origin; no marine organisms or burrows were observed, and the petrography revealed no pyrite. Sedimentary structures are uncommon, mostly small-scale ripple drift at the top of silty laminae.

Macroscopic Seam Characteristics

Sample Z-28 contains bands of clarain, durain, and vitrain (fig. 4 and 6). Clarain made up 75 percent of the seam; the high percentage of vitrinite made it appear bright. The durain also had vitrinite lenses throughout.

Petrography of Sample Z-28 and Inferred Environment

The clarain makes up thick sections of the bed and appears quite homogeneous, with concentrations of dull-luster material or mineral matter. In the lower parts of the seam, section III, vitrite and clarite make up 65 and 27 percent of the total composition, respectively. This consistency requires an environment that is stable for a long period of time. The exinite maceral is mostly resinite with sporinite subordinate; cutinite is rare. Stratigraphically up the seam, the vitrite is greatly reduced in all the clarain bands. This is due to the presence of disseminated material that reduces the homogeneity of the microlithotype, indicating a contribution of material from other plant parts. Despite this, the overall environment remained unchanged. This environment, represented by the

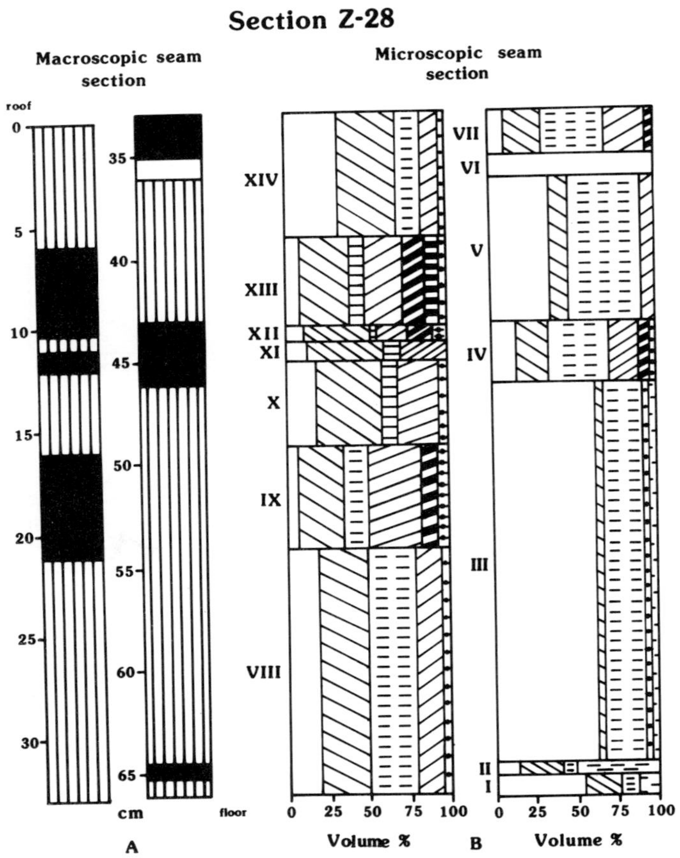

Figure 6. Macroscopic and microscopic section for sample Z-28. Roman numerals refer to individual sections discussed in the text. See figure 4 for explanation of symbols.

clarain, is a forest-moor with woody plants and trees, and a water table not high enough to drown the mainly terrestrial flora.

The durain is characterized by a low vitrinite content and a marked increase in clarodurite and durite in several places. The exinite maceral group is represented almost entirely by resinite. Fusinite is the most abundant inertinite maceral, and some mineral matter is admixed with the coal. Because of the high vitrinertite, duroclarite, and clarite content, from a petrographic standpoint, the durain could almost be clarain. The physical condition of the material is what distinguishes this sample. Much of the material shows signs of weathering and (or) transport, such as rounded edges, most likely from subaerial oxidation. The durain represents a forest-moor environment with a reduced water table.

Sample Z-28-2 BP and AP

The two samples with the designation Z-28-2 BP and AP are from the same 7-feet (2 m) thick seam at the Arness-McGriffin mine. This is the main seam 30 feet (9 m) above the Point Lookout Sandstone. The AP and BP denote above and below a 5 cm thick, highly carbonaceous, fine-grained sandstone parting with a thin clay on top. A second parting of similar composition lies beneath the lower coal sample, and provides a break point to mark the bottom of the sampling. The samples were separated at the parting only for convenience. Together the samples total 2 feet (61 cm) of coal.

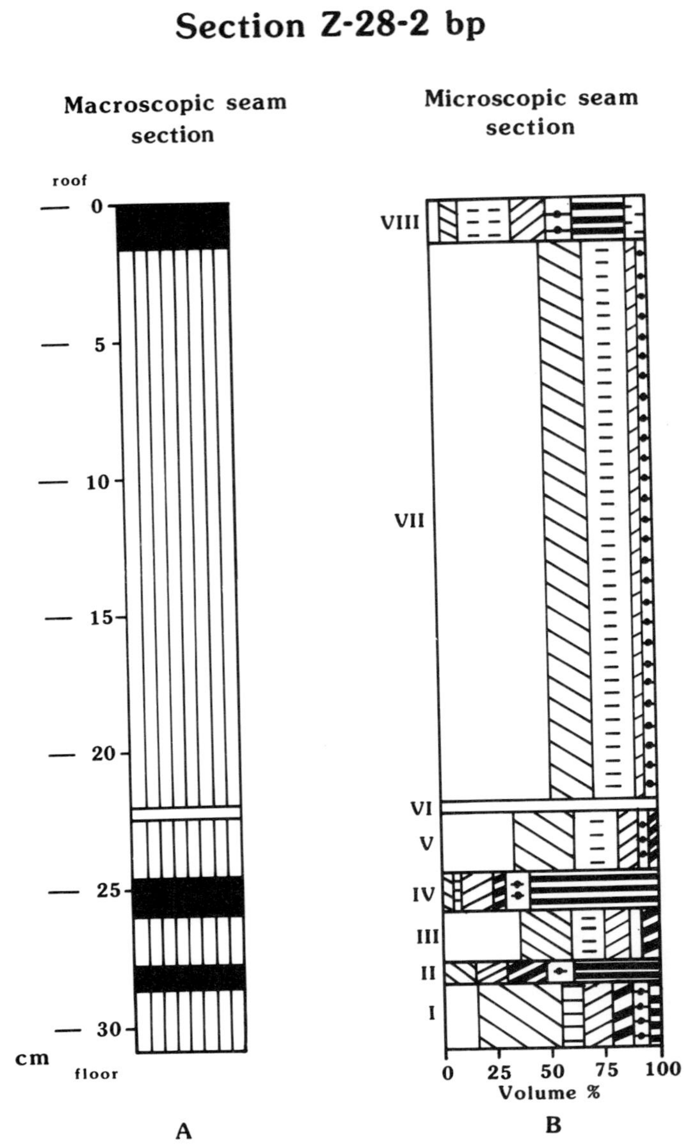

Figure 7. Macroscopic and microscopic section for sample Z-28-BP. Roman numerals refer to individual sections discussed in the text. See figure 4 for explanation of symbols.

The remainder of the seam was inaccessible due to flooding of the mine pit. The lateral extent of the seam was not discerned due to soil cover.

The associated lithologies are the same as the previous sample, Z-28, except that this bed is directly overlain by a channel sandstone 18 feet (5 m) thick. Laterally, beyond the channel sandstone, are siltstones and very fine grained sandstones in thin to medium thickness beds with mostly small-scale ripple drift as the sole sedimentary structure. This sequence was also interpreted as a fresh-water environment.

Macroscopic Seam Characteristics, Sample Z-28-2 BP

Figure 7 shows the lithotype composition of sample Z-28-2 BP. The column is composed mostly of clarain with several small durain and vitrain bands. The durain has a dull luster and is free of mineral matter. The clarain is bright to semi-bright, with few, but distinct, vitrain lenses in them.

Petrography of Sample Z-28-2 BP and Inferred Environment

The lower part of the bed, sections I through IV (fig. 7), is characterized by several changes of lithotypes. The clarain has a very different microscopic composition, possibly because of the influx of material that made up the parting below the sample. Sections I and III have much vitrinite in them but the proportions of exinite and inertinite are also high. The petrography reflects the macroscopic appearance; the clarain has no distinct banding, and the individual macerals are widely dispersed. Sporinite is the most common exinite maceral; resinite is uncommon. The fusinite has a weathered appearance. For these reasons, the lower clarain, sections I and III, represent a mixed forest and reed-moor environment, one where woody plants and trees coexist with or are very near a shallow water area inhabited by herbaceous plants.

Sections V, VI, and VII show more consistency in their composition, most likely from a long, uninterrupted growing term. The high vitrite, clarite, and vitrinertite content, the good microbanding, and the relatively unweathered appearance all suggest the flora was dominated by woody plants. These lithotypes represent a forest-moor facies with wet conditions.

The extreme heterogeneity of the durain in sections II, IV, and VIII, the low proportion of vitrite, the high proportion of sporinite, and the lack of microbanding indicate deposition where the organic material was transported and accumulated subaqueously, probably from fallout from suspension, in an open-water or open-moor facies. The clastic material in the last section, VIII, was related to the incursion which formed the parting above while agitating the accumulated peat.

Macroscopic Seam Characteristics, Sample Z-28-2-AP

Sample Z-28-2 AP, from above the parting, is overlain by a channel sandstone. It is not apparent that the channel removed any of the peat when it swept through the swamp; there is minimal scour into the coal, and the bottom of the sandstone is devoid of any coal spar or clasts. The thickness of the coal is constant for the length of the exposure, about 150 feet (45 m).

The only lithotypes identified in the coal blocks were clarain and durain (fig. 7). The clarain, with numerous vitrain lenses is bright; the durain was dull due to inertinite and detrital clays.

Petrography of Sample Z-28-2 AP and Inferred Environment

The clarain in this sample suggests long periods of stability in the swamp. Sections II, IV and VI (fig. 8) show good homogeneity, with vitrite and vitrinertite dominating. Exinite macerals are uncommon in all three bands, with resinite predominant over sporinite. Fusinite, in both small and large pieces, is the most common inertinite maceral group representative. This material is dispersed throughout rather than being concentrated in lenses or individual bands. This assemblage is indicative of a forest-moor facies.

The durite bands, sections I, III and V, are unique. The lowermost band, section I, shows the effects of flooding in the the peat swamp. The band is mostly clay with little organic material. The remainder of the material, vitrite and fusite, increase towards the top of the band, but the mineral matter also persists. The length of time for the deposition of the clays cannot be estimated. The presence of the clays in the clarain

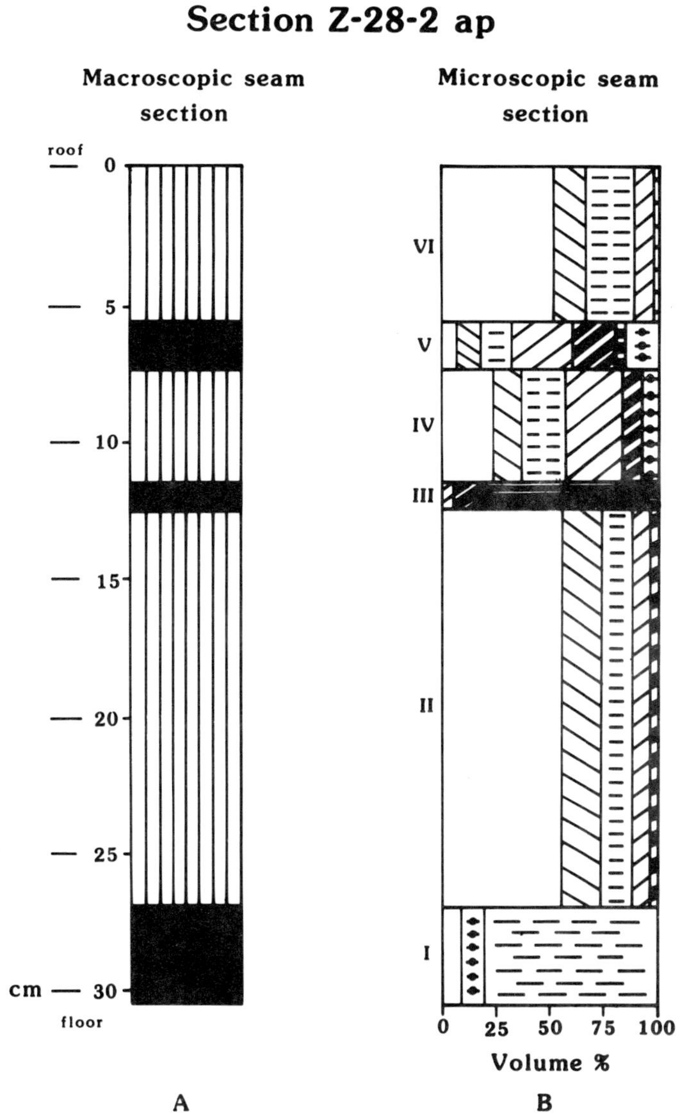

Figure 8. Macroscopic and microscopic section for sample Z-28-AP. Roman numerals refer to individual sections discussed in the text. See figure 4 for explanation of symbols.

band above suggests a longer deposition time. The durain bands were probably a forest-moor facies with a reduced water supply leading to partial oxidation of the plant material.

SUMMARY OF THE PETROGRAPHY

1) The bright coal intervals, comprising the microlithotypes vitrite, vitrinertite, clarite, and duroclarite represent relatively wet conditions where the water table was high and the peat accumulated under shallow water. This condition was probably a forest-moor environment in a terrestrial zone, at the high-water mark (Hacquebard and Donaldson, 1969).

2) The less-bright coal intervals, those with distinct lenses of vitrinite in relatively dull bands, generally comprise the microlithotypes duroclarite, clarodurite, and small percentages of fusite and vitrinertite. These microlithotypes together represent environmental conditions that were dry and at least intermittently exposed to subaerial oxidation. This exposure gives the vitrinite a weathered appearance. These coals suggest

a forest-moor environment but with a lower water table.

3) The dull-luster coals, which are uncommon in this study, contain little vitrite and high proportions of clarodurite, durite, and duroclarite often with mineral matter, either clastics or clays. Microbanding in the lithotype is generally lacking. This probably results from deposition in a subaqueous environment which corresponds to an open-moor environment.

4) There is no sedimentological evidence for a marine influence in the examined coals; this is supported by the absence of pyrite in the samples.

The petrography indicates that the swamp conditions were affected most strongly by changes in the water level which resulted in the conditions fluctuating between wet and dry. Such conditions in a fresh-water environment would most likely be due to seasonal water supply changes or flooding.

HYDROCARBON POTENTIAL

There are no petrographic analytical data, either standard or microlithotype, from the Menefee Formation in the literature to which to add or compare these analyses and interpretations. Also, there is a dearth of data related to methane from the Menefee coals. What follows is a summary from pertinent papers on the hydrocarbon generation potential from coals.

Bertrand and others (1985), after examining fourteen coals, of different rank and age from various worldwide locales, concluded that no clear correlation exists between petroleum potential, hydrocarbon generation, and bulk petrographic composition.

Stach (1982) noted that hydrogen-rich kerogen (from shale) is more sensitive to increasing temperature. With coal, the formation of methane begins at temperatures of about 80 degrees celsius in the subbituminous stage; by 110 degrees celsius in the bituminous stage, the generation is substantial (Van Heek and others, 1971). Similarly, Durand and others (1977) found that extracts and alkane yields were greatest from coals with vitrinite reflectances ranging between 0.8 and 1.2 percent Rm (mean random vitrinite reflectance). Comparison between the evolution of coals and the evolution of different types of kerogen showed that coals behave similarly to kerogen type III, although coal yields 5-10 times less extract. By comparison, coals generate more light hydrocarbons, especially methane. Also, the pyrolysis results of Bertrand and others (1985) demonstrate that coals have the potential to generate significant yields of liquid hydrocarbons, which on a weight basis, can be equivalent to or higher than yields from conventional source rocks.

The maturity of these coals was measured at 0.63 percent Rm, which is near the generally accepted threshold of the onset of hydrocarbon generation. Whether or not this number is significant in terms of methane generation from the Menefee coals remains to be discovered.

REFERENCES CITED

Anderson, J. A. R., 1964, The structure and development of the peat swamps of Sarawac and Brawny: Tropical Geography, v. 18, p. 7-16.

Bertrand, P., Behar, F, and Durand, B., 1985, Composition of potential oil from humic coals in relation to their petrographic nature: Advances in Organic Geochemistry, v. 10, p. 601-608.

Broughton, P. L., 1972, Petrology of the Estevan no. 3 lignite seam, southwestern Saskatchewan: Proceedings of the 1st Geological Conference on western Canadian coal, Information Series, no. 60, p. 185-201.

Cameron, A. R., and Birmingham, T. F., 1971, Petrographic and chemical properties of a lignite from Estevan, Saskatchewan: Geological Survey of Canada Paper 71-8, 20 p.

Campbell, C. V., 1971, Depositional model-Upper Cretaceous Gallup beach shoreline, Ship Rock area, northwestern New Mexico: Journal of Sedimentary Petrology, v. 41, p. 395-409.

Crelling, J. C. and Dutcher, R. R., 1980, Principles and applications of coal petrology: Society of Economic Paleontologists and Mineralogists Short Course notes No. 8, 127 p.

Durand, B., Nicaise, G., Roucache, J., Vandenbrouke, M., and Hagemann, H. W., 1977, Etude geochemique d'une serie de charbons: *in* Campos, R., and Goni, J. eds., Advances in Organic Geochemistry, 1975, Madrid, p. 601-632.

Fisk, H. N., 1960, Recent Mississippi River sedimentation and peat accumulation: Compte Rendu, 4th Carboniferous Congress, Heerlen, 1958, v. 1, p. 187-199.

Hacquebard, P. A., Cameron, A. R., and Donaldson, J. R., 1965, A depositional study of the Harbour seam, Sydney coalfield, Nova Scotia: Geological Survey of Canada Paper 65-15, 31 p.

Hacquebard, P. A, Birmingham, T. F., and Donaldson, J. R., 1967, Petrography of Canadian coals in relation to environment of deposition: Symposium on the Science and Technology of Coal, Ottawa, Canada, 41 p.

Hacquebard, P. A., and Donaldson, J. R., 1969, Carboniferous coal deposition associated with flood-plain and limnic environments in Nova Scotia: Geological Society of America Special Paper No. 114, p. 143-191.

Mackowsky, M.-Th., 1968, European Carboniferous coalfields and Permian Gondwana coalfields: *in* Murchison, D. and Westoll, S. eds., Coal and Coal Bearing Strata, American Elsevier, New York, 418 p.

Marchioni, D. L., 1980, Petrography and depositional environment of the Liddell seam, Upper Hunter Valley, New South Wales: International Journal of Coal Petrology, v. 1, p. 35-61.

Overbeck, F., 1950, Die moore Niedersachens: *in* Geologie und Laggerstatten Niedersachens, 4, Bremen-Horn, 122 p.

Pawlewicz, M. J., 1984, Stratigraphy, environments of deposition, and petrography of selected coals from the Upper Cretaceous Menefee Formation, near Durango, Colorado: Colorado School of Mines unpublished Masters Thesis, 112 p.

Seyler, C. A., 1954, Letter to the nomenclature subcommittee, *in* International Handbook of Coal Petrolography (1963): Centre National de la Recherche Scientifique, 15 Quai Anatole-France, Paris, France.

Smith, A. H. V., 1968, Seam profiles and characters: *in* Murchison, D. and Westall, T. S., eds., Coal and Coal Bearing Strata, American Elsevier, New York, 418 p.

Spackman, W., Riegel, W., and Dolsen, C. P., 1969, Geological and biological interactions in the swamp-marsh complex of southern Florida: *in* Geological Society of America Special Paper No. 114, p. 1-35.

Stach, E., 1982, Textbook of Coal Petrology: Gebruder Borntraeger, Berlin, 427 p.

Stopes, M. C., 1919, On the four visible ingredients in banded bituminous coal: Proceedings of Royal Society of London B, v. 90, p. 470-487.

———, 1935, On the petrography of banded bituminous coals: Fuel, v. 14, p. 4-13.

Taylor, M., 1981, Preparation and analysis of coal seam data utilizing paleoenvironment modeling, Hazard no. 7 coal, eastern Kentucky: International Journal of Coal Geology, v. 1, p. 213-233.

Teichmuller, M., 1950, Zum petrographischen Aufbau und Werdergang der Weichbraunkohle (mit Berucksichtigung genetischer Fragen der Steinkohlenpetrographie): Geologisches Jahrbuch, v. 64, p. 429-488.

Teichmuller, M., and Thompson, P. W., 1958, Vergleichende mikroskopische und chemishe Untersuchungen der wichtigsten Fazies-Typen im Haupfloz der niederrheinischen Braunkohle: Fortschr. Geologie Rheinland und Westfalia, v. 2, p. 573-598.

Theissen, R., 1919, Structure in Paleozoic bituminous coals: U. S. Bureau of Mines Circular, v. 117, 296 p.

Ting, F. T. C., and Spackman, W., 1971, The coal lithotype concept and seam profile: Compte Rendu, Septieme Congres International de Stratigraphie et de Geologie du Carbonifere, Krefeld, August 23-28, p. 307-311.

Van Heek, K. H., Juntgen, H., Luft, K. F., and Teichmuller, M., 1971, Aussagen zur gasbildung in fruhen inkohlungsstadien auf grund von pyrolyseversuchen: Erdol und Kohle, vol. 24, p. 566-572.

Wanless, H. R., Baroffio, J. R., and Trescott, P. C., 1969, Conditions of deposition of Pennsylvanian coal beds: Geological Society of America Special Paper no. 114, p. 105-142.

Zapp, A. D., 1949, Geology and coal resources of the Durango, area, La Plata and Montezuma counties, Colorado: U. S. Geological Survey Oil and Gas Investigations Preliminary Map 109.

Aerial view east with the San Juan Mountains east and north of Pagosa Springs, Colorado on the skyline. The prominent ridges dipping south into the San Juan Basin and snaking east are the Mesaverde Point Lookout Formation sands, left, and the Pictured Cliffs Formation sands, right. The Pine River, center of photo flowing left to right, bisects an anticlinal feature which extends into the San Juan Basin. The position of the Pine River on top of the structure and generally parallel to the south plunge may be controlled by localized faulting. Photo by Bruce Conrad.

Well-Log Determination of Ash Content in Fruitland Formation Coals, Southern Ute Indian Reservation, Southwest Colorado

Stephen E. Prensky[1]

ABSTRACT

A preliminary investigation of coal-ash estimation from well-log data for Fruitland Formation coals on the Southern Ute Indian Reservation of southwest Colorado indicates that it may be feasible to use density well-log data for coal-ash estimation provided sufficient core data are available for calibrating the log results. The gamma-ray log cannot be considered reliable for estimation of coal-ash in the study area due to the inconsistent presence of radioactive trace elements in the shales, siltstones, and coals. Two conditions precluded an accurate estimation of ash content at the present time: a) Only one core and only a very limited number of laboratory proximate analyses were available for developing a reliable ash-correction for the existing suite of well logs, and b) Discrepancies between laboratory- and log-determined ash values suggest that there may be some problems with the core-log depth correlation. Core penetration rate was used for depth correlation of core- and well-log data in lieu of other data. A core gamma log would improve the depth match in general and in the consolidated shaly intervals in particular. Recent information indicates that rank of Fruitland coal seams may vary within the same borehole and additional data from these seams are needed. The simple, linear, shale (ash)-coal density model utilized here may provide only a gross approximation of ash content and more sophisticated models, based on additional core data, would improve the accuracy and reliability of the ash estimates.

INTRODUCTION

This report is a preliminary investigation into the feasibility of using geophysical well-log data for estimating coal-ash content of Upper Cretaceous Fruitland Formation coals on the Southern Ute Indian Reservation. The ultimate goal of this and associated projects is the estimation of coal-quality for use in determining coal-bed methane production capability and the recoverable coal-bed methane resources on the Southern Ute Indian Reservation.

Wells were initially selected for study and analysis (table 1) on the basis of Fruitland-coal core recovery. Due to the competitive nature of coal-bed methane development in the northern San Juan basin, only core data from the Ladd Petroleum Company North Cox Canyon No. 2-3 well, cored at the request of the Southern Ute Tribal Energy Office, were available for this study. Figure 1 shows the location of the study wells within the Southern Ute Reservation.

FRUITLAND FORMATION COALS

The entire study area on the Southern Ute Indian Reservation (figs. 1 and 2), falls within the Ignacio Blanco gas field in the northern part of the central San Juan Basin. Figure 2 shows the location of the study area within the structural setting of the San Juan Basin. The geologic structure of the study area and in particular the structural nose of the Ignacio Blanco field (on which the majority of the wells were drilled) is shown in figure 3. The regional geology of the San Juan Basin in general and the geology of the Upper Cretaceous Fruitland Formation in particular, have been discussed by Fassett and Hinds (1971), Rice (1983, and this volume), Choate and others (1984), Meissner (1984), and Fassett (1987, and this volume).

Most of the following information on the geometry of the Fruitland Formation (including the coal beds) and its relationships with adjacent formations is from Fassett and Hinds (1971) and Fassett (1987, and this volume). The Fruitland Formation consists of interbedded sandstone, siltstone, shale, carbonaceous shale, carbonaceous sandstone, and coal, and ranges in thickness from 300 to 500 feet with individual coal beds ranging in thickness up to 40 feet. The contact with the overlying Upper Cretaceous Kirtland Shale is gradational and difficult to locate in the subsurface. The underlying Upper Cretaceous Pictured Cliffs Sandstone consists of massive sandstone and lower interbedded sandstones and shales. The Fruitland Formation represents nonmarine, lower coastal-plain deposition that occurred behind (shoreward of) the Pictured Cliffs marine shoreline. The coals occur in the lower part of the formation in close association with the Pictured Cliffs Sandstone and the thickest coal beds occur in northwest-trending bodies paralleling the Pictured Cliffs shoreline. Ash

[1] Stephen E. Presnsky
U.S. Geological Survey
P.O. Box 25046, MS 971, DFC
Lakewood, Colorado 80225

Craig Wandry of the U.S. Geological Survey, Branch of Petroleum Geology, Denver, generated the map in figure 1 and the three-dimensional surface projection in figure 3. I wish to thank J.E. Fassett, B.E. Law, J.W. Schmoker, C.W. Spencer, and R. Zech for reviewing the manuscript and providing valuable suggestions.

Table 1. List of study wells; well locations are shown on figure 1.

Well location no. on fig. 1	Well Name	Location: Section	T.N.	R.W.	API number	Year compl.	Logs run	Digitized interval log depth, ft	Surface elevation (KB, ft)	Fruitland Formation log depth (ft)	Fruitland Formation log elev. (ft)	Highest Fruitland Coal log depth (ft)	Pictured Cliffs Sandstone log depth (ft)	Pictured Cliffs Sandstone log elev. (ft)
Amoco														
1	So. Ute Tribal B 1	NW¼NW¼3	33	6	05-067-06688	1984	DL/DN/S	3,500-3,870	7,451	3,880	3,571	none	3,950*	3.501*
2	So. Ute Tribal C 1	SW¼SW¼21	33	6	05-067-06687	1984	DL/DN/S	2,800-3,300	6,732	3,135	3,597	none	3,254	3,478
3	Ute Gas Unit AB 1	NE¼SW¼16	33	7	05-067-06827	1986	DL/DN/S-NA	2,000-2,600	6,488	2,402	4,086	2,480	2,583	3,905
4	So. Ute Tribal D 1	SW¼SE¼17	33	7	05-067-06741	1984	DL/DN/S	2,200-2,650	6,583	2,433	4,150	2,427	2,594	3,989
5	Ute Gas Unit AC 1	NE¼SE¼18	33	7	05-067-06829	1986	DS/DN/S	2,050-2,670	6,550	2,422	4,128	2,448	2,596	3,954
6	Ute Gas Unit AA 1S	NE¼SW¼20	33	7	05-067-06797	1986	DS/DN	2,200-2,850	6,614	2,617	3,997	2,611	2,778	3,836
7	So. Ute Tribal F 1	SE¼SE¼28	33	7	05-067-06830	1986	DS/DN/S	2,100-2,550	6,351	2,330	4,021	2,410	2,503	3,848
8	H. Pearson G.U. A 1	SW¼NW¼20	34	7	05-067-06846	1986	DS/DN	2,450-3,000	6,822	2,716	4,106	2,725	2,922	3,900
9	G. Beebe G.U. A 1	NW¼NW¼25	34	7	05-067-06851	1986	DS/DN	2,400-2,930	6,749	2,674	4,075	2,723	2,886	3,863
10	So. Ute Tribal A 1	SW¼SE¼36	34	8	05-067-06361	1980	DS/DN/S	2,600-2,900	6,607	2,612	3,995	2,606	2,807	3,800
11	A.W. Mason G.U. A 1	SW¼NW¼20	34	9	05-067-06849	1986	DL/DN/CN	2,100-2,850	6,608	2,425	4,183	2,700	2,840	3,768
12	Fed. Land Bank C 1	SW¼NW¼36	34	9	05-067-06818	1986	DS/DN/S/CN	2,350-3,100	6,706	2,686	4,020	2,717	3,081	3,625
Ladd Petroleum														
13	N. Cox Canyon 2-3	NW¼SE¼3	32	10	05-067-06901	1987	DS/DN	2,500-3,300	6,602	2,810	3,792	2,968	3,209	3,393

*Available logs not run to TD of 4,502 ft

Log Abbreviations	Traces digitized
DL — dual induction laterolog	gamma, SP, deep induction, laterolog 8
DS — dual induction focused log	gamma, SP, deep induction, focused log
DN — combination borehole compensated formation density and neutron	gamma, caliper, bulk density, correction, neutron porosity
S — borehole compensated sonic log	gamma, caliper, delta T (transit-time)
CN — cased-hole spectral gamma-ray log	total gamma, potassium, thorium, uranium
NA — not available	

content ranges from 8 to 30 weight-percent and there is a general increase in ash content eastward across the basin. Hydrocarbon production from the Fruitland Formation consists of non-associated gas from lenticular sandstones and coals (Rice, 1983).

The top of the Fruitland Formation was identified by a carbonaceous shale which was used in well-to-well correlation for examining structural and stratigraphic relationships between the Fruitland Formation and the underlying Pictured Cliffs Sandstone. Individual lithologic units within the Fruitland Formation are often difficult to correlate, even between closely spaced wells, due to lithologic variability and discontinuity. This is particularly the case in the study area where most wells do not reach the Upper Cretaceous Huerfanito Bentonite Bed (named in Fassett and Hinds, 1971) of the Lewis Shale. This bentonite is a basin-wide time-plane recognizable in the subsurface on geophysical logs (Fassett and Hinds, 1971). The Fruitland Formation bed depths in table 1 sometimes represent a relative position within the Fruitland Formation rather the the tops of the carbonaceous shales. Table 1 lists both the log depth as well as elevation (log depth minus kelly-bushing elevation). Stratigraphic and structural relationships are presented in cross-sections A-A′, B-B′, and C-C′ (figs. 4-6).

Rice (1983) reported a pattern of increasing coal maturation (vitrinite reflectance) from south to north in the basin. The study area falls within the subituminous to medium volatile bituminous zone of Rice (1983). N. Bostick (1987, personal communication) analyzed cores from the Ladd Petroleum North Cox Canyon No. 2-3 well (well 13, table 1) and measured an average vitrinite reflectance of R_o = 1.05 percent, placing the coals in the high volatile bituminous A category of the ASTM coal rank classification (ASTM Standard, 1983). Fassett and Hinds (1971) mapped ash content within the basin using data obtained from proximate analyses of coal samples. However, they only had two samples available from within the area under consideration in this report. They found that in general, the ash content of Fruitland Formation coals is variable but relatively high in comparison with other western coals.

APPLICATION OF WELL LOGS TO COAL IDENTIFICATION AND EVALUATION

The use of well-logs for coal identification and coal-quality analysis stems from the unique responses of geophysical logging tools to coal. Figure 7 is an example of typical well-log response to coal and associated lithologies. The well-log coal responses vary with coal rank, due to changes in hydrogen content, the fixed carbon-volatiles ratio, and density; these changes accompany thermal maturation and compaction and affect the porosity-reading logging tools (neutron, sonic, and density) and the resistivity-reading logging tools. The

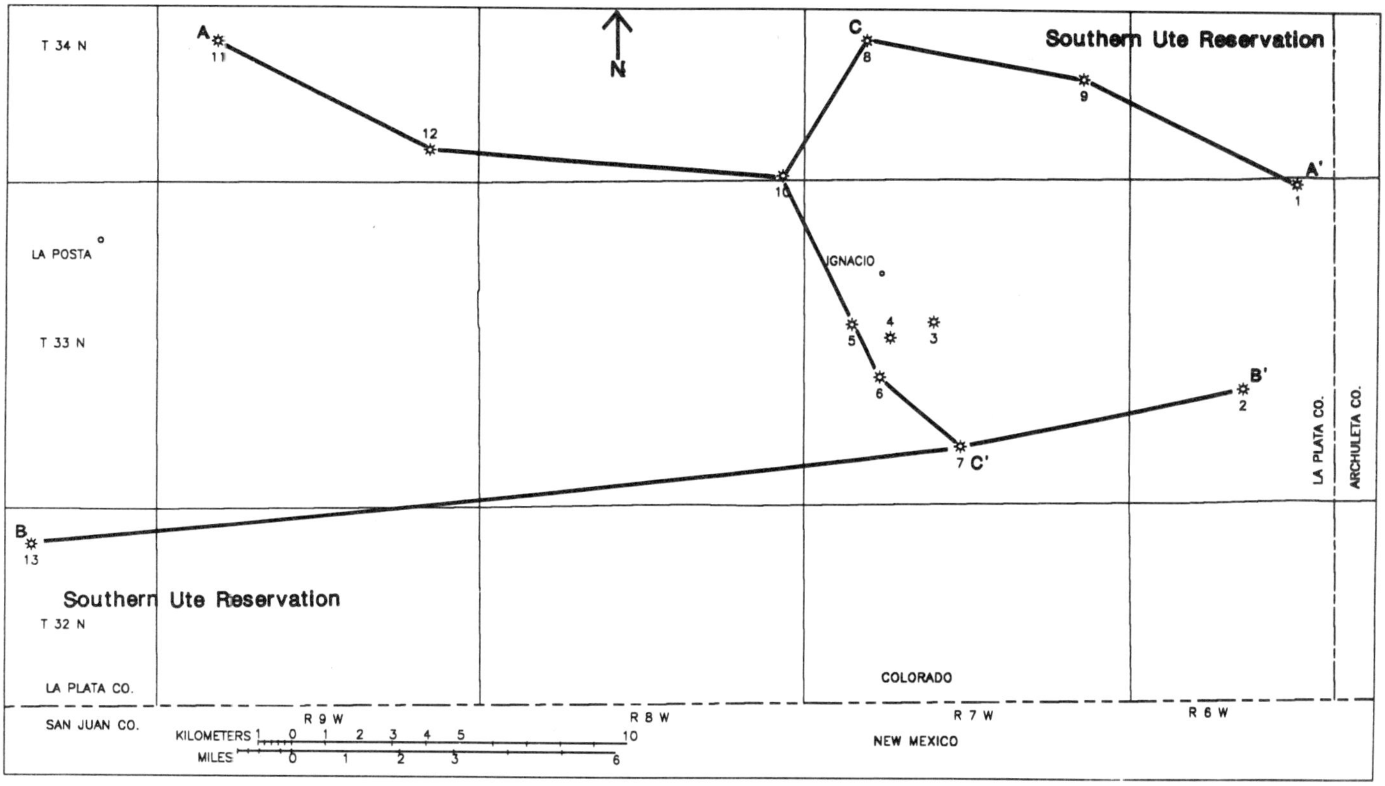

Figure 1. Map showing the location of study wells on a portion of the Southern Ute Indian Reservation and locations of geologic cross-sections A-A′, B-B′, and C-C′. Numbered wells are identified on table 1.

ranges of values for various lithologies as determined from well logs are shown in table 2. The ranges in values for coal include variations arising from differences in coal rank. Figure 8 shows the relationship between acoustic travel-time, bulk density, and coal rank. The presence of radioactive isotopes in the coal results in the range of gamma-ray API units.

The advantages and limitations inherent in the use of various well-logging parameters for coal identification and quantitative analysis have been reviewed by Hoffman and others (1982), Kern and others (1983), and Daniels and others (1983). The well-log responses of Fruitland coals have been discussed by Fassett and Hinds (1971), Kern and others (1983), Jones (1985), and Mullen (this volume).

WELL-LOG DIGITIZATION

The standard log suite that was run in the study wells consisted of dual induction laterolog/focused log, and combination borehole compensated density and neutron logs. An acoustic log was not run in every well. Although the well logs for all the wells listed in table 1 were originally recorded on digital magnetic tape at a 0.5 ft interval, the tapes were not available for this study. Consequently the available well-log data for these wells were manually digitized from analog prints at a 0.5 foot increment for computer processing. The variable quality of the analog data resulting from distortion in the copying process (from microfilm and microfiche) combined with operator error are areas of potential inaccuracies in the recorded values and in their depth locations.

These logs were obtained using standard oil field logging tools as opposed to specialized coal logging tools. Table 1 summarizes the types of logs available for each well as well as the curves and depth intervals digitized. Figure 9 presents the well-log data for the Ladd Petroleum well (well 13, table 1). The original bulk density curve was not available for 6 wells (wells 1, 3, 7, 10, 11, 13, table 1); the density porosity curve for each of these wells was digitized and then converted to bulk density. Because the maximum density porosity recorded on these logs was 70 percent, the converted bulk density values for these 6 wells were limited to minimum density values of 1.5 gm/cc resulting in the flattening of the bulk density curve seen in figure 9 and in various crossplots (figs. 11-15). Stratigraphically, the digitized interval included the lower shale member of the Kirtland Shale, the Fruitland Formation, and generally reached total depth (TD) within the uppermost part of the Pictured Cliffs Sandstone. Well-log data from the resistivity and acoustic logs were depth-shifted, as needed, based on visual inspection, to agree with the density gamma-ray trace which was chosen as the reference trace. This gamma-ray trace was then corrected for the effects of borehole size and mud weight.

Table 2. Well-log response in coal and associated lithologies (Suau, 1981).

Tool	Units	Coal	Shale	Sandstone
Density	gm/cc	1.25-1.4	2.5-2.8	2.4-2.7
Neutron	Phi %	25-65	25-40	1-15
Sonic	msec/ft	110-140	70-100	55-80
Gamma-ray	API	2-20	150-200	10-30

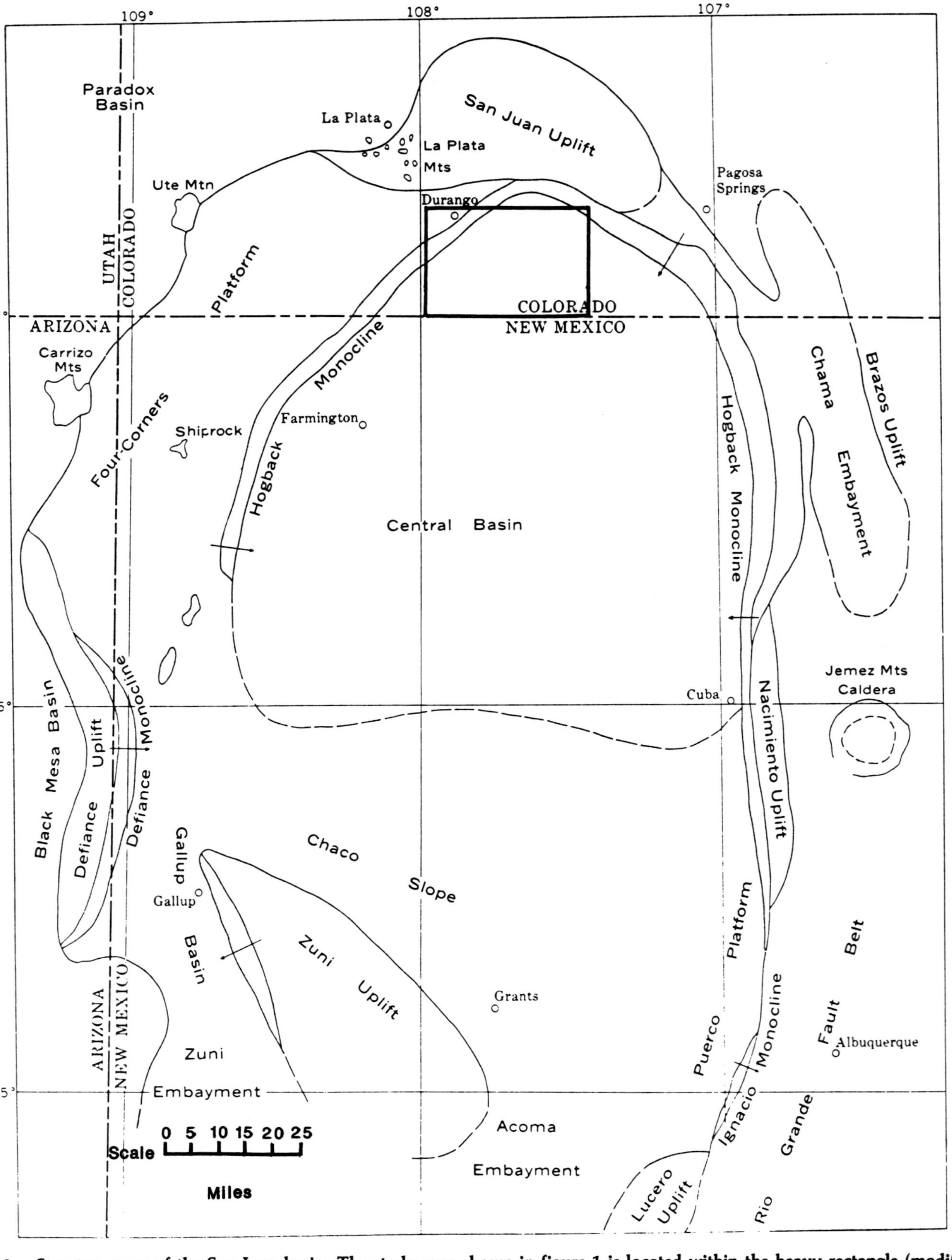

Figure 2. **Structure map of the San Juan basin. The study area shown in figure 1 is located within the heavy rectangle (modified from Fassett and Hinds, 1971).**

DATUM MEAN SEA LEVEL, CONTOUR INTERVAL 50 FEET, VERTICAL EXAGGERATION x15

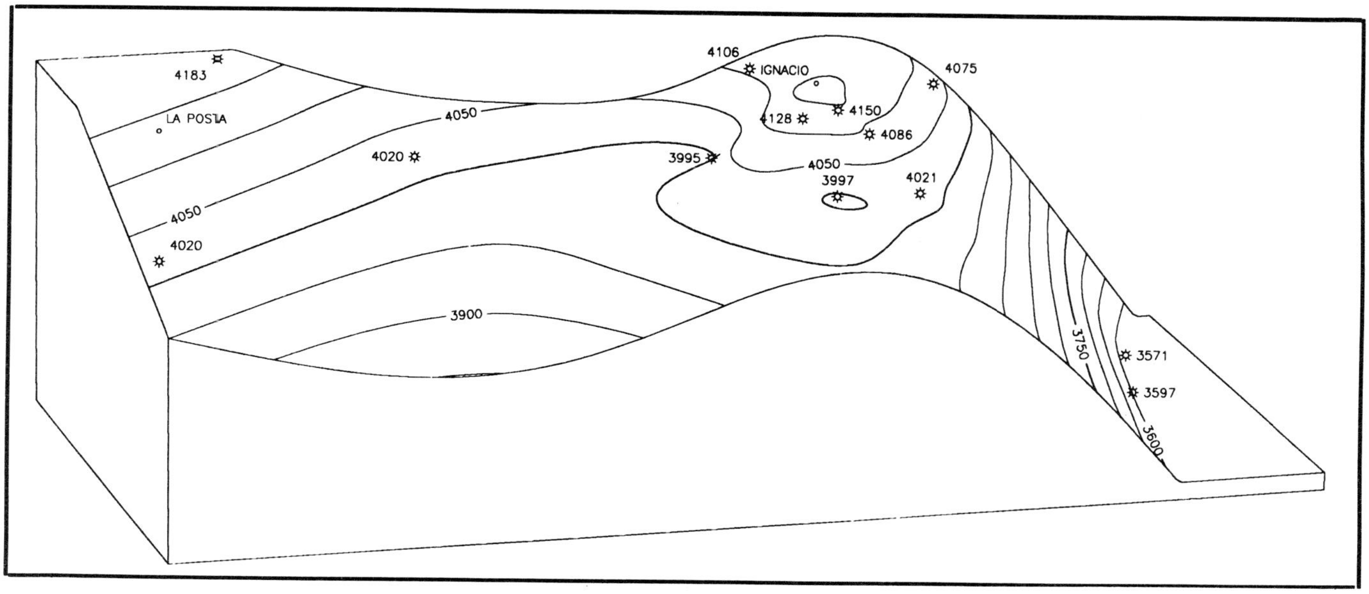

Figure 3. Structure contour map on top of the Fruitland Formation (depths listed in table 1) presented as a three-dimensional surface projection. Datum is mean sea level, contour interval is 50 feet. Due to the rotation of the projection and the resulting distortion of the original plane surface, no horizontal scale is given.

CORE DATA

A 30 foot continuous core was cut within the Fruitland coal zone in the Ladd Petroleum North Cox Canyon No. 2-3 well (well 13, table 1) at the request of the Southern Ute Tribal Energy Office. Core was recovered for the interval 3,086-3,118 feet. Core recovery and sample quality were good except in a few coaly intervals where the presence of trapped methane gas within the core resulted in an explosive discharge of the core from the core barrel which in turn resulted in the breakup of many of the thicker coaly intervals into small fragments; the exact location and orientation of which are uncertain. A detailed core gamma-ray log was not made. Such a log might have resolved the original make-up of the core and provided a means for accurate depth correlation with the open-hole logs run in the well.

Coal Analysis

The coaly intervals of the Ladd Petroleum North Cox Canyon No. 2-3 well were sampled by USGS geologists and five samples were sent to a commercial laboratory for proximate and ultimate analyses. The results are summarized in table 3, on which the equivalent log-depths, derived from the procedure discussed below, are indicated.

Core-Log Depth Matching

Differences in depths which arise during the coring and logging operations must be adjusted. Usually the core data are shifted to agree with the well-log data. The preferred means for correlating depths is via gamma-ray curves (core and open-hole log) as mentioned above. Since a core gamma-ray log was unavailable, depth correlation and core-depth shifting was based on the drilling penetration-rate log for the core.

SOURCES OF ERROR

Core-Log Depth Match

An assumption implicit in the depth-matching process, discussed above, where disparate data are involved is that variations in both parameters represent changes in the same variable, namely, lithology. However, variations in the gamma-ray count may result from changes in radioactive isotope (uranium or potassium) distribution that may be unrelated to lithologic change. The resulting depth match is shown in figures 9 and 10. An accurate depth match is essential for reliable quantitative analysis.

Data Sampling Increment

The different sampling increments of the well logs (0.5 foot), the drilling penetration-rate data (1 foot), and the core sampling (0.1 foot) leave room for error in the depth match and subsequent quantitative analyses.

Well Logs

The maximum vertical resolution of a logging tool, also called the intrinsic vertical resolution, is a function of its geometry, that is, the spacing between source and detector. The source-detector spacing selection is always a compromise because of the desire for fine vertical resolution (small spacing) and maximum depth of investigation (large spacing). The effective resolution is commonly less than the intrinsic value due to the effects of thin beds, shoulder effects of highly resistive beds adjacent to lower resistivity beds, and to the undersampling that arises with a coarse data-sampling increment (McCall and others, 1987). The vertical resolution for some standard petroleum well-logging tools such as those

Table 3. Results of core-coal analysis for the Ladd Petroleum North Cox Canyon No. 2-3 well (well 13, table 1).

Sample no.	Depths (ft) Core	Log equiv.	Proximate Analysis Moisture %	Ash %	Volatile matter %	FC %	Ultimate Analysis H %	C %	N %	S %	O %	Ash %
1	3,101-03	3,098-100										
As-received			4.38	42.60	18.90	34.12	3.72	44.23	0.83	0.39	8.23	42.60
Dry			—	44.55	19.76	35.69	3.38	46.25	0.87	0.41	4.54	44.55
Dry ash-free			—	—	35.63	64.37	6.10	83.41	1.57	0.74	8.18	—
2	3,103-05	3,100-02										
As-Received			9.67	14.10	23.44	52.79	5.42	66.01	1.33	0.66	12.48	14.10
Dry			—	15.61	25.95	58.44	4.80	73.07	1.48	0.73	4.31	15.61
Dry ash-free			—	—	30.75	69.25	5.69	86.58	1.75	0.87	5.11	—
3	3,105-07	3,102-04										
As-received			9.24	36.71	19.28	34.77	4.25	42.87	0.92	0.65	14.60	36.71
Dry			—	40.44	21.24	38.32	3.55	47.23	1.01	0.72	7.05	40.44
Dry ash-free			—	—	35.66	64.34	5.96	79.30	1.70	1.21	11.83	—
4	3,108-09	3,105-06										
As-received			3.17	46.64	19.38	30.81	3.33	40.77	0.84	0.43	7.99	46.64
Dry			—	48.16	20.02	31.82	3.08	42.11	0.87	0.45	5.33	48.16
Dry ash-free			—	—	38.62	61.38	5.94	81.23	1.68	0.87	10.28	—
5	3,086	3,083										
As-received			0.91	44.31	21.63	33.15	2.90	44.25	0.80	0.31	7.43	44.31
Dry			—	44.72	21.83	33.45	2.83	44.66	0.81	0.32	6.66	44.72
Dry ash-free			—	—	39.49	60.51	5.12	80.79	1.47	0.58	12.04	—

Abbreviations

FC — Fixed carbon C — Carbon O — Oxygen H — Hydrogen S — Sulfur N — Nitrogen

used for logging the wells studied in this report, is given in table 4.

By comparison, the standard tools used by the coal-logging industry (gamma-ray, density, neutron, resistivity) have a greater vertical resolution, ranging from 0.5-1.5 feet, due to reduced source-receiver spacings (Samworth, 1974; Lavers and Smits, 1976; Fishel and Mayer, 1979; Brom and Driedonks, 1981).

Differences in the vertical resolution of different logging tools can adversely affect the accuracy of depth matching of different log traces. Such differences can be seen in figure 9. The spontaneous potential (SP) and the deep resistivity log data are obviously providing less detail on shale breaks than the focused log, bulk density, neutron, or gamma-ray. The use of the SP and deep resistivity curves is not recommended for quantitative analysis.

Table 4. Examples of intrinsic vertical resolution of various logging devices (Kern and others, 1983; Serra, 1984; McCall and others, 1987).

Dresser Atlas Tool	Intrinsic vertical resolution (ft)	Schlumberger Tool	Intrinsic vertical resolution (ft)
Dual induction electrolog	3.5	Deep induction (ILD)	3.5
Dual induction focused	3.5	Spherically focused log (SFL)	2.5
Dual laterolog	1.5	Dual laterolog (DLL)	2.0
Minilog	0.16	Microlog	0.5
Spontaneous potential (SP)	4.0	Spontaneous potential (SP)	6-10.0
Dielectric	0.66	Electromagnetic propagation (EPT)	0.16
		Laterolog-8 (LL8)	1.2
		Gamma-ray	8-12.0
		Compensated formation density (FDC)	1.5
		Lithodensity tools (LDT) density	1.5
		Compensated neutron log (CNL)	1-2.0
		Borehole compensated sonic (BHC)	2-6.0

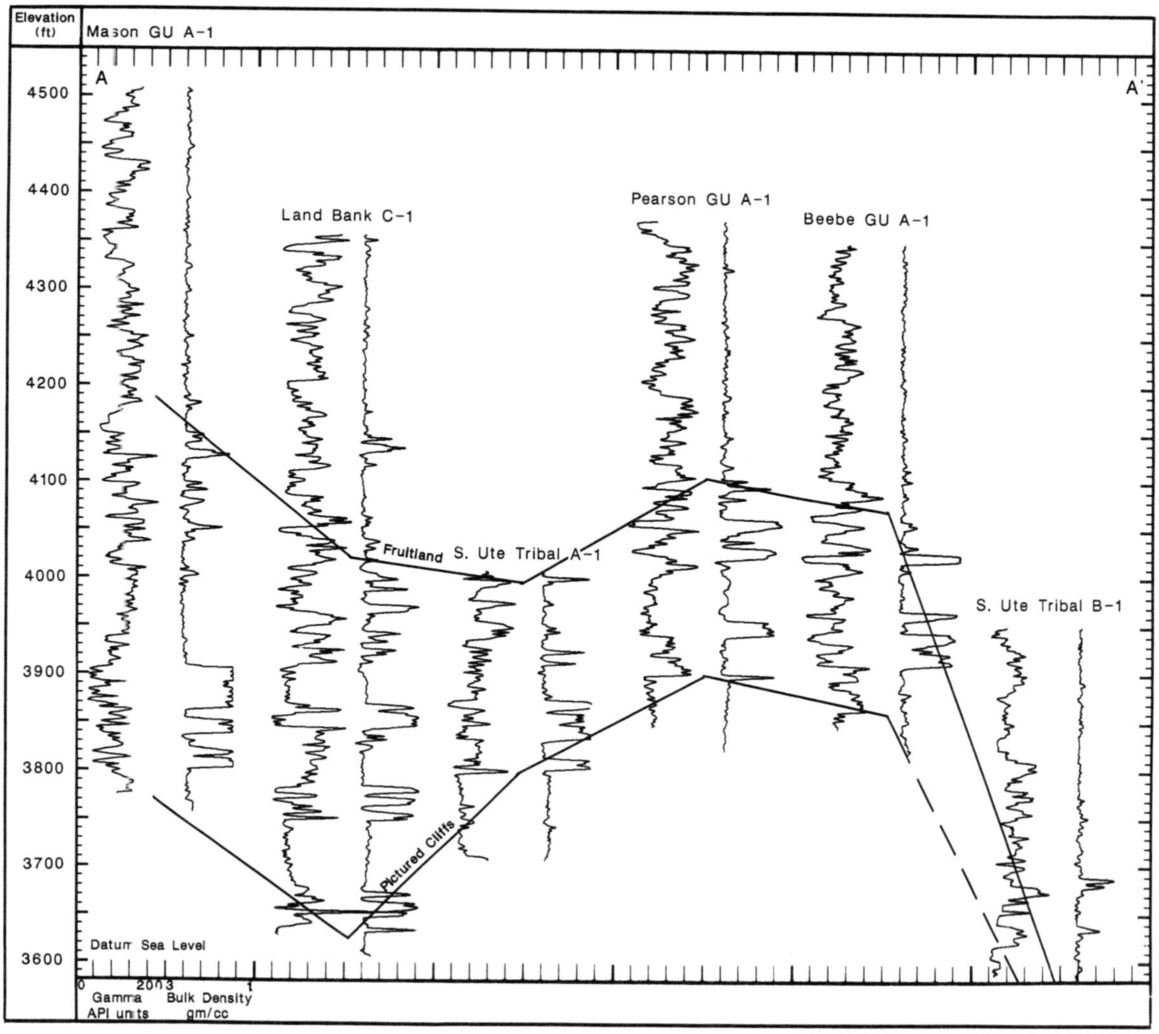

Figure 4. East-trending structural cross-section A-A′ showing well-log correlations of the top of the Fruitland Formation and the top of the Pictured Cliffs Sandstone. The wells used on this section are listed in table 1; the trace of the cross section is shown on figure 1.

Recent advances in computer processing by the logging service companies have resulted in the ability to enhance the vertical resolution of log data run at the standard 0.5 foot sampling rate (Galford and others, 1986; Flaum and others, 1987; McCall and others, 1987). However, the algorithms used in computing the results require the raw logging data, available only on the original logging tapes, as input.

Because nuclear events are random, their measurement is statistical. To reduce the effects of statistical variation on the lithologic response of the gamma-ray, density, and neutron logging tools, these data are filtered (averaged). Prior to the advent of digital recording, an electronic filter called a time-constant was used in the logging sonde. The modern logging units which record data directly in digital form (including all the well logs used in this study) use mathematical averaging algorithms. A large time-constant or too-rapid logging speed may result in reduced bed resolution and (or) shift the data upward in the hole, which are undesirable effects that can affect the maximum amplitude recorded. The selection of time constant and logging-speed represents a compromise between a need for maximum lithologic information with a minimum of statistical variation and the economics of rig-time used while logging (Hoffman and others, 1982). Differences in the filtering algorithms used by the various logging service companies also introduce variations which may affect quantitative analyses when data recorded by different service companies are compared and combined.

Finally, differences in tool calibration affect the log reading. Normalization of the data can be used to eliminate some of these differences. However, due to time constraints, the data used in this study were not normalized.

RESULTS

Coal Characterization on Well Logs

Well-log characteristics of the Fruitland coals (table 5) were determined from visual inspection of crossplots of composite data for the entire group of study wells (figs. 11-15). Data in each of these figures are presented in four panels labeled A-D. Data were plotted both for the total interval digitized (which includes the lower shale member of the Kirtland Shale, the Fruitland Formation, and the topmost part of the Pictured Cliffs Sandstone) on panel A, and for the Fruitland Formation alone on panel B. For those wells where the density-porosity

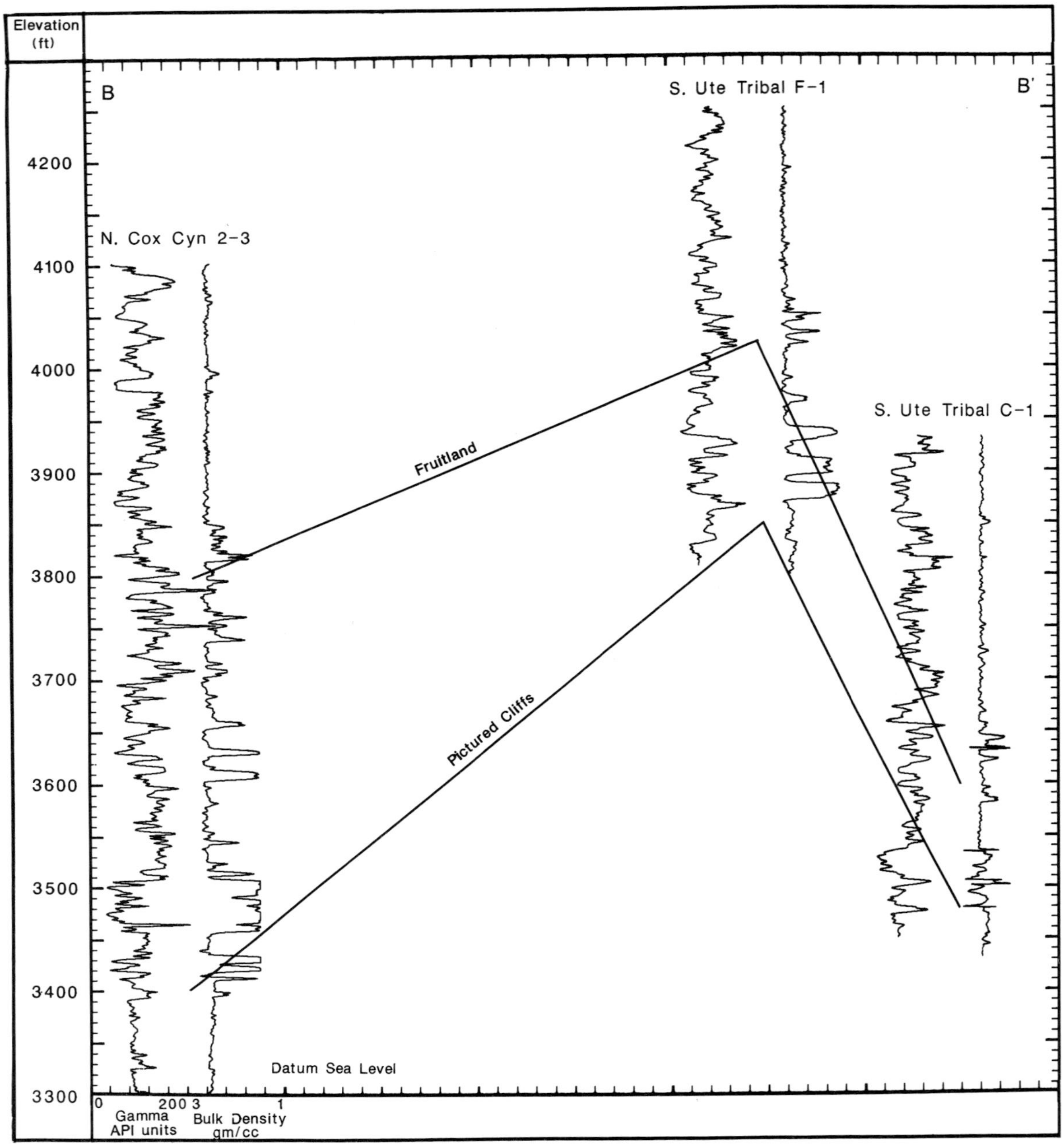

Figure 5. East-trending structural cross-section B-B′ showing well-log correlations of the top of the Fruitland Formation and the top of the Pictured Cliffs Sandstone. The wells used on this section are listed in table 1; the trace of the cross section is shown on figure 1.

curve was digitized, the maximum recorded porosity was limited to 70 percent (1.495 gm/cc), the limit of the backup curve. This truncation results in a data artifact on these crossplots. As a means of reducing or eliminating this artifact, an additional pair of crossplots were generated using a subset of the data plotted on panels A and B and consisting only of those wells where an original bulk density trace was available. The results are presented on panel C, total digitized interval, and on panel D, Fruitland Formation. Additional artifacts resulted from the absence of a backup trace on some neutron and density logs. On figure 11C the major lithologic endpoints; coal, sandstone, and shale are indicated for reference. The scatter of points reflects varying proportions of these

Table 5. Well-log characteristics of Fruitland Formation coals (field-wide basis).

Tool	Values	Units
Gamma-ray	15-45	API units
Density	1.25-1.35	gm/cc
Deep resistivity	150-300	ohm-m
Focused resistivity	1,000-2,000	ohm-m
Sonic	125-130	microsec/ft
Neutron	55-60	percent

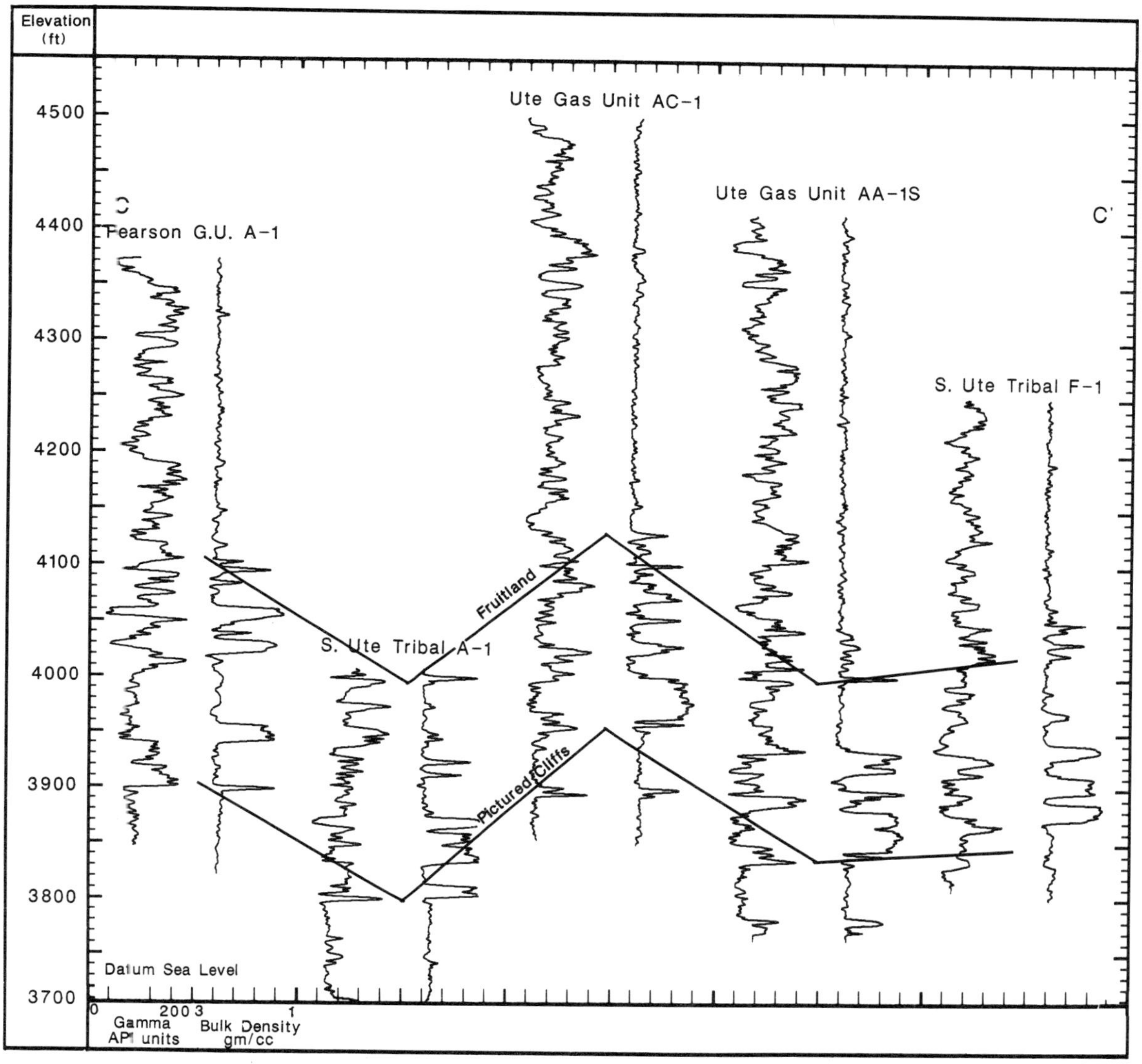

Figure 6. South-trending structural cross-section C-C′ showing well-log correlations of the top of the Fruitland Formation and the top of the Pictured Cliffs Sandstone. The wells used on this section are listed in table 1; the trace of the cross section is shown on figure 1.

three components as well as the presence of some calcareous material.

The field-wide values in table 5 assume the rank and log response for the different Fruitland coals to be identical and that these readings are for pure coal (carbon). However, this assumption is probably incorrect or a gross oversimplification because: a) the data in figures 11-15 are a composite of several wells and the data therefore reflect lithologic variation in addition to the sources of error and statistical variation discussed previously, b) B. Law (1987, personal communication) has recently measured the vitrinite reflectance (R_0) of the different Fruitland coal seams within a single well and found significant variation ranging from R_0 = 0.98-1.30 percent, while Jones (1985) reports laboratory-determined bulk density values of 1.45 gm/cc and 1.76 gm/cc for closely spaced (1.5 foot) Fruitland coal samples. Consequently, additional data on coal-rank and laboratory-determined ash content are necessary in order to assess the accuracy of data in table 5 and for making any corrections that may be necessary prior to a more detailed log-analysis. Differences in rank would explain, at least in part, the spread of data values in table 5. Nevertheless, the values in table 5 can serve as an approximation of the coal lithology endpoint values, that is, pure coal, for input to log-analysis models for lithology and coal quality pending more exact information.

Well-Log Estimation of Ash Content

The logs used most in coal analysis are gamma-ray, density, and the focused-resistivity log, primarily because their vertical resolution allows the identification of thin coals. Daniels and others (1983) have compared the problem of estimating coal ash-content from well logs to the shaly-sandstone problem in petroleum log analysis. In a two-lithology system of coal and argillaceous rocks under ideal conditions (the absence of uranium and pyrite in the coal) the coal can be treated as a three-component system of carbon, ash, and moisture for the purposes of well-log evaluation.

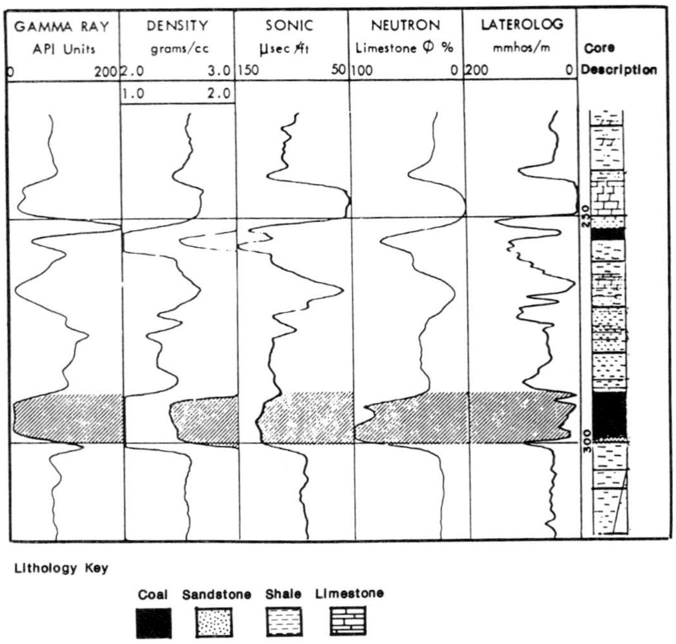

Figure 7. **Chart showing typical well-log responses of coal and associated lithologies. Depths in feet (modified from Bond and others, 1971, by permission).**

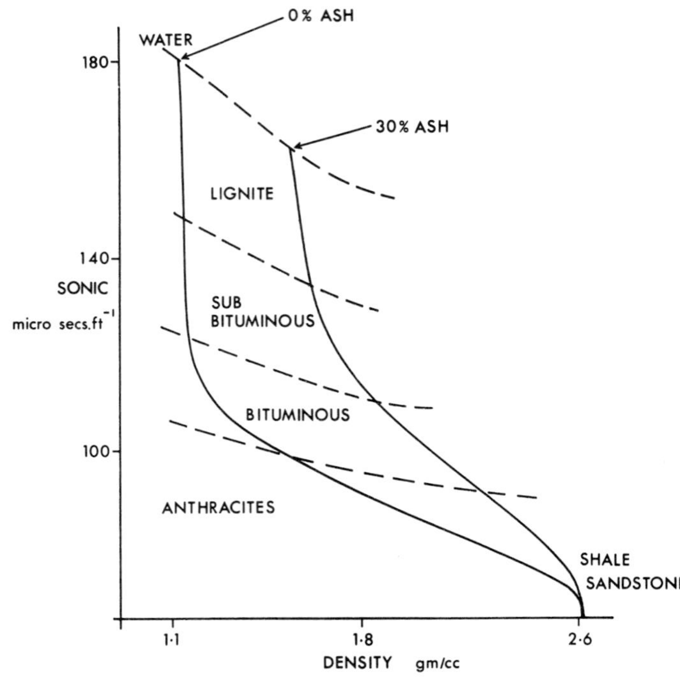

Figure 8. **Graph showing the relationship between acoustic log travel-time, bulk density, and coal rank (from BPB, 1981, by permission).**

Due to the variability inherent in neutron, sonic, and resistivity log measurements, these logs are generally considered less reliable than gamma-ray and density log measurements for estimation of coal quality in general and ash-content in particular (Daniels and others, 1983; Kern and others, 1983).

Estimation of Ash Content—Natural Gamma-ray Log

Due to the normally low radioactivity of coal, the gamma-ray log can be used for estimating ash content under specific circumstances: a) Where the rocks above and below a coal seam are mudstone or siltstone and it is assumed that these lithologies represent 100 percent ash, and, b) Where it can be determined that the gamma radioactivity is due primarily to associated clay minerals and to the presence of radioactive trace elements (uranium or potassium) and a linear relationship exists between ash content and gamma-ray value (Reeves, 1971). (This relationship assumes that non-radioactive kaolinite is not present in significant amounts.) This ideal situation is not exactly the situation in the study area.

To determine whether or not the gamma-ray log can provide reliable results in the northern San Juan Basin it is first necessary to examine the relationship between total gamma count, clay (shale), and coal. Figures 17 and 18 present data from cased-hole spectral gamma-ray logs run in two of the study wells, the Amoco Mason Gas Unit No. 1 and Amoco Federal Land Bank C No. 1 (wells 11 and 12, table 1). Bulk-density logs and open-hole total gamma logs are also presented. To identify which of the three radioactive components; uranium, potassium, or thorium, had the most influence on the total gamma curve, linear-fit plots of total gamma versus the individual components were made for each zone and for each well.

In the Mason well (fig. 17) there is good correlation between total gamma and both potassium (reduced major axis (RMA) r = 0.937) and thorium (RMA r = 0.935) in the Fruitland Formation and also in the overlying Kirtland Shale section while the correlation with uranium is not very good (RMA r = 0.708-0.780). This implies that the clays in the shale are the primary source of gamma radiation and that the gamma-ray log might possibly be used for estimating coal-ash in this well. The opposite is true for the Land Bank well. In this case there is an excellent correlation between total gamma and uranium in the Fruitland Formation (RMA r = 0.989) and only a fair correlation with thorium (RMA r = 0.730) and potassium (RMA r = 0.672). However, comparison of the cased-hole total gamma log with the open-hole equivalent (fig. 18) shows that the high gamma and uranium readings developed only after the first log was run. Note how the low natural radioactivity of the coals seen in the open-hole gamma log between 2,730-2,910 feet, is reversed on the cased-hole log where the coals show up as radioactive zones. The dramatic increase in coal radioactivity is probably the result of inflow of uraniferous formation waters through the production perforations. In any case, the correlation between clay (shale) and total gamma count is not sufficient for obtaining reliable and accurate coal-ash estimates from the cased-hole gamma-ray log in this well. The open-hole log may be more useful but there is no way to verify the results at this time.

Due to the wide range in variability of radioactive trace elements, particularly uranium, both areally and with depth, and to the lack of sufficient spectral data for determining the contributions of these trace elements within the study area, it is tentatively concluded that the gamma-ray log would not provide consistently reliable or accurate estimates of coal-ash in this area.

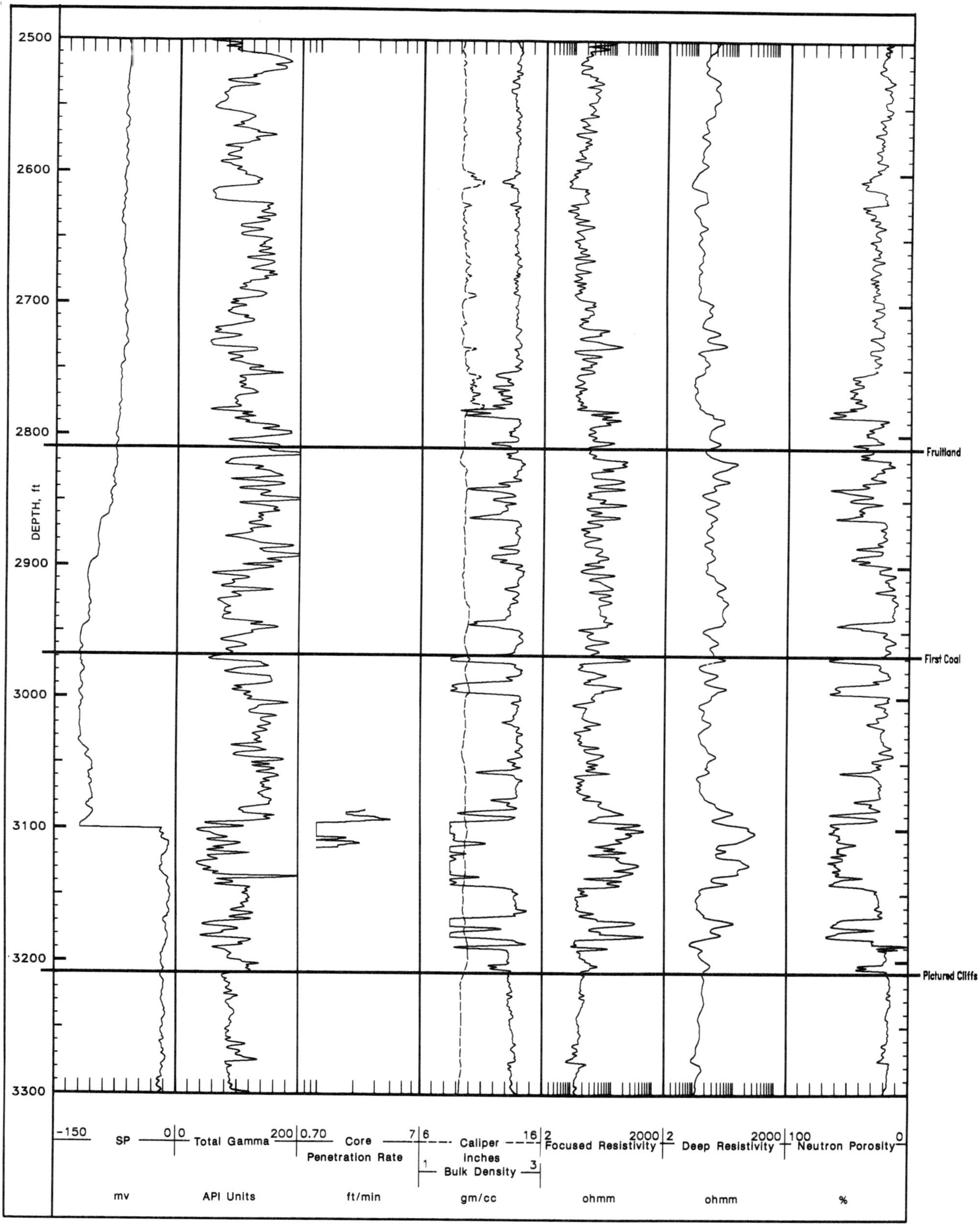

Figure 9. Plot of well-log data for Ladd Petroleum North Cox Canyon No. 2-3 Well (well 13, table 1) and core penetration-rate. The indicated stratigraphic markers are listed in table 1. Penetration rate and resistivity are plotted on logarithmic scales.

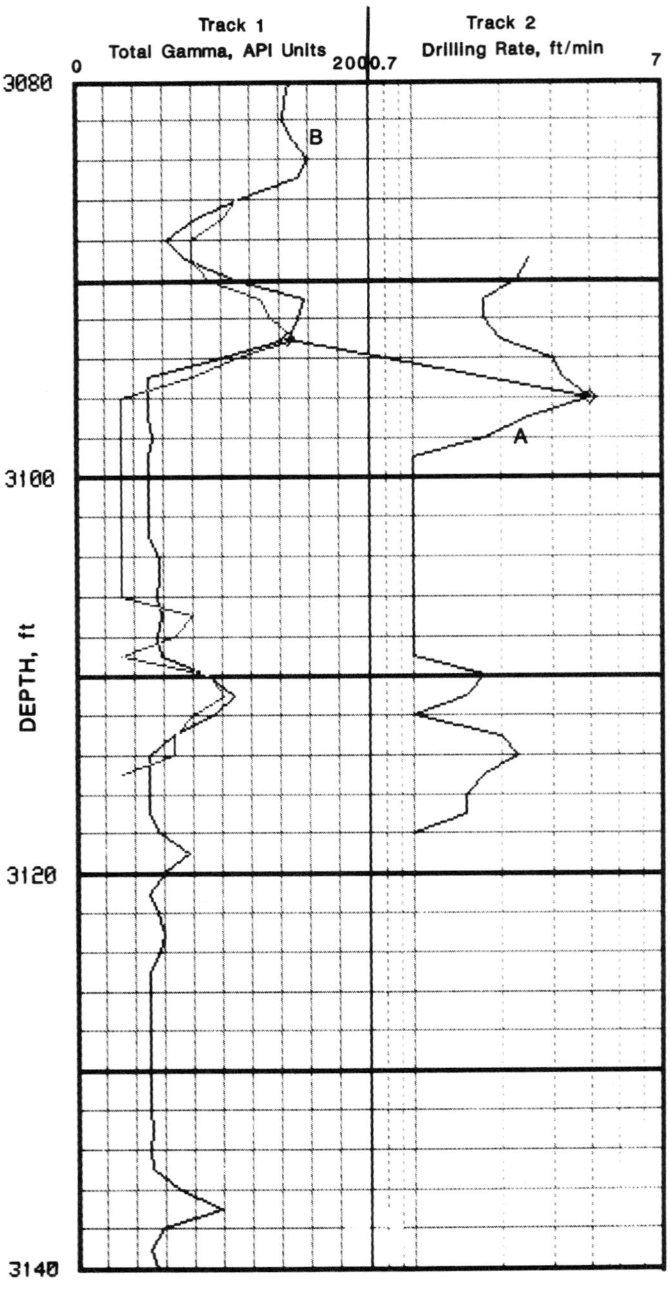

Figure 10. Core-log depth correlation and core-depth shift for the Ladd Petroleum North Cox Canyon No. 2-3 well (well 13, table 1). Trace A is core penetration rate in feet per minute on a logarithmic scale. Trace B is corrected open-hole gamma-ray curve from the density tool in API units. The depth increment used for the match is 1.0 foot. The depth match in track one required a 3-feet-upwards shift of core data in track two.

Estimation of Ash Content—Density Log

Because the bulk density of coal is very low; unique among the lithologies present in the typical depositional environment, the bulk density log can provide by itself, definitive identification of coal and is therefore the preferred log for coal-ash estimation. Crossplots of bulk density, particularly from high resolution tools and laboratory-determined ash content, show a linear relationship in the absence of significant amounts of pyrite and these data provide the most accurate means for estimating coal ash-content (Tixier and Alger, 1967; Reeves, 1971; Lavers and Smits, 1976; Agostini, 1977; Dyck and McKenzie, 1979; Norris and Thomas, 1980; and Daniels and others, 1983). Discrepancies between log-derived density values and coal-quality parameters can be ascribed to several factors which have been summarized by Daniels and others (1983) as follows: a) Inaccurate determination of core-sample depth, b) A core sample that is not representative of the effective interval measured by the logging probe, c) Variations in borehole conditions that cannot be corrected for by density compensation algorithms, d) Variations in the density of the ash component of the sample. Examination of the caliper versus bulk density plot (fig. 16) indicates that coaly zones (low bulk density) generally, do not exhibit preferential hole washout. Therefore, the bulk-density relations which can be adversely affected by borehole washout are generally reliable.

The simplest quantitative approach, and one that is generally capable of providing reasonably good results, assumes density values for 100 percent coal and 100 percent ash (shale) and a linear relationship between these end points (Norris and Thomas, 1980; BPB, 1981). Laboratory-determined values are preferred, although in their absence, individual well or field-wide values determined from crossplots, such as those in table 5, can be used. All results should be calibrated against laboratory-determined ash content, that is, ash-corrected, when possible, to ensure maximum accuracy. More sophisticated models which employ solution of simultaneous equations require, in addition to the laboratory ash values, laboratory-determined ash and coal grain-densities.

The ash content values presented in tables 3 and 6 represent two-foot samples (except for the sample at 3,083 feet which is a 0.25-foot-thick sample). For comparison with log data (table 6), the midpoint depth for each sample was chosen as the representative depth for the ash-content values. The log data, which are on a 0.5-foot sampling increment, were averaged for the equivalent core-sample interval.

A linear regression of these points gives a very poor fit (RMA $r = 0.547$), however, when the sample at 3,083 feet is excluded, the fit becomes excellent (RMA $r = 0.953$). The high correlation coefficient is probably an artifact, due in part to the original core sampling and to the lack of data for the low and high ends of ash-content. Specifically, a) the sample at 3,083 feet was very thin (4 inches) and the log response is affected by the adjacent shales and therefore not totally representative of the coal, and b) the density value for the sample at 3,101 feet is probably not correct (see note in table 6 above).

Mullen (this volume) presents an empirically derived bulk-density ash algorithm, based on data published by Fassett and Hinds (1971). Ash estimations using this algorithm are closer to the laboratory-derived values and are presented in table 6 for comparison.

The poor agreement between laboratory determined coal-ash content and the well-log estimates, including those using the Mullen (this volume) algorithm, are most likely due to a poor depth match between the log and core data. No appreciable improvement in the agreement of the data is noted when the density values for coal and ash are varied as shown in

Table 6. Comparison of core and well-log values. Coal-ash values from core are from table 3; data are averaged and depth is the mid-point of the core sample. [Assumed density, in gm/cc, of coal and ash, respectively: A = 1.30, 2.55; B = 1.30, 2.65; C = 1.40, 2.55; and D = 1.40, 2.65]

Sample no.	Log depth (ft)	Core ash (dry wt %)	Density log (gm/cc)	Log ash (dry wt %) A	B	C	D	Mullen ash (dry wt %)
1	3,099	44.6	1.51	16.0	9.3	15.4	8.6	31.3
2	3,101	15.6	1.50*	15.4	8.7	14.8	8.0	31.1
3	3,103	40.4	1.52	17.0	10.2	16.1	9.4	32.4
4	3,105.5	48.2	1.56	21.1	14.2	19.5	13.1	35.0
5	3,083	44.7	2.52	98.0	96.5	89.6	88.9	97.4

*The true log value may be less; bulk density was derived from density porosity which went off scale at a maximum of 70 percent (1.495 gm/cc).

Mullen ash; using the algorithm of Mullen (this volume): $Volume_{ash} = 64.94 \times (\text{bulk density}) - (66.27)$.

columns A-D on table 6.

The absence of a strong continuum of data between the shale and coal endpoints in the data crossplots (figs. 11-15) may be an indication that the linear model assumed in this investigation does not adequately reflect the in-place conditions.

CONCLUSIONS

Estimating ash content from density-log data may be feasible in the study area; however, two conditions precluded an accurate estimation in this study: a) Only one core and only a very limited number of laboratory proximate analyses were available for developing a reliable ash-correction for the existing suite of well logs, and b) Discrepancies between laboratory- and log-determined ash values suggest that there may be some problems with the core-log depth correlation. A core-gamma log would improve the depth match in general and in the shaly intervals in particular. Indications of variations in coal rank between seams within the Fruitland Formation in the same well require evaluation of additional core data from these seams. Due to difficulties in depth correlation resulting from differences in core and well-log sampling increments, laboratory analyses of smaller and contiguous core samples for the entire core would be helpful. The simple, linear, shale (ash)-coal density model utilized here may provide only a gross approximation of ash content and more sophisticated models, in combination with more core data, would improve the accuracy and reliability of the ash estimates.

REFERENCES

Agostini, A., 1977, Correlation of high resolution density log counts and ash content of coal: Australian Society of Exploration Geophysicists Bulletin, v. 8(2), p. 26-31.

American Society for Testing Materials (ASTM), 1983, Standard classification of coals by rank, ASTM designation D388-82, *in* Gaseous fuels, coal and coke: American Society for Testing Materials, Book of Standards, v. 505, 5 p.

Bond, L.O., Alger, R.P., and Schmidt, A.W., 1971, Well log applications in coal mining and rock mechanics: Transactions of the Society of Mining Engineers AIME, v. 250, December, p. 355-362. Reprinted in 1986, *in* Buchanan, D.J., and Jackson, L.J., eds., Coal geophysics: Society of Exploration Geophysicists, Geophysics reprint series no. 6, p. 28-35.

BPB Instruments Ltd., 1981, Coal interpretation manual: BPB Instruments Ltd., Loughborough, England, 100 p.

Brom, R.W.C., and Driedonks, F., 1981, Applications of petrophysical logging in the evaluation of coal deposits, *in* 22nd annual logging symposium transactions, paper KK: Society of Professional Well Log Analysts, Houston, 29 p.

Choate, R., Lent, J., and Rightmire, C.T., 1984, Upper Cretaceous geology, coal, and the potential for methane recovery from coalbeds in San Juan Basin, Colorado and New Mexico, *in* Coalbed methane resources of the United States: American Association of Petroleum Geologists, Studies in Geology Series, no. 17, p. 185-222.

Daniels, J.J., Scott, J.H., and Liu, J., 1983, Estimation of coal quality parameters from geophysical well logs, *in* 24th annual logging symposium transactions, paper KK: Society of Professional Well Log Analysts, Houston, 20 p.

Dyck, J.H., and McKenzie, C.T., 1979, Estimating certain analytical properties of lignite in Saskatchewan from gamma-gamma borehole density logs, *in* 7th formation evaluation symposium transactions, paper Z: Canadian Well Logging Society, Calgary, 16 p.

Fassett, J.E., 1987, Geometry and depositional environments of Fruitland Formation coalbeds, San Juan Basin, New Mexico and Colorado—Anatomy of a giant coalbed methane deposit, *in* 1987 Coalbed methane symposium [Tuscaloosa, Alabama, November 16-19], proceedings, GRI-8712: Gas Research Institute, Chicago, p. 19-35. Later published in 1988, *in* Fassett, J.E., ed., Geology and coalbed methane resources of the northern San Juan Basin, Colorado and New Mexico: Rocky Mountain Association of Geologists Guidebook.

Fassett, J.E., and Hinds, J.S., 1971, Geology and fuel resources of the Fruitland Formation and Kirtland Shale of the San Juan Basin, New Mexico and Colorado: U.S. Geological Survey Professional Paper 676, 76 p.

Fishel, K.W., and Mayer, R., Jr., 1979, Extremely high resolution density coal logging techniques, *in* Argall, G.O., Jr., ed., Coal exploration 2 [2nd international coal exploration symposium (Denver, 1978) proceedings]: Miller Freeman Publications Inc., San Francsico, p. 490-504. Also published

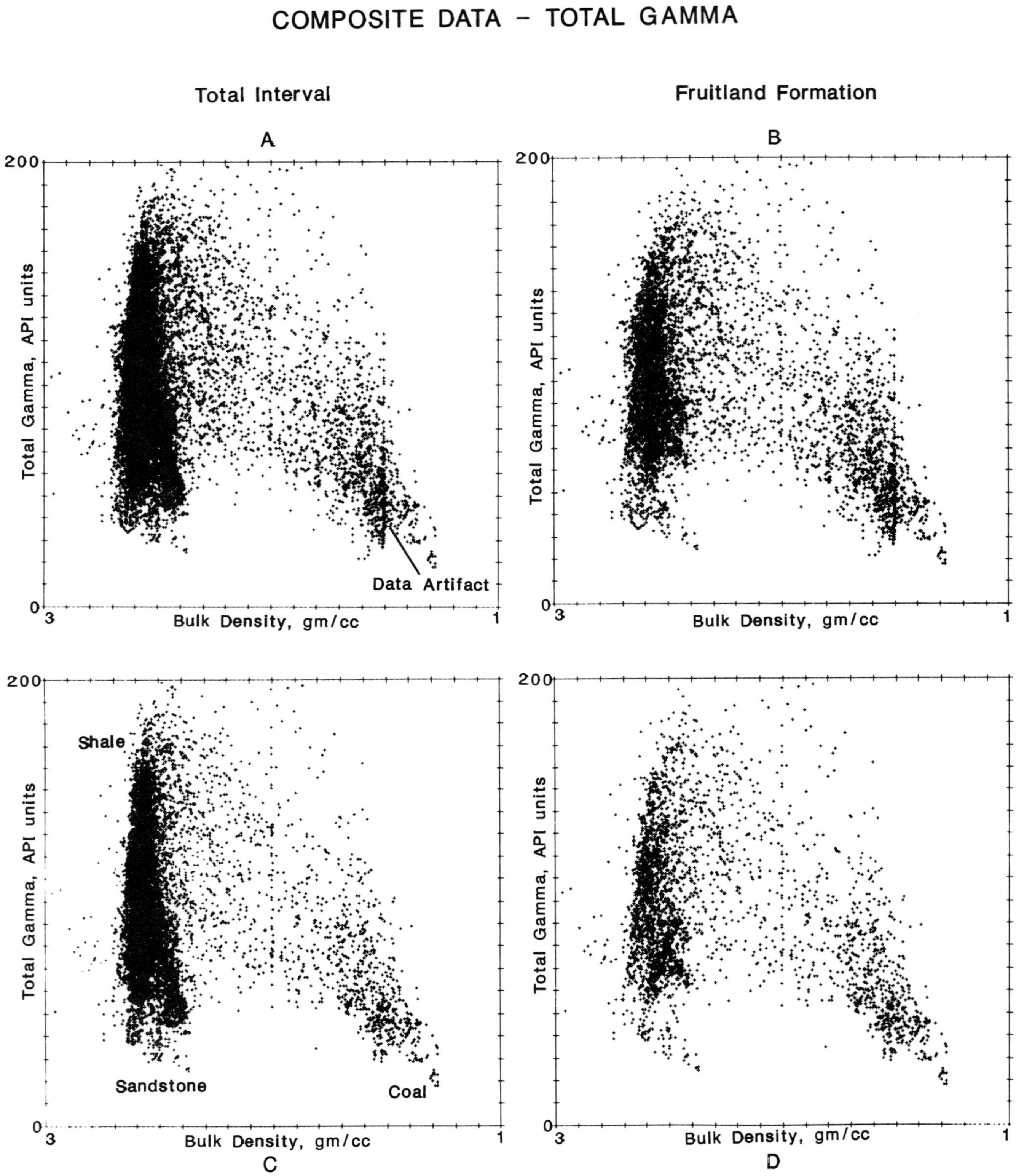

Figure 11. Gamma-ray (corrected) versus bulk density, composite data (13 wells). A, Data for the total digitized interval for all wells; B, Data for the Fruitland Formation, where present; C, Data for the total digitized interval for all wells where the original bulk density trace was available; D, Data for the Fruitland Formation for wells with an original bulk density trace. Data artifacts (panels A and B) result from the absence of a backup curve on the density log for some wells where density porosity is greater than 70 percent.

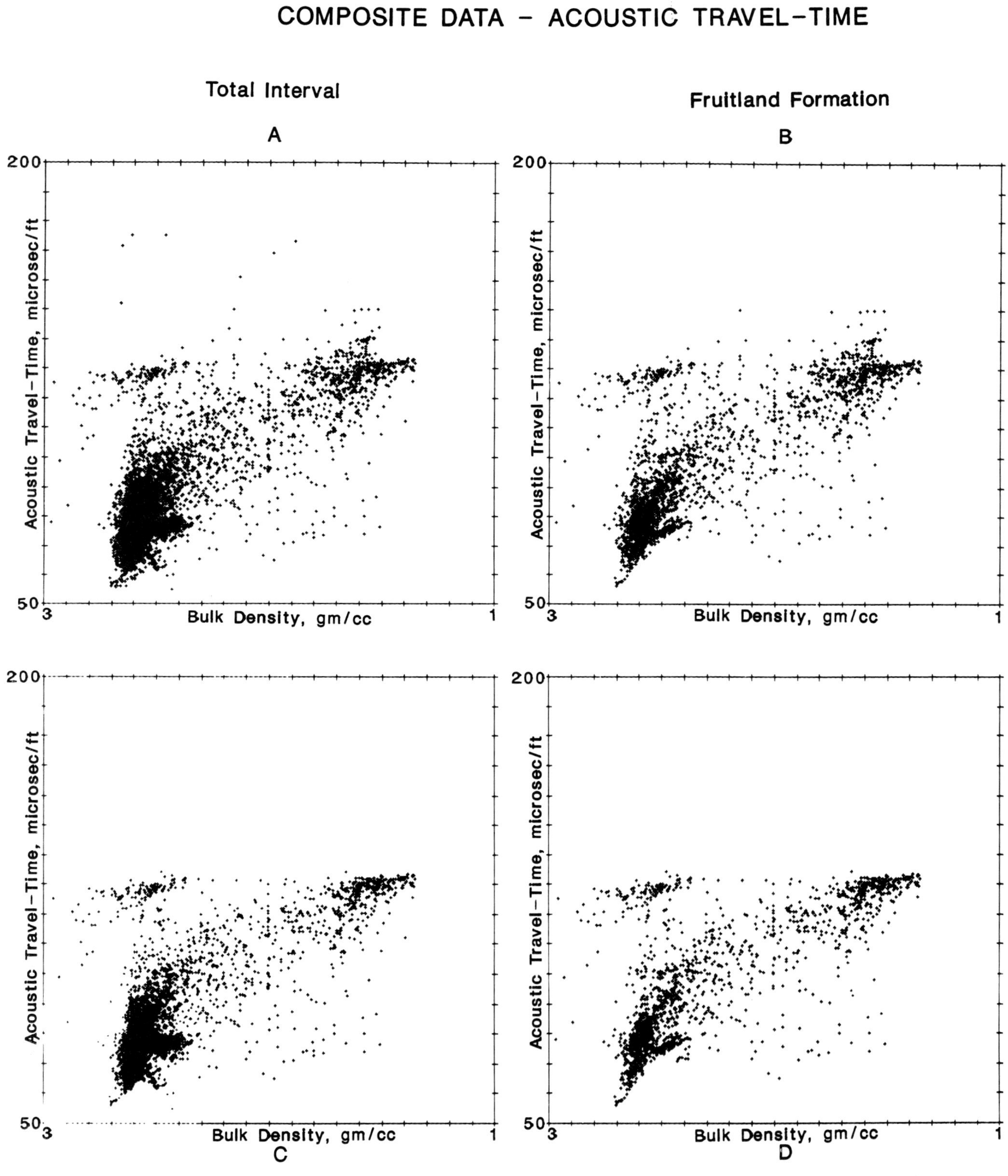

Figure 12. Sonic travel-time versus bulk density, composite data (8 wells). A, Data for the total digitized interval for all wells; B, Data for the Fruitland Formation, where present; C, Data for the total digitized interval for all wells where the original bulk density trace was available; D, Data for the Fruitland Formation for wells with an original bulk density trace.

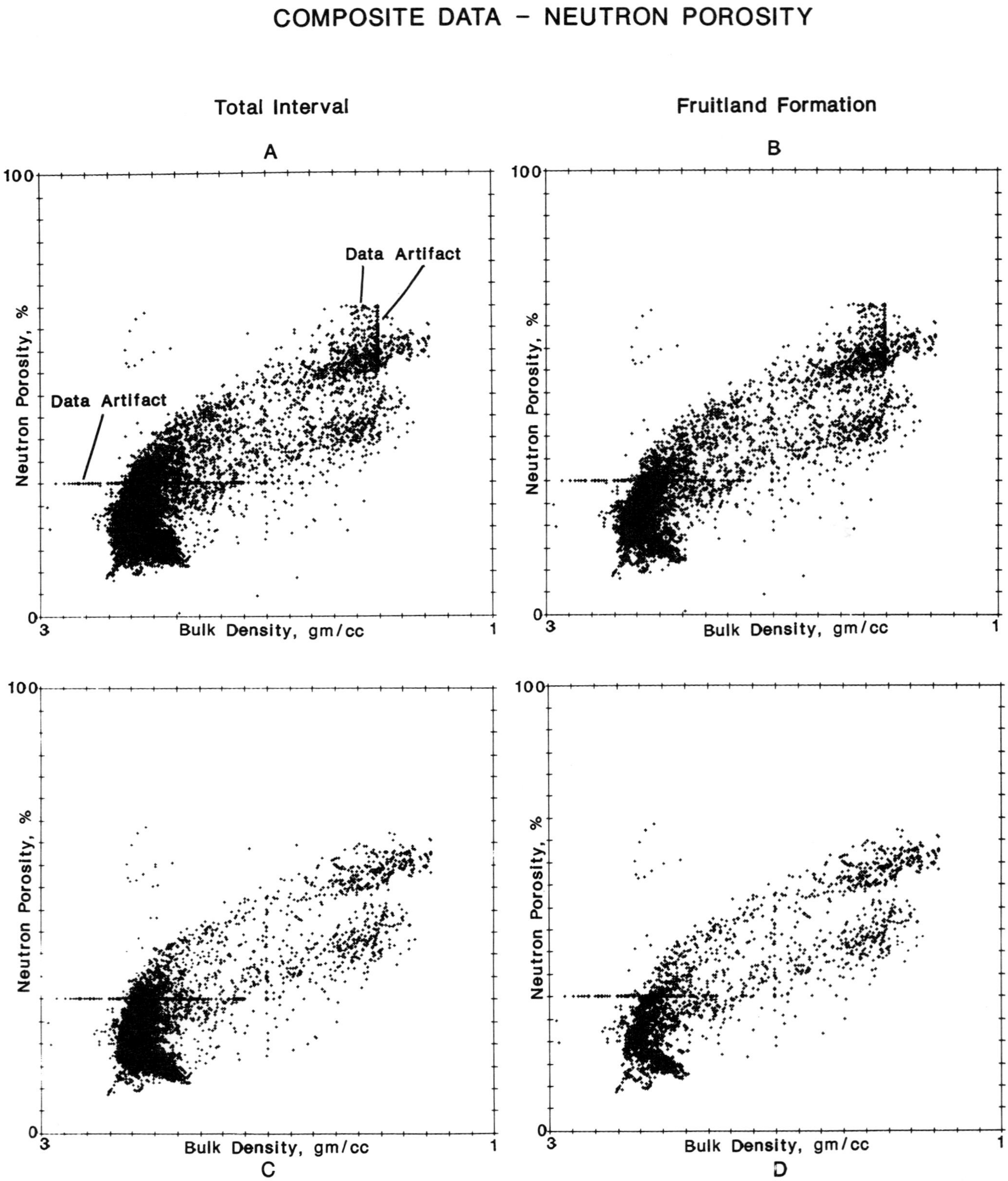

Figure 13. Neutron porosity versus bulk density, composite data (12 wells). A, Data for the total digitized interval for all wells; B, Data for the Fruitland Formation, where present; C, Data for the total digitized interval for all wells where the original bulk density trace was available; D, Data for the Fruitland Formation for wells with an original bulk density trace. Data artifacts (panels A-D) result from the absence of backup curves on the neutron and density logs for some wells where neutron porosity is greater than 30 percent and density porosity is greater than 70 percent.

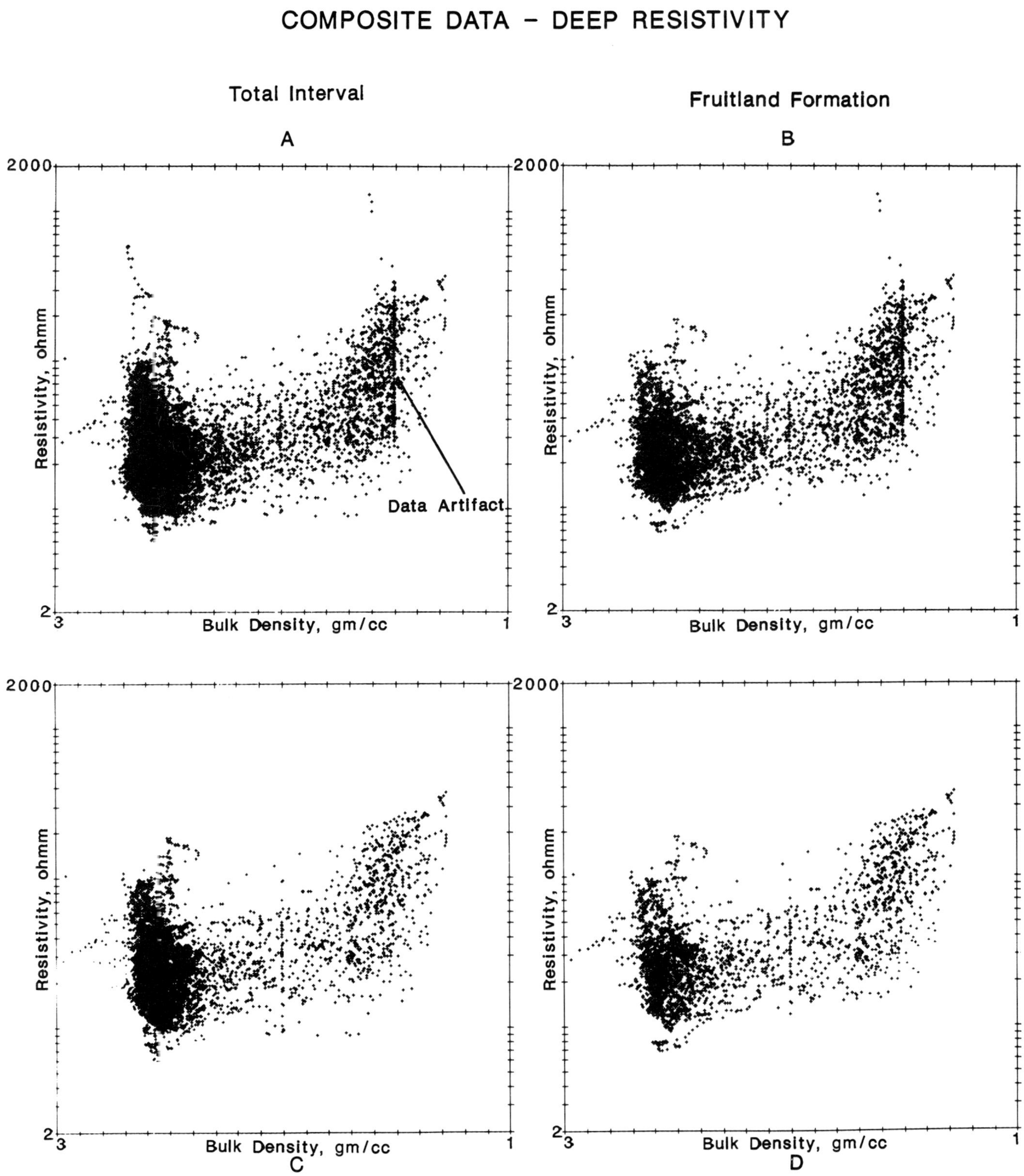

Figure 14. Deep induction log resistivity versus bulk density, composite data (13 wells). A, Data for the total digitized interval for all wells; B, Data for the Fruitland Formation, where present; C, Data for the total digitized interval for all wells where the original bulk density trace was available; D, Data for the Fruitland Formation for wells with an original bulk density trace. Data artifacts (panels A and B) result from the absence of a backup curve on the density log for some wells where the density porosity is greater than 70 percent.

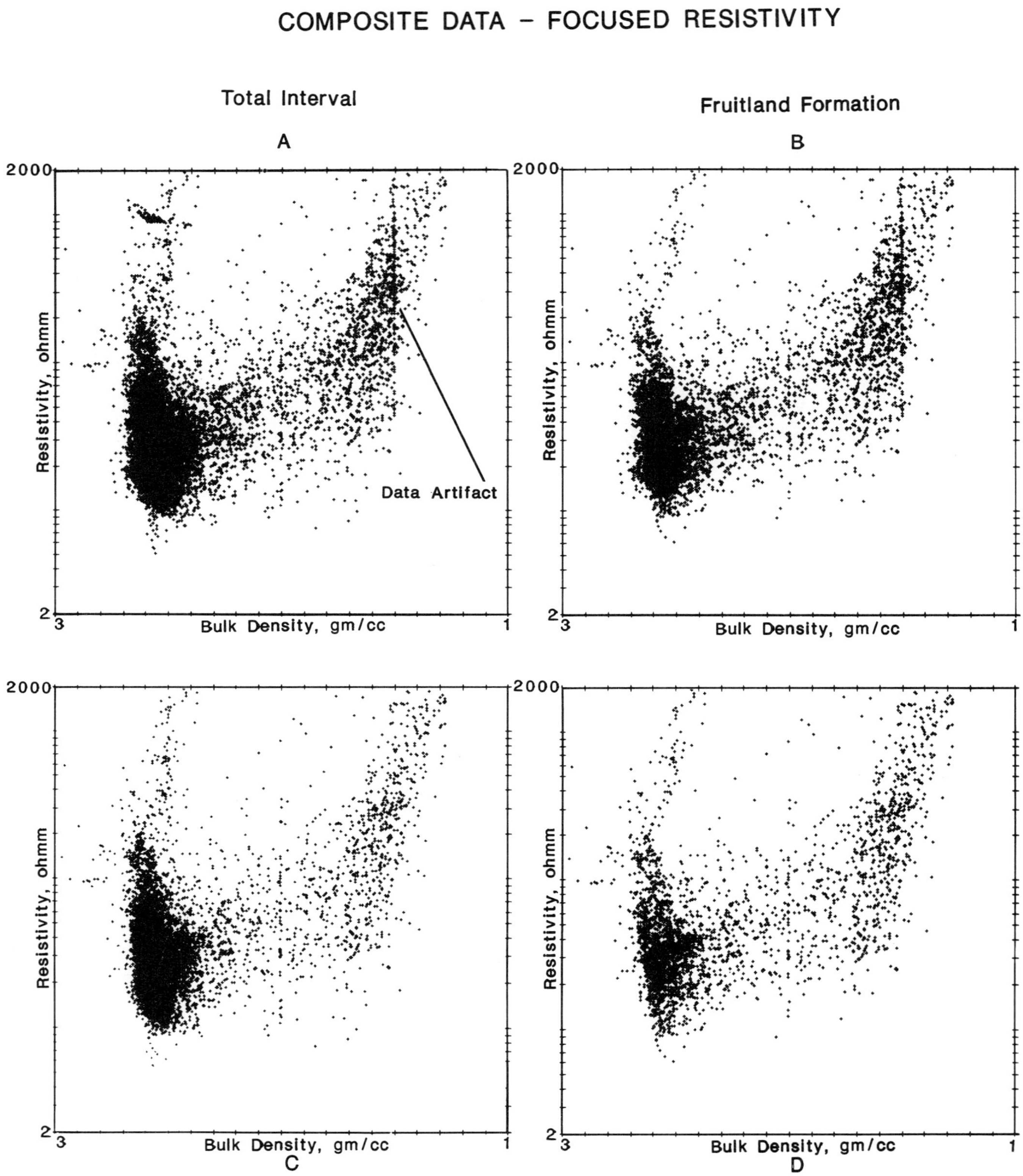

Figure 15. Focused resistivity versus bulk density, composite data (13 wells). A, Data for the total digitized interval for all wells; B, Data for the Fruitland Formation, where present; C, Data for the total digitized interval for all wells where the original bulk density trace was available; D, Data for the Fruitland Formation for wells with an original bulk density trace. Data artifacts (panels A and B) result from the absence of a backup curve on the density log for some wells where the density porosity is greater than 70 percent.

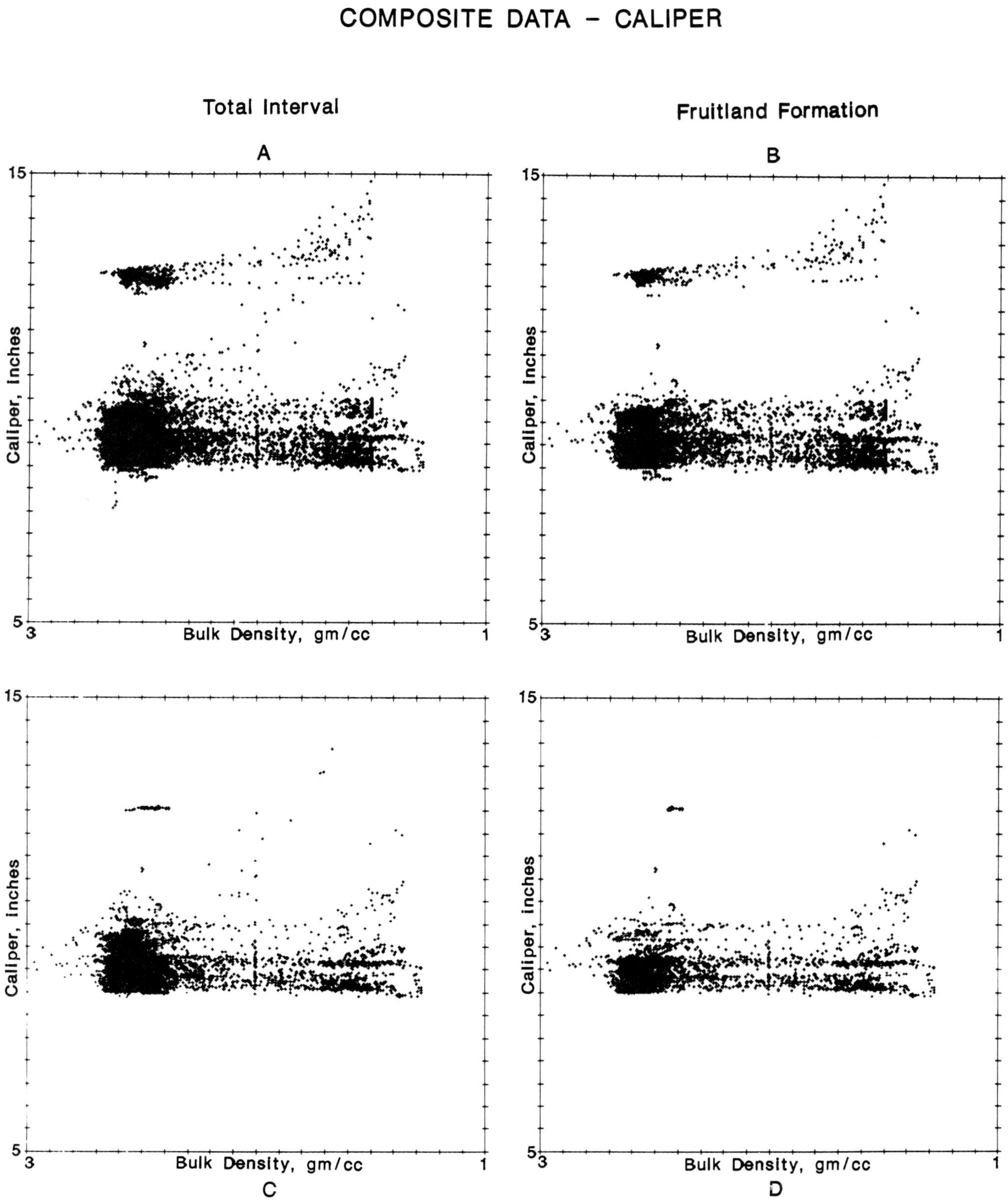

Figure 16. Caliper versus bulk density, composite data (13 wells). A, Data for the total digitized interval for all wells; B, Data for the Fruitland Formation where present; C, Data for the total digitized interval for all wells where the original bulk density trace was available; D, Data for the Fruitland Formation for wells with an original bulk density trace. The upper data field in A is an artifact due to a different bit size of 12.25 inches for well 12 versus 8.75 inches for the other 12 wells (table 1).

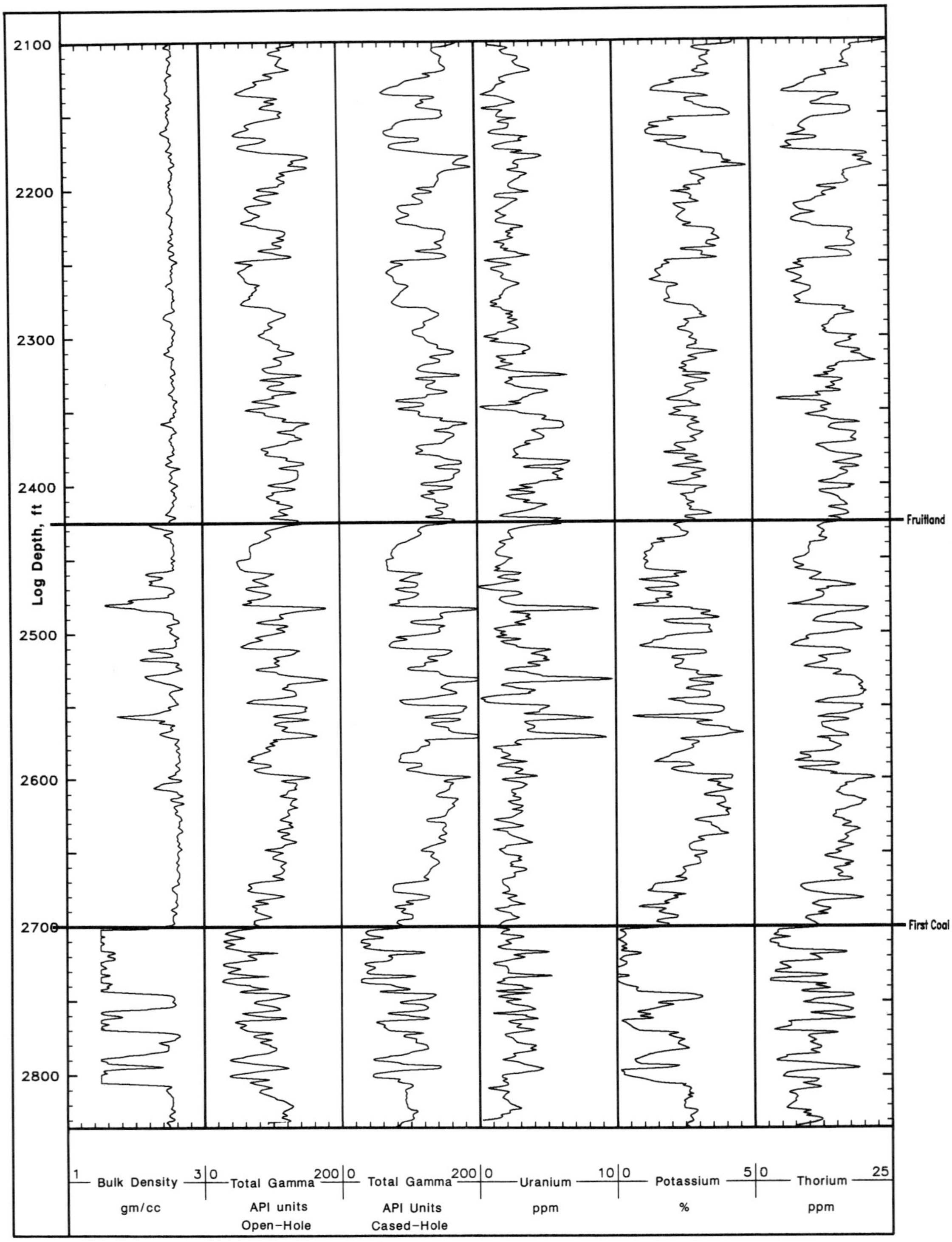

Figure 17. **Well-log display of the cased-hole spectral gamma-ray log, open-hole total gamma-ray log, and bulk density logs from the Amoco Mason Gas Unit A No. 1 well (well 11, table 1) showing variations in the gamma count of Fruitland Formation shales and coals due to fluctuations in the three radioactive trace elements (potassium, thorium, uranium). Large excursions of the total gamma trace are due primarily to changes in the uranium component.**

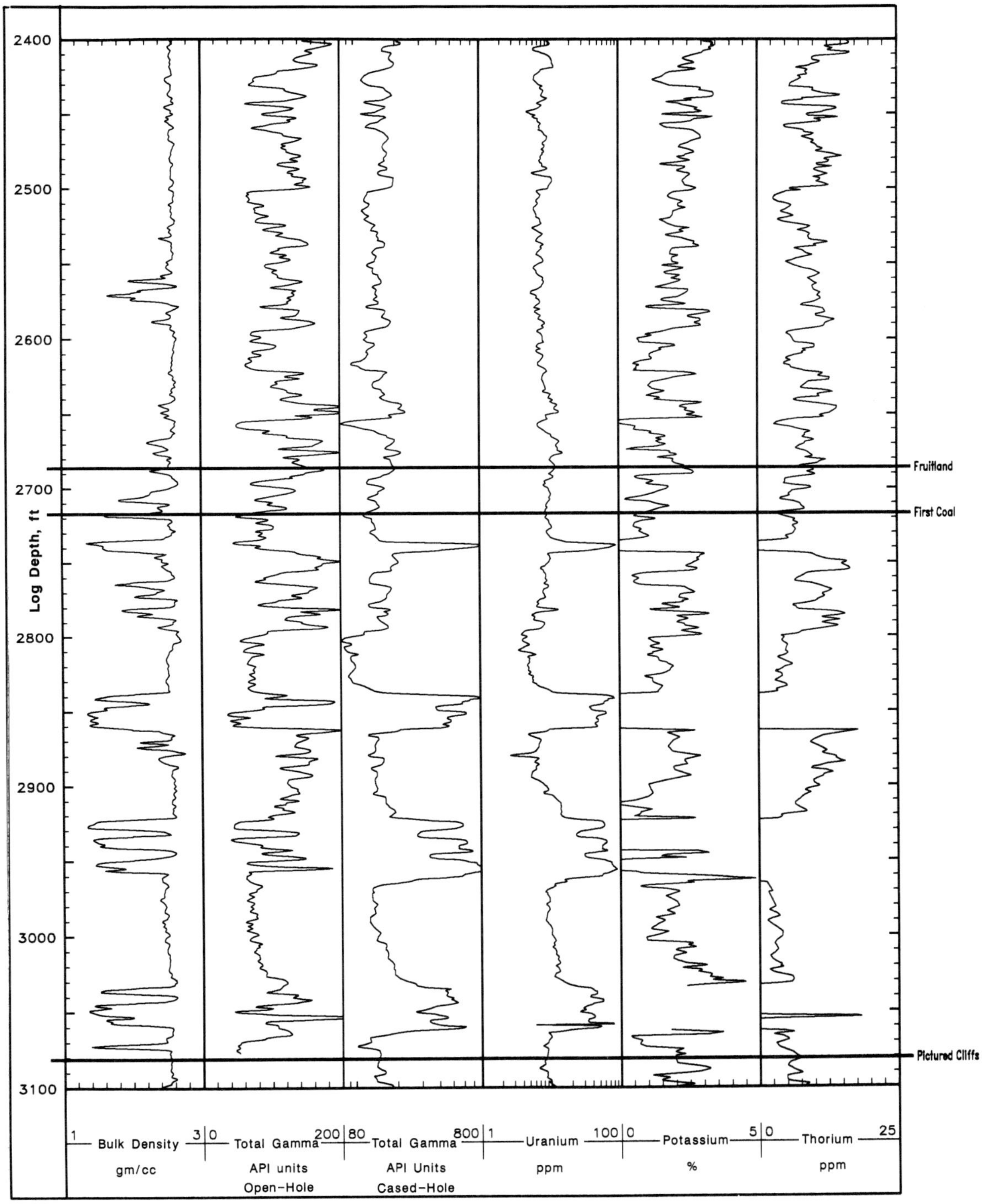

Figure 18. Well-log display of the cased-hole spectral gamma-ray log and open-hole gamma-ray log, and bulk-density logs from the Amoco Federal Land Bank C No. 1 (well 12, table 1) showing variations in the gamma count of Fruitland Formation shales and coals due to fluctuations in the three radioactive trace elements (potassium, thorium, uranium). Comparison of the open- and cased-hole total-gamma logs indicates that the high uranium values occurred after casing was set and are probably related to the flow of uraniferous ground waters through the production perforations. Total gamma and uranium are plotted on logarithmic scales.

in 1979 as, Mayer, R., Jr., and Fishel, K.W., Applications of electric logs in Applachian coal exploration, *in* 6th European logging symposium transactions, paper DD: Society of Professional Well Log Analysts, London Chapter, 18 p.

Flaum, C., Galford, J.E., and Hastings, A., 1987, Enhanced vertical resolution processing of dual detector gamma-gamma density logs, *in* 28th annual logging symposium, proceedings, paper M: Society of Professional Well Log Analysts, Houston, 25 p.

Galford, J.E., Flaum, C., Gilchrist, W.A., Jr., and Duckett, S.W., 1986, Enhanced resolution processing of Compensated Neutron Logs: Society of Petroleum Engineers Paper 15541, 13 p.

Hoffman, G.L., Jordan, G.R., and Wallis, G.R., 1982, Geophysical logging handbook for coal exploration: The Coal Mining Research Centre, Edmonton, Alberta, Canada, 270 p.

Jones, A.H., 1985, Methane production characteristics of deeply buried coal-bed reservoirs: Gas Research Institute, Chicago, Report GRI-85/0033, [prepared by TerraTek, Inc., under contract no. 5081-214-0577], 176 p.

Kern, J.W., Sinha, A.K., and Sookram, V.A., 1983, Geophysical techniques for coal-bed evaluation, final report: Gas Research Institute, Chicago, Report 83-0055 [prepared by Dresser Petroleum Engineering Services under contract no. 5081-214-0587], variously paginated.

Lavers, B.A., and Smits, L.J.M., 1976, Recent developments in coal petrophysics, *in* 17th annual logging symposium transactions, paper S: Society of Professional Well Log Analysts, Houston, 19 p. Reprinted in 1976 *in* Muir, W.L.G., ed., Coal exploration [1st international coal exploration symposium, (London, May 18-21), proceedings]: Miller Freeman Publications Inc., San Francisco, p. 129-152. Later published in 1977, *in* The Log Analyst, v. 18(1), January-February, p. 6-16.

McCall, D.C., Allen, D.F., Culberston, J.S., 1987, High-resolution logging—The key to accurate formation evaluation, *in* Formation evaluation and reservoir geology, volume sigma: Society of Petroleum Engineers Paper 16769, p. 283-298.

Meissner, F.F., 1984, Cretaceous and Lower Tertiary coals as sources for gas accumulations in the Rocky Mountain area, *in* Woodward, J., Meissner, F.F., and Clayton, J.C., eds. Hydrocarbon source rocks of the greater Rocky Mountain region: Rocky Mountain Association of Petroleum Geologists, p. 401-431.

Mullen, 1988, Coal-bed methane reserves calculation for the Fruitland Formation based on geophysical log evaluation, San Juan basin, New Mexico and Colorado, *in* Fassett, J.E., ed., Geology and coal-bed methane resources of the northern San Juan basin, Colorado and New Mexico: Rocky Mountain Association of Geologists Guidebook.

Norris, J.O., and Thomas, R., 1980, An in situ coal quality prediction technique: Society of Petroleum Engineers Paper 9467, 9 p.

Reeves, D.R., 1971, In-situ analysis of coal by borehole logging techniques: Canadian Institute of Mining and Metallurgy [CIM] Bulletin, v. 64(706), p. 67-75. Reprinted in 1971, CIM Transactions, v. 74, p. 61-69. Reprinted in 1986, *in* Buchanan, D.J., and Jackson, L.J., eds. Coal geophysics: Tulsa, Society of Exploration Geophysicists, Geophysics Reprint Series no. 6, p. 36-44.

Rice, D.D., 1983, Relation of natural gas composition to thermal maturity and source rock type in San Juan Basin, northwestern New Mexico and southwestern Colorado: American Association of Petroleum Geologists Bulletin, v. 67, p. 1199-1218.

Rice, D.D., Threlkeld, C.N., Vuletich, A.K., and Pawlewicz, M.J., 1988, Identification and significance of coal-bed gas, San Juan Basin, northwestern New Mexico and southwestern Colorado, *in* Fassett, J.E., ed., Geology and coal-bed methane resources of the northern San Juan basin, Colorado and New Mexico: Rocky Mountain Association of Geologists Guidebook.

Samworth, J.R., 1974, The radiation density log applied to the resolution of thin beds in coal measures, *in* 3rd European formation evaluation symposium, transactions, paper R: Society of Professional Well Log Analysts, London Chapter, 11 p.

Serra, O., 1984, Fundamentals of well-log interpretation, volume 1—The acquisition of logging data: Elsevier, Amsterdam, Developments in Petroleum Geology Series no. 15A, 423 p.

Suau, J., 1981, Logging methods for coal exploration: Bulletin des Centres de Recherchese Exploration-Production Elf-Aquitain, v. 5(2), p. 621-633.

Tixier, M.P., and Alger, R.P., 1967, Log evaluation of non-metallic mineral deposits, *in* 8th annual logging symposium transactions, paper R: Society of Professional Well Log Analysts, Houston, 22 p. Later published in 1970, Geophysics, v. 35(1) p. 124-142. Reprinted in 1971, Well Logging: Dallas, Society of Petroleum Engineers, Reprint Series no. 1, p. 368-390. Reprinted in 1986, *in* Openhole well logging: Society of Petroleum Engineers, SPE Reprint Series no. 21, p. 397-419.

Log Evaluation In Wells Drilled For Coal-bed Methane

Michael J. Mullen[1]

ABSTRACT

Methane from coal beds was for years a nightmare for the mining industry. Today, however it is adding additional valuable gas reserves on acreage already held by production. Coal beds constitute a new reservoir for the oil and gas industry to explore, however, this new reservoir does not lend itself readily to conventional well-log analysis. This study describes new wire-line logging procedures for coal-bed methane wells and new techniques for evaluating wire-line logs in the coal beds of the Fruitland Formation in the San Juan Basin, New Mexico and Colorado.

INTRODUCTION

Today coal beds present a new type of reservoir for the oil and gas industry to explore. Coal beds do not lend themselves readily to conventional well-log analysis. Until recently, the evaluation of coal beds for methane production has been done on a "try it and see what happens" basis, which has often been very costly. However, new log-analysis techniques have been developed and are being used to high-grade prospects, thus increasing the chance for a successful well.

Finding coal beds on a well log is just the first step in looking for coal-bed methane. Questions that are generally asked when evaluating a coal-bed prospect are: "How much gas will the well make?", "Is the coal fractured?", and "How much methane does the coal contain?"

There is only limited information in the literature on coal-bed methane to help log analysts answer these questions. The San Juan Basin of New Mexico and Colorado has seen a large number of wells drilled and recompleted for coal-bed methane in the past four years. This study presents the latest wire-line logging and evaluation techniques that are being used. The logging techniques are applicable wherever coal beds are being evaluated. Fruitland Formation coal beds investigated in this study are predominantly high- to medium-volatile bituminous coal. For coal beds of different rank, these evaluation techniques should be tested for their accuracy before being used extensively.

LOGGING PROCEDURE

The suite of logs normally run to evaluate a coal bed includes the spectral density* log (SDL), the dual-spaced neutron II* log (DSNII), and the dual induction log with gamma ray and spontaneous-potential logs. SDL data are used in determining coal quality, bed thickness, and methane content. DSNII data are used for gas detection in adjacent sandstones. The short-guard curve on the dual induction log is used for coal bed thickness measurements, the deep induction for coal-bed correlation, the spontaneous potential (SP) for indication of permeable zones, and the gamma ray for correlation.

*A copyrighted trademark of the Halliburton Company

[1] Senior Sales Engineer
Welex, A Halliburton Company
P.O. Box 170
Farmington, NM 87499-0170

The author wishes to express his thanks to the companies and individuals who provided log information, core data and time to discuss the ideas presented.

Coals are relatively easy to identify and correlate on well logs. The bulk density measurement from the spectral density log readily flags coal beds because the surrounding shales have a density of 2.65 g/cm³ and coal has an average density of 1.5 g/cm³. The short-guard curve from the dual induction log usually shows sharp increases in resistivity at coal-bed boundaries and usually reads more than 50 ohm-m. The deep-induction curve shows a more gradual increase in resistivity at coal-bed boundaries due to its large vertical resolution. The DSNII porosity curve is usually similar to the SDL porosity curve with both reading very high porosity, often 50 percent through coal beds. The gamma-ray curve usually indicates less than 70 API units in coal. Standard vertical resolution of the SDL, DSNII, and gamma ray is 2 feet.

If enough rat hole is provided, which is usually the case when logging at an intermediate casing point, all of these logs may be run stacked, on one trip in the hole. This saves rig time for logging and eliminates possible depth discrepancies that could occur if the tools were run separately. Typically, the well is logged by first making two runs (called repeat runs) over the coal beds, and then making the main logging run. A repeat run is normally recorded with a standard presentation to locate the coal beds. A logging speed of 30 ft/min is recommended when running nuclear tools.

In coal beds, partings or shale stringers may be thinner than the standard vertical resolution of the tools, thereby masking features that are seen in the core. High-resolution processing of the SDL measurement is being used to reduce the vertical resolution to 1 foot. Partings 1 foot thick will be accurately identified, but thinner partings may still be masked. A logging pass with the SDL tool is recommended at 18 ft/min using high-resolution processing to record the maximum detail in the coal beds. The main logging run should then be recorded with standard vertical resolution.

Figure 1 is a comparison between the standard vertical resolution pass and the high-resolution pass. What look like subtle differences in coal quality on the standard pass become identified as partings or carbonaceous shale layers on the high-resolution pass (see point A). The high-resolution pass showed excellent correlation to the core; it identified the partings and gave a more accurate bulk-density measurement in these thin coal beds than did the standard log.

The accuracy of the SDL tool at low bulk densities is checked, prior to logging, by inserting the tool in a water tank and recording the bulk density for several minutes, (see fig. 2). On the log shown, water reads 1.07 g/cm³ instead of 1.0 g/cm³. This discrepancy, which only affects bulk density

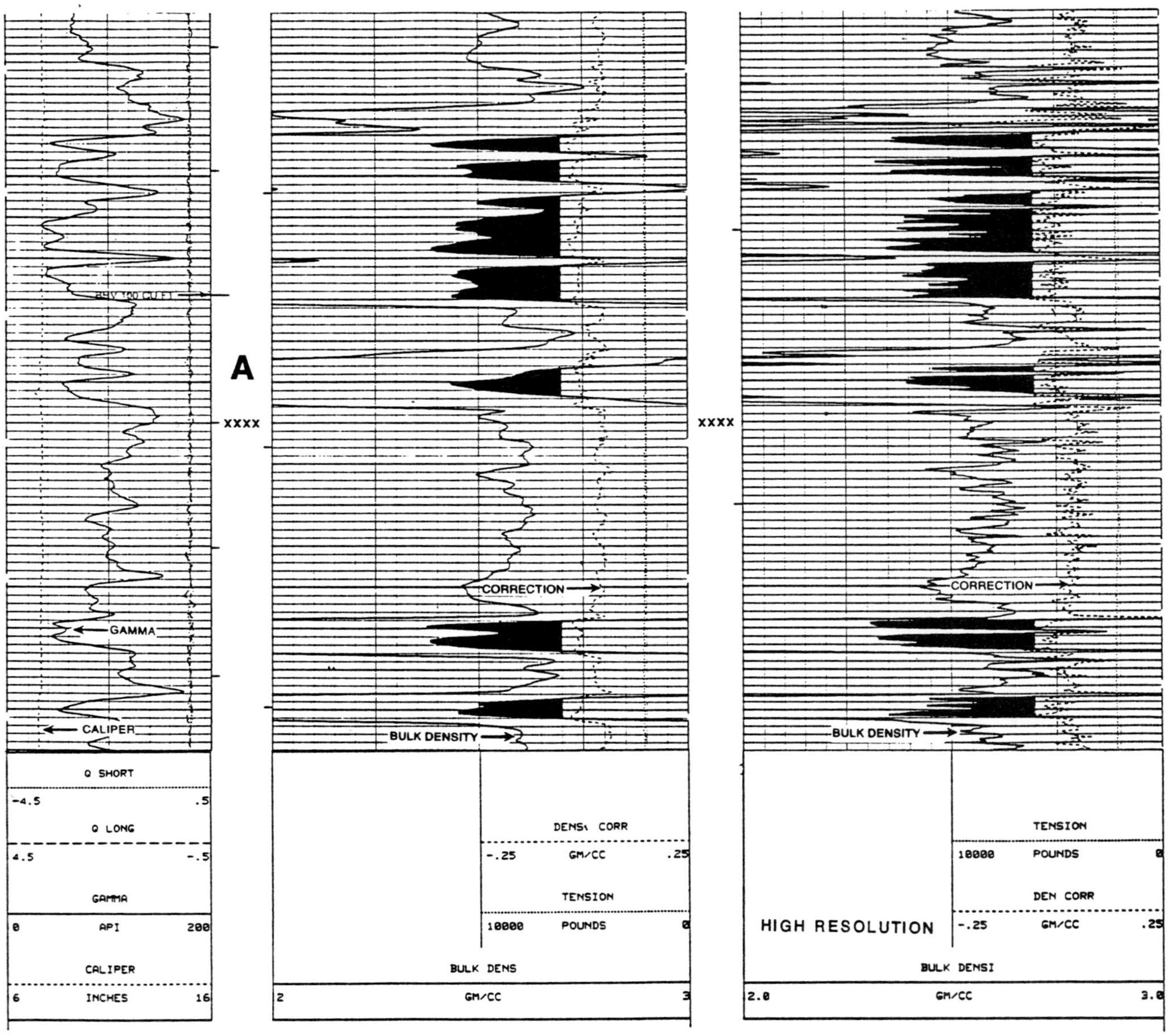

Figure 1. Comparison of standard and high resolution SDL Processing; high resolution is on right. Vertical-scale grid is in 2-foot increments.

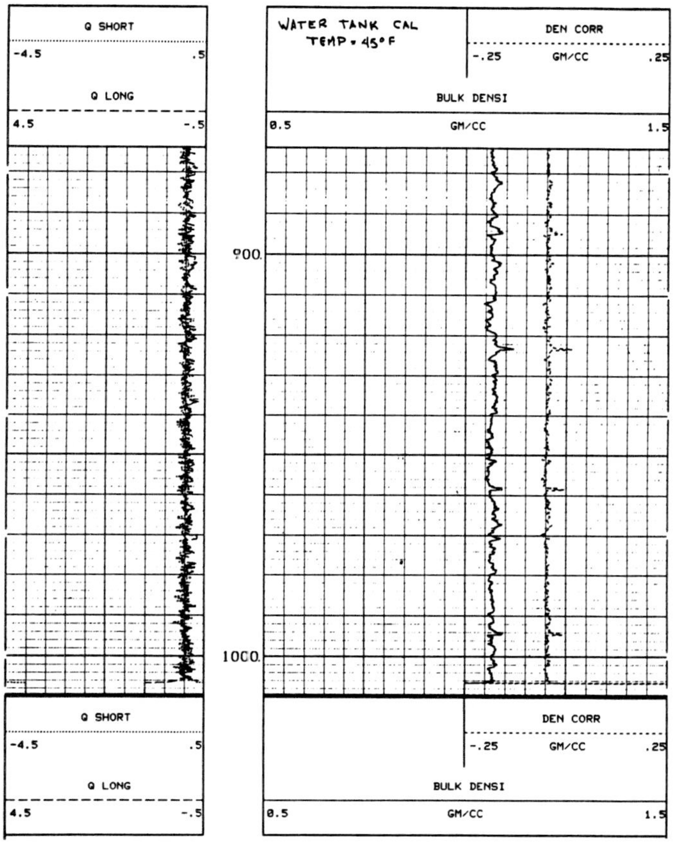

Figure 2. SDL bulk density recorded in the water tank.

readings from 1.0-1.3 g/cm³, is corrected by post-logging processing. For most coal beds in the San Juan Basin, this correction is insignificant since very few coal beds have a bulk density below 1.3 g/cm³. This correction has now been incorporated in real-time logging software.

With older generation density logging tools, long filter lengths were needed to reduce statistical variation. This leads to a question in high-resolution processing: "Are we looking at statistical variation or actual changes in the rock?" Figures 3 and 4 demonstrate the repeatability of the SDL tool with high-resolution processing. The two passes are nearly direct overlays. Thus, the high-resolution processing is displaying actual changes in bulk density, not just statistical variation.

With shorter vertical resolution, coal-bed boundaries are often very quick events on the SDL porosity curve. These bed-boundary changes become difficult to see on the log (fig. 4). Figure 5 shows the coal-bed evaluation format being recommended to record SDL bulk density and dual induction data on 5 in/100 ft scale. This portrayal is a playback from real-time data and is attached to the standard log presentation. With this presentation, the log analyst can more easily identify and analyze coal beds. A 10 in/100 ft grid (fig. 6) is also useful in accurately measuring coalbed thickness and showing detail within the coal beds. High-resolution processing must be done at the time the well is originally logged. It is not possible to take an old density log recorded with a 2 foot resolution and by processing, reduce its resolution to 1 foot.

FRACTURE OR CLEAT DETECTION

Conventional fracture detection methods are frequently ineffective in coal beds. The primary tools used for fracture detection today are the acoustic tool and the dipmeter. The acoustic frac finder is difficult to interpret in a formation, such as a coal bed, that rarely supports shear or stonley waves and is also laminated. The dipmeter frac finder is difficult to interpret in wells that have an irregular borehole, which is often the case through coals.

Most coal-bed methane wells are drilled with over-balanced mud programs to avoid blowouts. This practice usually forces mud cake into any open fractures or cleats in the coal. When fresh muds are weighted with barite, the mud cake packed in the fractures will also contain barite. The photoelectric curve (Pe) on the spectral density log is very sensitive to the presence of barite. The Pe response in barite is 266 where the Pe response in coal is about 0.3. Although the response in barite muds is somewhat less than 266, this Pe contrast is often enough to highlight the open fractures packed with mudcake. Based on experience, barite muds with weights greater than 11 lb/gal are needed to use the Pe curve as a fracture indicator. The log shown in figure 7 was recorded in 11.5 lb/gal barite-weighted mud. The Pe values are all shifted higher than the normal values expected (1.8 for sandstone, 3.5 for shale, 5 for limestone) because of the presence of barite in the bore hole. Large increases in Pe value are noticed in each coal bed below depth A. These increases are interpreted to represent large (or many) open fractures filled with barite mud. This well was completed naturally without stimulation and produced more than 1,000 MCFPD.

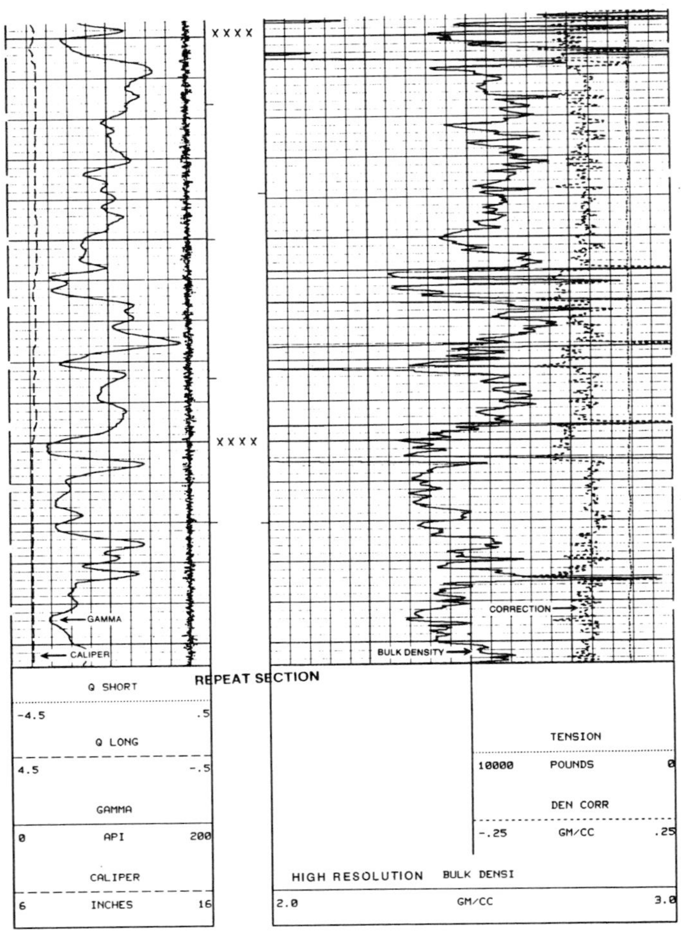

Figure 3. **High-resolution SDL curves through coal-bearing rocks. Depths marked by Xs are 100 feet apart.**

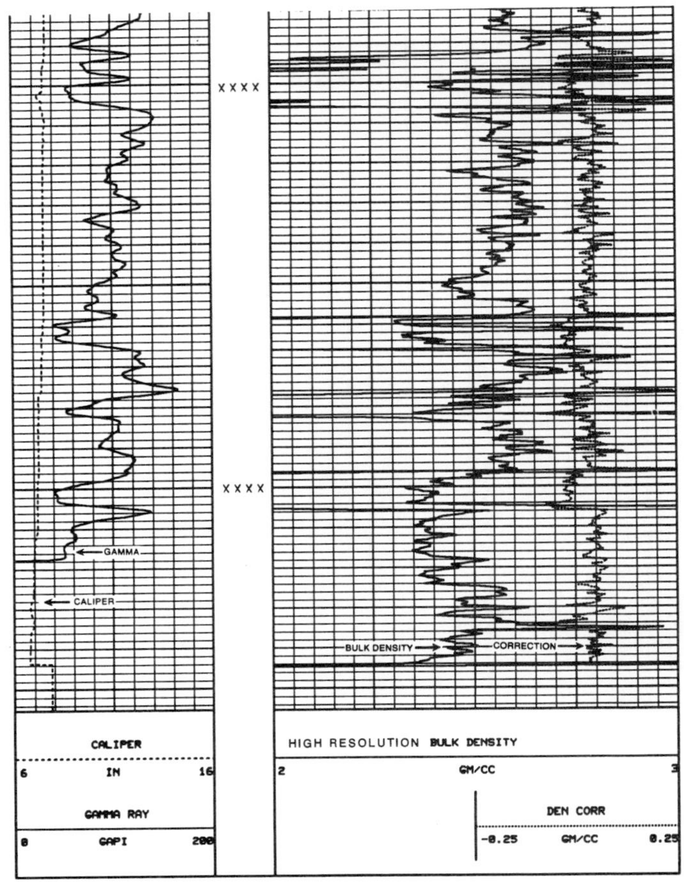

Figure 4. **High-resolution SDL curves repeat section in a second pass through the same coal-bearing interval shown on figure 3. Depths marked by Xs are 100 feet apart.**

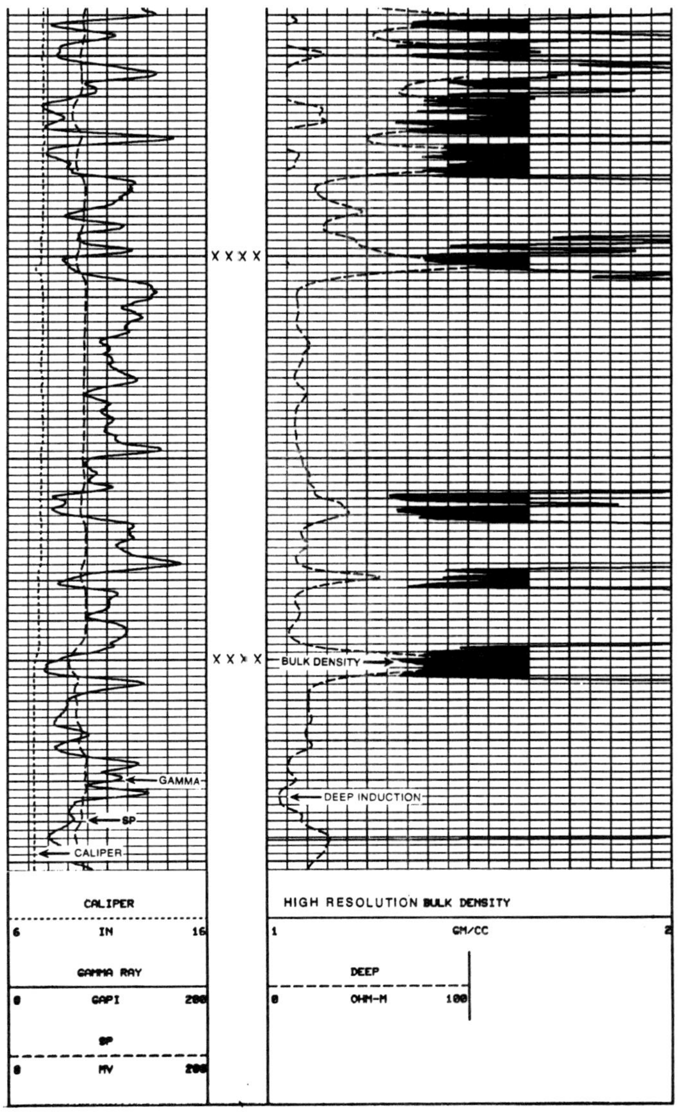

Figure 5. **Coal analysis presentation on an SDL porosity curve run at a scale of 5 in/100 ft (original scale). Depths marked by Xs are 100 feet apart.**

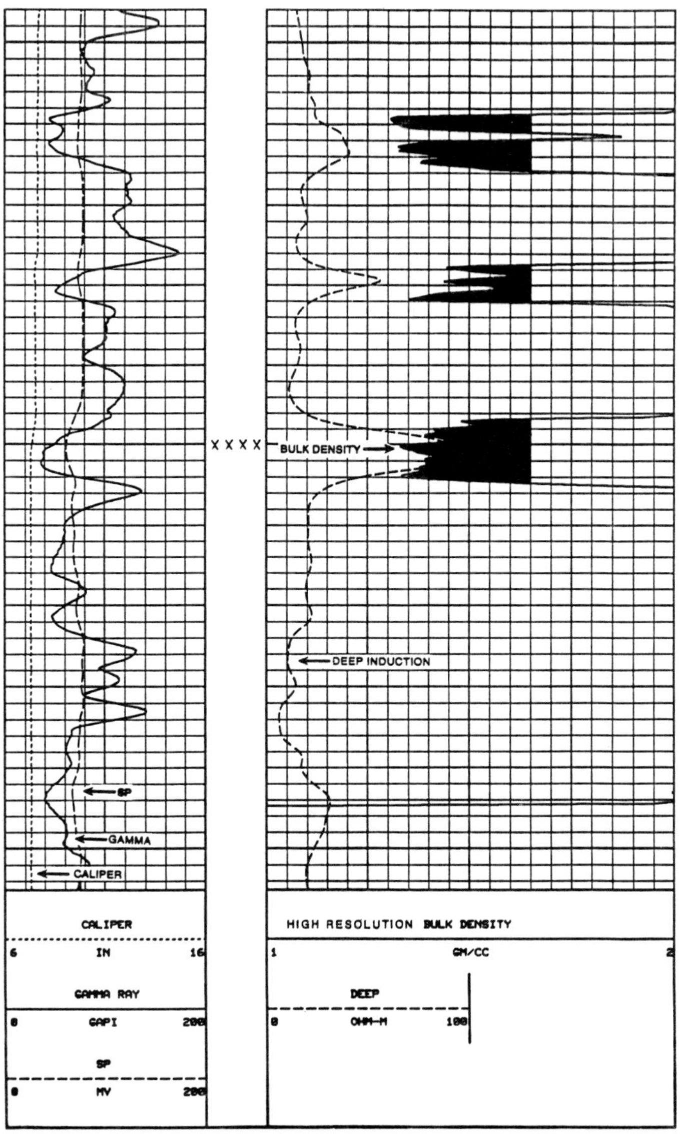

Figure 6. **Coal analysis presentation on an SDL porosity curve run at a scale of 10 in/100 ft (original scale). Vertical-scale grid is in 2-foot increments.**

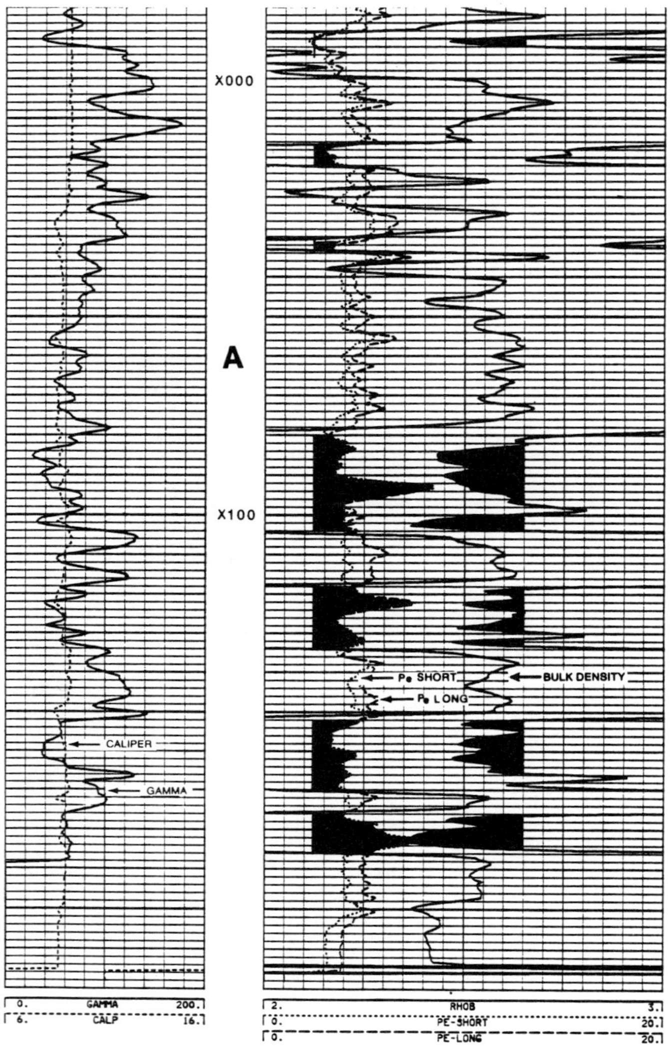

Figure 7. **SDL photoelectric (Pe) curve in coal beds; coal cleats are filled with barite-weighted drilling mud. Heavy lines on vertical-scale grid are in 10-foot increments.**

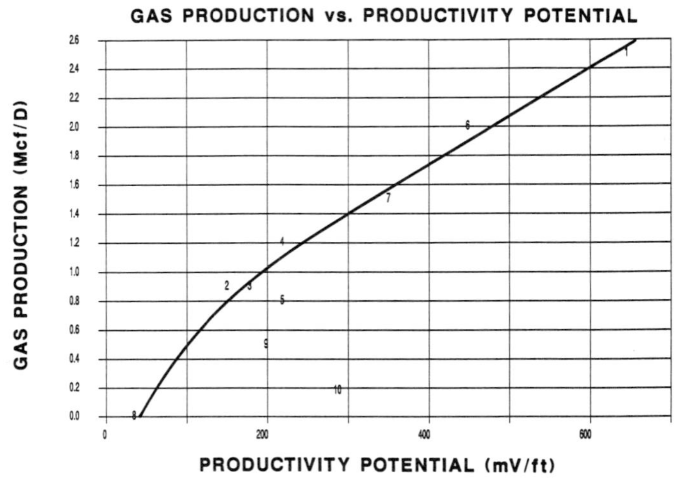

Figure 8. **Observed relations between the productivity potential as determined by log analysis and gas production for coal beds for ten wells. Mcf/D on left axis means million cubic feet per day.**

PRODUCTIVITY ANALYSIS

Estimating productivity of gas and water from a coal bed is dependent upon several factors: completion techniques, dewatering, producing practices, and time. Most coal-bed methane wells exhibit a production incline curve. Therefore, the more water lifted from the coal, the more gas the coal will produce. With time, the water production decreases and gas production increases.

It was noticed that the better-producing wells had more SP development in the coals than the poorer producing wells, but the magnitude of the SP did not correlate very well to production. Therefore, a new unit was derived to quantify productivity. The product of the maximum SP deflection through the coal bed and the thickness of the coal bed seemed to correlate to production. The millivolt-feet (mV-ft) product was chosen as the productivity-potential unit. The correlation of gas production to productivity potential is shown in figure 8.

Table 1 lists the ten wells used to construct figure 8, which indicates the relationship between millivolt-feet and gas production. Most of the wells in this group have been producing for from one to two years; therefore, significant dewatering has occurred. More information is needed on the 0-100 mV-ft range to better define the low end of the curve.

Figure 9 is well 1 of table 1. There are six coal beds shown totaling 77 feet of coal with 647 mV-ft productivity potential. This well makes nearly 2,500 MCFPD. In contrast, figure 10 is well 8 of table 1. This well has 36 mV-ft productivity potential from the coal beds. The well builds up to 880 psi but dies readily and produces little gas. There is very little permeability in the coal beds in this well.

To effectively estimate production from the productivity potential, a good quality SP log is essential. A poor quality SP log can mask or enhance subtle SP deflections in coal beds and thereby lead to an invalid analysis. A good SP log from an offset well can be used in place of a questionable SP log if the need arises.

Table 1. Coal thickness and SP response for the logs used in figure 8.

Well	Coal thickness (in feet)	SP	mV-ft	Production
1	5	6 mV	30	
	3	4 mV	12	
	25	9 mV	225	
	16	8 mV	128	
	18	9 mV	162	
	10	9 mV	90	
TOTAL	77		647	2,500 MCFPD, 500 BWPD
2	9	8 mV	72	
	8	10 mV	80	
	2	0 mV	0	
	5	? mV	?	
TOTAL	24		152+	900 MCFPD
3	12	15 mV	180	900 MCFPD, 25 BWPD
4	20	11 mV	220	1,200 MCFPD
5	20	11 mV	220	800 MCFPD, 50 BWPD
6	30	15 mV	450	2,000 MCFPD, 50 BWPD
7	16	22 mV	352	1,500 MCFPD, 50 BWPD
8	8	3 mV	24	
	5	0 mV	0	
	3	0 mV	0	
	4	3 mV	12	
TOTAL	20		36	No Production
9	3	0 mV	0	
	16	12 mV	192	
	2	2 mV	4	
	2	2 mV	4	
TOTAL	23		200	500 MCFPD, 30 BWPD
10	4	0 mV	0	
	20	8 mV	160	
	10	5 mV	50	
	10	7 mV	70	
	10	1 mV	10	
TOTAL	54		290	180 MCFPD, 30 BWPD

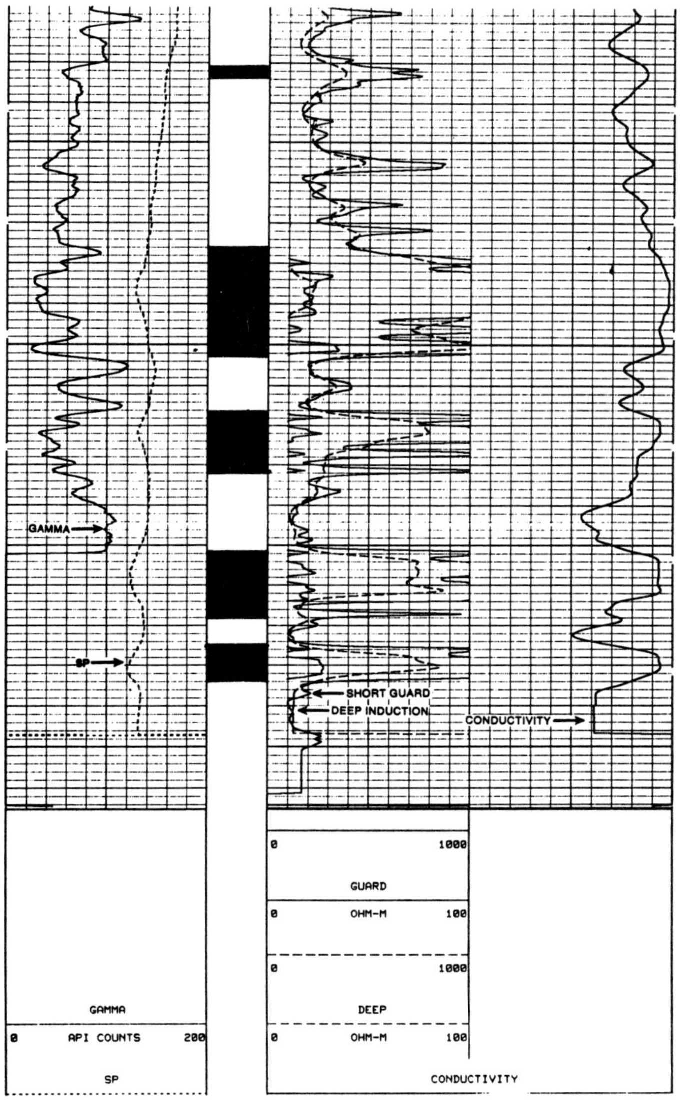

Figure 9. SP curve response in a well with a high rate of coal-bed methane production. Vertical-scale grid is in 2-foot increments.

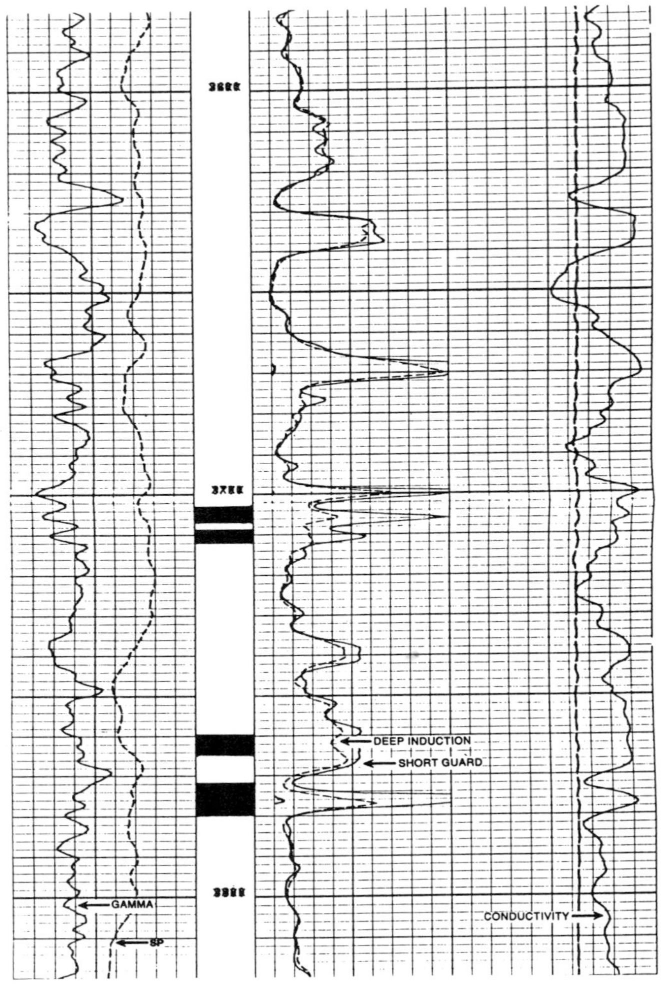

Figure 10. SP curve response in a well with low-permeability coal beds. Depths marked with Xs are 100 feet apart.

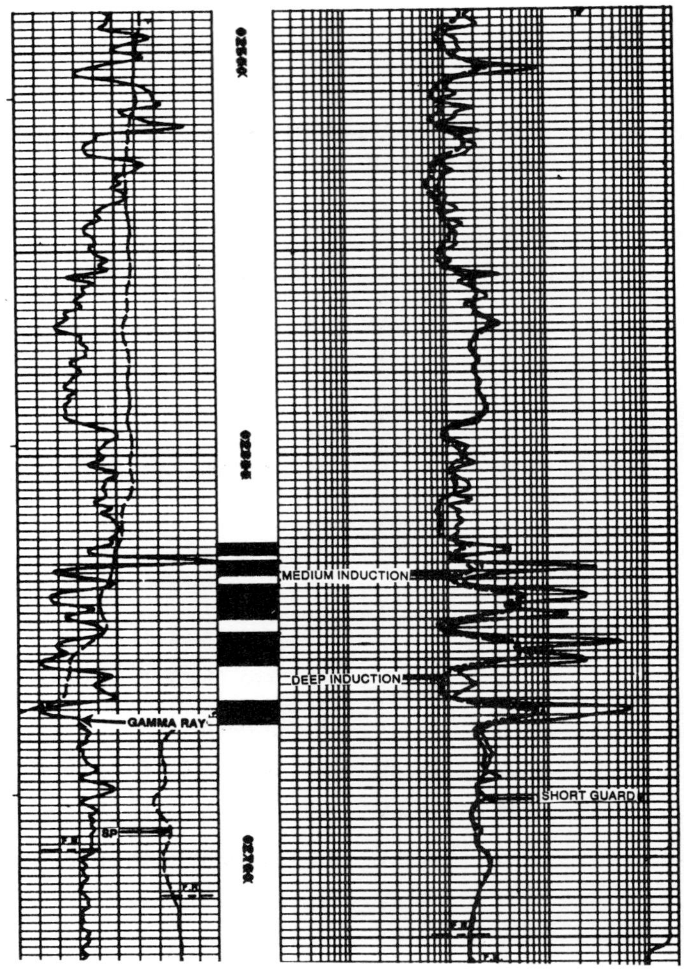

Figure 11. SP curve response through coal beds showing distortion of the SP curve due to cathodic-protection currents from an offsetting well.

Cathodic protection can also affect SP logs. Figure 11 is an example of a +90 mV SP shift caused by cathodic protection of an adjacent well. The SP deflections are seen superimposed on the shifting baseline. The productivity potential calculated on this well may be in error as some of the SP response could be masked. Again, an SP log from an off-setting well could be substituted if one could be found that had been run prior to the institution of cathodic protection in the area.

The SP response can also be influenced by reservoir and drilling fluids. Mud resistivity varies slightly from well to well but formation water can vary in salinity regionally which could cause an interpretation problem; mud and formation water salinity should be considered when making an analysis. Reservoir pressure can also influence the SP response, increasing or decreasing the deflection from the baseline.

Figure 12 is an example of a Fruitland coal section that has been altered by igneous intrusives. The deep induction curve erratically ranges from zero to infinity. The SP measurement is responding to non-conductive beds. Coal identification from the spectral density log is not possible in this well. This well makes about 30 MCFPD and no water. This example shows that quantitative evaluation of coal-bed methane potential by log analysis is not yet possible in wells where igneous material is present with the coals.

The productivity-potential correlation was derived in the northern San Juan Basin. Qualitatively, for other regions, higher productivity-potential values could indicate areas of higher gas production.

COAL-RANK ANALYSIS

Coal rank is usually determined from a proximate analysis measured from a coal sample; the analysis lists the percentages of moisture, ash, fixed carbon, and volatile matter in the coal sample, as received. The percentages of fixed carbon and volatile matter are also calculated on a moisture- and ash-free (DAF) basis (Geer and Yancey, 1968).

Proximate analysis requires a sample of coal, which requires coring in a new well. When re-evaluating old wells, it is not possible to obtain a sample of coal for analysis. Therefore, a method of determining the proximate analysis from wire-line logs must be used if coal rank information is desired. A relationship between log measurements and coal analysis was developed using an available, San Juan Basin data base. Using proximate analysis information from published data (Fassett and Hinds, 1971, p. 58-60), algorithms were derived to relate

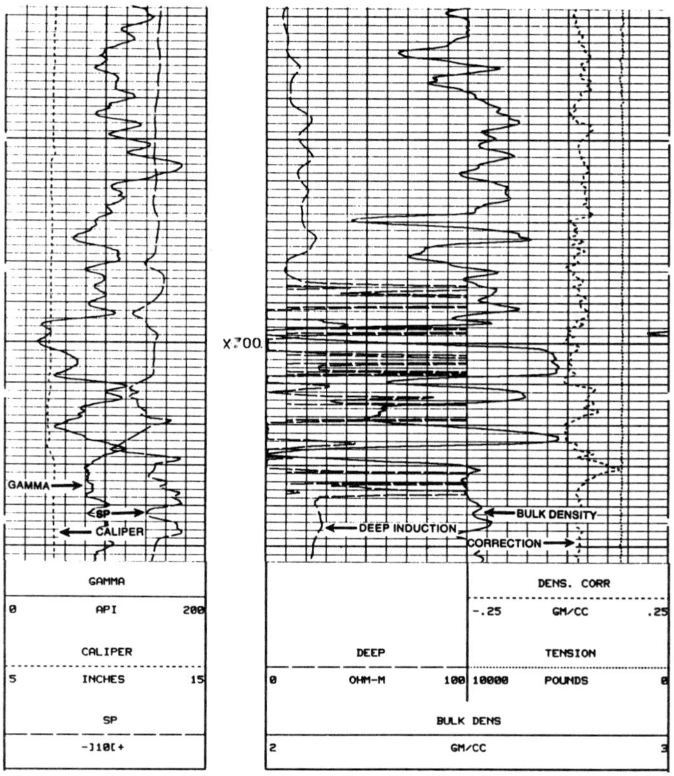

Figure 12. SDL curves through a section of the Fruitland Formation where the coals have been intruded by igneous rocks.

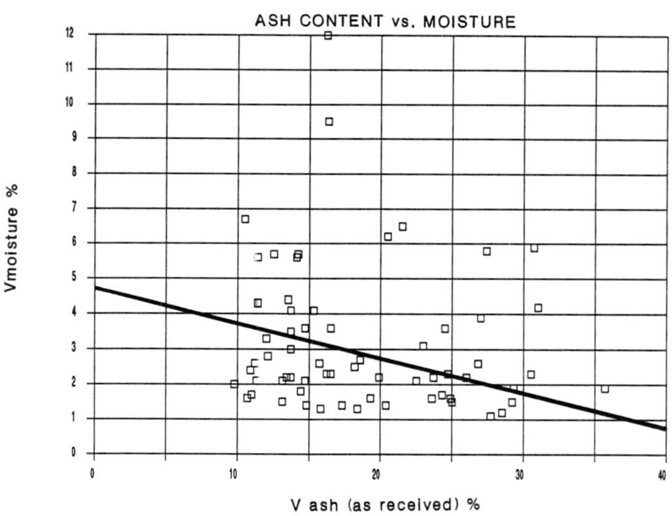

Figure 13. Relationship of Vash to Vmoisture (data is from Fassett and Hinds, 1971).

ash content to moisture, fixed carbon as received, and fixed carbon DAF for the northern San Juan Basin.

Vmoisture = −0.10 × (Vash) + 4.61,(see fig. 13);

Vfixed carbon = −0.517 × (Vash) + 51.2,(see fig. 14);

Vfixed carbon DAF = 0.75 × (Vash) + 42.3,(see fig. 15).

A relationship between bulk density and ash content was derived empirically as Vash = 64.94 × bulk density −66.27 (modified from Lavers and Smits, 1976). Volatile matter is determined by subtracting the percentages of ash, moisture,

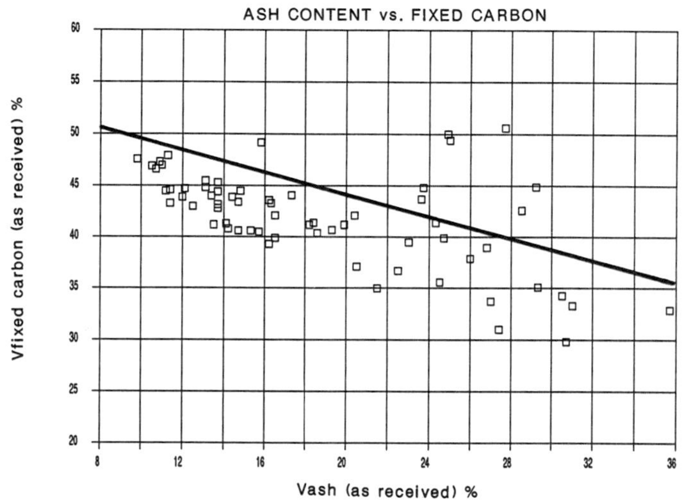

Figure 14. Relationship of Vash to Vfixed carbon, as received (data is from Fassett and Hinds, 1971).

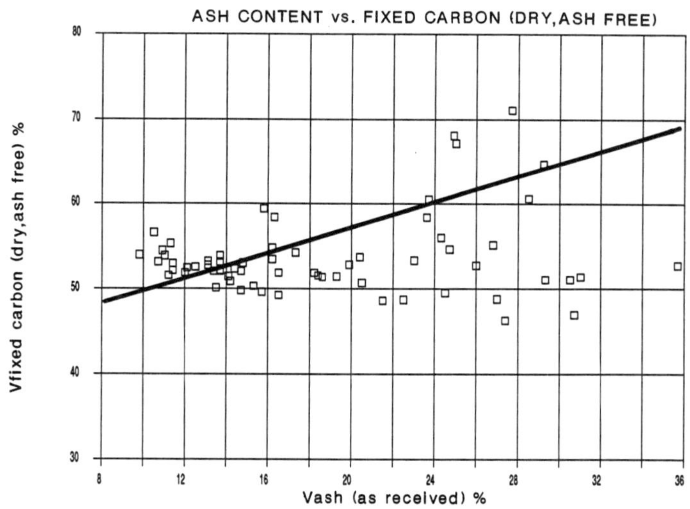

Figure 15. Relationship of Vash to Vfixed carbon (DAF) (data is from Fassett and Hinds, 1971).

and fixed carbon from one.

Vvolatile matter = 100 − Vash-Vmoisture-Vfixed carbon.

With the use of these algorithms, wire-line logs, specifically the SDL, can be used to estimate the proximate analysis (Lavers and Smits, 1976, Lientur, 1976, Norris and Thomas, 1980, and Peeters and Kempten, 1977). A comparison of log-calculated to measured proximate analysis is not releasable at this time. Preliminary results, however, show log-calculated values to be very representative of actual core analysis. More actual core data is needed to further refine these algorithms.

Appendix I is a Hewlett-Packard HP-41 calculator program the author wrote to calculate log-derived proximate analysis values. These same computations can also be made continuously by hand-digitizing old logs or using the digital tape recording of newer logs and processing them by computer. Figure 16 shows the results of this type of continuous processing from a well in the northern San Juan Basin. A digital listing with half-foot intervals is available to assist the log analyst in reading computed values accurately (not shown here).

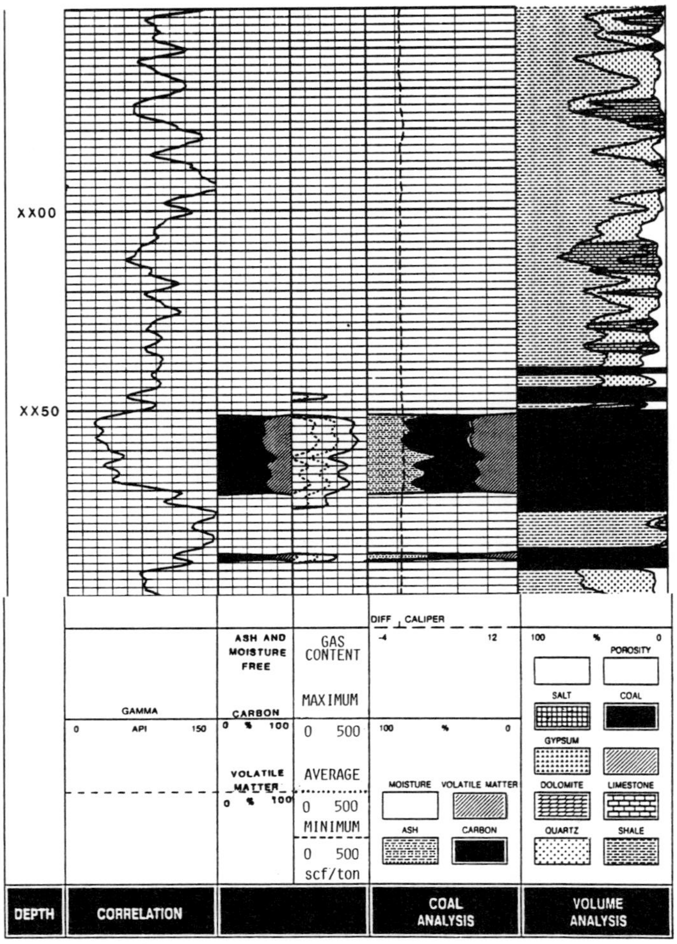

Figure 16. Graphic depiction of coal analysis based on continuous computation of log-derived values. Vertical scale is in 2-foot increments.

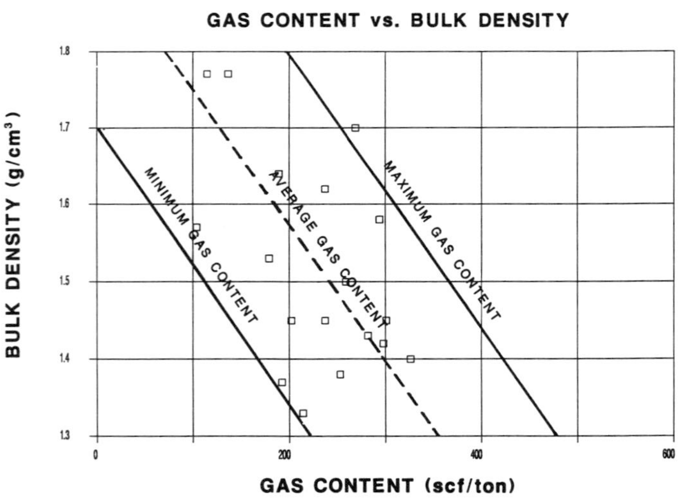

Figure 17. Observed relations of coal-bed bulk density and measured gas content.

GAS CONTENT

The gas content in coal is usually given as cubic feet of gas per ton (ft^3 gas/ton). If the amount of gas per ton of coal is known, reserves can be calculated as follows:

Table 2. Comparison of gas content computations.

Depth	Measured gas content ft^3 gas/ton	Average* gas content calculated ft^3 gas/ton	Coal quality** gas content ft^3 gas/ton
3,705	189	164	405
3,760	137	93	370
3,775	104	202	427
3,010	253	305	454
3,030	214	332	468
3,052	202	267	437
3,080	179	223	417
3,120	192	310	461
3,110	259	240	426
3,170	294	196	406
3,226	300	267	443
3,255	297	283	452
3,200	281	277	448
3,222	236	174	397
3,268	326	294	458
3,326	268	131	378
3,445	236	267	450
AVERAGE	233	236	429

*From figure 17
**From appendix I

Gas in place = Gas content × bed thickness in feet × coal tonnage per acre-ft × well spacing.

There are approximately 1,800 tons of coal per acre-foot of coal (Geer and Yancey p. 4-18). For example, in a 10 foot thick coal bed containing 250 ft^3 gas/ton over a well-spacing area of 160 acres, gas resources are calculated, as follows:

Gas in place = 250 × 10 × 1,800 × 160 = 720,000 MCF.

Desorption tests are commonly performed to measure the gas content of a coal bed by placing coal samples in a chamber and measuring the gas released as the pressure declines over time. The coal sample is then weighed and the results are expressed in cubic feet of gas per ton of coal (Choate and others, 1984, and Meissner, 1984). In older wells it is not possible to directly measure the gas content in the coal because the coal is behind casing.

A method of estimating gas content from wire-line logs was developed based on actual measured gas contents. Measured gas content was observed to correlate to the bulk density measurement within a range as shown on figure 17. The equations developed from this observation relating measured bulk density to methane content are as follows:

Maximum gas content = −542 × bulk density (g/cm^3) + 1178 ft^3/ton.

Average gas content = −542 × bulk density (g/cm^3) + 1053 ft^3/ton.

Minimum gas content = −542 × bulk density (g/cm^3) + 919 ft^3/ton.

The gas contents computed from the density log are shown on table 2 next to the values obtained from actual measured samples; these values are seen to be statistically close. The average gas content computed from the bulk density log using the equations above for the 17 samples was 236 ft^3 gas/ton and the average gas content measured from samples was 233 ft^3 gas/ton. When evaluating coal beds, a direct measurement of gas content is desirable but it is not always practical or cost effective to obtain. Therefore, wire-line logs can be used to give acceptable results for gas content.

Another method for computing the gas content of coal was proposed by Kim (1977, p. 1) based on coal-quality information from proximate analyses. Because the proximate analysis can be computed from logs with good accuracy, this method was investigated in the San Juan Basin. Gas content values were found to average 429 ft³ gas/ton for the wells in table 2. The proximate analysis and gas content were calculated using the program shown in appendix II. This method of calculating gas content appears to yield high values, thus its use is not recommended without further investigation.

CONCLUSION

This study has shown that log analysis can be a useful tool in evaluating unconventional reservoirs such as coal beds. Different logging procedures using high resolution processing with the spectral density log are recommended to improve the vertical resolution of the density measurement and thereby improve the accuracy of bed thickness determination and coal quality analysis in new wells. To predict coal-bed productivity, the relation between productivity potential and sustained coal-bed gas production has been utilized. Gas content in coal can be calculated from wire-line log measurements to approximate direct measurements from coal chips or cores.

Log analysis using these techniques can give the well operator or exploration geologist valuable insight into Fruitland coal-gas production potential by using existing well logs where no core data or production data exist. As new well-log and production data become available, the correlations presented here can be refined to make log analysis more accurate in coal beds.

APPENDIX II Gas Content from Coal Rank Data (from Kim, 1977).

$$V = 32 \times (1 - Vmois - Vash) \times .75 \times \{ko \times (.096h)^{no} - .14 \times \frac{1.8h}{100} + 11)\}$$

Where

V = Gas content of coal in ft³ gas/ton
Vmois = Volume Moisture
Vash = Volume Ash
ko = .8 (Vfc / Vvm) + 5.6
no = .39 − .013 ko
h = Depth of coal seam in meters
Vfc = Volume Fixed Carbon as received
Vvm = Volume Volatile Matter as received

Appendix I Coal Quality Program (C Q) for HP-41 Calculator

```
1:36AM 08/07
01•LBL "CQ"
02 ADV
03 "DEPTH="
04 PROMPT
05 ARCL X
06 AVIEW        Input
07 3.281
08 /
09 STO 07
10 "RHOB="
11 PROMPT
12 ARCL X
13 AVIEW
14 STO 00 ________
15 64.94
16 *
17 66.277      VAsh
18 -
19 STO 01 ________
20 -.045
21 *
22 3           Vmois
23 +
24 STO 02 ________
25 RCL 01
26 -.517
27 *           VFC
28 55
29+
30 STO 03 ________
31 RCL 02
32 +
33 RCL 01
34 +           VVM
35 100
36 -
37 CHS
38 STO 04 ________
39 RCL 01
40 .75
41 *     VFC DAF
42 42.3
43+
44 STO 05 ________
45 100
46 -    VVM DAF
47 CHS
48 STO 06 ________
49•LBL A
50 RCL 01
51 "VASH"
52 XEQ "OUT"
53 RCL 02
54 "VMOI"
55 XEQ "OUT"
56 RCL 03
57 "VFC AR"   Output
58 XEQ "OUT"
59 RCL 04
60 "V VOL AR"
61 XEQ "OUT"
62 RCL 05
63 "V FC AMF"
64 XEQ "OUT"
65 RCL 06
66 "V VOL AMF"
67 XEQ "OUT" ________
68 RCL 00
69 -542
70 *    Gas Content 1
71 1053       (Local)
72 +
73 "DSOR 1"
74 XEQ "OUT" ________
75 100
76 RCL 02
77 -
78 RCL 01
79 -    Gas Content 2
80 .24        Ref. 2
81 *
82 STO 08
83 RCL 07
84 .00252
85 *
86 1.54
87 +
88 STO 09
89 RCL 03
90 RCL 04
91 /
92 .8
93 *
94 5.6
95 +
96 STO 10
97 -.013
98 *
99 .39
100 +
101 STO 11
102 RCL 07
103 .096
104 *
105 RCL 11
106 Y↑X
107 RCL 10
108 *
109 RCL 09
110 -
111 RCL 08
112 ¢
113 "DESOR 22"
114 XEQ "OUT" ________
115 GTO "CQ"
116•LBL "OUT"
117 SF 21
118 "⊢="          Print
119 ARCL X
120 AVIEW
121 RTN
122 END
```

Example Output
DEPTH=3,052.00
RHOB=1.46
VASH=28.54
VMOI=1.72
VFC AR=40.25
V VOL AR=29.50
V FC AMF=63.70
V VOL AMF=36.30
DSOR 1=261.68
DESOR 2=371.89

Register Assignment
00 Bulk Density (RHOB)
01 V Ash
02 V Moisture
03 V Fixed Carbon
04 V Volatile Matter
05 V Fixed Carbon DAF
06 V Volatile Matter DAF
07 Depth
08 Scratch
09 Scratch
10 Ko
11 No

REFERENCES

Choate, R., Lent, J., Rightmire, C.T., 1984, Upper cretaceous geology, coal, and the potential for methane recovery from coal beds in San Juan Basin-Colorado and New Mexico, *in* Coalbed methane resources of the United States: American Association of Petroleum Geologists Studies in Geology Series no. 17, p. 185-222.

Fassett, J. E. and Hinds, J. S., 1971, Geology and fuel resources of the Fruitland Formation and Kirtland Shale of the San Juan Basin, New Mexico and Colorado: U.S. Geological Survey Professional Paper 676, 76 p.

Geer, M. R., and Yancey, H. F., 1968, Plant performance and forecasting cleaning results in Coal Preparation, Chapter 18: American Institute of Mining and Mechanical and Petroleum Engineers, p. 18-1—18-29.

Kim, A. G., 1977, Estimating methane content of bituminous coal beds from absorption data: U.S. Bureau of Mines Report of Investigations 8245.

Meissner, F. F., 1984, Cretaceous and lower tertiary coals as sources for gas accumulations in the Rocky Mountain area, *in* Woodward, J., Meissner, F. J., and Clayton, J. L., eds., Hydrocarbon source rocks of the Rocky Mountain Region: Rocky Mountain Association of Geologists Guidebook, p. 401-431

Norris, J. D., and Thomas, R., 1980, An in-situ coal quality prediction technique: Society of Petroleum Engineers, Paper 9467.

Peeters, M., and Kempten, N. H., 1977, Wire-line logging for coal exploration in Australia: Society of Professional Well Log Analysts, The Log Analyst, v. 17, no. 3, p. 24-29.

Lavers, B. A. and Smits, L. J. M., Recent developments in coal petrophysics: Society of Professional Well Log Analysts, 17th Annual Logging Symposium Transactions, p. S1-S19.

Lientur, S. ,W., 1976, Log evaluation of sub-bituminous coals in Magallanes, Chile: Society of Professional Well Log Analysts, 17th Annual Logging Symposium Transactions, p. K1-K33.

Completing, Equipping, and Operating Fruitland Formation Coal-Bed Methane Wells in the San Juan Basin, New Mexico and Colorado

William F. Clark[1]
and
Tom Hemler[2]

ABSTRACT

This paper is a review of drilling, completing, equipping and producing procedures used by operators in the San Juan Basin for Fruitland Formation coal-bed methane wells. Drilling fluids for both open- and cased-hole completions are compared as well as perforating and stimulating techniques. Coal-gas wells have special problems requiring design modifications of production equipment. Artificial lift methods such as the pump jack, gas lift, and progressive cavity pumps, which are in operation on Fruitland coal wells are discussed. Water disposal is a major problem for most Fruitland coal wells. Three disposal methods, evaporative units, evaporative pits, and disposal wells are evaluated. Finally, engineering challenges like zone identification and well spacing concerns are reviewed.

INTRODUCTION

Even though the first San Juan Basin Fruitland Formation coal-bed methane well was developed over thirty-five years ago, the best completion technique for coal-bed methane wells has still not been clearly established. Completion and operation of Fruitland coal-gas wells begins with a decision; to drill and complete to minimize damage or to drill and complete to maximize control, and then repair damage done. This basically translates into choosing between an open-hole completion and a cased-hole completion of the coal zones and is a matter of cost versus return.

An open-hole completion assumes a larger hole, gravel pack or liners, less controlled stimulation, if any, production of coal fines, and perhaps a high incidence of clean-out and other workovers. Cased completions assume a smaller hole, perforations, controlled stimulations to bypass formation damage, minimal coal-fines production, and moderate clean-out and repair workovers. This paper has been prepared to assist operators who must answer questions like: Should the completion be open or cased hole?, What kinds of drilling fluids will be required?, Should the well be completed natural or should it be stimulated? In addition, current production equipment design and artificial lift methods for coal-gas wells are reviewed. Engineering challenges such as water disposal and well spacing are outlined in the final sections. Although not intended to be a highly technical analysis, this paper presents a discussion of past and current operating procedures. We have not concluded that there is a best way to develop Fruitland coal-gas wells, but we have outlined some of the problems encountered by experienced operators.

[1] William Clark
Petroleum Engineer
Blackwood and Nichols Co, Ltd.
P. O. Box 1237
Durango, CO 81302

[2] Tom Hemler
Geologist
Carbon River Energy
1725 8th Ave., N.
Seattle, WA 98109

OPEN-HOLE COMPLETIONS

Defined as "no cement on the formation", open-hole coal-bed methane wells have proven so far to be the best producers. Typically, 7-inch or larger casing is set above Fruitland Formation coal zones and then some type of fluid is used to drill out. Completing "barefoot" (no casing through the open-hole section) is best suited when dealing with a single coal zone or closely grouped coal zones. When the open-hole well bore penetrates several coal zones extending over 100 to 300 feet, then some type of liner is usually employed.

Slotted liners and gravel pack completions have been tried on several Fruitland coal-gas wells over the past eight years and they have had mixed results. Back flushing of the pack or liner is usually necessary because of plugging. Coal fines, which have a wide range of size, easily bridge in the gravel or liner slots and can severely restrict the flow of gas and water.

Once a liner is installed through the coal zones of interest, it is usually perforated with a high density pattern of 4 to 12 shots per foot. For better control, one operator initiated running liners with predrilled 1 inch holes spaced 12 per foot. Temporary plugs were screwed into the liner's holes while it was being run and set in the well bore. Then the plugs, which extend into the inside of the casing, were milled off. This resulted in a liner with many large perforations at a reasonable cost. In some cases, predrilled or perforated liners set through the coal zones can plug with coal fines. To clean out an open-hole perforated liner plugged by coal fines, one operator washed the holes with water at high rates through a packer-like tool on the tubing. Another technique to clean up a plugged open-hole well is to pump 50 percent quality foam and then flow the well back.

CASED-HOLE COMPLETIONS

Cased completions of Fruitland coal zones are commonly 7-inch, K-55, 23-pound casing set through the coal zones or topset, and secondly 5½-inch, K or J-55, 15.5-pound casing set through and perforated. Wells exist with 4-inch or 4½-inch casing, but they are not common. In situations of low water volumes and low fines production, or in areas of dry-gas production, smaller casing design is both operationally and economically effective. Larger sized casing allows for greater volumes of water and fines to be pumped. More importantly, perhaps, using larger casing removes any potential limiting-strength factor in stimulation design allowing the operator to frac down the casing. Overall, the decision on casing size is an economic one with 5½-inch casing set in an 8¾-inch or 7⅞-inch hole being most common in cased-through completions; 7-inch casing set in an 8¾-inch hole is very common in topset-completions with liner or with gravel pack. Gas and fluid is produced up the tubing and gas may or may not be produced up the annulus.

Many different cementing systems have been used on Fruitland coal-gas wells with varying degrees of success and cost effectiveness. To minimize formation damage, light-weight cements were achieved with nitrogen foam or glass bubbles added to the slurries. Separation problems and occasionally stimulation problems were encountered when light-weight cements were used. A typical cement for a Fruitland coal-gas well would be 50/50 Pozmix with 2 percent gel, 10 percent Calseal (gypsum), and usually some gilsonite for lost circulation control with a weight of 13.5 to 14.0 pounds per gallon.

If an operator elects to do a cased-hole completion of a Fruitland coal-gas well, then some care and expense should be part of the design to protect the permeability of the cleat systems that naturally occur in the coal zones. Cement and drilling mud invasion should be a concern. It has not adequately been demonstrated that a chemical interaction between cementing fluids and the coals occur, but it has been suggested that clay in the Fruitland coals does swell in the presence of water.

DRILLING FLUIDS

Determining what fluid to expose the coal zones to is a major decision for an operator. Drilling fluids are selected to control formation pressures and to minimize formation damage during drilling and completion operations. Increasing the complexity of this decision is the stability of the coal zones when exposed to different hydrostatic pressures.

Perhaps similar to unconsolidated sandstone formations, coal zones usually flow coal fines when subjected to a low hydrostatic pressure. Drilling through the coal zones with natural gas, air, or air-mist, typically causes an intense flow of coal fines in the returns. Large chunks of coal, up to several inches in diameter, have been observed. Some operators complete their wells this way knowing their coal zones are significantly washed out or perhaps even a cavern has been formed.

In contrast, some operators maintain a high hydrostatic pressure to limit the flow of coal fines. Typically a gel mud, polymer mud, or heavy salt-fluid is used. It has not been clearly established which technique protects the permeability of the natural cleat system best.

Correctly weighted gel or polymer type muds provide good pressure control and hole cleaning capabilities with a strong possibility of formation damage. Whereas heavy fluids containing zinc chloride or magnesium chloride provide good pressure control and minimize formation damage. However, heavy fluids are quite expensive and somewhat difficult to work with because of their corrosive nature.

A typical mud program for cased-hole wells consists of either a gel- or starch-based mud, or a low solids non-dispersed system to the top of the Fruitland. To drill through the coal zones, the mud weight is usually increased to 10 pounds per gallon or higher, depending upon the geologic location. If the coal zones in an area are known to be primarily gas saturated, then a higher mud weight is required compared to an area where the coal zones are primarily water saturated.

Increasing the mud weight just before drilling into the Fruitland permits maximizing the penetration rate and should provide adequate protection from blow-outs. Great care must be taken to avoid an excessively overbalanced condition. Overweighted mud could permanently seal the natural permeability of the coal zones which would also seriously affect stimulation behavior. Some operators add potassium chloride to their mud systems to lessen the possibility of formation damage from the mud filtrate.

Completing a Fruitland coal-gas well open hole is usually done to reduce formation damage as much as economically feasible. Table 1, a drilling fluid comparison, presents the general relationship of pressure control, potential for formation damage, and expense. These factors need to be balanced to determine the best operating procedures for drilling and completing wells.

Table 1. Drilling fluid comparison.

Fluid	Pressure control	Formation damage	Expense
1) Weighted mud	Good	Possible	Moderate
2) Heavy fluids	Good	Unlikely	High
3) Produced water	Fair-good	Unlikely	Low
4) Natural gas or air	Poor	Unlikely	Moderate
5) Air mist	Poor	Unlikely	Moderate
6) Foam	Fair-good	Unlikely	Moderate-high

Air-mist or stiff foam have been used as alternative drilling fluids to mud, air, or natural gas. It is usually difficult to control formation pressures with air-mist fluid during drilling and completion. While foams can control the pressures, the cost generally exceeds that of muds. Foam has also been used to clean out open-hole wells which were plugged or bridged with coal fines.

Certain operators have been fairly successful using produced coal-formation water to drill and complete open-hole coal-bed methane wells. Continuously pumping produced water or alternating between water and gas down the drill string causes the well to flow in a heading manner, unloading slugs of gas, water, and coal fines. This well bore flushing technique tends to keep the coal fines in upward motion.

Produced water is also used to circulate out sand and coal fines after a cased-hole well has been stimulated. Once the hole is clean, rods and a pump are installed and the well is put on production. Produced water is popular in both cased- and open-hole completions because it will not damage the formation and it is inexpensive.

Overpressured Fruitland coal-gas wells are capable of

unloading a column of produced water. Well control is achieved by maintaining back pressure with a snubbing unit. When actually drilling or cleaning out the well bore, back pressure is not maintained. The snubbing unit enables the operator to hold back pressure on the well during bit trips or while running a liner.

When using fluids like natural gas or air for drilling and completing operations, the formation will frequently flow massive slugs of coal fines. In this situation the field personnel should be advised not to sharply over pull the drill-string weight. Such an action could cause the coal fines to bridge or pack around the drill string. It is better to keep the drill string rotating while only pulling up 5 to 10 points (thousands of pounds), then slack off and repeat until the drill string is free. This method of keeping the bit down in the flow has prevented numerous fishing jobs.

PERFORATING

Given a cased completion and the presence of some fluid used during drilling, two stimulation plans need to be considered to open the formation: 1) A perforating plan that allows maximum stimulation and clean up; and 2) A perforating plan that allows maximum productivity of the formation. Perhaps the most common technique is four jet shots per foot (JSPF) over the entire coal zone with hole sizes of 0.38 to 0.50 inches. Usually there is some separation between the coals and the top of the Pictured Cliffs Sandstone. The intent is: 1) To provide maximum surface area for production which minimizes damage from drilling; 2) Maximize formation control in terms of fines and pressures; and 3) To allow controlled stimulations.

In wells that have a single thick coal zone (20 feet or greater), some operators have gone to a limited-entry approach of 2 JSPF over part of the total zone. This approach is somewhat cheaper and may be as effective as 4 JSPF in achieving effective stimulations. Available data is limited, but does suggest that high break-down pressures can be encountered in both scenarios. Factors such as acidization, time before stimulation, and fracturing strength of the coal could be more significant than the number of perforations.

Casing slotting which is used in the Black Warrior Basin, has not become popular in the San Juan Basin. This is perhaps due to the cost of slotting or to the effectiveness of other perforating techniques. Slotting in terms of stimulation would create a well bore that would behave as an open hole and most likely would limit the overall-treatment effectiveness. Slotting could have value over perforations in naturally-completed wells.

STIMULATION

Given a cased Fruitland coal-bed completion, stimulation is almost a necessity, and occasionally should be considered for open-hole completions. Stimulations typically use cross-linked gelled water or 70 percent nitrogen foam. Non-propped fracs have been tried with both gel and foam. There is no apparent increase in effective permeability associated with either fluid due to the high closure-stresses of the Fruitland coals. Straight water fracs with low sand concentrations of up to two pounds per gallon at suitable rates, while not common, appear to be effective and appear to increase effective permeability.

At this time, a very common stimulation is a cross-linked gelled water frac at rates above 30 barrels per minute (BPM) and greater than 50,000 pounds of 10/20- or 12/20-mesh sands; volumes of 100,000 gallons and 100,000 pounds are quite common in thick zones. The intent is to create a wide vertical fracture of as great a length as possible, perhaps at concentrations of 6 pounds per gallon or higher. Break-down pressures can be quite variable. Foam is a more expensive medium that generally requires higher horsepower for injection of comparable volumes. It is unclear as to whether there is a difference in effectiveness of foam and cross-linked gelled water jobs. The tendency towards larger jobs and even massive hydraulic fracs of 400,000 pounds or more of sand suggests that gelled water-jobs are becoming the most common stimulation. Massive fracs that stay vertical appear to be more effective in creating frac length, but the resulting increased gas-production rates may not be enough to be cost effective.

At larger volumes and higher pressures the tendency to flow back sand can be very pronounced, and at least one operator has gone to resin-coated sands to seal the area around the well bore, preventing flow back and preserving permeability. This is an operational and economic decision dependent upon volumes and pressures anticipated.

Stimulation in open-hole completions in thick coal zones and multiple coal zones tend to be less effective in that they enter only parts of zones or single zones and flow back excessive fines and sands regardless of fluid and sand volumes. Open-hole completions result in much less control of the formation than is necessary for multiple completions and production operations.

PRODUCTION EQUIPMENT

Regardless of the type of completion, Fruitland coal-gas wells are usually troublesome to produce because of coal fines. Completed coal zones can be unstable resulting in production of various-sized coal fines in conjunction with the production of methane-gas and water. Coal fines, ranging in size from dust to chunks, create problems which require some design modifications to typical oil field production equipment.

Design considerations and field operation of wellheads, separators, filters, dehydrators, and evaporative units are discussed in the following sections. This summary is based on personal experience and conversations with equipment manufacturing representatives, production engineers, and field personnel.

Wellheads

Fruitland coal-bed methane wells in the San Juan Basin are typically overpressured. In the northern part of the San Juan Basin numerous drill-stem tests and producing wells indicate a pressure range for the Fruitland from 108 percent to 140 percent of hydrostatic pressure (K. C. Bowman, written communication). These percentages translate into surface shut-in pressures of 1,400 to 1,700 pounds per square inch gauge (psig).

Wellheads with a 2,000 psig rating are acceptable, however, a 3,000 psig rating is recommended to handle the abrasiveness of the coal fines when moderate flow rates are expected. Some operators who have high-volume wells (2 to 4 million CFPD and (or) 1,000 to 2,000 BWPD), use 5,000-psig-rated wellhead equipment.

Flow lines are typically 2-inch, however, for high volume wells, 3-inch flowlines are necessary and schedule 80 pipe is desirable to provide some protection against abrasion. Full-opening ball valves for flow control will be less likely to plug or cut out.

Separators

Gravity separation, production equipment's best friend, has trouble dealing with coal fines. Microscopic-sized coal dust stays entrained in the gas stream and must be filtered out. Large pieces or chunks of coal, ⅛ inch to 1 inch or greater, frequently plug or cause malfunctions in the separator's liquid dump valves and other controllers. Medium-sized coal fines usually are gas saturated and float at the gas-water interface in the separator requiring extra settling time.

Determining the best method to deal with the Fruitland's pressure is a primary element in surface-equipment design. If one elects to use standard-rated production separators with a maximum working pressure of 1,000 psig, then pressure control is necessary. Pneumatic motor valves controlled by a pressure-sensing element provides the necessary flow control for safe operations.

When working with 2-inch flow control valves the frequency of plugging can be high. Motor valve trim, seat, and (or) stem, can erode quickly with high flow rates and (or) coal-fine production. Keeping a separator's safety controls in working order can be a challenge. Use of 3-inch flow control valves results in less plugging, however the expense can be significant. The initial cost of a 3-inch motor valve is five to eight times higher than the 2-inch model and replacement parts follow suit.

Experience obtained from operating pressure-control equipment on coal-gas-well separators has caused some operators to increase the maximum working-pressure rating on their units. It is approximately 25 percent more expensive for a separator with a 1,440-psig rating than the 1,000-psig standard model. Increasing the maximum working pressure to 1,700 psig causes a price hike of 35 percent over the standard model. However, higher initial costs can be offset through lower maintenance expense and less downtime for repairs. These factors combined with the increased safety margin makes the higher-pressure-rated approach worth considering.

Most Fruitland coal-gas wells produce gas, water, and coal fines requiring two-phase separators. Because the gas-to-liquid ratios on these wells are usually low, vertical separators work best. Also, flowing Fruitland coal-gas wells tend to produce in a heading or slugging manner and a vertical separator should have large liquid capacity to handle this.

In the event a Fruitland well makes some condensate or oil, a two stage separator is recommended. A high-pressure vessel separates the gas from the liquids and then a low-pressure vessel separates the oil from the water. Condensate from Fruitland completions and commingled Fruitland-Pictured Cliffs completions is usually high in paraffin content which requires special handling, such as heaters in the stock tanks.

There are many factors to consider when sizing a separator for a Fruitland coal well: Obviously the production rate of gas and water are important, however other factors such as slug capacity, retention time, and operating pressures play a significant role in design. Floating coal fines in the water phase requires additional retention time for separation. Increasing the retention time from 1 minute (typical for oil separators with non-foaming conditions) to 3 minutes, decreases the liquid capacity by 67 percent.

Overall, it is recommended that a coal separator be sized with conventional design parameters and then increased to the next commonly available pipe size. In other words, if a design analysis requires a 20-inch-diameter by 10-foot-long vessel, then the size should be increased to a 24-inch by 10-foot size. If a 24-inch diameter is calculated then use 30-inch pipe. Optionally, increasing the length of the vessel also provides an economical way to achieve greater capacity, particularly when designing for high maximum working pressures. Once the initial gas head from a Fruitland well is produced, usually fairly quickly, warm formation water, 70 to 80 degrees F at the surface, prevents any freezing problems. A water bath around half of the separator vessel provides sufficient heat. Multiple-pass flow coils in a heated water bath are usually not necessary.

Separator control equipment, when exposed to coal fines and slug flow, can malfunction. A backup system for the liquid level controller and dump valve will decrease operating problems. A secondary-level controller mounted above the primary controller with the output signal sent to both the primary and secondary dump motor valves is good insurance to prevent liquid carry over.

Motor valves which dump the separator's liquid should have large trim to prevent plugging. Also snap action is preferred over throttling because of the coal fines. Some operators are using a motor valve designed for paper mill slurries that has a rotating ball action. A preferred design would have a full-opening isolation valve upstream of the liquid dump valve for convenient maintenance.

Conventional mist-extractors in coal-gas-well separators can plug with coal fines. This problem is decreased by installing a vein-type mist extractor followed by a conventional wire-mesh type. One design permits the mist extractor to be back flushed through a mounted nozzle system. If plugging is a severe problem then the increased expense of a removable head is justified.

Filtration Units

Small coal fines in the gas stream and separated-water system require filtration. Produced gas from the separator should go through a cartridge-type filter to prevent downstream problems in the dehydrator, compressor, and (or) meter. Filtration of produced water is usually required to prevent damage to a disposal well.

Gas filters can be free standing units with sufficient working pressure rating. One separator design incorporates the filter in the separator's water bath to prevent freezing and to allow for convenient maintenance. Replacement frequency of the filter cartridges depends on a coal-gas well's average production of fines and the occurrence of liquid carrying over from the separator.

Competency of filters can be stated in nominal or absolute terms. An absolute filter is rated to stop all particles larger than the specified size, that is, a 10 micron absolute filter stops all particles larger than 10 microns in size. A nominal competence is an arbitrary rating determined by the manufacturer based on the weight percent of removed particles. A test of a 1-micron nominally rated filter showed that no 1-micron-sized particles were stopped, while particles up to 50 microns were passed.

San Juan Basin operators working with Fruitland coal-gas wells are using 20- or 25-micron nominal filters for gas. Produced water is cleaned up with 10-micron absolute or 2-micron nominal filters before it is injected into a disposal well. Filtration procedures increase the cost of operating coal-gas wells but this should be recouped through less downtime.

Dehydrators

Dehydration of coal-bed methane is the same as conventional glycol-based systems developed for gas wells. Design parameters that might be different are maximum operating pressure, inlet gas temperature, and carbon dioxide content. Glycol absorption of water vapor works best at low gas temperatures (65 to 70 degrees) and higher operating pressures.

In the preceding discussion of separator design, pressure-control alternatives were outlined. If the production-facility design specifies that the separator is to handle pressure control, then a standard 1,000-psig, maximum-working-pressure dehydrator unit would be appropriate. In a design case where the separator has no pressure control, the dehydrator design should incorporate pressure control to protect the gathering facilities in a worst-case scenario.

Use of a compressor on a coal-gas well for gas-lift operations or for lowering the wellhead pressure can significantly increase the flowing temperature of the gas. Glycol-absorption efficiency decreases as the temperature increases. An overall analysis of production mechanics will insure that the maximum water content allowed by the gas-purchase contract will not be exceeded.

Another topic in most gas-purchase contracts is the maximum allowed level of carbon dioxide in the produced gas. Coal-gas wells produce varying amounts of carbon dioxide and treatment facilities are sometimes necessary. If the gas analysis shows more than 10 percent carbon dioxide, then the use of some stainless steel elements in the dehydrator should be considered.

Evaporative Units

There are several evaporative units in service on San Juan Basin Fruitland coal-gas wells. Basically, evaporative units boil produced water which releases steam to the atmosphere and concentrates the salts in the residual fluid. Handling produced water through an evaporative unit is expensive and has several operational problems. However, in some cases it may be the best alternative for an operator.

Cost of an evaporative unit depends on the size; approximate costs are: 20 barrels of water per day (BWPD) unit—$5,000; 50 BWPD unit—$8,000; 100 BWPD unit—$12,000. Fuel usage and maintenance costs depend on several factors and the following discussion is general in nature. Metering of consumed fuel gas used by evaporative units averages 0.600 MCF to vaporize one barrel of water; at a price of $2.00 per MCF, the fuel cost is $1.20 per barrel of water evaporated.

Operating an evaporative or steamer unit can be very expensive and time consuming. The buildup of mineral deposits on the fire tube and walls of the containing vessel significantly decreases efficiency and can cause control equipment to malfunction. Solutions to these problems include: 1) A time controlled flushing cycle to remove all or most of the residual fluid; 2) An aeration system to agitate the bottom of the vessel before draining off the residual fluid; 3) A holding tank in which the water is preheated with steam before entry into the evaporating vessel; 4) Dual float control valves to shut down the burner; and 5) A regular washing program to clean the vessel's interior.

Several of the approximately twenty evaporators built and used in the San Juan Basin have malfunctioned, requiring major repair work. Collapsed fire tubes, melted floats, and other disabling events have occurred. Regular inspection and a good maintenance program is clearly necessary.

Total operational cost of evaporative units is estimated to be $2.00 to $3.00 per barrel of water evaporated when capital, maintenance, and fuel expense are considered. For wells with remote locations and water production less than 100 BWPD, preferably less than 20 BWPD, an evaporative unit is probably the most economic choice.

Disposition of the residual fluid is a potential problem because of its high total dissolved solids (TDS). This fluid ranges from 100,000 to 300,000 parts per million TDS at discharge temperature depending on the flushing program. The volume of produced water is reduced approximately 90 percent by the evaporative unit. Trucking the reduced volume of residual fluid to a disposal site is more efficient, however the solid-mineral precipitates may require special handling.

ARTIFICIAL LIFT

Most Fruitland coal-bed methane wells are not capable of continuous fluid flow, resulting in the need for some type of artificial lift. Methods currently employed include sucker-rod pumps, cavity-type pumps, and gas lift systems. These artificial-lift methods are discussed primarily from an operational point of view. Limited data prevents an analysis of the average lifting cost of each method.

Pump Jack

The most common lift equipment is a sucker-rod pump with pumping units of 160, 300, and 640 size (peak torque rating) and 64- to 200-inch stroke lengths. Water productivity knowledge or estimates are necessary to configure a pumping system capable of pulling down the fluid level to the Fruitland. A ring type plunger pump, with a tolerance for solids, placed in the tubing at or below the producing formation has proven cost effective by minimizing plunger wear and maximizing sustained pumping. A sloppy pump with undersized pump rings results in a tolerance for solids without significantly decreasing efficiency. A fairly common assembly would be a stainless steel plunger with undersized plastic rings.

Some accommodation for gas locking through a gas anchor, charger valve, or equalizer valve may be necessary along with proper pump spacing. The size of the plunger depends upon the water volumes. Common sizes are 1.75- or 2.0-inch plungers. Ring-type plunger pumps with properly sized pumping units and sufficient stroke are capable of pumping 2,500 BWPD. More common are units capable of pumping 600 BWPD or less.

Fruitland coal wells equipped for sucker-rod pumping have a somewhat established workover pattern. Wells in the early phase of production need workovers for problems related to abrasion caused by sand and coal fines working on plungers, barrels, and valves. It is fairly common for pumping coal-gas wells to have two or more workovers in the first year.

In the later phase of more stable production (when the gas/water ratio has increased) workovers are more typical of gas wells. These workovers usually involve rod and tubing wear and more pump repair due to abrasion. Long-term problems such as scaling, corrosion, and rod and pump wear, mark the final phase of production; usually when the coal-gas wells are significantly dewatered. The use of sloppy pumps with cased completions has proven effective in minimizing workover frequency as have gravel packs and certain types of liners in open-hole completions. However, high volume producers might expect two workovers in the first year and one workover every two years after that for the first five years of well life.

Cavity Type Pumps

As an alternative to sucker-rod pumping, the progressive-cavity pump is currently being field tested on several Fruitland coal-gas wells. These pumps are highly tolerant of solids, and the turning of the rods in the tubing usually does not create excessive tubing wear. However, in the presence of high volumes of gassy fluid and deviated holes, tubing and pump wear can be excessive.

Under normal operating conditions, the ring-type plunger pump and the progressive-cavity pump are very comparable. Initial cost of a progressive pump system is significantly cheaper, approximately 70 percent of the cost of a beam-pumping unit for a well 3,000 feet deep. The rotor turning within the stator in a progressive-cavity pump, especially where high gas volumes are present, can create too much friction causing stator burnout and (or) excessive pump wear. Careful placement of the pump below the perforations or the deepest coal zone may reduce the problem.

Changing the rotation frequency of the rod string allows the cavity pump to handle a wide range of fluid volumes. Power failure, either electric or gas engine, can cause the coal fines in the tubing to settle and perhaps lock up the rotor. Some operators use a standing valve on the bottom of the tubing to prevent the liquid from falling back.

As more experience is gained with progressive-cavity pumps, they could become as popular as pump-jack systems on water-producing Fruitland coal-gas wells. If electricity is readily available the progressive pump system should be evaluated.

Gas Lift

Though not as popular as pump jacks, gas-lift systems are currently in operation on several Fruitland coal-gas wells in the San Juan Basin. Most of these installations are single-well systems where produced gas is recirculated to keep the well flowing. Conventional gas-lift systems usually involve a packer and gas-lift valves installed on the tubing.

"Poor boy" gas lift eliminates the gas-lift valves and packer. Compressed gas flows into the casing annulus, and then enters the bottom of the open ended tubing, to help the formation gas lift the water to the surface. This approach works fairly well on strong wells which are capable of unloading themselves. For the marginal wells, unloading by swabbing may be necessary to establish continuous flow. Operational problems with the compressor or other production equipment may cause the well flow to stop. This possibly is the biggest disadvantage of the poor boy approach.

A fluid-filled well can be unloaded with a packer and gas lift valves and a suitable compressor. However, a temporary gas supply may be needed. Once steady tubing flow is achieved then either the casing pressure or the flowing tubing pressure is monitored. When the monitored pressure is above the desired level then a motor valve allows the excess or produced gas to flow into the sales line. Maintaining the casing pressure and flow up the tubing is the first duty of the recirculated compressed gas with any excess being sold in a controlled manner.

When dealing with high-volume wells or excessive coal-fine production, it is at times desirable to circulate gas down the tubing. The larger tubing-casing annulus would be less likely to plug. Another gas-lift technique is the casing-lift system which is based on annular flow. Gas lift valves flow compressed gas from the tubing into the annulus. This would be particularly useful in completions where the coal zones are 100 to 200 feet apart, and a packer above the uppermost perforation is less likely to keep the lower zones producing.

Recirculating gas-lift systems are initially cheaper than a comparable beam-pumping system. There are no pumps to wear out or rods to pull, therefore workover expenses should be significantly less. If gas-lift valves are installed in a well, it is recommended that the bottom valve be wire-line retrievable for maintenance and performance optimization. Most operators need more experience with gas-lift systems to determine the cost effectiveness for use on coal-gas wells.

WATER DISPOSAL

Disposal of produced water is the largest single operational problem and cost encountered in Fruitland coal-bed methane development. Fruitland coal-gas wells in the San Juan Basin produce water in a wide range of amounts; some wells make less than 10 BWPD and a few wells make more than 2,000 BWPD. Coal-gas wells usually produce higher volumes of water initially with water volumes decreasing over time. This diminishing water production results in changing water disposal requirements over well lives. Produced water needs to be handled and disposed of in accordance with relevant Federal, State, and Tribal regulations.

There are many water-disposal methods available and the following is a discussion of evaporative pits and disposal wells. Evaporative units (see Production Equipment section) are in some cases a reasonable disposal technique. Produced water from Fruitland coal-gas wells ranges in composition from fresh to 20,000 parts-per-million TDS. Coal water contains predominantly sodium bicarbonate which disassociates into sodium and bicarbonate. These compounds form scale and can create land use problems.

Evaporative Pits

Initial coal-water disposal efforts in the San Juan Basin centered on evaporative pits sized to handle produced volumes of several wells based on the evaporation rate with seasonal storage adjustments. Historically, a 1- to 2-acre pit is capable of handling several hundred to approximately 1,000 barrels of water per day with year round evaporation.

The cost of an evaporative pit including purchase of surface lands as well as construction and liner costs totals approximately $50,000. Significant operating problems of vandalism, hydrogen sulfide gas pollution, and leakage have occurred. Fencing, bacteriacide treatment, and monitoring can lessen these daily operating problems. When an evaporation pit is abandoned, the concentrated brine and scale require special handling.

Disposal Wells

Disposal wells are an attractive alternative to surface pits in that they minimize vandalism, hydrogen sulfide problems, and do not require large surface-area ownership. Additionally, depending upon the receiving formation, disposal-well capabilities usually far exceed produced water volumes and can continuously operate without seasonal fluctuations. However, disposal wells can cost as much or more than a producing well, depending on the depth and need for stimulation.

Water-disposal wells have been completed in many formations from the Tertiary Ojo Alamo Sandstone down to the Jurassic Entrada Sandstone. Existing wells in the San Juan Basin are injecting water into the Ojo Alamo Sandstone, Pictured Cliffs Sandstone, Mesaverde Group, Morrison Formation and Entrada Sandstone. Capacities range from 500 to 600 BWPD at 600 psig into the Ojo Alamo, to over 10,000 BWPD at 2,500 psig with a cumulative volume of approximately 1,000,000 barrels of water into the Entrada. The tendency of these wells to pressure up over time may be a function of the plugging from fines resulting in decreasing capacity.

Disposal wells must meet mechanical and quality criteria administered either by the Environmental Protection Agency (EPA) or its administering agency within a state. Quality criteria may make certain zones unsuitable for injection because of water incompatibility or the disposal zone's inability to take large volumes of water. Administratively and mechanically the disposal well route may be difficult to pursue, however it may be the best solution for high-volume disposal.

Disposal into shallow zones such as the Ojo Alamo Sandstone or Farmington Sandstone Member of the Kirtland Shale may be restricted by quality-criteria problems or limited capacities. Although usually capable of accepting high water volumes, deep disposal wells into the Entrada Sandstone are wildcat in nature and are very expensive to drill and complete. Less-deep zones of the Pictured Cliffs Sandstone and Mesaverde Group will meet quality and capacity requirements for most operators and may be the most cost effective disposal zones provided these zones are not capable of commercial hydrocarbon production. Stimulation of these shallower zones will usually improve daily capacity at minimal cost.

ENGINEERING CHALLENGES

Determining the best procedures to develop Fruitland coal-bed methane wells is a challenge we all face. Clearly, the Fruitland coals contain a tremendous gas resource and through diligent work the people involved will be able to establish optimum reserves. Both the operators and regulatory agencies need to make difficult decisions with limited information to establish proper Fruitland coal-gas development. Two of the challenges are formation identification and well spacing.

Formation Identification

Determining which pool a well's hydrocarbons are coming from can be a problem in some areas of the San Juan Basin because of the intertonguing of the Fruitland Formation and the Pictured Cliffs Sandstone. In areas where the Pictured Cliffs is not capable of commercial production, or when wells are clearly completed only in the Fruitland coals, then zone identification is not a problem. However, some wells completed only in the Pictured Cliffs sandstones, can produce gas having a strong coal-bed methane signature. This could be the result of the stimulation treatment not staying in the intended zone. In other wells there appears to be production of a mixture of coal-bed and sandstone-reservoir gas and water.

Effective reservoir management is difficult in these mixed cases. To be able to keep meaningful production records, correct formation identification is necessary. Pools such as the Los Pinos Fruitland-Pictured Cliffs,South, in northern New Mexico, which are handled as down-hole commingled pools, need to be reevaluated and perhaps redefined.

In some areas of the San Juan Basin, one possible scenario would be for an operator to drill to a specified depth; complete and test the well; and finally present to the regulating agency, information which identifies the well as capable of producing from the Fruitland or the Pictured Cliffs or both. Well spacing and zone identification are integral matters which must be addressed harmoniously to insure the best development.

Well Spacing

Because of the site specific nature of Fruitland permeability, spacing should be flexible enough to allow down-spacing in the less permeable areas. Spacing-units of 320 acres would appear to be the largest that would not penalize most operators and still allow prudent development in non-unit situations. Unitized operations could include owner-agreements regarding well spacing perhaps as large as 640 acres without penalizing any mineral owner. This large spacing is not reasonable for prudent development of non-unitized lands.

The directional-permeability phenomenon in coal also suggests larger drill-site-spacing blocks; 320 acre-blocks are desirable, with downsizing allowed. The ultimate number of wells will depend upon economic recovery of the maximum amount of gas at minimum cost. The engineering challenge will be to use geologic and engineering data to locate wells in irregular patterns dictated by permeability zones, which vary. Perhaps these zones are influenced by the cleat system and they may change from well to well in a dynamic situation. These factors will require an understanding of the direction and extent of permeability in a field area concurrent with development.

CONCLUSIONS

Completion and operation of Fruitland coal-gas wells begins with a single decision; to drill and complete the well to minimize damage or to drill and complete the well to maximize control and then repair the damage done. An open-hole completion assumes a larger hole, gravel pack or liners, less controlled stimulation, if at all, production of coal fines, and a high incidence of clean-out and other workovers. Pumps and production equipment must be capable of handling the coal fines. Cased completions assume a smaller hole, perforations, controlled stimulations to bypass formation damage, minimal fines production, and limited clean-out and repair workovers.

The use of produced water as a drilling fluid is popular in both cased- and open-hole completions because of low cost and no damage to the formation. Regardless of completion type, the engineering challenge is to find the best production equipment and operating procedures for each well. Large separators and filtration units are necessary to handle the produced coal fines. Usually there are large volumes of produced water which should be properly disposed of at the cheapest price. Water disposal methods include evaporative units, evaporative pits, and disposal wells. With future Fruitland coal-bed methane development, the need for water disposal will increase and this issue will receive more attention and regulation by State and Federal agencies.

Horizontal Drainhole Drilling Techniques Used in Rocky Mountain Coal Seams

Terry L. Logan[1]

ABSTRACT

Methane gas contained in coal represents a potential new gas supply. Transformation of this large gas-in-place resource to reserves will depend on developments of modified and (or) new completion methods to achieve long-term economic production. Coal reservoirs typically have very low matrix permeability with gas being desorbed from the matrix to the well bore through a system of natural fractures or cleats. Due to low permeability, coal reservoirs usually require hydraulic fracture stimulation to achieve economic production. Investigations have shown that hydraulic fractures usually propagate parallel to the maximum stress or face-cleat direction and therefore may not adequately access an anisotropic reservoir. A more effective stimulation technique may be a horizontal borehole placed perpendicular to the maximum permeability direction. This paper discusses the short-radius and medium-radius directional drilling techniques that have been used to successfully place horizontal drainholes into coal seams.

INTRODUCTION

Coal seams in the contiguous United States have been estimated to contain as much as 400 trillion cubic feet (TCF; 11 Tm^3) of pipeline quality natural gas (Byrer and others, 1984). A large portion of this resource lies within the borders of five western states: Colorado, New Mexico, Utah, Wyoming, and Washington. Commercialization of this large-potential new gas supply will depend on the developments of modified and (or) new completion methods to achieve long-term economic production.

Gas is held in the coal reservoir in three possible ways: 1) As adsorbed methane molecules on the surface of micropores, 2) As free gas within fractures or pores, and 3) As dissolved gas in formation water (Decker, 1986). Of these ways, the first accounts for the majority of the retained gas volume. The free gas contained in the natural fractures or cleats within the coal constitutes a smaller fraction of the resource, however, the cleat system provides the primary flow channels or permeability of the coal. The cleat system usually contains two sets of closely spaced, vertical, natural fractures. The primary set of natural fractures are face cleats, a less defined set of perpendicular fractures are butt cleats. Gas flow depends on the natural fracture cleat system and diffusion within the coal matrix.

Coal reservoirs typically have low matrix permeability and therefore usually require stimulation to achieve economic production. Hydraulic fracturing is the most common stimulation technique used. Investigations of the orientation of hydraulic fractures in anisotropic coal reservoirs have shown that the fracture usually propagates parallel to the face cleat (maximum stress and permeability direction) and, therefore, may not provide the most effective access to the reservoir (Logan and others, 1986). A more effective stimulation technique may be a horizontal borehole placed perpendicular to the face cleat direction, thereby providing maximum access to primary flow channels. Reservoir tests and modeling studies of horizontal boreholes drilled perpendicular to the face cleat direction have shown that they provide the highest gas production potential (Ertekin and others, 1986).

This paper will discuss the two horizontal drilling techniques; short-radius and medium-radius, that have been used to successfully place horizontal wells into coal seams in the Rocky Mountain area. Short-radius horizontal drainhole drilling technology was first successfully demonstrated in 1986 at the Gas Research Institute (GRI)/Resource Enterprises, Inc. (REI) Deep Coal Seam Project site located in the Piceance Basin, Colorado. Medium radius horizontal drainhole drilling technology was first successfully demonstrated in 1987 at the Department of Energy (DOE)/GRI Rocky Mountain No. 1 project site located in the Hanna Basin, Wyoming. (A conversion table for the SI-to metric units is located on the last page of this text.)

HORIZONTAL DRAINHOLE DRILLING

Background

Horizontal drilling into coal seams is not a new idea. In-seam horizontal drilling which involves drilling into an outcropping seam at the coal face in an underground mine has been used for a number of years to degasify coal seams immediately prior to mining. The only successful pilot project currently being conducted in the Rocky Mountain area is at the Soldier Canyon Mine, Carbon County Utah. A total of 40 horizontal boreholes representing a total of over 65,000 feet (19,812 m) have been drilled into a 16-foot thick (4.9 m) coal seam at a depth of 2,000 feet (609 m) (Schwoebel 1986). Pipeline quality gas is collected using an underground pipeline net-

[1] Terry L. Logan
Manager, Operations
Resource Enterprises, Inc.
2516 Foresight Circle, No. 7
Grand Junction, CO 81505

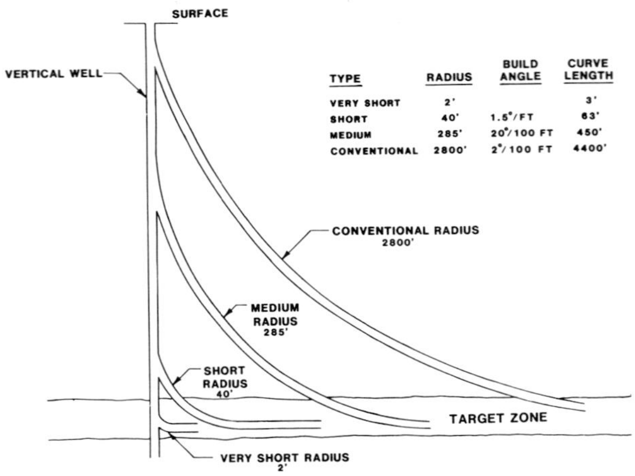

Figure 1. Horizontal drainhole drilling techniques.

work. Commercial quantities of gas have been sold continuously since 1983.

Horizontal drainhole drilling from a vertical wellbore is also not a new idea. In fact, the first patent for equipment to place a horizontal drainhole from a vertical well was granted in 1891. The first truly lateral holes were drilled in 1929 at Texon, Texas using equipment developed and patented by John Zubin. Since that time horizontal drilling technology has been significantly improved. Currently, there are four different techniques available for drilling horizontal drainholes, as shown on figure 1, and outlined as follows:

Type	Radius	Technique
Very short-radius	1 to 2 feet (.3 −.61 m)	High pressure focused water jet capable of drilling from 100 to 200 feet in length.
Short-radius	35 to 45 feet (10.7 −13.7 m)	Curve drilling guide with flexible articulated drill pipe, entire drill string is rotated from the surface.
Medium-radius	300 to 500 feet (91.4 −152.4 m)	Special designed steerable mud motor used with compressive drill pipe.
Conventional	1,800-2,800 feet (549 −853 m)	Conventional directional drilling equipment used, very long curve length of 2,800 feet to 4,400 feet needed to be drilled before horizontal drilling.

The primary difference between conventional or extended-reach directional drilling and the shorter-radius techniques is build angle. The build angle for conventional directional drilling is approximately 0.03 degrees per foot in comparison to the short-radius build angle of 2 degrees per foot. This means that less drilling and therefore, less cost is required using the shorter-radius techniques when compared to conventional directional drilling. One of the principal benefits of a horizontal drainhole is that it can be oriented to take direct advantage of the natural fracturing of the coal. Production tests performed on in-seam horizontal boreholes have shown that boreholes oriented perpendicular to the natural fracturing system or face cleat direction, produce more gas than holes oriented in other directions.

Short-Radius Horizontal Drainhole Drilling

Short-radius horizontal drainhole drilling in deeply buried coal was first successfully demonstrated in 1986 at the GRI/REI Deep Coal Seam Project site located in the Piceance Basin Colorado (Logan and others, 1987). The objective coal bed at the project site was the Cameo "D" seam which is part of the Cameo coal group of the Upper Cretaceous Mesaverde unit. The coal seam is at an average depth of approximately 5,500 feet (1,676 m) and is 16 to 20 feet (4.9 to 6 m) thick. The coal is of medium-volatile bituminous rank with a vitrinite reflectance of 1.3 percent R_0. Gas content of the coal is 257 cubic feet per ton (0.88 cc/Kg) which represents an in-place resource of 5.9 billion cubic feet (BCF) (167 million cubic meters; Mm^3) of methane gas per section. The reservoir is under-pressured at 1,900 psi with a total coal porosity of 4.1 percent. Permeability of the coal varies between 0.0003 to 0.008 millidarcy (md) with a permeability anisotropy determined from core tests of 12:1 (face cleat versus butt cleat).

The vertical portion of the 3 Deep Seam (3DS) well was designed and drilled specifically for optimal placement of a drainhole. The well was located between two existing wells in order to have: 1) Good stratigraphic control, 2) Presence of an overlying competent sandstone for stability of the short-radius curve, and 3) Presence of competent strata for placing the whipstock and packer in the open-hole interval above the coal. The vertical portion of the well was drilled to a depth of 5,645 feet (1,721 m), 105 feet (32 m) above the target coal seam. The hole was cased with 7⅝-inch (19.4 cm) casing and cemented in place. A 6¾-inch (17.2 cm) hole was then drilled and cored. Two 3-inch (7.6 cm) oriented cores were cut to verify the east-west orientation of the face cleat direction and to collect core for mechanical property tests to project future stability of the horizontal hole.

Triaxial stress tests on the coal core along with in-situ-stress-test data were used to determine mud weight required to maintain hole stability and to prevent formation fracturing. Results indicated that the horizontal hole would remain stable only within a very narrow bottom-hole pressure window of 150 psi (1,034 KPa). The mud-weight range required to maintain horizontal borehole stability and prevent fracturing was 10.0 ppg (1.19 g/m³) to 10.2 ppg (1.21 g/m³). Previous work has indicated a potential correlation between borehole stability and coal friability (Hardgrove, 1932). The coal seam grindability index from 3DS averaged 97 indicating that the coal is weak and friable, and horizontal borehole stability could be a problem. Therefore, the curve radius for the drainhole was placed primarily in the competent overlying sandstone.

The short-radius technique was selected for the 3 Deep Seam drainhole because at the time there was much more experience with the technique than with the other drainhole drilling techniques. In addition, the coal reservoir was anticipated to be partially water saturated which would require dewatering to allow the gas to desorb from the coal. The short, 35-to 45-foot (10.7 −13.7 m) radius would allow the hydrostatic pressure of water above the coal seam to be kept lower than with a medium-radius curve. The disadvantage of the short-radius drainhole drilling technique is that real-time surveying is not available, and therefore directional control can become a problem. The wellbore location is determined by removing the entire drill string to obtain a survey, this is time consuming and expensive.

The basic procedure for placing a short-radius drainhole consists of six steps as follows:

1. Set and orient whipstock.
2. Drill a 6½-inch (16.5 cm) OD, 3 foot (.9 m) long pilot hole.
3. Drill the short-radius curve with the curve drilling guide.
4. Survey the curve.
5. Drill a 6¼-inch (15.9 cm) OD horizontal drainhole.
6. Survey the drainhole.

The first step involves setting and orienting a deflection device or whipstock on a packer. It is critical that the whipstock is set at the correct depth and orientation. A surface readout directional orientation device was used to insure that the whipstock was set correctly. Previous drainhole drilling experience indicated that the curve typically drifts to the right approximately 10 to 15 degrees, due to the clockwise rotation of the bit. Therefore, the whipstock is offset to the left of target to allow for the curve drift while drilling.

Once the whipstock was set, the curve was initiated by slowly drilling a 3-foot (.9 m) long pilot hole with a conventional drill bit on flexible, articulated drill pipe rotated from the surface using a power swivel. This insured that the hole had been initiated properly off the whipstock. The remainder of the curve was drilled using the curved drilling guide which consists of a conventional 6½-inch (16.5 cm) drill bit and a pre-stressed or bent guide into which a flexible driveshaft is placed (fig. 2). The outer portion of the curved drilling guide does not rotate and therefore forces the bit to drill the predesigned 35-45-foot (10.7-13.7 m) radius curve. The drill bit was rotated from the surface using a power swivel via the inner driveshaft in the curved drilling guide and the flexible drill string above. Drilling fluid was circulated through 1-inch (2.54 cm) plastic tubing, sealed, and placed inside the curved drilling guide and flexible drill string. Once the curve had been completed, the curve was surveyed. This was accomplished by removing the drilling assembly from the well and running a multi-shot survey camera into the well on the end of flexible non-magnetic drill pipe. The current survey technique was time consuming, costly, and accelerated borehole deterioration by frequent tripping in and out of the wellbore.

The results of the curve survey indicated the actual radius drilled was 32 feet (9.7 m) which was 7 feet (2.1 m) less than the 39 feet (11.9 m) that the curve drilling guide had been designed to drill. This put the end of the curve approximately 4 feet (1.2 m) above the top of the coal seam. The shorter-than-design radius curve may have been caused by the very

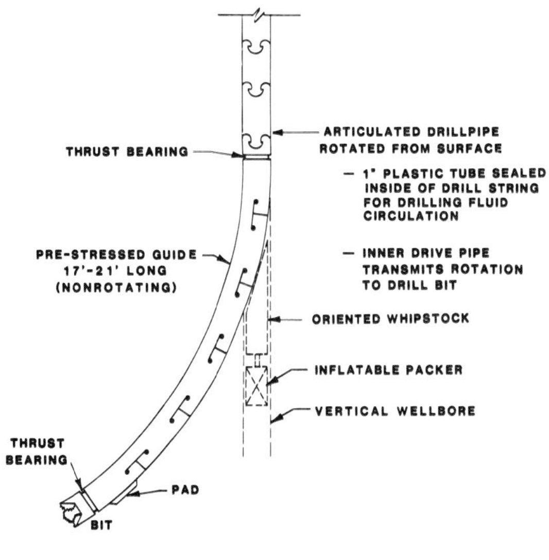

Figure 2. Short-radius curve drilling guide.

competent, hard sandstone and shale drilled with the curved drilling guide (CDG) above the coal. An in-gauge hole was actually drilled with the CDG when the tool is designed to build the proper angle in a hole that is slightly oversized.

Several unsuccessful efforts were made to drop into the coal seam using different horizontal drilling assemblies as shown on figure 3. This drainhole was abandoned and a new drainhole initiated oriented 180° to the first drainhole.

After drilling approximately two thirds of the new curve, a rigid drilling mandrel (RDM) was used to extend the current angle. Utilization of the RDM before reaching the end of curve was conceived and first used in this drainhole to assure coal seam penetration. After 10 feet (3.04 m) of hole was drilled with this technique, the curved drilling guide was again used to finish the curve. Upon entering the coal, gas shows increased to levels almost twice those previously encountered.

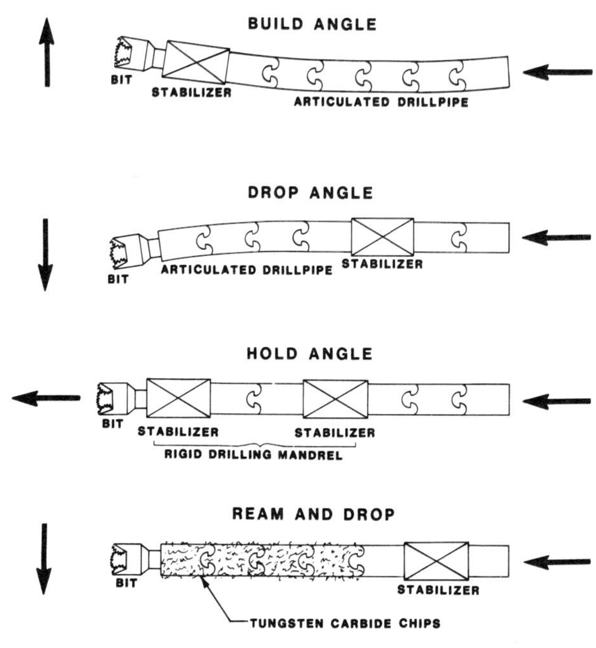

Figure 3. Horizontal drilling assemblies.

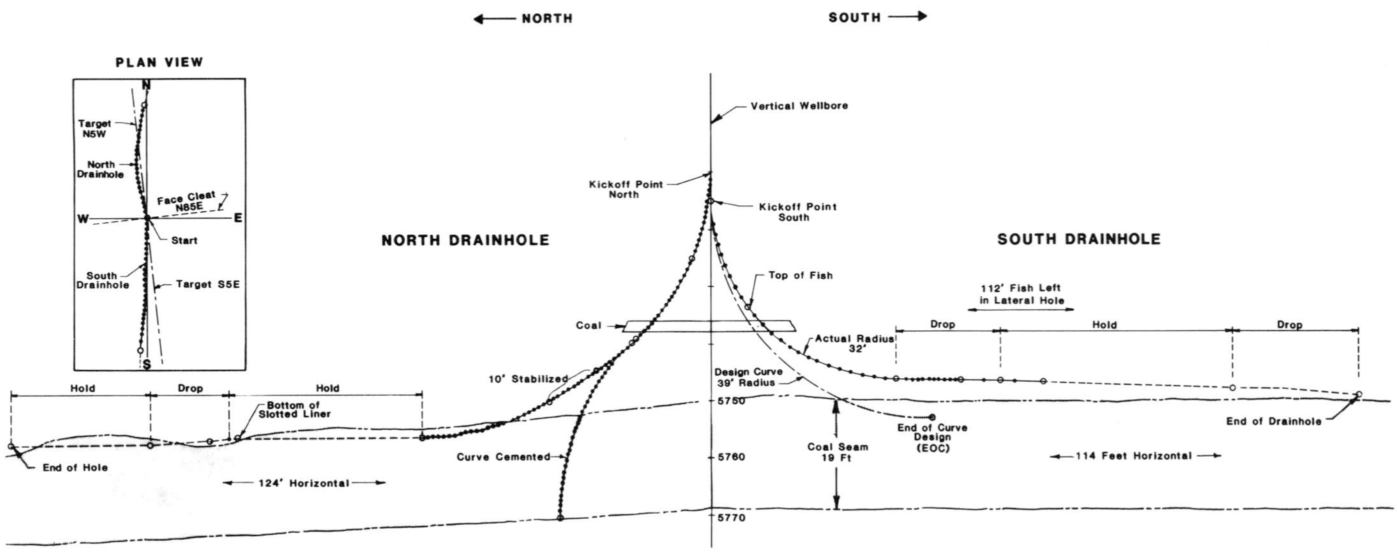

Figure 4. Three deep seam short-radius drainhole configuration.

A total horizontal displacement of 124 feet (37.8 m) was achieved in which a total of 73 feet (22.2 m) of coal was penetrated. To assure that the drainhole remained open for future well testing and production, a 4½ inch (11.4 cm) slotted liner was installed in the 6¼ inch (15.9 cm) horizontal drainhole.

Hole stability had been anticipated to be the major potential problem, however, this was not a problem due to detailed pre-drainhole planning and design. The major problems that were encountered and the solutions used are itemized as follows:

Problem	Solution
Hitting the target while drilling the short-radius curve.	Drill a tangent to the curve using a stabilized assembly.
Drainhole inclination control.	Use aggressive dropping and reaming horizontal drilling assemblies.
Excessive torque due to swelling shales.	Use special mud to stabilize shales.

The primary limitation of the short-radius drilling technology is the lack of real-time downhole survey capabilities. Generally, the problems encountered during this operation could have been identified earlier with real-time survey capabilities. The current multi-shot survey technique requires removal of the entire drill string and can be time-consuming, expensive, and cause borehole deterioration. It is hoped that the solution to this surveying problem is under development.

The final drainhole wellbore configuration is shown on figure 4. A total of 314 feet (95.7 m) was drilled in 207 hours, representing an average penetration rate of 1.5 feet per hour (45.7 cm/hr). The majority of the footage (241 feet; 73.5 m) was either in the curve or horizontally in the shale above the coal where penetration rates were kept low to help maintain directional control. The average penetration rate in the 73 feet (22.3 m) of drilling while in the coal was 8 feet per hour (2.4 m/hr).

TABLE 1. Short-radius horizontal well hours by operation.

3 Deep Seam well

Activity	Total hours	Percent of total time
Tripping	241	35
Drilling	207	30
Make-up tools	73	11
Ream	28	4
Set liner	27	4
Waiting	22	3
Cement operations	21	3
Circulate and condition	20	3
Survey reading	16	2
Set whipstocks	14	2
Rig repair	10	1
Miscellaneous	4	1
Total well	683 (28.4 days)	100

Survey

—Total time associated with surveying was 114 hours
—17 percent of total time associated with surveying
—8 surveys were run, average time per survey 14 hours

Drilling

—Total feet drilled: 314 feet (73 feet of coal)
—Average penetration rate: 1.5 feet per hour

The total drainhole operation required 28.4 days to complete, which included both north and south drainholes, subsequent abandonment of the first drainhole, cementing the reversed curve, and re-drilling the second drainhole. This is not representative of the time required to drill a problem-free drainhole. Analysis of the time required for each operation, as shown on table 1, indicates 17 percent (114 hours) of the total 28.4-day operation was associated with surveying. The aver-

TABLE 2. Short-radius cost by service (in dollars).

Drilling rig	$133,880
Directional equipment	155,430
Drilling fluid and water	86,657
Well site geology	10,079
Cement	7,008
Liner and hanger	40,221
Equipment rental and power swivels	39,575
Drill bits	12,090
Wireline and orientation	7,651
Supervision and facilities	5,932
Lost tools	22,925
Total well cost	$521,448

Average daily cost: $17,554 (not including lost tools)

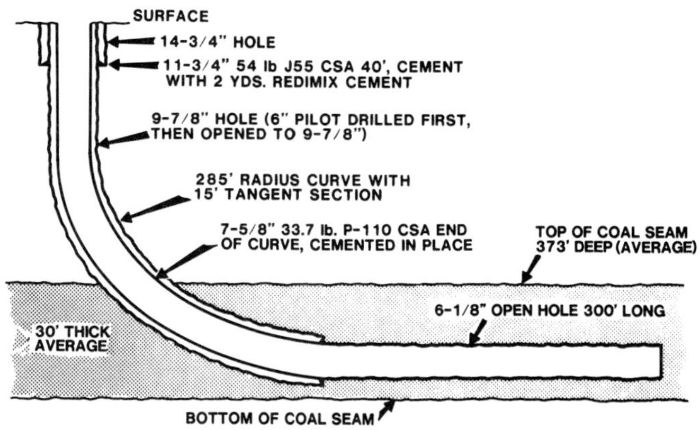

Figure 5. Basic horizontal wellbore diagram.

age time associated with each of the eight surveys was 14 hours. The average daily cost for all services during drainhole drilling was $17,554 per day, as shown on the cost analysis, table 2. Therefore, a substantial cost of the drainhole operation could be reduced if surveying time can be reduced.

Medium-Radius Horizontal Drainhole Drilling

Medium radius horizontal drainhole drilling is accomplished by turning a wellbore from vertical to horizontal in a radius of approximately 285 (87 m) feet representing a build rate of 20 degrees per 100 feet (30 m). Medium-radius horizontal drainhole drilling technology was first successfully demonstrated in a coal seam at the Rocky Mountain No. 1 Site in the Hanna Basin, Wyoming during spring 1987 (Logan, in press). The objective of the Rocky Mountain No. 1 project was to place a system of horizontal and vertical wells into a shallow coal seam for underground coal gasification (UCG). This project was funded by the Morgantown Energy Technology Center of the U.S. Department of Energy and an industry consortium headed by the Gas Research Institute and operated by Stearns-Rogers. Three horizontal wells were successfully drilled into the Hanna coal seam using special design mud motors, each well had approximately 300 feet (91 m) of open-hole horizontal section in the coal seam.

Rocky Mountain No. 1 was divided into two separate parallel UCG modules strategically placed so that they would not interact. The extended linked well (ELW) module was completed first with one horizontal well, PW-1. The second module completed was the controlled retracting injection point (CRIP) which was completed with two horizontal wells, CIW-1 and CPW-1. The relative location of the wells within each module was most important.

The Hanna coal seam at the project site contains three separate carbonaceous shale partings which were successfully used during drilling as markers to help determine the actual location of the horizontal hole in the coal seam. The target for all of the wells was the lower coal-seam interval. Due to the complex geology at the site, not only in the coal, but in the overlying sandstones and shales, detailed geologic analysis of cuttings while drilling was essential to help determine the actual well path in relationship to the well target. A three-dimensional, computer-processed, geologic grid was used during the planning and well-design phase. This tool, although expensive, was effective in identifying potential geologic anomalies which could affect the placement of the horizontal wells.

The Hanna coal seam is at an approximate depth of 363 feet (111 m) with an average thickness of 30 feet (9 m). Coal quality varies vertically and somewhat laterally across the site but averages high volatile C bituminous. Average proximate analyses for the coal are 8.8 percent moisture, 27.3 percent ash, 32.0 percent volatile matter, 31.9 percent fixed carbon, and approximately 8,600 Btu. Although the coal seam was shallow at this project site, the drilling procedure, well design, services used, and costs are applicable for placement of medium radius horizontal drainholes in much deeper coal seams.

The basic horizontal drilling plan for the wells at the project site is shown on figure 5. Due to the shallow depth of the wells, and therefore inability to get adequate weight on the drill bit around the medium-radius curve, a pull-down drill rig capable of applying 20,000 pounds (9,070 Kg) weight on the drill string in addition to collar weight was used. The pull-down capabilities were also required to push the casing into place around the curve. The basic drilling procedure consisted of rotary drilling a 14¾-inch (37 cm) surface hole to approximately 30 feet (9 m) and setting and cementing 11¾-inch (30 cm) casing. Then a 6⅛-inch (16 cm) hole was rotary drilled using a water-based drilling fluid to the kick-off point (KOP) at a depth of approximately 50 feet (15 m).

The KOP was determined based upon a curve radius of 300 feet (91 m) to the end of the curve at the target depth. A 300-foot (91 m) medium-radius curve was then drilled using the angle building assembly which consisted of a 6-inch (15 cm) bit (6⅛-inch on CRIP wells), 4¾-inch (12 cm), 20 degree per 100 foot (30 m) mud motor, and 3½-inch (9 cm) compressive service drill pipe, as shown on figure 6. A water-based gel mud fluid was used to drill the curve in PW-1 and a water-based potassium chloride fluid was used to drill the CRIP well curves. Hole azimuth and inclination were measured every 5 feet (1.5 m) while drilling using a downhole directional orientation tool (DOT) placed in a section of non-magnetic drill pipe located approximately 43 feet (13 m) behind the drill bit. The well was also surveyed with a magnetic multi-shot survey instrument which was pushed into the well after the drilling tools had been removed. The magnetic multi-shot data were used to plot the actual well path and to verify the DOT real-time survey data. The DOT survey data were consistently 1 to

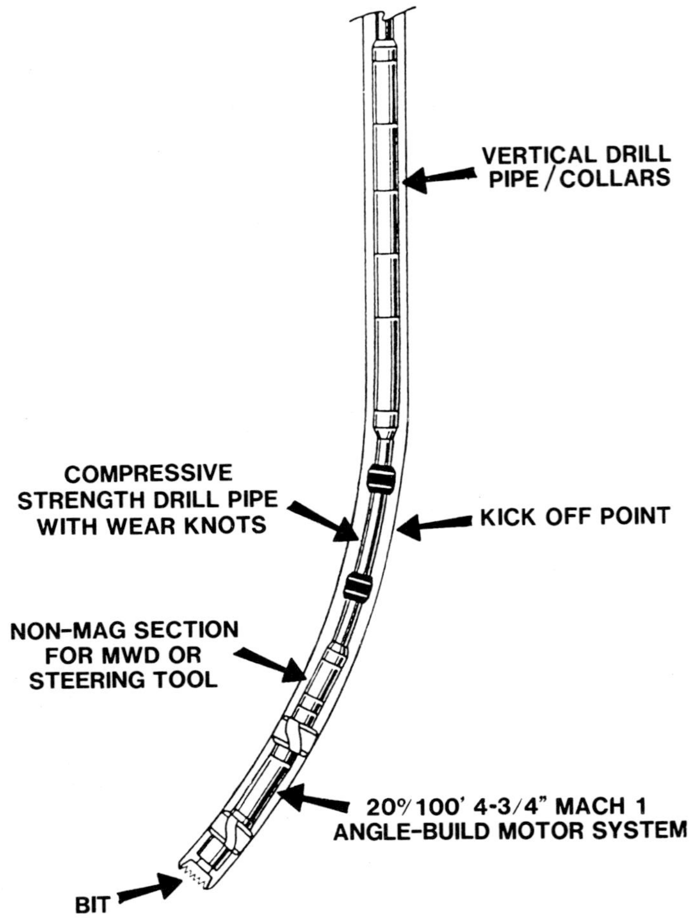

Figure 6. Medium-radius angle building assembly.

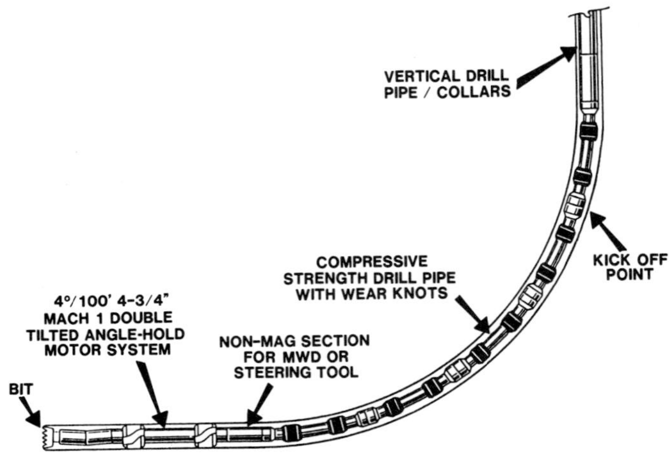

Figure 7. Horizontal drilling assembly.

3 degrees greater in inclination than the more accurate multi-shot data.

The 20 degrees per 100 foot (30 m) mud motor was designed to drill a 285-(87 m) foot radius curve which would cause the end of the curve to be 15 feet (4.5 m) above the target. Therefore, a tangent section with a vertical drop of 15 feet (4.5 m) was designed to be placed in each curve. This allowed build rate flexibility while drilling the curve. The actual angle build rate was 21.7 degrees per 100 feet (30 m). Therefore, the tangent sections had been placed earlier in each curve section. Once the curves were drilled, a 9⅞-inch (25 cm) hole opener was used to enlarge the 6 1/8-inch (15 cm) hole using rotary drilling. An average 475 feet (145 m) of hole was opened in each curve requiring approximately 30 hours per well. The 9 7/8-inch (25 cm) hole was then cased with 7 5/8-inch (20 cm) casing with bow spring centralizers strategically placed on the casing. Two horizontal wells were successfully cemented with a high temperature calcium aluminate cement containing 40 percent silica while the remaining horizontal well was successfully cemented with class G cement containing 30 percent silica. Once the cement had hardened, a 6 1/8-inch (15 cm) diameter horizontal hole was drilled using the horizontal drilling assembly which consists of a 6 1/8-inch (15 cm) tricone bit; 4 3/4-inch (12 cm), double tilted, 4 degrees per 100 foot (30 m) mud motor; and compressive service drill pipe as shown on figure 7. Azimuth and inclination were monitored as before using a DOT while drilling and at critical points using a magnetic multi-shot. Because the DOT is placed approximately 43 feet (13 m) behind the bit, the only nearly real-time technique available to determine the actual location of the drill bit is from analysis of drill cuttings. Samples of drill cuttings were collected and analyzed at least every 5 feet (1.5 m) while drilling and proved to be an effective tool to help determine the actual drill-bit location. A 6 1/8-inch (15 cm) diameter horizontal section approximately 300 feet (91 m) long was placed in each of the three wells.

The first of the three horizontal wells drilled was Process Well No. 1 (PW-1) located in the extended link module (ELW). This well was completed with a 6 1/8-inch (15 cm) diameter horizontal open hole 292 feet (89 m) in length at an average inclination of 88 degrees as shown on figure 8. This well was completed in 288.0 hours (12 days) and at a total cost of $253,633.

The PW-1 well was drilled first to use the experience and information gained to fine-tune the drilling procedure prior to drilling the more difficult horizontal wells in the CRIP module which had more stringent targets. Several changes to the drilling procedure were made after drilling PW-1 which included:

1. To help reduce the clay balling around motor stabilizer and thus increase penetration rate; drill a larger hole in the curve using a 6⅛-inch bit rather than a 6-inch bit.
2. Drill the curve with a shale inhibitive potassium chloride drilling fluid rather than the fresh-water-based fluid which was causing clay swelling and slow penetration rates.
3. Removing the wireline side entry sub used by the DOT and replacing it with a wireline packoff on a swivel to reduce DOT surveying time.

Drilling time and cost were significantly reduced on the first well in the CRIP module, CIW-1. This well was completed with 306 feet (93 m) of horizontal hole in the coal seam, as shown on figure 9. CIW-1 was completed in 202 hours (8.4 days) at a cost of $139,099 which was 3.6 days faster and $114,534 cheaper than the first well, PW-1. The final horizontal well, CPW-1 was completed with 313 feet (95 m) of open horizontal hole in the coal seam and on target, as shown on figure 10.

This well was completed in 256 hours (10.7 days) at a total cost of $164,726. A detailed time analysis breakdown and cost

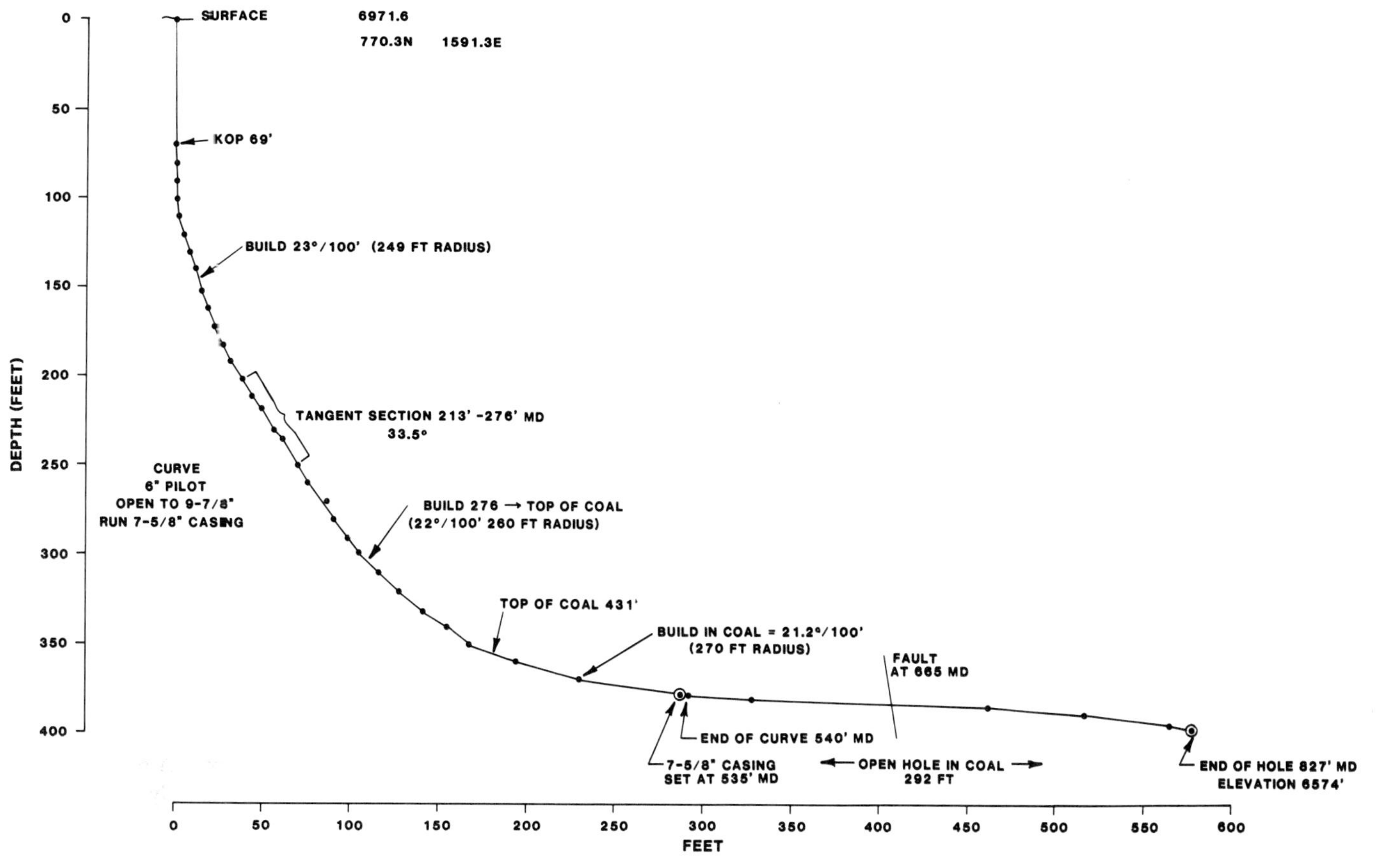

Figure 8. Medium-radius well PW-1.

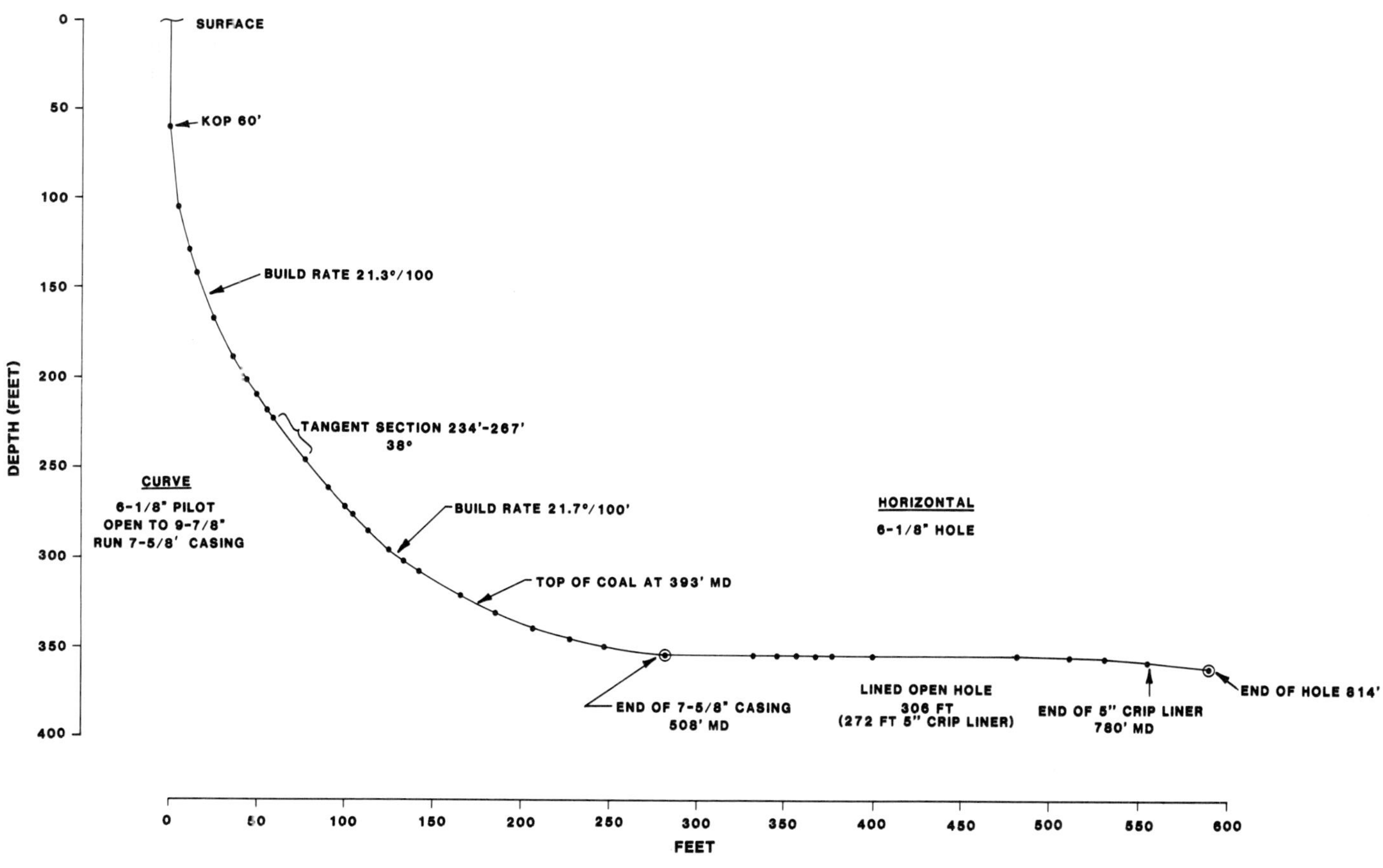

Figure 9. Medium-radius well CIW-1.

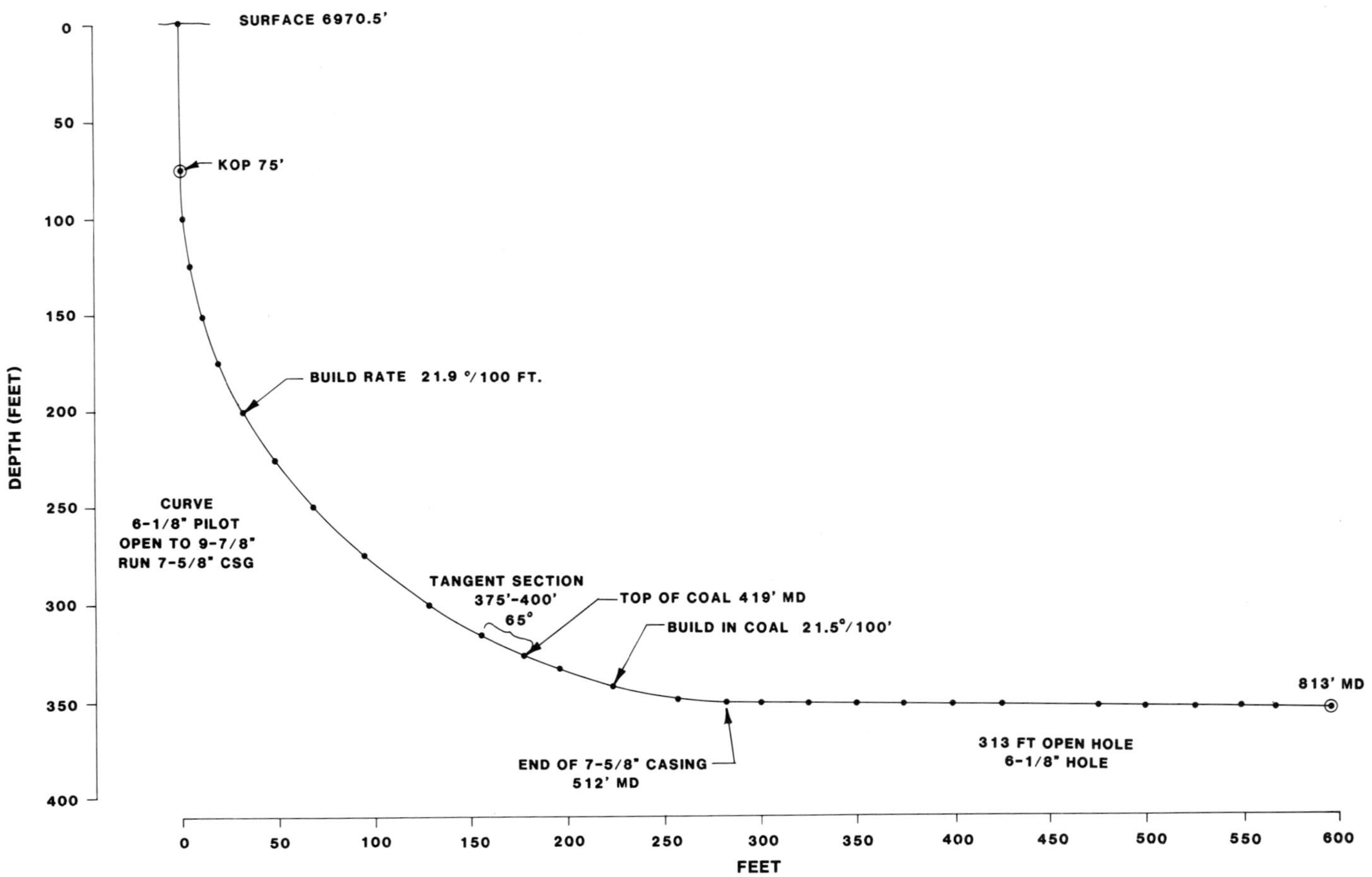

Figure 10. Medium radius well CPW-1.

TABLE 3. Medium-radius horizontal well hours by operation.

	Wells		
	PW-1 (%) hours (%)	CIW-1 (%) hours (%)	CPW-1 (%) hours (%)
Drilling	85 (30)	39 (20)	52 (20)
Tripping	43 (15)	27 (13)	48 (19)
Ream	27 (9)	33 (16)	37 (15)
Circulate	3 (1)	5 (2)	3 (2)
Survey	37 (13)	15 (8)	26 (10)
Repair	33 (11)	12 (6)	5 (2)
Casing and cement	14 (5)	33 (16)	11 (4)
Wait on cement	24 (8)	15 (8)	29 (11)
Rig up	4 (2)	5 (2)	5 (2)
Drill cement	6 (2)	6 (3)	7 (3)
Mix mud	2 (1)	4 (2)	5 (2)
Wait	3 (1)	1 (1)	14 (5)
Move rig	7 (2)	7 (3)	14 (5)
Total hours	288 (100) (12 days)	202 (100) (8.4 days)	256 (100) (10.7 days)

TABLE 4. Medium-radius horizontal well cost by service (in dollars)

	Wells		
	PW-1	CIW-1	CPW-1
Drilling rig	72,988	37,039	40,765
Directional equipment	106,203	65,371	88,728
Drilling fluid	7,832	7,632	6,773
Well site geology	6,162	10,444	4,116
Drilling supervision	9,563	4,986	5,904
3D geology model	28,930	1,230	2,310
Casing and tools	6,520	6,520	6,520
Cement	15,435	5,877	9,610
Total well cost:	$253,633	$139,099	$164,726
Average daily cost:	$ 21,136	$ 16,559	$ 15,394
Cost per foot: (Footage)	$ 306 (827)	$ 171 (814)	$ 203 (813)

breakdown for each medium-radius horizontal well is shown on tables 3 and 4. Each of the three horizontal wells were successfully drilled, cased, and completed on time and within budget.

SUMMARY

Horizontal drainholes have been successfully placed into coal seams in the Rocky Mountain area using both short-radius and medium-radius directional drilling techniques. The cost to place a horizontal drainhole using either technique is approximately $17,600 per day. The disadvantage of the short-radius technique is its lack of accurate azimuth and inclination control. The medium-radius technique currently has had limited use but has the capabilities necessary for placing long horizontal wells on target. With increased experience, the cost of this technique will become less. The success of these horizontal drainhole drilling projects required a team effort approach with fully integrated expertise in well design, field supervision, drilling, cementing, and well site geology.

SI-METRIC CONVERSION FACTORS

pounds, lb	lb × 0.4535	= Kg (kilogram)
feet, ft	ft × 0.3048	= m (meter)
volume, ft^3	ft^3 × 0.0283	= m^3 ($meter^3$)
cf/ton	cf/ton × 0.0044	= cc/Kg
inch	in × 0.0254	= meter
pressure, psi	psi × 6.8947	= KPa (kilopascal)
barrel, bbl	bbl × 0.1589	= m^3 ($meter^3$)
lb per gal, ppg	ppg × 0.119	= g/cc^3

REFERENCES

Byrer, C. W., Moroz, T. H., and Covatoh, G. L., 1984, Production potential for coalbed methane in U.S. basins: Society of Petroleum Engineers Paper 12832.

Decker, A. D. and Seccombe, J. C., 1986, Geologic parameters controlling natural gas production from a single deeply buried coal reservoir in the Piceance Basin, Mesa County, Colorado: Society of Petroleum Engineers Paper 15221.

Ertekin, T., Sung, W., and Schwerer, F. C., 1986, Production performance analysis of horizontal drainage wells for the degasification of coal seams: Society of Petroleum Engineers Paper 15453.

Hardgrove, R.M., 1932, Grindability of coal, Transactions of the American Society of Mechanical Engineers, v. 54, p. 37.

Logan, T. L., Seccombe, J. C., and Jones, A. H., 1986, Hydraulic fracture stimulation and openhole testing of a deeply buried coal seam in the Piceance Basin, Colorado: Society of Petroleum Engineers Paper 15251.

Logan, T. L., Schwoebel, J. J., and Horner, D. M., 1987, Application of horizontal drainhole drilling technology for coalbed methane recovery: Society of Petroleum Engineers Paper 16409.

Logan, T. L., in press, Application of medium radius horizontal drainhole drilling technology for underground coal gasification, Hanna Basin, Wyoming, Proceedings of the 13th Annual Underground Coal Gasification Symposium, Laramie, Wyoming, August 24-26, 1987.

Schwoebel, J. J., 1987, Coalbed Methane: Conversion of a liability to an asset: Mining Engineering Journal, April volume, Preprint no. 86-16.

Textbook meanders in the valley of the Animas River, River of Lost Souls, just north of Durango, Colorado. San Juan Mountains south of Silverton, Colorado on the skyline. Pennsylvanian system, Hermosa Group, form the steep slopes at the valley head. Photo by Bruce Conrad.

Stress-Dependent Permeability and Porosity of Coal

Chester R. McKee, Amar C. Bumb, and Robert A. Koenig[1]

ABSTRACT

Relationships have been derived for permeability, porosity, and density as a function of effective stress and depth. The primary formation parameter is pore compressibility and formulas are derived for both constant and variable pore compressibility. The grains are assumed to be incompressible. The Carmen-Kozeny equation is assumed valid and changes in porosity are proportional to changes in total effective stress. These relationships fit laboratory and field data. Four laboratory core tests on coal are used to illustrate the applicability of the permeability and porosity theory. In cases in which the correlation coefficient was applicable, it was above 0.95. A number of field permeability data points from the Piceance, San Juan, and Black Warrior basins were fitted to the theory. The correlation coefficient ranged from 0.85 to 0.92 in these cases. Only one of the coal samples required a variable pore compressibility. Further improvements to the derived relationships will be required to include the effects of structurally enhanced permeability.

INTRODUCTION

Economical coal-bed methane production depends on four important coal seam characteristics—gas pressure[2], gas content, coal seam thickness, and permeability. Without the necessary gas content or coal thickness, a project obviously cannot be successful. An extensive body of information exists for selecting a site with some degree of confidence based on gas content and coal thickness (National Petroleum Council, 1980; Wicks and others, 1987a; Lent, 1979, 1982; Wicks and others, 1987b; Tremain and Toomey, 1983; and Rightmire and others, 1984). However, unexpected variations in coal gas content can occur.

Information on coal permeability has been more difficult to obtain. For most coal-bed methane wells, permeability tests either have not been conducted or, if they have, the results remain proprietary. Consequently, very few permeability values have been reported in the literature (Way and others, 1984a). In 1982, the Gas Research Institute (GRI) began to pursue extensive permeability testing and data collecting to assess the permeabilities of coal-bed methane sites. A program was developed to acquire the necessary information from operators and from wells of opportunity using simple but effective slug testing methods (Way and others, 1984b). Far more information was available in the form of specific capacity data, which relates well-flow rate to drawdown; permeability can be estimated from this data by using available information or assuming depths to water. In this manner, more than 40 permeability data points were calculated for unstimulated wells in the San Juan, Piceance, and Black Warrior basins (Way and others, 1984a).

While some scatter was evident, the data indicated that permeability decreased with depth. Some initial success was achieved by drawing empirical straight-line correlations between depth and permeability of log-log plots and led to the search for a more fundamental theory to explain the data and correlations (Way and others, 1984a).

In view of the advances in reservoir engineering research, it is surprising that stress-dependent permeability has not been more extensively studied. As natural gas reservoirs are found at ever greater depths, understanding stress-dependent permeability will become essential, because under large drawdown, reduced permeability can lower the production from a stress-sensitive reservoir.

The work of Phil Randolph and Bob Rose at the Institute of Gas Technology (IGT), which documented the strong stress-dependence of core-sample permeability in laboratory experiments, was the basis for establishing a theoretical relationship of permeability reduction as a function of increasing stress or depth. These and other laboratory studies have shown that coal permeability decreases by one to two orders of magnitude when effective stress is increased by 2,000 psi (13.8 MPa), (Dabbous and others, 1974; Somerton and others, 1974; and Rose and Foh, 1984). Furthermore, stress-sensitive permeability and porosity have been observed in experimental studies on various reservoirs (Jones, 1975; Thomas and Ward, 1972). As

[1] In-Situ Inc.
210 South Third Street
P.O. Box I
Laramie, Wyoming 82070-0920

[2] Most coal seams are initially water saturated prior to production. If gas is present in solution, formation gas pressure is defined as the pressure which is in equilibrium with the measured gas content in solution. If gas is absent, gas pressure is determined by drawing the well down until gas first appears in solution. The bottom-hole water pressure on the coal seam is equal to the formation gas pressure. In either case, the gas pressure so determined must be in equilibrium with the coal's desorption isotherm.

The authors wish to express their thanks to Chuck Brandenburg, D'Arcy Horner, Leo Rogers, and Dick Schraufnagel of GRI for their comments and assistance. We are grateful to Phil Randolph of IGT for bringing this problem to our attention; his review of the concepts was very helpful. We express appreciation to Fred Schwerer, formerly of USX, Vello Kuuskraa of Lewin and Associates, and Ted Way of In-Situ for their thoughtful and constructive criticism. We also want to thank Arfon Jones of Terra Tek for bringing J. B. Walsh's work to our attention, Greg Bell of Terra Tek for helpful discussions, and Lindsay Holichek for her editing and preparation of the manuscript. This research was supported by GRI Contract No. 5082-214-0729.

a result, a number of empirical permeability-stress relationships have been proposed, most of which show a strong decrease in permeability with increasing effective stress which is normally proportional to depth (Jones, 1975; Ostensen, 1983; and Schwerer and Pavone, 1984).

OBJECTIVES

These efforts of the workers listed above laid the foundation for the theoretical relationship between stress-dependent permeability and porosity for coals. McKee and others (1986) developed the first theoretical equations using matrix compressibility as a fundamental property. The theory of McKee and others fits both laboratory and field data, and is of limited use because maximum closure stress, a function of matrix compressibility, is inherent and limits the depth to which the results apply (McKee and others, 1986). New theoretical equations are developed here for permeability and porosity in terms of pore compressibility. This formulation generally eliminates the need for variable compressibility in the range of interest, and therefore is simpler to use and is not limited by maximum stress. The resulting formulae are shown to fit both laboratory and field data, but their practical significance is that they provide a simple, usable tool to establish favorable areas of permeability for gas production from coal beds. However, the theory and correlations using the effective stress gradient can only indirectly account for structurally enhanced or reduced permeability, hence, the role of the geologist is still paramount in selecting those geologic factors leading to the best production prospects.

THEORY

Appendix A details the mathematics of the stress- or depth-dependent permeability derivations; major results are given in equations (A-6), (A-9), (A-10), and (A-15). Stress-dependent pore compressibility, which is required for some cores, is given in equation (1). The mathematically oriented reader may find it convenient to review these equations before proceeding to the discussion of results. Conceptually, the theory is very simple. It assumes the solid grains are incompressible and that the change in porosity is due to compression of the void space which is assumed to depend only on effective stress. Permeability is obtained by substituting the expression for porosity into the Carman-Kozeny equation for fractures. The tacit assumption is that the effective porosity is proportional to total porosity. The variation of density with effective stress is a result of the porosity relation.

COMPARISONS OF THEORY WITH LABORATORY DATA

The literature does not contain as much field data as laboratory data on stress-dependent porosity and permeability for coal. Hence, the theory first will be compared with core-test permeability data from three basins. Rose and Foh (1984) performed four tests on coal. Net effective stresses ranged from 150 to 2,400 psi (1.03 to 16.9 MPa).

Permeability Data

Constant Pore Compressibility

In reviewing the data from the first three samples, constant pore compressibility fit the permeability/effective-stress data very well. The first sample shown in figure 1 involved laboratory permeability vs. effective stress for a Pittsburgh seam coal sample obtained from a depth of 300 ft (91.4 m) in the U.S. Bureau of Mines test mine in the Appalachian Basin (Rose and Foh, 1984). The tabular data were reported in millidarcies (md) to one decimal point and ranged from 0.2 to 1.8 md (2 to 18 $\times$ 10^{-4} μm^2). While this does not create a problem with linear plots, on log plots it can show as scatter, particularly at low permeability ranges. To report all data as two significant digits, a range is shown with the data reported to only one significant digit. Figure 1 shows the Pittsburgh seam data on a semi-log plot with a least-squares straight line through the data. The correlation coefficient is 0.96 which is indicative of a good fit. The least-squares resulted in a constant pore compressibility of 1.87 $\times$ 10^{-3}/psi (0.271/MPa), and a permeability of 1.56 md (1.54 $\times$ 10^{-3} μm^2) at 150 psi (1.03 MPa).

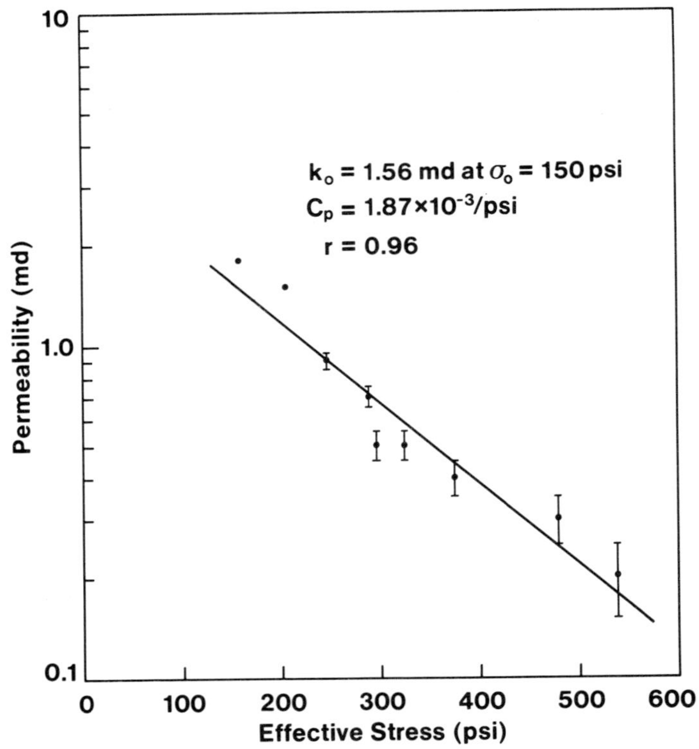

Figure 1. Laboratory-measured permeability versus effective-stress data and match with theoretical curve for Pittsburgh coal mine sample from depth of 300 ft, Northern Appalachian Basin.

The second sample in figure 2 used laboratory permeability vs. effective stress for a coal sample from the top seam of the Menefee Formation in the King Mine, San Juan Basin (Rose and Foh, 1984). For this core test, fluid pressure was kept at approximately 310 psi (2.14 MPa) while confining pressure was varied. As shown in figure 2, permeability data and theoretical values obtained from least-squares matching agreed quite closely for a constant pore compressibility of 1.34 $\times$ 10^{-3}/psi (.194/MPa), and an initial permeability of 35.8 μd (35.3 $\times$ 10^{-6} μm^2), at an arbitrarily selected initial effective stress of 250 psi (1.72 MPa). The permeability of this sample is more than two orders of magnitude smaller than that of the first sample and an excellent curve fit was obtained by assuming constant pore compressibility. A correlation coefficient of

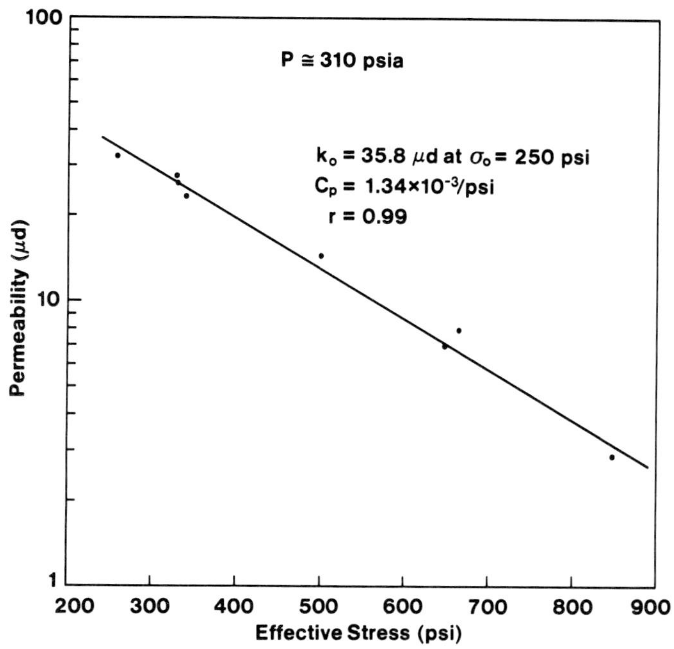

Figure 2. Laboratory-measured permeability versus effective-stress data at constant fluid pressure and match with theoretical curve for Menefee Formation (Mesaverde Group) coal outcrop sample, San Juan Basin, Colorado.

0.99 justifies constant pore compressibility.

The third sample in figure 3 used laboratory permeability data for a Cameo seam coal sample from a depth of 2,766 feet (843 m) in the Coors 1-23 well, Piceance Basin, Colorado (Rose and Foh, 1984). Effective stress was obtained by subtracting fluid pressure from confining pressure (equation (A-12), with s = 1). Data were obtained by holding confining pressure approximately constant at 2,615 psi (18 MPa). Almost all the data points fall on a straight line justifying the use of constant pore compressibility (fig. 3). The least-squares match resulted in a pore compressibility of 7.76×10^{-4}/psi (.112/MPa) and an initial permeability (k_o) of 273 nanodarcies (2.69×10^{-7} μm^2) at an arbitrarily selected initial effective stress of 1,000 psi (6.89 MPa). Again, the correlation coefficient of 1.00 and excellent fit to data (fig. 3) justify constant pore compressibility. Table 1 compares the pore compressibility of various samples.

Variable Compressibility

In the preceding section, three cases were shown where constant pore compressibility was a good assumption for the range of data in each case. However, Somerton and others (1974) and Scorer and Miller (1974) have shown that compressibility is unlikely to remain constant over the entire stress history. Instead it will approach the rock (or grain) compressibility as effective stress increases, because small asperities on the cleat face resist closure and must be deformed or crushed before further reduction in porosity is achieved. The result is a departure from linearity (constant pore compressibility). For the first three core samples, compressibility can essentially be assumed constant which implies that the material is initially relatively consolidated. A laboratory study of the relationship between relative compressibility and effective stress may be useful in understanding this phenomenon, provided a size effect (small size of laboratory core vs. large size of field core) is incorporated. The reason for doing so will be discussed in the following sections.

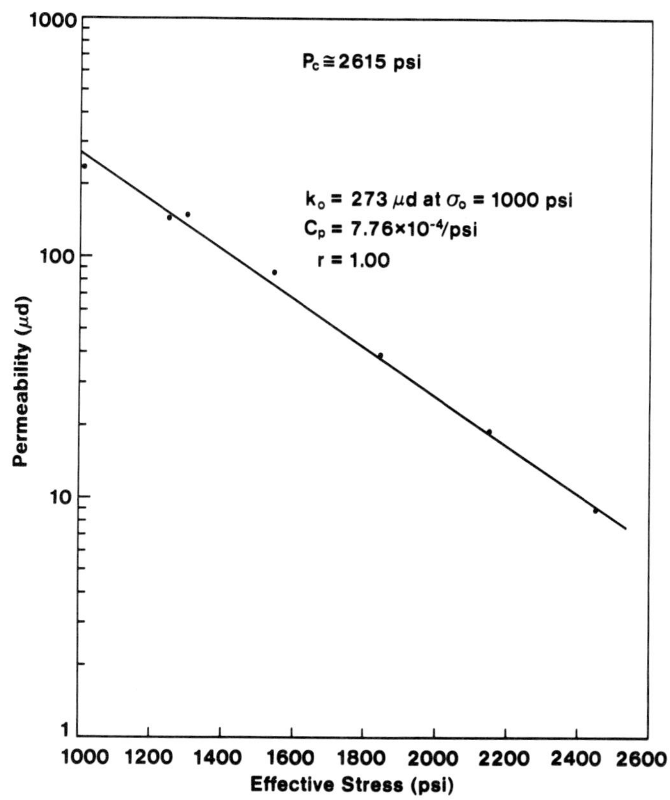

Figure 3. Laboratory-measured permeability versus effective-stress data and match with theoretical curves for Cameo coal-core sample from depth of 2,766 feet, Piceance Basin, Colorado.

TABLE 1. Pore compressibilities obtained by fits to experimental permeability/porosity and effective stress data.

Basin	Formation	Data type	Pore compressibility (psi^{-1})
Appalachian	Pittsburgh seam	Permeability	1.87×10^{-3}
San Juan	Menefee	Permeability	1.34×10^{-3}
Piceance	Cameo (2,766 ft)	Permeability	7.76×10^{-4}
Piceance	Cameo (2,767 ft)	Permeability	1.8×10^{-3}*

* initial pore compressibility shown for variable cases

As a general rule, compressibility declines with increasing stress, although on some samples it may increase (Scorer and Miller, 1974). Because a stress-dependent compressibility can be sample-specific, a general function is unavailable. The theoretical curves fit the laboratory values well when these samples were assumed to have exponentially declining compressibilities. Note, however, that because compressibility is variable, average pore compressibility ($\overline{C}_p$ in equation (A-5)) also is variable.

For exponentially declining pore compressibility, $\overline{C}_p$ is given by:

$$\overline{C}_p = \frac{1}{\sigma - \sigma_0} \int_{\sigma o}^{\sigma} C_0\, e^{-\alpha[\sigma - \sigma_0]}\, d\sigma$$

$$= \frac{C_0}{\alpha \Delta \sigma} \left[1 - e^{-\alpha \Delta \sigma} \right] \quad (1)$$

where α defines the rate of decline of pore compressibility as effective stress increases; C_0 is the pore compressibility at initial effective stress, σ_0; and $\Delta\sigma$ is the increase in effective stress from σ_0. For variable pore compressibility, $\overline{C}_p$ is not constant but simply represents the average pore compressibility over the stress interval $(\sigma - \sigma_0)$. If σ changes, the average $\overline{C}_p$ over the interval also will change. A core sample from the Piceance Basin displayed exponentially declining pore compressibility.

Laboratory permeability data for the Piceance Basin Cameo seam coal from a depth of 2,767 feet (843.3 m) are shown in figure 4 (Rose and Foh, 1984). Using the parameter $\alpha = 2.9 \times 10^{-3}$/psi (.4/MPa) in equation (1), the solid line in figure 5 yields an excellent fit to the Piceance coal sample data. While the sample in figure 4 was taken only 1 foot lower than the sample used in figure 3, the samples show different characteristics. This indicates that core samples, especially in fractured media, are not of sufficient size to constitute a relative elementary volume. In figure 3, constant pore compressibility gave an excellent approximation; however, the data in figure 4 require variable pore compressibility. This may be due to different sample orientations in the test equipment (the plug axis was perpendicular to the bedding planes for figure 3, while in figure 4 the plug axis was perpendicular to the face cleat) or just the local nature of the core and the effect of small sample size. The net effective stress was calculated using s = 1 in equation (A-12), where this assumption may not be true.

Although only a few cases have been included here, this theory fits other permeability data. The only other way to verify the theory is to analyze porosity data; unfortunately, porosity measurements as a function of effective stress are limited. However, as will be discussed in the following section, such additional data are useful in verifying the theoretical soundness of these concepts.

COMPARISON OF THEORY WITH FIELD DATA

This section compares the theory of stress-dependent permeability with field permeability and porosity measurements. The search for such field permeability data targeted the three most active coal-bed methane areas—Piceance Basin in Colorado, San Juan Basin in New Mexico and Colorado, and Black Warrior Basin in Alabama. In a recent GRI report, Way and others (1984a) reported the information acquired from 53 contacts with private companies and state and Federal agencies involved in coal-bed degasification, geologic, and hydrologic investigations.

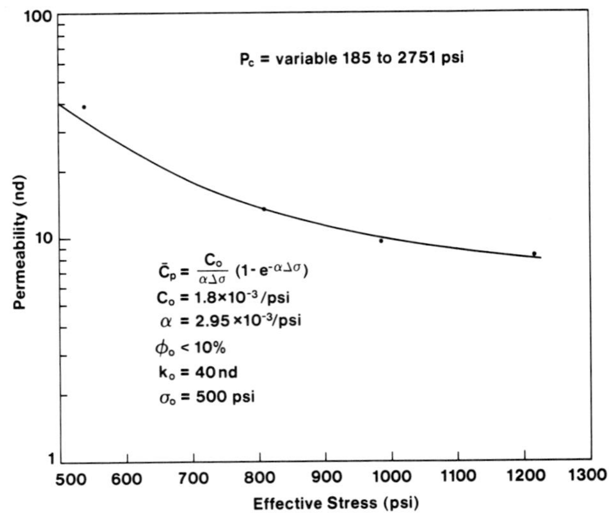

Figure 4. Laboratory-measured permeability versus effective-stress data and theoretical match using variable compressibility for a Cameo coal sample from a depth of 2,767 feet, Piceance Basin, Colorado.

Lithostatic Conditions

Overburden materials have an average specific grain density of approximately 2.65 and a porosity of about 20 percent, which gives an estimated stress gradient of 1.005 psi/ft (22.73 kPa/m)[3]. (Note that the commonly used value for this gradient is 1 psi/ft (22.63 kPa/m) and that for fresh water is 0.4335 psi/ft (9.81 kPa/m)). For those coal seams that are at or near hydrostatic pressure, the effective stress gradient is 0.572 (that is, 1.0005—0.4335) psi/ft (12.9 kPa/m), in which case the effective stress (in psi) at any depth, d (in feet), for lithostatic conditions is given simply by:

$$\sigma = 0.572d \quad (2)$$

For some data, information about effective stress and (or) water levels was available. This information was used to calculate equivalent lithostatic depth, the depth at which effective stress would equal the effective stress at lithostatic conditions (equation (2)). When information on effective stress and (or) water level was lacking, it was assumed that lithostatic conditions existed. If the average effective-stress gradients for these points differed from 0.572 psi/ft (12.9 kPa/m), compressibilities obtained by fitting equation (A-9) to the data would change inversely with the stress gradient. For with other parameters held constant, an actual in-situ effective-stress gradient (lower than the lithostatic assumption of equation (2)) would result in a higher in-situ permeability at a given depth than that predicted by equation (A-9).

Permeability Data

Hydrologic data were sparse and few data points were obtained from carefully conducted well tests. In most cases, permeabilities were estimated from static water level, water production, and productivity index data. (As pressure in the

[3] gradient $= [p_g (1 - \phi) + p_w \phi]$
$= [2.65(1 - 0.2) + 1(0.2)]$
$(62.4 \text{ lb/ft}^3)(1 \text{ ft}^2/144 \text{ in}^2)$
$= 1.005$ psi/ft

reservoir is lowered, gas begins desorbing until two-phase flow occurs. To avoid relative permeability effects typical of this flow regime, only the initial production phases were analyzed.)

For consistency, only data from unstimulated wells in coal seams have been used because hydraulically stimulated wells may yield larger, possibly scale-dependent, permeability values, which have been observed in fractured media by other investigators (Hsieh and others, 1983). A fractured medium such as coal exhibits small fractures (at spacings of centimeters) between larger fractures at spacings of several to tens of meters, which in turn are fault bounded at intervals of hundreds or even thousands of meters. A localized permeability measurement in an unstimulated well would have a probability of intersecting only the more numerous small-scale fractures. Large hydraulic fractures will have a higher probability of intersecting the larger-spaced, larger-scale natural fractures. Because permeability varies with the square of fracture aperture, permeability can increase as larger natural fractures are encountered. Therefore, with a sufficiently large hydraulic fracture, one can expect to attain a constant permeability value consistent with the intersection frequency of the main fractures in the coal. Interference tests conducted at GRI's Rock Creek site with wells spaced from 40 to 200 feet (12 to 61 m) apart have shown from zero to 75 percent increase in permeability over single-well tests (Koenig and Stubbs, 1986).

Hydrologic data by nature contain experimental errors. Furthermore, in low-permeability formations, several tests did not measure beyond the wellbore storage region, and, hence, only upper bounds on permeability could be inferred. The limits are indicated on figures 5, 6, and 7 by data points at the end of a range joined by a line. Open symbols indicate assumed lithostatic conditions where water level and (or) effective stress information was unavailable. Closed (filled) symbols represent data for which water level-effective stress information was available and used to calculate the equivalent lithostatic depth. The initial curve-fitting parameters (depth, permeability) and calculated pore compressibilities for the three basins are summarized in table 2.

McKee and others (1986) noted that permeabilities of the San Juan and Piceance Basin coals showed similar trends and compressibility values. The agreement between permeability values for the San Juan and Piceance basins could possibly be attributed to their similarities in geologic age and regional depositional settings. Also, due to the limited number of data points for these basins, they are plotted on the same figure (fig. 5), which shows that permeabilities from the San Juan and Piceance basins follow the same general trend. With depth (stress) dependent compressibility, the rate of change generally decreases with increasing depth; however, due to scarcity of data and lack of information on depth-dependence of compressibility, a constant value for pore compressibility is used for the San Juan and Piceance basins. Moreover, the porosity of these coals is less than 10 percent, so equation (A-10) is used. Figure 5 shows a good data fit to theory. If compressibility varied, one could expect a better fit to the data, that is, permeability decreasing more gradually with increasing depth and eventually becoming constant. Equation (A-9) accounts for this effect by incorporating the function for exponentially declining pore compressibility in equation (A-5).

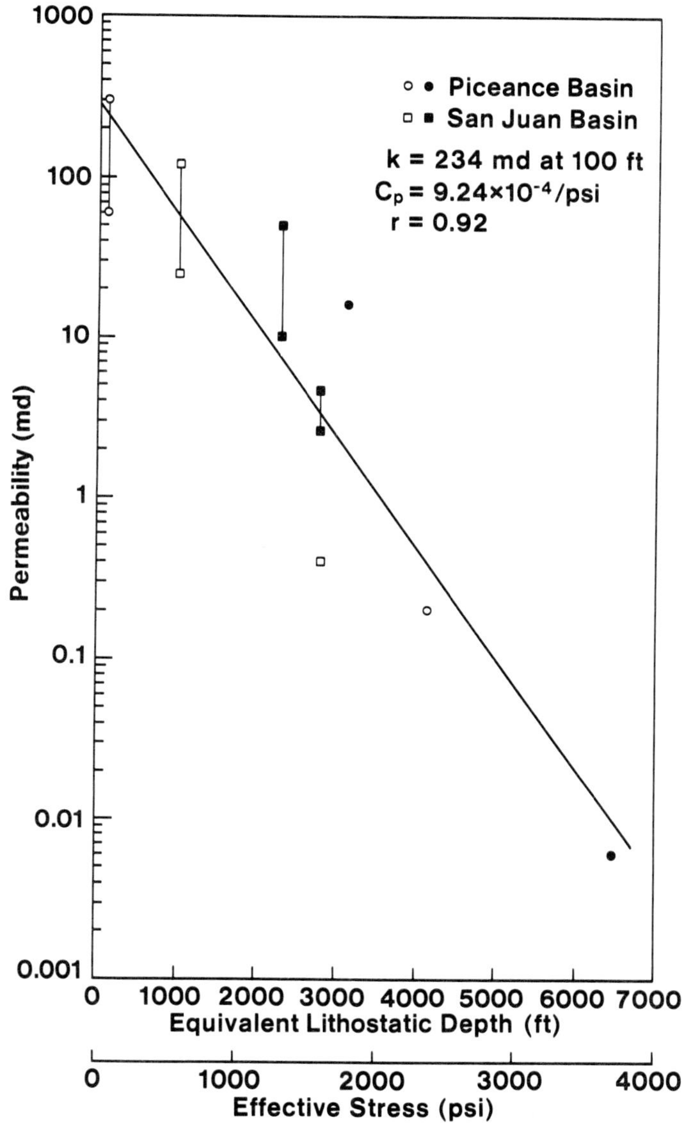

Figure 5. Semi-log plot of field-test derived and estimated permeabilities with equivalent lithostatic depth (and effective stress) and match with theoretical curve for coal seams in the Piceance and San Juan basins.

TABLE 2. Comparison of curve-fitting parameters for coal seams in the San Juan, Piceance, and Black Warrior basins.

Basin	Initial conditions: Depth (ft)	Initial conditions: Permeability (md)	Pore compressibility (psi^{-1})	Coefficient of correlation
Piceance & San Juan (fig. 5)	100	234	9.24×10^{-4}	0.92
Black Warrior (fig. 6)	100	186	1.87×10^{-3}	0.87
Piceance, San Juan, & Black Warrior (fig. 7)	100	109	8.78×10^{-4}	0.85

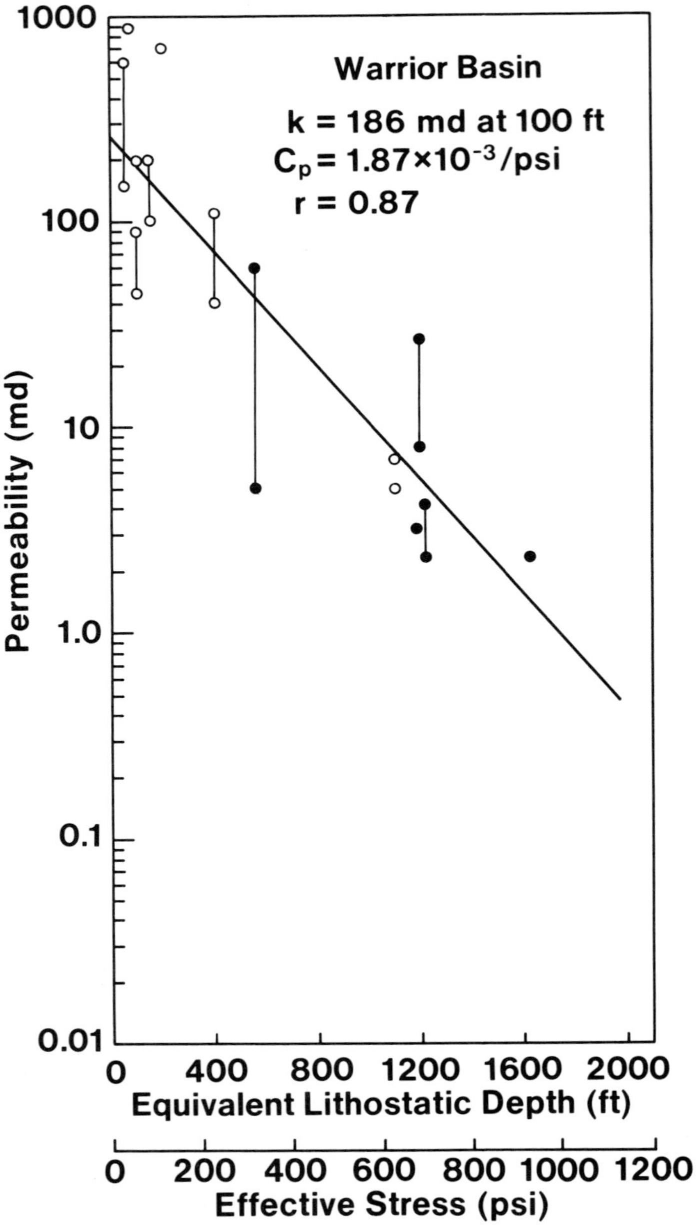

Figure 6. Semi-log plot of field-test derived and estimated permeabilities with equivalent lithostatic depth (and effective stress) and match with theoretical curve for coal seams in the Black Warrior Basin.

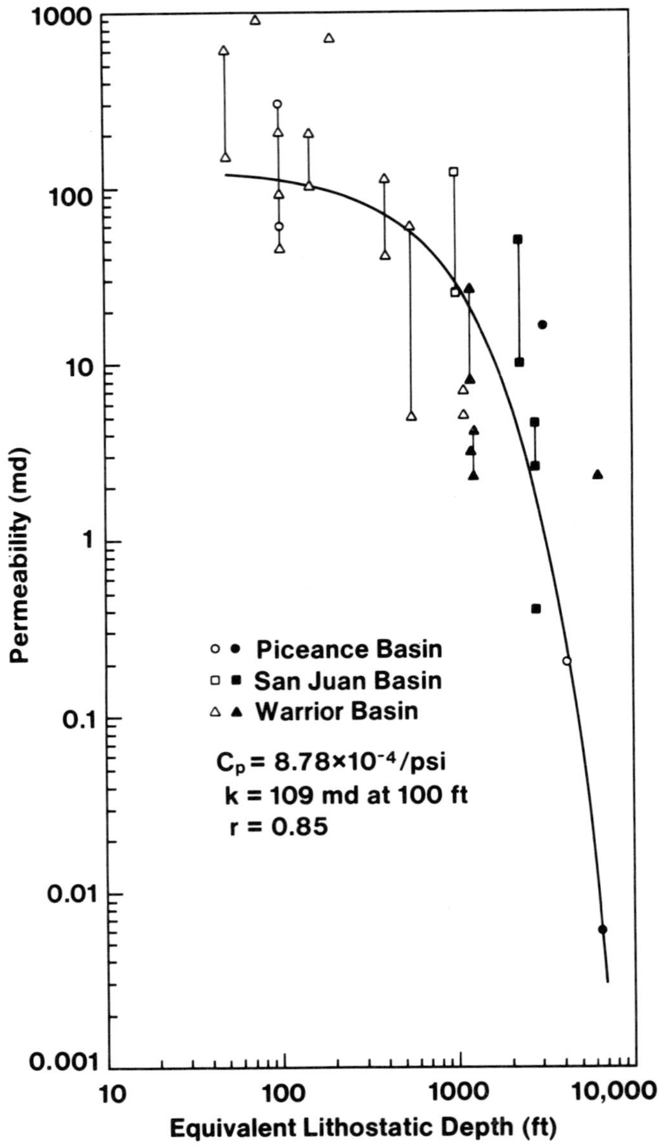

Figure 7. Log-log plot of field-test derived and estimated permeabilities with equivalent lithostatic depth (and effective stress) and match with theoretical curve for combined data from coal seams in the Piceance, San Juan, and Black Warrior basins.

Figure 6 shows similar comparisons of theory and permeability estimates for the Black Warrior Basin, excluding permeability data at known anomalies. The anomaly in the Black Warrior Basin represents a highly fractured well on the campus of the University of Alabama; high permeability (250 md (.246 5m2)) has made this well difficult to dewater and, accordingly, gas production has been low. For the Black Warrior Basin, pore compressibility is twice that for the San Juan and Piceance basins.

Figure 7 shows a log-log plot for all the coal permeability results obtained to date from the Piceance, San Juan, and Black Warrior basins, except for known anomalies. A good correlation of theory with the data is seen in the figure, and the pore compressibility for all data combined is nearly the same as that for individual basins. Again, constant pore compressibility was used as the simplest assumption pending additional data. Figure 8 compares constant pore compressibility against exponentially declining pore compressibility. The curve for constant pore compressibility is the same (same parameters) as in figure 7. For variable pore compressibility, initial compressibility was selected such that average pore compressibility between 100 feet and 5,000 feet (just before permeability becomes constant) is 8.78×10^{-4}/psi (0.127/MPa), the same as in figure 7. The constant defining rate of decline of pore compressibility, α, was taken as 1×10^{-3}/psi (0.145/MPa). Figure 8 shows that with exponentially declining pore compressibility, permeability will eventually become constant. However, the current field data available for these basins do not suggest an exponentially declining pore compressibility.

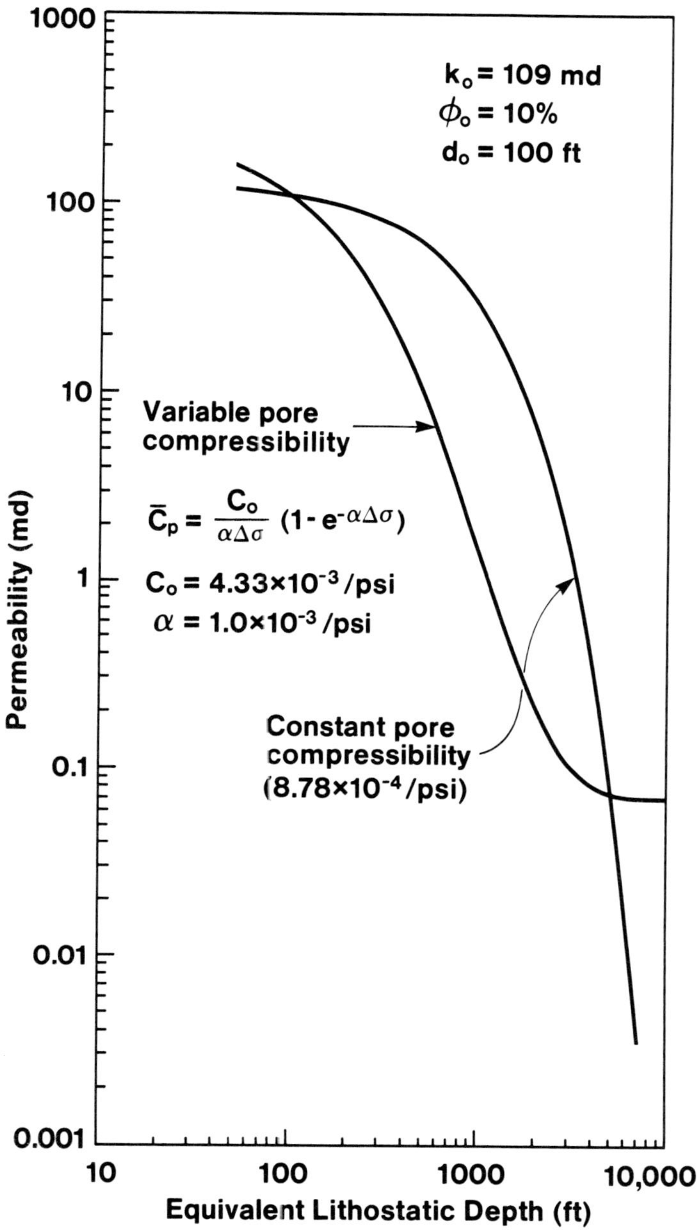

Figure 8. Comparison of constant with variable pore compressibility correlations for stress-dependent permeability.

COMPARISON OF FIELD AND LABORATORY-MEASURED CORE COMPRESSIBILITIES

It is noteworthy that pore compressibilities given in tables 1 and 2 are at least one order of magnitude higher than those measured using stress-strain response on core samples. For Fruitland Formation coal from a depth of 3,085 feet (940 m) in the San Juan Basin, Jones and others (1983) reported core matrix compressibilities of approximately 1 to 2.5 × 10^{-6}/psi (1.45 to 3.63 × 10^{-7}/kPa). In the Piceance Basin, Seccombe (personal communication) noted 3.8 × 10^{-6}/psi (5.5 × 10^{-7}/kPa) for the D Coal Seam (Mesaverde Group) at the Deep Seam Coal Project site. Finally, for the Black Warrior Basin at Rock Creek, Diamond (personal communication) reported a range of approximately 1.6 to 2.7 × 10^{-6}/psi (2.3 to 3.9 × 10^{-7}/kPa).

The porosity of these coals is approximately 4 percent implying mechanical-response pore compressibilities in the range of 2.5 × 10^{-5}/psi (3.63 × 10^{-6}/kPa) to 9.5 × 10^{-5}/psi (1.38 × 10^{-5}/kPa). The arithmetic average coal core (table 1) and basin pore (table 2) compressibilities determined from stress-dependent permeability theory are 1.45 × 10^{-3}/psi (2.10 × 10^{-4}/kPa) and 1.22 × 10^{-3}/psi (1.77 × 10^{-4}/kPa), respectively. These two permeability-response core and basin pore compressibilities are within 19 percent of each other, and are 22 times higher than the average mechanical-response pore compressibility of 6 × 10^{-5}/psi (8.7 × 10^{-6}/kPa). The implication is that the active porosity in the mechanical-response tests is the total porosity, whereas only the effective porosity to flow is responding in the permeability-effective stress tests. Assuming permeability scales are the cube of porosity (equation (B-7)), only $22^{-1/3}$ or 35.7 percent of the porosity is effective or active for fluid flow. Hence, with 4 percent porosity, only 1.43 percent is effective for purposes of permeability response on pore compressibility. Coal at Rock Creek at a depth of 1,450 feet (442 m) has a matrix compressibility of 2 × 10^{-5}/psi (2.9 × 10^{-6}/kPa); when divided by the effective porosity (0.0143), it yields a pore compressibility of 1.4 × 10^{-3}/psi (2.03 × 10^{-4}/kPa) which is in very good agreement with both the laboratory and basin permeability-response pore compressibilities. More work is required to establish the above hypothesis; however, if correct, it suggests that stress-dependent core permeability tests would be capable of determining in-situ pore compressibility values for use in reservoir modeling and in interpretation of well test pressure response.

PRACTICAL APPLICATION OF PERMEABILITY AND POROSITY CORRELATIONS

An important consideration in estimating site permeability from these correlations is the strong influence of local geology. In particular, structural deformation (folding and faulting) frequently causes secondary or fracture permeability. Stresses other than lithostatic stress can also enhance or reduce permeability. Although the theory presented here does not account for fracturing due to such processes, some of these phenomena can be modeled using an effective stress gradient other than lithostatic (0.572 psi/ft (12.9 kPa/m)) as shown in the section discussing field permeability data. Stresses less than lithostatic can occur for several reasons. First, because effective stress is the difference between total stress and fluid pressure (equation (A-12)), overpressure of gas or water can lower it. Second, formation buckling or uplifting can dilate fractures, thereby lowering effective stress and enhancing permeability. Third, a more competent unit in the stratigraphic section can bear a disproportionate amount of load and reduce the stress on underlying units. Many workers have used Heim's rule, which states that stresses in rock tend to approach lithostatic because of creep (Jaeger and Cook, 1979). If creep is not significant, however, then the horizontal stress will be less than the total stress by the ratio $\nu/(1-\nu)$, where ν is Poisson's ratio.

Under GRI sponsorship, Terra Tek Inc. has been working to determine optimum design conditions that will improve hydraulic fractures. One product of that study has been downhole stress gradients of 0.264 psi/ft (5.97 kPa/m) for the Brookwood unit and 0.194 psi/ft (4.39 kPa/m) for the Cedar Cove unit in the Black Warrior Basin, compared to an effective stress gradient of 0.572 psi/ft (12.9 kPa/m) where the measured total stress gradient approximates 1 psi/ft (22.6 kPa/m) in the Rock Creek and Oak Grove areas. Therefore, even in the same basin the critical factors determining permeability with depth may vary appreciably and it is important to obtain in-situ stress conditions for use with these correlations.

The theory's success to this point evidently has been attributable to two factors: (1) applicability of the Carman-Kozeny equation to fractured media; and (2) continuity of preservation of fractures and coal-pore geometries from shallow to greater depths. The theory fits both available laboratory and field data.

SUMMARY AND CONCLUSIONS

Relationships have been developed from fundamental principles regarding permeability and porosity changes in coal and other porous materials as a function of effective stress. The well-known Carman-Kozeny relation can be used to describe permeability. Because solid matrix can be considered incompressible, all volume changes due to compression were attributed to pore spaces. These simple assumptions were used to develop formulas for permeability and porosity as a function of depth.

The developed relations agreed closely with permeability and porosity-vs-stress data determined from laboratory core measurements. Although one of the four examples required exponentially declining compressibility to fit the data, the other samples could be fitted using constant pore compressibility. Based on these results, constant pore compressibilities were used in fitting field permeability-vs-depth data for the Piceance, San Juan, and Black Warrior basins. Surprisingly, all three basins showed close fits for pore compressibility. Mean pore compressibility ($\overline{C}p$) for Black Warrior (1.87×10^{-3}/psi (0.271/MPa)) is twice that of the San Juan and Piceance basins (9.24×10^{-4}/psi (0.134/MPa)).

Although the theory is capable of using either a variable matrix or pore compressibility, more data must be obtained to more precisely define variable compressibility. The effective stress gradient also was identified as being critical to determining permeability at a given depth. The theory predicts that a lower effective stress gradient will result in higher permeability.

The reader is cautioned that the theory discussed does not account for structurally enhanced permeability; when applying the theory to field cases, local effective stress gradients must be used rather than assuming lithostatic conditions.

EXPLANATION OF SYMBOLS

A = cross-sectional area (L^2)
a = aperture (L)
b = solid matrix thickness between two fractures (L)
C_m = bulk matrix compressibility ($M^{-1} LT^2$)
C_p = pore volume compressibility ($M^{-1} LT^2$)
C_i = intrinsic compressibility ($M^{-1} LT^2$)
$\overline{C}_p$ = average pore compressibility ($M^{-1} LT^2$)
d = depth (L)
k = permeability (L^2)
$\overline{l}$ = mean crack length (L)
n = number of fractures per unit cross-sectional area (L^{-2})
n_v = number of fractures per unit volume (L^{-3})
p = pore (fluid) pressure ($ML^{-1} T^{-2}$)
s = constant relating change in pore pressure to change in effective stress
T = tortuosity
v = volume of solids (L^3)
w = fracture height (L)
α = a rate constant defining rate of decline of compressibility as effective stress increases ($M^{-1} LT^2$)
Δz = change in vertical direction (L)
Δp = pressure drop ($ML^{-1} T^{-2}$)
σ = effective stress ($ML^{-1} T^{-2}$)
σ_t = total stress ($ML^{-1} T^{-2}$)
$\Delta\sigma$ = increase in effective stress ($ML^{-1} T^{-2}$)
ϵ = void ratio
ϕ = porosity
p = specific density
ν = Poisson's ratio

Subscripts

o = value at initial condition
w = water
g = grain

APPENDIX A

Porosity-Permeability Effective Stress Relationships

When the compressibility of individual grains in solids is negligible compared to the change in porosity, the following relationship holds for the change in porosity (ϕ) with respect to change in compressive (effective) stress (σ), which is derived by considering the volume of solid grains: ($V_s = A\ \Delta z(1 - \phi)$) constant (Hantush, 1964):

$$d\phi = -C_m(1 - \phi)d\sigma \qquad \text{(A-1)}$$

where C_m, bulk matrix compressibility, is related to pore compressibility (C_p) via porosity as follows:

$$C_m = \phi C_p \qquad \text{(A-2)}$$

Equation (A-1) assumes that all the stress relief is due to pore space comprising the effective interconnected porosity, ϕ, which is the basis of the commonly accepted storage coefficient used in hydrology and petroleum reservoir engineering. Combining equations (A-1) and (A-2) yields:

$$d\phi = -\phi(1 - \phi)C_p\, d\sigma \qquad \text{(A-3)}$$

Integrating equation (A-3) over the stress range yields the following porosity-effective stress relationship:

$$\epsilon = \frac{\phi}{1 - \phi} = \frac{\phi_o}{1 - \phi_o} \exp\left(-\int_{\sigma_o}^{\sigma} C_p d\,\sigma\right) \qquad \text{(A-4)}$$

where ϵ is void ratio. Somerton and others (1974) and Scorer and Miller (1974) observed that bulk and pore matrix compressibilities are stress dependent. Average pore compressibility ($\overline{C}_p$), introduced to reduce the complexity of the following equations, is defined as:

$$\overline{C}_p = \frac{1}{\sigma - \sigma_o}\int_{\sigma_o}^{\sigma} C_p d\,\sigma \qquad \text{(A-5)}$$

so that equation (A-4) then becomes

$$\epsilon = \frac{\phi}{1-\phi} = \frac{\phi_o - \bar{C}_p \Delta\sigma}{1-\phi_o} \quad \text{(A-6)}$$

where $\Delta\sigma$, the change in effective stress, equals $\sigma - \sigma_o$. The remaining derivations in this section are still valid for variable compressibility using equation (A-6), since $\bar{C}_p$ can vary as a function of stress. Equation (A-6) shows that if pore compressibility is constant, than a plot of log and void ratio versus effective stress should result in a straight line. Solving for porosity in equation (A-6) results in:

$$\phi = \phi_o \frac{e^{-\bar{C}_p \Delta\sigma}}{1 - \phi_o\left(1 - e^{-\bar{C}_p \Delta\sigma}\right)} \quad \text{(A-7)}$$

Based on the Carman-Kozeny equation (see appendix B), the following relationship between permeability and porosity can be written:

$$k \text{ varies as } \frac{\phi^3}{(1-\phi)^2} \quad \text{(A-8)}$$

It is unnecessary for the fracture surface to be smooth because the tortuosity factor (T) is included in equation (A-8), as shown in appendix B, equation (B-7). This equation also was used by McKee and Hanson (1975) to explain explosively created permeability. Combining equations (A-7) and (A-8) then:

$$k = k_o \frac{e^{-3\bar{C}_p \Delta\sigma}}{1 - \phi_o\left[1 - e^{-\bar{C}_p \Delta\sigma}\right]} \quad \text{(A-9)}$$

For most deeply buried, fractured media, porosity is on the order of one percent, and the denominator term can be approximated to unity to give (Reiss, 1980):

$$k = k_o e^{-3\bar{C}_p \Delta\sigma} \quad \text{(A-10)}$$

This simplification introduces an error less than the scatter in the experimental data. Schwerer and Pavone (1984) observed this simplification. Equation (A-10) shows that, with small porosity, a semi-log plot of permeability versus effective stress (log k vs. σ) results in a straight line if pore compressibility is nearly constant. The derivation for matrix compressibility as the fundamental parameter is given by McKee and others (1986).

Dependence of Permeability on Fluid Pressure

If the water pressure is assumed to act effectively throughout the elemental volume, Hantush notes that a change in fluid pressure results in an equal and opposite change in effective stress (Hantush, 1964),

$$dp = -d\sigma \quad \text{(A-11)}$$

Walsh (1981) then gave the following general relationship between effective stress (σ), total stress (σ_t), and pore (fluid) pressure (p):

$$\sigma = \sigma_t - sp \quad \text{(A-12)}$$

He assumed and justified that s was a constant for linearly elastic materials, and also rocks. A similar factor was also used by Brandt (1967). Integrating equation (A-12) over the stress range gives:

$$\Delta\sigma = \sigma - \sigma_o = s(p_o - p) = s\Delta p \quad \text{(A-13)}$$

where Δp is pressure drop, while $\Delta\sigma$ is the increase in effective stress, and total stress, σ_t, due to the overburden is taken as constant. Substituting equation (A-13) into equations (A-6) and (A-9) yields the desired pore pressure permeability and porosity relationship:

$$\frac{\phi}{1-\phi} = \frac{\phi_o}{1-\phi_o} e^{-s\bar{C}_p \Delta p} \quad \text{(A-14)}$$

$$k = k_o \frac{e^{-3s\bar{C}_p \Delta p}}{1 - \phi_o\left(1 - e^{-s\bar{C}_p \Delta p}\right)} \quad \text{(A-15)}$$

or, in the simplified form of equation (A-10):

$$k = k_o e^{-3s\bar{C}_p \Delta p} \quad \text{(A-16)}$$

Dependence of Density of Effective Stress

The specific density, p, of a formation can be written in terms of specific grain density, p_g, and formation porosity as:

$$p = p_g(1-\phi) \quad \text{(A-17)}$$

Porosity, ϕ, can be substituted from equation (A-7) to yield specific density as a function of effective stress:

$$p = \frac{p_g(1-\phi_o)}{1 - \phi_o\left(1 - e^{-\bar{C}_p \Delta\sigma}\right)} \quad \text{(A-18)}$$

Solving equation (A-17) for ϕ as a function of p gives:

$$\phi = \frac{p_g - p}{p_g} \quad \text{(A-19)}$$

Therefore:

$$\frac{\phi}{1-\phi} = \frac{p_g - p}{p} = e^{-\bar{C}_p \Delta\sigma} \quad \text{(A-20)}$$

Equation (A-20) indicates that a semi-log plot of $(p_g - p)/p$ versus effective stress would yield a straight line for constant pore compressibility.

APPENDIX B

Derivation of Permeability-Porosity Relationship for Fractures

The Carman-Kozeny permeability equation for fractures (Carman, 1937) as derived by McKee and Hanson (1975) is:

$$k = \frac{\phi^3}{12\,T(nw)^2} \quad \text{(B-1)}$$

where T is tortuosity, n is the number of fractures per unit area, and w is fracture height. The unit model used in calculating fracture permeability is shown in figure 9. Since porosity can be expressed as:

$$\phi = naw \quad \text{(B-2)}$$

where a is aperture. Therefore:

$$k = \frac{\phi\, a^2}{12T} \quad \text{(B-3)}$$

However, porosity can also be expressed as:

$$\phi = \left(\frac{a}{a + b}\right) w \quad \text{(B-4)}$$

where (a + b) is the cleat spacing, and b is the solid matrix thickness between the two fractures. Solving equation (B-4) for a as a function of ϕ:

$$a = \frac{b\phi}{w(1 - \phi/w)} \quad \text{(B-5)}$$

Substituting equation (B-5) into equation (B-3) yields:

$$k = (b^2/12Tw^2)\,\frac{\phi^3}{(1 - \phi/w)^2} \quad \text{(B-6)}$$

For continuous fractures, fracture height w is unity, which gives:

$$k = (b^2/12T)\,\frac{\phi^3}{(1 - \phi)^2} \quad \text{(B-7)}$$

Since pore compressibility is much higher than the solid matrix compressibility, b can be assumed to be constant thus:

$$k \text{ varies as } \phi^3/(1 - \phi)^2 \quad \text{(B-8)}$$

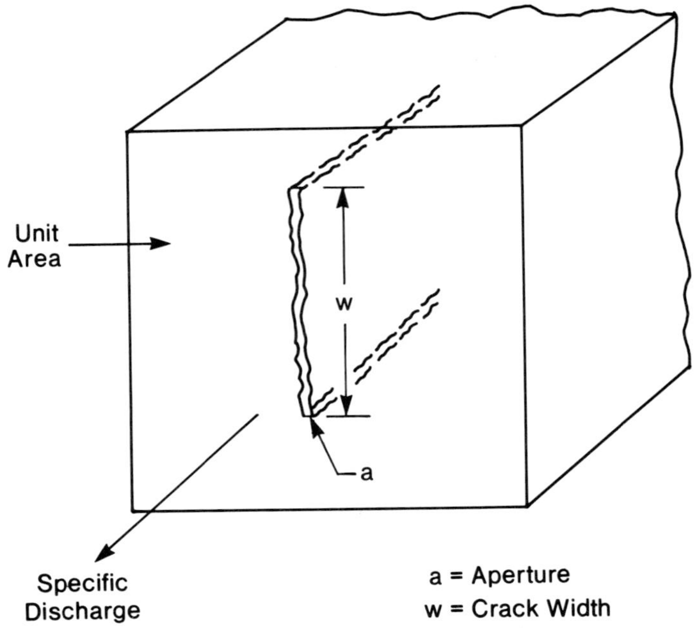

Figure 9. Diagrammatic representation of unit model used in Carman-Kozeny calculation of fracture permeability.

REFERENCES CITED

Brandt, H., 1967, Compressional wave velocity and compressibility of aggregates of particles of different materials: American Society of Mechanical Engineers Journal of Applied Mechanics, v. 34, no. 4, p. 866-872.

Carman, P. C., 1937, Fluid flow through granular beds, Transactions of the Institute of Chemical Engineers, v. 15, p. 150-166.

Dabbous, M. K., Reznik, A. A., Taber, J. J., and Fulton, P. F., 1974, The permeability of coal to gas and water: Society of Petroleum Engineers Journal, v. 14, no. 6, p. 563-572.

Hantush, M. S., 1964, Hydraulics of wells, *in* Chow, V. T., ed., Advances in Hydroscience, 2 v., Academic Press, New York City, v. 1, p. 281-432.

Hsieh, P. A., Neuman, S. P., and Simpson, E. S., 1983, Pressure testing of fractured rocks—A methodology employing three-dimensional crosshole tests: U.S. Nuclear Regulatory Commission Report NUREG/CR-3213, 176 p.

Jaeger, J. C., and Cook, N. G. W., 1979, Fundamentals of Rock Mechanics, 3rd ed., Halstead Press, New York City, 111 p.

Jones, A. H., Ahmed, U., Bush, D. D., Holland, M. T., Kelkat, S. M., Owen, L. B., Tibbits, G, and Bowman, K. C., 1983, Methane production characteristics of deeply buried coalbed reservoirs: Gas Research Institute Annual Report, GRI-83/0017.

Jones, F. O., 1975, A laboratory study of the effects of confining pressure on fracture flow and storage capacity in carbonate rocks: Journal of Petroleum Technology, no. 27, p. 21-27.

Koenig, R. A., and Stubbs, P. B., 1986, Interference testing of a Coalbed Methane reservoir: Society of Petroleum Engineers Paper 15225.

Lent, J., 1979 (revised 1982), San Juan Basin report—Early Tertiary geology, coal, and the potential for methane recovery from coalbeds in Colorado and New Mexico: Draft Report prepared for U.S. Department of Energy Morgantown Energy Technology Center, TRW Inc., Denver, Co.

McKee, C. R., Bumb, A. C., Way, S. C., Koenig, R. A., Reverand, J. M., and Brandenburg, C. F., 1986, Using permeability-vs-depth correlations to assess the potential for producing gas from coal seams *in* Quarterly Review of Methane from Coal Seams Technology, v. 4, no. 1, p. 15-26.

McKee, C. R., and Hanson, M. E., 1975, Explosively created permeability from single charges: Society of Petroleum Engineers Journal, v. 15, no. 6, p. 495-501.

National Petroleum Council, 1980, Coal seams, *in* Unconventional Gas Sources. v. 2: National Petroleum Council, 99 p.

Ostensen, R. W., 1983, Microcrack permeability in tight gas sandstone: Society of Petroleum Engineers Journal v. 23, no. 6, p. 919-927.

Reiss, L. H., 1980. The reservoir engineering aspects of fractured formations: Gulf Publishing Co., Houston.

Rightmire, C. T., Eddy, G. E., and Kirr, J. N., eds., 1984, Coalbed methane resources of the United States, American Association of Petroleum Geologists Studies in Geology, no. 17, 378 p.

Rose, R. E., and Foh, S. E., 1984, Liquid permeability of coal as a function of net stress: Society of Petroleum Engineers Paper 12856.

Schwerer, F. C., and Pavone, A. M., 1984, Effect of pressure-dependent permeability on well-test analyses and long-term production of methane from coal seams: Society of Petroleum Engineers Paper 12857.

Scorer, J. D. T., and Miller, F. G., 1974, A review of reservoir rock compressibility and its relationship to oil and gas recovery: Institute of Petroleum Paper IP 74-003, London.

Somerton, W. H., Soylemezoglu, I. M., and Dudley, R. C., 1974, Effect of stress on permeability of coal: U.S. Bureau of Mines Open-File Report 45-74, Contract H0122017, 56 p.

Thomas, R. D., and Ward, D. C., 1972, Effect of overburden pressure and water saturation on gas permeability of tight sandstone cores: Journal of Petroleum Technology no. 24, p. 120-124.

Tremain, C. M., and Toomey, J. J., 1983, Coalbed methane desorption data: Colorado Geological Survey Open-File Report 81-4, 514 p.

Walsh, J. B., 1981, Effect of pore pressure and confining pressure on fracture permeability, International Journal of Rock Mechanics and Mining Science and Geomechanics Abstracts, v. 18, no. 5, p. 429-435.

Way, S. C., Bumb, A. C., Koenig, R. A., McKee, C. R., Reverand, J. M., and Ward, W. P., 1984a, Hydrologic characterization of coal seams for methane recovery, activity 1 topical report—Hydrologic data base for Piceance, San Juan, and Warrior Basins: Gas Research Institute Final Report GRI-83-0064, 59 p.

Way, S. C., Koenig, R. A., Reverand, J. M., and Ward, W. P., 1984b, Hydrologic characterization of coal seams for methane recovery, activity 2 topical report—Hydrologic constraints and test procedures: Selected methods for well test analysis, v. 2,: Gas Research Institute Final Report GRI-83/0065.2, 108 p.

Wicks, D. E., McFalls, K. S., and Kuuskraa, V. A., 1987a, A geologic assessment of natural gas from coal seams in the Warrior Basin, Alabama: Gas Research Institute Topical Report GRI-86/0272.

Wicks, D. E., McFalls, K. S., and Kuuskraa, V. A., 1987b, A geologic assessment of natural gas from coal seams in the Piceance Basin, Colorado: Gas Research Institute Topical Report GRI-87/0060.

The San Juan Smelter located at the base of Smelter Mountain just south of Durango processed ores from the surrounding mountains. Over a million dollars worth of ores, including gold, silver, copper and lead, had been processed by 1888. (Smith, D.A., 1980) The physical plant ceased operations in 1963 when the Vanadium Corporation of America halted uranium milling and permanently closed the plant. Large radioactive tailing piles were left and as of 1988 the piles are being relocated to a repository in Bodo Canyon south of Durango. Photo courtesy of the Durango Public Library.

Improved Hydraulic Fracturing Strategy for Fruitland Formation Coal-Bed Methane Recovery San Juan Basin, New Mexico and Colorado

John Ely[1]
Stephen A. Holditch[1]
Ronald H. Carter[1]

ABSTRACT

Many different completion techniques have been applied to wells producing gas from coal seams. One of the most successful techniques for enhancing production from coal-bed methane is hydraulic fracturing. The Gas Research Institute, numerous operating companies, and several oil-field service companies have been conducting research trying to learn more about producing methane gas from coal seams. Several hundred wells have been completed in coal seams in the San Juan and the Black Warrior Basins. There are certain similarities between the application of hydraulic fracturing in coal seams and conventional elastic reservoirs, but there are also certain areas where distinct and important differences occur. This paper presents a strategy that can be applied to the completion and stimulation of wells producing from coal-seam reservoirs.

INTRODUCTION

Over the past several years there has been increasing interest in the production of methane gas from coal seams. Coal-bed methane is now an important energy source in parts of the United States such as the Black Warrior and San Juan Basins (Lambert and others, 1987, p. 81-82; Schraufnagel, and others, 1987, p. 83-92; Wicks and others, 1987, p. 131-140, Kelso and others, 1987, p. 119-130; Choate and others, 1984, p. 185-222; and Fassett, 1987, p. 19-38).

Hydraulic fracturing of wells is a stimulation method that has been widely and successfully applied for over 35 years. Most of the research and application of hydraulic fracturing has been targeted towards elastic reservoirs, such as sandstones, siltstones, and shales, or to carbonate reservoirs, such as limestones or dolomites. The mechanical behavior of these rocks can be described quite well using equations derived for a porous-elastic medium. In addition, virtually all fracturing techniques, fluids, additives, and propping agents have been developed to stimulate sandstone or carbonate rocks.

After a fracture treatment, a well is opened to produce the fracture fluids and the oil and gas stored in the pore space of the reservoir rock. Soon after the fracture treatment, the flow rate from the well will be maximized and will then begin to decline. The well is produced until the flow rate declines to the economic limit. The amount of gas produced will be controlled by the volume of free gas that is stored in the pore space in the formation.

Only recently has the attention of the petroleum industry been directed towards the completion of wells and the production of gas from coal seams. Compared to our knowledge of sandstone reservoirs, very little is known about completing, fracturing, and producing coal-seam gas reservoirs. Current research funded by Gas Research Institute (GRI) and other organizations is helping to develop a better understanding of how to produce gas from coal seams.

It is known that the mechanical properties of coal seams are much different from those of sandstones or limestones (Jeffrey and others, 1987, p. 113-116). The effective modulus of coal is much less than of other rocks, often resulting in the creation of wide fractures. It is also common to observe multiple parallel fractures or fractures that propagate in different directions (Diamond and Oyler, 1987).

The production of gas from coal seams is also much different from what is normally observed from sandstone reservoirs. Many times there is no free gas in the coal. The cleat (or fracture) system can be 100 percent saturated with water. All of the gas is adsorbed on the surface of the coal. To produce gas, one must produce the water from the cleat system and dewater the coal near the well bore. As the water is produced and the reservoir pressure is reduced, gas will begin to desorb and will eventually be produced.

Therefore, for many coal-seam wells, the gas flow rate increases with time (Byrer and others, 1987, p. 821-834). Maximum gas-flow rates may not be observed for months or even years after production is initiated. Even when free gas is present in the coal cleat system under initial conditions, desorption plays a major role in long-term gas recovery and economics.

If there is a similarity in what we are trying to accomplish with coal-seam fracturing compared to conventional fracturing, it would be compared to hydraulic fracturing of naturally fractured carbonate reservoirs. In such reservoirs, there is little or no matrix permeability and virtually all of the productive capability of the formation is controlled by the natural fracture system and the gas-in-place that exists in the natural fractures. To properly stimulate naturally-fractured carbonate

[1] S. A. Holditch & Associates
900 S.W. Parkway East, Suite 200
College Station, Texas 77840

reservoirs, one must interconnect the natural fracture system to the well bore. This, in fact, is what must be accomplished in hydraulic fracturing of coal-seam reservoirs. We must interconnect the cleat system, where the natural permeability of the reservoir exists, to the well bore. If the well bore is not in communication with the cleat system, little or no productivity will be seen.

One large dissimilarity between carbonate reservoirs and coal reservoirs is the low effective modulus or high compressibility of the coal-seam reservoir. As the modulus of a reservoir decreases, the created fractures will be much wider and shorter. If the effective modulus of the coal seam is too low, it may prevent the creation of deep-penetrating hydraulic fractures.

PRE-FRACTURE FORMATION EVALUATION

As with conventional reservoirs, it is important that we determine the critical formation properties prior to designing the completion and hydraulic fracture stimulation treatment. The three most important sources of data are cores, logs, and well tests.

The core can be used to determine the type of coal that exists in a particular location and the gas content of the coal. The core can also be used to determine the desorption-isotherm relationship that controls how much gas will be transported from the coal into the cleat system. Other properties such as Young's modulus, Poisson's ratio, and bulk modulus can be measured if sufficient core is recovered.

A full suite of logs run during the drilling operation can also be important to the fracture treatment design. The logs will help pinpoint the location and the thickness of all the coal seams that are potential completion candidates. As will be described below, the completion and stimulation strategy involved with completing a series of thin coal seams can be substantially different from attempts to complete a single thick coal seam. The logs also provide information that can be used by geologists to map the areal extent of coal beds.

Finally, pre-fracture well tests are needed to determine the initial pressure in the coal seam, the desorption pressure, and the permeability of the coal. If a coal seam is water saturated, then an injection-falloff test is usually the best way to determine the coal permeability. If a coal seam is initially gas saturated in the cleat system, then a conventional pressure-buildup test will provide the necessary results.

Many coal seams are initially saturated with water. In some circumstances, the pressure at which gas begins to desorb in significant amounts is less than the original pressure of the water in the cleat system. The pressure when gas first begins to desorb from the coal into the cleat system is called the desorption pressure. In water-saturated coals, it is sometimes necessary to perform pressure-drawdown tests at a series of bottom-hole flowing pressures in an effort to determine the desorption pressure. In some circumstances, artificial lift or swabbing units must be used to maintain the desired bottom-hole pressure.

An engineer cannot simply design a fracture treatment on a coal seam based upon limited data. To successfully complete and stimulate a coal reservoir, data from core analyses, log analyses, and well test analyses must be used to properly evaluate the formation. It is extremely important that the thickness of the coal seams, the initial pressure in the coal seams, the desorption pressure of the coal seam, and the permeability of the coal seam be known within reason, if one wishes to properly complete and stimulate a particular interval.

COMPLETION AND STIMULATION STRATEGY

To produce gas from coal seams, a hydraulic fracture treatment is normally required. The type of fracture treatment designed is a function of the depth and thickness of a coal seam. To better explain how the treatment design may vary, the following three scenarios are commonly observed: Scenario I involves a shallow coal seam where a horizontal fracture will be created; Scenario II involves a series of thin coal seams in a depth range where a single, planar, vertical fracture will be created; and Scenario III involves a single, thick coal seam where the hydraulic fracture will be confined entirely in the coal and a complex fracture system (multiple vertical or T-shaped fractures) is created. For these three scenarios, the following strategies should be used to complete the well and design the hydraulic fracture treatment.

General Strategy (Applies to All Three Scenarios)

The hydraulic fracturing of a coal seam is mainly a process used to interconnect the cleat system of the coal to the well bore. Without a well developed cleat system, creating a long fracture will not be profitable. If long fracture lengths cannot be created and propped, then some low permeability coals will not be economical to produce at current gas prices. For extremely low permeability reservoirs, we must create long conductive fractures to achieve acceptable flow rates. Because of the nature of coal and the inability to achieve long propped-fracture lengths in certain coals, there will be some coal seams with permeability so low that they cannot be produced at commercial flow rates.

Bottom-hole pressure during production should be minimized to accelerate desorption of natural gas. We need to take this into account in our fracture designs, making sure that we have wide, conductive, propped-fracture systems that can withstand large stress gradients.

Scenario I

In this scenario, involving shallow coal seams with horizontal fractures, individual horizontal fractures will have to be created using either limited entry methods or mechanical diversion. Linear fluids with a moderate size pad volume should be used to fracture-treat the interval. The bottom-hole treating pressure will be in excess of 1 psi/ft. Complex (multiple) fracture systems may be created if bottom-hole pressures increase substantially during the treatment.

Scenario II

This scenario involves a single vertical fracture through a series of thin coal seams. In this case the excess fracture pressure caused by the fluid moving down the fracture will be only 200 to 500 psi, which is far less than the 1000 to 2000 psi experienced when Scenario III occurs. Fracture propagation gradients should be less than 1 psi/ft in Scenario II.

Under these conditions a single, planar, vertical fracture cutting through several coal seams will be created; therefore, conventional 3-dimensional fracture design models can be used to estimate fracture dimensions. Leakoff is usually not a problem and normal pad volumes of 30 to 35 percent can be used to fracture-treat the coal. Viscous, delayed cross-link fluids (or foams) should be used to provide adequate proppant transport and to minimize proppant settling.

Scenario III

This scenario involves a single, thick coal seam where a complex hydraulic fracture is contained entirely in the coal. High injection rates must be used during the fracture treatment to combat high leakoff.

In addition to large injection rates, high viscosity and bridging-fluid-loss additives may be required to combat loss of fluid efficiency. In the San Juan Basin, operators have combated loss of fluid efficiency with an accelerated injection rate but for some completions high injection rates are not feasible and in those wells control of leakoff becomes a necessity. Because of the low Young's modulus, the high formation compressibility, and the presence of complex fracture systems, a hydraulic fracture will seldom penetrate more than a few hundred feet from the well bore in a single thick coal seam.

Obviously, there could be other scenarios that do not fall exactly within the three discussed above, however, it has been our experience that the vast majority of the coal seams being completed and stimulated currently in the United States can fit within one of the above scenarios.

THE WELL COMPLETION

Because of the extremely low values of Young's modulus for coal (relative to bounding formations) it is critical that proper isolation of the coal seam be achieved. In many cases, coal will wash out during drilling, which results in an enlarged bore hole through the coal seam. In areas where the coal seams are overpressured and the coal tends to spall during drilling operations, mud weights can be increased to maintain the competence of the coal seam. Loss of mud and (or) cement into the coal seam can also occur during drilling or cementing operations. Bridging-fluid-loss additives can be used in the cement and the mud to eliminate the loss of large quantities of fluid into the natural fracture (cleat) system of the coal. In areas where underpressured coal seams exist, the use of new light-weight, high strength, cements placed across the coal seam has achieved adequate isolation of the coal from the overlying and underlying rock layers.

The inability to achieve good coal-bed isolation can cause total failure of a completion attempt. In the San Juan Basin, Fruitland coals and the Pictured Cliffs Sandstone are sometimes adjacent. It is often very difficult to fracture treat just the coal seam. If the Pictured Cliffs and the Fruitland coal have equal access to the fracturing fluid, it is difficult to determine which zone will accept most or all of the fluid.

In other areas of the United States, such as the Black Warrior Basin of Alabama, where mining back into coals that have been hydraulically fractured has occurred, it has been observed that if a fracture is initiated in the coal-bed, the fracture remains contained within the coal. Although some of this experience involved depths that are considerably shallower than in the San Juan Basin, much of the knowledge concerning fracture behavior can be extrapolated to the San Juan Basin.

Field experience has indicated that it is desirable to initiate fractures within the coal seam. If fracture initiation starts below or above the coal, it is possible that the majority of the fracture treatment will not enter the coal seam. If, however, initiation starts within the coal, a very wide, very complex fracture system will usually propagate and remain contained within the coal.

Open-hole completions, slotted-liner completions, and a few gravel-pack completions have been tried in the San Juan Basin. Most of the time, these open-hole completions have encountered production problems with severe hole instability. Large workover costs are often necessary to clean out coal fines from the well bore. Cased-hole, perforated completions are preferred by most operators.

FRACTURE FLUID SELECTION

Virtually every type of fracturing fluid offered by a service company (other than alcohol or oil-based fluids) has been used in the stimulation of coal seams. Fracturing fluids have varied from straight water (which is used extensively in the Black Warrior Basin) to complex, cross-linked fluids. The cross-linked fluids are predominant in the San Juan Basin. There has been extensive use of foam and cross-linked foam fracturing fluids in the San Juan Basin. To state what particular type of fracturing fluid is best for all coals is an impossibility because the depths, thicknesses, and numbers of the coal seams vary.

In low-productivity areas where thin seams with minimal gas in place are the norm, economics becomes a major factor in fluid selection. The use of high-cost polymers and cross-linking agents can reduce the possibility of payout of a fracturing treatment in commercially marginal areas, no matter how successful the frac job. In areas where thick coal seams in overpressured areas exist, the use of a complex, high viscosity, cross-linked fracturing fluid carrying large volumes of proppant may turn out to be the most profitable fluid.

Experience in naturally fractured carbonate reservoirs has taught us that extremely high leakoff will occur and can be the most serious design problem. If high leakoff occurs, the potential for screenout is extremely high. If low fluid efficiency is to be expected, the treatment should be designed for high injection rate, high viscosity, and bridging-fluid-loss additives should be used. Experience in the San Juan Basin has taught operators to use 40 to 50 percent pad volumes, cross-linked fracturing fluids, and pumping rates exceeding 40 bbls/min. These techniques have allowed placement of large quantities of sand at relatively high concentrations in Fruitland coal beds.

Throughout the San Juan Basin, many different types of fracturing fluids have been used. At the present time, a typical fracturing treatment would consist of a 30 to 40 lb cross-linked guar-gum system. As a bridging-fluid-loss additive, 100-mesh sand can be used but there is a great deal of controversy concerning the effectiveness of the 100-mesh sand. Silica flour and other bridging-fluid-loss additives have been used with varying degrees of success. A typical treatment for a 40- to 50-foot-thick coal seam interval in the San Juan Basin would include 150,000 to 300,000 lbs of sand pumped at 40 to 50 bbls/min with an excess of 40 percent pad volume.

Post-fracture production from a typical coal-bed methane well in the San Juan Basin can vary from 50 to 1,000 MCFPD. The expected level of post-stimulation production often dictates the amount of money that can be spent on the fracturing fluid.

PROPPANT SELECTION

There has been some controversy concerning the type and concentration of proppants needed to properly stimulate a coal seam (Jones and others, 1987). The purpose of a fracture treatment in a coal seam is to place proppant in a fracture that will interconnect the cleat system within the coal to the well bore. With this in mind, large size (10-20 mesh) proppants are not necessarily optimum. Smaller size (20-40 or 16-30 mesh) proppants can provide adequate conductivity for most coal reservoirs. Given comparable conductivities, the smaller proppants are preferred due to their ability to be transported by the fracturing fluid more efficiently as compared to the larger (10-20 or 8-12 mesh) proppants.

Substantial productivity increases have been obtained through the use of very small proppants (100-mesh sand) (Perlman, oral communication with S. A. Holditch, 1985). Significant sand production has been observed from many of the wells stimulated in this manner, but the fact remains that effective fracture stimulation was achieved through the use of extremely small proppants.

Because of the complex nature of the fractures in coal and because of the relative softness of coal, we feel that high concentrations of proppant are necessary. Proppant concentrations in excess of 5 to 6 lbs/gal are commonly pumped in San Juan Basin fracture treatments. Wide, packed fractures are required for maintaining conductivity because of the softness of the coal and the potential for embedment and the production of fines.

There has also been a considerable quantity of curable-resin-coated sand pumped during the past few years. The consolidated sand pack has been successful in some areas, but in other areas it has been found that it is not required and the wells will produce gas and water without the production of fines. A typical application of the curable-resin-coated sand has consisted of a tail-in of 30,000 to 50,000 lbs of resin-coated proppant for a 200,000 to 300,000 lb treatment. For deeper coal-seam wells, resin-coated sand may be needed throughout the treatment to take advantage of its higher strength properties, as well as the control of fines production.

CONCLUSIONS

Although there are many areas where fracturing-technology transfer is applicable from sandstone and limestone stimulation, significant differences occur which require the development of new technology. The low effective modulus of elasticity of coal results in wide fractures and limited fracture length. Successful coal seam hydraulic fracturing is primarily due to interconnection of the fracture with the natural cleat system of the coal. As with conventional reservoirs, it is critical to ascertain formation properties such as permeability, gas in place, and elastic properties prior to attempting hydraulic fracture stimulation. Fracture design for coal seams will vary greatly based on: In situ stress of the coal and surrounding zones, thickness of the coal seam, and presence of multiple layers. Due to high leakoff potential, high injection rates and (or) bridging-fluid-loss additive may be required. Fluid and proppant selection depend upon specific coal-seam lithology. Key factors are the need for fracture penetration and sufficient fracture conductivity to interconnect the coal cleat system.

REFERENCES CITED

Byrer, C. W., Moroz, T. H., and Covatch, G. L., 1987, Coalbed methane production potential in U. S. basins: Journal of Petroleum Technology, July, p. 821-834.

Choate, R., Lent, J., and Rightmire, C. T., 1984, Upper Cretaceous geology, coal, and the potential for methane recovery from coal beds in San Juan Basin—Colorado and New Mexico, *in* Rightmire, C. T. Eddy, G. E. and Kirr, J. N. eds., Coalbed Methane Resources of the United States: American Association of Petroleum Geologists Studies in Geology Series, no. 17, p. 185-222.

Diamond, W. P., and Oyler, D. C., 1987, Effects of stimulation treatments on coal beds and surrounding strata: evidence from underground observations: U. S. Bureau of Mines Report of Investigations, no. 9083, United States Department of the Interior.

Fassett, J. E., 1987, Geometry and depositional environments of Fruitland Formation coal beds, San Juan Basin, New Mexico and Colorado: Anatomy of a giant coal-bed methane deposit: Proceedings of the 1987 Coalbed Methane Symposium, Tuscaloosa, Alabama, November 16-19, 1987, p. 19-38.

Jeffrey, R. G., Vandamme, L., Hinkel, J. J., and Horner, D. M., 1987, Propagation of fractures in coal: Equivalent modulus and parallel fracture effects: Proceedings of the 1987 Coalbed Methane Symposium, Tuscaloosa, Alabama, November 16-19, 1987, p. 113-116.

Jones, A. H., Bell, G. J., and Morales, R. H., 1987, Hydraulic fracture design rationale for recovery of methane from coal seams: Gas Research Institute, final report GRI-87/0016, Chicago, IL.

Kelso, B. S., Decker, A. D., Wicks, D. E., and Horner, D. M., 1987, GRI geologic and economic appraisal of coalbed methane in the San Juan Basin: Proceedings of the 1987 Coalbed Methane Symposium, Tuscaloosa, Alabama, November 16-19, 1987, p. 119-130.

Lambert, S. W., Niederhofer, J. D., and Reeves, S. R., 1987, Multiple coal seam completion experience in the Deerlick Creek field, Black Warrior Basin, Alabama: Proceedings of the 1987 Coalbed Methane Symposium, Tuscaloosa, Alabama, November 16-19, 1987, p. 81-82.

Schraufnagel, R. A., Lambert, S. W., Stubbs, P. B., Dobscha, F. X., and Boyer, C. M., II, 1987, The Rock Creek Field Laboratory—A project update: Proceedings of the 1987 Coalbed Methane Symposium, Tuscaloosa, Alabama, November 16-19, 1987, p. 83-92.

Wicks, D. E., McFall, K. S., and Zuber, M. D., 1987, An analysis of the coalbed methane resources of the Warrior Basin, Alabama: Proceedings of the 1987 Coalbed Methane Symposium, Tuscaloosa, Alabama, November 16-19, 1987, p. 131-140.

Tiltmeter Mapping and Monitoring of Hydraulic Fracture Propagation in Coal: A Case Study in the Warrior Basin, Alabama

George A. Gazonas, Chris A. Wright and M. D. Wood[1]

ABSTRACT

Significant advances in the use of tiltmeter technology for hydraulic fracture mapping have been made during the past decade. Improvements have progressed in instrument design, site preparation, array processing to reject noise, and models for interpretation of the signal. These improvements have increased the precision of determining the orientation, dimensions, and positions of hydraulic fractures; they have also allowed the use of this technology in a greater variety of formations, at different depths, and at different rates and volumes of fracture treatments.

The relatively complex response of coal to hydraulic fracture stimulation makes real-time monitoring and mapping of the fracturing necessary. Real-time mapping provides timely information for comparison of fracture design with actual execution and offers an opportunity to modify the stimulation process as new conditions are revealed. Complex fracturing has been observed in a wide variety of materials as expected at shallow depths. The cleats (joints) of coal can enhance the overall tendency for discontinuities to compound the complexity of fracture response to stress gradients at shallow depths. The coal-cleat system may lead to pathologies, such as fracture growth out-of-zone and growth of a fracture out of the original plane. These induced fracture pathologies may deviate so far from actual design that dehydration of the slurry, leading to screenout and possible loss of the well, cannot be avoided unless a premonitory signal can be established in real-time data; such signals would allow steps to be taken to modify injection rates and slurry concentrations to mitigate risks.

A case study of a coal-bed fracture mapping job completed in Alabama is used to illustrate most of these points. In particular, a statistical test (F-test for the equality of variances) is used as a criterion for justifying an increase in source complexity when a single-fracture source does not adequately describe the observed surface-deformation field.

INTRODUCTION

Real-time mapping of hydraulic fracture stimulations in coal is of major importance in situations where job execution can deviate substantially from job design. Timely transmission of critical information can facilitate beneficial changes in injection parameters to minimize undesired results. It is the purpose of this paper to discuss general procedures, benefits, and risks associated with real-time fracture mapping. The discussion is not comprehensive but instead focuses on practical aspects of detection and description of what appears to be a widespread problem in hydraulic fracturing of coal—namely, generation of dual fractures during fracture stimulation.

No distinction is made between coal and other material in which this phenomenon has been observed. The discussion is restricted to a "blind" analysis of tilt data from an array of high-gain (nanoradian) surface tilt sensors and does not include the more realistic and desirable situation of integrated interpretation of treatment records and other reservoir data. The text is intended to be more of a descriptive real-earth discussion than an exhaustive theoretical discussion on each of the topics covered.

The first section of this paper outlines the theory and principles for tiltmeter hydraulic fracture mapping. The next section outlines a method for real-time monitoring of fracture growth, with emphasis on detection of changes in fracture orientation. Finally, a case study is discussed which illustrates most of the major points outlined above.

[1] George A. Gazonas
Chris A. Wright
M.D. Wood
Hunter Geophysics
2344 Walsh Avenue, Building F
Santa Clara, California 95051

The work discussed in this paper includes results described in Hunter Geophysics' in-field reports on coal-seam hydraulic fracture mapping to C.M. Boyer II of United States Steel Corporation. The authors wish to thank the Rock Creek site personnel and GRI for their support of this project. The field work was subcontracted under GRI contract 5083-214-0847.

TILTMETER FRACTURE MAPPING

Theoretical Models

Theoretical displacement, tilt, and strain of the Earth's surface associated with seismic, volcanic, and other fracture sources is well established in the literature. Press (1965) discusses the dislocation modeling approach for seismic sources and provides an extensive bibliography of other more basic dislocation modeling. The dislocation model herein employed is based on the Davis (1983) solution.

Theoretical models output surface deformations (solution to the forward problem) as a function of the source, or the subsurface fracture characteristics such as location, orientation, and dimensions of the source. The dislocation model presently being used has eight fracture-source parameters; dip, strike, length, width, height, depth, and the x and y positions of the fracture center relative to the wellbore. The theoretical deformations predicted by the model are relatively insensitive to the material properties of the medium and demonstrates its great utility for a variety of rock types.

Changes in source parameters produce changes in magnitude and direction of tilt at each site in a surface array of tiltmeters. This is the principle which allows a solution to the problem of determining source parameters from surface tilt measurements. If some knowledge of the surface tilt field is known, then solution of the inverse problem yields estimates of the source parameters. The method used to solve the inverse problem is discussed in a subsequent section.

Data Acquisition

An array of tiltmeters is installed in shallow (20 ft. deep) boreholes surrounding the injection well. Tilt response is uniformly sampled at all instrument locations. Solution precision is a strong function of quality of tilt data and the number of locations where tilt data are recorded.

Detection of high-resolution tilt requires an excellent detector and excellent coupling of the detector to the earth in a minimum noise environment (Wood and Allen, 1973). Commercial applications compromise these ideal observatory type characteristics. Instruments currently being used record tilt in two orthogonal directions. Signals recorded at each site are embedded in a broad spectrum of local, regional, solid earth tidal, meteorological, and planetary noise. It is therefore important to be able to obtain noise records prior to the start of the fracture treatment (injection) of sufficient duration to define the spectrum of noise.

In real-time applications, a background noise trend is least-squares fitted, using an appropriate model (Ljung, 1987), up to time of the injection. At that point, coefficients in the least-squares fit are fixed and are used to predict the noise trend during the time of injection. The signal is then defined as the difference between observed data and predicted noise trend. A signal tilt vector is then constructed for each site from the orthogonal signal channels at each site.

Fracture Solution

After obtaining a set of tilt vectors from an array of instruments, the goal is to determine the nature of the source that produced the observed tilt vectors. Using the dislocation model, a set of theoretical tilt vectors can be calculated at instrument locations for any given set of source parameters. Tilt is defined as the spatial derivative of the vertical displacement field (fig. 1). Graphically, the solution to source parameters is found when there is a best fit match between observed and theoretical tilt vectors (fig. 2).

The solution for source parameters that best fits the observed tilt field is obtained by a numerical minimization technique. An error functional is defined as the sum of the

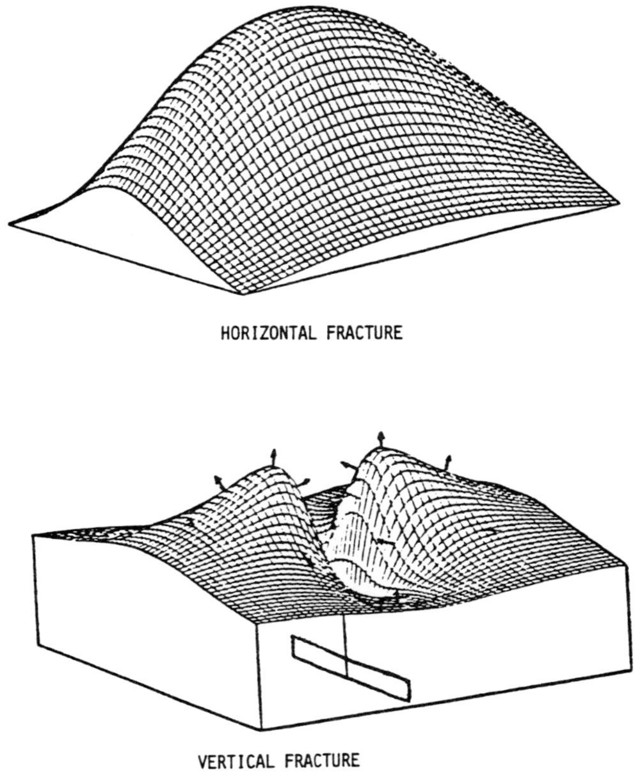

Figure 1. Vertical displacement fields induced above vertical and horizontal buried dislocations (adapted from Davis, 1983). Tilt is defined as the spatial derivative of the vertical displacement field.

squares of magnitudes of vector differences between observed and theoretical tilt vectors at each instrument location. The error functional, being a function of source parameters, is minimized by iterating over the possible range of source parameters. Ranges of possible source parameters can be constrained if specific information is known about the formation being stimulated or if conclusions can be drawn from other monitoring data (pressure analysis, temperature logs, radioactive tracers, etc.). Source parameters that give a minimum value of the error functional are considered best estimates of the actual fracture parameters.

It is important to simultaneously minimize the error functional with respect to all source parameters, rather than serially for each parameter, since the variation of the error functional for each parameter is dependent on the other parameters. The permutation of all possible combinations of eight parameters requires a practical assignment of the range and increment to be tested for each parameter and the rate of convergence that will give an acceptable computation time. The method of steepest descent facilitates these goals. This method is reviewed by Press and others (1986). It can be shown that this procedure will find the same minimum and is, therefore, equivalent to calculation of all possible permutations of values for the same parameters, if the error surface is "smooth" to at least second order derivatives. The procedure is rapid and output is the optimum value and local convergence sensitivity for each parameter.

It is illustrative to obtain slices through the 8th order error surface to show the relative behavior of convergence to the

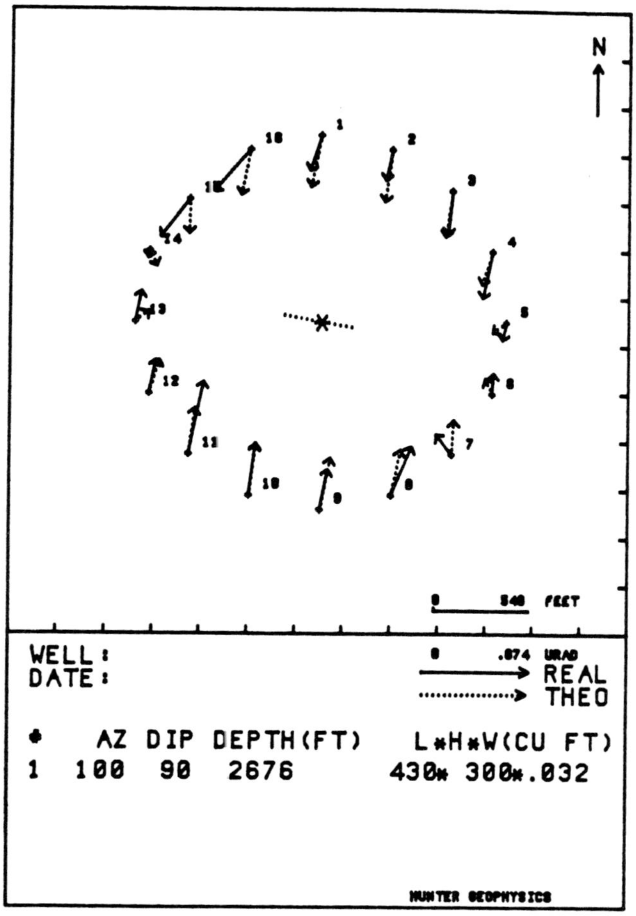

Figure 2. Overlay of observed (solid lines) and theoretical (dotted lines) tilt vectors for a vertical hydraulic fracture and the associated fracture solution (depicted by the dotted line through the wellbore).

best solution for each parameter. The error surface for orientation is useful to present as an error matrix of dip vs. strike (fig. 3). The error function for volume is obtained by varying fracture dimensions while keeping all other parameters fixed (fig. 4). Illustrations such as these assist in visualizing the convergence to the best fit obtained from the eight-dimensional method of steepest descent.

The model has highest sensitivity to orientation of the fracture. Azimuth and dip of a fracture can usually be precisely determined. The magnitude of the tilt signal is proportional to source volume and inversely proportional to the cube of the source depth. The least sensitivity of the model is to the linear dimensions of the fracture or how the fracture volume is partitioned. Volume partitioning between width and area is much more sensitive than partitioning of the area between height and length.

Complex Fractures

In some cases the error functional minimization yields a solution that still has a significant error between observed and theoretical tilt vectors. This can occur due to a very low signal-to-noise ratio, where there is great uncertainty in measuring the surface tilt field. In cases where expected tilt-signal magnitudes are very small or environmental noise is extremely

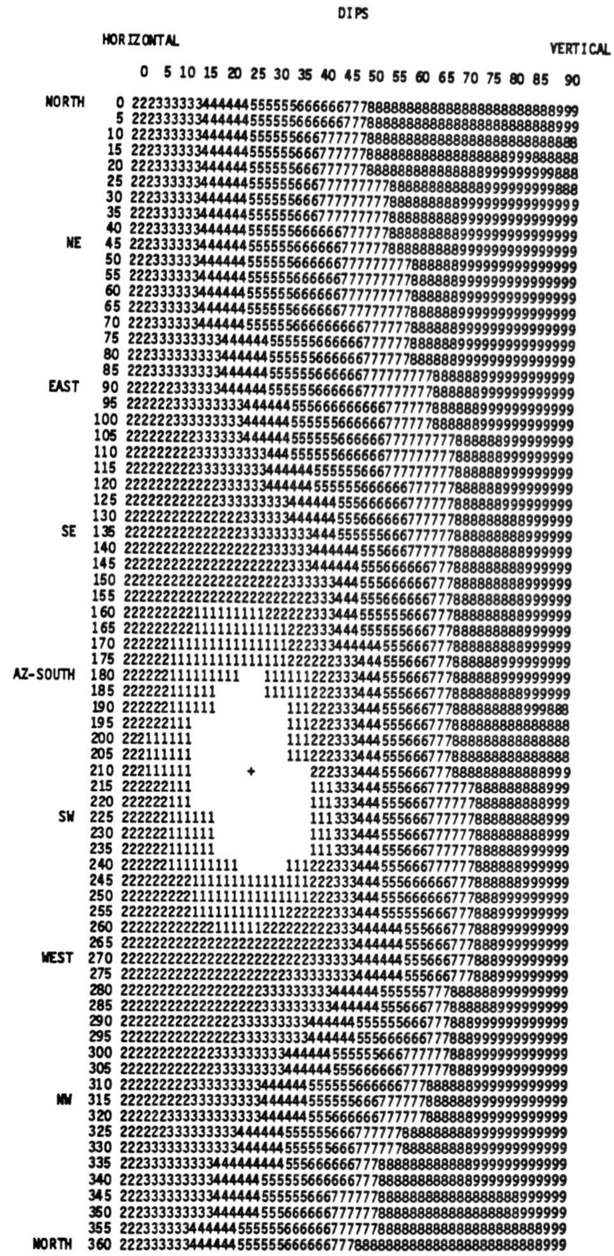

Figure 3. Contoured error surface for the P1A stimulation in azimuth-dip space. The surface is obtained by fixing all other fracture parameters and calculating the mean-squared error between observed and theoretical tilts at 5 degree increments in orientation over all sites. The minimum error occurs at the + sign at azimuth N. 30 E. and dip 25 northwest. Subsequent logarithmic contours are depicted by the number 1 through 9 in the plot.

high across the whole array, the solution for fracture source parameters may not be possible. In the case of poor fit due to poor signal-to-noise ratios, tilt error vectors will represent, by definition, noise. There is, however, another possible scenario:

Array processing techniques can determine whether a group of error vectors across an array are in some fashion coherent, spatially low frequency, or are spatially random (measurement noise). Details of this are omitted due to complexity of the

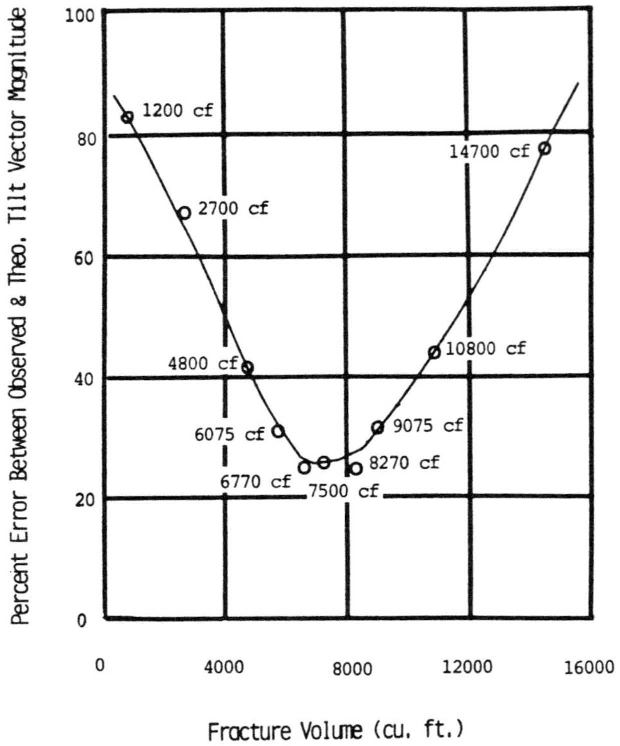

Figure 4. Parabolic error surface of the mean-squared error versus volume obtained by varying the dimensions of the single fracture and keeping all other fracture parameters fixed.

argument. The result is that in some cases high residual errors at the end of a best-fit minimization are not due to poor data quality but instead are due to insufficient model accuracy. In these cases, the remaining error vectors are not random, but display a regular pattern across the array. Experience has shown that if the single fracture model is expanded to include more complex fracture structures, then a quality of fit more commensurate with observed signal-to-noise ratio is possible. It has been found that fracture growth can change in orientation during a stimulation, or even grow in two planes simultaneously.

More than 1,000 fracture jobs have been monitored with tiltmeters with the bulk of them executed at depths of less than 3,000 feet within the past 48 months. A very large percentage of these jobs indicate that fracture complexity is far more widespread than is commonly acknowledged. The role of stress in determining fracture orientation is well established. Fracture stimulations alter the magnitude of principal stresses and in those cases where the difference between the intermediate and least principal stress is reversed in sense, incremental fracture growth will occur out of the original fracture plane. When fracture growth breaks out in another plane, (documented instances can be found in Evans and others, 1982; and Smith and others, 1985) often times there is an abrupt change in slope on the raw tiltmeter records on all (or many) channels in the array. The abrupt slope break can be detected and interpreted in real-time to indicate if and when fracture growth initiates in a second plane. In some instances, however, abrupt slope changes are not present on the raw tilt records, yet alternative evidence clearly indicates that a multiple fracture system has developed. This can occur in formations where the in-situ principal stresses do not appreciably vary with depth, or when one fracture component is significantly larger than the other. This paper discusses a simple but widespread condition of fracture growth in more than one plane to illustrate the value of real-time fracture mapping.

Real-time Monitoring of Fracture Growth

Real-time mapping of hydraulic fracturing is valuable in those operations that can deviate substantially from fracture design and possibly lead to pathologies such as screenout and loss of a well. Fracture complexity such as out-of-zone growth, growth in an orientation other than the original design, growth out of a single fracture plane leading to dual fractures, and unexpected leakoff are serious deviations from fracture design that surface tilt measurements are especially sensitive to. In many of these situations it may be possible to recognize premonitory conditions that can be altered by changes in injection rates and slurry concentrations to recover control sufficiently to produce a more desirable result, even if the original design cannot be restored. A grid of high resolution tilt sensors can contribute basic information, which together with other real-time data, serves to constrain the interpretation of each output parameter to a useful result.

Fracture growth in two planes, especially in orthogonal planes, can seriously threaten the successful completion of a stimulation. Difference in fracture aperture between a vertical and a horizontal fracture can cause dehydration of the slurry in one fracture, leading to premature screenout at undesirable locations in the fracture system. Screenout is not an instantaneous process, and if tilt sensors can detect the onset of new fracture growth out of the single plane of the fracture before dehydration causes bridging and screenout, options can be preserved for job completion.

Tilt vectors are quite sensitive to changes in orientation of new fracture growth. A horizontal fracture will have a four-fold greater leverage on the free surface (i.e. tilt signal) than a vertical fracture of the same volume and surface area at the same depth. Moreover, given the inverse cube relation of signal amplitude with depth, tilt is a sensitive detector of changes in orientation and depth. Thus, out-of-zone growth and growth out of a single fracture plane are pathologies that tilt data can address in real-time.

Changes in fracture orientation can appear on raw tilt records as abrupt changes in slope, or time-rate of change-of-tilt. If a fracture grows in a vertical plane and then changes orientation and some fraction, or all, of the injected fluid starts to flow into a horizontal fracture plane, the output slope of tilt data can change (fig. 5). The complex fracture propagation process can then be analyzed by temporally partitioning the tilt data (fig. 6).

For the first part of the job, the fracture was clearly oriented vertically, and for the latter part of the job, the fracture changed to a horizontal orientation. By the time of shut-in, the tilt component due to the horizontal fracture had almost entirely swamped the vertical tilt component. In fact, the fracture may have been diagnosed as horizontal if one only looked at orientation of tilt vectors at completion of the injection. Both the abrupt slope change, which was seen on all channels, and the resulting volume discrepancy point to the creation of a complex fracture structure. Real-time analysis is

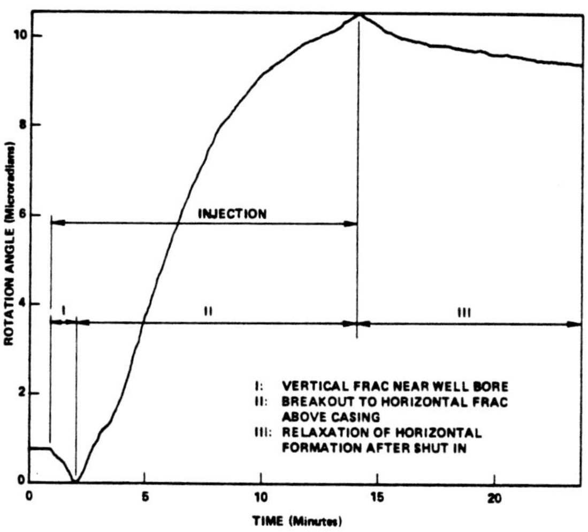

Figure 5. Three stages of fracture formation (from Wood and others, 1983).

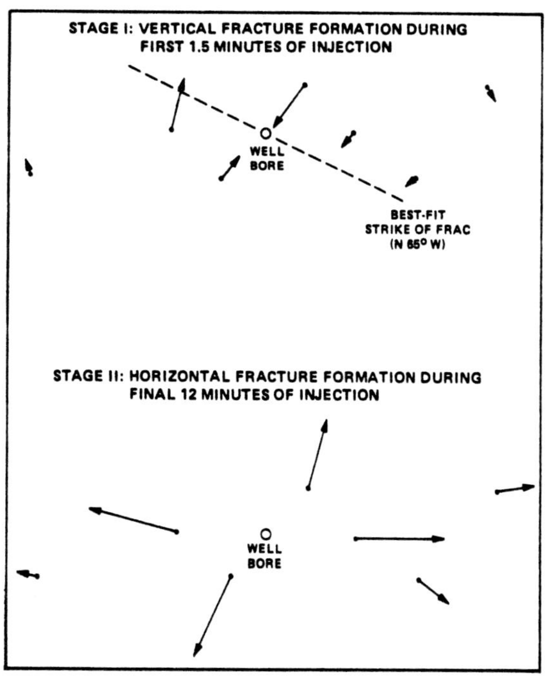

Figure 6. Measured tilt profiles during stages I and II of fracture formation (from Wood and others, 1983).

a type of path-dependent analysis that helps to resolve such ambiguities.

A real-time monitoring system would interpret a change in slope of tilt data across the whole array of instruments as a change in fracture orientation or fracture growth rate. If only one instrument had an abrupt slope change, this could be diagnosed as noise since it was not seen across the whole array. Once a change in orientation is detected, the fracture model would expand to include complex fractures, and monitoring would continue, unless it was decided to cease injection or modify treatment parameters.

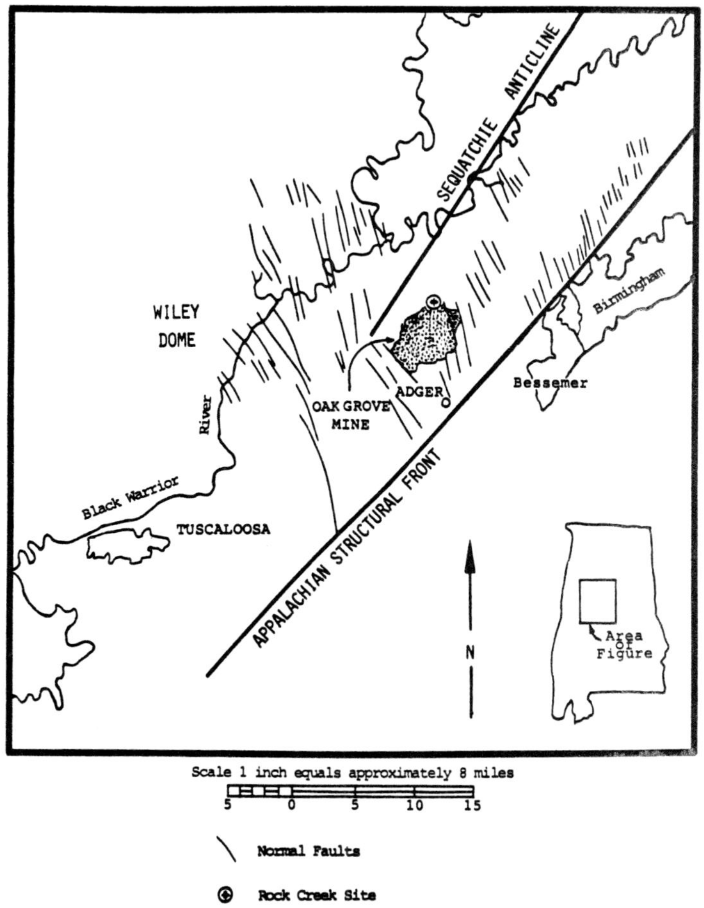

Figure 7. Map showing the location of the Rock Creek site and regional geologic structure of the southeastern Warrior Basin (redrawn from Camp and others, 1985).

COAL-BED FRACTURE MAPPING: A CASE STUDY

Geologic Setting

The Rock Creek site is located approximately 15 miles west of Birmingham, Alabama within the Warrior Basin and is bounded on the west by the Sequatchie Anticline and on the east by the Appalachian fold and thrust belt (fig. 7). The Warrior Basin of Alabama and Mississippi is structurally much less deformed than the Appalachian fold and thrust belt although numerous northwest-striking normal faults appear to be related to the Appalachian orogeny. Strata at the site reside within the Coalburg syncline and dip less than one degree to the south. Multiple coal seams at the Rock Creek site are in the Lower Pennsylvanian Pottsville Formation and primarily consist of interbedded sandstones, siltstones, and shales (Camp and others, 1985). Oriented cores were taken from monitor wells M1A (Pratt Group at 480 feet), M1B (Mary Lee Group at 1,050 feet), and M1C (Black Creek Group from 1,250 to 1,440 feet) (fig. 8). These coal groups contain an orthogonal set of joints; a preponderance of the face cleats have an average trend of N. 61 E. and butt cleats with an average trend of N. 142 E. (fig. 9).

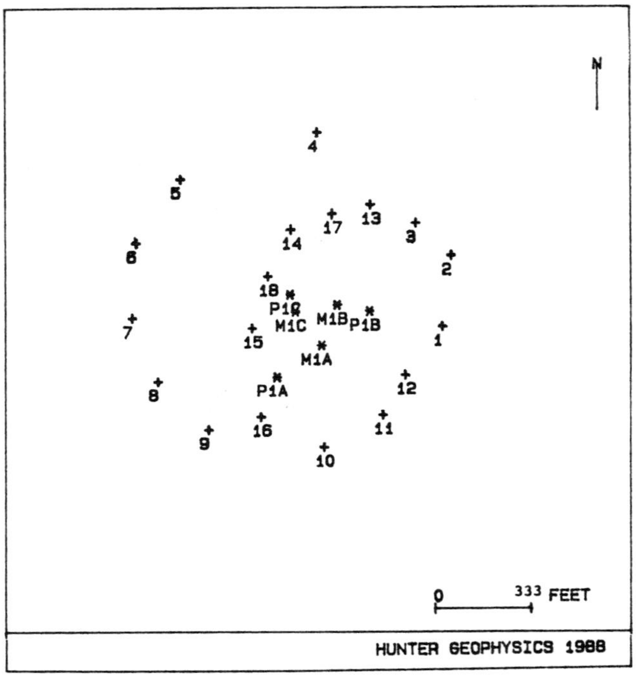

Figure 8. Map showing the location of monitor wells M1A, M1B, M1C; production wells P1A, P1B, P1C; and 18 tiltmeter sites.

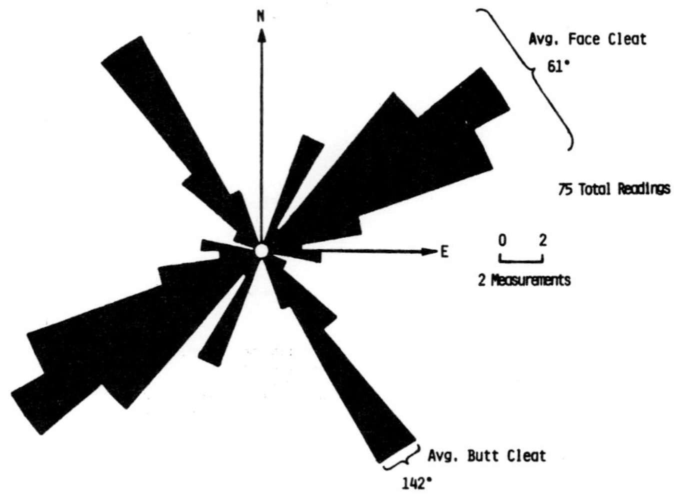

Figure 9. Composite rose diagram of average face and butt cleat orientations measured from oriented cores taken from monitor wells M1A, M1B, and M1C (redrawn from Boyer and others, 1984).

Coal Seam Stimulations

Three hydraulic-fracture stimulations were performed in multiple coal seams at the GRI/USX Rock Creek site in June and July of 1986. Methane production wells P1A, P1B, and P1C were stimulated at depths which ranged from 480 feet to 1,500 feet in the Pratt, Mary Lee-Blue Creek, and Black Creek coal groups (fig. 10). The stimulations consisted of a water prepad followed by a gel pad and a sand-laden cross-linked gel (Nielsen and Hanson, 1987). Injection rates ranged from 15 to 20 barrels per minute (bpm). Willis and others (1987) give a summary of fluid rheologies, pressures, and rates employed on these jobs. In addition, their paper provides the results of a fracture simulator which outputs in real-time, estimates of fracture dimensions, net fracture pressure, and treatment pressure. However, the precision in their estimates of fracture parameters is not given in their paper.

In a subsequent section, we summarize the results of a geophysical inversion technique which provides estimates, as well as the precision in the estimates, of fracture-source parameters. The accuracy in the predictions of modeling efforts can only be verified with direct observation of the fracture (for example with mine-back experiments).

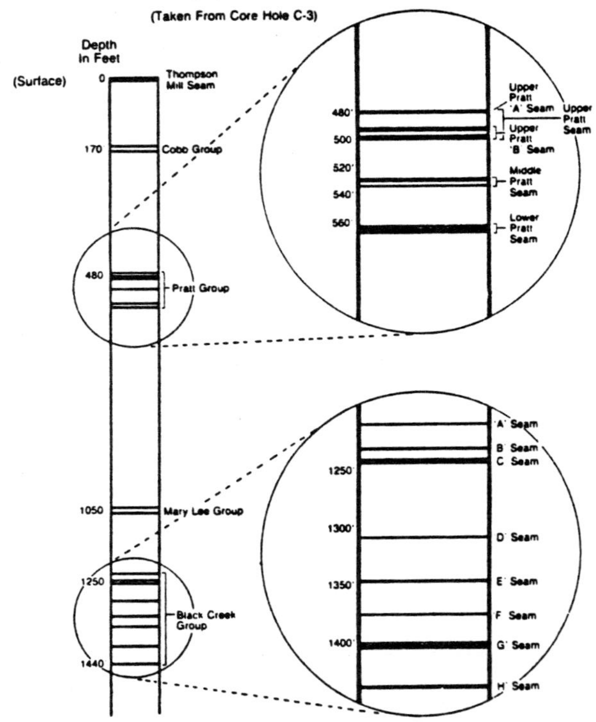

Figure 10. Stratigraphic column of Rock Creek site coals (from Schraufnagel and others, 1987).

Tiltmeter Measurements

Hunter Geophysics installed an array of 18 tiltmeters around wells P1A, P1B, and P1C in a configuration designed to optimize the expected tilt signals for each of the three stimulations (fig. 8). Maximum theoretical surface tilts for both horizontal and vertical fractures in the opening mode occur at a radial distance from the injection point which is equal to about 40 percent of the depth to the fracture center. Tilt data were continuously recorded at a rate of one sample per minute from a day before each stimulation until a few hours after shut-in of the wells. A re-analysis of the tilt data for each stimulation corroborates the results that were obtained in the in-field analyses. The fracture source parameter solutions and total-tilt vectors (accumulated from breakdown to shut-in) for each stimulation appear in figures 11a, 11b, and 11c.

The surface tilts which resulted from the stimulations created signals that were much larger than the resolution limits of the tilt sensor (5 nanoradians). Tilt amplitudes for the total injection interval (accumulated from formation breakdown to

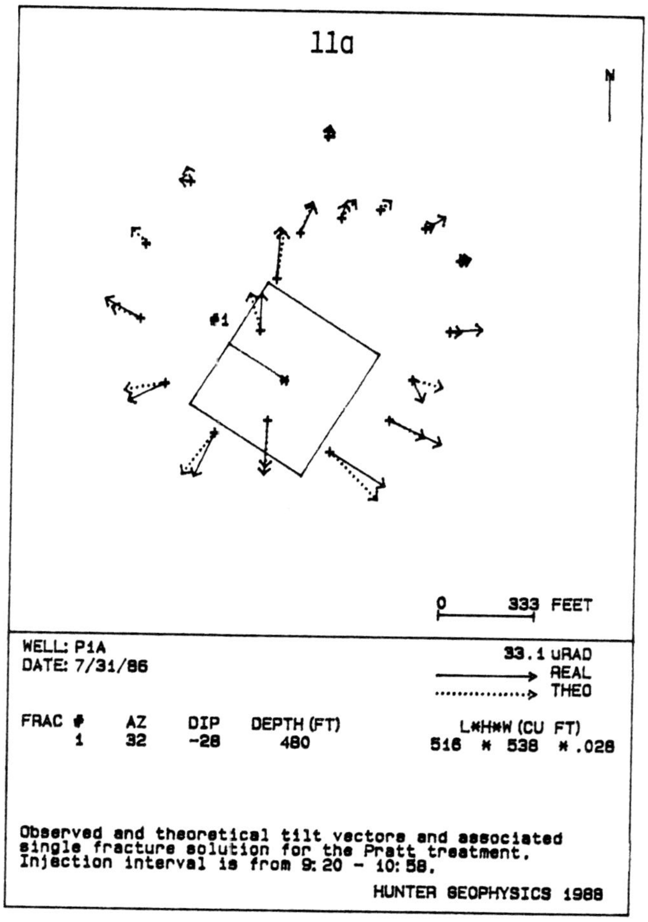

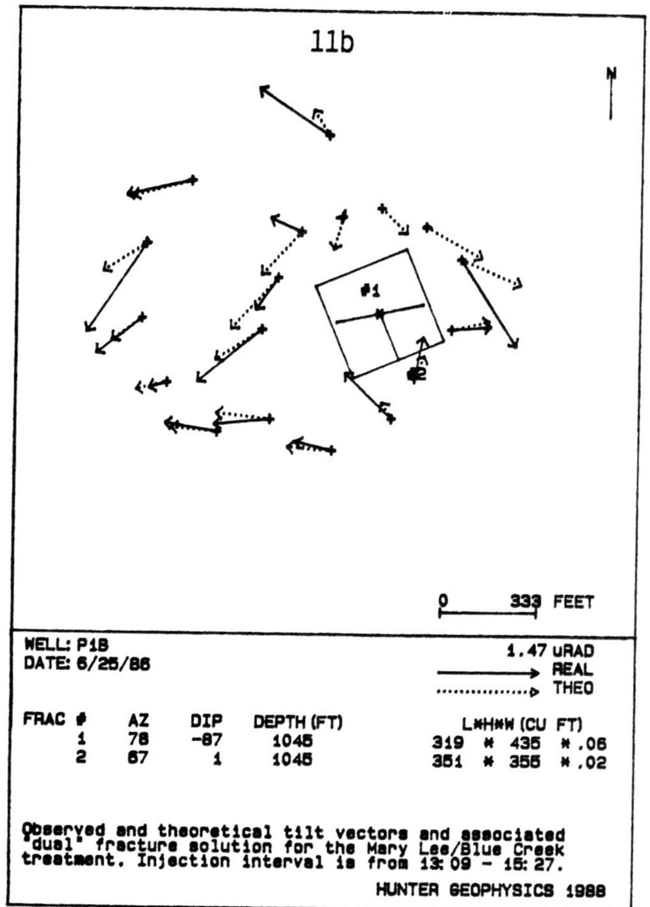

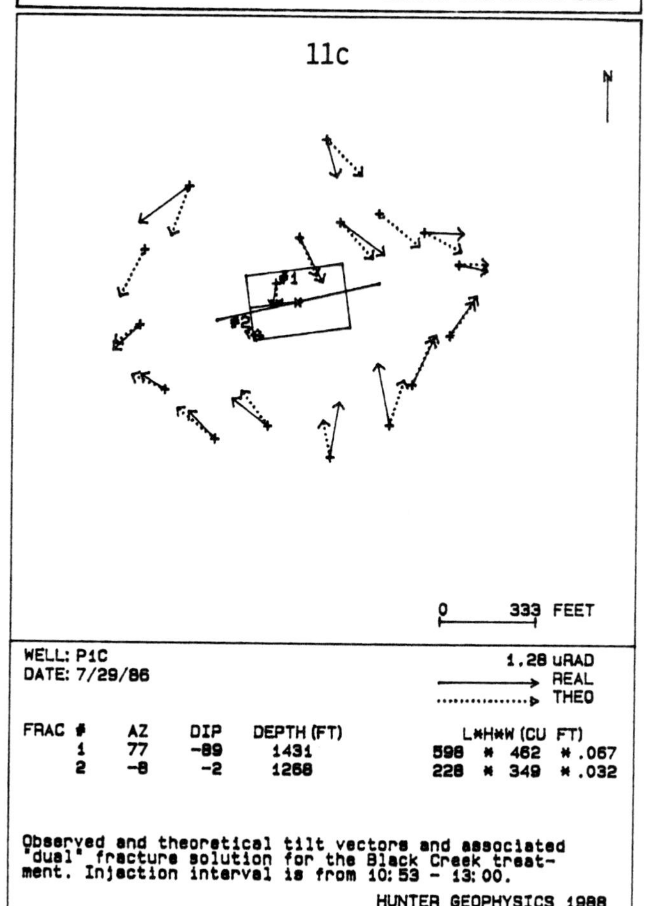

Figure 11. Overlay of observed (solid lines) and theoretical (dotted lines) and associated fracture solutions for the a) P1A—Pratt stimulation, b) P1B—Mary Lee-Blue Creek stimulation, and c) P1C—Black Creek stimulation.

shut-in) for each of these wells averaged 12.1 ± 6.3, .87 ±.45, and .58 ±.21 microradians respectively. The tilts induced by the P1A stimulation produced a radial pattern typical of a horizontally-oriented fracture at depth. In contrast to the P1A stimulation, the surface tilt patterns induced by the P1B and P1C stimulations are distinctly different and are both typical of vertically-oriented fractures at depth.

Tilt reversals were observed on many of the raw data channels for the P1A and P1B stimulations at sites in close proximity to the well and we attribute these changes to horizontal fracture propagation beneath the site, and to the formation of a dual fracture source. A few unexplained tilt reversals are also observed at some of the more distant sites for the P1A stimulation. The raw tilt data for the P1C stimulation are all relatively linear.

Interpretation of the Tilt Measurements

Well P1A—Pratt Group

The observed tilt pattern produced by the Pratt stimulation of well P1A can be modeled with an equidimensional, sub-horizontal dislocation with dimensions (in feet): length × height × width = 516 ± 83 × 538 ± 83 × .028 ± .005 (fig. 11a). The standard uncertainties of the estimated fracture parameters are determined by the method described by Press and others (1986). The fracture strikes N. 32 E. ± 11, and dips 28 ± 7 degrees northwest. Tilt data from sites 6 and 13 were deleted from the inversion data set since a large percentage of the solution error could be attributed to the mismatch between observed and theoretical tilt vectors for these sites. In addition, the fracture center is fixed at the wellbore and the fracture depth is fixed at the injection depth (approximately 480 feet). The solution provided is the best in a least squares sense and can be visualized in azimuth-dip space with a mean square error plot (fig. 3). The minimum error occurs at the + sign with subsequent logarithmic contours depicted by the numbers 1 through 9 in the plot.

Several attempts to model the observed tilt vectors using a more complex dual fracture source (a source consisting of both vertical and horizontal fracture components) did not significantly reduce the overall solution error. In fact, the ratio of the mean square errors (variances) for the dual and single fracture solutions is near unity, and hence there is no justification based on tilt data alone to increase the source complexity by admitting a vertical fracture component in the solution. The question as to whether a dual fracture solution is significantly better than a single fracture solution can be evaluated with an F-test for the equality of variances. In order to test for the equality of variances we can form the statistic $F = var1/var2$ which follows an F-distribution (Mendenhall, 1975). In this equation, the quantities var1 and var2 are the solution-error variances, or the mismatch between observed and theoretical tilt vectors, for the dual and single fracture solutions, respectively. Tabulated values of the F-distribution for various significance levels and degrees of freedom are obtainable from most introductory statistic textbooks.

Nielsen and Hanson (1987), incorporating pressure and well-logging data, provide an alternative interpretation of the tilt data with a solution that admits both vertical and horizontal fracture components. In their solution, the vertical fracture trends N. 73 E. and consists of less than 10 percent of the total injected slurry volume. The horizontal fracture comprises greater than 90 percent of the total fracture volume and has migrated up dip and to the northwest by approximately 110 feet. The method used to solve the inverse problem is not given in Nielsen and Hanson (1987), nor is any reference made to the precision of their estimates of the fracture source parameters. Willis and others (1987) provide real-time predictions of fracture length and net fracture pressure for the P1A stimulation, yet the precision in their estimates of the fracture parameters is not provided in their paper. In addition, their formalism is not general enough to provide precise estimates of fracture orientation.

It is interesting to note that gamma ray and acoustic televiewer logs reveal the presence of two vertical wellbore fractures in the Upper Pratt coal seams, yet there is no evidence of horizontal fractures in these seams. This could be attributed to the tools' inability to differentiate between the acoustic properties of a sand-filled horizontal fracture and partings in the coal seams. The orientations of the vertical fractures as determined by the acoustic logs is N. 78 E. and N. 93 E.; the fractures are oblique to the butt and face cleat orientations (fig. 9). Our analysis of the tilt data indicate that it is likely that the vertical fractures were limited to the vicinity of the wellbore and acted as fluid conduits to the dominant horizontal fracture that was created during the stimulation. Further evidence for the presence of horizontal fractures in the Pratt group comes from the results of in-situ stress measurements in well M1A. The measurements indicate that horizontal fractures can form in the Pratt group, at depths of 480 feet, after only 15 gallons of water are injected (at 5-6 gpm) into the formation, since fracture re-opening pressures during successive injection cycles are nearly equal to the overburden stress (Boyer and others, 1985). Hydraulic stimulation at greater depths in adjacent wells P1B and P1C created vertical fractures which volumetrically contributed 77 and 88 percent, respectively, to the total fracture volume.

Well P1B—Mary Lee-Blue Creek Group:

The observed tilt pattern produced by the Mary Lee-Blue Creek stimulation of well P1B can be modeled by a dual fracture source with a vertical component with dimensions (in feet): length × height × width = 319 ± 269 × 435 ± 238 × .060 ± .066. The vertical fracture component strikes N. 78 E. ± 4.5, and dips 87 ± 2 degrees northwest. One might have expected the azimuth of the vertical fracture to parallel either the butt or face cleat orientations in the coal (fig. 9). However, mining back into hydraulically induced fractures in coal has shown that the fractures have a stair-step pattern and follow oblique orientations, relative to butt and face cleat orientations (Diamond, 1987). The horizontal fracture component has dimensions (in feet): 351 ± 278 × 355 ± 282 × .020 ± .021. The horizontal fracture azimuth is N. 67 E. ± 360 and dips 1 ± 7 degree southeast (fig. 11b). The large standard uncertainty in fracture azimuth is expected, since azimuth becomes undefined as the fracture dip approaches zero. The standard uncertainties in fracture orientation are normally within 5 degrees. Tilt data from sites 3 and 13 were deleted from the inversion data set because a large percentage of the solution error could be attributed to the mismatch between observed and theoretical tilt vectors for these sites. In addition, both fracture centers are fixed at the wellbore and the fracture depths are fixed at the injection depth (approximately 1,045 feet). The vertical fracture component contributed 77 percent of the total modeled fracture volume. Additional evidence for the existence of complex fracture geometries in coal has been confirmed with direct observation of the fractures in mine-back experiments (Diamond, 1987).

The observed tilt vectors were initially inverted using a single fracture source model. The residual error was relatively large and the inversion proceeded by assuming a dual fracture source model. The ratio of solution error variances improved by a factor of 2.1, which is statistically significant at the 95 percent confidence level. A comparison of figures 11a and 11b illustrates the distinctive change in the surface tilt pattern with the introduction of a vertical fracture component.

Well P1C—Black Creek Group

The observed tilt pattern produced by the Black Creek stimulation of well P1C can be modeled by a dual fracture source with a vertical component with dimensions (in feet): length × height × width = 598 ± 245 × 462 ± 165 × .067 ± .026. The vertical fracture component strikes N. 77 E. ± 2.3, and dips 89 ± .6 degrees northwest. The horizontal fracture component has dimensions (in feet): 228 ± 135 × 349 ±

202 × .032 ± .018. The horizontal fracture azimuth is N. 8 W. ± 108 and dips 2 ± 3.8 degrees southwest (fig. 11c). Tilt data from sites 6 and 13 were deleted from the inversion data set since a large percentage of the solution error could be attributed to the mismatch between observed and theoretical tilt vectors for these sites. In addition, the fracture center depth for the vertical component is fixed at the injection depth (1,431 feet) and the fracture depth for the horizontal component is fixed at 1,268 feet. The vertical fracture component contributed 88 percent of the total modeled fracture volume.

The observed tilt vectors were initially inverted (as for the P1B stimulation) using a single fracture source model. The residual error was relatively large and the inversion proceeded by assuming a dual fracture source model. The ratio of solution-error variances improved by a factor of 4 (even though the horizontal component only contributes 12 percent of total volume), which is statistically significant at the 95 percent confidence level, and so we are justified in admitting the more complex fracture source model.

SUMMARY and CONCLUSIONS

The outline provided in this paper of tiltmeter fracture-mapping technology and the specific discussion of three separate stimulations in a coal formation demonstrate how it is used to obtain valuable real-time information of fracture-source parameters. Although the existing tiltmeter mapping is not optimized and is continually improving, it has been successfully applied to several hundred jobs in tar sands, diatomite, shale, and coal.

The geophysical inversion of surface tilt data provides estimates, as well as standard uncertainties, of fracture source parameters which include fracture azimuth, dip, depth, fracture center coordinates, and fracture dimensions which include length, height and width. The solution of the inverse problem using synthetic tilt data is unique, and since real data is contaminated with noise, the solution for fracture parameters often provides only fuzzy estimates of fracture dimensions. The standard uncertainties of the estimated azimuth and dip of single and dual fracture structures are usually within five degrees, whereas fracture dimensions are less precisely determined. However, given the sensitivity of tilt data to overall volume change, and the general condition that fracture closure following shut-in involves width changes with relatively little change in length or height, estimates of area and width can be made with models that also admit pressure data. Moreover, using pressure data to distinguish confined height versus length from area is determined by tilt with a model that jointly inputs pressure, rate, and tilt data.

Three hydraulic fracture stimulations were performed in multiple coal seams at the GRI/USX Rock Creek site in June and July of 1986. Methane production wells P1A, P1B, and P1C were stimulated at depths of 480 feet, 1,045 feet, and 1,431 feet, respectively, in the Pratt, Mary Lee-Blue Creek, and Black Creek coal groups. Fracture source parameters for each of the stimulations were obtained using geophysical inversion techniques. The tiltmeter data indicate that dual fracture sources (vertical and horizontal components) are required to best-fit the observed tilt patterns induced by the two deeper stimulations in the P1B and P1C wells. The direct observation of T-shaped fractures in the Blue Creek coal seam at USX's Oak Grove mine (Diamond, 1987) supports our interpretation for the creation of dual fractures during the P1B stimulation. The vertical fracture azimuths were found to trend approximately N. 78 E. and at oblique angles to the face cleat orientations (N. 61 E.) observed in oriented cores taken from adjacent monitor wells. An independent interpretation of the tilt data by Nielsen and Hanson (1987) essentially corroborates our interpretation for the creation of dual fractures during the P1B and P1C stimulations.

Inversion of the observed tilts induced by the stimulation of the shallowest P1A well revealed the presence of a large sub-horizontal fracture. Vertical fractures may have formed as a result of the stimulation but are volumetrically insignificant. Additional evidence for the presence of horizontal fractures and limited vertical fractures in the Pratt group comes from the results of in-situ stress measurements in well M1A. Pressure data indicate that horizontal fractures form through rollover after only 15 gallons of water are injected into the formation, since fracture re-opening pressures during successive injection cycles are nearly equal to the overburden stress (Boyer and others, 1985).

Gamma ray and acoustic televiewer logs revealed the presence of two vertical wellbore fractures in the Upper Pratt coal seams yet there is no evidence of horizontal fractures in these seams. This could be attributed to the tools' inability to distinguish between a horizontal sand-filled fracture and partings in the coal seams. The orientations of the vertical fractures as determined by the acoustic logs are N. 78 E. and N. 93 E., and they are oblique to the butt and face cleat orientations in the coal. Our analysis of the tilt data indicates that it is likely that the vertical fractures were limited to the vicinity of the wellbore and acted as fluid conduits to the dominant horizontal fracture that was created during the stimulation.

Further evidence for the absence of a significant vertical fracture component in the Pratt group is supported by the observation that post-stimulation methane production is poorest in the P1A well, although relatively low initial reservoir pressures could also explain this phenomenon (Schraufnagel and others, 1987). If our interpretation is correct, alternative stimulation designs should be developed which contact more of the seams in the shallow Pratt group.

The determination of fracture-source parameters by any method should include the precision of the estimated parameters. A precise comparison of the values of the same source parameter provided by different methods (seismic, borehole televiewer, and other logging data) was severely limited in this paper by the failure to provide, in the references cited, values for the precision in the estimated parameters. Non-uniqueness is a fact of inversion for the study of physical processes that are remotely sensed. No single geophysical tool will give equal quality determinations of source parameters; therefore, the trend of future analyses will be to integrate the determinations from a variety of complementary tools to constrain the solution to the least-error, most plausible condition. Inverting these data with a more comprehensive real-earth model of the processes will then be justified.

REFERENCES CITED

Boyer, C. M., II, Briscoe, F. H., Camp, B. S., Dobscha, F.X.,

Malone, P.G., Militzer, M.R., and Stubbs, P.B., 1984, Rock Creek methane from multiple coal seams completion project: Gas Research Institute Quarterly Report, November-January.

Boyer, C.M., II, Briscoe, F.H., Camp, B.S., Dobscha, F.X., Hirko, N.M., Malone, P.G., Militzer, M.R., Pavone, A.M., Schwerer, F.C., and Stubbs, P.B., 1985, Rock Creek methane from multiple coal seams completion project: Gas Research Institute Quarterly Report, May-July.

Camp, B.S., Briscoe, F.H., Malone, P.G., and Boyer, C.M., II, 1985, Rock Creek methane from multiple coal seams completion project, *in* Special data package preliminary geologic report: Gas Research Institute, Chicago, Illinois.

Davis, P.M., 1983, Surface deformation associated with dipping hydrofracture: Journal of Geophysical Research, v. 88, no. 87, p. 5826-5834.

Diamond, W.P., 1987, Characterization of fracture geometry and roof penetrations associated with stimulation in coalbeds, *in* Proceedings—The 1987 Coalbed Methane Symposium, The University of Alabama, Tuscaloosa, Alabama, November 16-19: Gas Research Institute, The University of Alabama, and the U.S. Department of Labor, sponsors, p. 243-253.

Evans, K., Holzhausen, G., and Wood, M.D., 1982, The geometry of a large-scale nitrogen gas hydraulic fracture formed in Devonian shale: An example of fracture mapping with tiltmeters: Society of Petroleum Engineers Journal, v. 22, no. 5, p. 755-763.

Ljung, L., 1987, System identification: Theory for the user: Prentice-Hall Inc., Englewood Cliffs, N.J.

Mendenhall, W., 1975, Introduction to probability and statistics, 4th edition: Duxbury Press, Massachusetts, 460 p.

Nielsen, P.E., and Hanson, M.E., 1987, Analysis and implications of three fracture treatments in coal at the USX Rock Creek site near Birmingham, Alabama, *in* Proceedings—The 1987 Coalbed Methane Symposium, The University of Alabama, Tuscaloosa, Alabama, November 16-19: Gas Research Institute, The University of Alabama, and the U.S. Department of Labor, sponsors, p. 105-112.

Press, F., 1965, Displacements, strains and tilts at teleseismic distances: Journal of Geophysical Research, v. 70, no. 10, p. 2395-2412.

Press, W.H., Flannery, D.P., Teukolsky, S.A., and Vetterling, W.T., 1986, Numerical recipes: Cambridge University Press, Cambridge, England.

Schraufnagel, R.A., Lambert, S.W., Stubbs, P.B., Dobscha, F.X., and Boyer, C.M., II, 1987, The Rock Creek field laboratory—A project update, *in* Proceedings—The 1987 Coalbed Methane Symposium, The University of Alabama, Tuscaloosa, Alabama, November 16-19: Gas Research Institute, The University of Alabama, and the U.S. Department of Labor, sponsors, p. 83-91.

Smith, M.B., Ren, N.K., Sorrels, G.G., and Teufel, L.W., 1985, A comprehensive fracture diagnostics experiment, Part II, Comparison of seven fracture azimuth measurements: Society of Petroleum Engineers Paper 13894.

Smith, M.B., Ren, N.K. Sorrels, G.G., and Teufel, L.W., 1986, A comprehensive fracture diagnostics experiment, Part 2-Comparison of fracture azimuth measuring procedures: Society of Petroleum Engineers Production Engineering, v. 1, no. 6, p. 423-431.

Willis, R.M., Wright, T.B., Buharali, A.M., and Ely, J. W., 1987, Monitoring and analysis of coal-bed fracturing operations using the GRI mobile test and control facility, *in* Proceedings—The 1987 Coalbed Methane Symposium, The University of Alabama, Tuscaloosa, Alabama, November 16-19: Gas Research Institute, The University of Alabama, and the U.S. Department of Labor, sponsors, p. 319-323.

Wood, M.D., and Allen, R.V., 1973, Methods for prediction and evaluation of tidal tilt data from borehole and observatory sites near active faults: Philosophical Transactions of the Royal Society of London, v. 274, no. 1239, p. 33-36.

Wood, M.D., Parkin, C.W., Yotam, R., Hanson, M.E., Smith, M.B., Abbott, R.L., Cox, D., and O'Shea, P., 1983, Fracture proppant mapping by use of surface superconducting magnetometers: Society of Petroleum Engineers SPE/DOE paper 11612.

A Review of the Physical and Mechanical Properties of Coal with Implications for Coal-Bed Methane Well Completion and Production

Arfon H. Jones[1], Gregory J. Bell[1] and Richard A. Schraufnagel[2]

ABSTRACT

Methane produced from coal seams is becoming an increasingly important commercial resource in the San Juan Basin of New Mexico and Colorado. However, coal seams in general are poorly understood with respect to their behavior as gas reservoirs. Gas is predominantly stored in an adsorbed state, as opposed to a compressed state. Diffusion of the desorbing gas will control early-time production. At later times, reservoir permeability controls the gas flows. In addition, coal seams are frequently saturated with water initially. Relative permeability effects commonly dictate the requirement for production enhancement by hydraulic fracturing or, alternatively, by drilling a drainhole.

There is a great need for improved knowledge of coal properties applicable to reservoir engineering and well completion in coal reservoirs. This paper addresses coal characteristics and, where applicable, their impact on well completion and gas production.

INTRODUCTION

Rocky Mountain Cretaceous coals are representative of common humic coals (formed form the remains of terrestrial plants) containing dominantly-pure type III vitrinitic kerogen (Meissner, 1984). Inertinite and exinite contents are typically of minor, but variable, amounts. In proceeding through the various rank-related stages of thermal metamorphism, these coals undergo progressive devolatilization. They lose only a small amount of volatile matter when progressing through the lower ranks, but lose substantial amounts when progressing through the high-volatile, medium-volatile, and low-volatile bituminous and various anthracitic ranks. Several investigators (Juntgen and Karweil, 1966; Karweil, 1969; Juntgen and Klein, 1975; and Hunt, 1979) have shown that the volatile matter, evolved as coals increase in rank, consists primarily of methane, carbon dioxide, nitrogen, and water. A generation model presented in Juntgen and Klein (1975) shows that: 1) the primary volume of water and carbon dioxide generation precedes the primary volume of methane generation during progressive metamorphism, and 2) nitrogen is a relatively small constituent when volatile matter drops below a critical level, approximately 46 to 35 percent (when coal rank increases above high-volatile C to A bituminous). The various models indicate a total methane generation capacity ranging from 800 to 1,920 standard cubic feet per ton (scf/ton) between the start of generation and when coal reaches 30 percent (dry, ash-free) volatile matter (equivalent to a vitrinite reflectance of 1.2 percent R_0).

Gas recovery is predominantly from coal reservoirs with rank in the range from high-volatile A bituminous to low-volatile bituminous. A better understanding of the physical and mechanical properties of coals within these ranks provides an insight into the mechanisms controlling production. Completions must be designed to allow for the larger fraction of the gas-in-place being desorbed at low reservoir pressures and at low relative permeability to gas. The physical, petrological, and mechanical characteristics of a San Juan Basin coal are provided here to form a basis for better understanding of the reservoir and of the requirements for hydraulic-fracture stimulations.

[1] Arfon H. Jones and Gregory J. Bell
Terra Tek, Inc.
400 Wakara Way
Salt Lake City, Utah 84108

[2] Richard A. Schraufnagel
Gas Research Institute
8600 West Bryn Mawr Avenue
Chicago, Illinois 60631

The work reported here was sponsored by the Gas Research Institute (Contract Nos. 5081-214-0577, 5084-214-1084, and 5087-214-1460). The authors express their gratitude to GRI for their support and for permission to publish these results.

Special thanks goes to Tiffany Gas Company for cooperating with research conducted on their Glover No. 1 well.

COAL ANALYSIS

Proximate and Ultimate Analysis

Proximate analysis establishes the volatile matter, fixed carbon, moisture, and ash content of the coal. The ultimate analysis is performed to determine the percentages of carbon, hydrogen, oxygen, nitrogen, and sulfur. As shown in figure 1, rank classification of coals with volatile matter contents greater than 31 percent are based on their moist, mineral-matter-free heating value. Ranks from medium-volatile bituminous through meta-anthracite are based on the weight percent volatile matter (VM) in the coal or fixed carbon (FC), (where FC = 100 − VM, percent) determined on a dry, mineral-matter-free basis (dmmf). VM range for medium-volatile bituminous

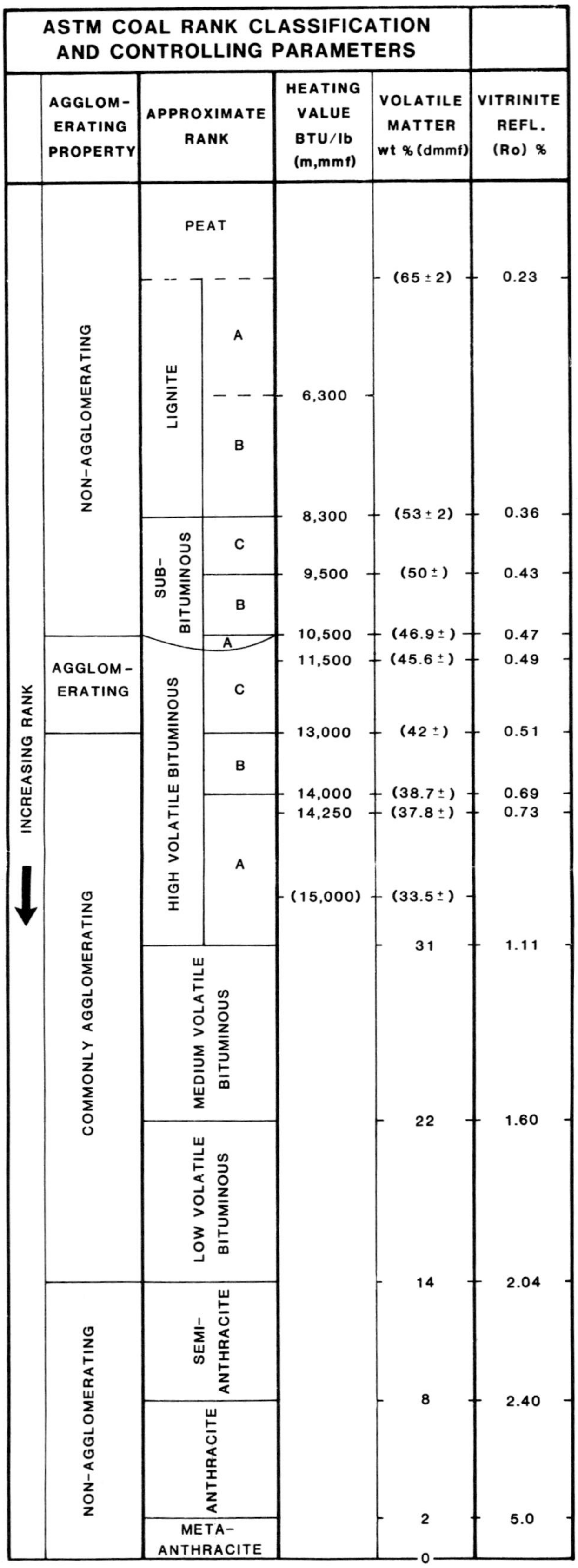

Figure 1. Coal rank classifications (ASTM Standard, 1983) and relation to vitrinite reflectance (modified from Meissner, 1984).

TABLE 1. Analysis of coal from the Glover No. 1 well.

Parameter	3,086.2 ft As received	3,086.2 ft Dry basis	3,088.2 ft As received	3,088.2 ft Dry basis
Proximate analysis: (percent)				
Moisture	2.19		1.23	
Ash	55.81	57.06	43.22	43.76
Volatiles	12.28	12.55	13.60	13.77
Fixed Carbon*	29.72	30.39	41.95	42.47
	100.00	100.00	100.00	100.00
Ultimate analysis: (percent)				
Moisture	2.19		1.23	
Carbon	36.02	36.83	48.72	49.33
Hydrogen	2.35	2.40	3.02	3.06
Nitrogen	0.86	0.88	0.96	0.97
Chlorine	0.11	0.11	0.13	0.13
Sulfur	0.29	0.30	0.42	0.43
Ash	55.81	57.06	43.22	43.76
Oxygen*	2.37	2.42	2.30	2.32
	100.00	100.00	100.00	100.00
Btu/lb	5,731	5,859	8,537	8,643
Air dry loss (percent)	1.12		0.51	
Total moisture (percent)	3.28		1.73	
Moist mm-free Btu/lb**	14,046		15,887	
Dry mm-free volatiles	20.60		19.23	
Dry mm-free fixed carbon	79.40		80.77	

*Determined by difference.
**Based on total moisture.

coal is from 31 to 22 percent and from 22 to 14 percent for low-volatile bituminous coal.

Table 1 summarizes the results of the proximate and ultimate analyses for coal samples retrieved from the Glover No. 1 well located in the northern San Juan Basin within the southeast quadrant of sec. 2, T. 32 N., R. 6 W., Archuleta County, Colorado. The VM value (dmmf) places the coal rank as low-volatile bituminous. Ash content of the coal is extremely high and is considered to reflect the high silt and clay particle content dispersed within the coal from periodic flooding of a marshy depositional environment during Late Cretaceous time. With an ash content of over fifty percent, the core sample from 3,086.2 feet is classified as a carbonaceous shale.

Maceral Composition

Macerals are the organic microscopic constituents of coals, analogous to minerals for rocks. They are classified into three major groups and many individual macerals as listed in table 2. The vitrinite group is derived from various decomposed woody tissues; the exinite (liptinite) group from spore and pollen coats, cuticles, resins, and other fatty secretions; and the inertinite group is derived mainly from partial car-

TABLE 2. **Maceral analysis of coal from the Glover No. 1 well.**

	Volume percent, mineral-matter free basis	
Maceral	3,086.2 ft	3,088.2 ft
Vitrinite	94.9	96.6
Pseudovitrinite	0.0	0.0
Exinite	0.0	0.0
Resinite	0.7	0.4
Semi-Fusinite	3.2	2.7
Semi-Macrinite	0.0	0.0
Fusinite	1.2	0.3
Macrinite	0.0	0.0
Micrinite	0.0	0.0
	100.00	100.00
Maceral group		
Vitrinite	94.9	96.6
Exinite (liptinite)	0.7	0.4
Inertinite	4.4	3.0
Vitrinite reflectance		
Mean maximum Vitrinite reflectance	1.22	1.22

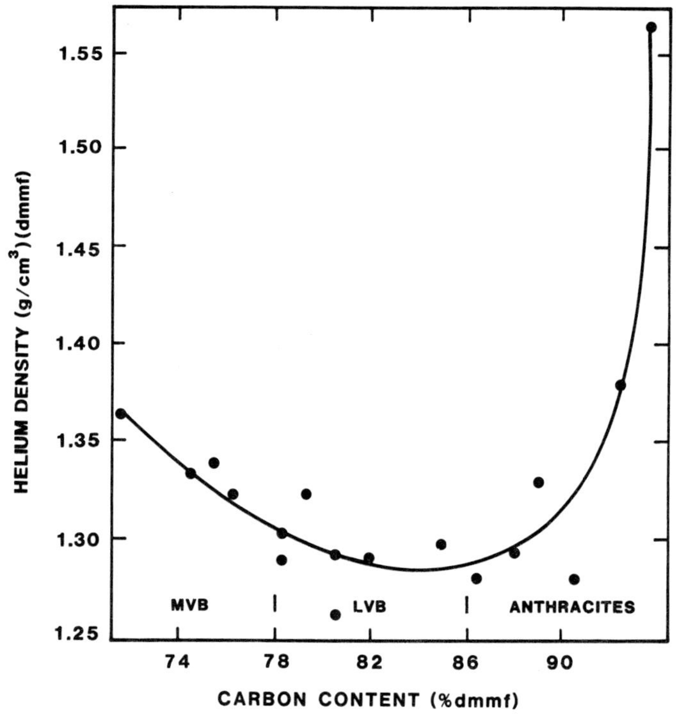

Figure 2. **Graph showing grain density for coal (from Nelson, 1979).**

bonization of the various plant tissues in the peat swamp stage (Ting, 1982). Progressive metamorphism or alteration will not change the maceral composition.

Maceral analysis for the two coal samples retrieved from the Glover No. 1 well are summarized in table 2. Their maceral content is predominantly vitrinite (approximately 95 percent) with minor amount of inertinite and exinite (in particular resinite). Snowden (1980) has shown the possible significance of resinite in the formation of heavy hydrocarbon gases and condensate.

Vitrinite Reflectance

A relatively accurate measurement of the degree of metamorphism or coalification is the reflectance of individual macerals and it is mostly performed on vitrinites (Ting, 1982). Vitrinites exhibit a wide range of reflectance in the coalification series, as shown in figure 1, ranging from 0.23 for lignite to 5.0 for anthracite. For high-volatile A bituminous coals, the range is from 0.69 to 1.11 percent R_0, with medium- and low-volatile bituminous coals ranging from 1.11 to 1.60 percent R_0 and 1.60 to 2.04 percent R_0, respectively. As shown in table 2, the vitrinite reflectance for coal samples recovered from the Glover No. 1 well was 1.22 percent R_0 which places the coal rank as medium-volatile bituminous, which is not in agreement with the proximate analysis. The discrepancy is due to the high mineral-matter content of the coal which tends to depress the vitrinite reflectance.

Coal Porosity

Coal density and porosity are normally measured with helium for two reasons: 1) it is the smallest inert molecule available and, therefore, has the best chance of penetrating the entire porosity of coal, and 2) helium adsorption on raw coals is not appreciable at room temperature to cause significant errors in density measurement. Since some porosities in coals may be closed to helium, the measured porosity and grain density will be lower bounds to the actual values.

Helium-determined grain densities of the 40 × 70 mesh fraction for various rank coals (Nelson, 1979) are summarized graphically in figure 2. The grain densities were found to: 1) decrease sharply with decrease in carbon content to 90 percent (semianthracite), 2) pass through a broad minimum in the carbon content range of 82-86 percent through the low-volatile bituminous rank, and 3) increase with further decrease in carbon content.

The same trend was observed by Ettinger and Zhupakhina (1960) for porosity of thirty-five coal samples taken from different seams. Carbon content for their samples varied from 51.4 percent to 98.1 percent on a dry, ash-free basis. Their results are shown in figure 3. There is a clearly defined minimum within the range of 70 to 80 percent carbon content (low-volatile bituminous to medium-volatile bituminous).

Test results for coal samples from the Glover No. 1 well are summarized in table 3. Average coal porosity was approximately 5 percent (coal porosities measured with water averaged about 20 percent lower). By comparison with figure 3, the data falls close to the upper limit of reported porosities for this rank of coal; this is not unexpected as the tests were performed on samples 1-inch in diameter by 2 inches long as compared to crushed coal reported in Ettinger and Zhupakhina (1960).

Based on transmission electron microscopy and small angle x-ray scattering analysis (Harris and Yust, 1976; Lin and others, 1978), the pore structures in coals are distinguished according to micropores (to 20Å diameter), mesopores (20Å to 500Å diameter) and macropores (> 500Å diameter). Low- and medium-volatile bituminous rank coals, such as found at the

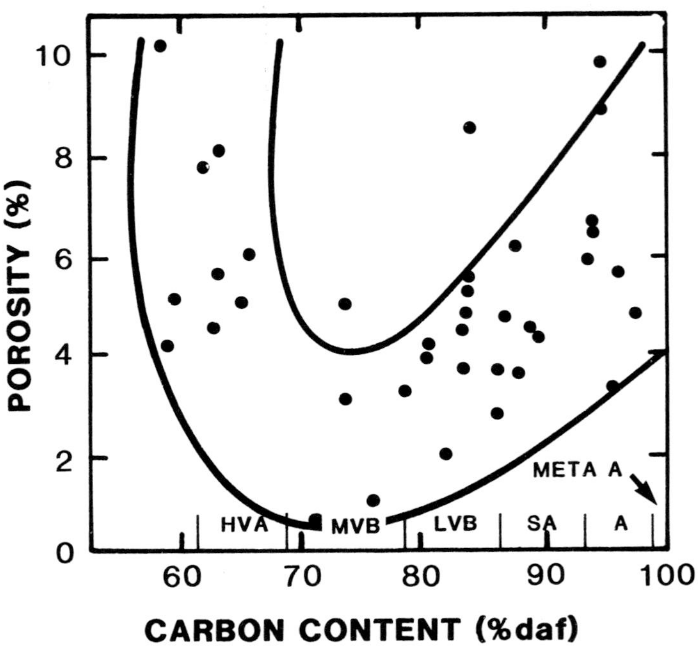

Figure 3. Graph of dependence of coal porosity on degree of metamorphism (from Ettinger and Zhupakhina, 1960).

TABLE 3. Coal porosity (helium); samples from the Glover No. 1 well.

Sample depth (feet)	Porosity (percent)
3,086.5	5.20
3,087.0	3.03
3,087.3	5.18
3,088.0	5.70

Glover No. 1 well, have porosities composed dominantly of micropores and mesopores; the pore volume is mainly associated with the mesopores and the surface area with the micropores. Such small pore structure will greatly influence capillary pressure.

Capillary Pressure

Capillary pressure data can be interpreted to provide information on pore size diameter as well as water retention characteristics. The latter is important to evaluate the degree of coal dewatering and, from that data, the maximum achievable effective gas permeability. Two coal samples recovered from a depth of approximately 3,085 feet at the Glover No. 1 well were tested; one sample had well-developed cleats that were clearly visible. All attempts to machine that sample into a right-circular cylinder failed. Consequently, an irregularly shaped sample was tested. The other sample was 1-inch in diameter by 1½ inches long. Both samples were dried in a humidity oven at 110°F until a constant weight was obtained.

The mercury injection method was used to determine capillary pressure after experimenting with the porous plate method and finding that unacceptable time was required to reach equilibrium at each saturation. Results of the tests, shown in figure 4, are based on porosity determined with helium. No correction for mercury compressibility was made in the data analysis.

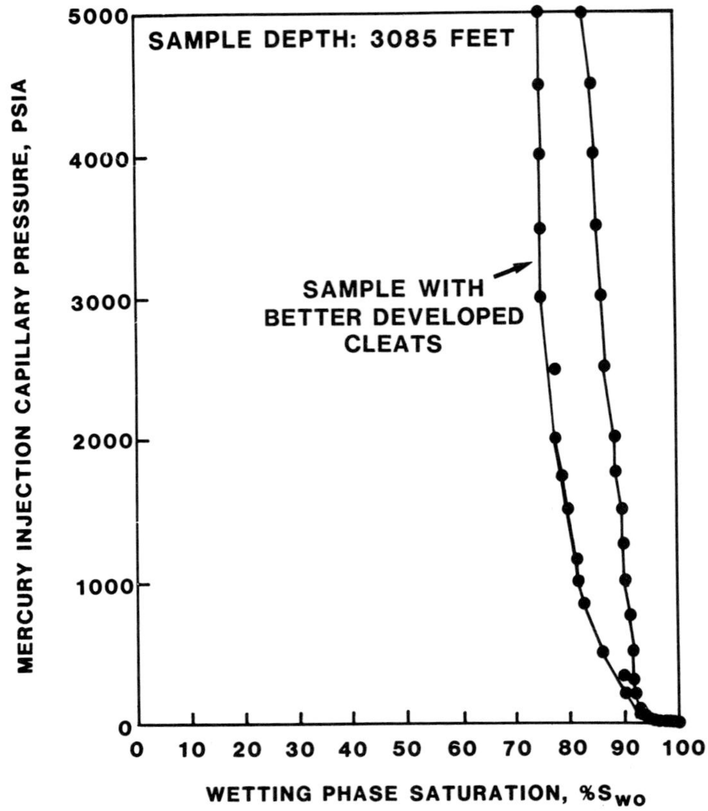

Figure 4. Mercury injection capillary pressure data for the Fruitland Formation coal (Glover No. 1 well).

The low, constant value for the mercury capillary pressure between 94 and 100 percent saturation (equivalent to a porosity of 0.3 percent for a 5 percent porosity sample), is considered to reflect the influence of the cleat fracture system (macropores). The other striking aspects of the results shown in figure 4 are the high irreducible water saturations of between 75 and 82 percent. Similar curves, but without the initial constant capillary pressure between 94 and 100 percent saturation, were observed for the Pittsburgh coal by Dabbous and others (1976) at high confining pressures (600 to 1,000 psia). The differences are thought to reflect cleat closure with increased confining stress in the Pittsburgh coal.

Pore throat size histograms derived from the two capillary pressure curves for the Glover No. 1 well samples are shown in figures 5a and 5b. These histograms indicate that the pore volume is dominantly in pores with throats less than 0.021 microns radius, that is less than 420Å diameter.

Surface Area

The small pore structure in coal provides a large surface area for storage of adsorbed gases. For example, the estimated surface area of one pound of coal is from 100,000 ft² to in excess of 1,000,000 ft². Coal containing 27 percent volatile matter can adsorb (at 77° F) up to 640 scf/ton (corresponding to an internal surface area of 460,000 ft²/lb with a monomolecular layer of methane molecules, 4Å in diameter). This storage is equivalent to a 30.1 percent gas filled porosity for a 3,000-foot-deep reservoir at normal pressure gradient

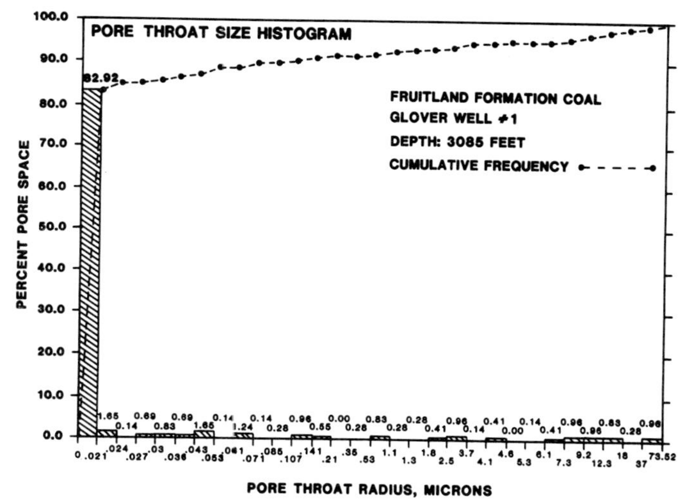

Figure 5a. Pore throat size histogram (sample without well-developed cleats).

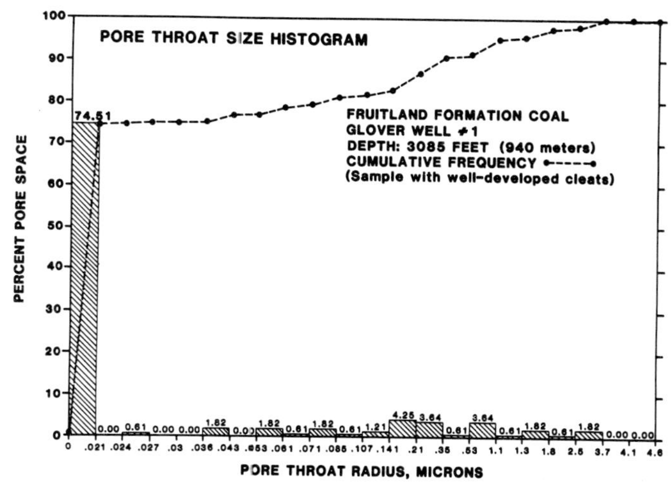

Figure 5b. Pore throat size histogram (sample with well-developed cleats).

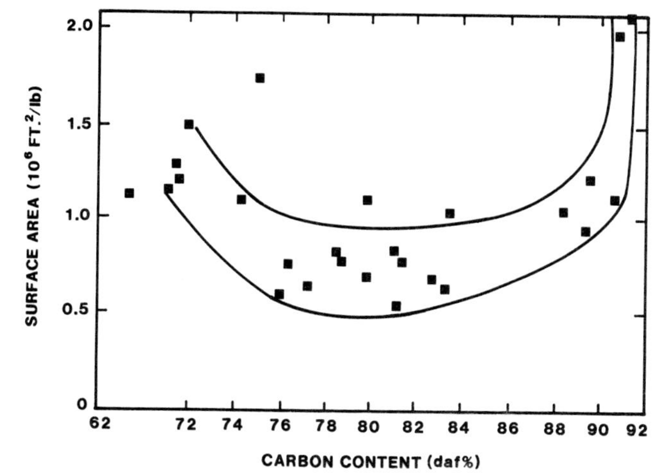

Figure 6. Graph of variation of CO_2 surface area of coals with carbon content (from Gan and others, 1972).

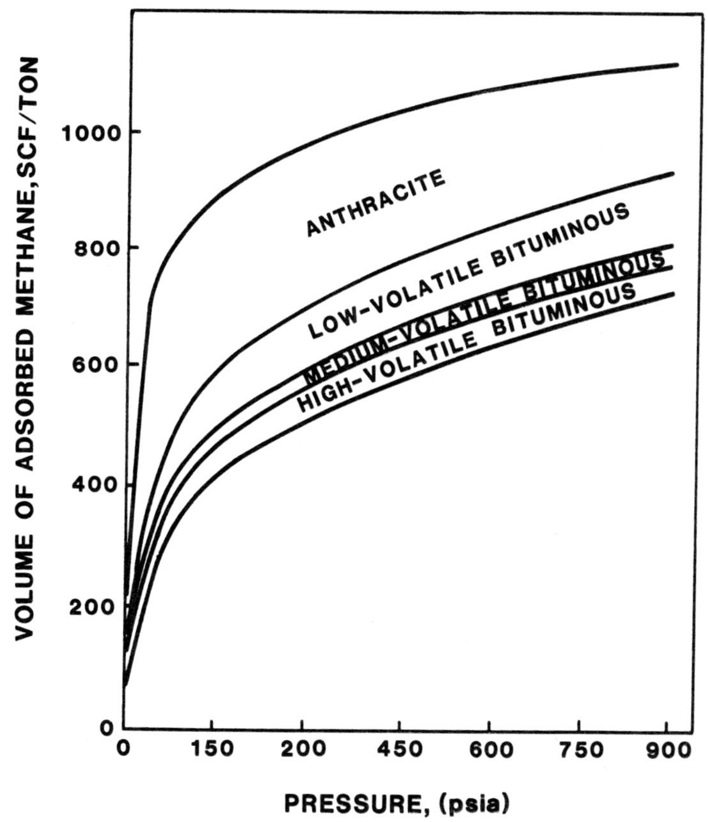

Figure 7. Graph of variation of methane adsorption isotherm with coal rank at 0° C (from Kim, 1977). The coal ranks represent the range of values between the bounding curves.

and a reservoir temperature of 100° F.

Surface area estimates are made from adsorption of various gases. To overcome low diffusion rate effects, tests are performed at the highest possible temperatures (Kini and Lahiri, 1960). CO_2 surface areas of coals versus carbon content are shown on figure 6 (from Gan and others, 1972). In general, the surface area decreases with a reduction in carbon content to about 83 percent. It remains more or less constant down to about 76 percent carbon content and increases thereafter. A data spread of 440,000 ft²/lb is shown for coal of the same carbon content. This is a much wider spread for the CO_2 areas than those reported by Marsh and Siemieniewska (1965).

Juntgen and Langhoff (1964) reported that the hydrocarbon surface area, calculated by the Langmuir theory, decreases with increasing volatile matter content of the coal and is expressed by:

$$A(10^6 \text{ ft}^2/\text{lb}) = 0.51705 - 0.010295 \times VM + 0.000104 \times VM^2$$

where VM is the percent of volatile matter. A medium-volatile bituminous coal with 31 percent VM will have a surface area of 298,000 ft²/lb and a semianthracite 441,000 ft²/lb at 8 percent VM.

COAL RESERVOIR PROPERTIES

Adsorption Isotherm

Adsorptive capacity of coal depends not only on the surface area but also on the gas pressure, temperature, and moisture level. Variation of the methane adsorptive capacity with pressure is shown by the adsorption isotherms illustrated in figure

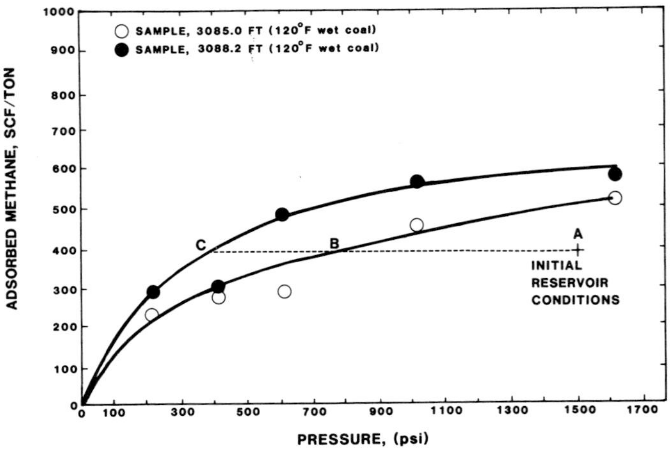

Figure 8. Adsorption isotherm for Fruitland Formation coal (Glover No. 1 well).

7. At low pressure, the adsorptive capacity increases almost linearly with pressure. As pressure continues to increase, the coal's ability to continue to adsorb gas decreases, eventually reaching a pressure level at which very little added gas can be stored. Higher rank coals, with larger surface areas, have the highest adsorption capacity.

Adsorption isotherms, for two crushed coal samples recovered from the Glover No. 1 well and tested to reservoir conditions (120° F, 1,500 psi), are shown in figure 8. In figure 8, as in figure 7, the adsorptive capacity increases rapidly with initial application of pressure and then the rate of increase gradually diminishes as gas saturates the accessible surface areas. For the coal cores retrieved from the Glover No. 1 well, the adsorptive capacities of the two samples at original reservoir pressure conditions were between 500 and 600 scf/ton. These can be compared to the 380 ± 10 scf/ton of gas-in-place as determined by core desorption. Since the coal will have generated a far greater quantity of gas during coalification, it implies that at some stage during burial history, a lower pressure and (or) a higher temperature condition existed in the reservoir which allowed the gas to migrate from its source.

Undersaturation of the coal with gas at initial reservoir conditions is most prevalent and has a strong bearing on well design, reservoir planning, and production operations. As can be deduced from figure 8, the onset of gas evolution occurs after reservoir pressure is dropped below 750 psi. In other words, no gas is desorbed between initial reservoir pressure A and point B (fig. 8). In a high permeability, thick-coal reservoir, a large diameter well will be required in order to pump at a sufficient rate to reduce downhole and reservoir pressures to achieve desorption. Otherwise, multiple pumping wells will be needed to successfully produce the reservoir. Under both scenarios, a large quantity of water must be disposed of.

For the reservoir in the vicinity of the Glover No. 1 well, figure 8 shows that even after reservoir pressures have been reduced to 400 psi on the adsorption isotherm OB (or 250 psi on adsorption isotherm OC) approximately 300 scf/ton of gas remains adsorbed, that is only 80 scf/ton of gas is desorbed by reducing the reservoir pressure by in excess of 1,000 psi. Dropping the reservoir pressure to 50 psi results in 300 scf/ton of gas being desorbed, in other words, to recover a large fraction of the gas in place, the reservoir must be produced to low

TABLE 4. Results of the nonlinear inversion for unstressed and stressed samples, early-time unipore and all-time bidisperse.

	Early time unipore and all-time bidisperse	All-time bidisperse
Test condition	D_a/a^2 (1/hr)	D_i/a^2 (1/hr)
Zero effective stress	0.210	0.015
Equivalent to initial reservoir conditions	0.129	0.0051

pressures. Therefore, in completing lower permeability reservoirs, long, high conductivity fractures or drainholes drilled in-seam will be needed to achieve commercially acceptable gas production.

Desorption Kinetics

Flow of adsorbing or desorbing gas in crushed coal is studied by using the non-steady-state diffusion of molecules either into or out of the coal particles. When the diffusion coefficient D is independent of concentration C and measurements are made at isothermal conditions, the equation of continuity for spherically symmetrical flow leads directly to Ficks's flow equation:

$$\frac{\partial c}{\partial t} = \frac{1}{r^2} \frac{\partial}{\partial r} \left(r^2 D \frac{\partial c}{\partial r}\right)$$

where r is the radius and t is time. Since the non-steady-state diffusion process is remarkably insensitive to particle shape, the radius of an equivalent sphere "a" may be used for coal particles of irregular shape.

Test results for coal-core samples (1-inch in diameter by 2-inches long) recovered from 3,087.0 feet depth at the Glover No. 1 well, rather than crushed coal, are shown in table 4. These tests were performed both at zero effective stress and at 350 psi effective stress. All tests were at reservoir temperature of 120° F. The results show that the early-time unipore diffusivity, D_a/a^2, decreases with increased confining stress. Both diffusivity values are much higher than the 0.0018/hr needed to match the observed gas flow during production tests and when compared with the use of diffusivity values of 0.0000036/hr and 0.018/hr by Price and Ancell (1978) in simulations of the Mary Lee and Pocahontas coals, respectively.

Airey (1968) found the unipore model was inadequate to describe the rate of diffusion near the end of the diffusion process. Accordingly, Smith (1982) adapted the bidisperse pore model with both macropore and micropore diffusivity to better describe the onset and completion of desorption. The approach outlined by Ruckenstein and others (1971), as detailed by Smith (1982), was used to analyze the same laboratory measurements as described above. Results are summarized in table 4. As with the unipore diffusivity, the micropore diffusivity decreases with confining stress.

Much remains to be understood about the diffusion of methane through solid coal. Fortunately, diffusion controls gas production at early times only. At the start of pumping, reservoir pressure in close proximity to the well will rapidly

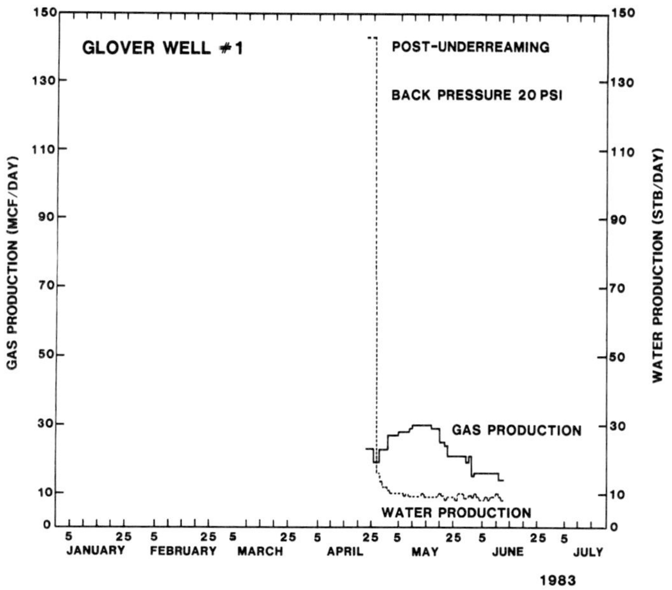

Figure 9. Production test graph following underreaming at the Glover No. 1 well.

reduce. Desorption kinetics will prevent the adsorbed gas from reaching equilibrium values on the desorption isotherm corresponding to the local reservoir pressure. With an increase of the drawdown radius, gas flow will eventually be controlled by Darcy flow at times beyond the observed maximum gas flow rate. This occurrence of a peak in the gas-flow rate is characteristic of the response of coal reservoirs after a rapid drawdown in water level. An example is shown from the production record for the Glover No. 1 well (fig. 9) following wellbore underreaming. As the radii of the reduced water pressure region enlarges, the rate of decline in reservoir pressure decreases. With a reduction in rate of pressure decline, diffusion no longer controls gas flow rate. Thereafter, gas production is governed by Darcy flow in the reservoir.

Reservoir Permeability

Coal-bed permeability is highly related to the cleats in coal. Cleats are essentially shrinkage cracks with orientation controlled by tectonic stresses at the time of fracture formation. Absolute permeabilities in the range from 0.1 to 250 millidarcies (md) have been attributed to various coal beds in the United States. Larger permeabilities are found near geologic structures and lower permeabilities occur at greater depths of burial. Permeability of deeply buried coal could be as low as those reported from other parts of the world. Australian coal permeabilities have been reported to range from 1 to 10 microdarcies, with European coal ranging from 1 to 50 microdarcies.

The presence of tectonically induced fractures will greatly influence permeability and will result in large differences in the permeability determined from well tests and on cores. A parameter that is often used to reflect the intensity of fracturing is:

$$\alpha = \frac{\text{permeability from drawdown/buildup or injection/fall-off}}{\text{matrix permeability for cores}}$$

Alpha (α) can range from about ten, when a few channels are open to flow, to values of a few thousand or more in the case of a highly fractured reservoir (Reiss, 1981). Laboratory-measured permeabilities of 0.113 md along the face cleats and 0.0045 md along the butt cleats were obtained at simulated in-situ stresses and temperatures on dried coal cores recovered from the Glover No. 1 well for depths of 3,087.0 feet and 3,087.3 feet, respectively. Field-measured permeability of 3.5 md was interpreted (fig. 10) during fall-off after injecting water at an average rate of 2 gallons per minute for eight hours. From these data $\alpha \approx 150$, suggesting that the coal contains an extensive fracture system.

In forecasting gas flow from coal reservoirs, two mechanisms leading to permeability changes during production must be included: 1) a decrease in permeability with decrease in reservoir pressure (that is, with an increase in effective stress) and 2) an increase in permeability as the coal shrinks with gas desorption.

Stress-Dependent Permeability

Rose and Foh (1984), Harplalani and McPherson (1985), and Gray (1987) observed large changes in permeability with change in effective stress on coal, as shown in figure 11. Schwerer and Pavone (1984) as well as McKee and others (1984) investigated the implication of observed data to well tests and water and gas production. Both studies showed that laboratory data predicts: 1) a nonlinear pressure change versus log of time response during well testing, 2) permeability would be dependent on production or injection rate, and 3) long-term production of water and gas would be significantly reduced from that expected on the basis of constant permeability. However, separate injection tests at the Glover No. 1 well, performed at different flow rates, showed consistent permeability (3.5 md), skin factor (-3.), and a linear response on the Horner plot. Figure 10 provides results for one such test. In addition, reservoir permeability deduced from a drawdown test was identical with the results of the injection tests.

At this time, there is no evidence that coal bed reservoirs exhibit the same stress-dependent permeability as observed in the laboratory. Presumably, the laboratory stress dependence observed results, at least in part, from damage during coring, relaxation fracturing within the recovered core, damage due to sample preparation, and damage when subjecting the sample to stresses and pore pressures during testing. A definitive explanation has not been published. [**Editors note: See the paper by McKee and others, this volume for more on stress-dependent permeability of coal.**]

Coal Shrinkage Associated Permeability

Although permeability changes as a result of gas desorption have not been demonstrated in the laboratory, data in the literature shows that carbon dioxide desorption has a definite affect on the linear dimensions of coal samples (Anderson and others, 1964; Lee, 1982; Hargraves, 1963; Reucroft and Patel, 1986). As summarized in table 5, significant length increases were observed in a CO_2 atmosphere at 20 psig, amounting to approximately between 0.10 percent to 0.44 percent after long time exposure; the lower-carbon-content coal achieved the higher degree of swelling.

Anderson and others (1965) noticed that the CO_2 adsorption mode is of the activated type and shows its presence as a

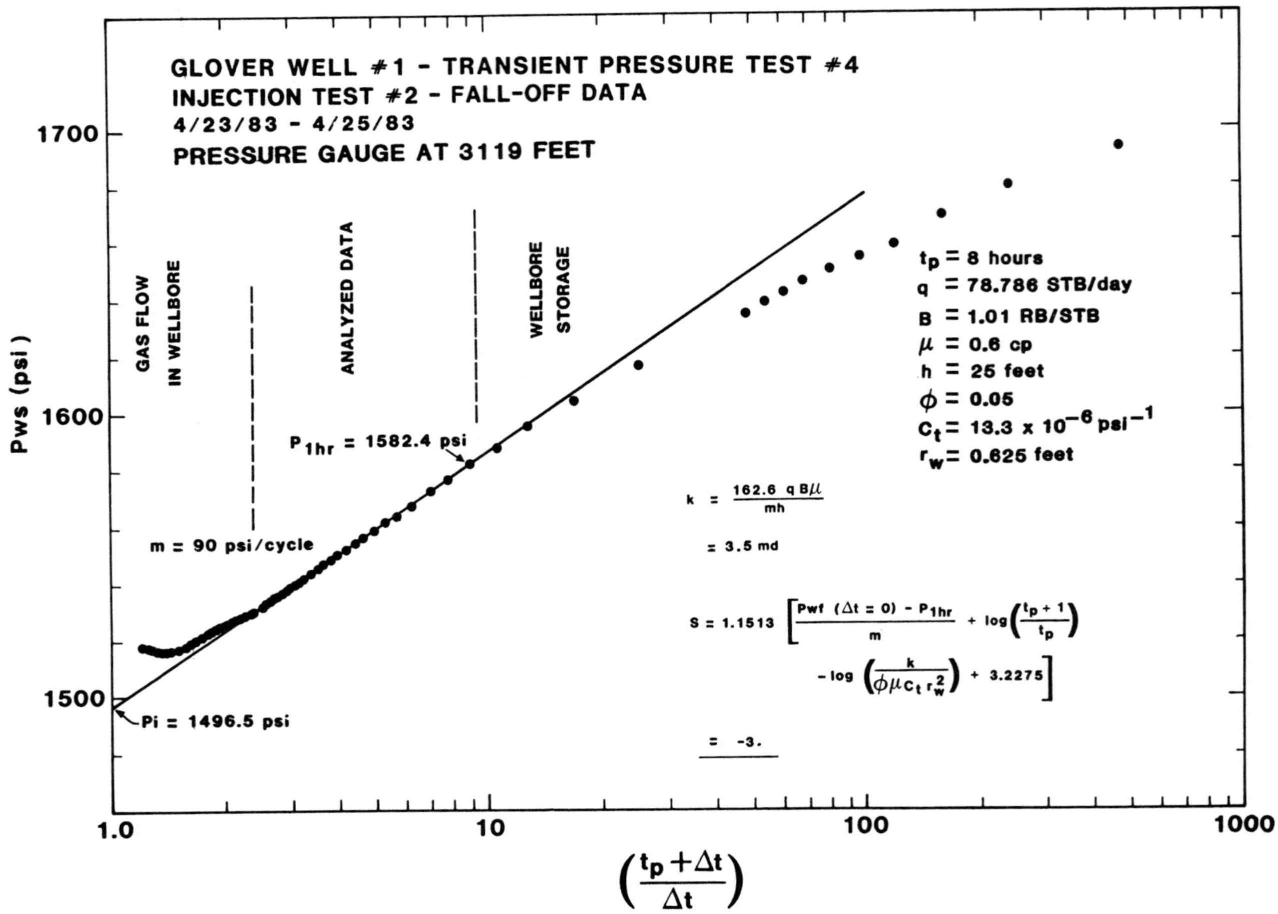

Figure 10. Horner plot for the fall-off data following the injection test prior to the six week production test of the Glover No. 1 well.

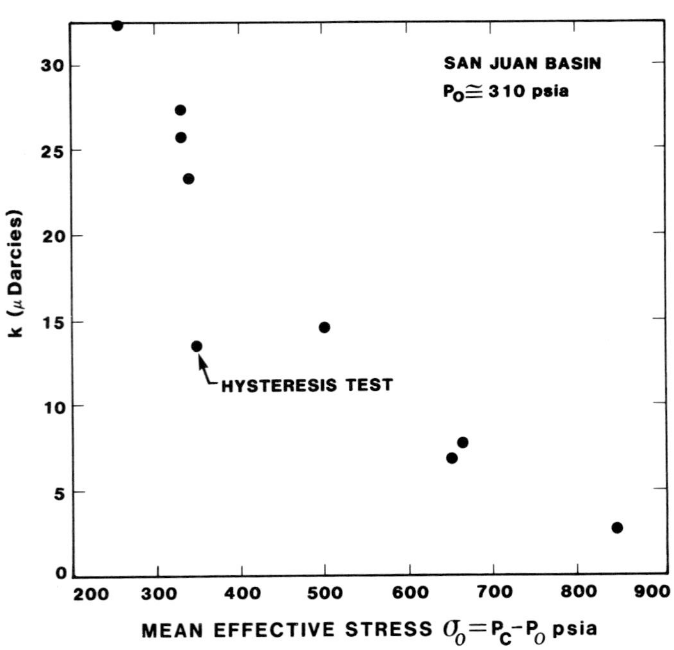

Figure 11. Permeability of a San Juan Basin coal with effective confining pressure ($\sigma_o = p_c - p_o$) where p_c is the confining pressure and p_o the pore pressure (from Rose and Foh, 1984).

TABLE 5. Swelling of coal on exposure in CO_2 atmospheres at 20 psig (from Reucroft and Patel, 1986).

C (daf) (percent)	Linear swelling (percent)
83.8	0.120
78.3	0.300
65.8	0.437

strong adsorption-desorption hysteresis. Only a small amount of methane adsorption-desorption hysteresis exists for United States coals (Bell and Rakop, 1986) which suggests that methane adsorption is only slightly activated. Therefore, coal swelling or shrinkage is expected to be much less for methane than for CO_2. It should be noted, however, that the CO_2 content of coal gas increases over time during gas production. Gas composition history for the Glover No. 1 well during the first fifteen months production history is illustrated in table 6. Therefore, there exists the potential for permeability to increase as CO_2 is desorbed from coal. This effect remains to be evaluated.

Relative Permeability

Downhole pressure history during a drawdown test at the

TABLE 6. Average coal gas composition (Molecular percent). Samples from the Glover No. 1 well.

Recovered during desorption tests

CH_4	C_2H_6	C_3H_8	C_4H_{12}	C_5H_{12}	O_2	N_2	CO_2	H_2	He	H_2S
99.50	—	—	—	—	<0.01	<0.01	0.20	—	—	—
99.50	—	—	—	—	<0.01	<0.01	0.20	—	—	—

Heating value = 1,004 Btu/scf

Recovered during well pumping

CH_4	C_2H_6	C_3H_8	C_4H_{12}	C_5H_{12}	O_2	N_2	CO_2	H_2	He	H_2S
98.50	<0.01	<0.01	—	—	<0.01	<0.01	0.99	—	—	—
98.50	<0.01	<0.01	—	—	<0.01	<0.01	0.90	—	—	—

Heating value = 992 Btu/scf

Recovered after 15 months production

CH_4	C_2H_6	C_3H_8	C_4H_{12}	C_5H_{12}	O_2	N_2	CO_2	H_2	He	H_2S
92.83	0.38	0.02	<0.01	<0.01	<0.01	<0.09	6.86	<0.01	<0.01	—
92.82	0.38	0.02	<0.01	—	<0.01	<0.09	6.54	—	0.06	—

Heating value = 944 Btu/scf

Glover No. 1 well is summarized in figure 12. Wellbore storage effects dominated for the first 4½ hours. Between 4½ hours and 9 hours the response was due to single phase flow. The increase in slope 9 hours into the test reflects the onset of unsaturated flow as gas bubbles began to form with gas desorption from the coal surfaces. Water was still the only phase flowing at this stage. As reservoir pressure was further reduced, gas bubbles grew larger until they connected and formed a continuous pathway to the wellbore. Two phase flow began at this point and gas entered the wellbore near the end of the test.

The large change in slope 9 hours into the test is consistent with the relative permeability behavior determined for coal as depicted in figure 13. The data predicts sharply-reduced water production with gas evolution.

The relative permeability to water was deduced from the capillary pressure curve for a sample "without well-developed cleats" shown in figure 4 using the relationship introduced by Purcell (1949):

$$k_{rw} = \frac{\int_0^{S_w} \frac{dS}{p_c^2}}{\int_0^1 \frac{dS}{p_c^2}}$$

where S is the water saturation and p_c the corresponding capillary pressure. Dabbous and others (1976) showed the calculated relative permeability to water to be in good agreement with the measured relative permeability for Pittsburgh coal. Relative permeability to gas, shown in figure 13, was determined by the stationary-fluid technique. The relative permeability to gas corresponding to the irreducible water saturation value will eventually control the rate of gas production. Figure 13 indicates the gas relative permeability will never exceed 0.1 at any stage during the producing life of the reservoir. At early production times, the gas relative permeability will be much lower. In view of the low relative permeability to gas (total field measured permeability was 3.50 md), a hydraulic fracture or a drainhole would significantly increase productivity.

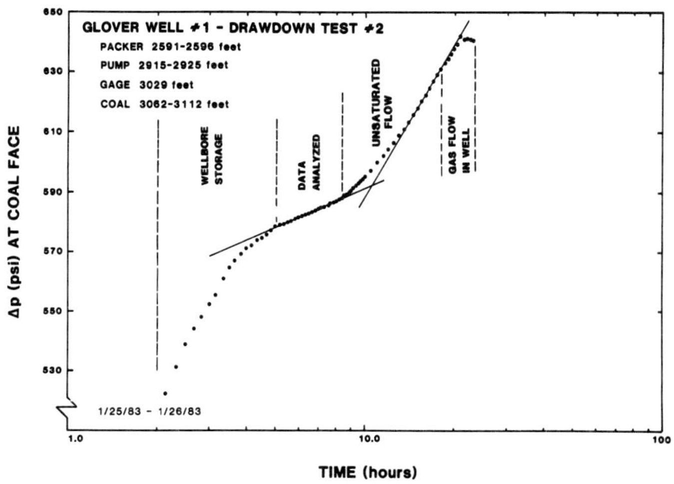

Figure 12. Semilog plot for drawdown test no. 2, Glover No. 1 well.

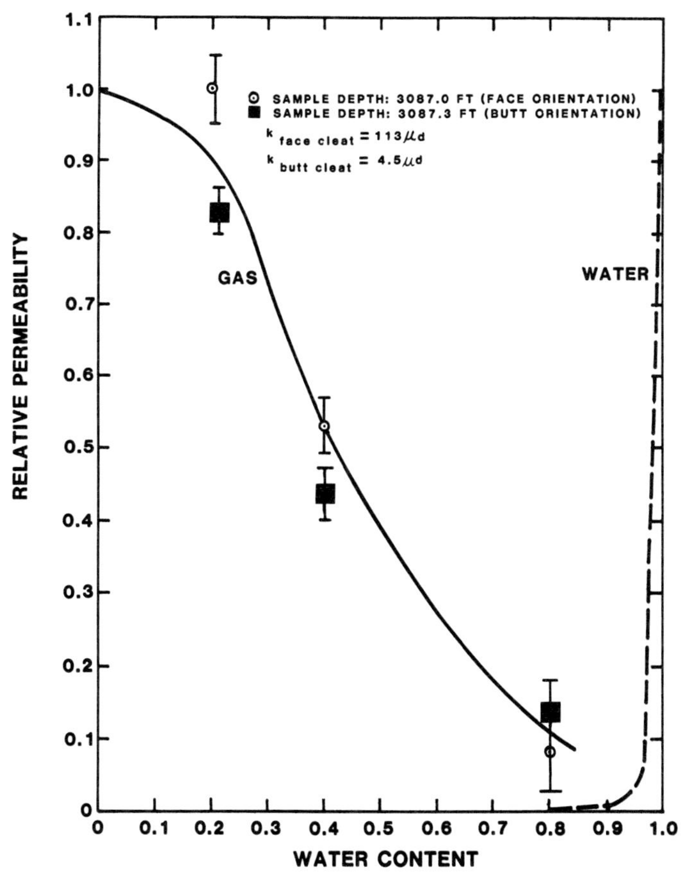

Figure 13. Relative permeability curves for the Fruitland coal, Glover No. 1 well.

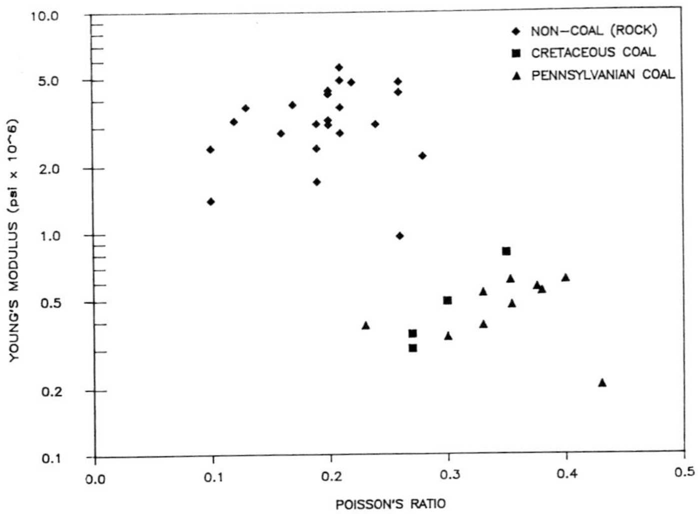

Figure 14. Mechanical properties of coal and non-coal rock.

TABLE 7. Critical stress intensity factor of coal under quasistatic loading, WLCT method (from Klepaczko and others, 1984)

Loading perpendicular to the bedding plane

Specimen no.	K_{Ic}(psi $\sqrt{}$ in)
1	154.0
2	173.2
3	90.0
4	142.7
Mean	140.0 ± 35.6

Loading parallel to the bedding plane

Specimen no.	K_{Ic}(psi $\sqrt{}$ in)
N1X	166.3
N4X	159.6
N5X	167.5
Mean	164.5 ± 4.3

PROPERTIES IMPORTANT TO HYDRAULIC FRACTURING

Mechanical Properties

A comparison of the static elastic properties (assumed isotropic) for several coals is shown in figure 14. There is not a great difference in Young's modulus or Poisson's ratio for the coals plotted. The Cretaceous-age coals (from the Piceance and San Juan basins) lie in the same range as the Pennsylvanian-age coals (from the Warrior and Appalachian basins). However, there is an obvious difference between the coals and non-coal rock properties. The elastic modulus for the non-coal rock ranges from 2.0×10^6 psi to 5.0×10^6 psi as compared to approximately 0.5×10^6 psi for coal. Similarly, the Poisson's ratio for non-coal varies from 0.1 to 0.28 with the coal ranging from 0.23 to in excess of 0.40.

The low elastic modulus implies large fracture widths for moderate net pressures (bottomhole treating pressures less the minimum in-situ stress) during hydraulic fracturing and low frictional pressure drop. Fracture tip opening will be proportional to the critical stress intensity factor, K_{Ic}, and inversely proportional to the elastic modulus. In such cases, the value used for the critical stress intensity factor is clearly important.

Stress Intensity Factor

Klepaczko and others (1984) determined the stress intensity factor for coal using the wedge-loaded-compact-tension (WLCT) specimen. Four WLCT coal specimens with the notch oriented perpendicular to the bedding plane, as well as three specimens, from a different coal batch, with the notch oriented parallel to the bedding plane were loaded to failure. No characterization was provided for the coals.

A summary of the quasi-static loading test results, reported in Klepaczko and others (1984), are provided in table 7. Because of the nature of the material, its stress intensity factor is relatively low. This low value for K_{Ic} implies the fracture opening will remain small. At high loading rates, however, the critical stress intensity factor was found to increase by over an order of magnitude.

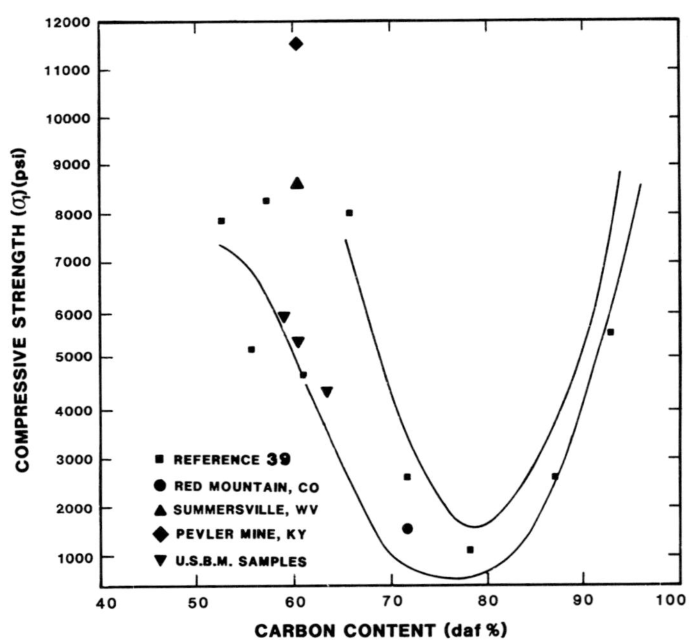

Figure 15. Comparison of unconfined compressive strength with volatile matter content (modified from Pomeroy and Foote, 1960). Reference 39, listed as a source on the graph, is Pomeroy and Foote (1960).

Compressive Strength

Coal is a weak, friable material that readily fails in compression. It is weakest at ranks ranging from high-volatile A bituminous to low-volatile bituminous as shown in figure 15 based on the data of Pomeroy and Foote (1960) along with results of recently tested coals at Terra Tek and the USBM (Costantino and Trettenero, 1983). Figure 15 shows that the unconfined compressive strength decreases to a minimum as the coal rank increases and the carbon content reaches 70 to 80 percent (daf). Strength then increases in the anthracitic ranks.

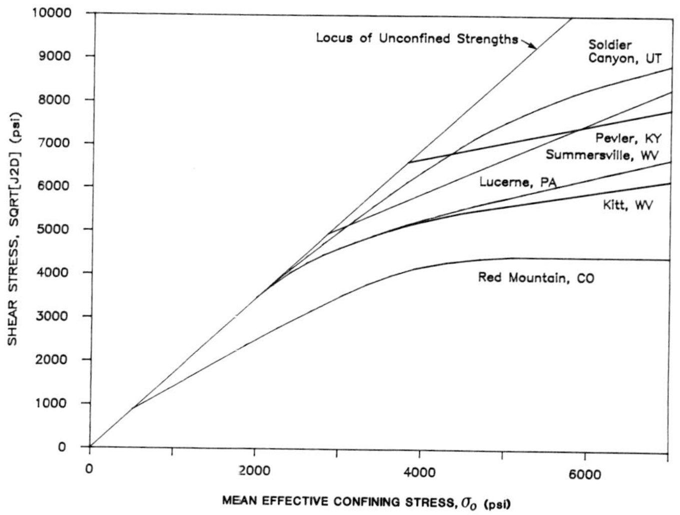

Figure 16. Coal failure envelopes (includes data from Costantino and Trettenero, 1983).

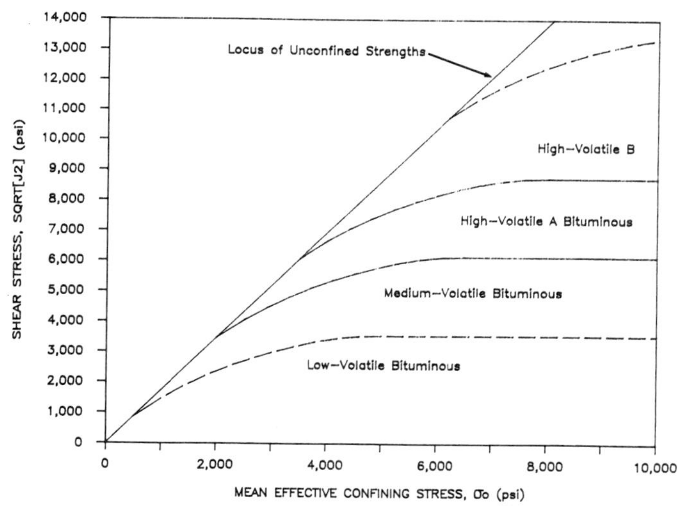

Figure 17. Generic coal failure envelopes.

Figure 16 compares the failure envelope for six coals tested in triaxial compression which includes data from Costantino and Trettenero (1983). The square root of the second invariant of deviatoric stress ($\sqrt{J2}$) has been used to represent the shear failure strength as a function of the mean effective confining stress. For axial stress σ_1 and confining stress σ_3, the following definitions apply:

$$\sqrt{J_2} = \frac{1}{\sqrt{3}} (\sigma_1 - \sigma_3)$$

The corresponding mean confining stress, (p_c) is:

$$p_c = \frac{(\sigma_1 + 2\sigma_3)}{3}$$

and the effective confining stress, σ_o

$$\sigma_o = p_c - p_o$$

where p_o is the pore pressure. In figure 16 the Red Mountain Unit coal is the weakest and also is the highest in rank. Although the Lucerne Mine and Kitt Mine samples are classified as being of different ranks (medium-volatile and high-volatile A bituminous, respectively) they are both near the medium-to high-volatile A borderline. These coals exhibit a similar strength which is greater than the mid-range medium-volatile bituminous Red Mountain coal. The Summersville Mine and Pevler Mine samples are high-volatile A to borderline high-volatile B bituminous and have the next highest strengths. The strongest coal is from the Soldier Canyon Mine which is mid-range high-volatile B bituminous in rank. Results from figure 16 were used to develop figure 17 which consists of generic failure envelopes for different ranges of coal rank. Figure 17 shows that the shear strength is a minimum for the low-volatile bituminous rank coal.

Coal weakness creates problems in well drilling and completion. Wash-out zones invariably exist, corresponding to the coal intervals. Failure of the coal during stimulation is considered to be the reason for the abnormally high treatment pressures (Jones and others, 1987). Mud weight must be carefully selected in drilling drainholes in highly stressed and (or) deeply buried coals to prevent sloughing and possible sticking of the drill pipe or well liner during placement. Too great a mud weight will result in enhanced fluid loss due to fracturing. Coal-fines damage to both downhole and surface equipment is commonly experienced when wells are first placed on production. These problems can be aggravated in regions having been subjected to tectonic activity.

SUMMARY

The physical and mechanical characterization methodology for conventional resources has been well-developed by the oil and gas industry. However, there has been less experience in determining the properties of coal as a gas-bearing formation. Material properties reviewed in this paper indicate that a different approach is needed for the completion of wells in coal and for the production of gas from coal reservoirs.

Gas storage is dominantly by adsorption in coal rather than by gas compression in the unfilled voids as in sandstone and limestone. A large fraction of the adsorbed gas is released at low pressures, whereas in conventional gas wells the gas produced is proportional to the pressure drop. To recover a large fraction of the initial gas-in-place, it is important to plan for gas production to low pressures.

Coal-gas reservoirs initially tend to be highly water saturated and water pumping is normally required in order to produce gas. With the large fraction of the pore volume residing in the micropores and mesopores, the irreducible water saturations are found to be greater than 70 percent, resulting in a low relative permeability to gas throughout the producing life of the reservoir. For this reason hydraulic fracture stimulations are effective for increasing gas production. Furthermore, high fracture conductivities are required for production to low reservoir pressures. Drainholes drilled in the coal seam should be equally effective.

High hydraulic-fracture treatment pressures are more the

rule rather than the exception. Research is being pursued to verify the mechanisms resulting in these high pressure treatments. One potential mechanism is failure of the weak friable coal at the wellbore and in the vicinity of the fracture tip during treatment resulting in fracture blocking. Also, coal failed during hydraulic fracturing, in the form of coal chips and fines, brought to the wellbore and pumped to the surface, creates problems with the operation of both downhole pumps and surface equipment.

Additional information on coal properties is required to aid in completion and reservoir engineering; for example: 1) the diffusion of gas from intact, water-saturated coal samples subjected to simulated in-situ conditions, 2) desorption of mixed gases from the adsorbed state (especially CO_2, methane, and the higher hydrocarbon gases), 3) the changes in pore compressibility with desorption of mixed-composition gases, and 4) the strength of coal previously subjected to high tectonic forces. Careful evaluation of coal sample preparation is needed. With improved knowledge of coal properties, better completion techniques and production of the coalbed methane resources will be achieved.

REFERENCES CITED

Airey, E. M., 1968, Gas emission from broken coal: An experimental and theoretical investigation: International Journal of Rock Mechanics and Mining Science, v. 5, p. 475-494.

Anderson, R. B., Bayer, J., Hofer, L. J. E., 1964, Equilibrium sorption studies of methane on Pittsburgh Seam and Pocahontas No. 3 Seam coal: Fuel, v. 63, p. 386-399.

Anderson, R. B., Bayer, J., and Hofer, L. J. E., 1965, Determining surface areas from CO_2 isotherms: Fuel, v. 64, p. 443-452.

Bell, G. J., and Rakop, K. C., 1986, Hysteresis of methane/coal sorption isotherms: Society of Petroleum Engineers Paper 15454.

Costantino, M., and Trettenero, S., 1983, The confined compressive strengths and Young's Moduli of three American coals: Journal of Applied Physics, v. 54, p. 76-82.

Dabbous, M. K., Reznik, A. A., Mody, B. G., Fulton, R. F., and Taber, J. J., 1976, Gas-water capillary pressure in coal at various overburden pressures: Society of Petroleum Engineers Journal, October, p. 261-268.

Ettinger, I. L., and Zhupakhina, E. S., 1960, Method of determining porosity of mineral coals: Fuel, v. 39, p. 387-392.

Gan, H., Nandi, S. P., and Walker, P. L., Jr., 1972, Nature of the porosity in American coals: Fuel, v. 51, p. 272-277.

Gray, I., 1987, Reservoir engineering in coal seams: Part I—The physical process of gas storage and movement in coal seams: Society of Petroleum Engineers Reservoir Engineering, v. 2, p. 28-34.

Hargraves, A. J., 1963, Instantaneous outbursts of coal and gas: unpublished Ph.D. Dissertation, University of Sydney, Sydney, Australia.

Harplalani, S., and McPherson, M. J., 1985, Effect of stress on permeability of coal: Quarterly Review of Methane from Coal Seams Technology, v. 3, p. 23-28.

Harris, L. A., and Yust, C. S., 1976, Transmission electron microscope observation of porosity in coal: Fuel, v. 55, 1976, p. 233-236.

Hunt, J. M., 1979, Petroleum geochemistry and geology: W. H. Freeman publishing company.

Jones, A. H., Bell, G. J., Morales, R. H., and Schraufnagel, R. A., 1987, The influence of coal fines/chips on the behavior of hydraulic fracture stimulation treatments, *in* Proceedings of the coalbed methane symposium, Tuscaloosa, Alabama, November 16-19, p. 93-102.

Juntgen, H., and Karweil, J., 1966, Gasbuildung und Gaspeicherung in Steinkohlenflozen: Part I und II: Erdol und Kohle, Erdgas, Petrochemie, v. 19, p. 251-258 and 339-344.

Juntgen, H., and Klein, J., 1975, Enstehung von Erdgas gus Kohligen Sedimenten: Erdol und Kohle, Erdgas, Petrochemie, Erganzungsband, v. 1, p. 52-69.

Juntgen, H., and Langhoff, J., 1964, Neue Ergebnisse zur Methansorption und Feiststruktur von Steinkohlen: I. Mechanismus der Methansorption und bei Raumtemperatur verfugbare innere oberflachen: Brennstoff-Chemie, v. 45, p. 28-32.

Karweil, J., 1969, Aktuelle Problem der Geochemic der Kohle, in Schenck, P. A. and Havernaar, I., eds., Advances in Geochemistry, Pergamon Press.

Klepaczko, J. R., Bassum, N. M., and Hsu, T. R., 1984, Fracture toughness of coal under quasi-static and impact loading: Engineering Fracture Mechanics, v. 19, p. 305-316.

Kim, A. G., 1977, Estimating methane content of bituminous coalbeds from adsorptive data: U. S. Bureau of Mines, Report of Investigation, RI 8245.

Kini, K. A., and Lahiri, A., 1960, Surface area of coal: Fuel, v. 39, p. 515-516.

Lee, Y. H., 1982, Methane recovery from coalbeds: Effects of monolayer capacity and pore structure on gas content: unpublished M.S. Thesis, University of New Mexico, Albuquerque, New Mexico.

Lin, J. S., Hendricks, R. W., Harris, L. A., and Yust, C. S., 1978, Microporosity and micromineralogy of vitrinite in a bituminous coal: Journal of Applied Crystallography, v. 11, p. 621-625.

Marsh, H., and Siemieniewska, T., 1965, The surface areas of coal as evaluated from the adsorption isotherms of carbon dioxide using the Dubinin-Polanyi equation: Fuel, v. 44, p. 355-367.

McKee, C. R., Bumb, A. C., and Bell, G. J., 1984, Effects of stress-dependent permeability on methane production from deep coal seams: Society of Petroleum Engineers SPE/DOE/GRI Paper 12858, p. 271-280.

McKee, C. R., Bumb, A. C., and Koenig, R. A., 1988, Stress dependent permeability and porosity of coal, *in* Fassett, J. E., ed., Geology and coal-bed methane resources of the northern San Juan basin, Colorado and New Mexico: Rocky Mountain Association of Geologists Guidebook.

Meissner, F. F., 1984, Cretaceous and Lower Tertiary coals as sources for gas accumulations in the Rocky Mountain area, *in* Hydrocarbon source rocks of the greater Rocky Mountain region: Woodward, J., Meissner F. F., and Clayton, J. L., eds., Rocky Mountain Association of Geologists, p. 401-431.

Nelson, J. R., 1979, Sorption of organic molecules into carbons, crosslinked polymers, and coals of varying rank: unpublished Ph.D. Dissertation, Pennsylvania State University, University Park, Pennsylvania.

Pomeroy, C. D., and Foote, P., 1960, A laboratory investigation of the relation between ploughability and the mechanical properties of coal: Colliery Engineering, v. 37, p. 146-154.

Price, H. S., and Ancell, K. L., 1978, The feasibility of methane production from coal: U. S. Department of Energy Report No. MERC/SP-78/1, p. 37-73.

Purcell, W. R., 1949, Capillary pressures—Their measurement using mercury and the calculation of permeability therefrom: Petroleum Transactions American Society of Mechanical Engineers, p. 39-48.

Reiss, L. H., 1981, The reservoir engineering aspects of fractured formations: Gulf Publishing Company.

Reucroft, P. J., and Patel, H., 1986, Gas-induced swelling in coal: Fuel, v. 65, p. 816-820.

Rose, R. E., and Foh, S. E., 1984, Liquid permeability of coal as a function of net stress: Society of Petroleum Engineers SPE/DOE/GRI Paper 12856, p. 253-259.

Ruckenstein, E., Vaidyanathan, A. S., and Youngquist, G. R., 1971, Sorption by solids with bidisperse pore structures: Chemical Engineering Science, v. 26, p. 1305-1318.

Schwerer, F. C., and Pavone, A. M., 1984, Effect of pressure-dependent permeability on well-test analyses and long-term production of methane from coal seams: Society of Petroleum Engineers SPE/DOE/GRI Paper 12857, p. 261-270.

Smith, D. M., 1982, Methane diffusion and desorption in coal: unpublished Ph.D. Dissertation, University of New Mexico, Albuquerque, New Mexico.

Snowden, L. R., 1980, Resinite—A potential petroleum source in the Upper Cretaceous/Tertiary of the Beaufort-Mackenzie sedimentary basin, *in* Miall, A. D., ed., Facts and Principles of World Petroleum Occurrence: Canadian Society of Petroleum Geologists Memoir, v. 6, p. 509-521.

Ting, F. T. C., 1982, Coal macerals, *in* Meyer, R. A., ed., Coal structure, Academic Press, p. 7-49.

The Pine River Store and Post Office serving the inhabitants of the Pine River valley. Buckskin Charlie, identified as no. 2, served as Chief of the Southern Utes. Photo courtesy of the Center of Southwest Studies.

The Durango & Silverton narrow guage railroad in Rockwood Gorge on its run from Durango to Silverton, Colorado. The same railroad bed is utilized for the scenic train ride between the two towns. Photo courtesy of the Center of Southwest Studies.

Disposal of Production Waters from Oil and Gas Wells in the Northern San Juan Basin, Colorado

Gerald L. Zimpfer[1], Eric J. Harmon[2], and Bradford C. Boyce[3]

ABSTRACT

Several different environmentally acceptable and economically sound brine conveyance and disposal alternatives are available for use in the northern San Juan Basin. These include truck or pipeline conveyance, direct use of produced water (with or without treatment), surface evaporation-pits, and underground injection wells.

Pipeline transport has fewer disadvantages than truck transport. A pipeline transportation system would be more difficult to implement because of required planning, engineering design, and high initial capital costs for equipment and construction. However, over the long-term, pipeline transportation is considerably less expensive on a unit cost basis, and provides less risk of environmental degradation.

On site, direct-use of brines without treatment is the least expensive water-disposal alternative, but would be limited to areas having relatively good quality water. Because use is limited by regulation to agriculture or wildlife use, treatment prior to discharge is normally too costly to justify.

Evaporation pits have relatively large land requirements, have high capital startup costs, and are frequently subject to environmental problems. However, they require a relatively small labor force, and passive pits have low energy consumption requirements. Injection wells also have relatively high startup costs, may require on-site brine storage, and may not be suitable for the disposal of small volumes. In general, for brine-disposal volumes below approximately 30,000 BWD per facility, injection-well disposal is less expensive than evaporation pit disposal.

Brine disposal on Indian reservations is subject to the jurisdiction of a variety of Federal, State and local agencies depending on land or mineral ownership, and the disposal method. No single disposal method is appropriate for all situations. Successful disposal systems must be designed on the basis of production or operational concerns, economics, environmental factors, and regulatory framework.

[1] G. L. Zimpfer
Yates & Auberle, Ltd.
303 East 17th Avenue
Suite 1060
Denver, Colorado 80203

[2] Eric J. Harmon
HRS Water Consultants, Inc.
200 Union Boulevard
Suite 400
Lakewood, Colorado 80228

[3] Bradford C. Boyce
Southern Ute Indian Tribe
Tribal Affairs Building
P. O. Box 737
Ignacio, Colorado 80202

This study was carried out under the direction of the Council of Energy Resource Tribes. The study was supported by funds from the Office of Federal Activities, U.S. EPA. The following persons provided valuable input to this report: D. Altman, ENSCO Corporation, Denver, Colorado; E. Bush, New Mexico Oil and Gas Conservation Commission, Aztec, New Mexico; M. Cook, Director of Energy Resources, Southern Ute Tribe, Ignacio, Colorado; D. Englishman, U.S. Bureau of Land Management, San Juan Resource Area Office, Durango, Colorado; R. Franks, Union Texas Petroleum Company, Farmington, New Mexico; J. Hudson, MECO Corporation, New Orleans, Louisiana; T. Pike, Indian Lands UIC Coordinator, U.S. Environmental Protection Agency, Denver, Colorado; W. Smith, Director, Colorado Oil and Gas Division, Denver, Colorado; and R. Young, Professor, Department of Agricultural Engineering, Colorado State University, Fort Collins, Colorado.

INTRODUCTION

Purpose

The coal-bed methane gas resource of the Fruitland Formation within the San Juan Basin is a major gas reserve currently undergoing expanded production. Proper management of the large volumes of coal-seam water produced along with the methane gas is crucial to the economic development of the resource. Prospect development is most successful when all relevant factors are considered and integrated into the prospect economics. Failure to fully develop and implement a comprehensive brine management program for Fruitland coal-bed methane development projects may unnecessarily increase expenditures and could result in failure of a project.

Proper brine management is also important in protecting tribal water resources during methane-gas resource development. The principal economic activities on the reservation: agriculture, recreation, and energy development, all require water for development. Improper brine disposal may have adverse effects upon surface and ground water quality. Proper brine disposal will protect the quality of water resources during additional development of methane reserves.

The Southern Ute Indian Tribe is concerned with the need to balance anticipated energy development with environmental considerations. The purpose of this report is to provide a brief description of the brine conveyance and disposal alternatives used in the San Juan Basin, to provide a preliminary economic assessment of the brine conveyance and disposal alternatives, and to describe the regulatory framework for brine management on the Reservation.

Location and Description of the Southern Ute Reservation

The Southern Ute Reservation, in southwestern Colorado, is a fifteen-mile by seventy-three-mile strip of land along the Colorado-New Mexico border. It covers parts of Archuleta, La Plata, and Montezuma Counties and adjoins the Ute Mountain Ute Reservation to the west, and the Jicarilla Apache Reservation to the southeast. The reservation is located within the Colorado Plateau physiographic province. The topography varies from high timber-covered mesas to fertile river valleys; elevations range from 6,000 to nearly 9,000 feet. Seven rivers cross the Reservation, the San Juan, Piedra, Pine, Navajo, Animas, Florida, and La Plata. The Reservation, which lies on the northern flank of the San Juan Basin, includes significant mineral resources including coal, oil, and natural gas. A combination of many factors in the geologic evolution of the San Juan Basin have resulted in the area becoming one of the largest known natural-gas reservoirs in the world.

Most of the Tribal members residing on the Reservation live in the immediate vicinity of the town of Ignacio or near the trading center of the Pine River (Los Pinos) Valley. The Tribal headquarters are located near Ignacio. The economic base for the Reservation includes oil and gas production, timber harvesting, agriculture, and recreation.

The governing body of the Southern Ute Tribe is the Tribal Council, elected by the popular vote of the people. A self-governing political entity, the Tribe operates under an approved constitution. The Chairman of the Tribal Council serves as administrative leader of the tribe. The Tribal Council, subject to restrictions contained in the Tribal Constitution and the U.S. Federal Code of Regulations, exercises a variety of rights and powers including the authority to enact and enforce Tribal codes or ordinances to protect the environment.

BRINE PRODUCTION AND MANAGEMENT

Current and Anticipated Brine Production

Although several natural-gas bearing formations underlie the Reservation, current activity is primarily concerned with the Fruitland Formation, which produces methane from coal layers (Butler, 1986). The depth of the Fruitland Formation varies from near or at the ground surface along the western and eastern edge of the Reservation, to a depth of over 4,000 feet in the south-central part of the Reservation, south of Ignacio. Ground water, varying from fresh to highly mineralized, occurs in the coal seams and in the sandstone layers which are interbedded with the coal layers in the Fruitland Formation (Brogden and others, 1977). Consequently, natural gas wells completed in the coal beds produce considerable volumes of ground water along with the natural gas produced.

Water quality in the Fruitland Formation varies depending upon geographical location. In the western and northwestern portions of the Reservation, Fruitland Formation water is fresh and has relatively good quality (total dissolved solids (TDS) concentration is less than 1,000 mg/l). As ground water migrates slowly to the south, downdip into the San Juan Basin, it becomes more highly mineralized (of poorer quality). Near the southern boundary of the Reservation, Fruitland Formation water has TDS concentrations of 10,000 mg/l to 15,000 mg/l. In the methane producing area on the Reservation, the TDS concentration of Fruitland brines is expected to be in the range 1,800 to 10,000 mg/l. (Although the water produced from methane-gas wells is a sodium bicarbonate type water with total dissolved solids concentrations lower than most oil field brines, production water from the methane-gas wells is called brine in this report.)

At the present time, approximately 100 coal-bed methane wells produce natural gas and, as a by-product, brine, from the Fruitland Formation. Natural gas production is approximately 25,000 MCFPD. Amoco Production Company is the largest producer of coal-bed methane on the Southern Ute Reservation, with over 90 percent of the present production (Smith, personal communication, 1987). Several smaller independent companies also produce coal-bed methane from the Fruitland Formation beneath Reservation lands. Brine production from methane wells varies from 14 BWD/well to about 1,000 BWD/well, with an average brine production of approximately 70 BWD/well. Therefore, total brine production from the 100 existing wells is about 7,000 BWD.

The Fruitland Formation within the Southern Ute Reservation contains an estimated 14.5 trillion cubic feet of recoverable methane (Kelso, this volume). Full development of the coal-bed methane resource may require an additional 1,300 methane production wells depending on well spacing, in addition to the 100 wells already in production, for a total of about 1,400 wells. Assuming an average brine production of 70 BWD/well, approximately 100,000 BWD of brine will be produced from Fruitland Formation coal-bed methane wells. (This is equal to approximately 4,600 acre-feet per year of brine which would be produced.) If 100 percent of this brine were to be disposed of by means of underground injection wells, approximately 20 injection wells with an average daily capacity of 5,000 BWD would be necessary. If greater volumes of brine are produced from the Fruitland Formation or if an average injection well has a capacity of less than 5,000 BWD, as many as 35 to 40 injection wells may eventually be needed.

The time period for such a large-scale increase of coal-bed methane production is difficult to predict. Given the presence of economic and regulatory conditions which are subject to change, it could take as many as 10 to 20 years to achieve full production from such a large gas reservoir. Tax incentives, due to expire in 1990, most likely will accelerate coal-bed methane production through 1988 and 1989 (Rountree, 1988). After that time, it is anticipated that production growth will slow somewhat due to greater industry regulation and expiration of tax incentives.

Development is expected to occur primarily in two areas of the Reservation. The first area is in the central to east-central portion of the Reservation, centered approximately three miles northeast of Ignacio, in an area of flat to rolling terrain which is relatively heavily populated. The area is extensively used for farming and ranching with canal-irrigated and sprinkler-irrigated crops and hay meadows. A second area anticipated to have an extensive initial increase of coal-bed methane production is a broad area encompassing the plateaus and canyons east and west of the Animas River valley. This area, which has relatively steep and broken terrain, is primarily used for natural gas production and for open-range cattle grazing. Except for the Animas River valley, the population density is low and there is little irrigated agriculture. This area probably would be developed for coal-bed methane production less intensively than the other area near Ignacio (Cook, personal communication, 1987).

Brine Management in the Northern San Juan Basin

Brine disposal practices in the northern San Juan Basin range from small passive evaporation ponds to 9,000-foot-deep disposal wells. The locations of underground injection wells and surface evaporation pits are shown on figure 1. Wells producing less than 5 BWD commonly have a small passive evaporation pit at the well location to evaporate the water produced. In other cases, brines are transported to commercial underground injection wells or evaporation pits for disposal. Brines are transported by trucking and by elaborate pipeline collection systems with constant monitoring of brine volume and chemistry on an individual-well basis.

Underground Injection Wells

In the Colorado portion of the San Juan Basin, there are currently five operating underground injection wells and several additional wells are being tested or proposed. Prior to 1985, one underground disposal well existed within the Ignacio Blanco gas field of Colorado. The well (Tenneco's Rea Unit No. 1, La Plata County, Colorado) is a commercial disposal facility injecting into the locally-gas-depleted Point Lookout Sandstone of the Mesaverde Group. Originally, this well was utilized for the disposal of conventional gas-sandstone-brines. Recently, because of increased gas and water production from the Fruitland Formation, this well has been used to dispose of brine from coal-bed methane wells.

In 1985, Amoco Production Company purchased a substantial acreage position from William Perlman, who previously had attempted to develop coal-bed methane resources in La Plata County, Colorado. In the three years subsequent to the purchase, Amoco has constructed two underground injection wells (the Salvador and McCaw subsurface injection facilities). A third Amoco injection well is currently being tested, and several additional injection wells are proposed.

Amoco's Salvador and McCaw injection systems dispose of the majority of the coal-bed brines produced in Colorado. Both injection systems utilize low pressure polyethylene or fiberglass pipes to collect brines from adjacent gas production wells. Brines are collected in holding tanks at the injection site, and filtered prior to injection.

The Salvador injection facility is located approximately three miles south of the town of Ignacio. The Salvador collection and injection system receives brine from 14 production wells connected to the injection facility by pipeline. The daily brine production for each well ranges from 10 to 400 barrels. Additional water is trucked to the Salvador injection site from outlying wells. Brine production from these outlying wells range up to 900 BWD. The Salvador well injects into the Bluff and Entrada Sandstones. The total water volume injected is slightly more than 1 million barrels (January 1, 1988). The maximum daily well injection rate is 4,800 barrels which is

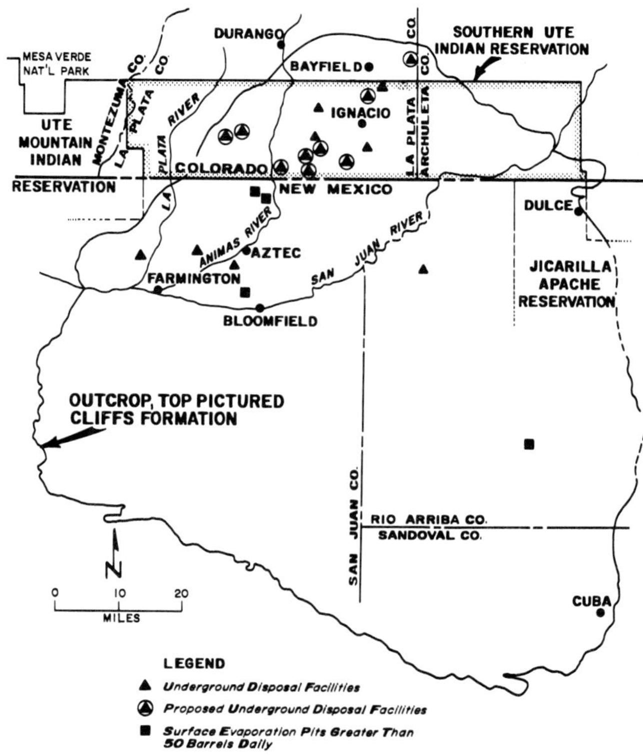

Figure 1. Index map of the San Juan Basin showing water disposal facility locations (Modified from Fassett and Hinds, 1971.)

limited by surface equipment.

The McCaw injection facility is located north of the La Plata County airport and adjacent to State Highway 172. The McCaw water collection system receives brine from 27 coal-bed methane production wells. Daily water production from each gas well ranges from less than 10 barrels to greater than 400 barrels. The McCaw well, injecting into the locally-gas-depleted Point Lookout Sandstone of the Mesaverde Group, has, as of January 1, 1988, disposed of approximately 1.5 million barrels of water. Maximum daily injection capacity is 9,600 barrels of water, the limit of the surface equipment.

Jay Magness, Inc., drilled and completed as a production and an injection well, the Echols-Ute 1-12 in NW ¼ sec. 12, T. 34 N., R. 7 W. in January of 1987. The Echols-Ute well injects Fruitland coal waters into the Point Lookout Sandstone of the Mesaverde Group which does not produce commercial gas in this area. Prior to obtaining approval for injection, Jay Magness, Inc. was required to demonstrate that the injection zone could not produce commercial quantities of hydrocarbons. The Point Lookout sandstone beds were proven to be water saturated at this well location and the minor volumes of gas could not be economically produced.

This well is unique in that it produces gas and associated water from the Fruitland coal seams while injecting the produced coal waters into the deeper Point Lookout (fig. 2). As of January 1, 1988, the mechanics of the well have been successful and the well has disposed of 47,200 barrels of water and produced 67,800 MCF cubic feet of Fruitland coal-bed methane gas.

In 1984, ARCO Oil and Gas began a pilot Fruitland coal-

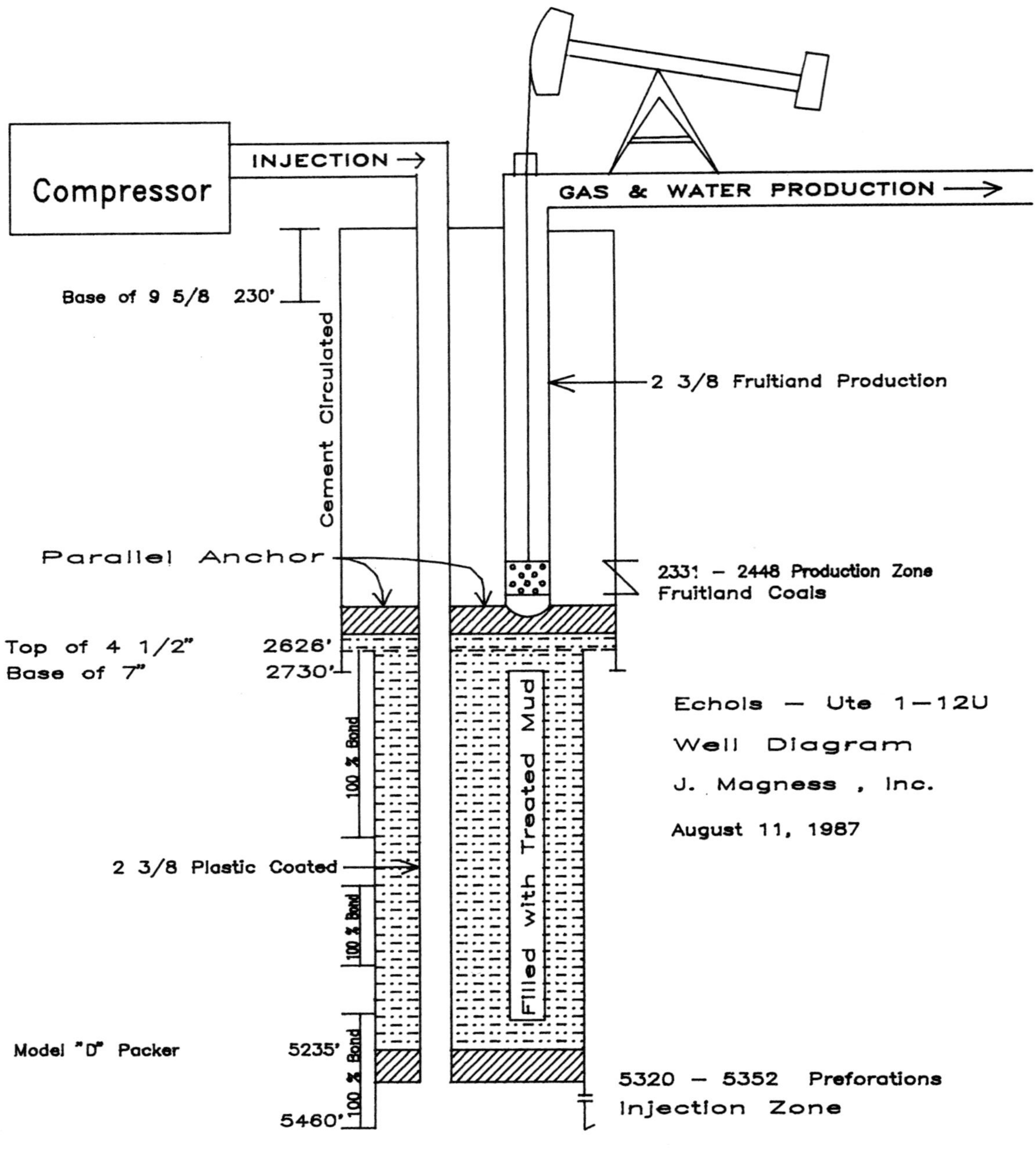

Figure 2. Schematic diagram of J. Magness combined production and disposal well.

degasification project drilling four Fruitland coal-bed test wells and an associated Pictured Cliffs Sandstone water disposal well in La Plata County, Colorado. The injection well was permitted in the fall of 1987. ARCO initiated the pilot project to resolve technical concerns prior to committing to a larger development program. ARCO's degasification project includes buried brine collection systems and will utilize various completion and stimulation techniques.

The San Juan Basin in northern New Mexico has approximately 23 operating underground injection wells (Bush, personal communication, 1988). Most of these injection well facilities are private operations associated with the disposal of brines produced from oil production in Northern New Mexico. Two injection wells are operated specifically to dispose of waters produced from Fruitland coal-gas production wells. Both of these injection wells are owned by Meridian Oil Company of Farmington, New Mexico.

The Meridian injection wells are the McGrath No. 4, San Juan County, New Mexico which injects brines into the Mesaverde Group, and the 30-6 unit No. 112, Rio Arriba County, New Mexico which injects into the Morrison and Entrada formations. Meridian drilled a third well, No. 58-A unit, to test the injection potential of the Ojo-Alamo Sandstone. The injection test indicated that the Ojo-Alamo has poor disposal potential. Meridian is in the process of abandoning this test well.

The McGrath No. 4 Mesaverde well has injected over one million barrels of brine and should continue to inject brine for many more years. Meridian's 30-6, No. 112 injection well, providing water disposal facilities for their 30-6 Fruitland coal-bed methane field, has injected over 800,000 barrels of water since its start up in 1986.

Evaporation Pit Disposal

Evaporation pits in Colorado are usually small, and handle low to moderate brine volumes depending on whether the system is passive or active. Frequently, a small evaporation pit associated with a single production well will receive 1 to 5 barrels of brine per day. Relatively poor evaporation rates and generally severe topographic relief limit the use of passive evaporation-pit disposal in Colorado.

Climate conditions in New Mexico are warmer and drier than in Colorado which results in higher evaporation rates. Data collected by the Arizona Public Service Company near Bloomfield, New Mexico shows an annual cumulative pan evaporation of 58.12 inches (Bush, personal communication, 1987). Monthly evaporation rates range from less than one-inch per month in December and January to almost nine inches per month in June and July. In addition, topographic relief in New Mexico is less severe, allowing construction of larger evaporation pits.

The New Mexico portion of the Northern San Juan basin contains several commercial surface evaporation pits, including the Basin Disposal Facility located outside Bloomfield, New Mexico and T.N.T. Water Disposal located in sec. 8, T. 25 N., R. 3 W., Rio Arriba County, New Mexico. In addition, Amoco Production Company and Union Texas Petroleum operate private evaporation pit systems; Amoco's evaportion pit is on their Cedar Hill coal-bed methane project located north of Aztec, New Mexico, and Union Texas's evaporation pit is located a few miles to the west of the Amoco operation.

Union Texas Petroleum constructed their private evaporative disposal system to handle produced waters from their adjacent gas wells. The evaporation pit, approximately 45,000 square feet, evaporates an average of 88 BWD utilizing a passive evaporation system. The evaporation pit was constructed such that conversion to an active system could be accomplished easily should water volumes justify the conversion.

COMPARISON OF BRINE CONVEYANCE AND DISPOSAL ALTERNATIVES

Brine conveyance is discussed separately from brine disposal. Two brine conveyance alternatives, truck transportation and pipeline transportation are considered. Once the brine is transported to a particular site for disposal, four alternatives are available: Direct use of the brine without treatment, direct use of the brine with treatment to improve water quality, evaporation pit disposal, and underground injection. The advantages, disadvantages, and cost of either brine conveyance alternative is independent of the brine disposal alternatives. Either brine conveyance alternative can be combined with any of the brine disposal alternatives.

Brine Conveyance Alternatives

Truck Transportation

Tanker trucks with a capacity of 7,000 to 9,500 gallons (approximately 167 bbl to 226 bbl) would collect the brine produced at each coal-bed methane production well, and convey it to a point of use, treatment, or disposal. Facilities and personnel needed by the operator (the Tribe, the gas-field operator, or some other private entity) would include trucks, haul roads, maintenance and storage facilities, drivers, managers, and a brine-storage tank or pond at each production well. Well-site storage would be necessary because gas production wells are commonly operated continuously to avoid harm to gas production and to maintain the mechanical integrity of the wells.

Land requirements for this conveyance alternative are comparable to the pipeline alternative. The trucking alternative would have a greater need for a continuous labor force of drivers and mechanics than the pipeline alternative. The primary advantage of truck transport of brine is that it can be implemented quickly at relatively little capital cost. There are several disadvantages to the truck-transport alternative (table 1). At the estimated level of full brine production from the Fruitland Formation of 100,000 BWD, an assumed average round-trip haul time of one hour for a 20-mile round trip, and a tanker volume of 200 bbl/tanker, brine transport would require 500 trips made per day, for a total mileage of 10,000 miles driven per day. This level of heavy-truck use may require significant investment in road maintenance and may result in several environmental and public health and safety problems (dust, noise, air pollution, degradation of driving safety for the residents of Tribal lands, and a potentially high risk of brine spills due to truck accidents). Another problem with trucking of brine is the potential for "midnight dumping", that is, unauthorized dumping of brine into irrigation canals, borrow pits along roadways, or into streams and arroyos. These problems could have an adverse effect on the overall quality of life for residents of the area.

Table 1. Brine conveyance alternatives comparison.

Advantages	Disadvantages
Truck transport	
Does not require a highly-trained labor force.	Requires a large labor force.
Could be implemented quickly.	Relatively high risk of environmental harm and safety concerns.
Would require relatively small capital outlay for startup.	Relatively high risk of unauthorized dumping.
	More expensive than pipeline transport over the long-term.
	Would require brine storage at each gas well.
Pipeline transport	
Less expensive than truck-transport over the long term.	Would require large capital outlay for startup.
Would not require brine storage at gas wells.	Would require relatively highly-trained personnel.
Relatively low risk of environmental harm and safety concerns.	Would require acquisition of rights-of-way.
Relatively low risk of unauthorized dumping.	

Pipeline Transportation

Alternatively, a system of buried pipelines could be used to convey brine from the Fruitland Formation production wells to the site or sites where the brine is to be used, treated, or disposed of. This alternative would require engineering design of a system of pipelines and pumping stations, as well as acquisition of rights-of-way for the pipeline corridors. Since a pipeline can accept brine on a continuing basis, well-site brine storage would not be required.

There would be several advantages to pipeline conveyance over tanker-truck conveyance (table 1). Although the potential for a brine spill due to leakage or pipe failure would exist, the risk would be relatively small provided that the pipeline and pumping-station system is properly designed, constructed and maintained. In addition, the potential problems of dust, noise, air pollution through diesel exhaust, and increased truck traffic would be minimized.

Initially, the pipeline alternative would be more capital-intensive in terms of planning, engineering, management, and investment for equipment and construction than the truck-transport alternative. (However, as later discussions of economics will show, it is less expensive than trucking on a unit-cost ($/bbl) basis over the long-term.) Another disadvantage of this alternative is the need for a relatively sophisticated control system, and the trained personnel to operate the pipeline system.

Brine Disposal Alternatives

Direct Use (No Treatment)

This alternative for disposal of coal-bed methane well brines would involve the use of the brine as a source of water supply for agricultural use (irrigation or stock watering) or wildlife. It assumes that the brine water quality is suitable for the intended use without treatment. Perhaps less than 25 percent of the produced water from the Fruitland Formation would be suitable for this alternative. In the southern half of the Southern Ute Reservation, the water quality in the Fruitland is anticipated to be unsuitable for most uses without prior treatment. However, in the northern half of the Reservation, where the water quality is better, direct use of the water should be considered. The total dissolved solids (TDS) concentrations of Fruitland Formation water in northern portions of the Reservation appear to be in the range of approximately 800 to 5,000 mg/l. (This assumes that use is limited by TDS and that there are no other water quality concerns with the Fruitland Formation water. The entities responsible for management of the irrigation systems may have additional water-quality requirements for water to be added to their system.) Water quality assessments would have to be made on a well by well basis to determine suitability for use without treatment.

In the major coal-bed methane development area centered near Ignacio, it may be possible to use water up to about 3,000 mg/l TDS for irrigation of crops or hay meadows, or for stock watering. This could, in part, be accomplished by use of the existing system of irrigation canals which supply the area.

In the less-populated plateau and canyon area centered near the Animas River, direct use of Fruitland Formation brines may be accomplished by construction of stock ponds to collect these waters or brines for livestock grazing in the area. Since there is almost no irrigated agriculture outside of the Animas River valley in this area, irrigation usage would require construction of new delivery systems. This would probably not be economical. Livestock can use water with a TDS concentration of 4,000 to 5,000 mg/l (McKee and Wolf, 1963), but cattle seldom thrive on water with a TDS concentration greater than 2,500 to 3,000 mg/l. Sheep can thrive on water with TDS concentrations up to 5,000 mg/l, provided that the water contains no other chemical constituents of concern.

Table 2. Brine disposal alternatives comparison.

Advantages	Disadvantages
Direct use (no treatment)	
Would require little capital outlay for startup. Operation and maintenance costs would be low.	Less than 25 percent of Fruitland brine may have suitable quality for direct use.
Unit cost of brine disposal is the lowest of the four disposal alternatives.	
Groundwater resources would be put to beneficial use.	
Direct use (with treatment)	
Groundwater resources would be put to beneficial use.	Would require large capital outlay for startup.
Could use wellhead natural gas as a power source.	Would require maintenance on a constant basis.
	Would require highly trained personnel.
	Some methods would require brine pretreatment.
	Would require disposal of brine concentrate or precipitate.
	Unit cost of brine disposal may be the highest of the alternatives studied.

The advantages and disadvantages for direct use without treatment are summarized in table 2. The primary advantage of direct use of brines without treatment is the low cost of this alternative. There would be little need for large-scale planning and management for direct use of the brines, and little maintenance cost except for occasional water-quality tests and maintenance on short lengths of pipeline, conveyance ditches, or stock ponds. Another potential advantage of direct use would be that valuable water resources could be put to beneficial use.

The primary disadvantage of direct use without treatment would be that a relatively small fraction of the anticipated brine volume could be used without treatment because of poor water quality.

Direct Use (with Treatment)

Direct use of brines with treatment would involve using any of several treatment methods to remove all or some portion of the dissolved chemical constituents in the brine making the water quality acceptable for agricultural or wildlife use. Common treatment methods for water with high total-dissolved-solids concentrations are: Distillation, ion-exchange (zeolite treatment), reverse osmosis, and dilution.

Distillation involves introduction of the brine into a sealed boiler where it is heated to the boiling point. The water vapor is then condensed, and released as purified water. Energy for heating the water can be provided by electrical power, diesel fuel, or on-site natural gas. (The latter source would be the most practical for use on Southern Ute Tribal lands.) Ion-exchange treatment involves pumping the brine through a bed or tank filled with a substance which will remove, by exchange of chemical ions, undesirable constituents from the brine. Zeolites, which comprise a family of naturally-occurring minerals, can be used as the ion-exchange medium (Schmidt, 1987). Reverse osmosis brine treatment is accomplished by pumping the brine through the molecular-size pores of a membrane. The membrane allows the water molecules to pass through, but does not allow the dissolved chemicals to pass through. Dilution is simply the process of mixing the brines produced by Fruitland Formation coal-bed methane wells with better quality water from some other source, resulting in water which is of suitable quality for the intended use.

With the exception of dilution, all of the treatment methods described above would have the disadvantage of requiring relatively high startup costs and constant maintenance to operate properly. The distillation and reverse osmosis processes result in a by-product consisting of a dry mineral precipitate or, more commonly in the case of reverse osmosis, a concentrated brine of perhaps one-tenth the volume but ten times the TDS concentration of the original brine. The reverse osmosis process has the added disadvantage that the membrane used must be operated with great care, and the input brine must be pre-filtered and monitored on a constant basis, so that no suspended solids enter the membrane and destroy its usefulness. Zeolite ion-exchange for brines which have TDS concentrations on the order of 5,000 to 10,000 mg/l or greater must be frequently backflushed, for perhaps one-third to one-half of the total operation time, to rid the zeolites of the accumulated chemicals from the brine. Such backflushing requires large volumes of relatively pure water to restore the zeolites to their original effectiveness (Reid and others, 1974; and Patton, 1974). The need for technical training by the operating personnel and constant monitoring of these processes is a distinct disadvantage. However, all of the treatment methods have the advantage that the end product is good quality water suitable for any beneficial use.

The dilution alternative is simple and requires little capital investment or operation and maintenance. However, an adequate supply of fresh water to mix with the brine is required to achieve a final TDS concentration which is suitable for the intended use. This alternative could be considered in the area near Ignacio, where perennial streams and an extensive irrigation-canal system may supply water for dilution. However, in the western portion of the Reservation, where the Animas and La Plata Rivers are the only perennial streams, this alternative may not be possible.

Evaporation Pit Brine Disposal

Disposal of brines in evaporation pits involves constructing and maintaining a shallow closed pit with a large surface area, introducing the brine into the pit, and allowing the brine to evaporate. An evaporation pit must be designed on impermeable soils, or constructed with a clay or impermeable-fabric lining, so that brine leakage out of the pit is minimized. In addition, the pit must be large enough to provide sufficient volume and surface area during winter when evaporation rates are low. After 10 to 20 years, depending on the brine concentration, the accumulated salt deposits must be removed from the evaporation pit. With a brine TDS concentration of 20,000 mg/l, which is probably the highest concentration expected from the Fruitland Formation brines, the accumulation would be about two inches after 20 years of continuous operation in a properly-sized evaporation pit (Haynes and Grubbs, 1969).

Table 3. Brine disposal alternatives comparison.

Advantages	Disadvantages
Evaporation pit brine disposal	
Less expensive than injection wells above approximately 30,000 BWD/facility.	Would require relatively large land areas. Use may be limited by topography and relatively low evaporation rates.
Low energy consumption.	Relatively large capital outlay for startup.
Relatively small labor-force needs.	Brine must be kept oil-free.
	Not well suited to disposal of small brine volumes.
	May have relatively high risk of environmental harm.
Deep-well injection brine disposal	
Relatively low risk of environmental harm.	Large capital outlay for startup.
Less expensive than evaporation pits up to approximately 30,000 BWD/facility.	Brine may require pretreatment.
	Not well suited to disposal of small brine volumes.
	May require on-site brine storage.

The advantages and disadvantages to using evaporation pits for brine disposal are summarized in table 3. A disadvantage of using evaporation pits is that relatively large land areas (usually several acres in extent) are required. Use of evaporation pits may be limited by topographic relief and by the relatively lower evaporation rates in the northern San Juan Basin. Also, evaporation pits are not suited to disposal of brines from just one or a few natural-gas wells because of economic considerations. As will be discussed in the economics section of this report, evaporation pits are best suited to disposal of larger volumes of brines collected from a number of production wells. Evaporation pits have the advantages of not consuming costly energy for operation, and of requiring relatively little maintenance, provided they are properly designed and constructed. Care must be taken with evaporation pits that the brine contains no oil. Even a small amount of oil will form a coating on the surface of the brine which will significantly reduce the evaporation rate of the brine. Historically, evaporation pits have been a common source of surface and ground water quality degradation as a result of embankment failure or seepage. Potential problems can be avoided, however, with proper design and construction of evaporation pits.

Passive evaporation systems may not be able to accommodate the large volume of brine commonly produced from methane-gas wells in the northern San Juan Basin. Active evaporation pits use a pump-and-spray system to enhance evaporation rates as much as five to ten times that of free surface evaporation rates. With an active system, evaporation rates will range from one to six percent of the pumped water volume (Franks, personal communication, 1987).

Active evaporation pits may offer several advantages over the passive evaporation pits discussed above. Use of active evaporation may enable the pit operator to reduce the surface area of the evaporation pit needed for disposal of a given volume of brine. Spray droplet size and local evaporation rates are the predominant factors controlling enhanced evaporation rates. Smaller droplet size, however, relies on smaller spray-head orifices, which may be susceptible to clogging if suspended material is present in the brine. A finer droplet-size spray may also result in a greater downwind travel distance of the sprayed brine and the accumulation of salt deposits in adjacent areas.

Underground Injection Brine Disposal

Disposal of brines is accomplished by injecting the brines back into the subsurface by pumping them into a deep well. An existing oil or gas production well which is no longer economic to operate as a production well can be used for disposal or a new well can be drilled and constructed specifically for the purpose of injection of brine into the subsurface.

A successful underground injection program depends upon the chemical compatibility of the injected brine and formation water, and upon the porosity and permeability of the receiving formation. Generally, brines are pumped into the selected geologic formation requiring an on-site source of energy. Brines can be disposed of in a well without pumping at very slow rates. However, the air bubbles entrained in the brine can become lodged at the wall of the borehole, effectively sealing the formation and preventing it from accepting the injected brine. Therefore, it is standard practice to pump the brine into the bore under pressure, in a surface and downhole piping system, which eliminates air bubbles and increases the rate at which brine can be injected into the receiving formation.

As with the evaporation pit method of disposal, injection wells are not particularly well suited to small-scale operations such as the disposal of brine from only one or a few production wells (table 3). At an average brine-production volume of 70 BWD/well, and at an average maximum brine-injection capacity of 5,000 BWD for each injection well, the ratio of coal-bed methane wells to brine-disposal wells would ideally be about 70 to 1 on the Southern Ute Reservation. A potential disadvantage of using injection wells is the need for pre-treatment of the brine prior to injection, to prevent plugging of the perforations in the casing of the injection well. It is also necessary to periodically clean injection well perforations of encrusted material which accumulates, even if pre-treatment is used (Davis and Hineline, 1986). Temporary storage or alternative disposal facilities would be required nearby while a specific injection well undergoes cleaning. An advantage of using injection wells for disposal of brines is that the brine is returned to a deep geologic zone in which the natural quality of the water is as bad, or, more likely, worse, than the injected brine. If the well is properly designed, maintained, and operated, there is little risk of surface water contamination with brine resulting from a leak or spill (Geraghty and Miller, 1980).

On the Southern Ute Reservation, the Mesaverde Group would most likely be used for brine disposal. These strata, consisting of interbedded sandstone, shale, and siltstone, are located at depths below that of the Fruitland Formation. The Mesaverde lies at depths of approximately 5,000 to 8,000 feet below ground surface beneath a majority of the Southern Ute Reservation. It would not be feasible to reinject the Fruitland-derived brines back into the Fruitland Formation coal-bed methane production zone. In order for a coal-bed natural gas reservoir to effectively yield large amounts of gas, the hydrostatic pressure of the formation must decrease. Removal of brine from the coal beds within the formation reduces the formation pressure, so that the natural gas production is stimulated. Injection of brine into gas-production zones would again raise the hydrostatic pressure of the formation, reducing gas production.

ECONOMIC ASPECTS OF BRINE DISPOSAL

The preliminary economic analysis of the brine conveyance and disposal alternatives also provides separate discussions of brine conveyance (truck transport or pipeline transport) and of brine disposal (direct use, treatment, evaporation, and underground injection). The cost of trucking or pipeline transportation must be combined with a particular disposal cost to obtain total disposal costs. The economic aspects of brine treatment and direct use are presented more generally than other conveyance and disposal alternatives. Complete economic analysis of all the treatment alternatives is beyond the scope of this study.

Unit costs in dollars per barrel of brine transported or brine disposed of are necessary to compare alternatives. Therefore, a computer program was developed which calculates unit cost ($/bbl) for each alternative over a range of brine volumes transported or disposed of. The component cost factors used to calculate unit costs are based on the methodology presented by Reid, and others (1974).

Several cost factors common to each alternative were included in the economic analyses. The cost factors common to each alternative included:

1. Quantity of brine to be conveyed or disposed of (variable, measured in BWD).
2. Estimated project life (20 years).
3. TDS concentration of brine (12,000 mg/l).
4. Discount interest rate (10 percent per annum) (Glenn, 1970).
5. Current ENR building-cost index (2564) (ENR Market Trends, ENR (formerly Engineering News Record) November 12, 1987, p. 78).
6. Unit electrical energy cost ($.075 per kilowatt-hour).
7. Land-acquisition cost ($5,000 per acre).
8. Contingencies (ten percent of capital cost).

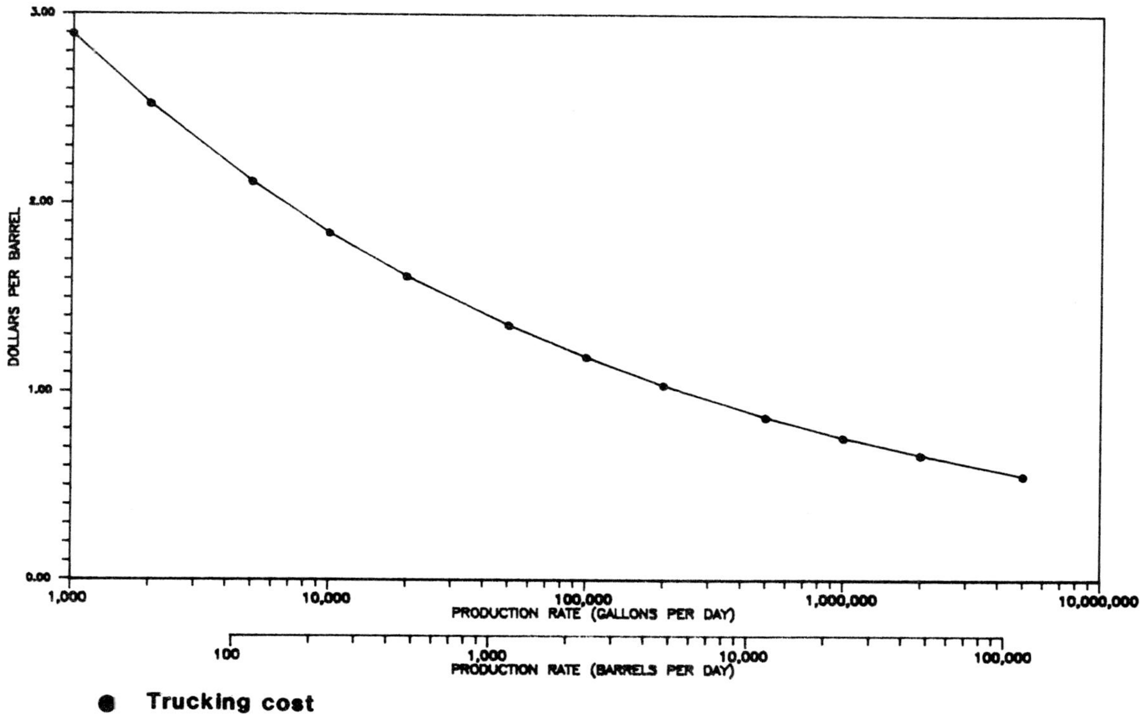

Figure 3. Graph of brine volume vs. trucking conveyance costs.

9. Engineering (ten percent of capital cost plus contingencies).
10. Interest on construction (1.625 percent of cumulative capital cost).
11. Interest on working capital (0.25 percent of all annual expenditures).

(Source for cost factors 8-11 was Reid and others, 1974.)

Brine Conveyance Alternatives

Truck Transport

A method for calculating unit cost for truck transportation was not included in Reid and others (1974). Therefore, based on discussions with representatives of Groendyke Transport Co. of Denver, Colorado, and B & N Water Trucks of Bayfield, Colorado; and information provided by the Southern Ute Tribe, an empirical mathematical relationship was developed to estimate trucking costs. This relationship takes into account the following factors which influence the unit cost ($/bbl) for truck transport of brine:

1. State and Federal highway taxes.
2. Licensing fees.
3. Fuel cost.
4. Cost of salaries for drivers, management, and support personnel.
5. Cost for purchase and periodic replacement of trucks and trailers.
6. Cost of maintenance and storage facilities for trucks.
7. Cost of management offices.
8. Cost of operation and maintenance.

Other factors which may be important in determining the costs of brine conveyance in a large-scale truck-transport operation (including the costs of rights of way for roads, road design and construction, and road maintenance) were not considered. In a detailed economic analysis of this conveyance alternative, these factors should be taken into account.

From ranges of costs provided by the sources identified above, and an assumed average tanker volume of 160 bbl (6,720 gallons) with an average haul distance of ten miles from production well to disposal facility, the relation between unit brine-transport costs and the volume of brine transported is described in the following equations:

$$C = 11.1\,(Q)^{-.195}$$

where C = unit cost of brine conveyance ($/bbl)
Q = volume of brine to be transported (gallons/day)

Figure 3 shows unit costs for truck transport versus brine volume as computed from this relation. Unit costs range from almost three dollars per barrel to less than one dollar per barrel, depending upon the volume transported. The economics-of-scale for the trucking conveyance alternative (and for most of the other conveyance and disposal alternatives) shows that the greater the volume of brine transported (or disposed of), the lower the unit cost.

Pipeline Transport

There are eight cost factors which make up the total unit-cost of pipeline conveyance of brine. These factors are:

1. Length of pipeline necessary.
2. Discharge elevation (relative to pump-station elevation).
3. Cost of pipe to be used.
4. Cost of pipe installation.
5. Pump and motor efficiency.
6. Cost of pump-station facilities.
7. Operation and maintenance.
8. Energy cost for pumping.

For the purpose of this analysis pipeline costs were based on three components; the related cost factors listed above were combined into: 1) Pipeline costs, 2) pump-station costs, and 3) energy costs. Total unit costs for pipeline transportation

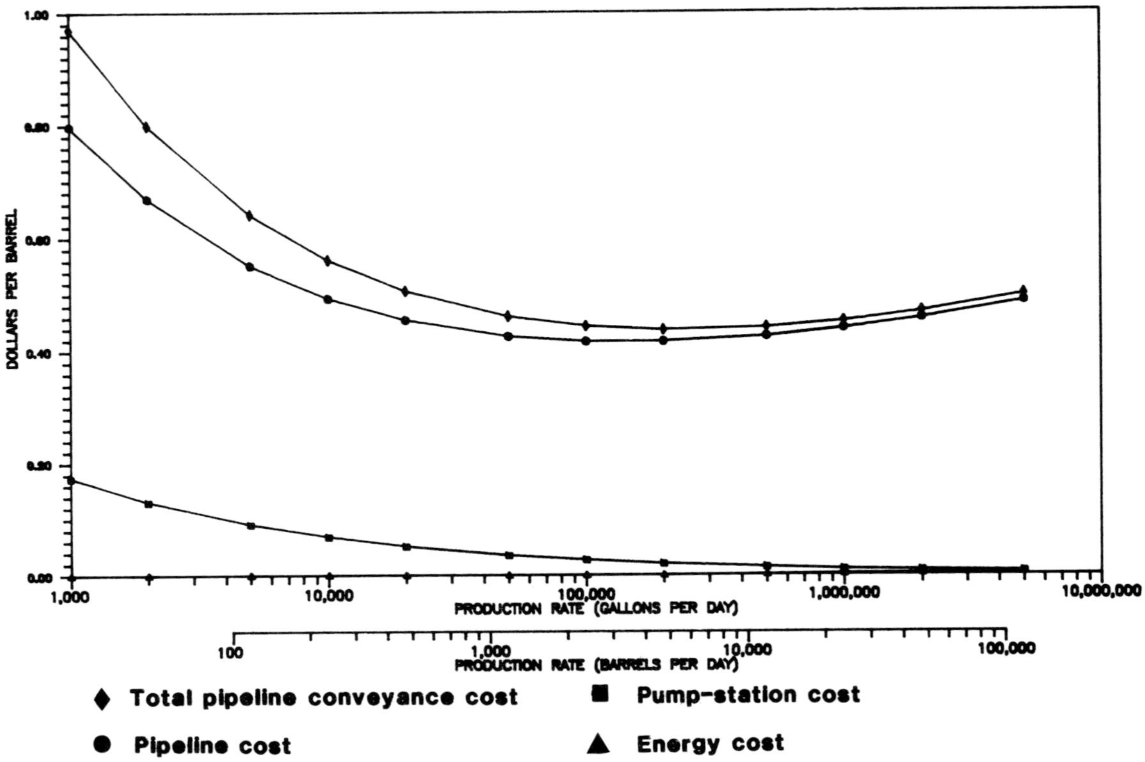

Figure 4. Graph of brine volume vs. pipeline conveyance costs.

ranged from approximately $0.44/bbl. to $0.97/bbl. The minimum unit cost occurs at a brine conveyance rate of approximately 5,000 BWD. At higher conveyance rates, the unit costs increase. This is due to the relatively high cost of the large-diameter pipelines necessary to convey large volumes of brine, as compared with the pipelines needed to convey volumes less than about 5,000 BWD. Figure 4 is a graph showing costs versus brine volume for pipelines. Comparison of figure 4 with figure 3 shows that brine transport using tanker trucks is considerably more expensive than pipeline conveyance, over the range of brine volume considered.

Brine Disposal Alternatives

Direct Use (No Treatment)

Brines having suitable water quality can be put to direct use either on site at the well location or they can be transported from the production well location to the point of use.

The unit cost for on-site use, such as stock-watering or irrigation near or at each methane production well, would be minimal, involving only the short lengths of ditch or pipeline necessary to deliver the brine to the point of use. Unlike the other disposal alternatives, direct-use cost generally is not a function of the volume of brine in use; there is no economy of scale to be considered. It is estimated that the cost to dispose of brine by direct, on-site use is less than $0.01/bbl. By way of comparison, irrigation water users in Colorado currently pay between $5.00 and $15.00/acre-foot (about $0.005 and $0.01/bbl) for irrigation water (Young, personal communication, 1987). For brine having a quality suitable for irrigation or stock watering, on-site direct use is the least costly disposal alternative.

Off-site direct use would require tanker truck or pipeline transportation to the point of use. The unit cost for off-site direct use of the brine (without treatment) can be considered as the conveyance cost. As shown in figures 3 and 4, this alternative is considerably more costly than on-site use.

Direct Use (With Treatment)

Brines having poor water quality may require treatment prior to discharge for direct use. Brine treatment options were discussed with treatment-industry representatives to determine a potential range of treatment costs.

Brine-distillation units used in the oil industry, generally are able to use any of several energy sources for the heat distillation process. The most practical energy source for the Southern Ute Reservation would be natural gas produced from the coal-bed methane production wells. Energy cost alone is a small portion of the total cost for any of the treatment processes; for distillation, it is estimated that the energy cost would be in the range of $0.15/bbl to $0.25/bbl. The total cost for distillation may be $5.00/bbl or more, including capital costs and the ongoing costs for operation and maintenance (Hudson, personal communication, 1987).

Although the instantaneous energy demand for the ion-exchange process (zeolite treatment) may be about one-half of the energy demand for distillation, the final unit cost per barrel of brine would be similar to the unit cost for distillation (Altman, personal communication, 1987). This is because of higher operation and maintenance costs for the ion-exchange system. A 30- to 50-percent downtime of the ion-exchange system is required for backflushing of the zeolite, and for disposal of the brine concentrate.

The unit cost for the reverse-osmosis treatment process would be even more expensive than either distillation or ion-

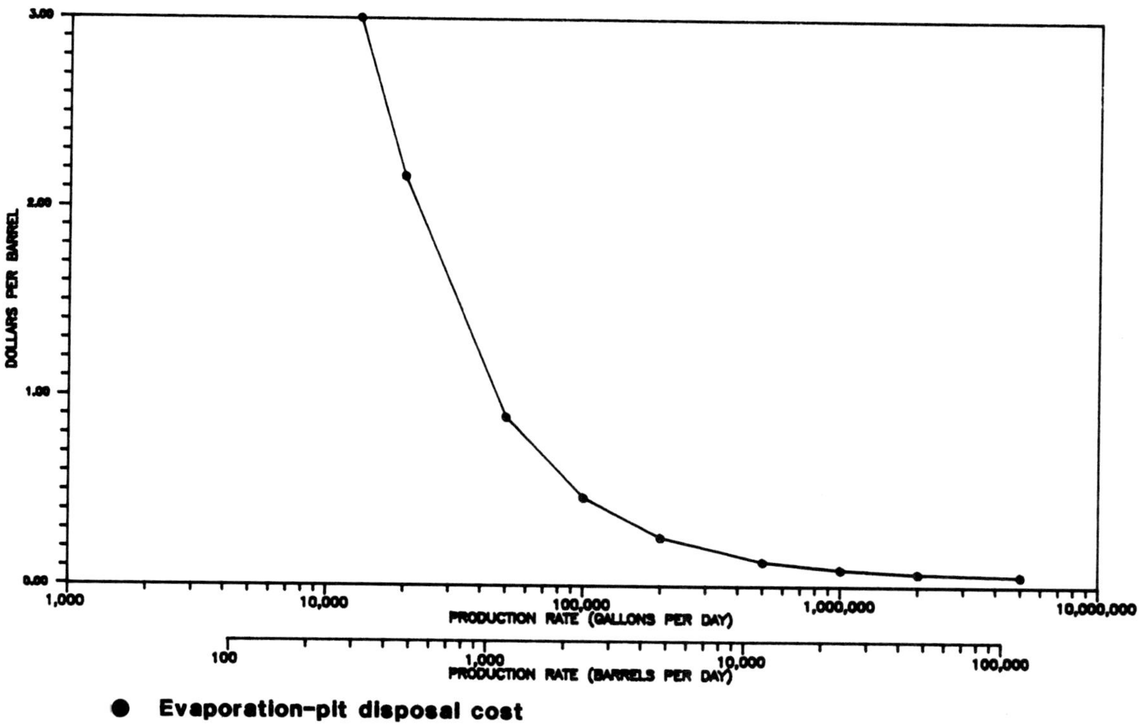

Figure 5. Graph of brine volume vs. evaporation-pit disposal costs.

exchange (Altman, personal communication, 1987), due to the high costs associated with providing highly-trained personnel to operate a reverse-osmosis system. Reverse-osmosis systems require continuous monitoring and maintenance, and pre-filtering of the brine may be needed to remove suspended solids.

The lowest-cost treatment method is dilution of the brine with better-quality water to obtain an acceptable water quality prior to use. This method, however, assumes that there is an economically-available supply of good-quality water with which to dilute the brine, and that some level of degradation of the good-quality water is acceptable.

There may be certain areas of the Southern Ute Reservation where both of these conditions can be met; however, it is doubtful that these can be met over the entire Reservation. The total unit costs for direct use with treatment can be obtained by adding treatment cost to the on-site or off-site direct use costs discussed earlier.

Evaporation Pit Brine Disposal

The cost estimate for using evaporation pits as a method of brine disposal includes consideration of several economic, climatologic, and engineering factors. The following assumptions were used in the analysis of the unit cost of brine disposal using evaporation pits:

1. 50-year, 24-hour rainfall (0.25 feet).
2. Net evaporation (36 inches/year).
3. Cost of clearing land ($500/acre).
4. Cost of topsoil stripping ($500/acre).
5. Cost of dike excavation ($5.00/cubic yard).
6. Design pit liquid depth (1.0 feet).
7. Twenty-year total residue buildup depth (0.12 feet).
8. Volume of dike material (7 cubic yards/linear yard).
9. Cost of pond liner ($0.50/square yard).
10. Cost of liner fill ($5.00/cubic yard).
11. Labor, material, and equipment for dirt work ($2.50/cubic yard).

(Items 1 and 2 are from National Oceanographic and Atmospheric Agency climatic records; items 3, 4, 5, 9, 10, and 11 are from ENR Market Trends, ENR, Nov. 12, 1987; and items 6, 7, and 8 are from Reid and others, 1974.)

In addition to the cost-component and design items listed above, the computer program developed to estimate unit costs for evaporation-pit brine disposal included operation and maintenance of the facility.

The disposal cost of brine in evaporation pits is a function of the volume of brine disposed of (fig. 5). For brine volumes less than approximately 10,000 gallons per day (gpd) per evaporation pit, unit costs are very high (greater than $3/bbl), and may not be competitive with brine treatment costs. Unit costs decrease significantly as the volume of brine disposed of in a pond increases. Evaporation pit disposal is most cost effective when few evaporation pits are constructed, and each pit is constructed to the largest volume possible. Comparison of evaporation-pit disposal costs (fig. 5) with underground injection disposal costs (fig. 6) shows that above a brine-disposal volume of approximately 30,000 BWD, evaporation pits are a less expensive alternative for disposal than underground injection.

Spray-enhanced evaporation pits may provide a more economically-competitive disposal method. The smaller pit surface areas needed for active evaporation would mean lower capital costs for land acquisition, land clearing, pit liner, and liner-cover fill. However, additional costs would be incurred for energy, piping and pumps, and operation and maintenance.

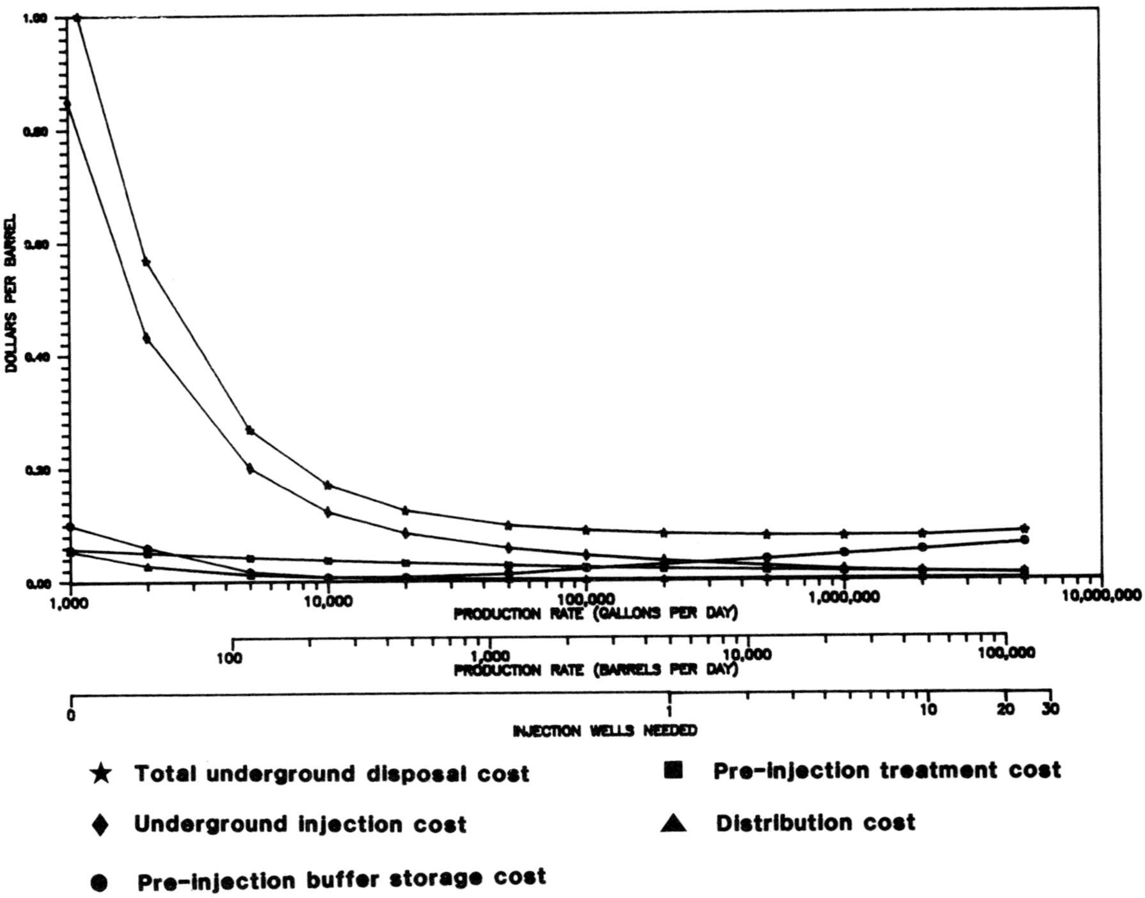

Figure 6. Graph of brine volume vs. underground injection disposal costs.

Underground Injection Brine Disposal

As with evaporation pits, there are several additional economic and engineering factors to be considered in analyzing the unit cost of brine disposal in underground injection wells. Geologic factors must also be taken into account. In addition to the common factors previously identified, the following factors were considered in the economic analysis of using injection wells for brine disposal:

1. Formation porosity (30 percent).
2. Formation permeability (0.05 Darcy).
3. Formation effective thickness (200 feet).
4. Formation hydrostatic pressure (1,500 psi).
5. Brine-injection capacity (5,000 BWD/well).
6. Depth to base of deepest fresh-water sand (2,000 feet).
7. Disposal well depth and cost (4,000 feet/well, $400,000/well).
8. Wellhead-equipment cost ($5,000/unit).
9. Cost of well-integrity testing ($20,000/well).
10. Cost of well-injectivity testing ($4,500/well).

(Items 1, 2, 3, 4, 5, and 6, are from the Southern Ute Tribe and Mr. Ed Demattio, Colorado Oil and Gas Conservation Commission; items 7 and 8 estimated from Hammershaimb and Kuuskra, 1984; and items 9 and 10 estimated from Nielsen and Aller, 1984.)

Other items involved in the cost analysis of disposal wells, such as casing and downhole tubing size, operation and maintenance costs, brine pre-treatment cost prior to injection, and cost of pumping for injection, have been estimated from Reid and others, 1974; Haynes and Grubbs, 1969, and from personal experience with similar projects.

The unit cost of brine-disposal in underground injection wells is relatively constant for volumes of brine above approximately 200 to 300 BWD (fig. 6). A minimum unit cost (approximately $0.10/bbl) is reached at approximately 30,000 BWD, above which the costs for deep-well brine disposal increase slightly due to increasing costs of pre-injection buffer storage. Buffer storage is necessary because injection wells require routine redevelopment and cleaning to maintain acceptable levels of injection. A temporary storage facility must be provided for a period of up to several days, to accept the brine which is produced on a continuous basis by the coal-bed methane production wells. Above approximately 30,000 BWD per facility, use of evaporation pits would be slightly less expensive than the use of disposal wells.

REGULATORY ASPECTS

The disposal of brines on Indian Reservations is a complex regulatory issue. Depending upon the disposal method selected and the land and mineral ownership, a number of State and Federal regulatory programs may be applicable to brine disposal (table 4). In the western United States, production water may be put to agricultural or wildlife use under provisions of the Clean Water Act. Evaporation pits are regulated under programs administered by the Bureau of Land Management (BLM), the Colorado Oil and Gas Commission, or the Colorado Department of Health and individual counties. Under-

Table 4. Regulatory framework for brine conveyance and disposal alternatives.

	Lease or land status	Regulatory agency	Comments
Brine Conveyance			
Trucking and piplines	Fee minerals	Colorado Oil and Gas Commission	Section 325 of Colorado Oil & Gas Commission regulations. Has not been applied to pipelines.
Trucking and piplines	Indian lands (Allotted and Tribal)	Bureau of Land Management	Removal of brines from lease must be approved.
Brine Disposal			
Direct use	Indian lands (All lands within Reservation boundary)	Environmental Protection Agency	NPDES Permits of Reservation issued by EPA. Subject to Agricultural and Wildlife Water Use Subcategory (40 CFR 435.50). Tribes eligible for program primacy.
Evaporation pits	Fee minerals	Colorado Oil and Gas Commission	Section 325 of Colorado Oil and Gas Commission regulations.
Evaporation pits	Indian lands (allotted and Tribal lands)	Bureau of Land Management	NTL-2B (January 1, 1976)
Evaporation pits	Commercial pits	Colorado Department of Health/County Commissioners	Subject to Colorado Solid Waste Regulations. County issues permit (Certificate of Designation) with technical assistance from the State. Enforcement by State.
Underground injection	Fee minerals	Colorado Oil and Gas Commission	State Class II UIC program approved April 1984.
Underground injection	Indian lands (All lands within Reservation boundary)	Environmental Protection Agency	Tribes now eligible for program primacy.

ground injection wells are regulated under the Safe Drinking Water Act under which the Environmental Protection Act (EPA), the Colorado Oil and Gas Commission, or the Southern Ute Tribe may have program primacy.

Brine Conveyance

The State of Colorado provides some regulation of brine transportation. Truck transporters of brine are required to register with the Colorado Oil and Gas Commission, and to file monthly transporter reports. These reports identify the operator from whom the brine is received, the volume transported, and the brine delivery point. Transport of brine by pipeline is not regulated by the Colorado Oil and Gas Commission (Smith, personal communication, 1988).

Removal of brine from Federal or Indian leases must be approved by the BLM regardless of who has jurisdiction over the actual disposal facility (evaporation pond or injection well). In addition, if a pipeline is used, the BLM must approve those portions of the pipeline on Federal or Indian leases. Trucking must be specifically approved as a method of removing brine from the lease. Once approved, however, the trucking of brine is unregulated (Englishman, personal communication, 1988).

Brine Disposal Alternatives

Direct Use

Discharge of brine is authorized under the Clean Water Act if the water has sufficient quality to be used for livestock watering or other agricultural uses. The agricultural and wildlife use subcategory (40 CFR 435 Subpart E) was established at the request of Western states to allow produced water to be discharged and put to beneficial use (EPA, 1986). The regulation is intended to be restrictive based upon the unique factors of prior usage in the region, arid climatic conditions, and the existence of low-salinity production waters. To qualify for discharge, the facility must be located west of the 98th meridian and the discharge must be used for agricultural or wildlife purposes.

The EPA has National Pollutant Discharge Elimination System (NPDES) permitting authority on Indian lands. However, permits are required so infrequently on Indian lands that the EPA may rely on State assistance in preparing the actual permit that would be issued by the EPA (Pike, personal communication, 1988).

Colorado is one of several states that have received program primacy under the Clean Water Act. The Colorado Department of Health has permitted 10 to 15 commercial brine disposal facilities to discharge brine under the wildlife and agricultural use subcategory (EPA, 1986). Discharge water quality limitations are based on EPA national effluent limitations (40 CFR 435, Subpart E), state receiving stream water quality standards, and biomonitoring results. Biomonitoring is now required by the State for new discharges to identify toxicity concerns. The results must be provided to the State so that treatment requirements can be determined prior to discharging.

Evaporation Pit Brine Disposal

The regulation of evaporation pits is subject to complex jurisdictional considerations although the regulations themselves are not particularly stringent. The Colorado Oil and Gas Commission and the BLM regulate evaporation pits associated with wells on specific Oil and Gas Leases regulated by those agencies. Commercial brine evaporation pits are regulated by the State of Colorado and the local county governments. Disposal of brine from Indian lands in commercial evaporation pits would require BLM approval.

The Bureau of Land Management has responsibilities over many aspects of oil and gas operations on Federal and Indian Lands. The BLM defines Indian lands to include allotted and Tribal lands and minerals (Englishman, personal communication, 1988). Indian leases do not include leases with private surface and minerals within the Reservation boundary. BLM responsibilities include the regulation of production wells, waste handling and disposal (including production water), and site reclamation, as well as any other areas not handled by the Bureau of Indian Affairs and the Minerals Management Service (EPA, 1986). Leases must comply with applicable Federal laws and regulations, with the lease terms, onshore oil and gas orders, and Notice-to-Lessees (NTLs).

The specific requirements for the disposal of brines are contained in Notice to Lessees and Operators of Federal and Indian Oil and Gas Leases, NTL-2B (January 1, 1976). This notice requires that all brines from Federal and Indian oil and gas leases be disposed of by: 1) Injection into the subsurface, 2) lined impoundments, or 3) by other acceptable methods. (The only other acceptable method identified in NTL-2B is unlined impoundments.) All disposal of brine from wells under BLM jurisdiction must be approved in writing regardless of the physical location of the disposal facility. On site or central collection pits must file application with the BLM. Before brine from wells on Indian leases can be removed to a commercial disposal facility, the lease operator has to demonstrate that the facility meets BLM requirements.

Disposal of brine into unlined impoundments is allowed under five conditions: the water has a total dissolved solids concentration less than 5,000 ppm, the brine is used for beneficial purposes, the brine is of better quality than potentially affected surface or ground water, the volume of brine does not exceed 5 barrels per day, or the discharge is permitted as a surface discharge under the NPDES.

The State of Colorado also regulates evaporation pits receiving brine in excess of 5 barrels per day on a monthly basis. The State bases its jurisdiction over evaporation pits upon land ownership (Smith, personal communication, 1988). On-site disposal facilities (facilities used to dispose of the brine produced from one or more oil and gas wells at the lease location from which it is produced), and central disposal facilities (those facilities receiving brine from several leases and operated by one or more oil and gas operators) are regulated by the Colorado Oil and Gas Commission. These impoundments are subject to Section 325, Disposal of Water Produced with Oil and Gas Operations or Other Oil Field Waste of the Colorado Oil and Gas Commission Regulations.

Commercial brine evaporation pits are not operated by the owners or operators of specific oil and gas leases. Commercial facilities are independently operated and receive water from production wells on a contract basis. Commercial brine disposal facilities are regulated by the Colorado Hazardous Materials and Waste Management Division and by county government under the Colorado Solid Waste Disposal Site and Facility Act. Jurisdiction is exercised on all operations within the State whether on private, State, Federal, or Indian land.

Commercial brine evaporation pits are regulated as waste impoundments subject to Section 6 of the Colorado Solid Waste Regulations. Waste impoundments include any pit or pond used for the storage, treatment, or disposal of nonhazardous solid or liquid waste. Under the State Solid Waste Act, the actual facility permit (called a certificate of designation) is issued by the county with technical assistance in permit review provided by the Colorado Department of Health.

Underground Injection Control

The Safe Drinking Water Act (SDWA) of 1974 required the EPA to develop a national program to prevent unsafe underground injection of fluids. The EPA was required to prepare minimum requirements for State underground injection control (UIC) programs and to enforce UIC programs if a state did not develop its own program. The SDWA classifies injection wells into five categories, depending upon the fluid injected. Class II injection wells include waters associated with oil and gas production (brines), fluids injected to enhance recovery of oil or natural gas, and the underground storage of hydrocarbons. States were given the option of developing full UIC programs or of developing programs limited to Class II wells. The SDWA also gave the EPA authority to enforce UIC programs on Indian lands.

On April 4, 1984, the EPA approved the State of Colorado UIC program for Class II wells. The EPA retained responsibility for the administration of the UIC program in Colorado for other classes of injection wells and for Class II wells on Indian lands in Colorado. Under the State of Colorado UIC program, the Colorado Oil and Gas Commission claims jurisdiction over non-Indian mineral leases within the Reservation boundary, and the State has issued several brine injection well permits on the Southern Ute Reservation.

The EPA also maintains jurisdiction on Indian lands, which means "Indian Country" as defined in 18 USC 151 (Pike, personal communication, 1988). Under this definition, Indian lands include all land within the limit of the Reservation (40 CFR 144.3). The EPA administered program implements the current Federal UIC minimum requirements unless a tailored program has been developed in response to concerns of the affected tribal government. (This national UIC program is described in 40 CFR parts 124, 144, and 146.) This is the same program that the EPA administers for most non-Indian lands.

Most tribes have not actively sought special programs for the regulation of their injection wells and the EPA intends to enforce the generic Federal UIC program on most Indian lands (52 FR 17685). The EPA has developed modified UIC programs for several tribes such as the Navajo, the Ute Mountain Ute, and some other Indian tribes in New Mexico and Oklahoma. Modified programs may be developed for other reservations in the future at the request of the tribe (52 FR 17685).

Therefore, the EPA claims jurisdiction on all lands within the boundary of the Southern Ute Reservation and the State of Colorado claims jurisdiction of operations associated with fee minerals within the Reservation boundary. As a result, the existing injection wells on the Reservation which were permitted by the State of Colorado are now required to also obtain permits from the EPA (Pike, personal communication, 1988).

To receive UIC program approval (40 CFR 145), state run UIC programs must include specific program provisions and must be at least as stringent as the equivalent Federal regulations. Therefore, the EPA and State UIC permit requirements are essentially identical. However, the permitting procedures are slightly different. The Colorado Oil and Gas Commission allows drilling and construction of the injection well prior to permit issuance if adequate design plans have been submitted. The intent of the Colorado process is to gather as much data as possible which might affect injection, prior to issuing final permit approval. The EPA permit procedure requires public notice and permit approval before drilling and well construction can begin. The EPA and the Colorado Oil and Gas Commission are currently negotiating a memo of understanding to coordinate future Class II UIC permit reviews on the Southern Ute Reservation.

The 1986 Amendments to the Safe Drinking Water Act specifically directed the EPA to promulgate a Federal UIC program for all Indian lands not already covered by an applicable UIC program. The amendments also authorize the EPA to treat tribes as states, by delegating primary enforcement responsibility for Public Water Systems (PWS) and Underground Injection Control (UIC) programs, and by providing grant and contract assistance to Indian tribes where appropriate (52 FR 28112). Tribes meeting the conditions set forth in the Act may receive authorization as the states do to enforce their own UIC or Class II UIC programs over lands within their jurisdiction. On July 27, 1987, the EPA published proposed regulations to allow Indian tribes to administer their own UIC programs. Final regulations are scheduled for publication in February, 1988, and would take effect 30 days after publication (Pike, personal communication, 1988).

SUMMARY

Several different environmentally acceptable and economically sound water disposal systems are currently in use in the northern San Juan Basin. No single disposal method is appropriate for all situations. Successful disposal systems will be designed on the basis of production and operational concerns, economics, environmental factors, and regulatory framework.

Of the brine-conveyance alternatives, pipeline transport has fewer disadvantages than truck transport. Truck transport would require a larger labor force, would provide greater environmental and safety risks, and would provide greater opportunity for unauthorized dumping. Both alternatives have comparable land requirements for roads or pipeline corridors. A pipeline transportation system would be more difficult to implement because of required planning, engineering design, and higher initial capital costs for equipment and construction than the trucking alternative. However, over the long-term, pipeline transportation is considerably less expensive on a unit cost basis, and would provide less risk of environmental degradation than truck transport.

Of the four brine-disposal alternatives considered, on-site direct use of the brines without treatment is the least expensive disposal alternative. However it is doubtful whether untreated brines can be used in more than about one-quarter of the coal-bed methane development areas because of poor water quality.

Direct use with treatment to improve chemical quality has the advantage that the end product is good quality water which can be used for almost any common usage. However, Federal regulations would limit usage to agricultural or wildlife use which may make treatment costs difficult to justify. Brine treatment methods involving distillation, ion-exchange (zeolite treatment), or reverse osmosis are heavily dependent on monitoring and maintenance on a constant basis by highly trained personnel. Operation and maintenance costs and the initial capital costs for purchase and installation of treatment systems coupled with use limitations, limit the use of brine treatment for direct use on the Southern Ute Reservation. Dilution was also considered as a brine treatment method; although this method would require little operation and maintenance costs, an adequate supply of good-quality water must be available at a given location for dilution. In addition, some level of degradation of good-quality water must be environmentally acceptable.

Several advantages and disadvantages are associated with evaporation pit and underground injection well brine disposal. Evaporation pits have relatively large land requirements, have high capital startup costs, and are frequently subject to environmental problems. However, they also require a relatively small labor force to operate, and passive pits have low energy consumption requirements. Underground injection wells also have relatively high startup costs, may require on-site brine storage, and may not be well suited to dispose of small brine volumes.

Evaporation-pit and injection-well brine disposal both appear to be more cost-effective than the maintenance-intensive treatment and direct use alternative discussed above. For brine-disposal volumes below approximately 30,000 BWD per facility, injection-well disposal would be less expensive per barrel of brine than evaporation-pit disposal. Above about 30,000 BWD per facility, the evaporation-pit alternative is slightly less expensive than the injection-well alternative because a larger diameter pipe is required.

Brine disposal on Indian reservations is subject to the jurisdiction of a variety of Federal, state, and local agencies depending upon land or mineral ownership, and disposal method. Evaporation pits associated with specific leases are regulated by the Colorado Oil and Gas Commission or the U.S. Bureau of Land Management. However, underground injection wells on the Reservation may be subject to both State and Federal jurisdiction. The Colorado Oil and Gas Commission claims jurisdiction over fee mineral leases within the Reservation boundary while the EPA claims UIC jurisdiction over all lands within the Reservation boundary. Until a memo of understanding between the two agencies can be developed, injection wells for brines produced from fee minerals within the Reservation boundaries are required to obtain both State and Federal permits.

The regulatory role of Indian tribes is changing. Recent amendments to the Safe Drinking Water Act (and more recently to the Clean Water Act) authorize the EPA to treat tribes as states and delegate program responsibilities to Indian tribes. The EPA is now developing programs to delegate UIC authority to qualifying Indian tribes.

Operators developing adjacent coal-bed methane resources should consider establishing central brine-disposal units. The potential savings of large volume disposal may be substantial, however difficult the coordination effort may be.

Future research and experience may provide cost effective alternatives, including water processing equipment allowing surface discharge of produced formation waters. Keeping informed of such research, improved technologies, and jurisdictional and regulatory issues is essential for operators involved in disposing of produced oil and gas field waters.

REFERENCES CITED

Brogden, R.E., Hutchinson, E.C., and Hillier, D.E., 1977, Availability and quality of ground water, Southern Ute Reservation, Southwestern Colorado: U.S. Geological Survey Open File Report 77-623, 29 p.

Butler, D.L., 1986, General surface- and ground-water quality in a coal-resource area near Durango, Southwestern Colorado: U.S. Geological Survey Water Resources Investigation Report 86-4073, 53 p.

Davis, K.E., and Hineline, T.L., 1986, Two decades of successful hazardous waste disposal well operation—A compilation of case histories *in* Proceedings of the International Symposium on Subsurface Injection of Liquid Wastes: National Water Well Association, p. 295-308.

Fassett, J.E., and Hinds, J.S., 1971, Geology and fuel resources of the Fruitland Formation and the Kirtland Shale of the San Juan Basin, New Mexico and Colorado: U.S. Geological Survey Professional Paper 676, 76 p.

Geraghty and Miller, Inc., 1980, Waste disposal effects on ground water: Premier Press, Berkeley, California, 512 p.

Glenn, B.P., 1970, A guide to using interest factors in economic analysis of water projects: U.S. Bureau of Reclamation, 103 p.

Hammershaimb, E.C., and Kuuskra, V.A., 1984, The economic potential of re-injection into geopressured aquifers: Journal of Petroleum Technology, v. 37, July, p. 1164-1170.

Haynes, C.D., and Grubbs, D.M., 1969, Design and cost of liquidwaste disposal systems: University of Alabama, Natural Resources Center Report 692, 90 p.

Kelso, B.S., and Wicks, D.E., 1988, A geologic analysis of the Fruitland Formation coal and coal-bed methane resources of the San Juan Basin, southwestern Colorado and northwestern New Mexico, *in* Fassett, J.E., ed., Geology and coal-bed methane resources of the northern San Juan Basin, Colorado and New Mexico: Rocky Mountain Association of Geologists Guidebook.

McKee, J.E. and Wolf, H.W., 1963, Water quality criteria: California State Water Quality Control Board Publication 3-A, Sacramento, California.

Nielsen, D.M., and Aller, L., 1984, Methods for determining the mechanical integrity of Class II injection wells: National Water Well Association, Dublin, Ohio, 263 p.

Patton, C.C., 1974, Oilfield water systems: PetroTech Ltd., London, England, 152 p.

Reid, G.W., Streebin, L.E., Canter, L.W. and Smith, J.R., 1974, Brine disposal treatment practices relating to the oil production industry: U.S. Environmental Protection Agency, Report no. 660/2-74-037, 275 p.

Rountree, R., 1988, Coal bed methane to gear-up: Western Oil World, v. 45, no. 3, p.7.

Schmidt, R.R., 1987, Zeolites: Their market potential and occurrence in New Mexico: New Mexico Bureau of Economic Geology Report of Investigations Number 1, 68 p.

U.S. Environmental Protection Agency, 1986, Wastes from the exploration, development, and production of crude oil, natural gas and geothermal energy, an interim report on methodology for data collection and analysis: Technical Report EPA/530-SW-86-051, Office of Solid Waste and Emergency Response, Washington D.C.

Production History of the San Juan 32-7 Unit No. 6 Well, Northern San Juan Basin, New Mexico.

Brent W. Hale and Carolyn H. Firth[1]

ABSTRACT

The San Juan 32-7 Unit No. 6 well, located in the San Juan Basin of New Mexico, has produced 1,364,000 MCF of gas from the Upper Cretaceous Fruitland Formation over the last 30 years with no decline in rates or reservoir pressure. The presence of 40 feet of coal within the producing interval and the gas composition indicate that this well produces from coal seams. The readjustment time in flow rates following pressure changes suggests a desorption process is taking place. The well has produced minimal amounts of water throughout its life and has required very little maintenance, demonstrating that coal-seam methane can be economically produced over an extended period of time.

INTRODUCTION

On July 15, 1953, San Juan Engineering Company staked a location for the San Juan Unit 32-7 Fruitland No. 6-17 well. This was the beginning of what is currently known as the San Juan 32-7 Unit No. 6 well operated by Northwest Pipeline Corporation. Our records give no indication that anything unusual was expected in drilling the well which is one of the oldest continuously producing coal-seam wells in the United States. Production records are very brief. They show no evidence of any significant operational problems. However, the production history has some unusual characteristics when viewed in the context of typical San Juan Basin gas production. These characteristics are better understood by relating the well's history to coal-seam methane production performance. The data and notes presented in the following sections document the nature of coal-seam methane production in this well.

DRILLING HISTORY

The San Juan 32-7 Unit No.6 well was spudded on July 23, 1953. The recommended geologic program called for the well to be drilled to the top of the Fruitland Formation and casing set so that drilling could be converted from mud to gas. A natural completion was planned. Approximately 150 feet of Fruitland section was to be drilled slowly to provide a safe means of detecting moisture, even though none was expected, and to safeguard against drilling into a large gas flow too rapidly.

The well was drilled to 3,055 feet and 7-inch OD, 23 lb J-55 casing was set on August 7, 1953. The August 12 drilling report shows drilling through sandstone and coal with 10.9 MCFPD of gas being produced as measured with a pitot tube. On August 13, drilling of sandstone, shale, and coal was reported with the first show of gas noted at a depth of 3,202 feet. Gas-flow rates increased between depths of 3,232 and 3,236 feet. Gas-flow measurement, through a two-inch line with a pitot tube and mercury manometer, was reported as 2,420 MCFPD. Drilling continued to a total depth of 3,240 feet leaving a 185-foot open-hole section. The well was shut in on August 15 and 16 with a 48-hour buildup pressure of 1,500 psig and a 10-day buildup pressure of 1,504 psig reported. At the end of a three-hour flow test, the well was producing 1,790 MCFPD.

COAL GEOLOGY

The No. 6 well is located 660 feet from the north line and 660 feet from the east line of sec. 17, T. 31 N., R. 7 W., San Juan County, New Mexico. The location is approximately four miles south of the synclinal axis of the San Juan Basin (fig. 1). The well is situated near the southwest edge of the thickest stratigraphic rise or vertical buildup of the underlying Pictured Cliffs Sandstone (Fassett and Hinds, 1971).

The Fruitland Formation is Upper Cretaceous in age and consists primarily of interbedded sandstone, siltstone, shale, carbonaceous shale, and coal. The sediments were deposited landward of the regressive Pictured Cliffs beach sand, in coastal swamp, floodplain, and deltaic environments (Fassett and Hinds, 1971). The well is in an area of intertonguing between the Fruitland Formation and the Pictured Cliffs Sandstone (fig. 2). Coals in this area are among the thickest in the San Juan Basin (Fassett, this volume).

Figure 1 shows the structural configuration of the Huerfanito Bentonite Bed of the Lewis Shale. This marker bed was named and described by Fassett and Hinds (1971). The map shows no significant structural anomalies within the vicinity of the No. 6 well. This suggests that structure has little or no influence on the well's production.

The open-hole section of the well includes approximately 40 feet of coal. Apparently, the most gassy coals are located below a depth of 3,200 feet, as the first significant gas show was reported at 3,202 feet. A seven-foot-thick clean sandstone between the depths of 3,203 feet and 3,210 feet may also be contributing to the gas production.

A northwest-oriented cross section is shown in figure 2. To

[1] Brent W. Hale and Carolyn H. Firth
Northwest Pipeline Corporation
P.O. Box 8900
Salt Lake City, Utah 84108-0899

The authors wish to thank Northwest Pipeline Corporation for permission to use company records in the preparation of this paper and to publish data about this well.

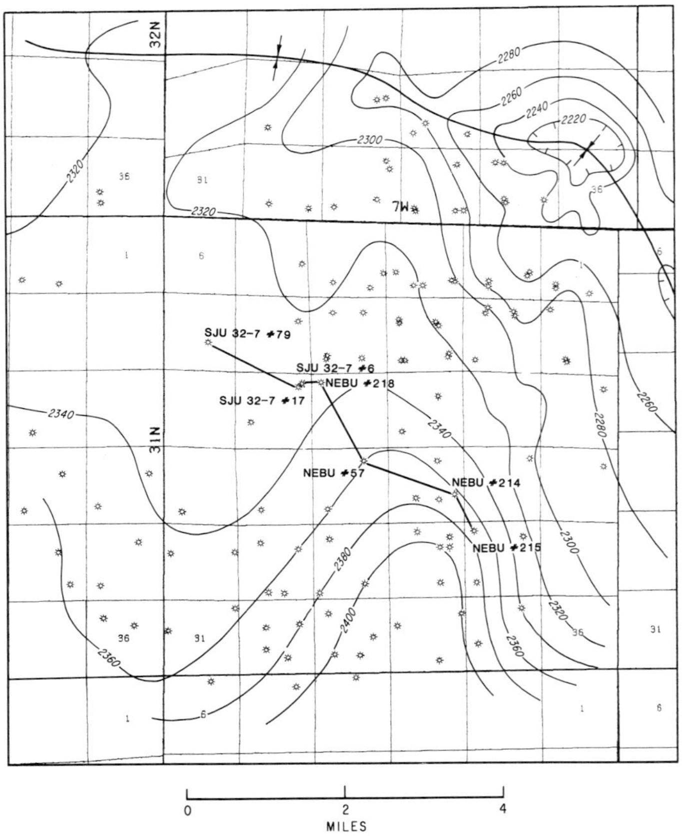

Figure 1. Structure contour map of the area around the San Juan 32-7 Unit No. 6 well, San Juan County, New Mexico. The line of the cross section illustrated on figure 2 is shown. Structure contours are drawn on the Huerfanito Bentonite Bed of the Lewis Shale; the structural axis of the basin is shown crossing the northern part of the map area. Contour interval is 20 feet.

facilitate correlation of coal beds, the northwest trend of the line of section was chosen to parallel the Pictured Cliffs Sandstone paleoshoreline. The San Juan 32-7 Unit No. 17 well was included in the cross section because of its proximity to the No. 6 well. These wells are only 325 feet apart.

The cross section shows that the thicker coals are more continuous than the thinner coals. The most apparently continuous coal bed extends through the entire four-mile length of the cross section. However, some coals in the No. 6 well may be only 1.5-2 miles long. Thinning and pinching out of coal beds is evident. Some coals are probably cut out by channel-sandstone deposits. For example, the coal at 3,200 feet in the No. 6 well is thinner than in the two immediate offset wells, and the sandstone bed directly below this coal is thicker in the No. 6 well than in the offset wells. The basal coal in the No. 6 well is absent in the Northeast Blanco Unit No. 218 well, but may correlate with the coal at 3,050 feet in the Northeast Blanco Unit No. 57 well. Floodplain deposits may have spread over the coal deposits for a period of time in the vicinity of the Northeast Blanco Unit No. 218 well.

Total Fruitland Formation coal thickness is indicated at the bottom of each log on the cross section and thicknesses range from 51 feet to 75 feet in wells completely penetrating the Pictured Cliffs Sandstone. If the No. 6 well had been drilled through the Pictured Cliffs interval, an additional 20 feet of coal would probably have been penetrated.

PRODUCTION HISTORY

The No. 6 well has produced since 1953 (New Mexico Oil and Gas Engineering Committee, 1953-1975) with no significant water production problems reported. Production notes during 1958 state that the well "does real good—cleans self" and "kicks self off about 1 per day." In 1962, a Nordstrom 2-inch wing valve was installed to blow the well down. In 1978, a water production rate of one barrel per day was reported. The well continues to produce gas with little to no water production except when the well is first produced following a shut-in. It unloads a large slug of water following each shut-in and then produces gas with little additional water production.

The well initially delivered during September 1953 and produced 15,339 MCF that month. There was little production until the Pacific Northwest Pipeline was placed in service during 1957. The well has produced continuously since then except for a market-related curtailment through 1972 and again beginning in 1983. Cumulative production through November 1987 is 1,364,000 MCF at a pressure base of 14.73 psia.

Figure 3 shows both annual production and monthly flow rates. In 1957 and 1958, production was high with volumes dropping in 1959 and then increasing steadily through 1976. Produced volumes cannot be directly converted to flow rates because of missing producing time before 1977. For the pre-1977 monthly data and for all annual data, the well was assumed to be flowing 100 percent of the time.

The annual production rates, shown in figure 3, are volumes divided by 365 days yielding minimum possible flow rates. The same procedure was used for monthly data up to 1977 where minimum rates are shown. Since 1976, when producing time records first became available, actual flow rates are shown. These flow rates are seen to increase during the 1983 to 1987 market decline, where minimum rates dropped during the 1959 to 1971 period for a number of months each year when production was curtailed. From 1972 to 1976, the No. 6 well produced almost continuously so that minimum reported rates are close to actual flow rates.

Production records show gas production rates in excess of 400 MCFPD for one month when the well initially delivered during 1953 and again during 1957. Except for these two months, the production rate has remained constant at about 150 MCFPD to 180 MCFPD.

The increase in rates seen during the 1983 to 1987 period is not stabilized flow. This indicates that the producing formation has relatively low permeability and significant porosity creating large pressure gradients and reservoir storage capacity. Both are necessary for flush production to occur.

Unchanging peak rates from the 1958 to 1976 period indicate that there has been little change in reservoir energy. Production rates have not declined. Increased rates during the 1980s are not stabilized and do not indicate any overall improvement in well performance.

DETAILED PRODUCTION HISTORY

Detailed production performance for 1977 and 1978 is seen in figure 4. From January to June, 1977, the rate drops from 184 MCFPD to 114 MCFPD. This reduction occurred with no

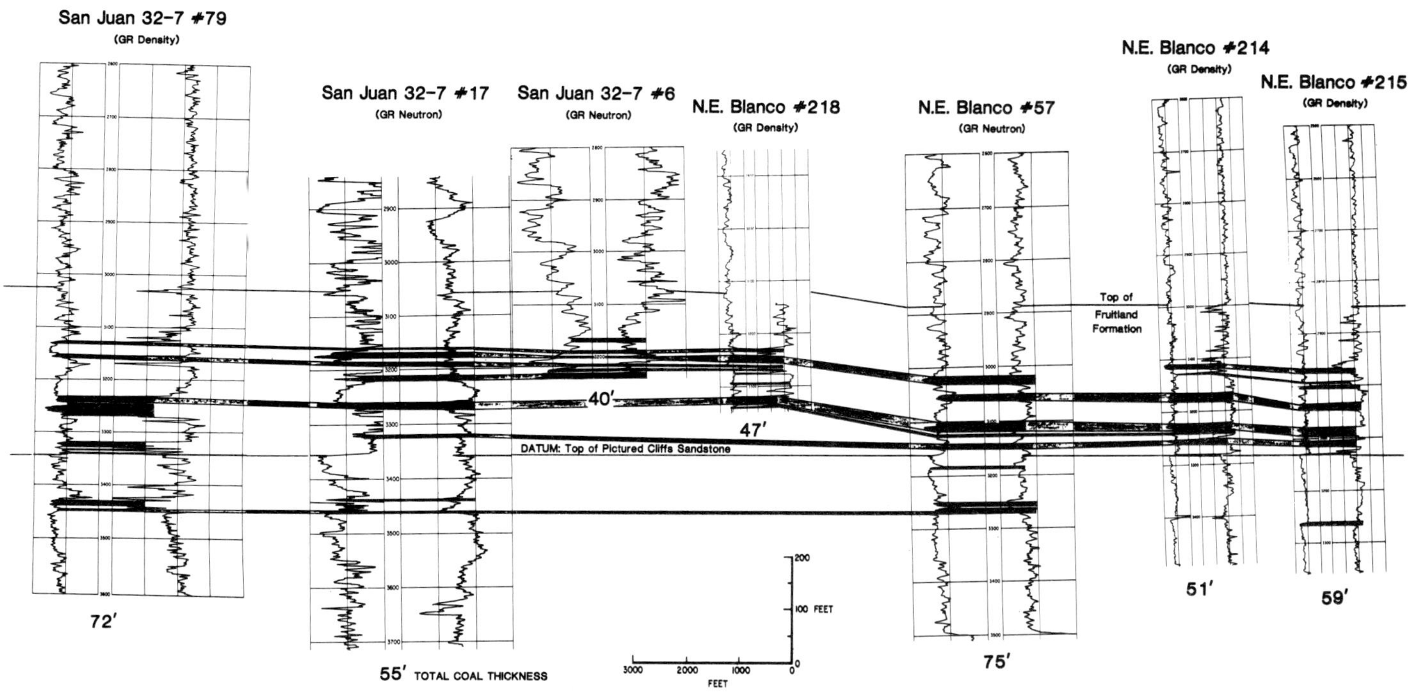

Figure 2. Northwest-trending stratigraphic cross section showing correlations of Fruitland Formation coal beds. Line of section is shown on figure 1.

change in operating pressures. Following the 1977 shut-in period of seven days, flush production is observed over a six month period. Figure 4 shows that reduced rates prior to the test were a result of increased bottom-hole pressures resulting from accumulating wellbore liquids with the well responding favorably to a blowdown to remove water.

A gathering system pressure spike during April and May, 1978 shows production performance is adversely impacted by increased wellhead pressure. The production-rate spike seen after the 32-day pressure-buildup test is similar to that seen in 1977. The first day rate is higher in 1978 only because measurement charts were manually read to show detailed short-term performance. The 1977 data show only seven-day averages which mask the first-day peak by combining it with lower rates later in the week.

1978 PRESSURE SPIKE

Figures 5 and 6 are graphs of flow rates and flowing pressures during the 1978 pressure spike; they show performance which is characteristic of a coal-seam desorption process. About April 1, pipeline pressures increased from 240 psig to nearly 400 psig. For 48 hours, the rate response is dominated by wellbore storage. Production rates dropped from 163 MCFPD to 115 MCFPD and then increased to 143 MCFPD. Over the next 30 days, rates gradually declined to a new stable rate of 80 MCFPD. The long-term rate response, controlled by methane desorption from the coal seam, is a gradually decreasing rate as reservoir pressure increases in response to higher wellhead pressures.

When pressure dropped back to the 250 psig range late in

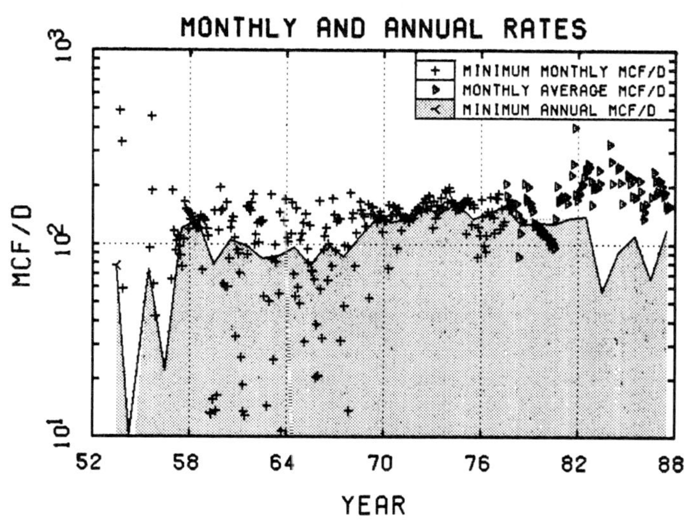

Figure 3. Graph showing the production history of the San Juan 32-7 Unit No. 6 well.

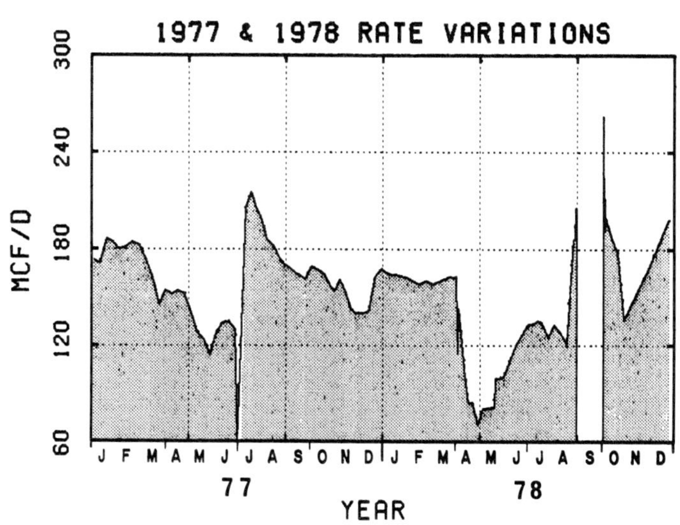

Figure 4. Graph showing detailed production from the San Juan 32-7 Unit No. 6 well in 1977 and 1978.

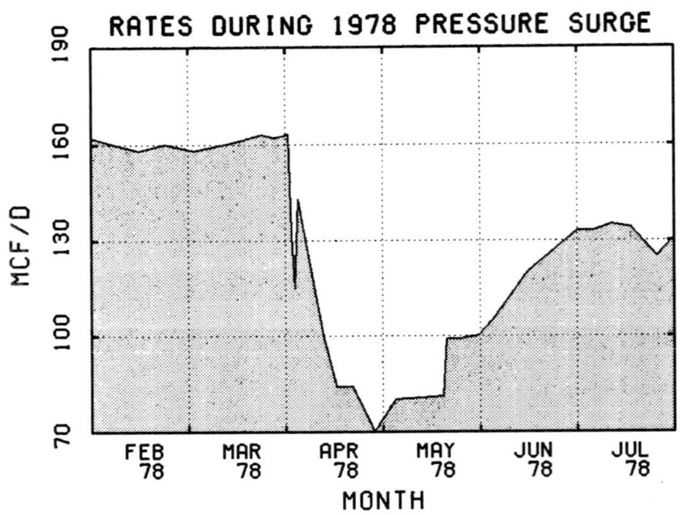

Figure 5. Graph showing flow rates for the San Juan 32-7 Unit No. 6 well in 1978.

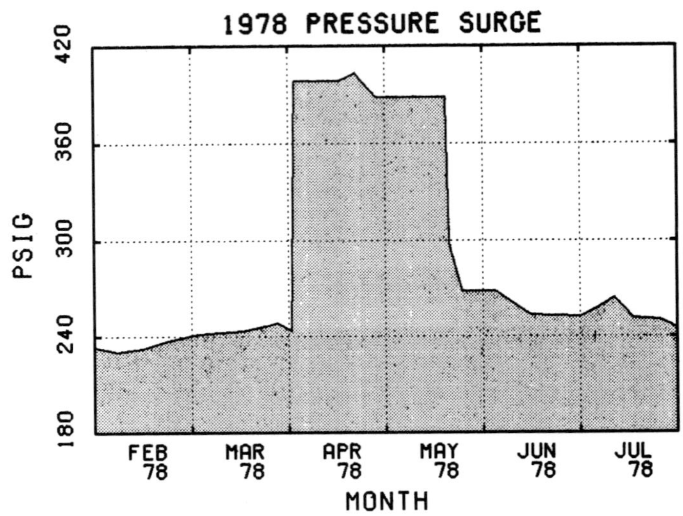

Figure 6. Graph showing flowing pressure for the San Juan 32-7 Unit No. 6 well in 1978.

May, there was an immediate storage related production spike from 80 MCFPD to 100 MCFPD. Over the next 60 days, as the desorption process readjusted to lower pressures, rates increased to 135 MCFPD as compared to the previous stable rate of 163 MCFPD. Both the lack of a significant production-rate spike and a lower stable production rate indicate that fluid levels increased in the wellbore during the pressure spike.

The time required for rates to readjust to new operating pressures, as seen in figure 5, is evidence that gas availability is both pressure and time dependent. As pressure increases, observed stable rates decrease showing the pressure dependency of gas production. However, rates change slowly over a 30 to 60 day period. This gradual change is related to the time required after a pressure change for methane desorption rates to readjust to new operating pressures.

32-DAY PRESSURE BUILDUP

In order to better understand the No. 6 well, an extended pressure-buildup test was conducted from September 1, 1978 to October 3, 1978. Pressures were recorded at the surface using dead weight gauges and chart recorders. The tubing pressures increased from 275 psig to 1,463 psig over a 32-day period. All data were converted to real gas pseudopressure (Al-Hussainy and Ramey, 1966) for analysis. Superposition (Carlslaw and Jaeger, 1959) was used to account for variable rates prior to the pressure buildup.

Figure 7 is a Horner plot of the buildup data. Wellbore storage dominates during the first 24 hours (Earlougher, 1977). The final flowing rate was 205 MCFPD with a pressure increase of 204 psi/hour at the surface during the first hour. At this rate, the storage volume is estimated to be 620 cubic feet which is close to the actual casing volume of 675 cubic feet.

A straight line after the initial wellbore storage period indicates radial flow of gas into the well. It is better seen in the expanded pressure scale of figure 8 than in figure 7. This Horner slope of −7,420,000 extrapolates to a pseudopressure of 223,000,000 or 1,502 psig (Hale, 1981). The change from

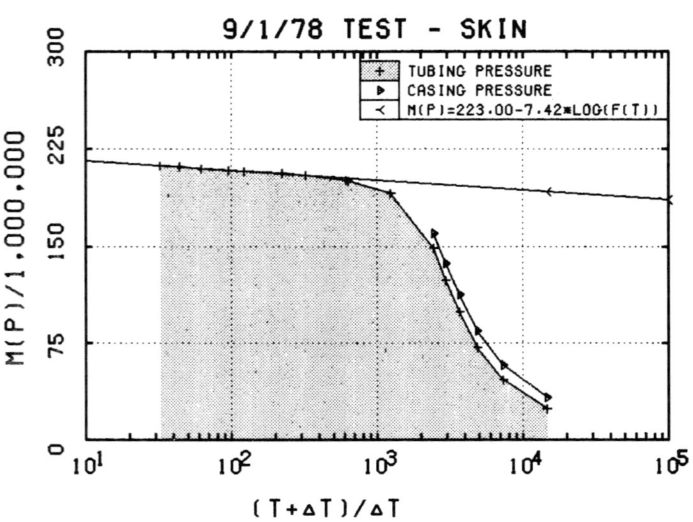

Figure 7. Horner Plot based on a pressure-buildup test on the San Juan 32-7 Unit No. 6 well, September 1 to October 3, 1978.

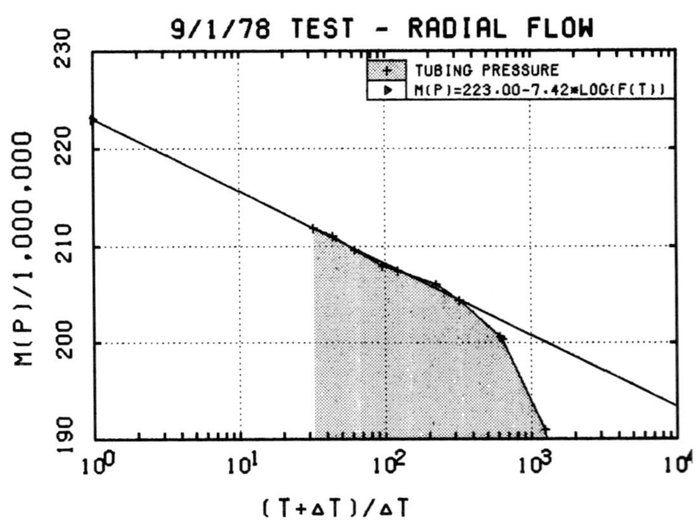

Figure 8. Expanded Horner Plot based on a pressure-buildup test on the San Juan 32-7 Unit No. 6 well, September 1 to October 3, 1978.

the 1,504 psig reported in 1953 is not significant. Based on the final flow rate of 205 MCFPD, the permeability-thickness product is 26.3 md/ft. Data used in the calculation are shown in table 1.

Analysis of wellbore damage shows that 85 percent of the pressure drop occurs in a damaged zone near the wellbore. The well's potential in the absence of damage is estimated to be 1,480 MCFPD. The pressure-drop data used in the calculation are listed in table 1. This potential is reasonably close to the initial potential of 1,790 MCFPD reported in 1953. At that time, the well appears to have been undamaged with damage occurring over the first two to three months of production. The relatively constant producing rate indicates that there has been little change in damage since 1953.

The most notable finding from the 1978 pressure-buildup test is the lack of measurable pressure depletion. This test shows that the current condition of the well is very near to the original condition of the well with the possible exception of wellbore damage.

GAS COMPOSITION

A number of gas analyses taken on this well are listed in table 2. They show very little change in the gas quality which is typical high-methane, coal-seam gas. The average gas composition during the last five years includes 4.97 percent carbon dioxide, 94.24 percent methane, 0.36 percent ethane, and small amounts of nitrogen and heavier hydrocarbon gases. The Btu content of the gas has varied from a low of 955 Btu/scf to a high of 991 Btu/scf with an average of 969 Btu/scf over the last ten years. The unchanging gas quality is evidence that the reservoir has changed very little, if at all, with 35 years of production. The gas composition is distinctly different from the average San Juan Basin gas well producing from a sandstone reservoir which is 87.87 percent methane, 5.85 percent ethane, and 1.50 percent carbon dioxide. [**Editors note: See paper by Rice and others in this volume for more on the chemistry of gasses from coal and sandstone reservoirs in the San Juan Basin.**]

The gas composition from the No. 6 well is very similar to gas composition from other Fruitland wells in the area. The offsetting Yager No. 3 COM, Yager No. 4, and NE Blanco Unit No. 201 wells are all near 94 percent methane with 5 percent carbon dioxide. The high methane content of the gas in these wells indicates that they produce from Fruitland coal seams as does the San Juan 32-7 Unit No. 6 well.

CONCLUSIONS

The San Juan 32-7 Unit No. 6 well produces from an interval dominated by coal seams as shown by well logs. It is close to the San Juan Basin synclinal axis but shows no apparent structural enhancement. Even though production records are brief and no detailed testing has been done on this well, available data indicate that this well produces from Fruitland coal seams. This conclusion is supported by the gas analysis which has the characteristics of coal-seam gas. The lack of pressure decline in the reservoir sets the well apart from other wells. The readjustment time following pressure changes suggests that a desorption process is taking place. Most importantly, this well demonstrates that coal-seam methane can be economically produced over an extended period of time. As the demand for natural gas improves, similar coal-seam wells can be expected to play an increasingly important role as part of the U. S. natural gas supply.

Table 1. Permeability and undamaged potential calculation; data is from a pressure-buildup test on the San Juan 32-7 Unit No. 6 well, September 1 to October 3, 1978.

KH = 1,640,000 Qsc T/Mr

Final rate	(Qsc) =	0.205 MMcf/d
Horner slope	(Mr) =	7,420,000 psi*psi/cp
Formation temperature	(T) =	580 deg R.
Permeability	(KH) =	26.3 md-ft
Formation pressure	=	1644 psia
Shut-in time	=	816 hours
Flow time	=	190,900 hours
Cumulative Production	=	997,000 Mcf
Gas Gravity	(air = 1.000) =	0.6006

AOF = Qsc M(P)//(M(P) −M(Pwf) − M(Pd))

Final rate	Qsc =	0.205 MMcf/d
Static pseudopressure	M(P) =	223,000,000 psi*psi/cp
Flowing pseudopressure	M(Pwf) =	14,200,000 psi*psi/cp
Damage	M(Pd) =	177,900,000 psi*psi/cp
Undamaged potential	AOF =	1.480 MMcf/d

Table 2. Gas composition history; San Juan 32-7 Unit No. 6 well.

DATE	N2	C1	C2	CO2	C3	iC4	nC4	iC5	nC5	C6+	BTU	
60/07/15	.04	94.93	.53	4.40	.07	.01	.02	-	-	-	955	(1)
65/09/09	.06	95.00	.32	4.49	.08	.02	.02	.01	tr	-	969	(1)
average	**.05**	**94.96**	**.42**	**4.45**	**.08**	**.02**	**.02**	**.01**	-	-	**962**	
82/02/10	.11	93.37	.39	5.88	.19	.03	.03	.00	.00	.00	961	(2)
82/09/23	.32	94.27	.42	4.75	.18	.03	.03	.00	.00	.00	970	(2)
83/02/18	.12	94.54	.28	4.86	.15	.03	.02	.00	.00	.00	970	(2)
84/01/04	.20	94.47	.25	4.89	.14	.03	.02	.00	.00	.00	968	(2)
84/02/03	.24	94.22	.50	4.78	.19	.03	.04	.00	.00	.00	972	(2)
84/09/06	.23	94.24	.30	5.03	.15	.03	.02	.00	.00	.00	967	(2)
85/02/28	.35	94.05	.40	4.96	.18	.03	.03	.00	.00	.00	968	(2)
85/08/02	.12	94.46	.29	4.94	.14	.03	.02	.00	.00	.00	969	(2)
86/02/25	.11	94.45	.31	4.96	.13	.02	.02	.00	.00	.00	968	(2)
86/10/23	.28	94.31	.36	4.86	.14	.03	.02	.00	.00	.00	967	(2)
87/03/10	.19	94.24	.50	4.74	.22	.04	.04	.01	.00	.02	974	(2)
average	**.20**	**94.24**	**.36**	**4.97**	**.16**	**.03**	**.03**	**.00**	**.00**	**.00**	**969**	

(1) Data from Dudley Rice, U.S.G.S., Denver
(2) Data from Northwest Company records

REFERENCES CITED

Al-Hussainy, R., and Ramey, H. J., Jr., 1966, The flow of real gases through porous media: Journal of Petroleum Technology, v. 18, no. 5, p. 624-636.

Carslaw, H. S., and Jaeger, J. C., 1959, Conduction of heat in solids, second edition: Oxford at the Clarendon Press, Oxford, England, p. 30-32.

Earlougher, R. C., 1977, Advances in well test analysis: American Institute of Metalurgical Engineers Monograph, v. 5, p. 10-13.

Fassett, J. E. and Hinds, J. S., 1971, Geology and fuel resources of the Fruitland Formation and Kirtland Shale of the San Juan Basin, New Mexico and Colorado: U. S. Geological Survey Professional Paper 676, 76 p.

Fassett, J. E., 1988, Geometry and depositional environments of Fruitland Formation coal beds, San Juan Basin, New Mexico and Colorado: Anatomy of a giant coal bed methane deposit, *in* Fassett, J. E., ed., Geology and coal-bed methane resources of the northern San Juan Basin, Colorado and New Mexico: Rocky Mountain Association of Geologists Guidebook.

Hale, B. W., 1981, Elliptical flow equations for vertically fractured gas wells: Journal of Petroleum Technology, v. 33, no. 12, p. 2489-97.

Hale, B. W., 1986, Analysis of tight gas well production histories in the Rocky Mountains: Society of Petroleum Engineers Production Engineering Journal, v. 1, no. 4, p. 310-322.

New Mexico Oil and Gas Engineering Committee, 1953-1975, Annual Reports, Volume II (northwest New Mexico), Hobbs, New Mexico.

Rice, D. D., Threlkeld, C. N., Vuletich, A. K., and Pawlevicz, M. J., 1988, Identification and significance of coal-bed gas, San Juan Basin, northwestern New Mexico and southwestern Colorado, *in* Fassett, J. E., ed., Geology and coal-bed methane resources of the northern San Juan Basin, Colorado and New Mexico: Rocky Mountain Association of Geologists Guidebook.

Buckskin Charley and his wife Emma Nailor Buck (Towee-Indian name). The photo was taken in front of the old Consolidated Ute Agency in Ignacio, Colorado. Buckskin Charley, Chief of the Southern Utes from 1913 to 1936 is dressed in his everyday attire. His wife Towee is also wearing her traditional everyday attire and wears a headdress as she was the first lady of the Head Chief. Colorado Historical Society photo.

The Ignacio Blanco Gas Field Northern San Juan Basin, Colorado

Clarence L. Harr[1]

ABSTRACT

Ignacio Blanco gas field is located in the northern end of the San Juan Basin, north of the Colorado-New Mexico State line and south of the steep monoclinal dip rimming the north end of the basin. The field is within the Southern Ute Indian Reservation and the 888 producing wells (December 1986) are located on both Southern Ute Tribal lands and interspersed fee acreage.

Stanolind Oil and Gas Company (Amoco Production Company) completed the Southern Ute Indian No. 1-B well on December 6, 1950 as the discovery which opened development of the field. The well was completed open hole in sandstones in the Upper Cretaceous Dakota Sandstone and the upper part of the Jurassic Morrison Formation for over 3,500 MCFPD. The Fruitland-Pictured Cliffs pool was opened in 1951, and the Mesaverde pool in 1952.

During its 37 year history, the Ignacio Blanco gas field has produced 815 billion cubic feet of gas from five primary reservoirs: The Lower Cretaceous Burro Canyon Formation, the Upper Cretaceous Dakota ("A") Sandstone, the Point Lookout Sandstone of the Mesaverde Group, the Pictured Cliffs Sandstone, and coal beds in the Fruitland Formation. Gas accumulation is controlled primarily by stratigraphy in the San Juan Basin. Other then on Ignacio anticline and the Bondad closure, structure is a secondary trapping component within the Ignacio Blanco gas field. Permeability variations, hydrodynamic effect, and natural fracture intensity locally modify gas accumulation and well deliverability.

Field production from 1981 to 1985 averaged 25.5 billion cubic feet of gas per year (BCFPY). Due to market curtailments only 17.8 BCF was produced in 1986. The calculated remaining recoverable gas reserve for the 888 producing wells in the field (December, 1986) is 595 BCF. Recent infill drilling has proven new additional reserves in the Morrison-Dakota pool. In December 1987, Northwest Pipeline Corporation recovered 16,500 MCFPD from a new, previously untested Morrison sandstone. Only 1 well per 22 square miles of the field area has penetrated deep enough into the upper Morrison to have adequately tested the gas potential of its lenticular sandstone reservoirs. Fruitland coal-bed methane resource estimates suggest a recoverable gas volume of 4 trillion cubic feet for the Ignacio Blanco field.

Existing production, additional conventional gas reserves, and the successful development of the coal-bed methane potential insures a long and economically successful life for the Ignacio Blanco gas field.

INTRODUCTION

Twenty-nine years after the first commercial gas discovery in the San Juan Basin near Aztec, New Mexico, Stanolind Oil and Gas Company (Amoco Production Company) completed the Southern Ute Indian No. 1-B as the first commercial well of the Ignacio Blanco gas field in the northern end of the San Juan Basin. (The well is presently operated by Northwest Pipeline and is now designated the Ignacio 33-7 No. 1-18.) It was completed open hole in sandstones in the Dakota Sandstone and the upper part of the Morrison Formation on December 6, 1950. Gas flows of 10,000 to 12,000 MCFPD were reported during testing. The well's initial potential was over 3,500 MCFPD.

During 1951 the Fruitland-Pictured Cliffs pool was opened and in the following year, 1952, the Mesaverde pool, Ignacio Blanco's prime producing zone, was discovered. In its 37 year history, the Ignacio Blanco gas field has produced over 815 billion cubic feet of gas (BCF). Production from the Dakota Sandstone and Mesaverde Group reservoirs will continue for another 20 years, however, at a decreasing rate. Recent successes in developing and producing coal-bed methane gas from the vast coal resources of the Fruitland Formation insure a long and prolific remaining life for the Ignacio Blanco gas field.

The Ignacio Blanco field is located in the northern end of the San Juan Basin, north of the Colorado-New Mexico State line and south of the steep monocline rimming the north end of the basin (fig. 1). Gas accumulation in the basin is con-

[1] Consulting Petroleum Geologist
Southern Ute Indian Tribe
West Gate Offices-Suite 207
2210 Hwy. 6 and 50
Grand Junction, Colorado 81505

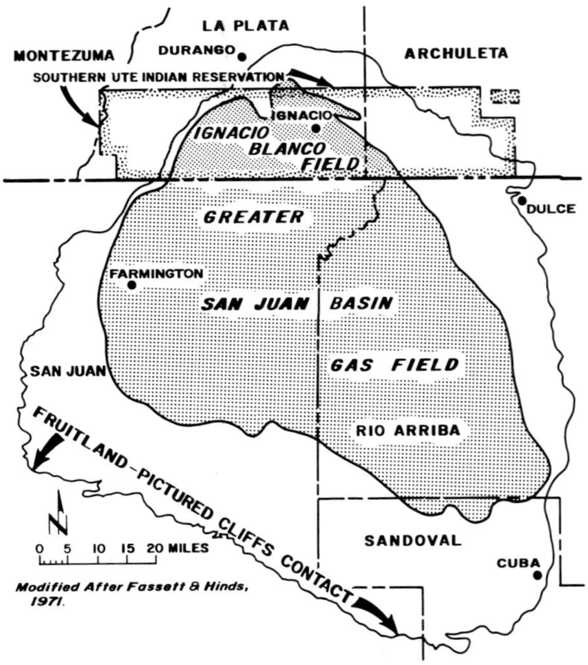

Figure 1. Index map showing the location of the: 1) greater San Juan Basin as outlined by the Fruitland-Pictured Cliffs contact; 2) greater San Juan Basin gas field; 3) Ignacio Blanco gas field; and 4) Southern Ute Indian Reservation.

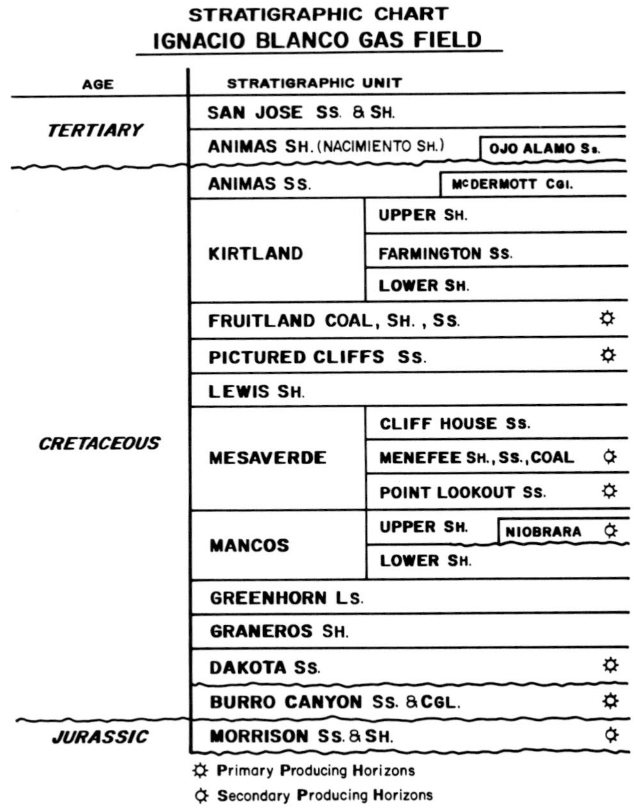

Figure 2. Stratigraphic chart of the Ignacio Blanco gas field. The primary and secondary producing horizons are indicated on the right side of the chart.

trolled primarily by stratigraphy. Other than on Ignacio anticline and the Bondad closure, structure is a secondary component. Permeability variations, hydrodynamic effect, and natural fracture intensity locally modify gas accumulations and well deliverability. The field's 888 producing gas wells (as of December 1986) are contained within the Southern Ute Indian Reservation and are located on both Southern Ute Tribal lands and interspersed fee acreage.

Five primary reservoirs have supplied all but a minor amount of the field's 815 BCF of cumulative gas production (fig. 2). The primary producing horizons are the Lower Cretaceous Burro Canyon Formation, the Upper Cretaceous Dakota ("A") Sandstone, the Point Lookout Sandstone of the Mesaverde Group, the Pictured Cliffs Sandstone, and coal beds in the Fruitland Formation. Production statistics for the Jurassic Morrison sandstones and the Cretaceous Burro Canyon sandstones are included with the Dakota Sandstone. The Fruitland and Pictured Cliffs reservoirs are designated by the Colorado Oil and Gas Conservation Commission (COGCC) as one producing pool.

Ignacio Blanco field production data from discovery to 1976 is from the COGCC files as presented by Bowman (1978a and 1978b) and Craney (1978). Production data from 1976 through 1986 is from the computer files of the Energy Resources Division, Southern Ute Indian Tribe.

STRUCTURE

Structure in the northern end of the San Juan Basin is dominated by 1) the steep monoclinal flexture which brings the Pictured Cliffs and Mesaverde sandstones to outcrop rimming the northern part of the basin and 2) the Ignacio anticline (fig. 3). Ignacio anticline, located approximately one mile southwest of the town of Ignacio, has 200 feet of structural closure and is asymmetrically steep to the northeast. The closure is approximately 2 miles wide and 9 miles long and is located on the northwest end of a southeast plunging structural nose which extends 18 to 20 miles down to and across the New Mexico State line.

The Bondad closure, in the center of T. 33 N., R. 9 W., exhibits close to 75 feet of relief, along an east-trending axis. A structurally weak, south plunging nose along the west side of T. 33 N., R. 9 W., the Bondad structure, and the Ignacio anticline form an arcuate structural trend which parallels the basin's monoclinal flexture.

The Spring Creek structural nose, which plunges south along the west side of T. 34 N., R. 6 W., is a relatively sharp feature, asymmetrically steep to the east. There is a poorly controlled, weak closure in the south half of T. 34 S., R. 10 W., and a north trending closure in the east part of T. 32 N., R. 11 W.

ANNUAL TOTAL FIELD GAS PRODUCTION

Discovery of Dakota gas on the Ignacio anticline, near the end of 1950, initiated the drilling in the Colorado portion of the northern San Juan Basin. However, it was the completion of the El Paso Natural Gas Co. 24 inch pipeline to California in 1951 (Matheny, 1978) and the Mesaverde gas pool discovery on Southern Ute Indian lands in 1952 that accelerated the development of the Ignacio Blanco gas field. During the 10 years from 1954 to 1964 an average of 46 commercial gas

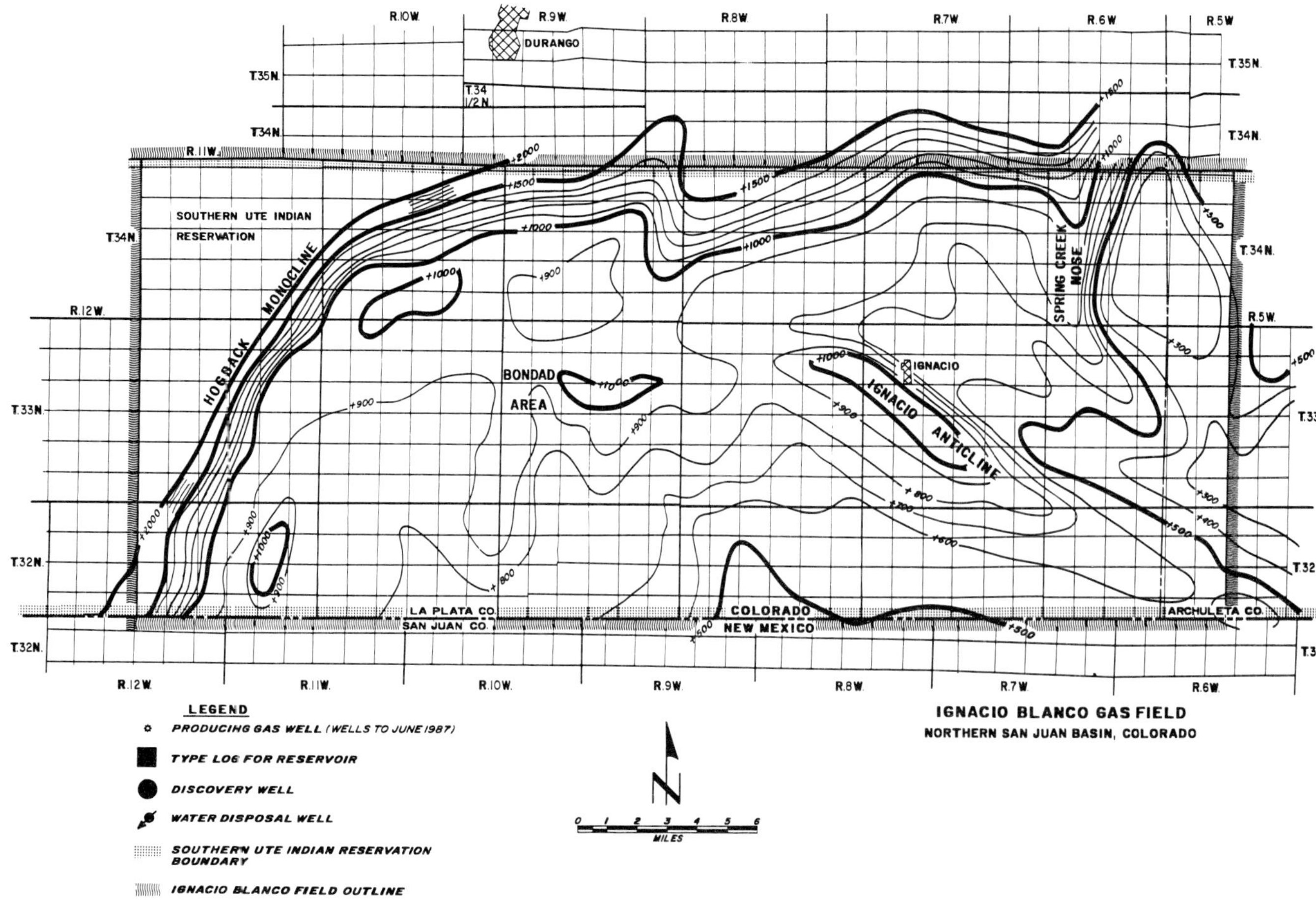

Figure 3. Structure map on the top of the Mancos Shale for the Ignacio Blanco field area. Contour interval is 100 feet. The significant structural features are labeled. Ignacio Blanco gas field outline is shown.

wells per year were completed and placed on stream. Peak annual production of 37.9 BCF was obtained in 1966. Production declined for the next 12 years until 1978. Field production and the number of producing wells per year are plotted on figure 4; the tabulated data are given in table 1.

The second Ignacio Blanco drilling boom started in 1977 when the average price of San Juan Basin gas went above $1.00 per MCF. As the price of oil headed towards $40 per barrel due to the 1973 oil embargo and gas prices followed, going to nearly $5 per thousand cubic feet, an average of 58 wells per year were drilled between 1977 and 1984. During 1980, 148 producing gas wells were completed, an all time high for the Ignacio Blanco field.

This increased drilling reversed the field's annual production decline. Gas production increased from 1976 to 1984 with only minor fluctuations, remained steady through 1985, and then dropped by 7.8 BCF (a 30 percent decrease) in 1986. This sharp decline resulted from curtailment of production due to the nationwide oversupply of gas, decline in price, deregulation and the general upheaval in the gas transmission business.

As of December 31, 1986, the Ignacio Blanco gas field has delivered a total of 814,540,729 MCF through 35 years of production. Calculated remaining recoverable gas for all of the field wells that were producing as of December 1986, is 594,553,699 MCF. Based on a normal year's production of approximately 25 BCF and a 7.1 percent decline, the field's

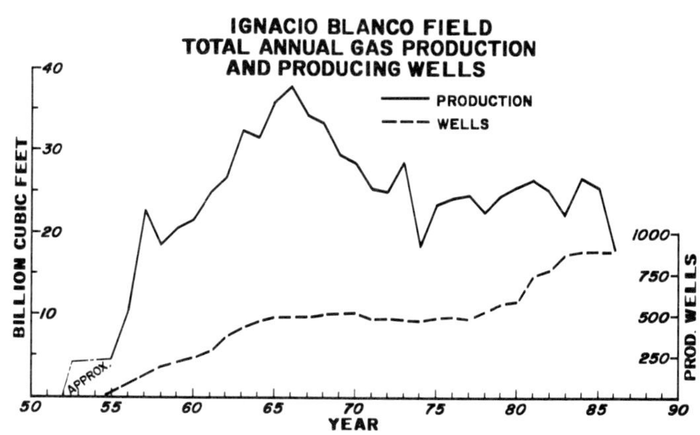

Figure 4. Graph of total annual gas production of the Ignacio Blanco field and number of producing wells per year from 1951 through 1986.

1986 proven reserves production life will extend beyond the next 20 years. Successful development of the Ignacio Blanco coal-bed methane potential will insure a long and economically successful life for the field.

DAKOTA POOL

Dakota pool gas is produced primarily from two field areas, as shown on figure 5. Ignacio anticline, a structural

Table 1. Ignacio Blanco field total annual gas production 1950-1986. This data is plotted in graph form on Figure 4.

Year	Producing wells	Annual production MCF	Cumulative production MCF
1950	Field discovered		
1951-54		ND	*13,923,622
1955	*56	* 4,501,223	18,424,845
1956	*48	*10,839,852	29,264,597
1957	*57	*22,701,174	51,965,871
1958	*26	*18,531,142	70,497,013
1959	*35	*20,966,362	91,4673,375
1960	*35	*21,587,017	113,050,392
1961	*88	24,918,151	137,968,543
1962	*45	*26,551,474	164,520,017
1963	*50	*32,277,785	196,797,802
1964	461	31,660,881	228,458,683
1965	487	35,823,972	264,282,655
1966	487	37,888,390	302,171,045
1967	491	34,035,265	336,206,310
1968	500	33,102,244	369,308,554
1969	503	29,642,569	398,951,123
1970	503	28,647,718	427,598,841
1971	*484	25,599,681	453,198,522
1972	*483	25,109,378	478,307,900
1973	*479	28,591,924	506,899,824
1974	472	18,237,424	525,137,248
1975	480	23,219,593	458,356,841
1976	487	24,209,961	572,566,802
1977	483	24,542,487	597,109,289
1978	513	22,323,695	619,432,984
1979	555	24,386,035	643,819,019
1980	588	25,612,103	669,431,122
1981	736	27,305,336	696,736,458
1982	783	25,059,728	721,796,186
1983	864	2,379,436	744,175,622
1984	892	26,934,398	771,110,020
1985	900	25,604,967	796,714,987
1986	888	71,825,742	814,540,729

*Numbers are near approximations based on incomplete state records.
ND—No data

accumulation in T. 33 N., R. 7 W., and the "A" sandstone trend, a stratigraphic accumulation extending from T. 32 N., R. 7 W. through the south half of T. 33 N., R. 9 W. have produced the bulk of the Dakota gas. Drilling depths in these areas range from 7,400 feet to 9,000 feet. Four separate stratigraphic units are included as a single entry in the Dakota pool statistics: The Dakota "A" and "B" sandstone units, as defined by Bowman (1978a), (figs. 6 and 7); the Burro Canyon conglomeratic sandstone, and the upper Morrison Brushy Basin sandstones are grouped as Dakota pool reservoirs by the Colorado Oil and Gas Conservation Commission.

Morrison Formation

The Brushy Basin Member of the Morrison Formation is primarily light colored, mostly gray, claystones and shales with interbeds of siltstone. Numerous white, green, red, fine- to coarse-grained fluvial channel sandstones are present and contain producible gas. Green and Pierson (1977) consider the Brushy Basin to be a low-energy fluvial and in part lacustrine deposit near the distal facies of a large alluvial fan.

Pan American Petroleum (Amoco) in 1953 drilled the Southern Ute Indian-B No. 6 well (sec. 17, T. 33 N., R. 7 W.) to a depth of 13,127 feet, plugged back, and completed in the Morrison. The well produced approximately 210,000 MCF in 3 years before its recompletion in the Dakota. Bowman (1978a) suggests the Morrison gas recovery was due to structural accumulation since the well is located near the crest of the Ignacio anticlinal closure.

Northwest Pipeline Corporation on December 11, 1987 recovered on a test of Morrison sandstone perforations, 7,856-64 feet and 7,866-84 feet (26 feet open), approximately 16,500 MCFPD. The Northwest Pipeline Ignacio 33-8 No. 22-13 well in sec. 13, T. 33 N., R. 8 W. was completed in the Morrison Formation, Burro Canyon Formation, Dakota Sandstone and the Mesaverde Group. After fracing all units and gauging 1,500 MCFPD from the upper 3 intervals, the bridge plug between the Burro Canyon Formation and a Morrison sandstone was drilled out and the well came in at nearly 20,000 MCFPD. This well is located on the southwest flank of the Ignacio anticline, 1 mile southwest and approximately 100 feet downdip from the depleted Morrison sandstone in the Pan Am Southern Ute-B No. 6 well. The proximity of the two wells, their structural relationship and the magnitude of difference in producing characteristics suggests positive separation of the reservoirs and stratigraphic accumulation of the Morrison gas.

More than 1,000 holes have been drilled in the Ignacio Blanco field, yet in an area that covers 550 square miles, fewer than 25 wells (1 well per 22 sq. mi.) have penetrated deep enough into the upper Morrison to have adequately tested the gas potential of these lenticular sandstones.

Burro Canyon Formation

While recognizing that disagreements exist regarding the correct name of the Lower Cretaceous sandstone in the San Juan Basin, the usage followed in this paper will be Burro Canyon Formation (Craig, 1981), (fig. 6 and 7). The Burro Canyon is a fluvial sandstone and mudstone deposit formed by a system of coalescing channels and floodplains. The formation as a whole was deposited over most of the northern San Juan Basin. However, due to the erratic distribution of the stream channels and subsequent pre-Dakota erosion, the Burro Canyon sandstone reservoir lacks any large scale pressure-fluid continuity. Structural arching such as at Ignacio anticline and the Bondad feature (figs. 3 and 5) is of assistance in trapping Burro Canyon gas, but the stratigraphic character of pinchouts and permeability variations within the sandstone reservoir has been a prime contributor to gas accumulation.

The Burro Canyon is a conglomeratic sandstone, light gray to white, very coarse grained to conglomeratic at or near the base and becoming medium- to fine-grained towards the top. Thicknesses range from over 100 feet to zero feet in local areas. The unit contains from 1 to 3 or 4 sandstone beds interbedded with dark gray mudstone and siltstone beds which are from 2 to 30 feet thick. A majority of the Burro Canyon

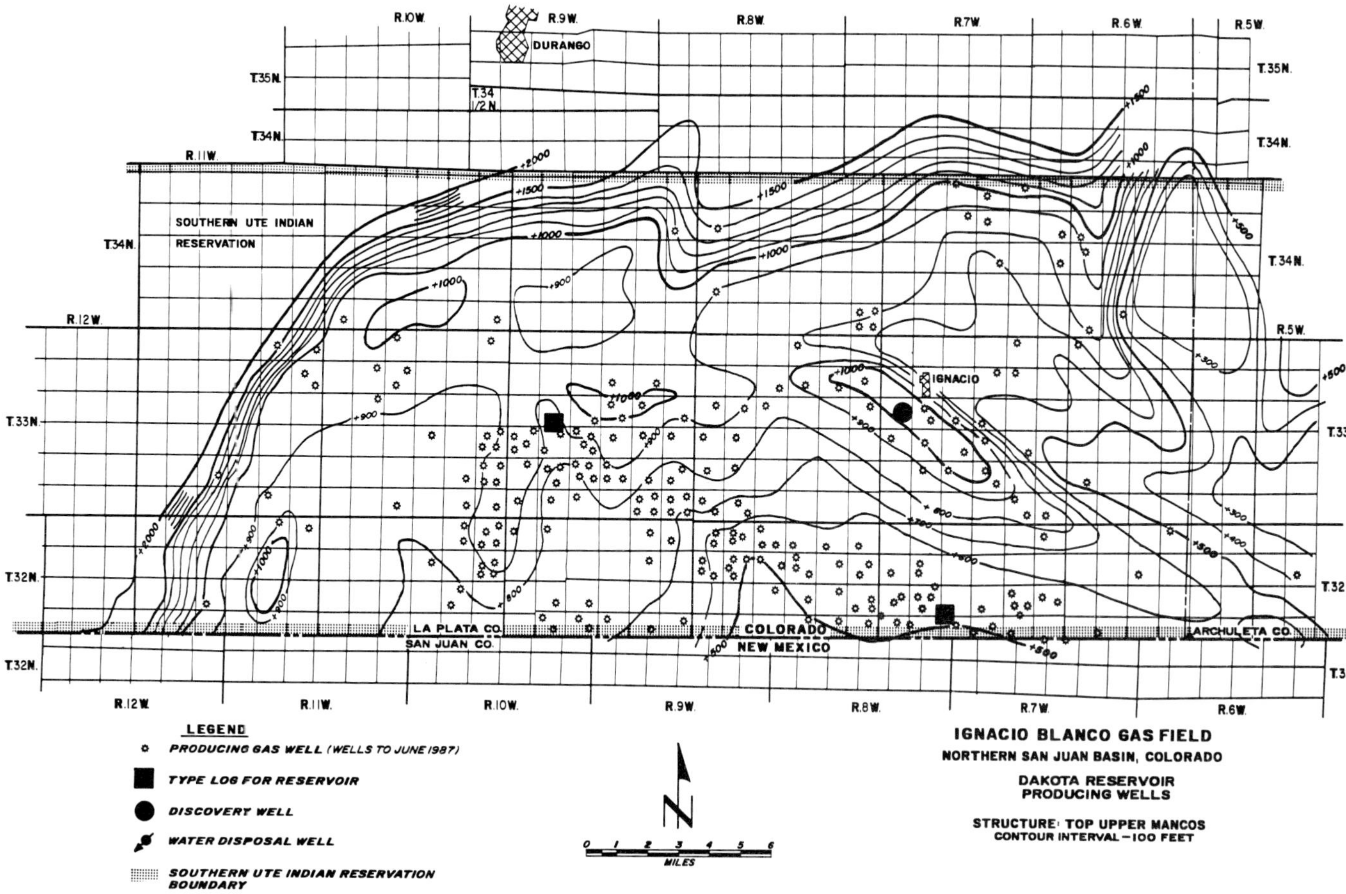

Figure 5. Map showing areal distribution of Dakota pool producing wells as of June 1987. Structure is on top of Mancos Shale.

tests penetrated only the top of the unit, leaving lower sandstones untested.

The type log for the Burro Canyon reservoir (fig. 6) is from the TXO Ute 17 No. 1 well located in the Bondad area in sec. 17, T. 33 N., R. 9 W. Two Burro Canyon sandstones separated by some interbedded mudstone and tight sandstone were encountered. The upper and lower sandstones were completed for a calculated absolute open flow (CAOF) of 3,300 MCFPD. Cumulative production as of December 1986, is 1.5 BCF with no reported water.

Sohio discovered gas in the upper part of the Burro Canyon Sandstone in their Southern Ute No. 2 well, sec. 16, T. 32 N., R. 7 W., adjacent to the New Mexico State line (fig. 7). There is no structural closure near this well (fig. 5), as is the case with other Burro Canyon producers. It is my contention that the meandering stream system of the Burro Canyon formed numerous individual stratigraphic traps which are presently spread across the relatively low dip of the northern San Juan Basin. All Dakota wells should penetrate through the Burro Canyon and the upper Morrison to test the potential for such up dip meandering-channel-sandstone gas accumulations.

Dakota Sandstone

The Dakota "B" sandstone (figs. 6 and 7), as defined in this paper and by Bowman (1978a), is the interval of interbedded carbonaceous shale and mudstone, occasional thin coal beds, and intermixed slightly conglomeratic sandstones which disconformably overlie the Lower Cretaceous Burro Canyon Formation and are moderately disconformable with the overlying marine Dakota "A" sandstone. The carbonaceous shales and coals are coastal swamp to lagoonal sediments. The sandstones are fluvial and generally siliceous and tight. Bowman (1978a) considers only those "B" sandstones that are highly fractured as contributors to Dakota production.

The Dakota "A" sandstone (fig. 7) is presently the prime contributor to Dakota pool gas production. Gas was trapped stratigraphically along a west-northwest-trending barrier island. Littoral marine currents working across a Dakota delta receiving sediment from the southwest, deposited the sand body nearly parallel to the southwest transgressing shoreline of the Mancos sea. Core data suggests a rapid change from the barrier sandstone to lagoonal dark, silty, organic shale on the south (Bowman, 1978a). The sandstone fairway extends west-northwest from T. 32 N., R. 6 W. through the lower half of T. 33 N., R. 11 W., and continues west up the Hogback monocline and through the Red Mesa field in T. 33 N., R. 12 W. Sandstone thicknesses vary from 5 feet to 35 feet. Producing gas wells shown on figure 5 clearly define this stratigraphic trend. The sandstone body is up to 3 miles wide and contains within it thicker northwest-southeast sub-trends localized along the fairway. Gas is trapped by local permeability barriers along the general sandstone trend.

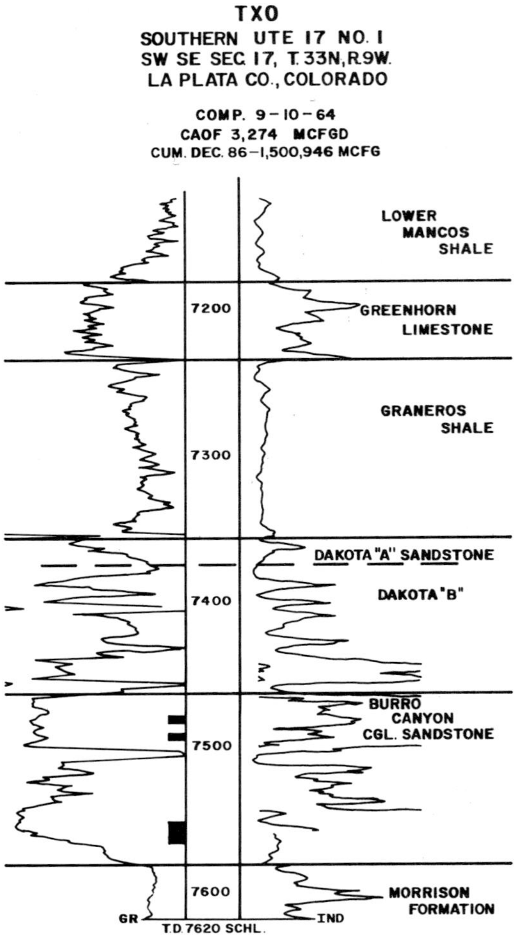

Figure 6. Type log of the Dakota pool stratigraphic units showing the perforated and producing Burro Canyon Sandstone reservoirs. The Dakota "A" sandstone is not developed in this well.

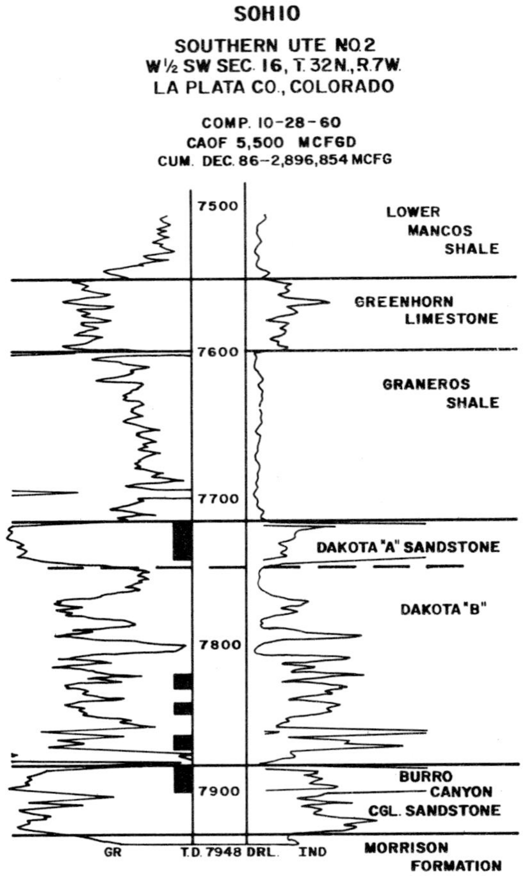

Figure 7. Type log of the Dakota pool stratigraphic units showing the perforated and producing Dakota "A" sandstone, Dakota "B" sandstones, and Burro Canyon sandstone reservoirs.

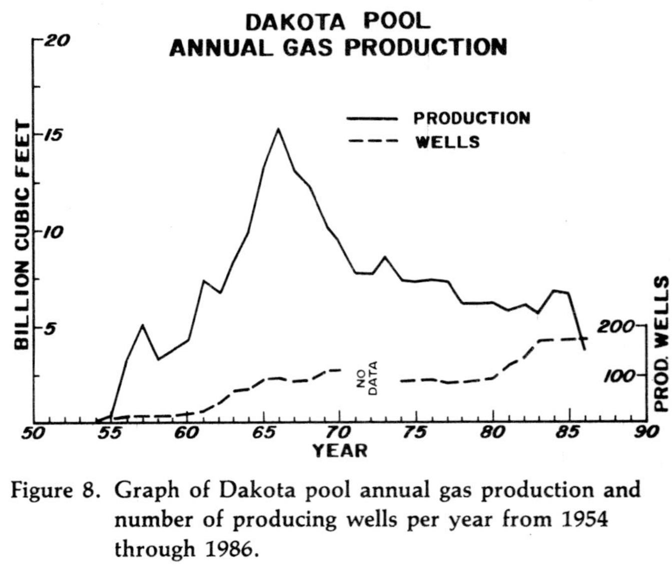

Figure 8. Graph of Dakota pool annual gas production and number of producing wells per year from 1954 through 1986.

Dakota Pool Production

Through 33 years of production, the four Dakota-pool reservoirs have produced 233,795,314 MCF. Figure 8 is a graph showing the annual Dakota pool production and the number of wells producing through the life of the pool up to 1986. Table 2 lists the data plotted on figure 8. Pool production peaked in 1966 at 15,097,972 MCF with 91 wells producing. Much of this peak production came from the structurally trapped gas in the Burro Canyon Formation. Drilling through the 1970s and the flurry of well drilling from 1981 to 1983, due to price increases, was conducted primarily along the Dakota "A" sandstone trend. The numerous wells drilled from 1981 to 1983 reversed the Dakota pool's annual decline, but only slightly (fig. 8). All new Dakota pool reserves added in the last 15 years, either Dakota "A", Burro Canyon, or Morrison, have been in stratigraphic traps.

MESAVERDE POOL

Southern Union Gas Company completed the Southern Ute No. 1 well in the Point Lookout Sandstone of the Mesaverde Group for an initial potential of 1,700 MCFPD on May 14, 1952. The well located in the SE1/4 SW1/4 sec. 15, T. 32 N., R. 11 W., La Plata County, Colorado was the Mesaverde discovery well for the Ignacio Blanco field Mesaverde pool (fig. 9). Southern Union drilled the well as an extension of the Blanco Mesaverde pool to the southeast, in New Mexico. The Southern Ute No. 1 well is located in the extreme southwest corner of the Ignacio Blanco Mesaverde pool, 12 miles southwest from the main northwest-trending production fairway.

Table 2. Dakota pool annual gas production 1954-1986. This data is plotted in graph form on Figure 8.

Year	Producing wells	Annual production MCF	Cumulative production MCF
1950	Field discovered		
1954	1	ND	ND
1955	3	364,444	364,444
1956	12	3,071,446	3,435,890
1957	13	$5,128,518	8,564,488
1958	14	3,326,553	11,890,961
1959	18	3,925,886	15,816,847
1960	20	4,383,209	20,200,856
1961	26	7,338,817	27,583,873
1962	41	6,566,296	34,105,169
1963	56	8,437,260	42,542,429
1964	73	9,973,500	52,515,929
1965	84	13,269,350	65,785,279
1966	91	15,097,972	80,883,251
1967	89	13,386,344	94,269,595
1968	95	12,171,897	106,441,492
1969	113	10,226,933	116,668,425
1970	113	9,529,107	126,197,532
1971	ND	7,876,908	134,074,440
1972	93	7,747,656	141,822,096
1973	ND	8,619,931	150,442,027
1974	84	7,330,778	157,772,805
1975	87	7,290,731	165,063,536
1976	90	7,405,801	172,469,337
1977	81	7,377,904	179,847,241
1978	83	6,053,419	185,900,660
1979	86	6,145,854	192,046,514
1980	92	6,279,847	198,326,361
1981	120	5,932,900	204,259,261
1982	137	6,193,344	210,452,605
1983	163	5,682,618	216,135,223
1984	165	6,934,748	223,069,971
1985	174	6,775,209	229,845,180
1986	172	3,950,134	233,795,314

ND—No data

Whether the intent was truly an extension attempt out of New Mexico, bold exploration, or serendipity, it initiated 34 years of drilling and production that has netted 597 producing wells and 527 BCF of gas as of December 1986. Sixty five percent of the Ignacio Blanco field gas has been produced from the Mesaverde pool. Drilling depths to the base of the Mesaverde, range from 5,500 feet to 6,500 feet.

The Mesaverde Group comprises three formations in the northwest San Juan Basin. The basal Point Lookout Sandstone, the middle Menefee Formation, and the upper Cliff House Sandstone (figs. 2 and 10). The three units represent regressive and transgressive cycles of the Cretaceous sea. In the northeast part of the basin, these three formations have not been subdivided on the outcrop because the Menefee Formation thins northeastward and the Point Lookout and Cliff House come closer and closer together in that direction and are difficult to differentiate on a geologic map. Thus, in the northeast these rocks are mapped as the undivided Mesaverde Group. (See Fassett, 1977, for a more detailed discussion of the geometry and environments of deposition of the Mesaverde Group.)

Point Lookout Sandstone

The Point Lookout Sandstone is the most regionally widespread regressive sandstone in the San Juan Basin (Molenaar, 1977). As the Mancos sea regressed to the northeast, in hesitating steps, thick sand deposits formed as barrier beaches and occasional bars in various places across the basin at northwest trending littoral still-stands of the shoreline. Each succeeding sand buildup was stratigraphically higher and farther northeast than the previous one (Fassett, 1977, fig. 2).

The coastal barrier sandstones of the Point Lookout, which make up the prime Ignacio Blanco Mesaverde gas-producing fairway, trend northwest from New Mexico through T. 32 N., R. 7 W. to the southern portion of T. 34 N., R. 10 W., a distance of approximately 23 miles. The sandstone corridor (with net sandstone thicknesses of 20 feet or greater) is generally 3 to 4 miles wide. Individual sandstone build-ups, irregularly located along the gross trend, attain thicknesses up to 50 or 55 feet of net sandstone.Point Lookout porosities range from 10 percent to occasionally 22 percent. The sandstones are light colored, very fine to fine grained, slightly argillaceous, and locally calcareous.

Reservoir sandstones were developed in 4 benches. The type log from the Jerome McHugh, Southern Ute No. 3 well located in sec. 20, T. 32 N., R. 9 W. (fig. 10) shows benchs "D" through "B" of the Point Lookout. Fluvial deposition of the Menefee Formation forced the marine coastal waters out of the immediate area of the McHugh Southern Ute No. 3 well prior to deposition of the "A" sandstone bench. In the southwest corner of the Ignacio Blanco field, the overlying continental Menefee Formation rests directly on the "D" bench, but climbs stratigraphically until it directly overlies marine siltstone and shale above the "A" sandstone bench in the northeast corner of the field.

Sandstone benches "B" and "C" are the best developed reservoir sandstones. Several of the wells completed in these intervals along the prime trend have individually produced over 8 to 9 BCF of gas in 30 years and have a calculated remaining reserve of over 1 BCF. Sandstone bench "D" is tight in most areas and needs natural fractures to produce.

Except within the Ignacio anticlinal closure, the sandstone reservoirs of the Mesaverde pool are stratigraphic gas accumulations. Lenticularity of the individual Point Lookout sandstone benches, permeability barriers due to increased shaliness or diagenetic clay boundaries, and termination of permeability-enhancing fracture fields are the most important trapping factors.

Menefee Formation and Cliff House Sandstone

The Menefee Formation, which overlies the Point Lookout (figs. 2 and 10), consists of interbedded carbonaceous shales, fine- to coarse-grained, kaolinitic sandstone, and generally thin coal beds. The unit is the coastal, continental-paludal and fluvial phase of the Point Lookout regression and the Cliff House transgression. Menefee gas is trapped primarily in the updip meanders of the fluvial channel sandstones and by kaolinitic clay permeability barriers within the sandstones. Some gas is also trapped within the coal beds. Bowman (1978) attributes

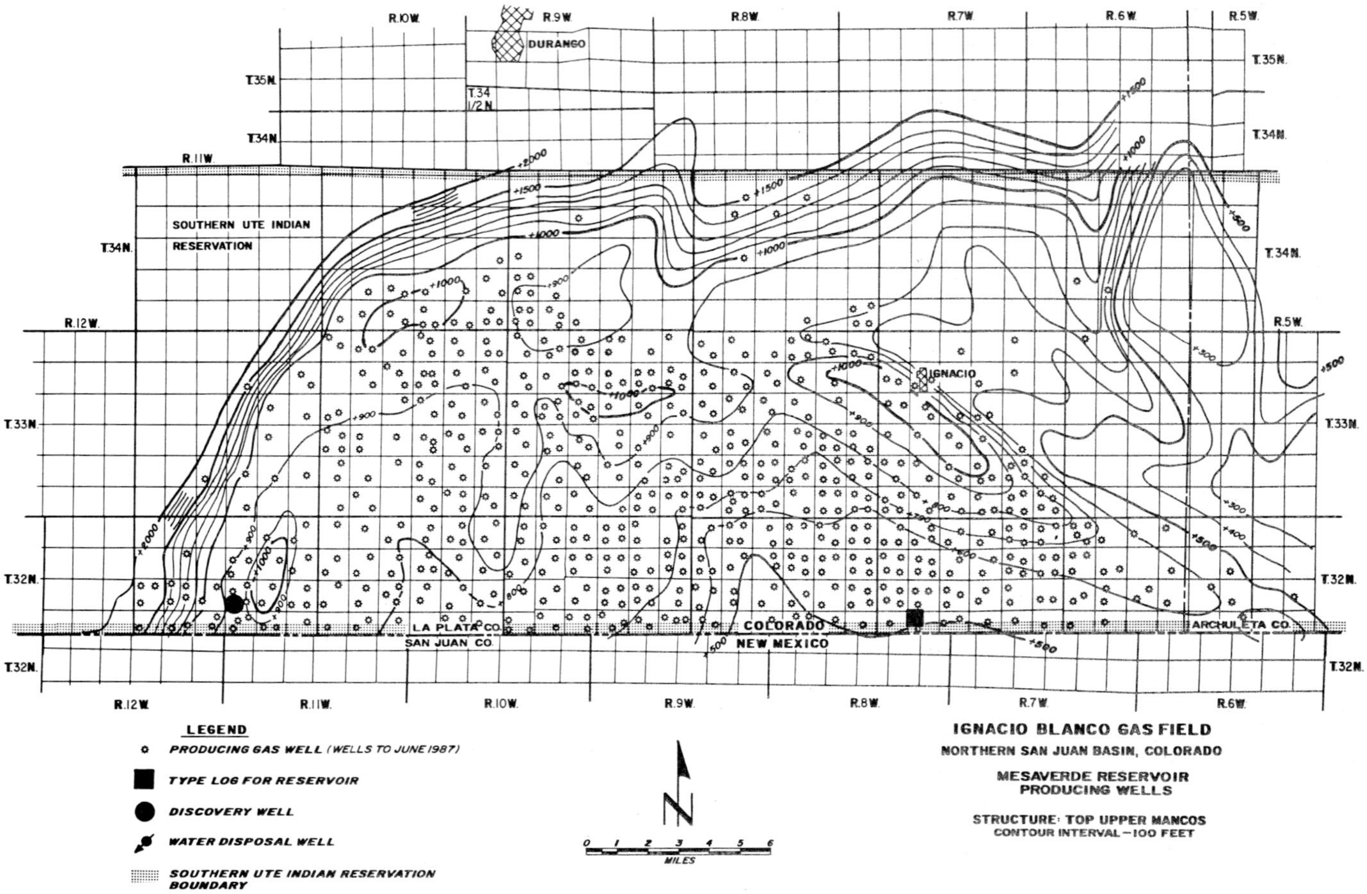

Figure 9. Map showing areal distribution of Mesaverde pool producing wells as of June 1987. Structure top of Mancos Shale.

only 10 percent of the Mesaverde pool gas production to the Menefee Formation.

Only a few wells, located in T. 32 S., R. 11 W., produce gas from the Cliff House Sandstone. The Cliff House is lensy and according to Fassett (1977) is not present at the top of the Menefee in many areas in the northern San Juan Basin. Where it is present, the Cliff House is generally highly argillaceous (fig. 10), but it occasionally cleans-up at the base. It is an insignificant contributor to the gas production in the Mesaverde pool.

Mesaverde Pool Production

Figure 11 is a graph of the annual Mesaverde pool gas production and yearly number of producing wells. Table 3 is the tabulated data plotted on figure 11. Because the Mesaverde produces 65 percent of the field's total gas production, both curves on figure 11 express trends similar to the field's total production shown on figure 4.

Production peaked in 1963 at 22,948,634 MCF from 330 producing wells. The second development boom of wells came in 1979 following the gas market price increases; annual production climbed by approximately 1 BCFPY. The decrease in production in 1986 is due to an oversupply of gas, price declines, and market curtailment.

FRUITLAND-PICTURED CLIFFS POOL

The COGCC in the early 1950's defined the Fruitland Formation and the Pictured Cliffs Sandstone as a single producing pool within the Ignacio Blanco gas field. Figure 12 and table 4 show the combined Fruitland-Pictured Cliffs gas production data and number of producing wells from discovery through 1986. The two units are individual reservoirs and should be designated as separate pools. Figure 13 and table 5 show the Pictured Cliffs gas production from 1976 through 1986. Fruitland reservoir gas production from 1976 through 1986 is shown on figure 14 and table 6. The production statistics from the combined reservoirs have been included in this paper to give historical continuity to the units involved. Only the production data from each individual reservoir is considered to be of technical importance.

Production and producing well data through 1976 is from the COGCC files as presented by Craney (1978). Production and well data from 1977 through 1986 is from the computer files of the Southern Ute Indian Tribe.

The Fruitland-Pictured Cliffs pool was opened by discovery of gas in the Pan American Petroleum (Amoco) Southern Ute No. D-1, located in sec. 7, T. 33 N., R. 7 W. on the Ignacio anticline (fig. 15). The well was completed September 4, 1951, at a plug-back depth of 2,470 feet to 2,615 feet. Craney (1978) stated that the completion was attempted using an open-hole nitroglycerin shot in the Pictured Cliffs Sandstone. The well was plugged back after making water from the Pictured Cliffs and completed in the lower half of the Fruitland Formation. On a 3-hour test the well produced 2,200 MCFPD of gas. Subsequent wells on the anticline were, however, completed in the Pictured Cliffs Sandstone.

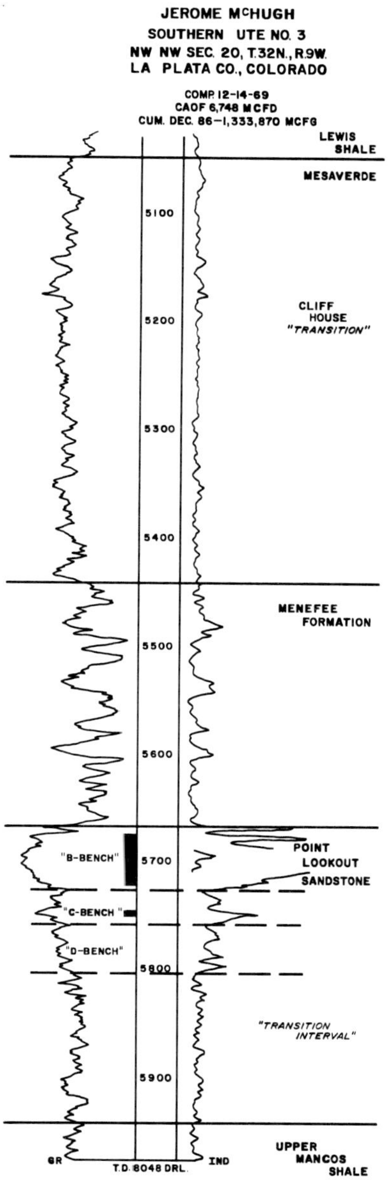

Figure 10. Type log of the Mesaverde pool stratigraphic units showing the perforated and producing "B" bench reservoir of the Point Lookout Sandstone. The "A" bench is absent by non-deposition and the "C" and "D" benches are shaly.

The pre-coal-bed methane nemesis of "if it's starting to kick water you're through" caused all of the Ignacio anticline Fruitland-Pictured Cliffs pool wells to be abandoned as well as numerous other similar pool wells throughout the field. Craney (1978) in his paper on the Fruitland-Pictured Cliffs pool indicated that there were 79 abandonments and 11 shut-in wells as of the end of 1976. Amoco's 1977 discovery of the Cedar Hills field in New Mexico, the first designated coal-bed gas field in the western U.S., initiated industry thought, if not action, regarding dewatering the Fruitland coal beds for degasification of the coal.

Prior to 1964, the COGCC Fruitland-Pictured Cliffs pool production records are incomplete. They do, however, indicate high early pool production rates with rapid decline and a short life for the producing wells. The short life of the wells as compared to Dakota or Mesaverde production can be attributed to depletion of the free (previously desorped) coal-bed methane of the Fruitland, contained in fractures within the gas cap of the Ignacio anticline, and subsequent encroachment of the coal-bed generated water.

Figure 12 and table 4 show 8 wells producing in 1974 with an annual production of only 85,159 MCF. Pictured Cliffs development from 1976 through 1983 and Fruitland coal-bed methane development in 1983 and 1984 increased the 1984 pool gas production to 3,532,612 MCF, it's highest since pool discovery. As of December, 1986 the cumulative production of the combined Fruitland Formation and the Pictured Cliffs Sandstone is 57,640,060 MCF.

Table 3. Mesaverde pool annual gas production 1954-1986. This data is plotted in graph form on Figure 11.

Year	Producing wells	Annual production MCF	Cumulative production MCF
1954	161	ND	ND
1955	8	136,779	136,779
1956	50	4,268,406	4,405,185
1957	97	14,822,656	19,227,841
1958	148	12,784,589	31,932,430
1959	176	15,247,924	47,180,354
1960	198	16,311,917	63,492,271
1961	227	16,687,443	80,179,714
1962	300	19,093,287	99,273,001
1963	330	22,948,634	122,221,635
1964	363	21,262,888	143,484,523
1965	379	22,174,514	165,659,037
1966	376	22,474,911	188,133,948
1967	383	20,390,874	208,524,822
1968	389	20,682,541	229,207,363
1969	380	19,180,503	248,387,866
1970	379	18,908,535	267,296,401
1971	ND	17,529,355	289,825,756
1972	ND	17,271,733	302,097,489
1973	ND	19,833,176	322,163,053
1974	380	15,821,487	337,984,540
1975	384	15,829,592	353,814,132
1976	384	16,313,192	370,127,324
1977	380	16,310,618	386,437,942
1978	389	14,730,821	401,168,763
1979	418	16,134,172	414,302,935
1980	443	16,949,884	434,252,819
1981	531	17,724,677	451,977,496
1982	551	16,178,380	468,155,876
1983	576	13,975,669	482,131,545
1984	587	16,467,038	498,598,583
1985	587	16,205,948	514,804,531
1986	597	11,773,996	526,578,527

ND—No data
SI—Shut in

PICTURED CLIFFS RESERVOIR

Pictured Cliffs gas development and production has shift-

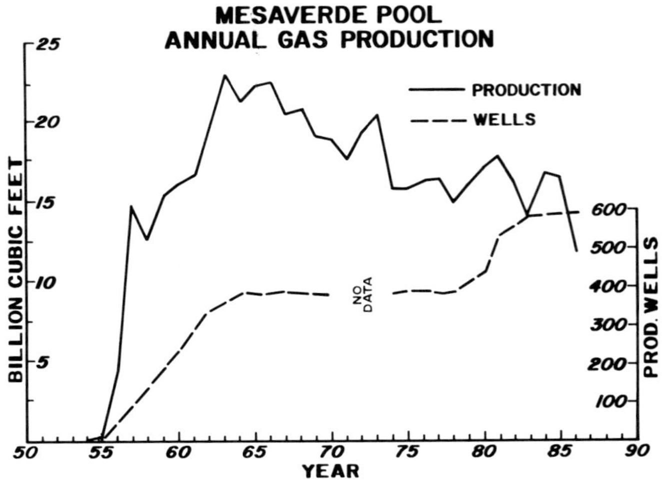

Figure 11. Graph of Mesaverde pool annual gas production and number of producing wells per year from 1954 through 1986.

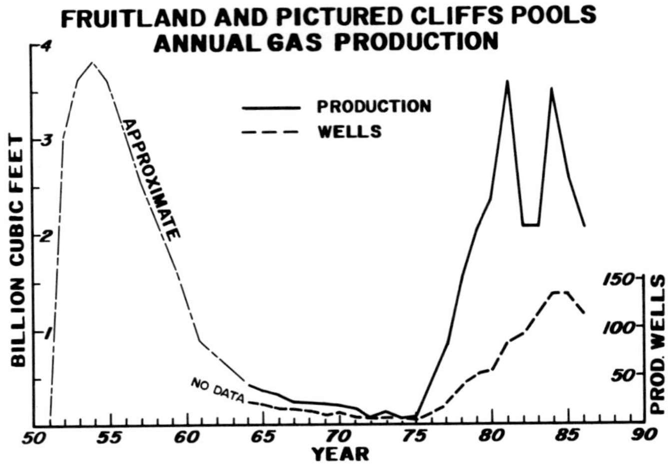

Figure 12. Graph of Fruitland-Pictured Cliffs pool annual gas production and number of producing wells per year from 1951 through 1986.

ed from the depleted structural accumulation at Ignacio anticline to stratigraphically trapped gas in two areas which exhibit negligible structural influence. The most active area centers around the northeast corner of T. 32 N., R. 9 W. (fig. 16).

ARCO's Southern Ute 32-9 No. 1-2 well, sec. 1, T. 32 N., R. 9 W. is the type log for the Pictured Cliffs Sandstone (fig. 17). The well was perforated across the Pictured Cliffs, which appears as a shaly sandstone reservoir yet was completed for 3,700 MCFPD. As of December 1986, the well has produced 442,917 MCF. Other wells in the immediate area are producing 200 to 300 MCFPD after 9 years of productive life.

Additional Pictured Cliffs development exists in the northwest corner of T. 32 N., R. 9 W. (fig. 16), and also scattered through the southern portion and into the northwest corner of T. 33 N., R. 10 W. The wells are in a rather loose northwest trending development pattern which is parallel to the sandstone reservoir's depositional trend.

The Pictured Cliffs Sandstone is a northwest-trending regressive coastal-barrier sandstone unit, deposited by the final northeast retreat of the Lewis sea (Molenaar, 1977). Directly overlying the Pictured Cliffs, the coastal-swamp coals and river and flood-plain shales and sandstones of the Fruitland Formation were deposited in a trailing effect as the Lewis sea regressed. Fassett and Hinds (1971) and Fassett (this publication) have demonstrated by geophysical log study that the thicker Fruitland coals develop along trend and shoreward (southwest) of the thick sandstones formed along the Pictured Cliffs shoreline still-stands.

The Ignacio Blanco Pictured Cliffs production corridor is confined to a position northeast of thick northwest-trending Fruitland coal and southwest of a thick buildup of Pictured Cliffs sandstone. The Pictured Cliffs pool gas production is undoubtedly controlled by local sandstone lenticularity and permeability barriers developed along the shore-side slope of the coastal-barrier. The sandstone is white, gray, very fine to fine grained, and locally kaolinitic. Boyce (personal communication) contends the higher yield wells are producing from individual thicker sandstone lenses which are less shaly and contain a minimum of clay. The cleaner sandstone lenses are

Table 4. Fruitland and Pictured Cliffs pool annual gas production 1951-1986. This data is plotted in graph form on Figure 12.

Year	Producing wells	Annual production MCF	Cumulative production MCF
1951	Pool discovered		
1952-58	ND	ND	26,673,622
1959	ND	1,792,552	28,466,174
1960-63	ND	ND	32,033,736
1964	25	424,493	32,458,229
1965	24	380,108	32,838,337
1966	19	315,507	33,153,844
1967	19	258,047	33,411,891
1968	16	247,806	33,659,697
1969	10	235,133	33,894,830
1970	11	210,076	34,104,906
1971	9	193,418	34,298,324
1972	10	89,989	32,629,099
1973	9	138,817	32,767,916
1974	8	85,159	32,853,075
1975	9	99,270	32,952,395
1976	13	490,968	33,443,313
1977	22	853,965	34,297,278
1978	41	1,539,455	35,836,733
1979	51	2,106,009	37,942,742
1980	53	2,382,372	40,325,114
1981	82	3,647,759	43,972,873
1982	91	2,688,004	46,660,877
1983	121	2,721,149	49,382,026
1984	136	3,532,612	52,914,638
1985	135	2,623,810	55,538,498
1986	115	2,101,612	57,640,060

ND—No data

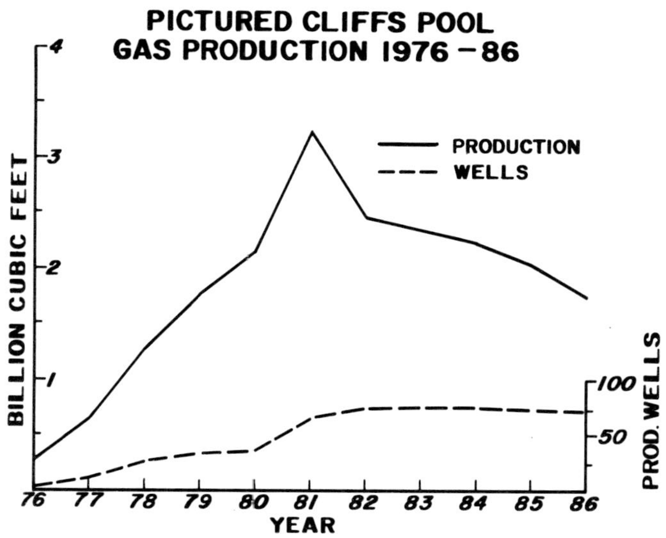

Figure 13. Graph of Pictured Cliffs reservoir annual gas production and number of producing wells per year from 1976 through 1986.

Table 5. Pictured Cliffs pool annual gas production 1976-1986. This data is plotted in graph form on Figure 13.

Year	Producing wells	Annual production MCF	Cumulative production MCF
1975			*6,590,469
1976	4	290,612	6,881,081
1977	10	645,169	7,526,250
1978	27	1,291,789	8,818,039
1979	34	1,787,947	10,605,986
1980	37	2,111,756	12,717,742
1981	67	3,204,667	15,922,409
1982	73	2,465,435	18,387,844
1983	76	2,355,847	20,743,691
1984	76	2,261,123	23,084,814
1985	74	2,011,616	25,016,430
1986	73	1,733,922	26,750,352

* Pictured Cliffs 1975 cumulative estimated at 20 percent of Fruitland-Pictured Cliffs 1975 combined cumulative production.

probably due to selected current winnowing along the coastal-barrier as it developed.

Pictured Cliffs Production

Figure 13 is a graph showing the Pictured Cliffs reservoir's annual production and number of producing wells from 1976 through December, 1986. Table 5 is the tabulated data through the same producing period. The 1975 Pictured Cliffs cumulative production is estimated at 20 percent of the combined Fruitland-Pictured Cliffs cumulative as carried in the COGCC files. The 20 percent Pictured Cliffs, 80 percent Fruitland division is based on assumed producing characteristics of the two reservoirs during the early production history at Ignacio anticline, which leaves the cumulative totals of the two individual units in question.

Ignacio Blanco field gas production from the Pictured Cliffs Sandstone had decreased to 290,612 MCF per year from only 4 wells in 1976. Additional development was initiated in 1976, and by 1982, 63 new wells had been completed. Production peaked at 3,204,667 MCF in 1981, took a rapid decline the following year then smoothed out into a gradual decline. As of December 1986, the Pictured Cliffs Sandstone has yielded 26,750,352 MCF.

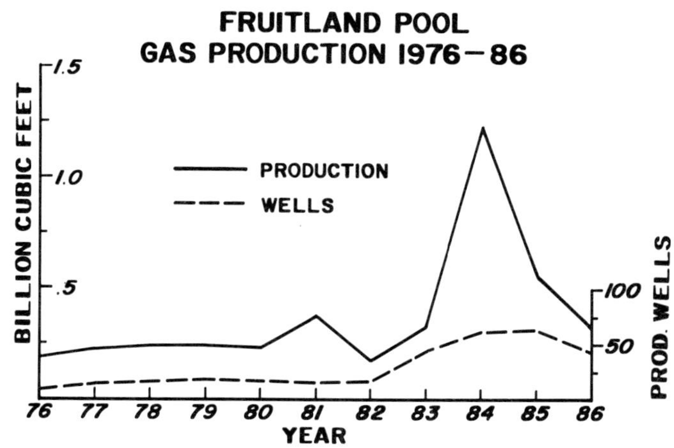

Figure 14. Fruitland reservoir annual gas production and number of producing wells per year from 1976 through 1986.

Table 6. Fruitland pool annual gas production 1976-1986. This data is plotted in graph form on Figure 14.

Year	Producing wells	Annual production MCF	Cumulative production MCF
1975			*26,361,876
1976	9	189,001	26,550,877
1977	12	202,226	26,753,103
1978	14	240,841	26,993,944
1979	17	243,043	27,236,987
1980	16	232,798	27,469,785
1981	15	375,001	27,844,786
1982	18	172,470	28,017,256
1983	45	321,826	28,339,082
1984	60	1,223,075	29,562,157
1985	61	558,993	30,121,150
1986	42	337,995	30,459,145

* Fruitland 1975 cumulative estimated at 88% of Fruitland-Pictured Cliffs 1975 combined cumulative production.

FRUITLAND RESERVOIR

Figure 15 shows the location of 132 coal-bed methane wells within the Ignacio Blanco gas field (within the boundaries of the Southern Ute Indian Reservation). The wells were drilled, but not necessarily producing, by June 1987. Seven additional wells are located north of the Ute line which divides T. 34 N. Ninety percent of the wells shown are operated by Amoco Production Company. The locations represent new coal-bed

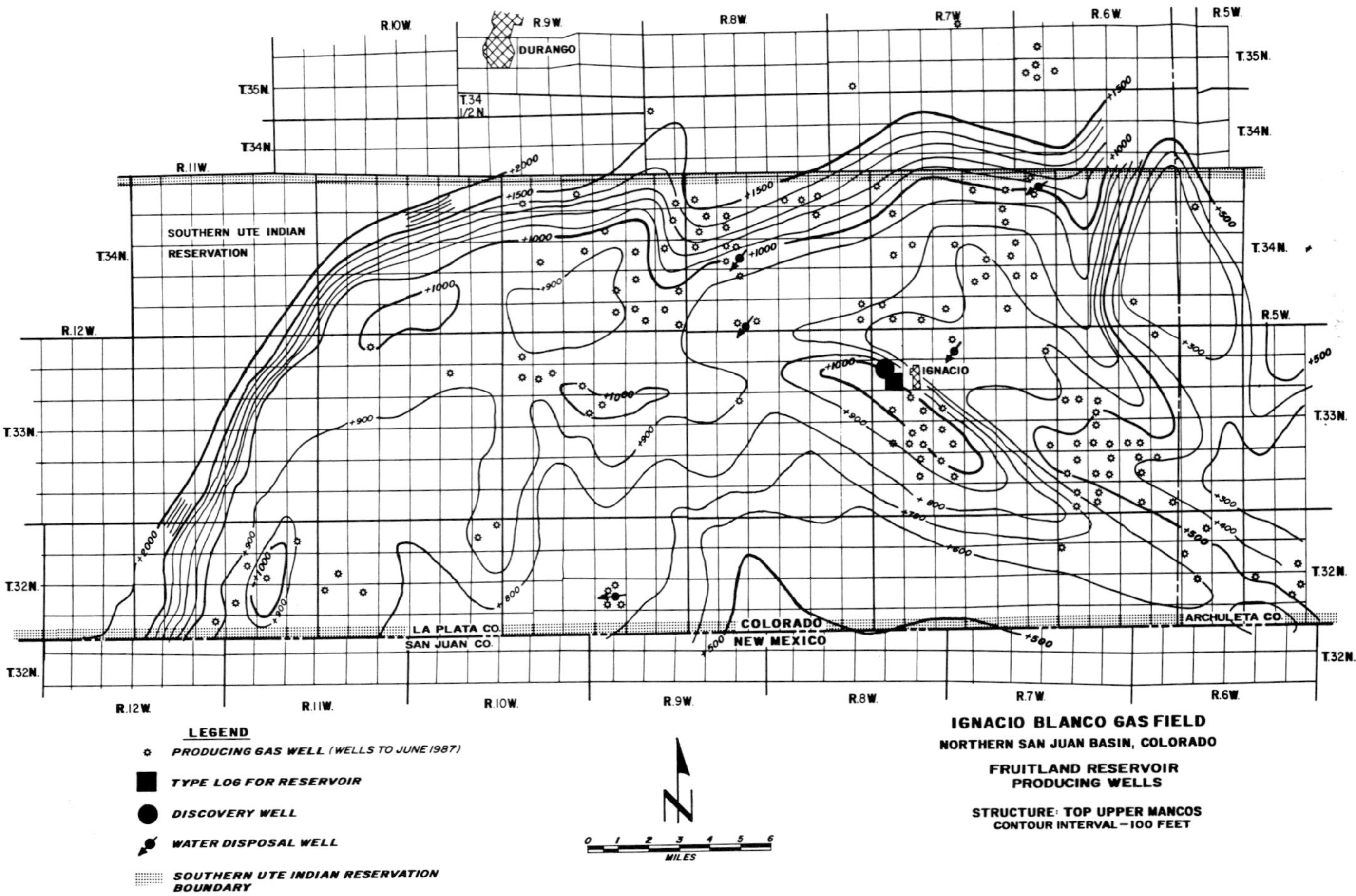

Figure 15. Map showing areal distribution of Fruitland reservoir producing wells as of June 1987. Structure top of Mancos Shale.

methane wells drilled between 1981 and June 1987.

The majority of the Fruitland wells are grouped in four separate areas of rather loose development patterns which will receive additional wells as production characteristics or lease obligations dictate. At present the wells are located on 320 acre spacing units with occasional 1-well infills. Fruitland drilling depths range from 1,500 feet to 4,000 feet and could be shallower if the basin's monoclinal flanks are drilled.

Three of the larger Fruitland coal development areas are located on structurally prominent flextures. The wells at Ignacio anticline are all new Fruitland coal-bed wells drilled subsequent to abandonment of the original Fruitland wells. The Fruitland coal development area centered around sec. 24., T. 34 N., R. 9 W., is situated on a structural nose which plunges southwest through the basal flexture of the Hogback monocline. Wells in the southwest corner of T. 33 N., R. 6 W., form a development area in a broad syncline paralleling the Ignacio anticline. The wells located in T. 34 N., R. 7 W. are so loosely spaced, I doubt that they represent a planned coal-bed development pattern.

The tight pattern of five wells located outside of the Ignacio Blanco field, in T. 35 N., R. 6 W., are positioned on the Spring Creek structural nose. The up-plunge contours of the nose were omitted.

There are presently five injector wells for Fruitland brine disposal in the field. Amoco has two, which are located in sec. 4, T. 33 N., R. 7 W. and sec. 20, T.34 N., R. 8 W. ARCO has an injector located in a 4 well pattern in sec. 15, T. 32 N., R. 9 W. Tenneco has 1 disposal well in sec. 32, T. 34 N., R. 8 W., and Magnus has an injector in sec. 12, T. 34 N., R. 6 W.

Within the Ignacio Blanco gas field as of March 1988 there were 138 completed Fruitland coal-bed methane wells, 14 drilling or near-completed wells, and 46 Fruitland permitted wells.

Fruitland rocks were deposited landward of the northeastward-regressing Pictured Cliffs Sandstone shoreline. The Fruitland conformably overlies the Pictured Cliffs Sandstone and is conformably overlain by the nonmarine Kirtland Shale (fig. 2). Thickness of the Fruitland Formation varies from 200 feet to over 500 feet in the Ignacio Blanco field area. The carbonaceous shale, interbedded siltstone, sandstone, and thick coal beds are evidence of a swamp and flood plain environment which trailed and interfingered with the retreating Pictured Cliffs Sandstone. Fassett and Hinds (1971), Fassett (this publication) and Sandberg (this publication) have documented in their comprehensive work on the Fruitland and Pictured Cliffs that the thicker coal beds develop along trend and shoreward (southwest) of the northwest-trending thick Pictured Cliffs sandstones.

Total thickness of the coal along the better developed Fruitland trends reaches 70 feet (Fassett and Hinds, 1971 and Fassett, this publication). Sandberg (this publication) has divided the Fruitland coals into three separate zones. The lower zone, immediately above the Pictured Cliffs, usually contains the

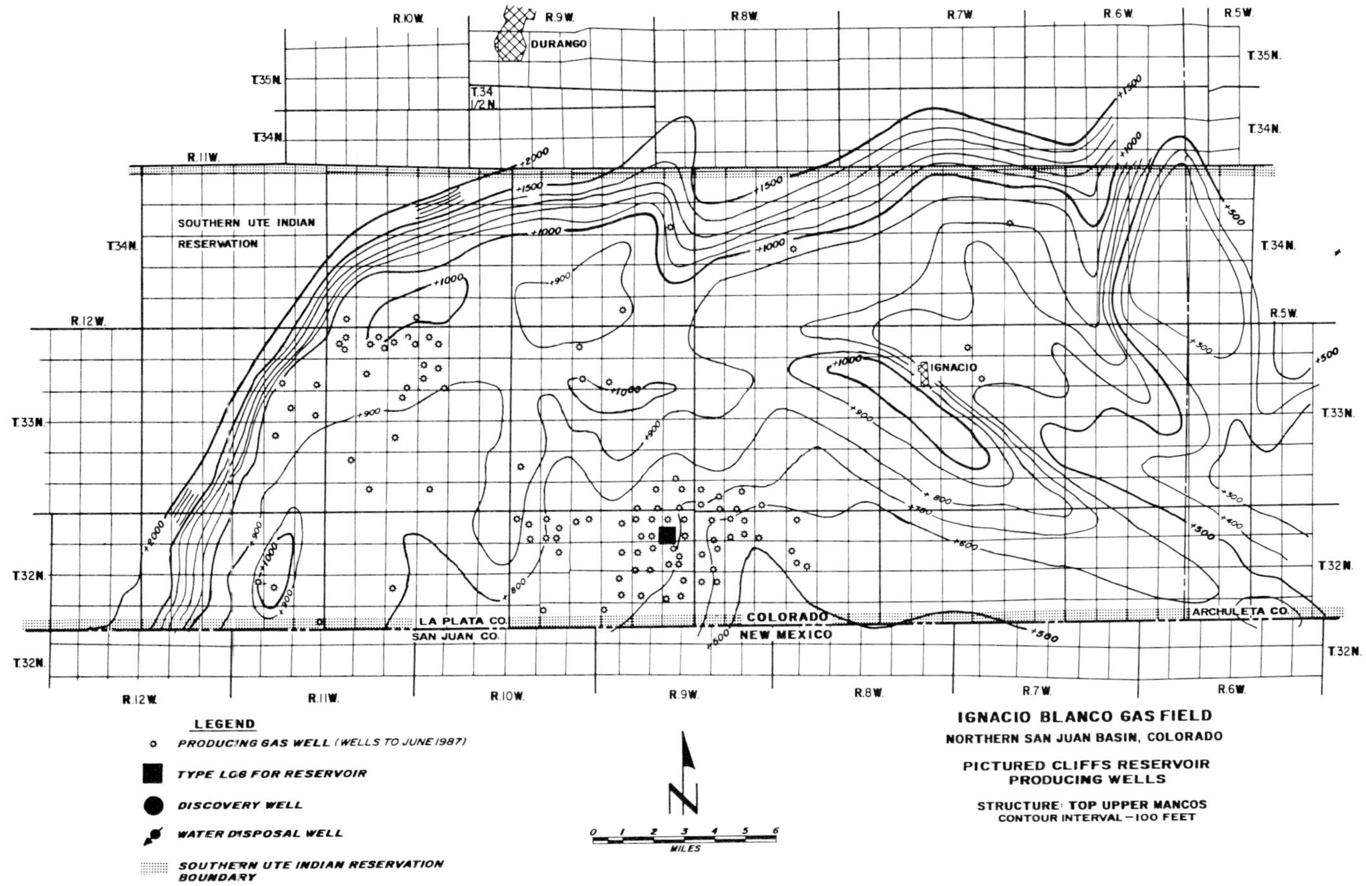

Figure 16. Map showing areal distribution of Pictured Cliffs reservoir producing wells as of June 1987. Structure top of Mancos Shale.

thickest coal beds. The type log in figure 18 used to illustrate the Fruitland reservoir shows a 27 foot coal bed of the lower zone. The well is Amoco's Southern Ute Gas Unit "AC" No. 1, sec. 18, T. 33 N., R. 7 W. Amoco completed the Fruitland coal-bed methane well March 20, 1987 for 76 MCFPD and 117 BWD.

Production

Early Fruitland production on the Ignacio anticline was from both sandstone and coal-bed reservoirs. Adequate data is not available to determine the quantity of gas produced individually from either the Fruitland sandstones, the Pictured Cliffs Sandstone, or from the Fruitland coal beds. Except for fracture permeability, the reservoir quality of the sandstones of both formations is exceptionly poor in the area of Ignacio anticline. Many of the early wells were completed with 7-inch casing set approximately 40 feet above the Fruitland Formation and completed open hole without stimulation (Craney, 1978). Such a completion would allow those Fruitland coals above the structurally controlled gas-water interface to produce.

Table 4, Fruitland and Pictured Cliffs pool annual gas production 1951-1986, shows that during the 8 year span from 1952 through 1959, 28 BCF was credited to the pool production, which at that time would be from the Ignacio anticline. I would speculate that during the first 8 years of pool production, the 28 BCF which is 50 percent of the 35 year life-time total of 58 BCF of pool production, came mostly from the Fruitland coal beds at Ignacio anticline prior to the early encroachment of water into the gas cap.

Based on the above sparse data and my assumptions, I credit the Fruitland reservoir(s) with 80 percent of the combined Fruitland-Pictured Cliffs pool 1975 cumulative production. The assigned 80 percent figure may be low. Eighty percent or 26,361,876 MCF was used in table 6 to initiate the annual Fruitland cumulative figures derived from the 1976 to 1986 Fruitland annual production figures. Table 6 offers a basis from which to show future growth in the Ignacio Blanco field of Fruitland coal-bed methane production volumes.

Individual well-production averages for the Fruitland are about 120 MCFPD and 130 BWD with the characteristic coal-bed gas production of decreased water and increased gas through extended production life. Gas production per well usually ranges from 10 to 600 MCFPD with some recent wells reporting initial tests at 1,000 to 3,000 MCFPD. Water production ranges from 5 to 700 BWD.

The number of producing Fruitland wells per year as shown on table 6 and figure 14 should not be confused with the number drilled per year. There was a very active coal-bed methane drilling program in the northern San Juan Basin from 1981 into 1985 with over 120 Fruitland coal-bed methane wells being drilled. However, due to implementing new completion techniques for a new unique gas resource, the need for dewatering, and the need for a water-disposal system, coal-

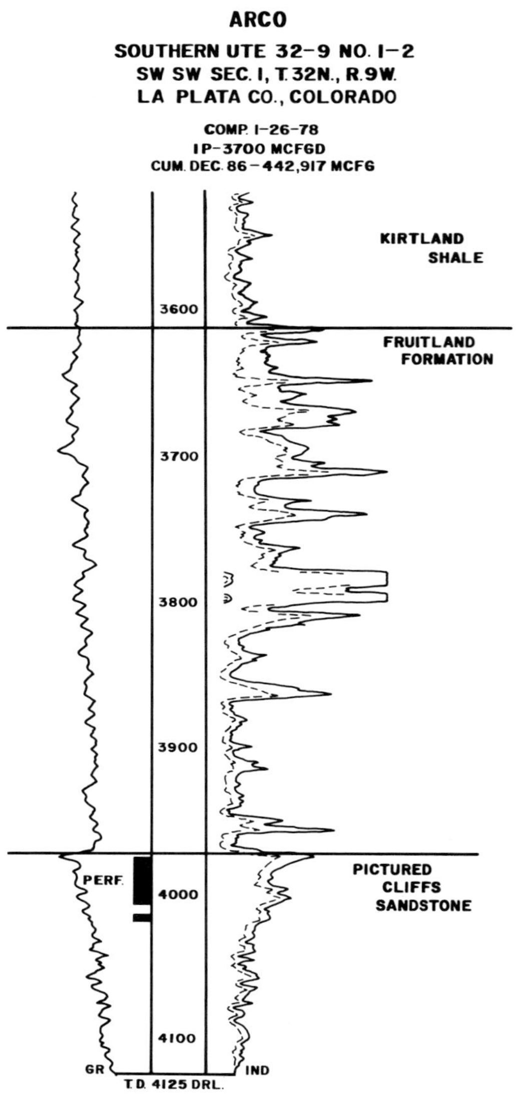

Figure 17. Type log of the producing Pictured Cliffs Sandstone reservoir showing the perforated interval.

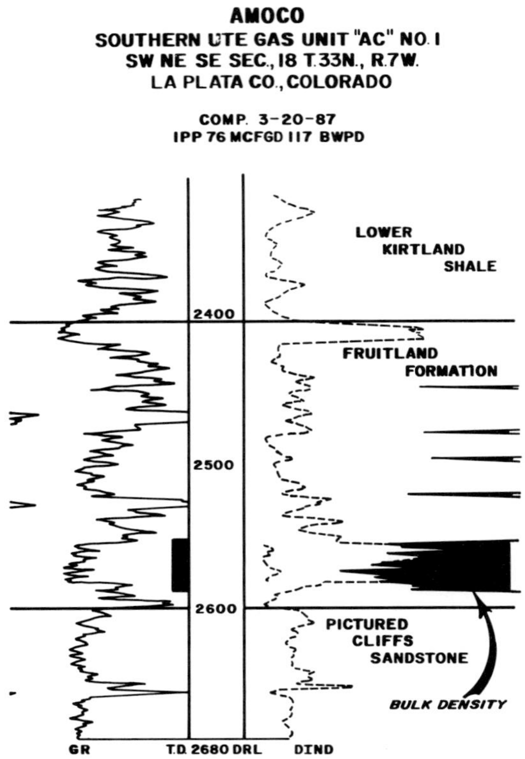

Figure 18. Type log of the Fruitland lower coal zone producing interval. Coal thickness is 27 feet.

bed methane wells are now slower to go on line, once drilled.

The number of producing wells listed on table 6 and shown by graph on figure 14 represent only those wells producing gas within that year. Twenty-seven new wells went on line in 1983 and 5 in 1984 which resulted in peak production of 1.2 BCF for 1984. The decline in gas produced in 1985 and 1986 was due to oversupply, decreased price, and market curtailment. Nineteen producers were shut in or at least inactive in 1986. As of December 1986, the Fruitland Formation has produced 30,459,145 MCF of gas.

SPACING REGULATIONS

The Ignacio Blanco gas field received spacing requirements under Order 3 and Order 112 in the early life of the field between 1952 and 1954. Under subsequent rulings, certain modifications were implemented. Present regulations set Dakota pool spacing at 640 acres with optional infill drilling on 160 acre units. Mesaverde pool wells carry a 320-acre spacing requirement with an optional 160-acre infill.

The Fruitland-Pictured Cliffs pool spacing order at the present requires that wells be drilled on 320-acre units, but allows for a 160-acre infill. The COGCC, Southern Ute Tribe, Bureau of Land Management, and Ignacio Blanco field operators are investigating the possibility of spacing and pool limit changes to insure the proper conservation and development of the Fruitland coal-bed methane resource in Colorado. (The State of New Mexico is also working with this group to try to avoid state-line changes in spacing or field-rule regulations.)

CONCLUSION

Many geologists and engineers have considered the Ignacio Blanco gas field (as well as all of the San Juan Basin) to be an essentially fully-explored hydrocarbon province with little additional new potential. Such pessimistic thinking has been applied to many other basins throughout the history of oil and gas exploration and development. "When will we ever learn?"

The saga of Fruitland coal-bed methane, the production of 16,500 MCFPD from a deeper sandstone reservoir in the Morrison Formation because someone drilled a little deeper, and recent oil discoveries in the southern part of the basin makes premature the use of the word "mature" in describing the hydrocarbon potential of the Ignacio Blanco gas field or the basin in its entirety. For the company and the person with the optimism and the courage, I'm sure there is more gas and oil to find and produce.

REFERENCES CITED

Bowman, K.C., 1978a, Ignacio Blanco Dakota (Gas), *in* Fassett, J.E., ed., Oil and gas fields of the Four Corners Area: Four Corners Geological Society, v. 1, p. 131-132.

Bowman, K.C., 1978b, Ignacio Blanco Mesaverde (Gas), *in* Fassett, J.E., ed., Oil and gas fields of the Four Corners Area: Four Corners Geological Society, v. 1, p. 137-139.

Craig, L.C., 1981, Lower Cretaceous rocks, southwestern Colorado and southeastern Utah, *in* Wiegand, D.L., ed., Geology of the Paradox Basin: Rocky Mountain Association of Geologists, p. 195-200.

Craney, D.L., 1978, Ignacio Blanco Fruitland-Pictured Cliffs, *in* Fassett, J.E., ed., Oil and gas fields of the Four Corners Area: Four Corners Geological Society, v. 1, p. 134-136.

Fassett, J.E., 1977, Geology of the Point Lookout, Cliff House and Pictured Cliffs Sandstones of the San Juan Basin, New Mexico and Colorado, *in* San Juan Basin III: New Mexico Geological Society 28th annual Field Conference Guidebook, p. 193-197.

Fassett, J.E., 1988, Geometry and depositional environment of Fruitland Formation coal beds, San Juan Basin, New Mexico and Colorado: Anatomy of a giant coal-bed methane deposit, *in* Fassett, J.E., ed., Geology and coal-bed methane resources of the northern San Juan Basin, Colorado and New Mexico: Rocky Mountain Association of Geologists Guidebook.

Fassett, J.E., and Hinds, J.S., 1971, Geology and fuel resources of the Fruitland Formation and Kirtland Shale of the San Juan Basin, New Mexico and Colorado: U.S. Geological Survey Professional Paper 676, 76 p.

Green, M.W. and Pierson, C.T., 1977, A summary of the stratigraphy and depositional environments of Jurassic and related rocks in the San Juan Basin, Arizona, Colorado. and New Mexico *in* Fassett, J.E., ed., San Juan Basin III: New Mexico Geological Society Guidebook, 28th Field Conference, p. 147-152.

Matheny, M.L., 1978, A History of the Petroleum Industry in the Four Corners Area, *in* Fassett, J. E., ed., Oil and Gas Fields of the Four Corners Area: Four Corners Geological Society, v. 1, p. 17-24.

Molenaar, C.M., 1977, Stratigraphy and depositional history of Upper Cretaceous rocks of the San Juan Basin area, New Mexico and Colorado, with a note on economic resources *in* Fassett, J.E., ed., San Juan Basin III: New Mexico Geological Society Guidebook, 28th Field Conference, p. 147-152.

Sandberg, D.T., 1988, Coal resources and coal-bed geometry, Fruitland Formation, Southern Ute Indian Reservation, Archuleta and LaPlata Counties, Colorado, *in* Fassett, J.E., ed., Geology and coal-bed methane resources of the northern San Juan Basin, Colorado and New Mexico: Rocky Mountain Association of Geologists Guidebook.

Chief Buckskin Charley, the last hereditary chief of the Utes, was named Chief of the Utes at Ouray's request. He was a sub-chief under Ouray for many years. Buckskin Charley helped his people through the difficult adjustment to reservation life. He was trusted and liked by the Anglos, and his advice was listened to with respect. He insisted all tribal festivals and ceremonies be carried out in traditional manners. He spoke english well, but preferred his native Ute language. Buckskin Charley wore anglo clothes only when he visited with them, wearing his native attire among his people. Buckskin Charley died on May 8, 1936. In this photo taken in the 1880s, Buckskin Charley is dressed in his traditional attire which his wife made for him. Colorado Historical Society photo.

Geology, Geochemistry, Reservoir Engineering, and Completion Methods at the Cedar Hill Field, San Juan County, New Mexico: A Field Study of Classic Coal Degasification Behavior

David Decker[1], Stephen J. Jeu[1], James D. Cooper[1], Donald E. Wicks[2]

ABSTRACT

The Cedar Hill coal degasification field was discovered by Amoco Production Company with the drilling and completion of the Cahn No. 1 well in May 1977. Since that time, the field has been in-filled with 16 production wells and 3 pressure monitor wells. A total of 8.6 billion cubic feet of coal-bed methane has been produced from the field through the end of 1987.

Production has been established from the basal Fruitland coal seam at a depth of approximately 2,800 feet. Varying in thickness from 4 to 26 feet, the basal Fruitland coal reservoir is laterally continuous throughout the field area. As determined from coal isopach maps and measured gas contents, the basal seam contains an estimated 313 million tons of coal and 89 billion cubic feet of gas in place. High permeability permits effective water removal and gas recovery in central portions of the field. The coal reservoir is water saturated and overpressured with a gradient of 0.58 psi/foot. Over time, wells have exhibited the classic coal-bed methane production profile: increasing gas production with decreasing water production.

Applied geochemical techniques have been effective at distinguishing bicarbonate-rich coal-derived water from chloride-rich water produced from sandstone reservoirs. Likewise, a low C2+ hydrocarbon composition with the presence of CO_2 identifies gas originating from coal. The combined use of gas and water chemistries verifies the origin of produced fluids from the Cedar Hill field as coming from the Fruitland coals.

Eight wells were initially completed open hole and later equipped with gravel pack assemblies or predrilled liners to stabilize the coal interval and inhibit the flow of coarse coal fines. The gravel packs were removed, where possible, and production appeared to increase. Cemented and cased wellbores that are perforated and hydraulically stimulated in the coals appears to be Amoco's preferred completion technique. This technique was utilized on wells drilled as replacements for wells originally completed open hole.

The Cedar Hill field is significant because: 1) Commercial levels of coal-bed methane have been established and a long production history is available, 2) the field serves as a case history to examine favorable geologic and reservoir characteristics that may be helpful in identifying other commercial coal-bed methane fields, and 3) many varied completion and stimulation techniques were employed documenting an evolution of drilling, completion, and production practices.

INTRODUCTION

An estimated 50 trillion cubic feet (TCF) (Kelso and others, 1987; Kelso and Wicks, this volume) of pipeline quality gas is present within Fruitland Formation coal reservoirs of the San Juan Basin. To cost-effectively exploit this large resource will require reliable exploration and engineering practices tailored to the special requirements of coal reservoirs. Examining an area where commercial production has been established provides insight into the origin of favorable reservoir properties and cost-effective drilling, completion, and production techniques. The Cedar Hill field is the only multiple-well coal-degasification field in the San Juan Basin with a relatively long production history. The level of productivity and the amount of available reservoir data make this field a logical candidate for a detailed coal-bed methane field study. This report relies exclusively on data available in the public record.

The Cedar Hill field is located in San Juan County, New Mexico (fig. 1). The field covers approximately sixteen-square-

[1] David Decker, Stephen J Jeu, and James D. Cooper
Resource Enterprises, Inc.
2516 Foresight Circle, Number 7
Grand Junction, Colorado 81505

[2] Donald E. Wicks
ICF-Lewin, Inc.
1090 Vermont Ave., NW
Suite 700
Washington, DC 20005

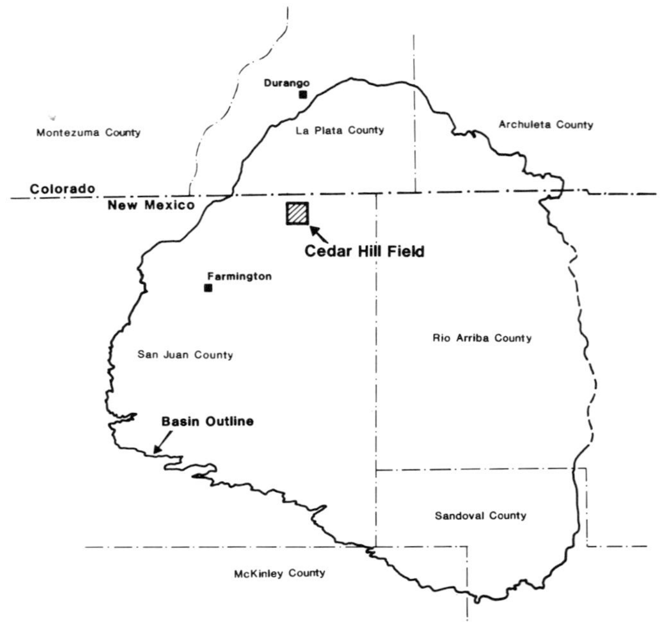

Figure 1. Map showing location of the Cedar Hill field in the San Juan Basin. Basin outline is defined by the outcrop of Fruitland coal seams.

miles and overlaps the Ignacio Blanco, Mt. Nebo, and Animas Chacra fields. In the field area, ninety-three wells produce from the Fruitland Formation, Pictured Cliff Sandstone, Mesaverde Group, Dakota Sandstone, and "Chacra interval" (fig. 2). Field development has been pursued by Amoco Production Company, Union Texas Petroleum, and Meridian Oil, Inc., as shown in figure 3. With the large number of Fruitland penetrations, an accurate inventory of coal reserves is possible along with detailed stratigraphic and structural analysis. This field study focuses on Amoco's efforts because of the long history of their degasification program.

GEOLOGY

In spite of a lack of publicly available core information, a geologic investigation was conducted to investigate critical geologic factors responsible for the highly-productive basal Fruitland coal found in the Cedar Hill field. The lack of core data obviously makes our conclusions somewhat speculative. Should core information become publicly available, it would be desirable to try to relate to reservoir quality and productivity these geologic parameters: thermal history, maceral composition, reservoir lithotypes, coal-forming environments and their possible impacts upon gas generation, migration, and porosity and permeability development during basin evolution. Specific geologic factors defining gas-in-place, reservoir distribution, reservoir pressure, and enhanced permeability are presented.

Fruitland Coal Thickness and Continuity

Both an upper Fruitland coal zone and the basal Fruitland coal seam are present within the lower 200 feet of the Fruitland Formation (fig. 4). The basal coal seam is the primary target in the field and immediately overlies the Pictured Cliffs Sandstone. This basal coal seam is commonly split into several individual coal benches by shale partings, yet the seam is

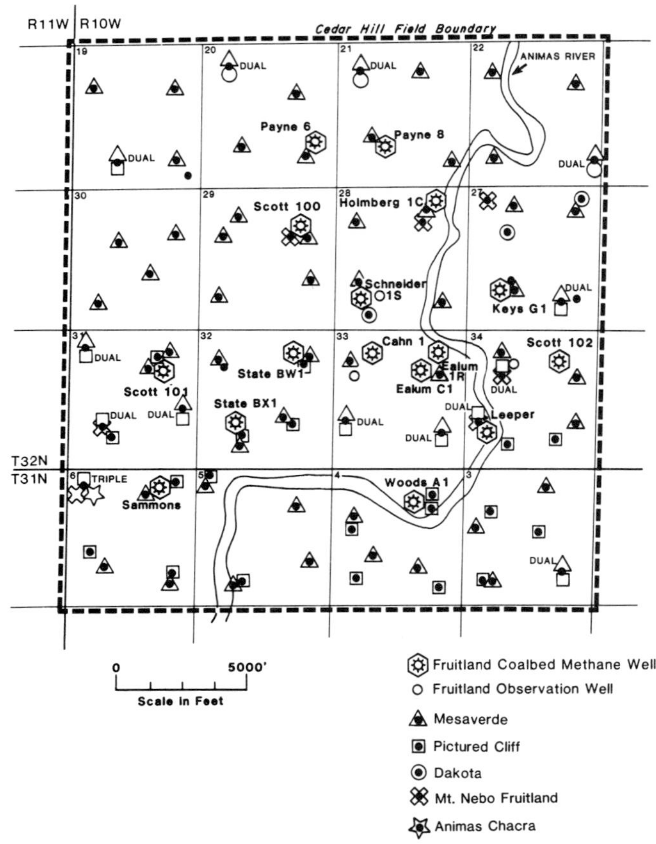

Figure 2. Production map of the Cedar Hill field area showing locations of all gas wells; the producing formation in each well is shown.

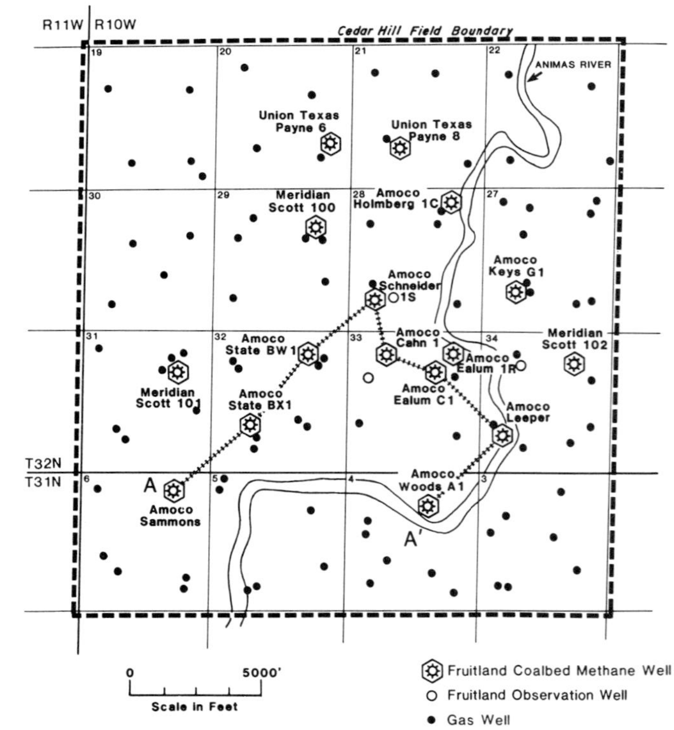

Figure 3. Base map of the Cedar Hill field locating Fruitland coal-bed methane production wells with operator's name. Line of cross section A-A' (figure 4) is shown.

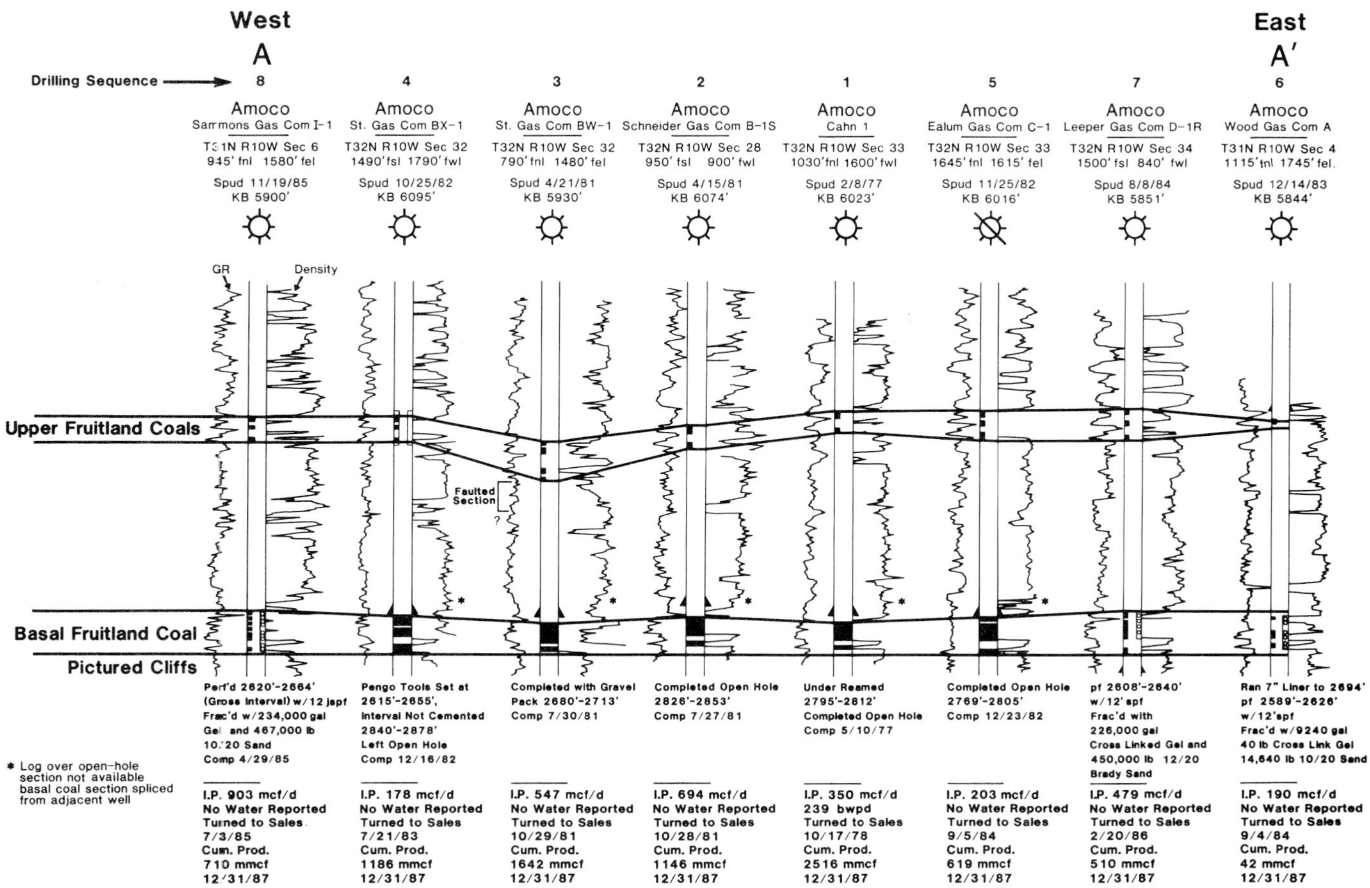

Figure 4. Annotated stratigraphic cross section for the Cedar Hill field including logistical information, completion, and production data. Each well has been correlated with respect to the top of the Pictured Cliffs Sandstone.

laterally continuous throughout the Cedar Hill field varying in thickness from 4 feet to 26 feet, as established by the 53 control points shown on figure 5. Geometry of the coal isopach suggests a northeast coal-forming trend with maximum coal thickness occurring within the central portion of the field. Interestingly, the Keys G1 with the thickest coal development and the Woods A2 having the thinnest coal development are the two poorest producing wells in the field. Apparently, there is not a simple relationship between coal thickness and productivity.

The upper Fruitland coal zone is a forty-foot thick interval of interbedded coal, shales, and sandstones in the upper part of the Fruitland Formation (fig. 4). Thin, multiple coal seams are typical of the coals in this interval. A total-coal isopach map of the Cedar Hill field (fig. 6) indicates that the upper Fruitland coal zone contains almost as much coal in-place as the basal Fruitland coal seam, although, density logs show that the coals in the upper zone have slightly higher density suggesting higher mineral (ash) content and thus lower gas contents. Amoco reportedly attempted a separate completion of the upper Fruitland coal in the State Gas Com B 1 well, however, the results of that test are not of public record. In Union Texas' Payne No. 6 and No. 8 wells, production is commingled from both coal zones, but the relative contribution of the individual zones is not known. The limited interest in completing the Upper Fruitland coals suggests that this zone may not be as attractive, in spite of the acknowledged gas-in-place, however, no firm evidence is publicly available to confirm this supposition.

Both the basal Fruitland coal seam and the upper Fruitland coal zone appear to be laterally continuous throughout the field area, yet gas production rates over the field area are not evenly distributed. This suggests that favorable coal-gas reservoir properties are not related simply to coal continuity or thickness.

Gas Content

The gas present in the coal is molecularly adsorbed on the coal's extensive surface area. Gas content measurements using standard gas desorption techniques typically consist of three components: lost gas, measured gas, and residual gas. Lost gas is the amount of gas desorbed from the coal during drilling, core or cuttings retrieval, and surface handling. Lost gas can be estimated using a number of different techniques, such as the Bureau of Mines or Williams and Smith (1984) methods. Measured gas is the amount of gas recovered once the core or cuttings have been placed in desorption canisters. After gas desorption is complete, the coal is crushed and any remaining gas liberated is considered residual gas. Residual gas is not thought to be recoverable, and therefore, is not commonly used for in-place gas determination. Generally, gas content data suffers from the lack of an industry standard for reporting and measuring gas-content data.

Reported gas contents for the Fruitland coal, as measured

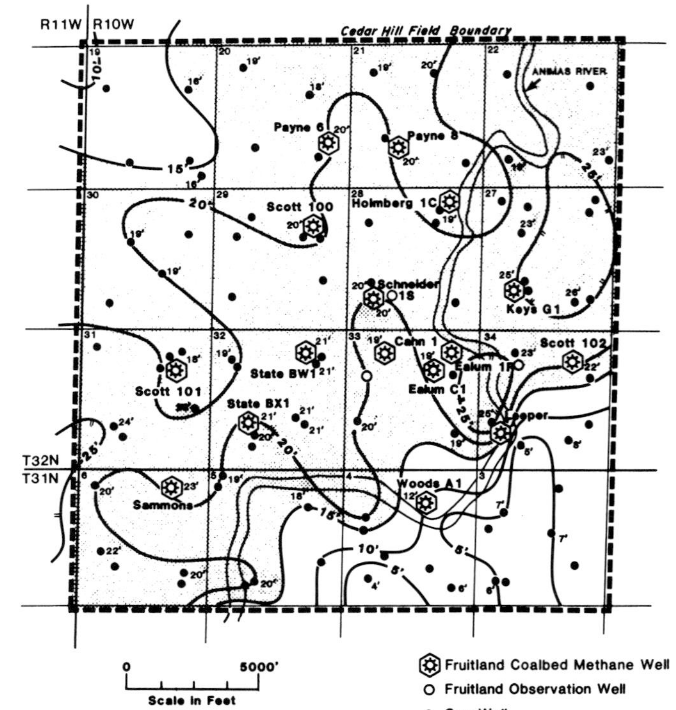

(Courtesy Amoco Production Co.)

Figure 5. Isopach map showing net coal thickness of the basal Fruitland coal seam. Common shale partings are not included in coal thicknesses. Coal thicknesses greater than 15 feet are shown by stippled pattern. (Modified from exhibit in NMOCD Case 8014.)

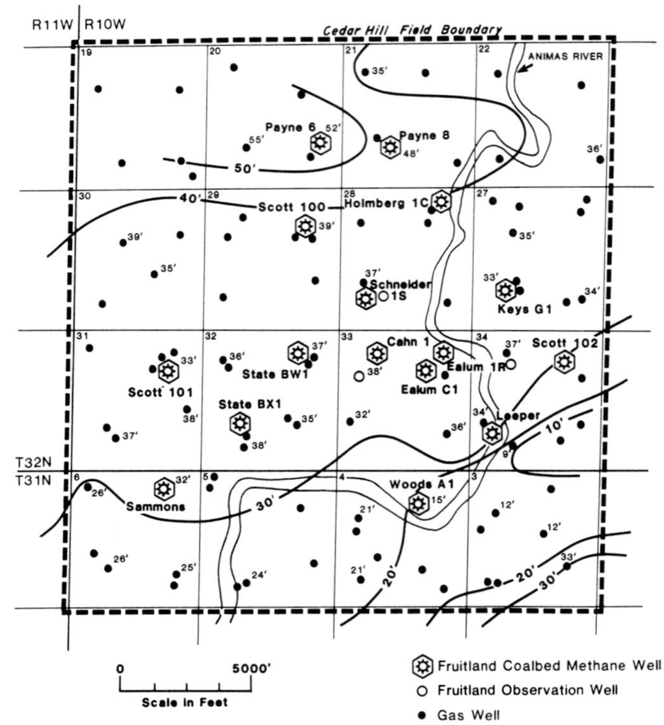

Figure 6. Isopach map of total Fruitland coal thickness in the Cedar Hill field. Coal thicknesses were determined from gamma ray and density logs. Individual coal seams less than 2 feet thick were not included in net coal thicknesses.

by Amoco using standard gas desorption techniques, are shown in table 1; an average value for the field is 284 SCF/ton. The method Amoco used to estimate lost-gas was not disclosed, however, the reported lost-gas volumes were generally less than amounts reported for other coal cores recovered from similar depths. No residual gas was reported. The gas contents reported by Amoco do appear to be consistent with the gas contents inferred from the work of Kim

Table 1. Gas content for Cedar Hill Field.

Well	Interval	Measured (scf/t)	Lost at surface (scf/t)	Total (as received) (scf/t)
Ealum B1	2785.7-2787.9	254	110	336
(core)	2783.5-2784.5	100	30	121
Cahn 1	2797.6-2800.5	233	35	267
(core)	2808.8-2810.2	325	31	330
Schneider B1S	2832-2834			245
(chips)	2836-2837			361
	2840-2842			376
	2847-2849			337
State BW1	2693-2694			292
(chips)	2696-2698			284
	2703-2704			284
	2709-2711			190
				avg. 284 scf/t

(1977) which describes the adsorptive capacity for coals of various ranks and for high volatile A bituminous coals as estimated from the coal rank maps of the San Juan Basin developed by Fassett and Hinds, 1971.

Gas In Place

The amount of coal-bed methane in place underlying the Cedar Hill field was estimated from the amount of coal reserves in place multiplied by the amount of gas present in the coal as estimated from Amoco's desorption measurements. Using coal volume calculations from the basal Fruitland coal isopach map (fig. 5), the average gas content from table 1, and average density for a high volatile bituminous coal, in-place gas was determined as shown below and listed on table 2:

$$\text{(coal volume)(coal density)(gas content)} = \text{gas in place}$$

In the Cedar Hill field, the basal coal seam contains an estimated 313 million tons of coal and 89 billion cubic feet of coal-bed methane in place.

Permeability

Coal itself is essentially impermeable. Cleats, the natural fractures which form as coal matures, provide the primary permeability network for desorbed gas to flow through the coal seam to the well. The degree to which these cleats are open and permeable depends on in situ stresses and coal compressibility (McKee and others, 1984, and this volume). Tectonic fracturing may provide additional flow paths and more importantly allow stress relief along structural flexures and

Table 2. In-place gas volume for basal Fruitland coal, Cedar Hill field.

Coal thickness	Coal resource (10'6 tons)	Gas resource (bscf)
25' - 20'	42.26	12.00
20' - 15'	171.56	48.72
15' - 10'	78.69	22.35
10' - 5'	16.68	4.74
5'	3.87	1.10
Totals	310.06	88.91
		5.26 bscf/mi 2

(or) shattered fault zones thereby contributing to areas of enhanced permeability.

A detailed structure contour map on the top of the Pictured Cliffs Sandstone (fig. 7) was constructed in an effort to identify permeability-enhancement features. Structurally, the Cedar Hill field is quite flat with a maximum of 80 feet of relief over a sixteen-square-mile area. Low relief and the high density of well control required structure contouring on a 20-foot interval to resolve structural elements of the field. Honoring all data, a complex, strike-oriented, northeast-trending series of parallel structural highs and lows terminate at a northwest-trending trough. Extensive fracturing may result from these low angle dip reversals, however, more realistically, the structural fabric may be dominated by high angle, low amplitude faults. Topography at the Cedar Hill field is dominated by the Animas River valley, Cox Canyon, and Ditch Canyon; features that may be the surface expressions of deeper fault trends. The sudden change in the course of the Animas River within central portions of the field may be surface expressions of subsurface faulting.

The wide range of gas and water production rates across the field seems to be related to areas of possibly-enhanced fracturing and stress relief. It is evident that the high-permeability areas have allowed effective and accelerated dewatering which has had a marked effect across a large portion of the field.

Classic Coal-Bed Methane Reservoir Behavior

The gas that is adsorbed onto the coal matrix is held there by reservoir pressure. The sorption isotherm, which describes the relationship between gas content and reservoir pressure, is quite non-linear which leads to two effects discussed below which are responsible for the classic coal-bed methane production profile; increasing gas production with decreasing water production.

For water-saturated coals, relative permeability changes help to restrict water production and improve gas flow as the coal seam is dewatered. The non-linear shape of the sorption isotherm requires, at the higher initial reservoir pressures, that a significant reduction in reservoir pressure occur before measurable amounts of gas are desorbed from the coal matrix. In a water-saturated reservoir, removal of water (dewatering) must occur to affect a reduction in reservoir pressure. As water is removed and the reservoir pressure declines, an increasing amount of gas is desorbed from the coal thereby generating a gas saturation in the cleats. As the gas saturation increases, the relative permeability to gas gradually increases and the

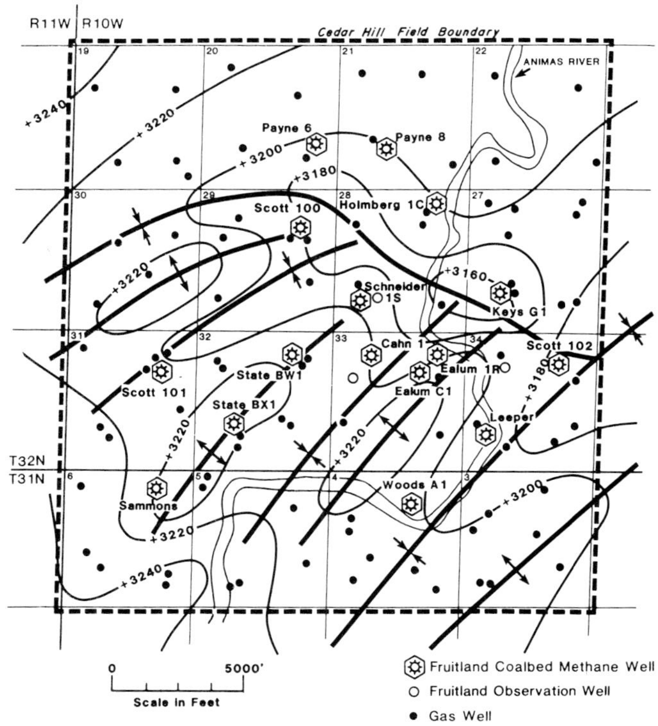

Figure 7. Structure contour map (feet above sea level). Datum is the top of the Pictured Cliffs Sandstone. Structural highs and lows may be related to faulting rather than folding.

relative permeability to water decreases. The shift in relative permeability results in a decreased water production and an increase in gas rates.

Because of the non-linear shape of the sorption isotherm, at lower reservoir pressures, ever-increasing amounts of gas will desorb for the same amount of pressure drop. Thus, as reservoir depletion occurs, the rate of gas desorption actually increases. This effect also contributes to the negative decline in gas production observed in most coal-bed methane wells.

Reservoir Pressure and Drainage

To quantify the performance of the coal reservoir, Amoco recompleted three non-commercial Pictured Cliffs wells to the basal Fruitland coal to monitor formation pressures (fig. 8). Amoco reported initial shut-in bottom-hole pressures and results from pressure transient tests from several wells as shown in table 3.

On May 10, 1977, the first producing well, the Cahn No. 1 was completed. Approximately one month after completion of the Cahn No. 1, two of the pressure monitoring wells, the Schneider B No. 1 and the Cahn No. 2, were recompleted. The data in the public record from the pressure-monitor wells begins in 1979, approximately two years after the Cahn No. 1 was completed. The lack of production and pressure data from the field from 1977 to 1979 hinders a quantitative reservoir history match.

The reservoir pressure measured at the Cahn No. 2 well in 1979 was 250 psi lower than the initial shut-in bottom hole pressure measured in the Cahn No. 1 well measured in 1977. The decrease in reservoir pressure suggests that production at the Cahn No. 1 caused significant pressure drawdown. As

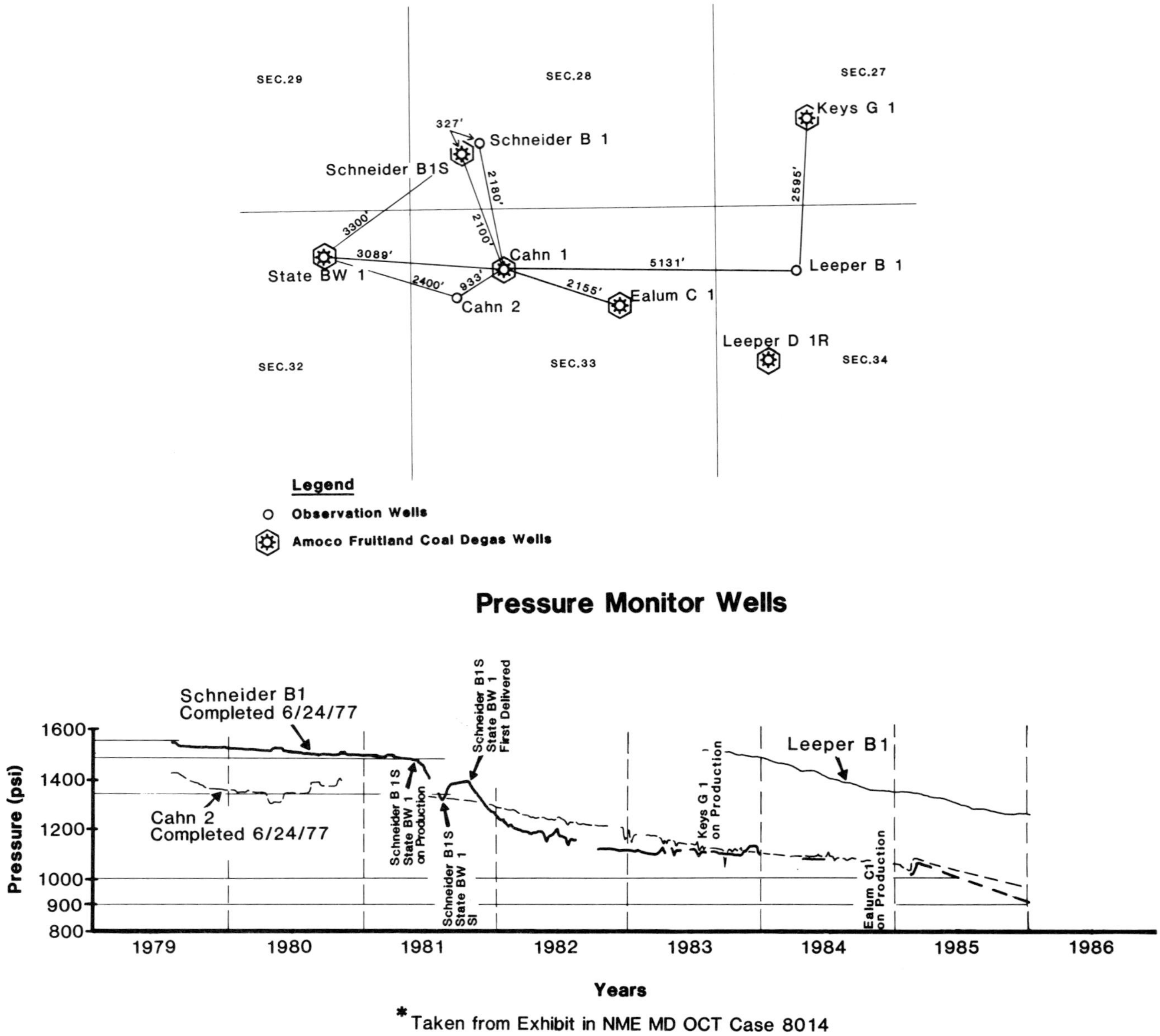

Figure 8. Map showing Amoco-established well pattern in Cedar Hill field and graph showing pressure changes in wells through time. Note that on wide spacing (Cahn No. 2 and Schneider B No. 1 wells) little interference has been observed over three years of production history. However, closely spaced wells (Cahn No. 1 and Cahn No. 2, Schneider B No. 1S and Schneider B No. 1) demonstrated rapid interference. The conclusions reached by Amoco and concurred with by NMOCD were: 1. 640-acre well spacing was inadequate for effective gas recovery, 2. 160-acre well spacing was considered over-developed, and 3. 320-acre well spacing was ideal for gas recovery.

production at the Cahn No. 1 continued, a constant decrease in reservoir pressure, as observed in the Cahn No. 2 and Schneider B No. 1 monitor wells, indicated a drainage radius at least 2,200 feet away from the wellbore. In effect, the single producing well significantly reduced reservoir pressure over a large area. The Cahn No. 1 apparently achieved widespread dewatering, as was evidenced in the low water production and high gas production from the wells drilled after the Cahn No. 1.

By the end of 1985, the Cahn No. 1, State Gas Com BW No. 1, and the State Gas Com BX No. 1 wells had resulted in a reduction of field reservoir pressure from 1,580 psi to 900 psi, yet combined gas production from the three wells continued to increase (figs. 9, 10), most likely the result of the strongly non-linear desorption behavior as discussed previously. Figures 11, 12, and 13 are production histories for the three wells studied.

Following the gas sales curtailments at the end of 1986, the Ealum C No. 1 and the Cahn No. 1 wells demonstrated higher water production rates and sharply reduced gas rates. Before the gas curtailments, both wells were flowing high volumes of gas and little water. Possible explanations included: Increased problems with coal fines, a redistribution of water within the highly per meable coal and an unfavorable shift in relative permeability, or instability of the open-hole completion. Amoco has drilled replacements for these two wells and they

Table 3. Cedar Hills field reservoir pressures.

Well name	Test zone	Date	Type of test	Reservoir pressure-psi
Cahn Gas Com No. 1 SW NE NW 33-32N-10W	Coal	5/77 ?	SI BHPBU	1562.00 1465.00
Schneider Gas Com "3" NE SW SW 28-32N-10W	No. 1S Coal	8/81	BHPBU	1362.00
State Gas Com "BW" No. 1 SW NW NE 32-32N-10W	Coal	12/82	BHPBU	1421.00
Keys Gas Com A2 NW SE NW 27-32N-10W	Coal	12/81	DST	1590.00
Keys Gas Com E1 SE NW NW 27-32N-10W	Fruitland	2/72	SI	1267.00
Holmberg Gas Com A1 SE NW NE 28-32N-10W	Fruitland	12/72	SI	1287.00
Scott No. 1 NW SE NE 29-32N-10W	Fruitland	5/52	DST (8)	1580.00 (Max)
Martinez Gas Com H1 NW SE SW 32-32N-10W	Pictured Cliffs	11/73	SI	1094.00
Valentine Gas Com B1 SW NE NE 32-32N-10W	Pictured Cliffs	12/76	SI	1040.00
Gardner Gas Com No. 1 SW NE SE 33-32N-10W	Pictured Cliffs	6/79	BHPBU	884.00
Leeper Gas Com C1 NE SE SW 34-32N-10W	Pictured Cliffs	3/77	SI	886.00
Scott No. 1 NW SE NE 29-32N-10W	Pictured Cliffs	5/52	DST	250.00

have been cased-hole completions. This suggests that a problem with the open-hole completions may have occurred and that substantial reserves remain to justify the cost of redrilling.

Water and Gas Geochemistry

Wells completed in the basal Fruitland coal seams appear to produce gas exclusively from that interval as established by water and gas analysis. Sampling of water and gas has proved to be a useful tool in identifying and quantifying the origin of produced gases and water from coal-bed methane wells.

Produced water was collected and analyzed from Fruitland sandstones, the Fruitland basal coal, and the Pictured Cliff Sandstone. The data have been segregated by anions and cations, normalized to milliequivalent/liter, and plotted on figure 14. Use of a Piper (tri-linear) diagram to determine the origin of produced water and the relative contribution of waters of a mixed origin was first utilized by Decker and others (1987). Conclusions from that work were that connate coal water is primarily a bicarbonate-rich water resulting from sulfate reduction occurring during early coal diagenesis; sandstone water is dominated by chlorides due to selective retention during post depositional, cross-formation fluid flow. Waters produced from the Fruitland basal coal and associated sandstones within the Cedar Hill field are consistent with these findings.

Water produced from the Fruitland coal completions are dominantly bicarbonate in anion composition and show little variation over nine years of production. High concentrations of total dissolved solids (TDS) in Fruitland coal water at the Cedar Hill field suggest that a sluggish hydrologic system was present throughout geologic time. Similar coal-water chemistries have been observed in other water saturated, overpressured, coal-bed methane fields. It is also interesting to note that water recovered from the Valentine B No. 1 well, a Pictured Cliffs Sandstone producer, displays a mixed origin. Production from both coal and sandstone is further supported by the gas composition of the produced gas.

Gas analyses from samples collected from Fruitland sandstones, the basal Fruitland coal, and the Pictured Cliffs Sandstone are shown in table 4. Also included are tabulated analyses of desorbed gas taken from the desorption canisters containing basal Fruitland coal cores and cuttings. Cedar Hill basal Fruitland coal gas can be distinguished from gas from other formations by: (1) C2+ less than 1 percent of total gas composition, and (2) CO_2 greater than 4 percent of total gas composition.

During nine years of production, the Cahn No. 1 well has not shown any changes in gas composition.

Drilling and Completion Methods

The Cahn No. 1, the discovery well for the Cedar Hill field, was drilled in February, 1977 specifically for coal-bed methane. Since that time, the Cedar Hill field has been developed with 16 production wells and 3 pressure monitor wells. An annotated cross-section (fig. 4) illustrates stratigraphy,

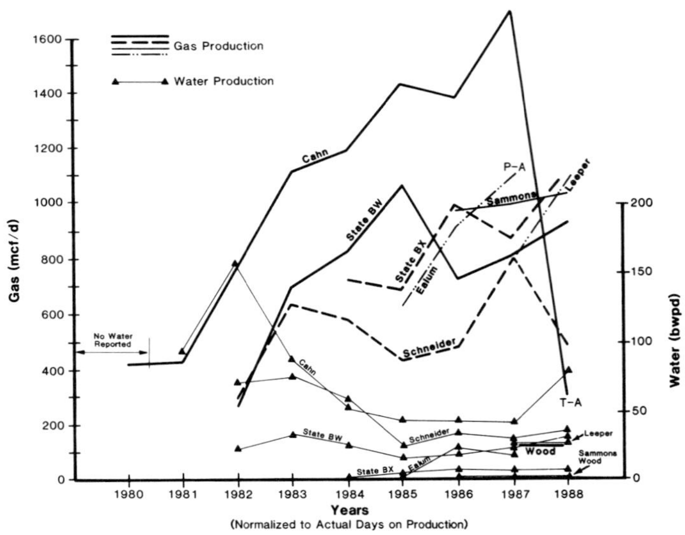

Figure 9. Graph of the average daily water and gas production per well over the first nine years of production history at the Cedar Hill field. By comparing gas and water production rates during 1979 through 1985 of the Cahn No. 2, State BW, and Schneider wells with reservoir pressure, on figure 8, the following general conclusion may be made: Gas rates increased and water rates decreased in response to declining reservoir pressure and change in relative permeability. (Note the increased water-production rates and the sharply reduced gas-production rates following production curtailment, December, 1986.)

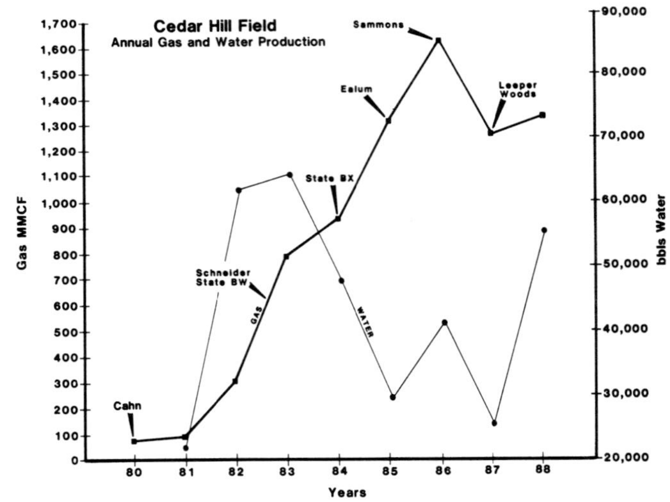

Figure 10. Graph of annual gas and water production for the Cedar Hill field. This plot shows an increase in gas production and decrease in water production through 1986. Gas curtailments during 1987 resulted in decreasing gas production and an increase in water production.

completion methods, and the sequential development of the field. The first five wells, the Cahn No. 1, Schneider B No. 1S, State BW No. 1, State BX No. 1, Keys G No. 1, and the Ealum C No. 1, were drilled and cased above the top of the basal coal. These wells were then deepened to expose the basal Fruitland coal without penetrating the underlying Pictured Cliffs Sandstone. To remove damage caused by drilling mud and coal fines, the open-hole section in the first three wells was under-reamed. It is suspected that problems with borehole stability and production of coal fines in several of the open-hole completions led to the installation of gravel packs and pre-packed screened liners. After several of these gravel packs were removed production increased, suggesting that gravel packs may plug with fines. More recent open-hole completions have been attempted by Meridian Oil Inc. on the Scott 100, 101, and 102 wells in which the coal zone was drilled with air and completed open hole with the installation of pre-drilled liners to provide wellbore stability.

The remainder of the Cedar Hill wells have been completed with casing and cement set through the coals. The coal intervals were then perforated and fracture stimulated. Lacking pre-stimulation and post-stimulation productions tests, conclusions regarding the most effective completion techniques are tentative. Some general observations are as follows:

1. Open-hole completions may be effective in coals which have adequate permeability.
2. Open-hole completions present problems with wellbore instability and coal fines production.
3. Gravel packs, while providing a stable wellbore environment for open-hole completions, may plug with coal fines. Predrilled liners may not plug as easily.
4. Cemented, cased holes that are perforated and fracture stimulated with large 10/20 mesh proppant appears to be Amoco's preferred completion technique. Cased-hole completions potentially offer several advantages: Stable wellbore environment, ability to complete multiple coal seams, provide a rathole for fines accumulation, and allows the placement of pumps in a sump below the coal seam.

Reservoir Properties

Reservoir properties for the Cedar Hill field were estimated using a single-well reservoir simulator by matching historical gas and water production from three wells. The Schneider, State BW, and Ealum wells were selected because of their close proximity to the original Cahn well in the field and their definable drainage area.

The three wells were history matched using COMET, a finite-difference reservoir simulator. A sample match of gas and water production rates for the Schneider well is shown in figures 15, 16, and 17. A good match was made of daily gas and water production rates over the 70 months for which production data were available.

The matching reservoir parameters for the three wells are listed as follows:

Well	Permeability	Porosity	Initial water saturation
	(md)	(%)	(%)
Schneider	50	2	93
State BW	36	1	90
Ealum	75	1	90

The significance of these values is discussed below:

High Permeability

The relatively high permeabilities for coal at depths of

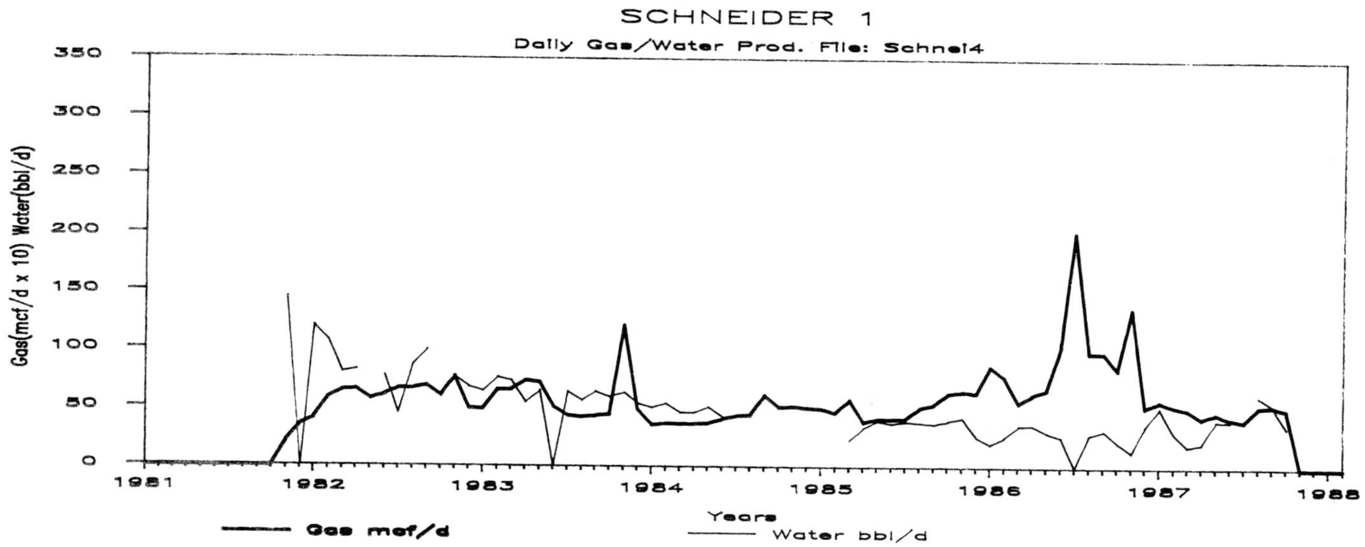

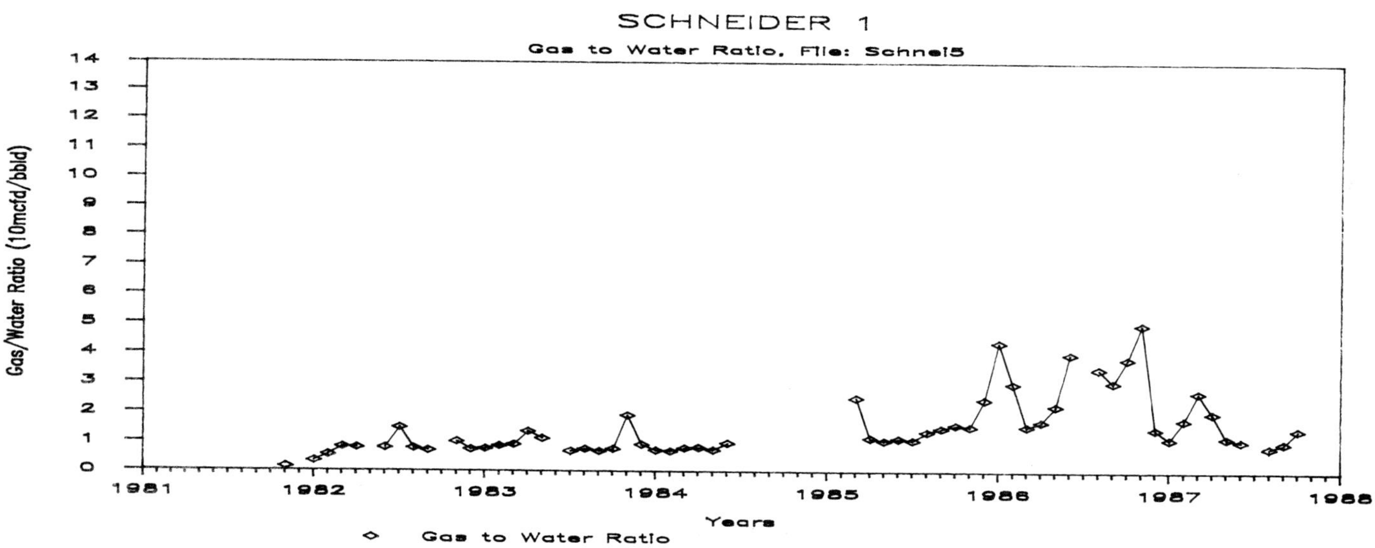

Figure 11. Normalized plot of gas and water production for the Schneider B No. 1S well from 1982 through 1988. Note the increase in the gas/water ratio in 1986 and a reversal in the trend for 1987. During 1987 a decrease in gas sales resulted in an increase in water production.

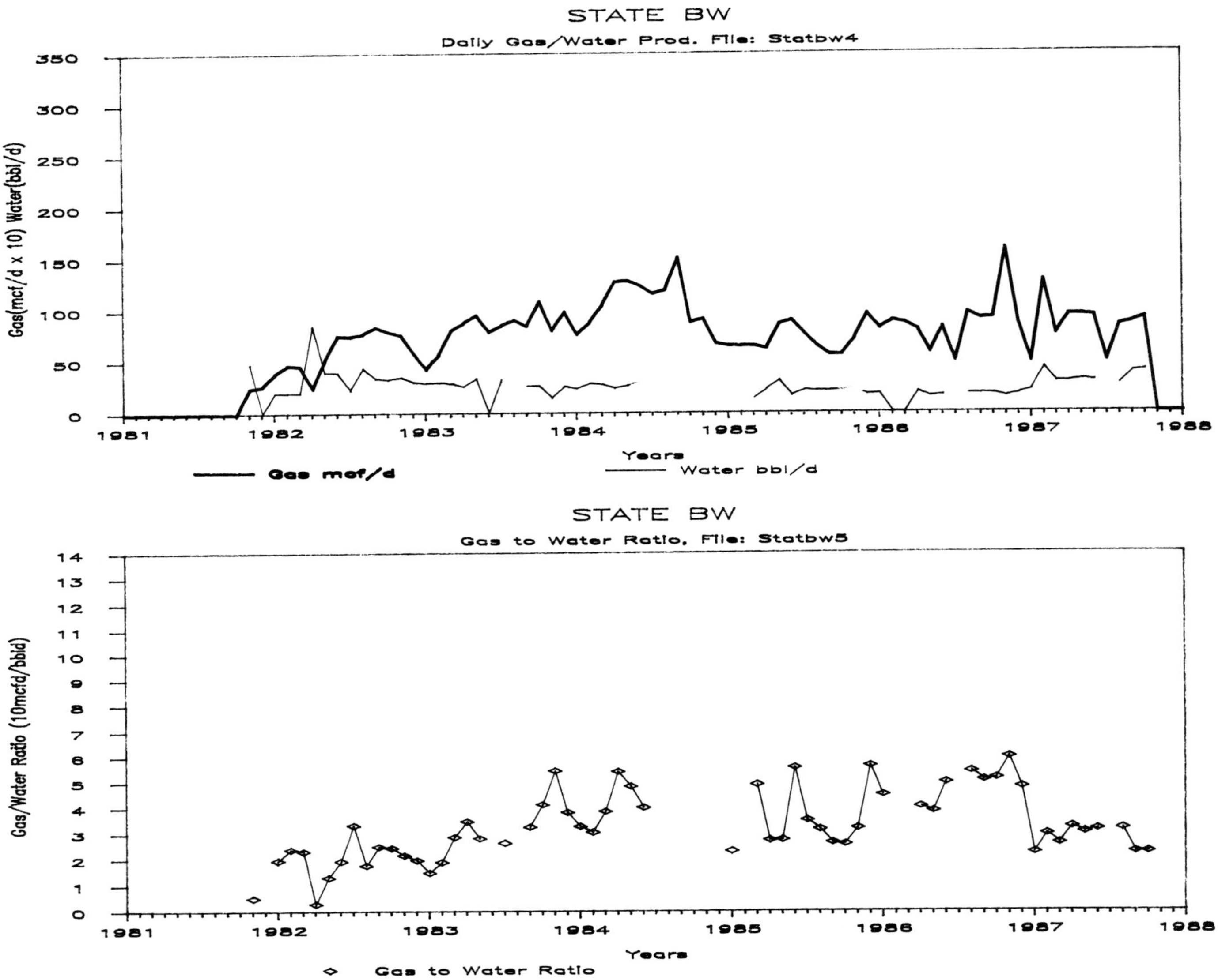

Figure 12. Normalized plot of gas and water production from the State BW No. 1 well from 1982 through 1988. Note the increase in the gas/water ratio in 1986 and a reversal in the trend for 1987. During 1987 a decrease in gas sales resulted in an increase in water production.

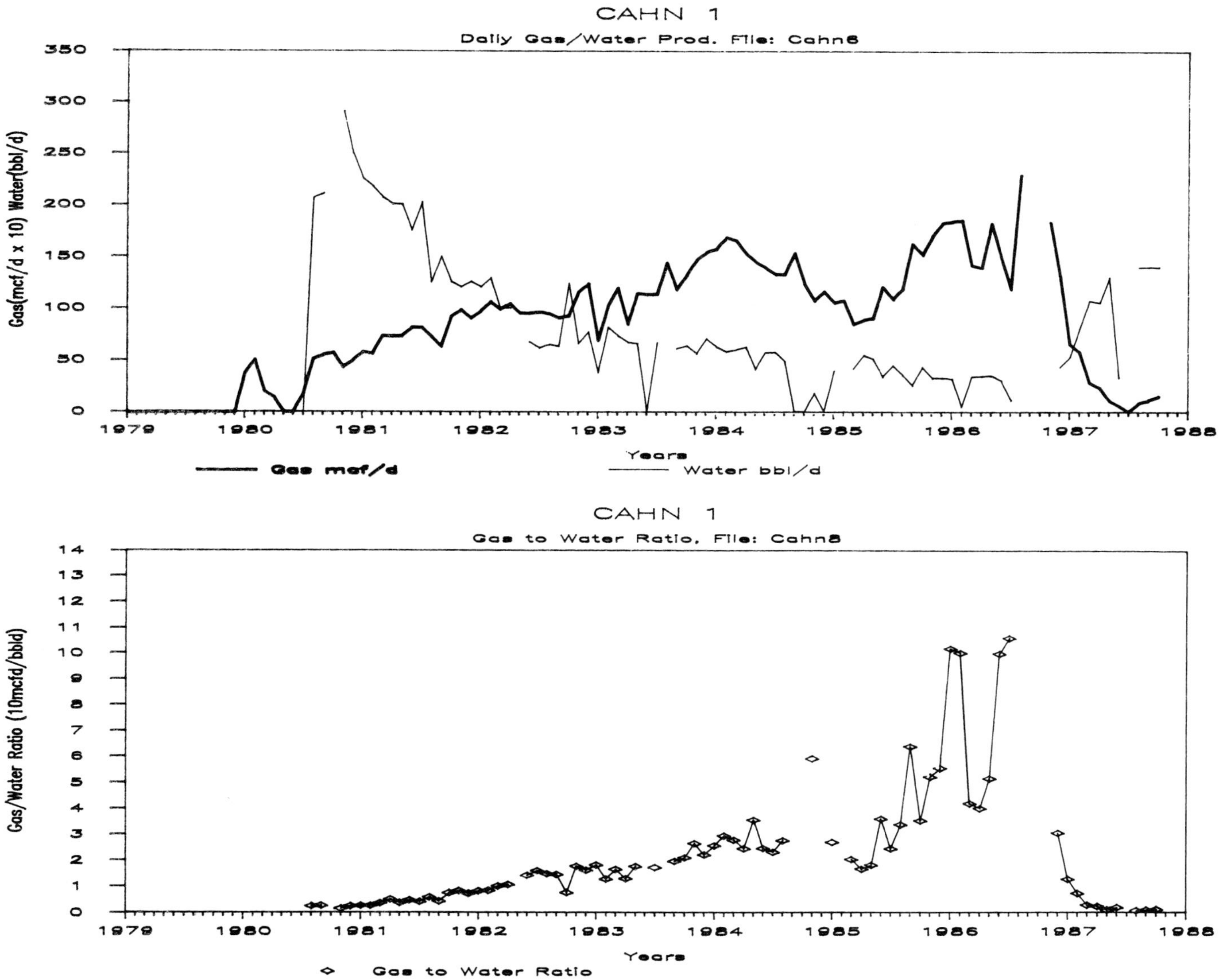

Figure 13. Normalized plot of gas and water production for the Cahn No. 1 well from 1979 through 1988. Note the increase in the gas/water ratio in 1986 and a reversal in the trend for 1987. During 1987 a decrease in gas sales resulted in an increase in water production.

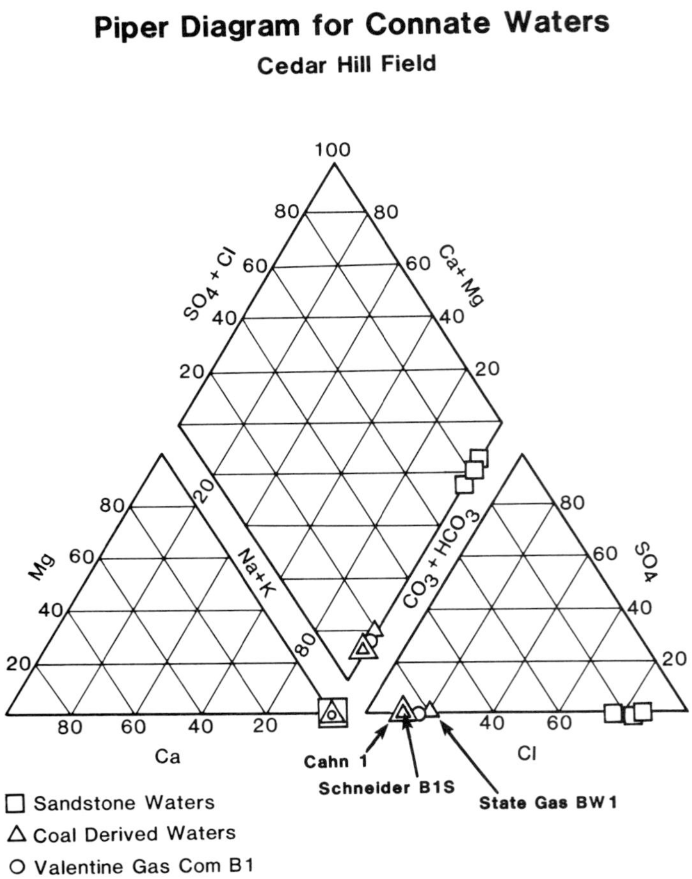

Figure 14. **Piper diagram plotting produced-water chemistry from Fruitland and Pictured Cliffs wells versus Fruitland coal wells. Note clear separation of sandstone-derived, chloride-rich water from a bicarbonate-rich coal water. Based on this plotting technique, produced water from the Valentine Gas Company B No. 1 well is of a mixed origin indicating commingling of fluids from sandstone and coal reservoirs.**

approximately 2,800 feet indicate that the Cedar Hill field is in an area of enhanced permeability. This may be due to low effective net stress as discussed earlier. These permeabilities are largely responsible for the high production rates exhibited by these wells.

Low Porosity

The effective porosities of the coal cleat system are fairly low (at 1-2 percent) for coal-bed methane wells in general, but are consistent with values found elsewhere in the San Juan Basin. Lower porosities generally result in shorter dewatering periods.

Initial Water Saturation

The initial water saturations were found to be less than 100 percent. These values were thought to be reasonable in light of the fact that the Cahn well had been producing for two years prior to the time that the Schneider or State BW wells came on line. These two wells came on line at approximately the same time, although there was a 3-percent difference in initial water saturations. This is probably due to the fact that the Schneider well lies down dip from the Cahn well while the State BW well is up dip. Therefore, it is to be expected that the area up dip would have been dewatered to a greater extent. The fact that these areas had less than 100 percent water saturation indicates dewatering support provided by the Cahn No.1 well which allowed them to produce gas almost immediately with much lower water production than observed at the Cahn No. 1 well.

These parameters could be investigated further once a three-dimensional, multi-well, finite-difference simulator for coal-bed methane has been created. Such a multi-well model would allow for the gravity effects of dewatering to be included and would provide a more accurate estimation of the matched parameters by modeling the initial conditions of each of the wells in a single, field-wide simulation.

Field Rules

On July 9, 1984 the State of New Mexico Energy and Minerals Department, Oil Conservation Division issued Order No. R-7588 in Case No. 8014, effective February 1, 1984, creating and designating the Cedar Hill-Fruitland Basal Coal Pool, with vertical limits comprising the basal coal bed of the Fruitland Formation as found from approximately 2,795 feet to 2,828 feet on the type log of the Amoco Production Company Schneider Gas Com "B" No. 1 well, SW1/4SW1/4, sec. 28, T. 32 N., R. 10 W., NMPM, San Juan County, New Mexico; with horizontal limits consisting of the following described lands: secs. 3-6 (all), T. 31 N., R. 10 W., and secs. 19-22 and 27-34 (all), T. 32 N., R.10 W., NMPM.

Order No. R-7583 in Case No. 8014 established, among other things, 320-acre standard spacing or proration units comprising any two contiguous quarter sections of a single section. The Order also requires the first well drilled or recompleted on every 320-acre unit in the Cedar Hill-Fruitland Basal Coal Pool to be located in the NE1/4 or SW1/4 of a single section, not closer than 790 feet to any outer boundary of the tract or closer than 130 feet to any quarter-quarter section inner boundary. Case No. 8014 was reopened on February 19, 1986 and Order No. R-7588-A was issued March 7, 1986, continuing Order No. R-7588 in full force and effect until further order of the Division.

NGPA Determination

Wells completed in the Cedar Hill-Fruitland Basal Coal Pool (upon application to the New Mexico Energy and Minerals Department, Oil Conservation Division and presentation of evidence at the Commission Hearing) qualify as coal-seam gas wells for purposes of Section 107, High Cost, Occluded Natural Gas Produced From Coal Seams, under the Natural Gas Policy Act of 1978.

SUMMARY

A thorough search and review of public records has brought to light important geologic, geochemical, reservoir, and completion data; this information allowed for an integrated analysis of the Cedar Hill field. Lack of core information and certain data during early stages of field evolution complicates interpretations but certain general conclusions can be reached.

The volume of gas contained within the basal Fruitland coal at the Cedar Hill field is small relative to other portions of the basin. However, commercial production has been achieved because of unusually high permeability present which should allow most of the in-place gas to be recovered by the existing

Table 4. Analyses of gas produced from three different reservoirs in the Cedar Hill gas field.

Well name	Date tested	Nitrogen	Carbon dioxide	Methane	Ethane	Propane	Isobutane	Normal butane	Pentane	Hexane +	Btu	Gr
				Basal Fruitland coal								
Cahn 1	2798-2801*		10.35	89.10	0.55						910.00	
	2809-2810*		9.43	90.04	0.55						919.00	
	3/77	0.00	7.33	92.32	0.35	0.00	0.00	0.00	0.00	0.00	939.00	0.6262
	7/79	1.38	6.39	92.02	0.20	0.00	0.00	0.00	0.00	0.00	934.00	0.6225
	9/80	0.05	5.68	93.62	0.48	0.12	0.03	0.01	0.00	0.01	963.00	0.6140
	8/82	0.00	6.39	93.30	0.29	0.00	0.00	0.00	0.00	0.00	947.00	0.6168
	8/82	0.00	6.38	93.38	0.22	0.00	0.00	0.00	0.00	0.00	947.00	0.6164
	9/82	0.43	5.30	93.16	0.67	0.25	0.05	0.05	0.03	0.05	971.00	0.6170
	11/82	0.00	6.63	93.17	0.16	0.04	0.00	0.00	0.00	0.00	945.00	0.6190
	11/82	0.14	6.37	93.30	0.16	0.03	0.00	0.00	0.00	0.00	946.00	0.6170
	11/82	0.00	6.33	93.37	0.18	0.03	0.00	0.00	0.00	0.00	947.00	0.6157
	8/86	0.13	5.38	93.91	0.30	0.15	0.03	0.03	0.01	0.06	967.00	0.6130
State BW	12/81	0.01	5.82	93.68	0.21	0.02	0.00	0.00	0.00	0.00	955.00	0.6128
	12/81	0.00	5.82	93.75	0.20	0.03	0.00	0.00	0.00	0.00	951.00	0.6104
	8/82	0.02	5.84	93.95	0.18	0.01	0.00	0.00	0.00	0.00	952.00	0.6113
	8/82	0.00	5.87	93.91	0.22	0.00	0.00	0.00	0.00	0.00	952.00	0.6116
	11/82	0.00	6.09	93.71	0.15	0.05	0.00	0.00	0.00	0.00	950.00	0.6138
	11/82	0.00	6.11	93.68	0.13	0.03	0.00	0.00	0.00	0.00	950.00	0.6140
	11/82	0.36	5.96	93.55	0.00	0.07	0.00	0.00	0.00	0.00	947.00	0.6133
Schneider B-1S	*	0.10	4.39	95.08	0.31	0.10	0.00	0.05				
	12/81	0.00	4.76	94.99	0.23	0.02	0.00	0.00	0.00	0.00	964.00	0.6011
	12/81	0.00	4.83	94.92	0.22	0.03	0.00	0.00	0.00	0.00	963.00	0.6018
	8/82	0.06	4.95	94.78	0.02	0.00	0.00	0.00	0.00	0.00	961.00	0.6028
	9/82	0.13	4.45	93.29	1.61	0.19	0.08	0.08	0.05	0.12	994.00	0.6150
	11/82	0.00	5.45	94.33	0.16	0.05	0.0	0.00	0.00	0.00	957.00	0.6078
	11/82	0.00	5.23	94.56	0.17	0.04	0.00	0.00	0.00	0.00	959.00	0.6055
	11/82	0.00	4.87	94.91	0.17	0.05		0.00	0.00	0.00	963.00	0.6022
Ealom CL	11/85	0.02	5.96	93.73	0.22	0.04				0.03	950.00	0.6130
State BX	11/85	0.01	7.00	92.71	0.24	0.04	0.00	0.00	0.00	0.00	939.00	0.6220
Leeper	11/85	0.02	6.99	92.71	0.23	0.04	0.00	0.00	0.00	0.01	939.00	0.6230
				Pictured Cliffs sandstone								
Sammons G No. 1	11/85	0.14	1.56	86.36	7.15	2.65	0.46	0.68	0.22	0.64	1.15	0.6710
Martinez Gas Com A1	9/77	0.15	1.37	89.07	5.83	2.12	0.34	0.54	0.32	0.26	1115.00	0.6440
NW SE SW 32-32N-10W	12/81	0.17	1.52	86.10	6.57	2.53	0.32	0.63	0.35	1.72	1184.00	0.6930
	11/82	0.33	1.70	87.99	5.74	2.20	0.32	0.57	0.30	0.85	1130.00	0.6625
Gardner Gas Com No. 1	12/81	0.21	1.74	89.44	4.75	1.60	0.11	0.48	0.27	1.10	1114.00	0.6524
SW NE SE 33-32N-10W	9/82	0.22	1.78	90.02	4.86	1.89	0.30	0.44	0.24	0.25	1095.00	0.6370
Valentine Gas Com B	9/79	0.12	5.91	93.70	0.00	0.07	0.00	0.00	0.00	0.03	950.00	0.6122
SW NE NE 32-32N-10W	12/81	0.25	5.77	93.26	0.42	0.09	0.03	0.04	0.04	0.10	964.00	0.6190
	9/82	0.20	5.56	93.21	0.54	0.28	0.05	0.07	0.04	0.05	971.00	0.6190
Leeper Gas Com C1 NE SE SW 34-32N-10W	10/80	0.15	1.49	90.16	5.09	1.93	0.33	0.42	0.25	0.18	1098.00	0.6340
				Fruitland Formation sandstone								
Keys Gas Com E1	3/77	0.09	1.09	95.06	1.54	0.82	0.23	0.46	0.48	0.23	1066.00	0.6070
NW NW 27-32N-10W	12/81	0.03	0.77	93.63	1.51	0.97	0.25	0.63	0.86	1.22	1125.00	0.6413
	12/81	0.00	0.80	93.36	1.57	0.90	0.08	0.68	0.99	1.62	1143.00	0.6531
	6/82	0.03	1.15	94.66	1.81	0.93	0.24	0.46	0.55	0.17	1070.00	0.6090
Holmberg Gas Com A1	12/81	0.00	0.18	92.85	2.11	1.49	0.16	0.76	0.69	1.76	1162.00	0.6555
NW NE 28-32N-10W	12/81	0.03	0.14	93.23	2.09	1.65	0.44	0.73	0.60	0.92	1131.00	0.6352
	9/82	0.03	0.19	94.48	2.11	1.62	0.45	0.54	0.39	0.19	1096.00	0.6100

*Gas analysis from basal fruitland coal gas desorption canister

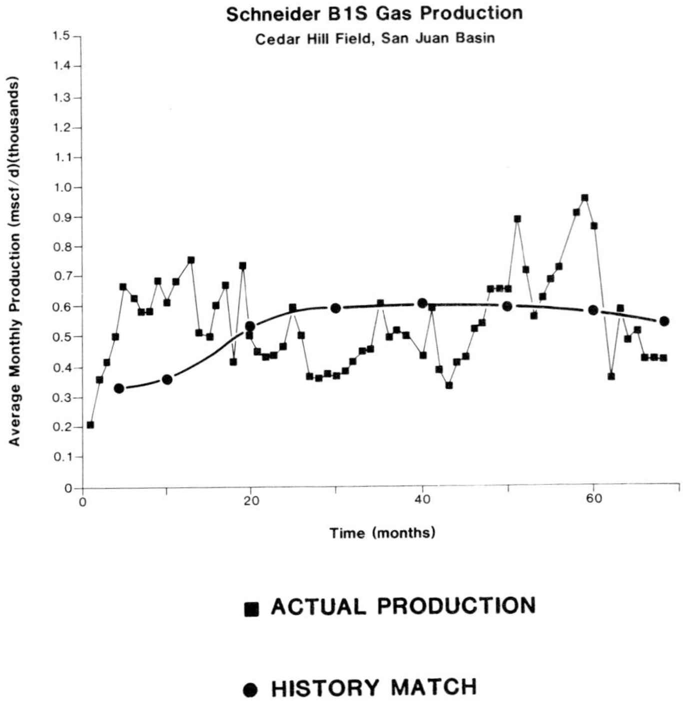

Figure 15. Graph showing comparison of actual average monthly gas production versus COMET model history match for Schneider B No. 1S well.

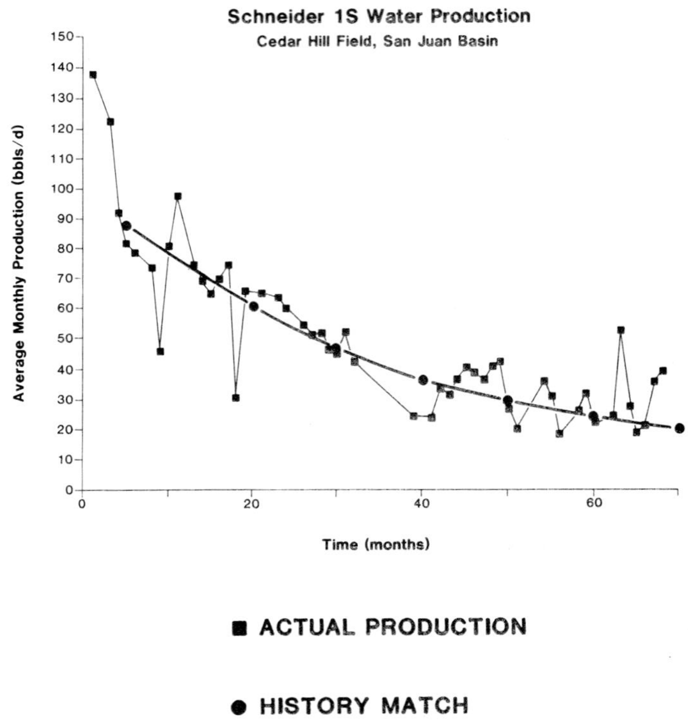

Figure 16. Graph comparing the actual average monthly water production versus COMET model history match for the Schneider B No. 1S well.

wells. There is no supporting data to relate production levels to completion practices or, separately, coal thickness. Rather, well deliverability is probably tied to structural influence that relieves in situ stress and maintains higher cleat permeability in the coals.

A consequence of shutting-in coal-bed methane wells, especially those completed open hole, may be increased water production, hole instability, and reduced gas production as evidenced by the production history of the Cahn and Ealum wells.

The understanding of geologic processes leading to the reservoir conditions found at the Cedar Hill field and the subsequent development of reliable predictive methods will hopefully result in the finding of additional, commercial coal-bed methane fields.

REFERENCES CITED

Decker, A. D., Klusman, R., and Horner, D.M., 1987, Geochemical techniques applied to the identification and disposal of connate coal water, *in* Proceedings of the 1987 Coalbed Methane Symposium, November 16-19, 1987, The University of Alabama, Tuscaloosa, Alabama: Gas Research Institute, The University of Alabama, and the U.S. Department of Labor, sponsors, p. 229-242.

Fassett, J. E., and Hinds, J.S., 1971, Geology and fuel resources of the Fruitland Formation and Kirtland Shale of the San Juan Basin, New Mexico and Colorado: U.S. Geological Survey Professional Paper 676, 76 p.

Kelso, B. S., Wicks, D. E., and Horner, D. M., 1987, GRI geologic and economic appraisal of coalbed methane in the San Juan Basin, in Proceedings of the 1987 Coalbed Methane Symposium, November 16-19, 1987, The Universi-

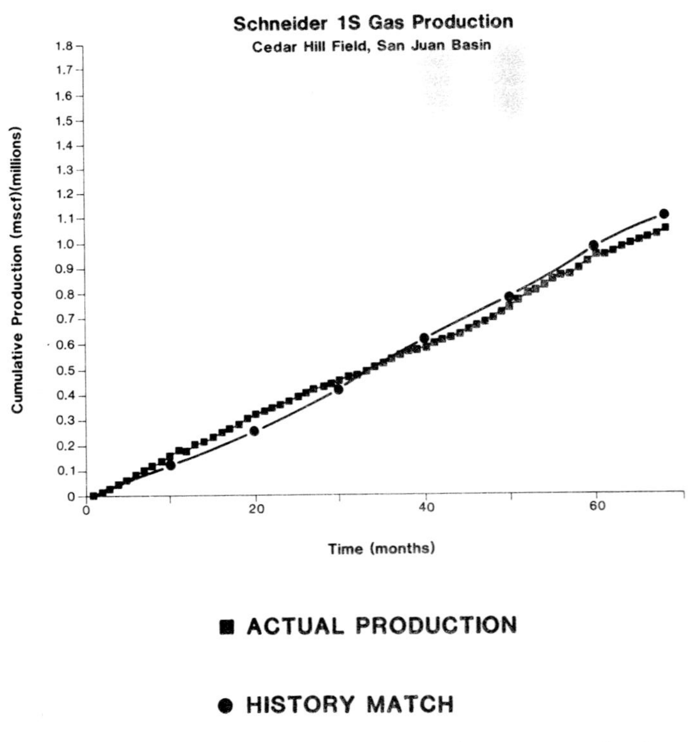

Figure 17. Graph comparing actual cumulative gas production versus COMET model history match for the Schneider B No. 1S well.

ty of Alabama, Tuscaloosa, Alabama: Gas Research Institute, The University of Alabama, and the U.S. Department of Labor, sponsors, p. 119-130.

Kelso, B. S., and Wicks, D. E., 1988, A geologic analysis of the Fruitland Formation coal and coal-bed methane resources of the San Juan Basin, southwestern Colorado and northwestern New Mexico, *in* Fassett, J. E., ed., Geology and coal-bed methane resources of the northern San Juan Basin, Colorado and New Mexico: Rocky Mountain Association of Geologists Guidebook.

Kim, A. G., 1977, Estimating methane content of bituminous coal beds from adsorption data: U. S. Bureau of Mines, Report of Investigations 8245, 22 p.

McKee, C. R., Bumb, A. C., and Bell, G. J., 1984, Effects of stress-dependent permeability on methane production from deep coal seams: Society of Petroleum Engineers SPE/DOE/GRI Paper No. 12858

McKee, C. R., Bumb, A. C., and Koenig, R. A, 1988, Stress-dependent permeability and porosity of coal, *in* Fassett, J. E., ed., Geology and coal-bed methane resources of the northern San Juan Basin, Colorado and New Mexico: Rocky Mountain Association of Geologists Guidebook.

Williams, F., and Smith, D. M., 1984, Lost gas determination method: Fuel, v. 63, No. 2, p. 251.

Outcropping Jurassic Entrada Sandstone is the prominent resistant massive bed in the center of the photo. This sandstone formation is utilized for disposal of produced Fruitland Formation coal waters in parts of the San Juan Basin. The foreground valley contains the Animas River with its meanders. The visible rock units range from Pennsylvanian to Cretaceous in age with the Dakota Sandstone capping the slope dipping to the south (left) in the center. La Plata Mountains background. Photo by Bruce Conrad.

White (left) and Theodosia Frost in the late 1860s, approximate age, 30s. Their dress is of traditional style and the suit White wears is one of many he owned. Mr. Frost served as Chief of Police for two years. The badge he wears was issued by the B.I.A. representing his position as an officer.

White and his wife lived in a tepee north of Ignacio, Colorado. They had numerous children with several who died in infancy and three known to survive. Mr. and Mrs. Frost died of an unknown disease which spread through the Utes in the late 1800s. The Frost family descendents include Clement Frost who served as Tribal Council Vice Chairman and Darlene Vigil, registration manager for this Guidebook Symposium. Photo courtesy of the Colorado Historical Society.

A Study of Coal-Bed Methane Production Trends as Related to Geologic Features, Warrior Basin, Alabama

F.H. Briscoe, B.S. Camp, L.K. Lottman and P.G. Malone[1]

ABSTRACT

Coal-bed methane has been produced in various parts of the Warrior Basin of Alabama since the late 1970's. Because there has been a significant amount of data available over a period of several years within this basin, it is possible to determine the relationship between geologic features and coal-bed methane production. A detailed case study was conducted in the Oak Grove coal degasification field, which consisted of mapping surface joints, identifying linear features from remote sensing data, and evaluating production data from the field. This paper briefly summarizes the results of that study and reveals that zones of high gas and water production are directly affected by geologic structure. The dip of the beds affects the manner in which several wells act collectively to dewater and therefore degas the coal seam. Joints mapped on the surface in the Oak Grove field and observed underground in the adjacent Oak Grove mine are thought to enhance production by increasing gas contributions from adjacent seams and (or) increasing the permeability of the coal seam. This conclusion is supported by recent field mapping and evaluation of other coal-bed methane fields throughout the Warrior Basin. Assuming other factors (depth, gas content, coal thickness) are constant, significant increases in gas production can be achieved by locating coal-degas wells on fracture systems defined by a combination of field mapping and remote sensing. A joint classification system was devised to interpret the affect of fracture patterns on production. Fractures allow conductivity both horizontally between wells and vertically between coal seams. Knowledge of all local structures, including dip of beds, folding, faulting, and joints can be used to positively influence dewatering of wells and optimize spacing of wells, thus increasing production. Based on Oak Grove production data, an alternative method of determining recoverable methane was also developed.

INTRODUCTION

Coal has unique physical and chemical properties which make conventional production methods for gas-sandstone reservoirs ineffective. In coal-bed methane, the coal seam itself serves as both the source rock and the reservoir. Many coal-bed methane reservoirs are found at pressures predicted by their depth from the surface (Rightmire, 1984). Much of the methane is adsorbed on the surface of the coal and held by reservoir pressure, usually the hydrostatic head of ground water. Dewatering causes a pressure decline, allowing the methane to desorb from the coal surface. This liberated methane can then flow through the pores and fractures of the coal. As dewatering continues, the rate of gas production generally increases, in contrast to production from sandstone reservoirs, in which production normally declines with time.

The coal seams in the Warrior Basin contain both water and gas. The removal of the water is a primary factor in the commercial recovery of methane. The initial amount of water which must be removed is directly related to certain geologic features such as natural fractures and dip of beds. Since these features allow communication vertically and horizontally, dewatering of coal-bed degas wells is enhanced by drilling multiple wells. Assuming all other factors are equal, multi-well fields yield higher initial gas production rates than single wells, since single wells do not lower coal-bed pressure or water saturation as efficiently or as rapidly. Placement of multiple wells at optimum spacing creates an interference pattern which efficiently dewaters the field and increases gas production (Hewitt, 1984). Data on gas and water production for the Oak Grove mine are extensive enough to study trends and their relationships to the geologic structure, which is well known in this area. Trends for the Oak Grove mine and Warrior Basin may have applications in other coal basins.

Although engineering and hydraulic fracturing techniques play key roles in the development of coal-bed methane wells, understanding of geologic features is critical to the development of an economically successful field. Upon completion of studies conducted in the Warrior Basin, it was determined that the geologic features which most influence methane and water production are structural in nature, specifically dip of beds and fracture zones. Numerous other geologic factors were considered, including original methane content, in-place reserves per well, coal thickness, coal cleat, and length of production. The original in-place reserves affect the upper and lower limits of the ultimate gas production but are not as influential in the

[1] GeoMet, Inc.
6200 Flintridge Road
Fairfield, Alabama 35064

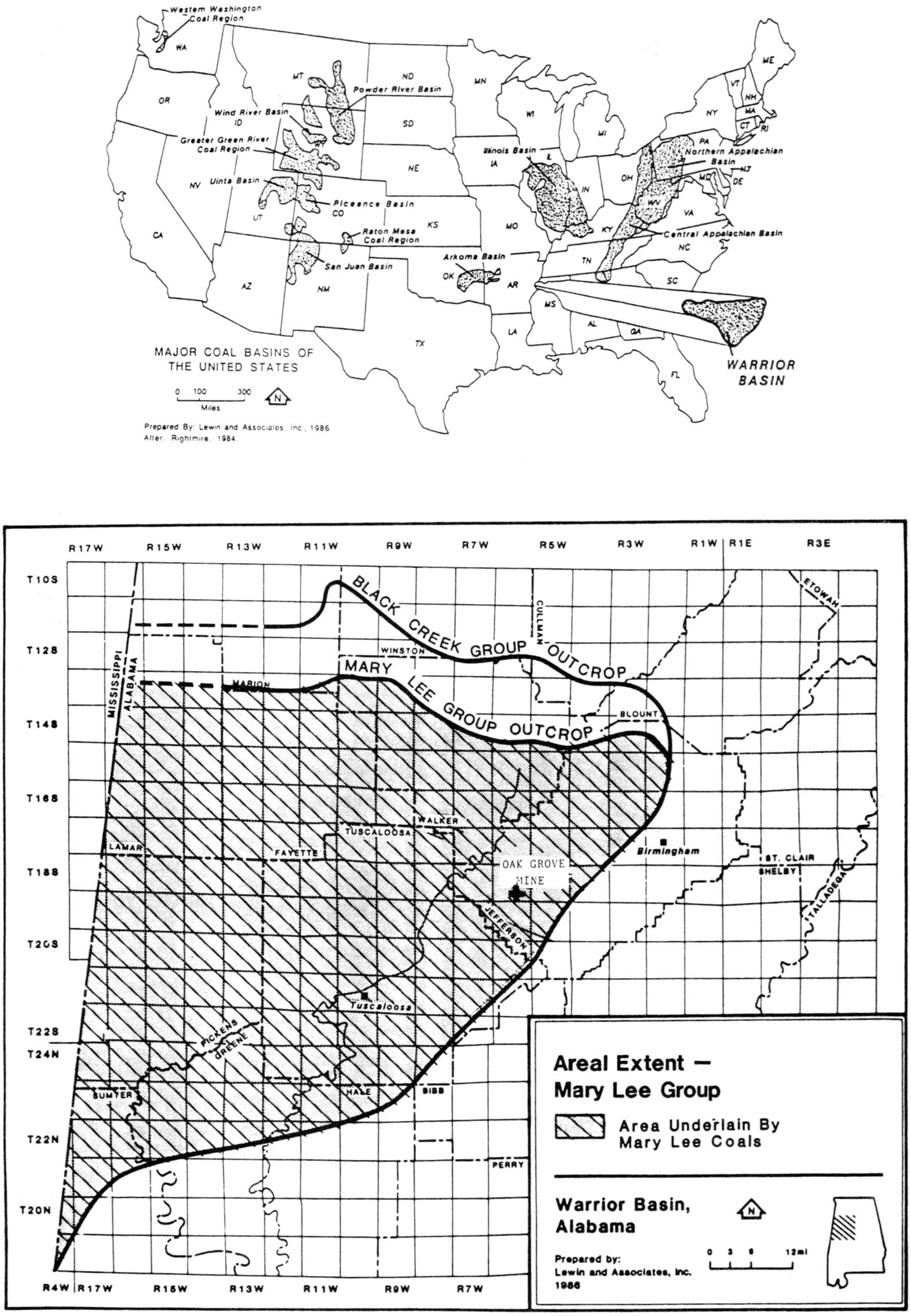

Figure 1. Index map showing the location of the Warrior Basin, Alabama, and the location of the Oak Grove mine within the basin.

early production as structure (Briscoe and Camp, 1983).

OAK GROVE CASE HISTORY

In 1977, USX Corporation (formerly U.S.Steel) in conjunction with the U.S. Bureau of Mines, developed a coal-bed degasification field. This was accomplished by utilizing vertical wells in advance of mining at USX's Oak Grove Mine in Jefferson County, Alabama (fig. 1). The initial field consisted of a 23-well grid, and was the first large-scale degasification field in the United States. The original intent of the field was to reduce gas emissions into the mine; thereby improving mine safety, reducing ventilation requirements, and allowing faster mining advance. Gas from the field was originally vented, but as natural gas prices increased, the gas was collected and sold. The results at the Oak Grove Mine and subsequent degasification fields in the Warrior Basin reveal that in addition to being a mining aid, degasification of coal beds is economically feasible as an independent venture unrelated to mining.

With emphasis on coal-bed methane as an economic resource, the need for a more thorough understanding of factors affecting production was realized. While the Oak Grove degas field served its original purpose of degassing the coal seam prior to mining, it also raised questions concerning the role geologic features play in controlling the rate and amount of gas produced from individual wells.

In September 1983, USX, the Gas Research Institute (GRI), and the U. S. Department of Energy (DOE) funded a study to analyze the geology of the area in and adjacent to the Oak Grove degasification field. USX geologists mapped and interpreted surface joints, and analyzed linear features from aerial photographs. The relationship between these features and the gas and water production from the Oak Grove degas field, as well as other parameters which potentially affect gas production were evaluated (Briscoe and Camp, 1983). Since this study, data from additional field mapping throughout the Warrior Basin and evaluation of production data from several other fields within the basin reinforce the findings of this study regarding the influence of structure on production trends.

The degasification target of the USX Oak Grove degas field is the Blue Creek seam, at an average depth of 1,000 feet. The Blue Creek seam of the Pottsville Formation, Pennsylvanian age is contained within the Mary Lee group, which consists of the New Castle, Mary Lee, Blue Creek, and Jagger seams (fig. 2). The Mary Lee seam occurs an average of 7.0 feet above the Blue Creek seam and separated from it by siltstone and silty shale. The Jagger seam underlies the Blue Creek seam by 5 to 15 feet. Although discontinuous in some areas of the Warrior Basin (McFall, 1986), the Jagger seam is laterally persistent throughout the Oak Grove 23-well grid area.

As the Oak Grove field was developed, it became apparent that production from coal seams both above and below the Blue Creek was occurring. Fracture zones within the Mary Lee group allowed conductivity between the overlying and underlying strata, including the coal seams. In the Oak Grove area, this study revealed that gas was being produced not only from the Blue Creek seam but also from the Mary Lee and Jagger seams (fig. 2). Production from these additional seams can be explained by the hydro-fracing itself and (or) by access via natural fracture systems.

Regional structure in the study area consists of gentle folds, numerous northwest trending normal faults, minor reverse

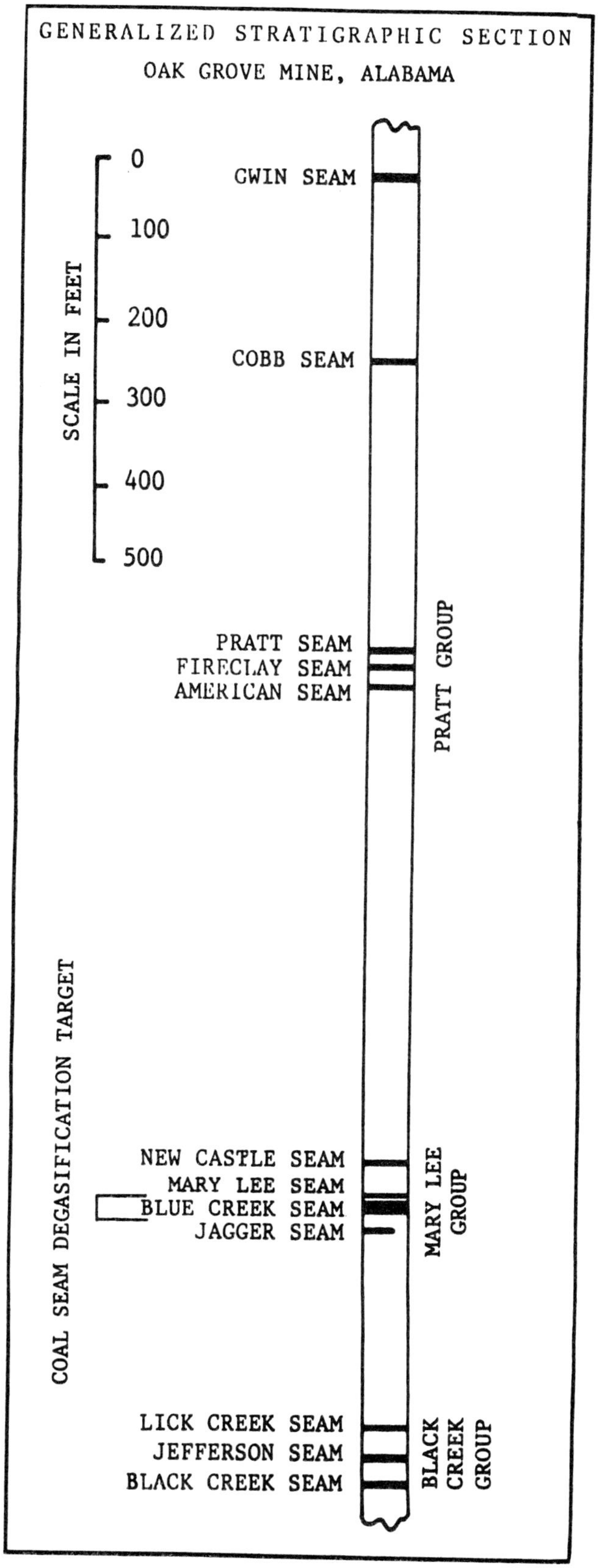

Figure 2. Stratigraphic section showing the major coal groups of the Pennsylvanian age Pottsville Formation, Warrior Basin, Alabama. The degasification target of the Oak Grove Mine is the Blue Creek seam within the Mary Lee group.

faults, and jointing (fig. 3). Ward (1977) established two major joint systems in the Warrior Basin; one system oriented parallel to the Appalachian structural front (40°-50°) and the second system perpendicular to the front (285°-305°). The normal faults generally form local sets of parallel, en echelon patterns. The Sequatchie anticline, a prominent structure of the Valley and Ridge Province striking northeast, dies out approximately three miles northwest of the study area. The Oak Grove degas field is bounded to the east by a major northwest-trending normal fault with up to 140 feet of vertical displacement (Briscoe and Camp, 1983). This fault is near the southern terminus of an en echelon fault zone. Structure within the mine and degas field is relatively simple, with small normal faults limited to seven feet of displacement, and with beds that generally dip to the south (fig. 4).

Evaluation of gas and water production histories of the Oak Grove field reveals great variability over time. Experimental production began in 1977 and most of the original 23 wells were producing by 1979. Oak Grove gas and water production data from 1982 to the present were obtained from production reports of the State of Alabama Oil and Gas Board. Data prior to 1982 were evaluated and showed similar trends but are not included in this report. Figure 5 shows gas production from the first quarter of 1982 through the third quarter of 1983. An additional 15 wells were added to the original 23-well degas grid in the fourth quarter of 1982. A trend of higher gas production near the center of the field is well-defined by production in the first quarter of 1982. This zone of higher gas production migrates to the south and southeast (fig. 5) through time.

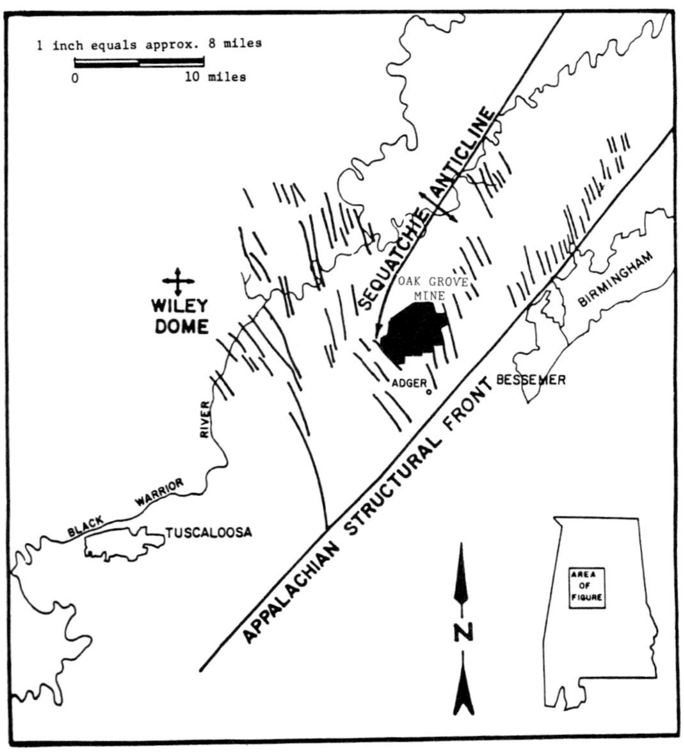

Figure 3. Regional structure map, Warrior Basin, Alabama. Major normal faults, the Sequatchie anticline, and the relative position of the Oak Grove mine are shown.

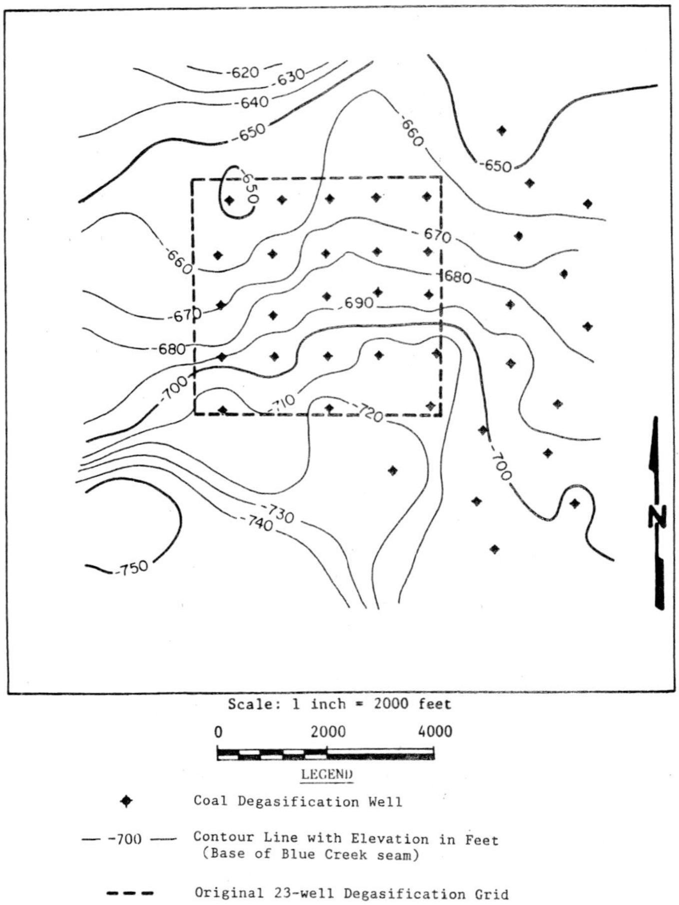

Figure 4. Structure contour map on the Blue Creek seam for the original 23-well Oak Grove degasification grid and the surrounding area. Regional dip of beds is to the south.

Figure 6 illustrates quarterly water production for the same time period. Water production trends indicate that the central and western portions of the field were extensively dewatered by the first quarter of 1982. Excluding the northwest corner of the field, the dewatering trend continues to migrate southward with time.

Both the gas and water production trends indicate the effect of the dip of the beds on the dewatering process. The lowering of the hydrostatic pressure is migrating in a southern, or down-dip direction. Therefore the position of an individual well relative to regional dip and to other wells within the field significantly impacts its production history.

This conclusion is reinforced by the production trends of the 15 wells added in late 1982. Dewatering and methane production of the additional wells were enhanced by the original 23-well grid. The location of the added wells relative to the original degas grid is shown on figure 7. Three distinct trends were recognized in the added wells. The northernmost wells, which are structurally the highest (up-dip) wells, yielded a peak production early, with a gradual decline (fig. 8). The central wells, closest to the original 23-well degas field, immediately yielded a very high methane production, benefiting from the dewatered wells to the west (fig. 9). These central wells subsequently underwent a relatively rapid decline. The southern wells, which were structurally the lowest (down-dip) wells,

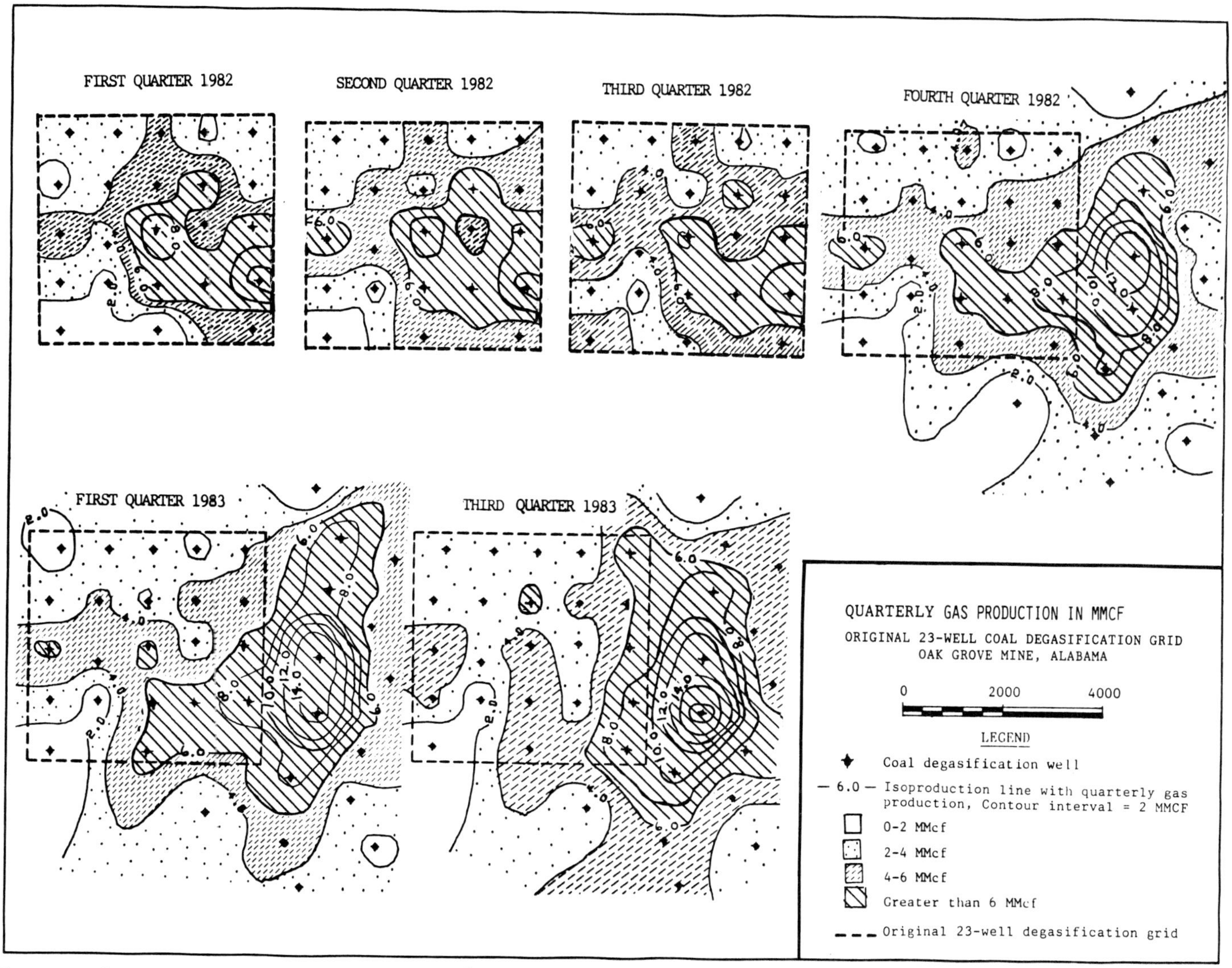

Figure 5. Quarterly gas production in million cfg, original Oak Grove degas field. Time sequence of production includes the period from the first quarter of 1982 to the third quarter of 1983. Higher gas production migrates to the south and east, through time.

reached a production peak as high as the wells to the north, but it took 18 months to reach maximum production (fig. 10). This data suggests the significant impact of the dip of beds on production rates of individual wells, as well as the effect of adjacent wells.

Two wells in the northwest corner of the original grid produced anomalously large volumes of water and low quantities of gas. This trend dramatically reversed when advancing mine activity penetrated a fracture system. This fracture system occurs near these two wells, and provides a major conduit to the Blue Creek coal reservoir in an up-dip direction. This connection to the reservoir up-dip inhibited the ability of these two wells to extensively depressurize the surrounding area. When the underground mine intersected the fracture zone, it cut off the water supply up-dip, thus allowing the wells to dewater the adjacent area and increase gas production (fig. 11).

BROOKWOOD CASE HISTORY

A similar evaluation of the relationships between gas and water production histories and geologic structure was conducted for the Brookwood degas field, developed by Black Warrior Methane Corporation in conjunction with Jim Walter Resources, Inc. This degas field lies approximately 15 miles southwest of the Oak Grove field (fig. 12) adjacent to the structural front. This area is structurally more complex than Oak Grove. Regional dip of the beds in this area is to the south. The Blue Creek seam is the degasification target as in Oak Grove; however, at the Brookwood field the seam occurs twice as deep as at Oak Grove, at an average depth of 2,000 feet.

Gas and water production data were analyzed for the Brookwood field from the fourth quarter of 1982 through the third quarter of 1985. As with the production trends at Oak Grove, the initially high gas producing zone in central Brookwood migrates over time southward (down dip). The highest water production is initially in the south, but as dewatering continues to affect the down-dip wells the southern area becomes progressively higher in gas production, until depletion occurs.

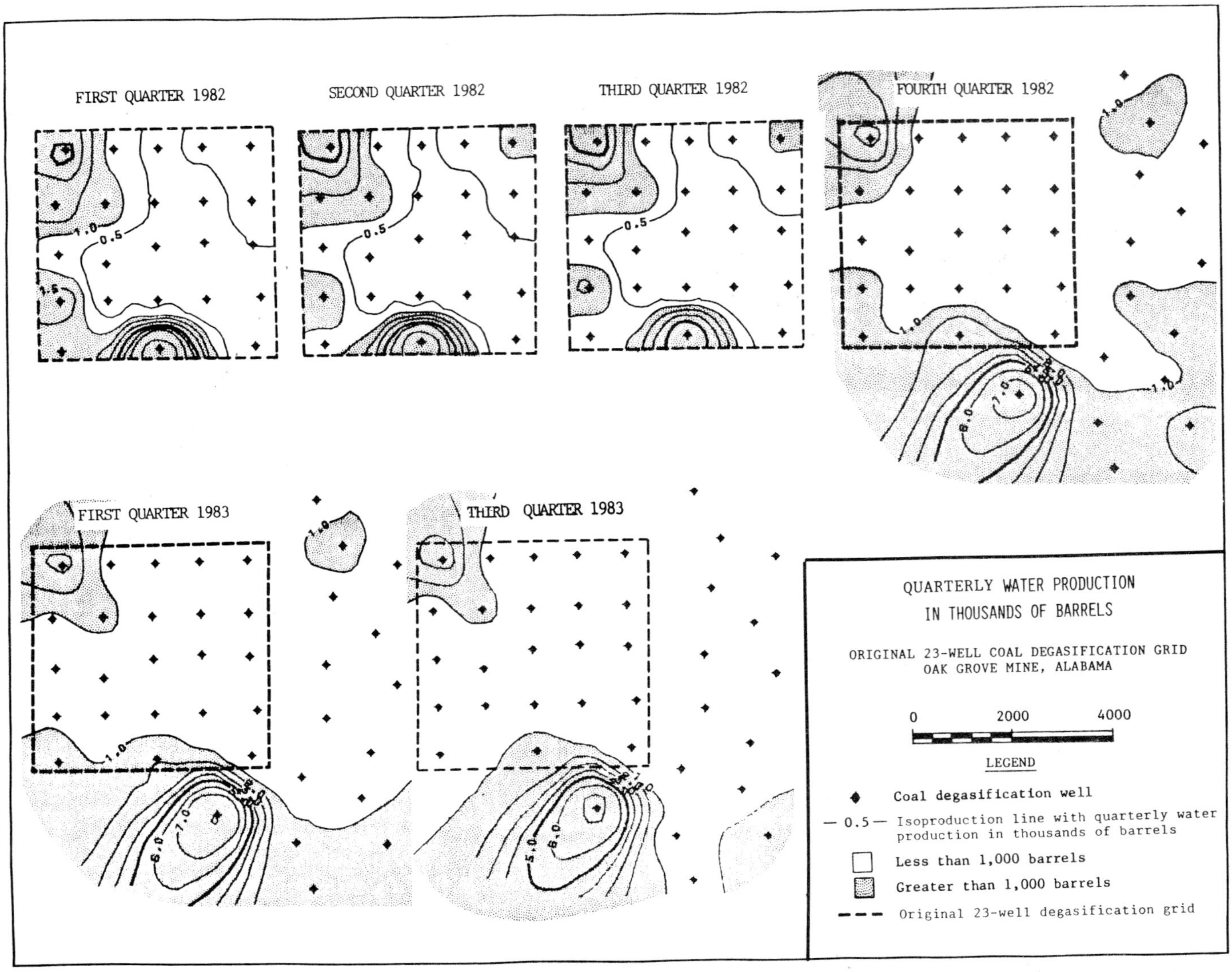

Figure 6. Quarterly water production in thousands of barrels, original Oak Grove degas field. Time sequence includes the period from the first quarter of 1982 to the third quarter of 1983. Water production rates are generally inversely related to gas production for the same period of time (fig. 5)

FIELD MAPPING-JOINT CLASSIFICATION

When used exclusively, conventional remote sensing yields inadequate definition of underground fracture systems for specific areas. A combination of mapping surface joint features and interpretation of remote sensing data provides an improved method of identifying zones of higher fracture intensity. In order to consistently define linear features and joint patterns in the Oak Grove field, as well as throughout the basin, a method evolved which establishes a joint classification system. The field mapping program which was devised classifies the strike, development, and spacing of joints observed in the field from all available exposures. Joints are then grouped into four categories based on joint plane development. The system is relatively simple in order to be consistent and reproducible. This data can be entered into a computer and used to generate fracture maps and rose diagrams.

Although somewhat subjective, used in conjunction with aerial photography this method defines fracture patterns so that they can be interpreted with respect to production. This method allows evaluation of large areas, as well as defining detailed structures within a specific area. Detailed study of production trends versus fracture zones in the Oak Grove field confirmed the importance of structure; wells connected to fracture systems, in the form of major linear features and (or) closely-spaced joints, yielded higher production (Briscoe and Camp, 1983). Wells within 200 feet of fracture zones obtained 25 percent greater methane production and 50 percent greater water production than wells in unfractured areas.

While data obtained from the Oak Grove degas field established the existence of subsurface communication between seams via fracture systems, field mapping in the Warrior Basin also revealed fracture avenues from subsurface coal seams to the surface, as evidenced by continuous streams of methane bubbles observed in water in creek beds.

Knowledge of local structure may optimize well spacing in order to increase methane production and minimize costs. Although conventional well spacings of 40 or 60 acres are commonly used in degas fields, optimum grid spacing for an individual field is highly dependent on fracture patterns. Since

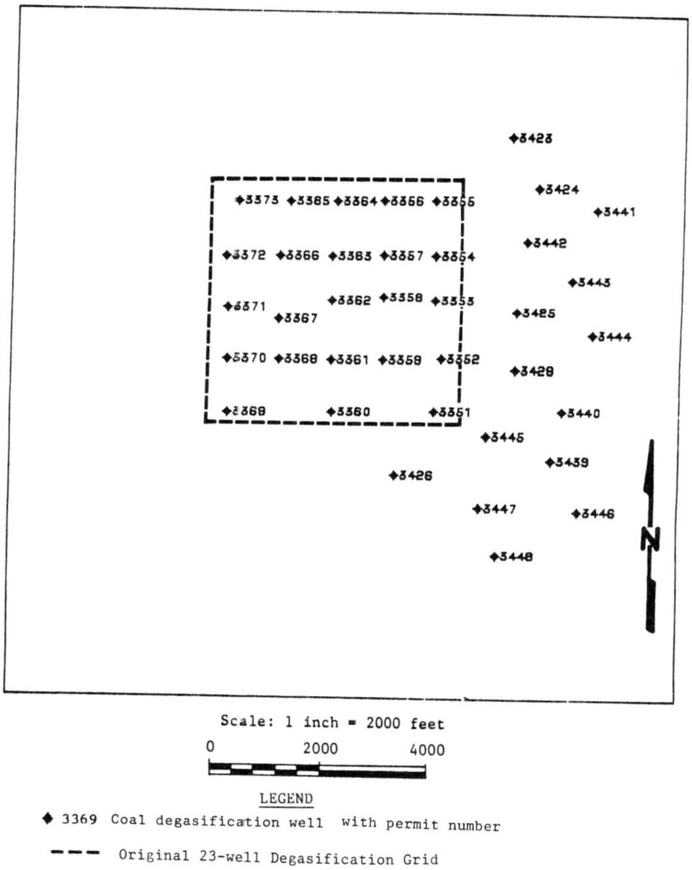

Figure 7. Location of the Oak Grove field degasification wells. The wells within the box represent the original 23-well grid, and the 15 wells to the east and south were added in the fourth quarter of 1982.

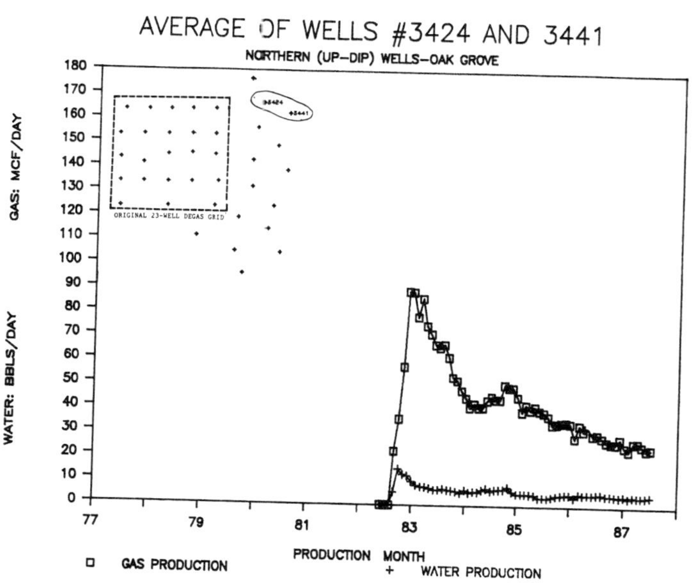

Figure 8. Production trends of two of the northernmost of the 15 added wells in the Oak Grove field. These structurally higher wells showed an early peak production, then steadily declined. These two wells are circled in the upper left corner of the map.

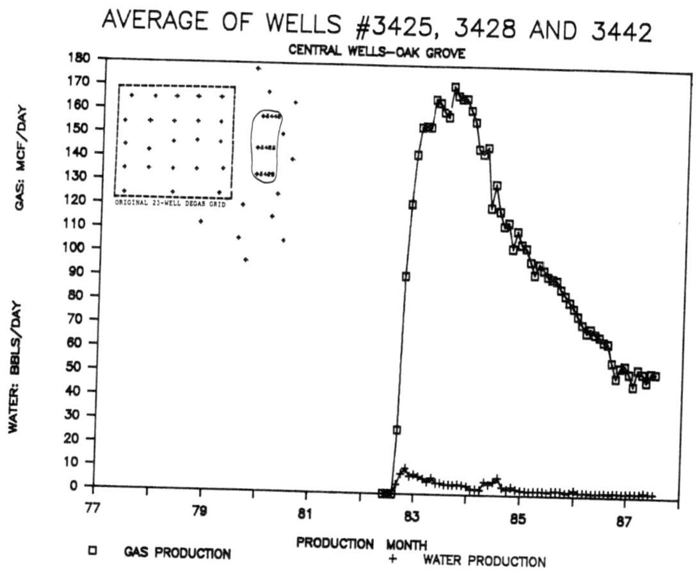

Figure 9. Production trends of three central wells, closest to the original 23-well degas grid. These wells benefited from the proximity of dewatering wells to the immediate west and yielded an immediate peak gas production followed by a relatively rapid decline. Location of these three central wells within the field is shown in the upper left corner of the map.

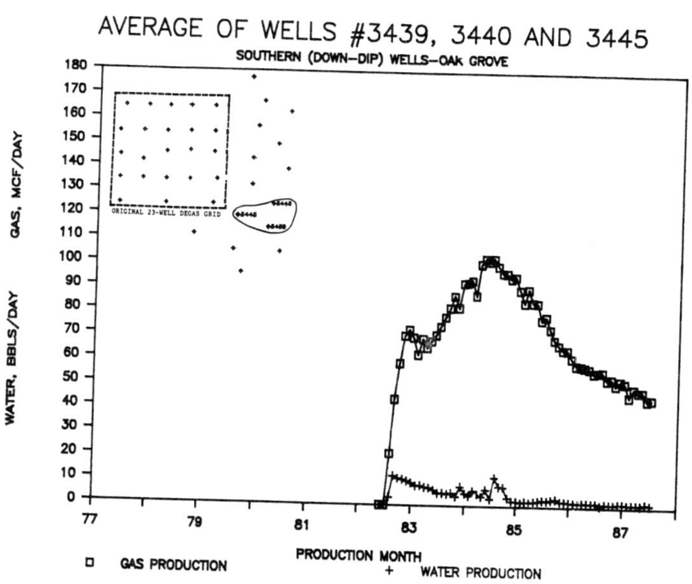

Figure 10. Production trends of the southernmost wells, structurally the lowest of the 15 added wells. These wells produced significant water initially, and peak production was obtained after a total of 18 months. Location of these three southern wells within the field is shown in the upper left corner of the map.

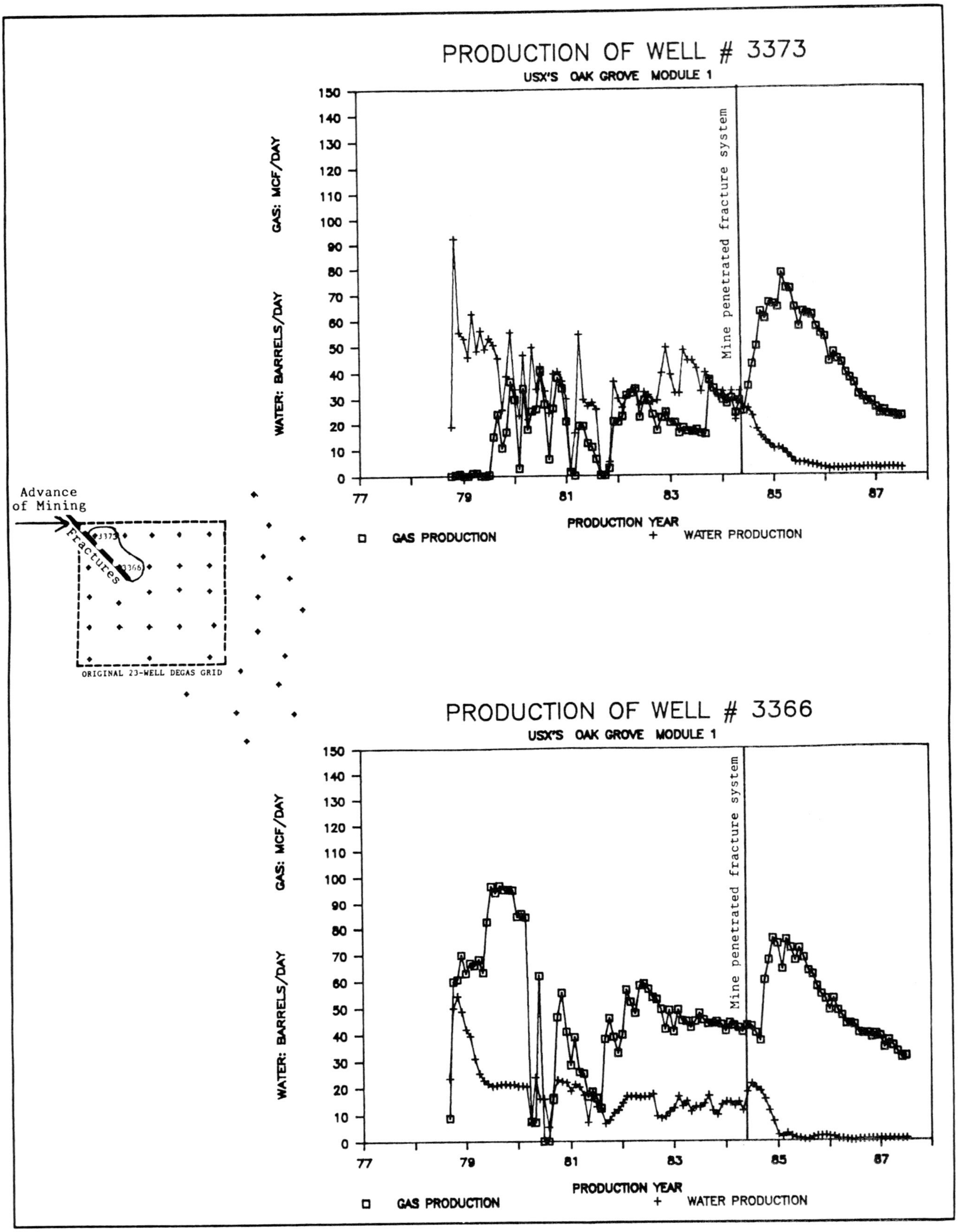

Figure 11. Production trends of two northwestern wells with anomalously low gas production. Relative locations of the wells, advance mining, and a major fracture system are shown. When mining activity encountered the fractures, a substantial decrease in water production and increase in gas production occurred in both wells.

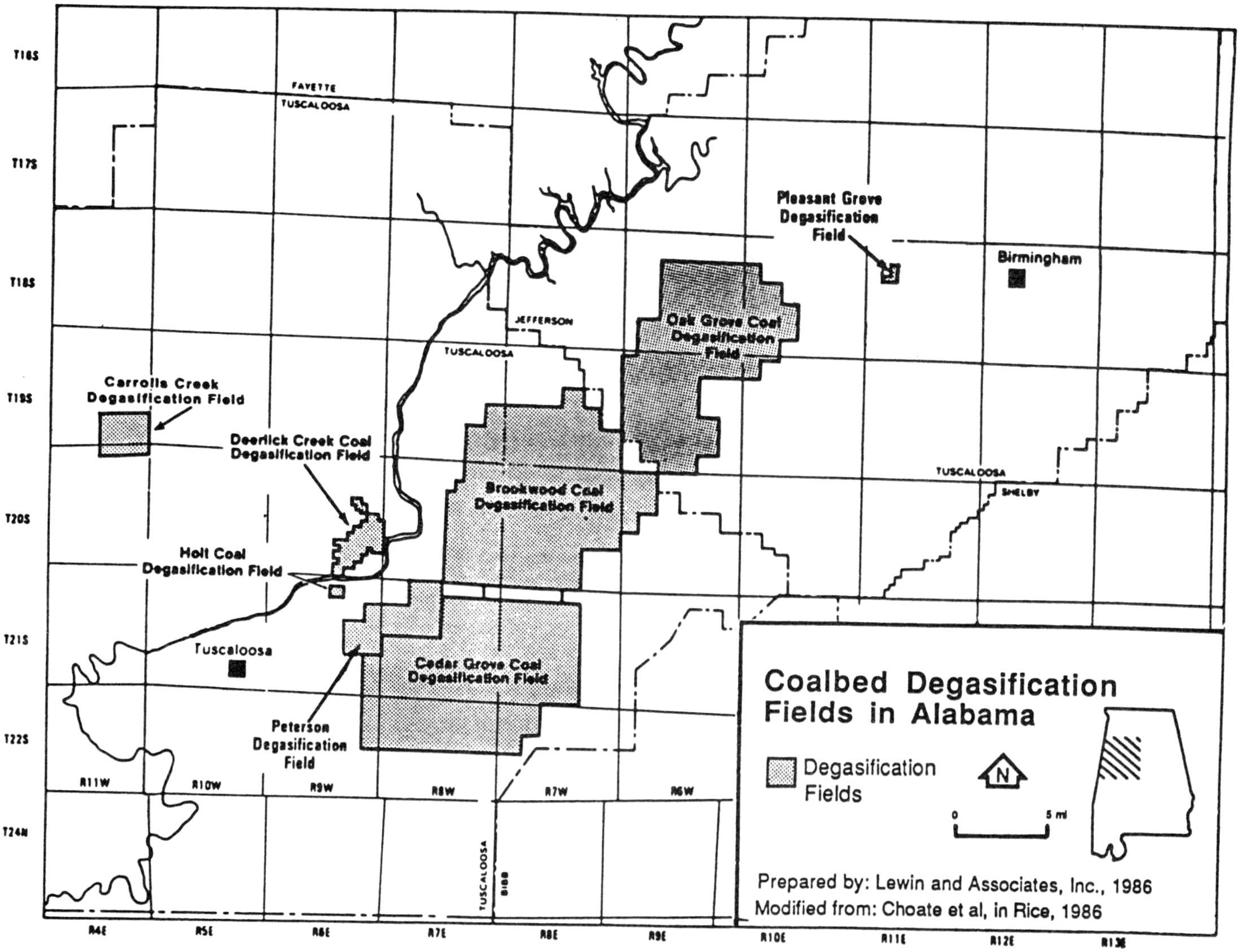

Figure 12. Index map of the current degasification fields within the Warrior Basin showing the relative positions of the Oak Grove and Brookwood fields.

fracture zones greatly increase the communication between wells horizontally and between seams vertically, it may be possible to increase well spacing in highly fractured areas without decreasing early production, yielding a longer production period. Structural highs such as anticlines and faults also may favorably affect production. A current, large scale degas project in the Warrior Basin has successfully utilized the method of joint mapping and structural analysis described above to optimize well placement and spacing, thus increasing production.

IN-PLACE VERSUS RECOVERABLE COAL-BED METHANE RESERVES

In addition to evaluating the influence of geologic factors on production, the problem of determining how much of the measured coal-bed methane is actually recoverable was studied. The method commonly used is based on the assumption that a percentage (commonly 50 percent) of the in-place gas can be recovered (Rightmire, 1984). This method can be extremely misleading, as it assumes that the initial gas volumes can be used to predict the final gas volume of the reserve when gas production has declined to a non-economic level. For example, if two coal seams contain in-place gas of 200 and 600 cfg/ton, a 50 percent reduction of the gas content of these two seams would yield final shut-in gas contents of 100 cfg/ton and 300 cfg/ton, respectively. Reducing methane values by an arbitrary percentage fails to compensate for initial variation; if the seam containing 200 cfg/ton can be reduced to 100 cfg/ton, then the seam containing 600 cfg/ton should also be reduced to 100 cfg/ton.

An alternative and more realistic approach was conceived based on data collected from the original Oak Grove degas field grid. Since coal thickness, original coal seam gas content, and production data are available for this area, this information provided a unique opportunity to determine how much of the known original methane was ultimately produced. Oak Grove is the only degas field in the Warrior Basin with a production history of sufficient length to study annual production rates versus in-place methane. Based on Oak Grove gas production decline curves, it was established that the in-place coal-seam gas content was reduced to approximately 100 cfg/ton. For a specific degas field the final value will be a function of well spacing, permeability, coal rank, and the price of gas. Using this lower threshold of 100 cfg/ton appears to be more accurate than assuming a percentage to calculate recoverable methane. Recoverable gas reserves for a seam containing 600 cfg/ton (in-place gas) are calculated as follows: Cubic feet of methane per acre = 600 cfg/ton − 100 cfg/ton × coal density × coal volume per acre. Through June 1987, the original in-place coal-gas content for the area of the 23 Oak Grove wells was reduced from 450 cfg/ton to 118 cfg/ton. Although the issue of recoverable methane is still being debated, production data indicate that using a 50 percent recovery formula yields unacceptably low estimates of gas volumes for wells currently being produced in the Warrior Basin.

REFERENCES CITED

Briscoe, F.H., and Camp, B.S., 1983, Oak Grove coal degasification field joint and linear study and their relationship to methane production: Department of Energy, unpublished report, 9 p.

Hewitt, J.L., 1984, Geologic overview, coal, and coalbed methane resources of the Warrior Basin—Alabama and Mississippi, *in* Coalbed Methane Resources of the United States, American Association of Petroleum Geologists Studies in Geology Series no. 17, p. 73-104.

McFall, K.S., Wicks, D.E., and Kuuskraa, V.A., 1986, A geologic assessment of natural gas from coal seams in the Warrior Basin, Alabama: Gas Research Institute, Topical Report contract number 5084-214-1066, 80 p.

Rightmire, C.T., 1984, Coalbed methane resource, *in* Coalbed Methane Resources of the United States, American Association of Petroleum Geologists Studies in Geology Series no. 17, p. 1-13.

Ward, W.E., 1977, Jointing in a selected area of the Warrior Coal field, Alabama: Unpublished Master's Thesis, University of Alabama, Tuscaloosa, Alabama, 61 p.

Durango's Main Avenue looking south with Smelter Mountain in the background. Note the vehicles to date the photo. Photo courtesy of the Durango Public Library.

Marketing San Juan Basin Gas

David M. Posner[1]

ABSTRACT

Marketing natural gas produced in the San Juan Basin of New Mexico and Colorado principally involves four gas pipeline companies with significant facilities in the basin. The system capacity, transportation rates, regulatory status, and market access of each of these companies is evaluated. Because of excess gas supplies available to these pipeline companies, producers can expect improved take levels and prices by selling gas directly to end users and utilities as opposed to selling gas to the pipelines for system supply. The complexities of transporting gas today suggest that the services of an independent gas marketing company may be beneficial to smaller producers with gas supplies in the San Juan Basin.

INTRODUCTION

Only a few years ago, petroleum industry scientists concentrated exclusively on the geological and engineering aspects of a potential natural-gas play. The only relevant marketing question concerned how much gas there might be in the new reservoir. Market conditions since then have dramatically changed the environment that explorationists face. Reserves in the ground are no longer enough. Today, one must determine the feasibility of delivering the gas to market, the relative attractiveness of alternative markets, and the regulatory constraints that are associated with different market options.

This paper describes the various markets available for San Juan Basin gas. Government regulations are discussed as they relate to selling and transporting San Juan Basin gas. Alternative approaches to marketing gas, including the role of independent marketers are also considered.

MARKETS FOR SAN JUAN GAS

Figure 1 is a map showing the major pipelines that are available in the San Juan Basin. There are four major pipeline systems that offer an outlet for San Juan gas. El Paso Natural Gas Company has the largest system, covering both the Colorado and New Mexico sides of the basin.

Northwest Pipeline Corporation also spans portions of both Colorado and New Mexico. Western Gas Supply Company (WestGas), an affiliate of Public Service Company of Colorado has gathering lines throughout the Colorado portion of the Basin. Sunterra Gas Gathering Company (formerly Southern Union Gathering Company), an affiliate of the Gas Company of New Mexico, covers the New Mexico portion of the Basin. The Sunterra and WestGas systems serve intrastate markets in New Mexico and Colorado, respectively. The El Paso system primarily serves markets in California and Nevada. The Northwest system serves markets in Washington and Oregon, and to a limited extent also Nevada.

In addition to these four major pipeline systems, there are several smaller systems that are being developed in the San Juan Basin. Greeley Gas Company is one example of a smaller system. The Greeley Gas system has the ability to redeliver gas to El Paso Natural Gas Company as well as supply gas directly to customers in the Durango area. Ladd Gas Marketing, Inc. as well as other producers and independent gathering companies have also built gathering facilities in the San Juan Basin to help minimize transportation costs and improve market access.

In selecting a market outlet for a gas well, several factors need to be considered in evaluating alternative pipeline outlets. These include: The proximity to the well and the investment required to make the pipeline connection; the markets available through alternative pipeline outlets; the regulatory status of pipelines relative to availability of open access transportation; the capacity constraints on the pipeline for receiving gas and redelivery of gas to alternative markets; the transportation rates charged to deliver gas to market; and the nature of competing gas supplies that are available to serve selected market areas. These factors are reviewed as they apply to each of the four major pipeline markets.

EL PASO NATURAL GAS COMPANY

The El Paso system in the San Juan Basin converges with the rest of the El Paso System in McKinley County, New Mexico, just south of the basin. The flow of gas in this part of the system is toward California. It is estimated that the El Paso system in the San Juan Basin has the capacity to move approximately 1.5 BCF (billion cubic feet) of gas per day.

As mentioned above, the primary markets for the El Paso system gas are within California. El Paso's system stops at the California border, where gas is redelivered to Southern California Gas Company (SOCAL) and Pacific Gas and Electric Company (PG&E). These two pipelines transport virtually all gas within California, either from intrastate production sources or redelivery of gas supplied from interstate pipelines such as El Paso. In addition to extensive retail utility systems, PG&E and SOCAL act as wholesale suppliers to smaller local distribution companies in California such as San Diego Gas and Electric. The complexities of gas marketing in California are discussed further below.

In addition to California, El Paso also is capable of delivering gas to Southwest Gas Corporation (serving Las Vegas and southern Nevada), and to the Gas Company of New Mexico. It is also possible to backhaul and exchange gas between El Paso and other interstate pipelines to access secondary mar-

[1] Ladd Gas Marketing, Inc.
370 17th Street, Suite 1700
Denver, Colorado 80202

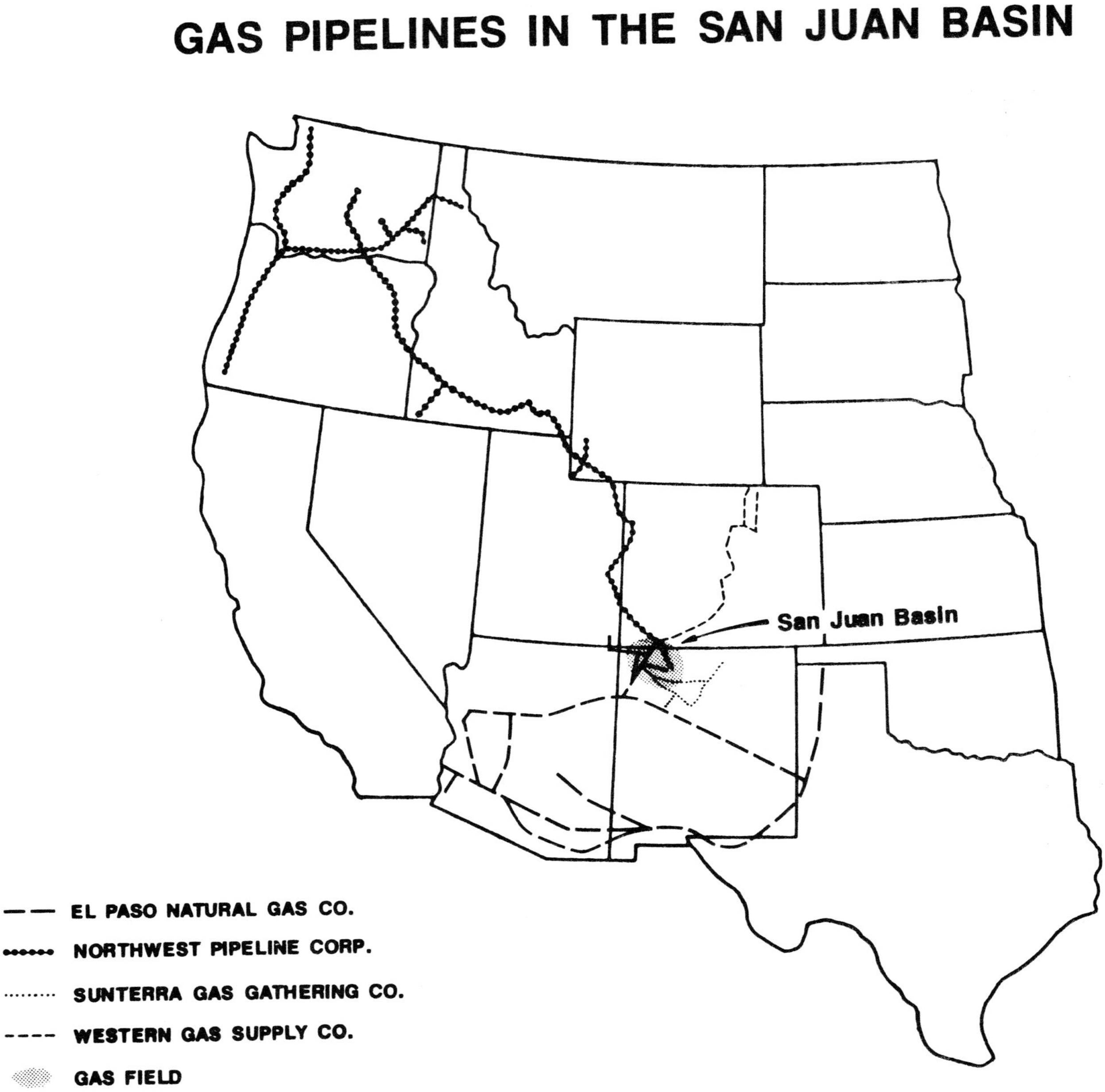

Figure 1. Map showing the major pipelines that carry gas out of the San Juan Basin of New Mexico and Colorado.

kets. In serving secondary markets, additional transportation charges often make it difficult to compete with gas directly available into the primary pipeline.

El Paso currently operates as an open-access transporter of gas under the Federal Energy Regulatory Commission (FERC) interim transportation regulations. Under these rules, El Paso will transport gas under Natural Gas Policy Act Section 311 transportation agreements, which must be on behalf of a local distribution company or intrastate pipeline company. El Paso is now in the process of completing its filing to become an open-access transporter of gas pursuant to FERC's permanent open-access regulations originally outlined in FERC Order 436 during the autumn of 1985 and subsequently modified by FERC Order 500 during late 1987. As an open-access transporter under FERC's permanent rules, El Paso will transport on a nondiscriminatory, first come, first served basis for any entity.

A producer evaluating connecting a new well to El Paso's pipeline theoretically has two options: a) Dedicate the gas to El Paso and sell the gas to El Paso under a traditional system supply contract; or b) Sell the gas under a direct sale arrangement to a third party marketing company, utility or end user. Under the second arrangement, the El Paso system would be used to transport the gas to the ultimate market destination.

In making the decision whether to sell gas to El Paso for system supply or use El Paso as a transporter, several considerations need to be addressed. An overriding consideration is the terms that are available from El Paso for a system-supply sale. The recent history of system-supply sales is not very good. Take levels have been low and highly seasonal and prices have been regularly adjusted to new "market out" levels that reflect current spot-market conditions. Recent experience with El Paso system-supply sales has motivated Ladd and other producers to release or terminate all system-supply contracts in order to have the flexibility to sell to other markets. While the possibility of selling gas to El Paso as system-supply sales remains an option, it is not currently a very attractive option because of the recent interstate pipeline contractual problems that are pervasive in the gas industry today.

In the past, one of the advantages of selling to an interstate pipeline was that the pipeline was willing to extend its gathering facilities directly to the well. In the current environment, it is usually a difficult and long process to get the pipeline to install new connections. More often than not, new connections are made either directly by the producer or by an independent gathering company.

The option of selling the gas through a direct-sale or transportation arrangement is, in most cases, the more favorable approach in today's environment, both for new production and for existing wells already connected to the El Paso system. The principal advantage of this approach is the flexibility to sell the gas to a party that has an immediate need. There are many ways to go about transportation sales, depending on the volumes of gas controlled and the time and expertise available. A producer can attempt to market gas directly to utilities and end users, but this takes a tremendous amount of time and expertise in all aspects of sales and transportation arrangements, as well as substantial volumes of gas to justify the costs involved. More often, producers will sell to a third party marketing company (including El Paso's marketing company) which then handles all aspects of the marketing and transportation. The duration of sales arrangements can vary from month-to-month to several years including provisions for periodic price redetermination.

The good news about transportation-type sales is their flexibility. The bad news is that transportation contracts have a lower priority than system-supply contracts relative to access to pipeline capacity. Ladd and many other producers have been selling gas in the San Juan Basin on the spot market through various transportation arrangements for more than two years. The gas has flowed continuously for that time until the extreme cold periods encountered at the end of 1987 in California, when deliveries were curtailed due to capacity constraints on the El Paso system coming out of the San Juan gathering system.

These curtailments illustrate that pipeline system-supply sales take priority over transportation arrangements. When the California utilities determine that additional gas supplies are needed due to a combination of higher demand and supplier non-performance, the utilities nominate additional gas purchases from El Paso under existing system-supply contracts. El Paso then is required to increase its field-gas purchasers under producer contracts, and in some instances recall gas supplies that have been previously released. This increase in sales volumes necessitates that transportation volumes be reduced to make pipeline capacity available. And so the cycle continues—more need for the California utilities to purchase El Paso system gas and then more curtailments of transportation gas—until virtually all transportation gas has been curtailed.

This curtailment cycle often means that end users that were purchasing gas directly from suppliers are forced to go back to system-supply purchase arrangements. Industrial customers with fuel oil capability are often forced to discontinue gas use. The recent history on transportation curtailments indicates that there will be interruptions during isolated periods, but on average, transportation sales will experience a good rate of flow.

El Paso's transportation rates for interruptible service are broken down into mainline-transmission charges and production-area charges. Maximum rates for mainline forward hauls range from a low of $0.141 per million Btu for deliveries to Texas to a high of $0.166 per million Btu for deliveries to California. El Paso reduces these rates by 50 percent for backhaul and selectively discounts forward hauls from time to time on a nondiscriminatory basis. Added to this mainline charge is a field-gathering charge of up to $0.23 per million Btu. The field-gathering charge applies to all wells not connected to the mainline system. In addition to these fees, El Paso applies a fuel retention charge of 5 percent (for California deliveries) for mainline hauls as well as an additional fuel retention fee for field gathering which is based on actual fuel requirements. Fuel-retention charges are simply another pipeline charge for transportation services. In essence, the pipeline redelivers less gas at the redelivery point than it receives into its system. In many cases, fuel-retention charges approximate the pipeline's fuel requirements for compression and line losses. Other small fees for the Gas Research Institute and miscellaneous FERC approved adjustments add approximately another $0.02 per million Btu to these transportation charges.

Another complexity of El Paso's transportation rates is the special category of displacement markets. Displacement markets include any market that is currently a sales customer of El Paso and is paying standby service charges to El Paso. Direct sales to all displacement markets are not subject to El Paso's production-area charges, such as gathering fees, and are assessed only a flat 5 percent production-area fuel charge. Spot sales to the major California utilities qualify as displacement markets. Non-displacement markets include most off-system transportation sales that involve redelivery to another interstate pipeline and any new customers within El Paso's traditional service area. El Paso is currently discounting production-area charges to non-displacement markets to $0.03 per million Btu but is assessing full fuel retention charges, which in the San Juan Basin typically run 15 percent.

Table 1 shows how El Paso's fees translate into net prices at the wellhead for different types of direct sales. A $2.30 per million Btu California border price nets back $1.91 per million Btu to the wellhead when selling to a displacement market. The same $2.30 per million Btu California border price nets only $1.68 per million Btu at the wellhead when selling to a non-displacement market.

The additional transportation charges for non-displacement markets represents one of the difficulties of marketing gas directly to end users in California today. Obviously, non-displacement markets need to be willing to pay a higher price at the California border than displacement markets to be able to compete for gas supplies. Regulatory constraints also currently inhibit direct sales to customers in California. The California Public Utilities Commission and the major California pipelines make it difficult to transport gas within California. Sometime in the future, it is expected that these regulatory constraints and rate differences will be resolved, and that direct sales to customers in California will be one of the most promising markets for San Juan Basin gas.

Direct sales arrangements are usually subject to wide price

Table 1. Producer net back prices on El Paso Natural Gas Co. (Based on January 1988 prices); prices are dollars per million Btu.

	Displacement Market	Non-Displacement Market
California border price	$2.30	$2.30
El Paso mainline charges		
transport fee	(0.18)	(0.18)
fuel retention @ 5 percent	(0.11)	(0.11)
Net mainline price	$2.01	$2.01
El Paso production-area charges		
Gathering & dehydration [a]	-	0.03
Fuel retention [b]	(0.10)	0.30
Net wellhead price	$1.91	$1.68

[a] As of January 1988, El Paso was discounting production-area charges for non-displacement markets.

[b] Fuel retention for displacement markets is 5 percent; non-displacement market fuel is 15.3 percent.

variations with changes in market conditions, unless a long term contract is used with some type of fixed price schedule. Over the last year, net wellhead prices in the San Juan Basin have gone from $1.55 per million Btu in January 1987 to a low of $1.30 per million Btu in August 1987 and then recently increased to the $1.90 per million Btu level as of January 1988.

Gas supplies from the Permian Basin in west Texas are the primary El Paso supplies competing with San Juan Basin gas. These supplies can also be delivered via El Paso to the California market. San Juan Basin supplies also compete with gas on Transwestern Pipeline, now part of the Enron Corporation. During the summer months, when capacity is readily available on El Paso's system, supplies from wells connected to other interstate pipelines in the Anadarko Basin are also available to the El Paso system for redelivery to California.

One other consideration in transporting gas on El Paso's pipeline, or any other interstate pipeline, is the consequences of FERC Order 500. In this order (and the subsequent FERC Orders 500B and 500C), any producer requesting interstate transportation services is required to provide to El Paso volume for volume take-or-pay credits on all gas transported. These credits can be applied by El Paso against any take-or-pay liability that El Paso has under any contract with a given producer. Producers that are still negotiating for take-or-pay settlements, or are in litigation, will need to think twice before signing an offer providing for take-or-pay credits. Many producers, including Ladd, have settled all significant take-or-pay problems, and are therefore willing to provide the necessary offer letters.

NORTHWEST PIPELINE CORPORATION

Northwest Pipeline has a significant gas-gathering presence throughout the San Juan Basin. The configuration of Northwest's market and supply areas, system constraints, and regulatory posture all present challenges for marketing San Juan Basin gas through Northwest. Northwest's principal markets are utility systems in Washington and Oregon. Northwest also hands off gas to other interstate pipeline companies, including El Paso and Colorado Interstate Gas Company (CIG).

Northwest's position with regard to providing open-access transportation has changed several times during 1987. For a few months, Northwest provided transportation under Natural Gas Policy Act Section 311 provisions. At this time, however, Northwest is a closed system. Transportation is only available pursuant to Section 7(c) of the Natural Gas Act. Section 7(c) transportation is far more restrictive than is Section 311. A specific FERC filing is required before Section 7(c) transportation can commence. It usually takes several months to file and receive approval from FERC for a Section 7(c) certificate; preparation for this filing can take a year or longer. Once a 7(c) certificate is granted by FERC, it can not be amended without making a new filing. In most cases, a Section 7(c) certificate is too restrictive to work effectively in marketing transportation gas, especially when the seller desires to have the flexibility to sell to multiple customers. Northwest states that it intends at some time in the future to become an open-access transporter, but the timing is uncertain.

With transportation essentially unavailable on Northwest, the other option is to sell gas to Northwest for system supply. Northwest is a pipeline that currently has substantial quantities of excess gas. The reserve life of the Northwest system, based upon an annual throughput of 400 BCF is now 28 years. With those quantities of reserves under contract, it is not difficult to see why Northwest is not buying new gas. Low takes on gas currently under contract and the unavailability of other transportation combine to create real problems for producers with gas already connected to Northwest.

Northwest's domestic supplies come from the San Juan Basin and the Rocky Mountain area. The San Juan Basin is at the southern end of the Northwest system (fig. 1). The transportation of San Juan Basin gas is subject to severe capacity limitations in the Idaho area as the gas flows to the Pacific Northwest. In actuality, most of the San Juan Basin gas is delivered off the system by Northwest to Colorado Interstate Gas, El Paso, and other pipelines under various exchange and sales arrangements.

Another important aspect of Northwest is that a large portion of its supplies are derived from Canada via West Coast Transmission Company. The availability of extensive Canadian supplies contributes to the problem of low takes and is likely to limit Northwest's need to acquire new San Juan Basin gas supplies.

Should Northwest at some time become an open-access transporter, it is difficult to anticipate the exact structure of its transportation rates. Under rates currently available to Section 7(c) customers, Northwest charges up to a maximum of $0.50 per million Btu for gathering plus 4.5 percent fuel retention in the San Juan Basin. Mainline charges under current rates are $0.235 per million Btu plus 0.5 percent fuel retention for each 100 miles.

Northwest has three programs that make it possible for producers to improve takes on gas already connected to the Northwest system. One program, referred to as the nomination program, enables producers to nominate gas that the producer desires to sell to Northwest. This gas must be competitive with other supplies available to comply with Northwest's least-cost purchase strategy. Gas nominated under this program generally experiences high levels of takes. The price to participate in the nomination program for January 1988 is $1.25 per million Btu. To be eligible to participate in this program, Northwest is likely to require that all take-or-pay or other contractual claims against Northwest be waived.

Northwest will also take some gas under contracts still in effect at a market-out price of $1.52 per million Btu. Take rates under the market-out price are expected to be less than the nomination program, and are likely to be in the 50-75 percent range depending on the specific contract.

A third approach to move San Juan Basin gas currently contracted for Northwest is to exchange gas from Northwest's gathering system to El Paso's gathering system. Because gas remains entirely on the Northwest gathering system, a Section 7(c) certificate for transportation is not required. The maximum rate of $0.50 per million Btu for gathering in January, has been discounted to $0.25 per million Btu by Northwest. Once the gas is delivered to El Paso's gathering system, it is subject to the normal El Paso gathering charges. Table 2 shows that the net price for sale to a displacement market off El Paso in January works out to be $1.58 per million Btu, substantially better than the Northwest nomination program price.

Table 2. Producer net wellhead price on Northwest Pipeline for El Paso redelivery (January 1988 prices); prices in dollars per million Btu.

El Paso net wellhead price:	$1.91
(For displacement market sale)	
Northwest gathering charge	(0.25)
Northwest retention @ 4.5 percent	(0.08)
Net wellhead price	$1.58

SUNTERRA GAS GATHERING

Sunterra Gas Gathering Company covers portions of the New Mexico side of the San Juan Basin. Sunterra gathers an average of 200,000 MCF per day from the San Juan Basin, and has a pipeline supply capacity of over 300,000 MCF per day. Sunterra's primary market outlets are the Gas Company of New Mexico, its parent company, and El Paso Natural Gas Co.

Like most other pipelines, Gas Company of New Mexico and Sunterra have more system supply gas under contract than needed. Therefore, take levels on sales for system supply have been very low during the past several years. Because of low takes, Ladd has released all of its production from existing contracts with Sunterra and the Gas Company of New Mexico. This gas is now being sold through transportation arrangements.

As an intrastate pipeline company, Sunterra is not subject to the same FERC regulations that an interstate pipeline company faces. Sunterra performs transportation services under established rates. The system gathering rate is currently $0.32 per million Btu.

A good example of how a direct sales arrangement works on Sunterra is provided by Ladd's current sale of gas to an end user in Albuquerque. Table 3 shows how the transportation charges are assessed. The wellhead price is $1.60 per million Btu after all transportation fees. The delivered price is $2.95 per million Btu, significantly below the sales price available to this industrial plant from the Gas Company of New Mexico under the appropriate rate category. In addition to a reasonable price, this approach offers predictable volumes and good take levels. There are numerous marketing companies working on direct sales on the Sunterra system.

Table 3. Illustration of direct gas sale on Sunterra (January 1988 price); price in dollars per million Btu.

Wellhead price:	$1.60
Sunterra gas gathering:	0.32
Gas Company of New Mexico:	
Transportation	$0.95
Fuel retention @ 4 percent	$0.08
Total delivered cost	$2.95

One problem with the Gas Company of New Mexico market is that it does not have a large gas requirement, especially in the summer months. Another option is to use the Sunterra System to redeliver gas to El Paso, which would make the markets discussed above available. With Sunterra's gathering charge, a January 1988 sale of this type would result in a net wellhead price of $1.53 per million Btu.

WESTERN GAS SUPPLY COMPANY

WestGas faces problems similar to other pipelines in the San Juan Basin—too much gas supply relative to market requirements and pipeline capacity available to transport gas. WestGas' market outlets for San Juan gas are Public Service Company of Colorado and a few smaller gas distribution companies along the Colorado front range area, and two interstate pipeline companies: El Paso and Northwest. WestGas faces severe capacity restrictions in moving gas out of the San Juan Basin to the Denver area, where it redelivers gas to Pubic Service Company of Colorado. The majority of gas moving out of the San Juan Basin on WestGas is delivered to WestGas markets in the Leadville and Buena Vista areas, which are relatively small markets. The most feasible haul from a capacity standpoint is redelivery from the WestGas gathering system to El Paso's gathering system. As an intrastate pipeline company, WestGas transports gas under established rates, which can be negotiated. The gathering fee on WestGas is $0.20 per MCF plus 2 percent fuel retention which would apply to a gathering system hand off from WestGas to El Paso. If the WestGas system is used for transmission to one of the markets off of the WestGas system, an additional fee of $0.36 per million Btu plus 2 percent fuel retention is incurred.

WestGas is not actively buying gas currently because of its over supply position. Takes in recent years have been so low that Ladd has been released from sales contracts for almost all of its gas connected to WestGas. The WestGas system can be used to redeliver gas to El Paso for subsequent sale to various direct-sale markets. Sales of this gas to an El Paso displacement market in January 1988 would net back to the wellhead $1.68 per million Btu. In a number of instances, producers and independent gatherers have laid new gathering lines from wells connected to WestGas and then connected these gathering lines directly to El Paso's main line. By taking this approach, the WestGas gathering fee is avoided.

In addition to redelivery of gas to El Paso via WestGas, it is possible to bid gas on WestGas to various utilities and end users served by WestGas. Capacity limitations on moving gas out of the San Juan Basin, however, make direct sales very difficult. WestGas itself, as a significant supplier of gas to various markets, also requests bids for spot gas from time to time. Bid prices to WestGas in the fall of 1987 were in the $1.20 to $1.40 per million Btu range.

CONCLUSIONS

Similar to any gas producing region of the U. S. today, marketing gas in the San Juan Basin poses substantial challenges. It is nearly impossible to take a passive approach to marketing San Juan Basin gas. Waiting for the pipelines in the area to purchase gas under the type of sales arrangements common during the late 1970s and early 1980s may take a long time. To be successful, one must aggressively pursue the most attractive markets, which will most likely be those accessed by El Paso.

In many instances, access to markets for new drilling will be accomplished most effectively by construction of producer-owned gathering lines, or small gathering systems built by independent gatherers that connect into the El Paso system. Construction of independent gathering systems will mean timely connection of gas to market, and can avert paying second gathering fees to pipelines such as WestGas and Northwest in order to access the El Paso Natural Gas Co. market. The Greeley Gas system, Ladd's gathering system, and other similar systems provide good examples of the feasibility of this approach.

There may be some cases where markets for San Juan Basin gas can be established through Sunterra, WestGas, or Northwest. These markets are likely to be limited in volume, highly competitive, and subject to capacity limitations and regulatory constraints.

While the El Paso Natural Gas Co. market today is dominated by large volume spot sales to California utilities, it is likely that during these next few years, direct sales to end users in California will become more common. The extent to which direct end-user sales will offer price advantages is dependent on how El Paso's rates and the rates charged by California pipelines and utilities evolve.

With the complexity of transportation and marketing gas today, many producers will be well served by using the services of an independent gas marketer. Using a marketer to aggregate gas volumes is often essential because of the current need for substantial volumes to bid for the California utility markets (many require minimum bid volumes of 10,000 MCF per day). The services of a marketer can be employed by a variety of methods, from selling the gas to the marketer to entering some type of agency arrangement to have the marketer represent the producer's interest in actively seeking the most favorable market for the producer's gas.

Non-Conventional Fuel Tax Credit

Peet M. Sööt, Ph.D., P.E.[1]

ABSTRACT

Coal-seam methane, along with certain other non-conventional fuels, is eligible for a tax credit. This production tax credit allowed coal-seam methane producers to receive $0.7526 per million Btu of gas sold during 1986. In 1987, this credit rose to $0.78 per million Btu. The tax credit is a very significant element of the economic analysis of current coal-seam methane projects. In today's spot market, gas prices are around $1.50 per million Btu. Allowing for costs of production, the gas producer will net more income from the tax credit than from the sale of the gas.

The Crude Oil Windfall Profit Tax Act of 1980 is the source of this tax credit. There were some minor changes made by subsequent legislation, but most of the tax credit has remained intact. Wells must be drilled by 1990 to qualify for the tax credit but the production from such wells is eligible for the tax credit until 2001. Projections have been made, showing that the tax credit should increase to $0.91 per million Btu for production in 1990 and $1.34 per million Btu in 2000. Variables which may decrease the tax credit from these projections are dramatically lower oil prices or general economic price deflation.

INTRODUCTION

The Crude Oil Windfall Profit Tax Act of 1980 (Tax Act) provided an incentive for production of coal-seam methane (CSM). This incentive was intended for times when a general decline of oil prices might lead to market non-competitiveness for unconventional gases. The current market place reflects such a condition. Until 1982, producers were generally receiving significant incentive prices for CSM, so there was little attention given this tax credit. With the current low prices for all natural gas, CSM producers should become aware of the significance of this credit.

The Tax Act placed the tax credit for coal-seam methane in Section 44D of the IRS Code. The 1984 Tax Reform Act changed this IRS Code section number from 44D to 29, but otherwise made no significant changes in the tax credit. Additional legislation in 1985 and 1986 made two changes in the wording of the law. One related to the tax credit not exceeding the estimated minimum tax for the taxpayer and the second change made gas sales between affiliated corporations eligible for the credit.

Production of fuels other than CSM can also utilize the tax credit. The list of qualified fuels includes: (1) Oil produced from shale and tar sands, (2) gas produced from biomass, geopressured brine, or Devonian shale, (3) liquid, gaseous or solid synthetic fuels from coal, (4) some processed wood fuels, and (5) steam produced from some agricultural byproducts. Gas produced from tight formations was initially a qualified fuel, but it was subsequently deleted by Congress. Some of these other fuels have specific rules associated with them and one should refer to the Tax Code for more details on these other fuels.

TAX CREDIT DEFINITION

The tax credit formula for CSM can be expressed in algebraic form as:

$$TC = (\$3/BOE) \times I \times PH.$$

Where TC = tax credit in dollars per million Btu, BOE × Barrel of oil equivalent, that is, 5.8 million Btu, I = Inflation adjustment factor which is based on the GNP (gross national product) implicit price deflator, and PH = Phaseout factor (always less than or equal to 1).

The phaseout factor is based on domestic crude oil prices, as follows:

$$PH = 1 - \left\{ \frac{\text{Domestic Oil Price} - \left(\frac{\$23.50}{\text{bbl}}\right) \times I}{\left(\frac{\$6}{\text{bbl}}\right) \times I} \right\}$$

The value of the tax credit for a given year cannot be calculated until March of the following year. At that time the IRS presents values for the inflation adjustment factor and the reference domestic oil price. These values are published in the Federal Register by April 1. Using these values, the formulas written above can be used to calculate the tax credit and to determine whether any phaseout may be applicable.

The decline in oil prices since 1981 has brought the tax credit into full effect. With high prices in 1980, the tax credit was phased out for that year. In 1981, the credit was only partially phased out and provided producers $0.25 per million Btu. The credit jumped to $0.66 per million Btu in 1982, as oil prices dropped sharply and no phaseout occurred. The credit has increased much more slowly since 1982. With no phaseout element, it only increases with general inflation. Figure 1 presents a graph of historic and projected values for the tax credit. Production during 1986 was eligible for $0.7526 per million Btu of tax credit.

This year's and near term future values of the tax credit are not difficult to predict. One only needs a forecast of the GNP implicit price deflator. This is a commonly predicted variable. Predicasts Forecasts (Cratcha, 1987) is a quarterly review of many economic forecasts. In the April 1987 issue, this quarterly had eleven forecasts listed from various publications for the

[1] Consulting Engineer
7675 S.W. Miner Way
Portland, OR 97225

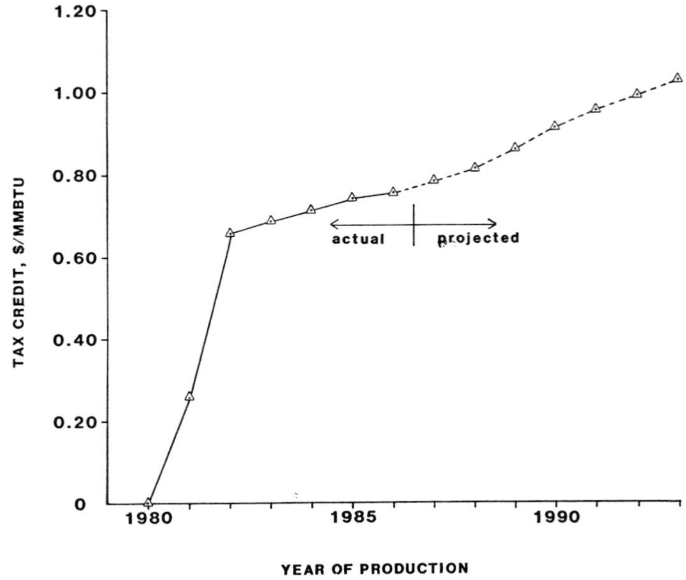

Figure 1. Graph showing coal-seam methane production tax credit from 1980 through 1986 and projected to 1993.

1986-87 annual GNP implicit price deflator. They varied from 3.0 to 4.5 percent, with an average value of 3.5 percent. The deflator for 1987-88 was predicted in 5 publications. The range was from 3.6 to 5.2 percent, with an average of 4.1 percent. These values were used to predict future values of the tax credit as listed in table 1. In 1987, the tax credit should be $0.78 per million Btu. It will rise to $0.91 per million Btu in 1990 and $1.34 per million Btu in 2000. These projections are based on the following assumptions: The GNP deflator forecasts are correct, the tax credit will not be repealed by Congress, and crude oil prices will stay below the phaseout values listed in table 1.

If crude oil prices rise above the listed values, then the credit would be partially, or totally phased out in the year in which the price rise occurs. The phaseout factor has not been an element in the calculation since 1981. It is not expected to play a role for several more years, if ever. Figure 2 is a graph showing the domestic crude oil reference price upon which the phaseout factor is based. The crude oil price would have to exceed $35.40 per bbl in 1987 before the phaseout factor would lower the tax credit from the projected $0.78 per million Btu level. This is quite unlikely. It has never been at that level. Even if prices rise sharply, as they did in the early 1980's, it would take years before those levels are reached. In 1990, crude oil prices would have to be above $41 per bbl before any phaseout occurred.

Since the actual reference price is not published until the following year, a current source for the reference price would be valuable. The U.S. DOE (Department of Energy) provides that source in the Monthly Energy Review published by the EIA (Energy Information Administration). Since 1981, the required reference price and the "Average Domestic First Purchase Price" as reported by the EIA have been the same. Prior to that date there were some differences due to price controls on new and old oil. By reviewing the crude oil price summary table in the Monthly Energy Review, one can calculate a moving average for the current year. These data are often several months old, but they are still better than having to wait until

Table 1. Coal-Seam Methane Tax Credit

Year of Production	Inflation Adjustment Factor	GNP Deflator	Oil Price, $/Bbl	CSM Credit, $/MMBtu	Oil Price Required for Phase-Out Partial	Total
1979	1.0	1.0	—	—		
1980	1.0896	1.0896	33.03	0.0	25.61	32.15
1981	1.1900	1.0921	32.19	0.2513	27.97	35.11
1982	1.2676	1.0652	28.50	0.6556	29.79	37.40
1983	1.3197	1.0411	26.19	1.0626	31.01	38.93
1984	1.3673	1.0361	25.88	0.7072	32.13	40.33
1985	1.4211	1.0393	24.08	0.7351	33.40	41.93
1986	1.4555	1.0242	12.66	0.7528	34.20	42.93
1987	1.51	1.035	—	0.78	35.40	44.43
1988	1.57	1.041	—	0.81	36.85	46.25
1989	1.66	1.056	—	0.86	38.91	48.84
1990	1.75	1.057	—	0.91	41.13	51.62
2000	2.59	1.04		1.34	60.90	76.43

Note: Recalculations may be in error due to rounding

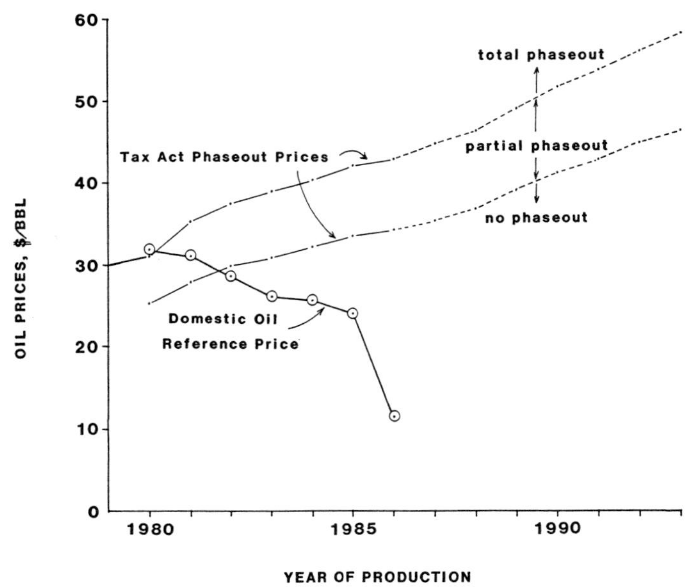

Figure 2. Graph showing phaseout factor for coal-seam methane tax credit for the years 1980 through 1986 and projected to 1991.

the following year. Knowing current spot prices for domestic oil will provide a more timely, albeit semi-quantitative indication of the Reference Price.

TAX CREDIT APPLICABILITY

The tax credit is only applicable for production from new wells on new property as defined by the Tax Act. Wells must have been drilled after December 31, 1979 and before January 1, 1990. The wells cannot be drilled into reservoirs where commercial gas was produced from coal seams prior to 1980. There are no definitions as to what is meant by "drilled". One might assume that it is the same as with other oil and gas drilling operations where spudding by a certain date is the equivalent of drilling. Production from valid wells will be subject to the tax credit until January 1, 2001.

This tax credit differs from other tax credits which have

been allowed by the Tax Code. The fuel tax credit must be taken in the year of gas production, it cannot be carried forward or backward into other tax years. Once taken, it is not subject to recapture, like investment tax credits. If the owner of coal-seam methane production does not have a current tax liability against which the credit can be taken, there are ways of reorganizing projects where the credit could be taken.

The benefits of the tax credit are apportioned among owners of the CSM production according "...to their respective interests in the gross sales...". This means that royalty interest owners are entitled to take the tax credit on their share of the gross sales. Some parties have questioned whether this would apply in the case where no record title to the gas production resides with the royalty owner. One must still consider the IRS position regarding "economic interest" in a property. Royalty owners that pass the economic interest criteria will most likely be considered as being entitled to the tax credit. This section of the Tax Code does not have any regulations to help one interpret the law. In such a case, tax counsel may have to be retained if an aggressive position is taken in the matter.

If the CSM gas is sold at incentive prices allowed under Section 107 of the Natural Gas Policy Act (NGPA) of 1978, then the tax credit may not be applicable. This is an academic consideration at the present time. CSM is currently being sold at the same prices as conventional gases due to market constraints. Those prices are not even near the levels allowed for conventional gases under Section 102 of the NGPA.

The following items are some more details of the Tax Credit which may be important in particular applications: The producer claiming the tax credit must sell the gas to an unrelated party (This was broadened under recent legislation. Now affiliated corporations are considered as unrelated parties for the purposes of the tax credit.); any grants or subsidies received from the Federal or State governments for a given project must be deducted from the tax credit; and only production within the United States is eligible for the tax credit.

TAX CREDIT INTERPRETATION

There are no regulations in force regarding this tax credit. No court rulings have been located addressing the credit. The only interpretive sources which one can reference are IRS Letter Rulings, Revenue Rulings, and General Counsel Memoranda. Even though Letter Rulings cannot be cited as precedent in non-originating cases, they do provide an indication of the IRS' current stance regarding some of the issues surrounding the credit.

Numerous Letter Rulings have been published by the IRS relating to this tax credit. Some involve tight-gas sands and other qualifying fuels. They all provide some guidelines for interpreting the credit. Several Letter Rulings address incentive price considerations. They generally conclude that producers can switch back and forth between accepting incentive prices and the tax credit, but one cannot double dip. Other Letter Rulings consider the allocation of the tax credit among partners of a partnership. All of the rulings and published memoranda should be studied if one is interested in a full picture of the tax credit and its interpretation.

REFERENCES CITED

Cratcha, George, ed., 1987, Predicasts Forecasts: Predicasts, Inc., Cleveland, Ohio, p. B23-B24.

View looking north over Durango from Smelter Mountain prior to 1956. Fort Lewis College had not been moved to the bluff on the right side of the photo where a small airplane landing strip is visible. Photo courtesy of the Center of Southwest Studies.

Coal-Bed Gas Development in the San Juan Basin: A Primer for the Lawyer and Landman

by
J. Hovey Kemp*
and
Kurt M. Petersen**

ABSTRACT

The recovery of coal-bed gas as an energy resource has become increasingly attractive due to recent technological advances and changing market conditions. The northern San Juan Basin, located in northwestern New Mexico and southwestern Colorado, has been and will continue to be the focus of coal-bed gas development. This new interest has brought two legal questions to the fore: the question of ownership of the coal-bed gas and the rights to develop that gas, and the question of competing rights when both the coal and the gas are to be developed. Although those issues are far from being resolved, the risks from a legal standpoint are ascertainable and, in most cases, should be manageable.

The legal issues surrounding ownership of the coal-bed gas must be analyzed separately for each type of land ownership in the San Juan Basin—fee, Federal, Indian, and state. Although the law is by no means settled or uniform on the ownership question, all coal-bed gas development to date has been done pursuant to the terms of oil and gas leases. It appears to be an acceptable business risk at this time for the owner of the oil and gas on fee, Federal, Indian (tribal or allotted), or state lands, or its lessee, to continue to develop coal-bed gas in this manner, provided there are no specific contrary grants in the chain of title. The magnitude of the legal risk, and the potential ownership challenge by the coal owner or its lessee, is directly related to the minability of the targeted coal seam—the greater the minability, the greater the chance for conflict.

The second legal issue and area of potential conflict involves the planned or concurrent development of both the coal and the coal-bed gas resources. The major areas of potential conflict are access to the target coal seam through shallower coal seams and the possible damage to the coal seam done by the development of the coal-bed gas. As above, the more minable the coal, the greater the legal risk and chance for conflict.

The key to anticipating disputes regarding coal and coal-bed gas rights lies with the early identification of possible conflicting claimants to those rights. Careful and extensive analysis of the land ownership records, with an eye toward any special interests in the coal, should identify the legal risks and the likelihood of disputes. If such disputes seem likely, advance negotiation is the key to minimizing the legal risks associated with coal-bed gas development.

* Attorney at Law
Davis, Graham & Stubbs
370 Seventeenth Street, Suite 4700
Denver, Colorado 80201
A.B. 1976 Harvard University;
J.D. 1980 Georgetown University Law Center.

** Attorney at Law
Davis, Graham & Stubbs
370 Seventeenth Street, Suite 4700
Denver, Colorado 80201
B.A. 1974 St. Olaf College;
Ed.M. 1980 Harvard University;
J.D. 1986 University of Denver.

EDITOR'S NOTE: The format for this paper is different from the other papers in this guidebook. Because the paper was primarily written for lawyers and landmen, it was decided to leave the paper in standard legal-paper format and not try to force it into our scientific-paper format. Having said that, I would urge all geologists, oil and gas company managers, and even engineers to read this paper carefully; it is clearly written and summarizes this complex problem well.

CONTENTS

I. INTRODUCTION.

Coal-bed gas,[1] the principal component of which is methane, was traditionally considered to be a dangerous nuisance rather than an energy resource in and of itself. Therefore, disputes involving the legal issues of ownership of and access to coal-bed gas seldom arose. Coal-bed gas was part of the workaday vocabulary of the coal operator (probably in a profane sense) and was not something to which natural gas producers gave much thought. Historically, coal mine operators vented coal-bed gas from coal seams during their mining operations for safety reasons; now, Federal law requires them to have stringent methane control programs. Such wasting of coal-bed gas by ventilation continued until recently without regard for the fact that, depending on the legal definition of coal-bed gas, the coal mine operator may have been venting into the atmosphere what was actually the valuable property of the owner of the gas rights.

The recovery of coal-bed gas as an energy resource is now more economically attractive. Because of a combination of improved economics and certain technical advances, the last five years have seen a marked increase in experimental and commercial coal-bed gas development projects in the Appalachian coal fields, which includes the Warrior Basin of Alabama, in the San Juan and Raton Mesa Basins of southern Colorado and northern New Mexico, in the Piceance Basin of western Colorado, in the Arkoma Basin of Oklahoma and in the Book Cliffs coal field of the Uintah Basin in Utah. As a result of these activities, increased attention is now being paid to the legal questions of ownership of and access to coal-bed gas found in Federal, Indian, state, and fee coal deposits. At the present time, answers to those questions are beginning to take shape but are far from being settled, which means that certain legal risks must be considered by the coal-bed gas developer.

The issues which are the subject of this paper have received substantial attention from legal commentators.[2] This article does not reiterate the legal debate among such commentators. Rather, it is meant to be a practical guide by which coal-bed gas developers may avoid disputes or costly litigation in this area. It is the goal of the authors to summarize the current state of the law surrounding the issues of ownership of and access to coal-bed gas, define the resulting legal risks, and discuss the means by which such risks may be mitigated.

This paper, which concentrates on the issues as they pertain to the northern San Juan Basin in New Mexico and Colorado, is organized as follows: Part II is a general discussion of coal-bed gas as an energy resource. Part III focuses on the legal issues of ownership of and access to coal-bed gas, as such issues pertain to Federal lands. Part IV considers these issues as they relate to Indian lands, specifically those of the Southern Ute Indian Reservation. Part V examines these issues as they relate to state-owned minerals. Part VI discusses such issues as they relate to fee lands; particular attention is paid to the holding of the Pennsylvania Supreme Court in *United States Steel Corp. v. Hoge, et al.*[3] and its possible impact on coal-bed gas development activities on fee lands in the western United States. Part VII discusses coal-bed gas activities as they relate to the jurisdiction of state oil and gas conservation agencies. Finally, Part VIII offers some conclusory thoughts and summarizes the steps necessary to mitigate the legal risks associated with coal-bed gas development projects.

II. THE NATURE OF COAL-BED GAS.

A. Scientific Definition.

Coal-bed gas is present in all coal beds and is formed by biochemical and physical processes during the conversion of accumulated plant material into coal. Of the mixture of gases termed coal-bed gas, methane constitutes 80 to 99 percent. Coal-bed gas is similar to natural gas in physical and chemical properties and can be interchanged with natural gas, added directly to natural gas pipelines, used as a chemical feedstock, converted to liquid natural gas, burned as a boiler or process-heating fuel, or used in gas turbines for generating electricity. The amount of coal-bed gas contained in a specific coal bed depends on a number of geologic factors, including the thickness and extent of the coal bed, the rank of coal, the thickness of overburden, and the hydrostatic pressure. Coal-bed gas is present in coal fractures and is adsorbed on the micropore surfaces of even the tiniest piece of coal; the gas is released when confining pressure is reduced. Although the quality of coal-bed gas varies, it is generally quite good; the heat of methane combustion ranges from 950 to 1,050 Btu per cubic foot. Coal-bed gas has few impurities; carbon dioxide and water vapor are the primary undesirable components; sulfur dioxide and hydrogen sulfide gases are not present even in the more sulfur-rich coals.[4]

[1] For purposes of uniformity, the term coal-bed gas is defined to mean all natural gas, including methane, which is or can be produced from a coal seam.

[2] For legal periodicals which contain detailed discussions of the issue of ownership of coal-bed gas, see Cohen, Farnell and Thompson, Legal and Regulatory Aspects of Coal-bed Gas Development, University of Alabama School of Mines and Energy Development, July 1984 (hereinafter, SOMED Research Report at ______); Farnell, *Methane Gas Ownership: A Proposed Solution for Alabama*, 33 Ala. L. Rev. 521 (1982) (hereinafter, Farnell at ______); Farnell, *Legal Constraints on Methane Gas Development* Monograph No. 5, Office of Energy and Environmental Law, University of Alabama Law Center (July 1981) (hereinafter, Monograph at ______); Bowles, *Coal-bed Gas: Present Status of Ownership Issue and Other Legal Considerations*, 1 E. Min. Law Inst. 7-1 (1980) (hereinafter, Bowles at ______); Comment, U.S. Steel v. Hoge, 82 W. Va. L. Rev. 1451 (1980); McGinley, *Legal Problems Relating to Ownership of Gas Found in Coal Deposits*, 80 W. Va. L. Rev. 369 (1978) (hereinafter, McGinley at ______); Craig and Meyers, *Ownership of Methane Gas in Coal Beds*, 24 Rocky Mtn. Min. L. Inst. 767 (1978) (hereinafter, Craig and Meyers at ______); Olson, *Coal-bed Gas: Legal Considerations Affecting Its Development as an Energy Resource*, 13 Tulsa L. J. 377 (1978) (hereinafter, Olson at ______); Williams, *On Leasing Gas from Coal Seams*, 47 W. Va. L. Q. 211 (1941). For two excellent legal articles that discuss (in part) the issue of access to coal-bed gas, see Lear, *Multiple Mineral Development Conflicts: An Armageddon in Simultaneous Mineral Operations*, 28 Rocky Mtn. Min. L. Inst. 79 (1983) (hereinafter, Lear at ______); Schissler, *Developmental Conflicts and Constraints Dealing With the Problem of Coexistent Estates*, 22 Rocky Mtn. Min. L. Inst. 203 (1979).

[3] 468 A.2d 1380 (Pa. 1983) (hereinafter referred to as *U.S. Steel*).

[4] The above description of the resource was based on various technical papers on coal-bed gas. Since this article is not meant to be a technical paper, the authors have elected not to present a list of citations to the many technical papers which have been prepared on the production of coal-bed gas by engineers, geologists and the like. For a complete discussion of technical and geological aspects of coal-bed gas development, the articles in this volume should be consulted. In addi-

In *U.S. Steel*, the seminal judicial decision in this area which will be discussed later, the trial court judge eloquently described coal-bed gas as follows:

> It is a gas. . . which. . . has. . . a close affinity for and association with coal seams. In its original state it permeates and penetrates the coalbed, is its alter ego, its constant companion, its geological handmaiden, and is sometimes viewed as its contumacious free-spirited bride, but more generally regarded as its ill-chosen bridesmaid. It is found with the coal when they come to mine it, stays with the coal as it leaves, and remains in the space after the mining has been done. Its past has been filled with peril and tragedy, its present is seen as having a modest commercial attractiveness, and its future as a fuel potential has become increasingly brighter.[5]

B. Coal-bed Gas and Mine Safety.

Because methane in air is extremely explosive at concentrations of about 5 to 15 percent, historically it has received the greatest attention as a coal mine safety hazard. The Federal Mine Safety and Health Amendments Act of 1977 includes stringent requirements regarding methane control in underground coal mines.[6] Mine ventilation is the usual means employed for coal-bed gas removal. However, research aimed at developing technologies for removal of coal-bed gas prior to mining has also been undertaken, largely under the aegis of the U.S. Bureau of Mines and, recently, the U.S. Department of Energy.

Principal methods utilized to remove methane from coal beds prior to mining include drilling vertical holes from the surface and horizontal holes from the bottom of shafts. One of the most heated technological controversies in the area of coal-bed gas surrounds the issue of whether removal of coal-bed gas by techniques such as hydraulic stimulation, otherwise known as hydrofracturing, causes damage to the structural integrity of the coal seam, thereby making it more hazardous and expensive to later mine the coal.

The removal of methane prior to mining is attractive because such removal enhances mine safety, reduces ventilation costs, and decreases atmospheric pollution; its recovery, if marketed, conserves an energy resource.

C. Estimate of Available Resources of Coal-bed Gas.

Coal-bed gas, which is currently produced or exhausted into the atmosphere through mine ventilation systems, is a potential resource which could greatly benefit the nation's need for energy. Estimates of coal-bed gas removed from all United States coal mines range from 200 to 250 million cubic feet per day; this translates into enough gas to heat 800,000 homes every day.[7]

The Bureau of Mines estimates the volume of gas contained in coal in the United States to be 300 trillion cubic feet,[8] equal to the entire amount of proven and currently recoverable reserves of conventional natural gas. Of this 300 trillion cubic feet of gas, 150 trillion cubic feet are now, according to the Department of Energy, commercially feasible for development.[9]

D. General Description of the San Juan Basin.

Most of the data on coal-bed gas resources has been obtained through research on bituminous coals in the East. Deep, high-rank coal beds are the most promising sites for the production of coal-bed gas. The Black Warrior Basin of Alabama has been the focus of perhaps the most intense coal-bed gas play in the East, with the San Juan Basin occupying that position in the West. Although 90 percent of the coal reserves in the Western States are subbituminous or lignite in rank (much lower than eastern coals), recent studies indicate that because of their great thicknesses, these lower-grade western coals contain large volumes of coal-bed gas.[10]

The San Juan Basin is located in northwestern New Mexico (San Juan, Rio Arriba, Sandoval and McKinley Counties) and southwestern Colorado (La Plata and Archuleta Counties) and includes parts of the Navajo, Southern Ute, Ute Mountain Ute, and Jicarilla Apache Indian Reservations. The basin is almost 100 miles long (north to south) and 90 miles wide (east to west) and extends over an area of about 7,500 square

tion, for a complete bibliography of such papers, the authors suggest that the Gas Research Institute in Chicago be contacted. A few of the more notable sources which the authors depended upon are the Gas Research Institute's Quarterly Review of Methane from Coal Seams Technology; *see also* Gas Research Institute, *Position Paper: Unconventional Natural Gas* (May 1981); National Petroleum Council *Unconventional Gas Sources: Coal Seams* (June 1980); Kuuskraa and Meyer, *Review of World Resources of Unconventional Gas* (1980); Potential Gas Agency, *Potential Supply of Natural Gas in the United States* (May, 1981); Deul and Kim, *Coal Beds: A Source of Natural Gas*, The Oil and Gas Journal (June 16, 1975); Gallo, *Report of The Ad Hoc Task Force on Methane Drainage to The Assistant Secretary—Resource Application*, Dept. of Energy (Dec. 14, 1977); Deul, *Natural Gas From Coalbeds*, The Natural Resource Council's Natural Gas From Unconventional Geologic Sources (1976); Kim, *Estimating Methane Content of Bituminous Coalbeds from Adsorption Data*, U.S. Department of the Interior, Bureau of Mines, Report of Investigations (USBM RI) 8245 (1977).

[5] Unreported opinion in Nos. 78-682 and 691 (C. P., Greene Co., Pa., September 29, 1980) at 20. After the court entered its adjudication and decree nisi on March 24, 1980, the plaintiff, United States Steel, filed exceptions. The final decree of the trial court dismissed those exceptions and dissolved an injunction previously entered in United States Steel Corp. v. Hoge, No. 78-682 (C. P., Greene Co., Pa., Jan. 9, 1978).

[6] 30 U.S.C. §§ 863, 877(h).

[7] Olson at 381.

[8] Deul and Kim, *Coal Beds: A Source of Natural Gas*, The Oil and Gas Journal at 47 (June 16, 1975). The 300 trillion cubic feet figure is apparently based on methane estimates in deep coal only.

[9] Gallo, *Report of the Ad Hoc Task Force on Methane Drainage to the Assistant Secretary—Resource Applications*, Dept. of Energy at (Dec. 14, 1977). This estimate may not be uniformly accepted, depending on matters such as the deregulation of gas prices in FERC.

[10] *See* Deul, *Natural Gas from Coalbeds*, in "Natural Gas from Unconventional Geologic Sources," The National Research Council at 195 (1976). In addition, the Colorado Geological Survey is in the process of completing substantial reports on their studies. To the best knowledge of the authors, two have been completed to date, covering the San Juan River Coal Region (Open-File Report 80-2) and the Raton Mesa Coal Region (Open-File Report 80-4).

miles. The Fruitland Formation, the primary target of coal-bed gas developers, contains in excess of 200 billion tons of coal throughout the San Juan Basin. Fruitland coal is distributed in a series of northwest-trending belts across the San Juan Basin, and is found at outcrops around the rim of the basin down to depths of more than 4,000 feet in the north central part of the basin.[11]

Interest in the coal-bed gas underlying the San Juan Basin has increased since Phillips Petroleum Company deliberately drilled the first coal-bed gas well into the Fruitland Formation in 1953.[12] To date, 243 Fruitland and Fruitland-Pictured Cliffs test wells and locations have been reported in the Quarterly Review of Methane from Coal Seams Technology. Recently published reports indicate that the coal-bed gas resource of the Fruitland is around 50 trillion cubic feet[13] which greatly exceeds the fourteen trillion cubic feet of gas produced to date from conventional sandstone reservoirs in the San Juan Basin. The development of new drilling, completion, and production techniques,[14] relatively shallow drilling depths of 2,500 to 3,500 feet costing $200,000 to $300,000 per well,[15] and the size of estimated reserves make coal-bed gas development in the San Juan Basin one of the brighter spots in an otherwise depressed oil and gas industry. One of the most promising areas for coal-bed methane is the northern San Juan Basin of Colorado and New Mexico and especially the Southern Ute Indian Reservation in southwestern Colorado. Speculation is that the Fruitland Formation in that area may be tested by as many as 200 wells in the next two years. In fact, Amoco Production Company alone may drill as many as 100 wells by January of 1989.[16]

E. Reasons for Lack of Development.

Despite its promise, development of coal-bed gas as an energy resource has not occurred on any large scale until quite recently. There are a number of reasons for this situation:

(1) Coal-bed gas wells initially (first five years) generally produce lower volumes of gas than do gas wells producing from conventional gas reservoirs. Thus, most natural gas producers have found that the value of coal-bed gas at previously regulated prices was too low to justify development compared to that of natural gas found in conventional reservoirs. The gas-tax credit, discussed by Soot elsewhere in this volume, however, changes the economics of coal-bed gas.

(2) Because gas production from coal is still a developing technology, the economics of producing coal-bed gas have not yet been finalized. The production of coal-bed gas requires pumping large amounts of water out of the coal, and the disposal of this water in accordance with state and environmental regulations is a problem in some areas.

(3) Until recently, most coal companies were not interested in, nor did they have the available capital to invest in the equipment necessary for commercial coal-bed gas production; they found it easier to vent unwanted gas during mining. Now that these companies have found that it is more economical and efficient to produce the hazardous gas before mining the coal, they have also found it profitable to sell the collected gas rather than vent it to the atmosphere.

(4) Coal-bed gas ownership and access to the resource have been clouded by certain unresolved and complex legal issues.[17]

Due to technical advances and changes in the world energy supply, the recovery and utilization of coal-bed gas now appears to be more feasible.[18] As a result, recent years have been marked by increased interest and development of the resource and, during this time, the legal risks associated with coal-bed gas development have been brought into clearer focus and are beginning to be addressed. Notwithstanding the newfound promise of coal-bed gas as an energy resource, the developer of coal-bed gas should evaluate these legal risks in conjunction with geologic and economic feasibility studies of the targeted coal-bed gas prospect. These legal risks and the steps that can be taken to mitigate them are discussed in detail below, as they pertain to Federal, Indian, state and fee mineral resources, with a particular focus on the San Juan Basin.

III. COAL-BED GAS ACTIVITIES ON FEDERAL LANDS.

This section focuses on the special problems the coal-bed gas developer may encounter and the procedural hurdles which should be anticipated where either the coal or the oil and gas, or both, are owned by the United States.

A. Introduction.

Where the United States owns the coal but not the oil and gas, one cannot unequivocally conclude that a fee oil and gas lease covers the gas found in the Federal coal deposit. Nor can one unequivocally conclude, where the United States owns all the minerals under a tract of land, that a Federal oil and gas lease covers the gas found in the Federal coal deposit. In part, this is because Congress has never specifically addressed the question of whether the right to extract coal-bed gas is part of a Federal coal estate. Thus, these issues have to be resolved by the construction of existing Federal statutes which (1) provide for the reservation of mineral resources to the United States and (2) authorize the leasing of those mineral resources.

The coal-bed gas developer, however, is not totally without guidance in this area. In fact, a thorough analysis of the issues surrounding the ownership of the coal-bed gas in the Federal

[11] Fassett and Hinds, Geology and Fuel Resources of the Fruitland Formation and Kirtland Shale of the San Juan Basin, New Mexico and Colorado; U.S.G.S. Professional Paper 676 (1971).

[12] The Phillips #6-17 well is located in T.31N., R.7W., San Juan County, New Mexico.

[13] See Kelso paper, this volume.

[14] Bowles at § 7.04[1].

[15] *Coal-bed Methane Romancing Operators*, Western Oil World, p. 19, November 1987.

[16] Amoco Production Company Press Release, February 12, 1987.

[17] See SOMED Research Report at v.

[18] It is suggested that the coal-bed gas developer get on the mailing list of the Gas Research Institute's quarterly journal, the Quarterly Review of Methane from Coal Seams Technology. In addition, there is a group called the Coal-bed Gas Forum which meets monthly in Denver to discuss coal-bed gas exploration and production problems. Attendance at such meetings by a coal-bed gas developer's technical personnel can assist in identifying and discussing new technologies and other technical matters pertaining to coal-bed gas.

situation results in a conclusion that the legal risks are probably manageable from a business standpoint insofar as ownership of the resource is concerned. Further, regarding the access question, the legal risks depend on whether a Federal coal lease has been issued. Here again, an analysis of the relevant statutes and regulations governing multimineral development can provide the coal-bed gas developer with good guidance.

On May 12, 1981, the Solicitor of the Department of the Interior, after examining the relevant Federal statutes, issued his Opinion that, first, a reservation of coal to the United States does not include coal-bed gas and, second, a Federal reservation of gas does encompass gas found in deposits of coal and, therefore, such gas is disposable as a gas under Section 17 of the Mineral Leasing Act.[19] Thus, according to the Solicitor, where the oil and gas and coal are federally owned, a Federal oil and gas lease covers gas found in coal seams and coal-bed gas operations, and production by the Federal oil and gas lessee is subject to the regulations governing normal oil and gas drilling and production operations. Where the oil and gas is owned in fee and the coal is federally owned, the Solicitor's Opinion disclaims ownership of the coal-bed gas, subject to the government's right to protect the federally owned coal resource from damage.

B. **Analysis of the Solicitor's Opinion.**

The Solicitor first considered the laws surrounding Federal reservations of coal in patents. Patents issued pursuant to the Act of March 3, 1909,[20] or the Act of June 22, 1910,[21] reserved to the United States "all coal" and the right to prospect for, mine, and remove the "coal deposits" underlying the patented lands. In examining the legislation and its history, the Solicitor found that, in the 1909 and 1910 Acts, Congress intended to, and did, reserve only the coal and not other minerals found in association with coal. Accordingly, the Solicitor concluded that all minerals other than coal, including coal-bed gas, passed to the surface owner at the time a patent was issued pursuant to the 1909 or 1910 Act.[22]

Next, the Solicitor considered coal-bed gas in the context of the question of Federal reservations of oil and gas. After reviewing the statutes and the legislative history, the Solicitor concluded that where coal-bed gas occurs in lands on which oil and gas were reserved to the United States in agricultural patents issued under the Act of July 17, 1914,[23] coal-bed gas is reserved to the United States.

In reviewing the Mineral Leasing Act of 1920[24] and its legislative history, the Solicitor found that the Mineral Leasing Act refers only to gas or natural gas, without any qualifying adjectives, thus supporting a nonrestrictive reading of those terms to include coal-bed gas.[25] The Solicitor noted that the standard Federal oil and gas lease grants to the lessee the right to drill for, extract, and dispose of all oil and gas, except helium, and that current Federal oil and gas lease forms do not reserve coal-bed gas from the granting clause. The Solicitor concluded that coal-bed gas is leasable under the oil and gas leasing provisions of the Mineral Leasing Act and, thus, drilling and production of Federal coal-bed gas are governed by the Oil and Gas Operating Regulations.

Conversely, the Solicitor found that the coal leasing provisions of the Mineral Leasing Act did not provide for a coal lessee's extraction of minerals related to or associated with coal.[26] The Solicitor concluded that those statutory provisions do not authorize a coal lessee's extraction of coal-bed gas, other than the ventilation of the gas required by mine safety and health laws. However, the Solicitor made it clear that nothing in his Opinion warrants title to any coal-bed gas deposit nor detracts from any coal lessee's obligation to comply with the methane ventilation provisions of the Federal and State laws relating to the safety and health of the nation's coal miners. The Solicitor also noted that the oil and gas lessee does not have a license to develop the coal-bed gas in a manner that would damage the coal or create hazards for subsequent coal mining operations. In addition, the Solicitor stated that the rights of an oil and gas lessee would be restricted to the rights not previously granted to the coal lessee.[27]

The authors are not aware of any recent or impending change in or challenge to the position taken by the Office of the Solicitor regarding the issues covered by the May 12, 1981 Opinion. It is cautioned, however, that the formal Opinion of the Solicitor is not binding upon the courts and can be revoked by a subsequent Solicitor who concludes differently. Although a court will not be bound by the Solicitor's interpretation of the Federal statutes in question, the current interpretation should continue to be persuasive in a judicial setting, since courts tend to defer to the judgment of the agency charged with implementing the laws they are interpreting, and since the current interpretation does not favor the interest of the United States. Accordingly, until the issue is settled by litigation, reliance on the Solicitor's Opinion involves some risk. This risk, however, seems minimal at the present time because states such as New Mexico have informally followed the Federal position, and because the Solicitor's Opinion has gone unchallenged for almost seven years. Thus, the position of the Solicitor enunciated in his May 12, 1981 Opinion is still good law within the Federal agencies having jurisdiction over mineral leasing activities on Federal lands.

C. **The 1981 Position of the USGS.**

Although the position taken by the Solicitor seems to have prevailed, other approaches to the coal-bed gas leasing dilemma have been investigated in the past and may still be implemented in the future. The position taken by the United States Geological Survey (USGS) in 1981 regarding the leasing of coal-bed gas provides a good example of another approach to the issue. At about the same time the Solicitor was issuing his Opinion, the USGS was considering formally recommending to the Secretary of the Department of the Interior that the

[19] *Ownership of and Right to Extract Coal-bed Gas in Federal Coal Deposits* (M-36935) (Memorandum of the Solicitor to the Secretary of the Department of the Interior) (May 12, 1981) 88 I.D. 538 (1981).

[20] 30 U.S.C. § 81.

[21] 30 U.S.C. §§ 83-85.

[22] 88 I.D. at 542.

[23] 30 U.S.C. § 121.

[24] 30 U.S.C. § 181, *et seq.*

[25] 88 I.D. at 544-49.

[26] *Id.* at 547, citing 30 U.S.C. § 201.

[27] *Id.* at 549; *see also* Lear at 147.

current Federal statutes governing the leasing of Federal mineral resources be charged so as to provide for combined leases covering coal and associated coal-bed gas. The proposed recommendation contemplated a statutory amendment similar to the change made to the Mineral Leasing Act in regard to combined hydrocarbon leasing.[28] This recommendation was not accepted at the time, and to the authors' knowledge has not been reproposed because the position taken by the Solicitor's Opinion apparently carried the day within the Department of the Interior.

D. Coal-bed Gas Development on Federal Lands.

Coal-bed gas found in Federal coal deposits may be explored for and produced under either a fee or Federal oil and gas lease, depending on who owns the non-coal minerals. The legal risks associated with developing the gas in the Federal coal deposit are fairly clear. First, there is the above-referenced risk that the Solicitor's Opinion, which currently governs the ownership issue, will be somehow overturned. This risk seems relatively remote at the present time, and should be viewed by developers as an acceptable business risk. Second, even if no coal lease has been issued, there is a risk that the Bureau of Land Management (BLM) will nonetheless move to protect its coal resource and inhibit the coal-bed gas developer's access to the resource. At least at present, this risk, too, seems manageable. Third, there is a legal risk that the Federal government has issued a coal lease and that either the BLM or the Federal coal lessee will oppose the oil and gas lessee's attempts to gain access to the coal to develop the resource. Here the degree of risk in proceeding with the development of the coal-bed gas must be assessed on a case-by-case basis, depending on the possibility of development under the coal lease.

E. Federal Units.

In general, our experience is that the BLM is very interested in cooperating with private companies in the development of coal-bed gas on Federal lands. As industry interest in coal-bed gas increases, several companies have formed Federal units to assist them in developing their coal-bed gas prospects and to protect correlative rights.[29] A wise coal-bed gas developer should look to the experience of these companies for guidance.

IV. COAL-BED GAS ACTIVITIES ON INDIAN LANDS.

Let us turn now to the legal risks involved when the coal-bed gas developer has targeted a coal-bed gas prospect involving Indian lands. For purposes of this paper, we will focus in particular on the Southern Ute Indian Reservation in southwestern Colorado (the Reservation).

A. Introduction.

This section discusses two issues: (1) the ownership of minerals on the Southern Ute Indian Reservation and (2) the right to develop the coal-bed gas under those lands. The ownership of Indian land and hence the ownership of the minerals beneath the land differs from other types of ownership within the United States. This fact affects all operational phases of mineral resource development on Indian land. Historically, mineral developers operated on Indian land in much the same manner as on Federal land, recognizing that title matters and document execution required more attention.[30] In recent years, as tribes have moved toward control of their own destiny vis-a-vis mineral development, these historical patterns have been changing. Moreover, with interest in the coal-bed gas of the San Juan Basin escalating, the heretofore unaddressed issue of coal-bed gas ownership on Indian land is coming to the fore.

B. Formation of the Southern Ute Indian Reservation.

To ascertain the nature of the ownership of the minerals under the Reservation, it is necessary to examine the actual mineral rights granted the Southern Ute Indian Tribe (the Tribe) and individual members of the Tribe in the documents forming the Reservation. The Reservation as it exists today is the result of a series of events that occurred from 1868 through the present. Although two previous treaties were made with various bands of what later became known as the Ute Indians,[31] the March 2, 1868 Treaty between the United States and Confederated Bands of Ute Indians (the 1868 Treaty) actually established the Ute Indian Reservation in Colorado.[32] At the time of its creation, the Reservation contained 14.7 million acres and covered the western quarter of Colorado.

In 1873 and again in 1880, the Ute Tribe ceded 3 million and 11.7 million acres, respectively, of their lands to the United States.[33] The 1880 agreement (the 1880 Agreement) ratified the terms of the 1868 Treaty and divided the Confederated Band of Ute Tribes into three groups, the Uncompahgre, White, and Southern Ute Indians.[34] Certain lands were set apart for each

[28] This information is based upon United States Geological Survey, *Secretarial Issue Document (Draft)* (June 3, 1981), a copy of which the authors obtained in 1982. To the authors' knowledge, this document was not finalized, submitted to the Secretary of the Department of the Interior, or publicly issued.

[29] An example of such a unit is the Cactus Valley Unit, BLM Contract No. 14-08-0001-20420, which unitized approximately 30,000 acres for the development of coal-bed gas in Pitkin and Garfield Counties, Colorado.

[30] Moore, *Mineral Development on Indian Lands—Cooperation and Conflict*, 28 Rocky Mtn. Min. L. Inst. 1 (1983) (hereinafter, Moore at ______).

[31] In the twenty years following the treaty of Guadalupe Hidalgo with Mexico in 1848, the United States entered into negotiations with various Indian tribes occupying the land acquired from Mexico. Two of those treaties were made with "Ute" tribes—one with the Utah Indians (December 30, 1849, 9 Stat. 984) and one with the Tabeguache Band of Ute Indians (December 14, 1864, 13 Stat. 673). Whatever may have been the interest of the Ute Indians in the land occupied by them in this period, the treaty of 1868 superseded the two previous treaties and established a reservation for the Ute Indians having the same status as any other Indian reservation in the United States.

[32] Treaty With The Ute, 1868, 15 Stat. 619, ratified July 25, 1868, proclaimed November 6, 1868. The status of this territory as a reservation has been uniformly recognized by the Congress, the Court of Claims (The Ute Indians v. The United States, 45 Ct. Cl. 440 (1910)), and the Department of the Interior. The Ute Reservation established by the 1868 Treaty is clearly within the definition of an Indian reservation enunciated by the Supreme Court in Minnesota v. Hitchcock, 185 U.S. 373, 389-90 (1902)—that an Indian reservation is created when from what has been done there results a certain defined tract appropriated to Indian purposes.

[33] 18 Stat. 36-40, c. 136 (1874), and 21 Stat. 199 (1880).

[34] The Act of June 15, 1880 was ratified by Congress as amended at 21 Stat. 199-205, c. 223 (1880); 22 Stat. 449, c. 61 (1883); 23 Stat. 22, c. 50 (1884).

group, with lands located along the La Plata River in southern Colorado designated for the Southern Utes. The 1880 Agreement also made provisions for allotments of land to individual Indians. These allotments were inalienable and nontaxable for twenty-five years and then were to be patented in fee, unless the President of the United States extended the period of restrictions.[35]

Approximately one year after the 1880 Agreement was signed, the White and the Uncompahgre bands of the Utes were moved onto the Uintah Reservation in Utah. The relocation of the Southern Utes onto their designated reservation, however, did not proceed as smoothly. The designated reservation was not suitable agricultural land and was incapable of supporting the Utes. Bills attempting to relocate the Southern Utes were introduced in Congress each year from 1886 through 1893, but failed to win support. In 1895, Congress passed the Act of February 20, 1894, locating the Southern Utes on the Reservation as originally defined in the 1880 Agreement.[36] The bill also provided that individual allotments of land were to be distributed to eligible Ute Indians who elected to take agricultural lands under the provisions of the 1880 Agreement. By April 1896, the allotments had taken place; 72,811 acres were distributed to 371 Southern Ute Indians.[37] Although those Indian allottees retained their interest in Tribal property, many of them conveyed away the surface and (or) mineral interest in their allotted lands to the Tribe or non-Indians after restrictions on the allotted lands were removed. Not all of the lands available to be allotted were in fact allotted, and the land remaining was opened to settlement by non-Indians on May 4, 1899.[38] Shortly thereafter, Congress passed the Ute Free Homestead Act of June 13, 1902,[39] which threw the unallotted lands not disposed of before the date of the Act open for free homesteading. This Act specifically contradicted the terms of the 1880 Agreement, which provided that the lands were to be sold for cash. That portion of the opened land which was not in fact settled became the subject of a great controversy after the Indian Reorganization Act (IRA) was passed in 1934.[40]

The dispute centered around the classification of the undisposed lands ceded by the Utes under the 1880 Agreement as remaining surplus lands, as defined in the IRA. If so classified, the undisposed lands were available for restoration to Tribal ownership. After determining that the lands were so classified,[41] in 1938 the Secretary of the Interior restored 222,016 acres of remaining surplus lands to the Southern Ute Reservation.[42] This restoration of lands also included the severed minerals, primarily coal, under approximately 500,000 acres, previously reserved by the United States.[43] The lands not restored to Indian ownership, but ceded under the 1880 Agreement, became the subject of numerous legal proceedings focusing on the provisions of the 1880 Agreement providing for cash payment for the land.[44]

Since the late 1930s, the Southern Ute Tribe has actively acquired both surface and mineral interests from Indian allottees and private landowners, and has also participated in a few private exchanges of land. Most of these acquisitions have been negotiated by the Tribe and all have earned the approval of the Secretary of the Interior. Currently, the Reservation covers approximately 618,000 acres, with the following divisions of ownership: Tribal surface acreage—304,000; Tribal oil and gas acreage—331,000; Tribal coal acreage—520,000; allotted surface acreage—6,800; fee surface acreage—312,000; and Federal government (Forest Service) acreage—49,000.

C. Mineral Ownership on the Southern Ute Reservation.

As the pages which follow discuss, a coal-bed gas developer interested in drilling for coal-bed gas on the Southern Ute Indian Reservation and in assessing the legal risks associated with such operations must come to understand the ownership of the minerals on the Reservation. Although the Southern Ute Indian Tribe or individual Southern Ute Indians own the Reservation lands, this ownership, whether by the Tribe or by an individual, differs from other types of land ownership in

[35] As part of the Agreement, the Utes were to receive the proceeds from the sale of the 11.7 million acres. Following the 1880 cession, some of the land was sold for cash but, contrary to the Agreement, some of the land was given away under the Ute Free Homestead Act of 1902, and some was withdrawn for national forests and other purposes. Since the United States had made no accounting to the Utes, in 1909, Congress authorized the Utes to sue in the Court of Claims for the proceeds from the sales and for the value of the lands withdrawn for public purposes and given away under the Homestead Act. Claiming that they were not fairly compensated for the sale of certain lands pursuant to the 1880 Agreement, the Ute Indians successfully brought suit against the United States for the equivalent of what they would have received if the land had been sold pursuant to the public land laws. The Ute Indians v. United States, 45 Ct. Cl. 440 (1910); The Ute Indians v. United States, 46 Ct. Cl. 225 (1911). The judgment covered the land disposed of from 1880 through June 30, 1910, and affected about 4.5 million acres.

[36] One-half of the Southern Utes, including the Weeminuche band under the leadership of Ignacio, disapproved of this agreement and the planned allotments. This band had initially wanted to, and did, move to San Juan County, Utah, but were forced to return to Colorado when Congress failed to approve their relocation. This band settled on the western end of the original Southern Ute Reservation and a subagency was opened for them at Navajo Springs in 1897. This was the beginning of the separation of the bands of Ute Indians, and after the turn of the century, the two divisions became known as the Ute Mountain Tribe and the Southern Ute Tribe and now they occupy separate reservations.

[37] Jefferson, Delaney and Thompson, *The Southern Utes* (1972), p. 42 (hereinafter, Jefferson at ______).

[38] *Id.* at 33-43, Proclamation of May 4, 1899.

[39] 32 Stat. 384 (1902).

[40] Indian Reorganization Act, 48 Stat. 984, ch. 576 (June 18, 1934), codified at 25 U.S.C. §§ 461-479.

[41] Restoration to Tribal Ownership of Ceded Colorado Ute Indian Lands, 56 I.D. 330 (June 15, 1938) and Restoration of Lands Formerly Indian to Tribal Ownership, 54 I.D. 559 (August 10, 1934).

[42] The Secretary of the Interior's action was affirmed by the Act of June 28, 1938, amended July 15, 1941, 52 Stat. 1210-1211, c. 776, § 6; 55 Stat. 593-594, c. 299.

[43] Minerals were reserved pursuant to many laws, some of which are the Acts of March 3, 1909 (30 U.S.C. § 81) (coal), June 22, 1910 (30 U.S.C. §§ 83-85) (coal), July 7, 1914 (30 U.S.C. §§ 121-123) (oil and gas), February 27, 1917 (30 U.S.C. §§ 86-89) (coal) and December 29, 1916 (43 U.S.C. §§ 291-301) (all minerals).

[44] Confederated Bands of Ute Indians v. United States, 112 Ct. Cl. 123 (1948). This suit specifically focused on the retention of interest by the Ute Indians in lands ceded to the United States by the 1880 Agreement, and taken by the United States by Section 6 of the Act of June 28, 1938 (100 Ct. Cl. 413).

the United States in two significant respects. First, no sale or other disposition (*e.g.*, lease) of land by any Indian or Indian tribe is valid unless it complies with certain statutory restrictions and is approved by the Federal government.[45] Second, Indians are considered wards of the United States government and as such they are to be protected.[46] This trust relationship is most clearly seen in the way title to much of the Indian lands is held; that is, by the United States, in trust for the particular Indian tribe.

As can be gleaned from the history of the formation of the Reservation, there are three major types of Indian mineral ownership on the Southern Ute Indian Reservation: (1) ownership by the Tribe; (2) ownership by individual Indian allottees or their successors in interest; and (3) Tribal ownership of coal or other minerals previously reserved by the United States, with the surface and possibly the oil and gas owned by non-Indian third parties.

1. Tribal Lands. Tribal land ownership is ownership in common. Equitable ownership is held by the tribe for the benefit of all members of the tribe; individual tribal members have no alienable or inheritable interest in the tribal lands. Tribal governments decide how to use the land and individual tribal members can control this use only to the extent of their participation in tribal government.[47]

Interests in real property have been acquired by the Southern Ute Indian Tribe in at least four ways: (1) by treaty; (2) by act of Congress; (3) by Executive Order; and (4) by purchase.[48] For most purposes, tribal rights created by these four mechanisms are "as sacred as the fee-simple title of whites."[49] Because the nature of the property interest granted by each method has been subject to some controversy in the past, the actual property rights conveyed as granted by each method are analyzed below.

a. Treaty and Congressional Action. Whether the 1868 Treaty and subsequent congressional action in 1880, 1895, and 1938 granted the minerals to the Ute Indians has not been specifically litigated. The 1868 Treaty, however, contains the initial grant language and is the basis for all future congressional actions. The actual grant language of the 1868 Treaty "set apart [the lands described] for the absolute and undisturbed use and occupation" of certain bands of Ute Indians. Identical language in a similar treaty between the United States and the Shoshone Indians has been interpreted by the Supreme Court as granting and assuring to the Indian tribe the peaceable and unqualified possession of the land in perpetuity, specifically including timber and minerals as constituent elements of the land.[50] Inasmuch as the 1868 Treaty contained no provisions specifically excluding minerals from the grant, the minerals underlying the Reservation lands passed to the Southern Ute Indian Tribe with the grant of Reservation lands. The 1880 Agreement served as the basis for the cession of the Tribal lands, as well as the subsequent restoration of certain ceded lands to the Tribe in 1938. The 1880 Agreement specifically affirms the terms of the 1868 Treaty, making the treaty's grant language applicable.

In sum, the Southern Utes were initially granted the entire surface and mineral interest in the lands of the Reservation, and they ceded portions of their land back to the United States. The Tribe retained both the surface and mineral estate in the lands not ceded. In 1938, certain portions of the ceded lands were returned to the Tribe, including both the surface and mineral estates in certain portions of land and only the mineral estate in other portions.

b. Executive Order. Certain portions of the Reservation were set aside by the Executive Order of February 7, 1879.[51] Courts have sustained the validity of reservations established in this manner and the power of the President to create them, based largely upon the acquiescence of Congress in such practices.[52] In addition, Congress has specifically authorized by statute the leasing of oil and gas on Executive Order reservations in the same manner as on congressionally created reservations.[53] By implication, then, the Executive Order reservations pertaining to the Reservation conveyed both the surface and the mineral interest under the subject lands.

c. Purchased Lands. Indian tribes have the capacity to enter into contracts for the acquisition of land, and the validity of such acquisitions has been recognized by both Congress and the courts.[54] The purchase of such land vests the tribe with an enforceable property interest, irrespective of the form in which title to the property is held. But, as with all transfers of property, the transferee can only obtain what the transferor owned, and nothing more.

[45] This restraint upon alienation has been imposed through Congressional adoption of a series of Trade and Intercourse Acts enacted between 1790 and 1834. The sections of these acts prohibiting unapproved land acquisitions from Indians are generally called Nonintercourse Acts by the Supreme Court. Oneida Indian Nation v. Oneida, 414 U.S. 661 (1974). *See generally*, F. Cohen, Handbook of Federal Indian Law 510-22, 618-28 (1982 ed.) (hereinafter, cited as Cohen at ______).

[46] This trust relationship was premised on the constitutional provisions granting to Congress the right to regulate commerce with the Indians, found in Article I, section 8, clause 3 of the U.S. Constitution, and on the treaty-making powers granted by Article VI.

[47] Cohen at 475-76.

[48] *Id.* at 472.

[49] Mitchel v. United States, 34 U.S. (9 Pet.) 711, 746 (1835). The exceptions to this rule are that: 1) aboriginal title may be extinguished by the United States without paying compensation as required by the fifth amendment, Tee-Hit-Ton Indians v. United States, 348 U.S. 272 (1955), and 2) title held pursuant to an unratified executive order may be revoked without requiring the payment of compensation under the fifth amendment, Sioux Tribe v. United States, 316 U.S. 331 (1942)

[50] United States v. Shoshone Tribe of Indians, 304 U.S. 111 (1938).

[51] Act of June 30, 1919, 41 Stat. 34, ch. 4, 41, codified at 43 U.S.C. § 150. Reservations were created by this method from 1871 until 1919, when Congress put an end to the practice.

[52] Arizona v. California, 373 U.S. 546, 598 (1963), decree entered, 376 U.S. 340 (1964). In addition to judicial recognition of the validity of creating reservations by Executive Order, Congress has acted in a number of ways to assure that title to many Executive Order reservations stands on the same footing as title to reservations created by statute. One such method has been congressional acknowledgement of this presidential power by enacting legislation which recognized and confirmed tribal title to a number of reservations created by Executive Order. *See generally* Cohen at 495.

[53] Act of March 3, 1927, 25 U.S.C. §§ 398a-398e.

[54] Cohen at 482.

d. Summary of Tribal Lands Situation. There is no doubt that the Southern Ute Indian Tribe owns the minerals and hence the coal-bed gas under the Tribal lands acquired by treaty, congressional action and Executive Order. The Tribe also owns whatever property rights it has purchased, which may or may not include the coal-bed gas, depending on the nature of the conveyance. The Tribe or one of its lessees has the right to develop the coal-bed gas under such lands. Mineral deposits on Tribal lands are leased under the authority of the Indian Leasing Act of 1938 [55] and the Indian Mineral Development Act of 1982.[56] The Southern Ute Indian Tribal Council has the authority, pursuant to the Southern Ute Indian Tribal Constitution,[57] to act on behalf of the Southern Ute Tribe in the execution of minerals agreements and leases. The nature of the Tribe's property interest in the land is the key to determining the legal risks associated with coal-bed gas development.

e. Coal-bed Gas Development on Tribal Lands. As the above discussion indicates, the developer's first step in evaluating possible legal risks associated with coal-bed gas development on Tribal lands is to thoroughly examine the manner in which the Tribe acquired its mineral interest. If the Tribe acquired the entire mineral interest in the land or the oil and gas rights, pursuant to the referenced grant of minerals in 1938, the Tribe has the right to develop the coal-bed gas. The legal risk in this situation is significantly affected by the fact that no Tribal coal leases, or minerals agreements covering coal, exist. To date, all coal-bed gas has been developed pursuant to the terms of the Tribal oil and gas lease[58] or minerals agreement[59] covering the property. The first legal risk associated with coal-bed gas development on Tribal lands concerns these leases and their execution—the risk that the lease or minerals agreement was not executed in accordance with Federal[60] and Tribal rules and regulations. This risk is straightforward and can be completely eliminated by careful examination of the existing lease and leasing procedure, or by careful compliance with the applicable rules and regulations when taking a new lease or minerals agreement.

The second legal risk involves the typical question faced by coal-bed gas developers—whether the oil and gas lessee or coal lessee has the right to develop the coal-bed gas. Because no coal leases exist on the Tribal lands and because the Tribe itself has not challenged an oil and gas lessee's right to develop the gas in the Fruitland coal, this issue has not been specifically addressed either by governmental edict or judicially as it pertains to Tribal lands. In addition, as was stated in our Federal discussion above, not much can be learned from the Federal leasing rules and regulations since they do not specifically address the potentially competing rights of an oil and gas lessee and a coal lessee in this situation. The definitions of gas in the Tribal and Allotted Oil and Gas Leases ("helium gas, carbon dioxide gas, sulfur gas, and all other natural gases...") and in a recently completed Southern Ute Tribe Minerals Agreement ("gas (being defined without limitation), and all other gaseous substances and associated hydrocarbon substances...") are easily broad enough to include coal-bed gas. In fact, one paragraph of a recently completed Southern Ute Tribe Minerals Agreement the authors reviewed specifically addresses the "development of the methane potential of the Fruitland Coals." The departmental approval required by Federal rules and regulations, coupled with the above-referenced language and the development of coal-bed gas under Tribal oil and gas leases or minerals agreements, would seem to indicate at least tacit, if not deliberate, policies of the Bureau of Land Management, Bureau of Indian Affairs (BIA), and Southern Ute Indian Tribe encouraging development of coal-bed gas pursuant to the terms of the Tribal oil and gas lease or minerals agreement, rather than a Tribal coal lease. The specific legal risk here is that this de facto grant of the right to develop coal-bed gas pursuant to the Tribal oil and gas lease or minerals agreement will be challenged. This possibility seems remote at the present time in view of the lack of Tribal coal leases and the present Tribal mineral leasing policy, and should be viewed by developers as an acceptable business risk.

One particular possibility, however, should be closely monitored. The Southern Ute Indian Tribe is currently faced with a dilemma. Because all coal-bed gas development to date has been done by the owner or lessee of the oil and gas rights, the Tribe has tacitly conceded that they do not have the right to develop coal-bed gas under lands where the Tribe owns only the coal. If the Tribe desired to develop the coal-bed gas in that Tribal coal, it might challenge this assumption. Such a challenge, however, seems highly unlikely in light of past and present leasing and development policies. That is, taking such a position would be grossly inconsistent with the Tribe having, in the past, leased land and collected royalties on so much of the Fruitland coal-bed gas under oil and gas leases.

Even if the Tribe were to continue allowing oil and gas lessees (whether Tribal or fee) to develop the gas from the Tribal coal, the possibility exists that the Tribe may move to protect its coal resource by either restricting access or development techniques or requiring compensation for the possible damage done to the coal by the development of the coal-bed gas, or both. Although there are no standard restrictions or provisions which compensate the Tribe for damage to the coal at the present time, the intensity of the development activity may accelerate the development of Tribal, BIA and (or) BLM policy addressing these issues in the future. Current research on the effect of coal-bed gas development on the coal resources is not definitive. The magnitude of this risk will eventually be determined by the results of this ongoing research. At present, at least with respect to the deeper Fruitland coals, this risk seems manageable, as the authors have seen no indication of movement by the Tribe in this regard.

2. Allotted Lands. Indian lands can also be owned by

[55] 25 U.S.C. §§ 396a-396g.

[56] 25 U.S.C. § 2101-2108.

[57] Southern Ute Indian Tribal Constitution, approved November 4, 1936, amended October 1, 1975.

[58] The form commonly used in the past was United States Department of the Interior, Bureau of Indian Affairs, Oil and Gas Mining Lease, Form S-157 (July 1964).

[59] Such minerals agreements are executed pursuant to the terms of the Indian Mineral Development Act of 1982, 25 U.S.C. § 2101 *et seq*.

[60] 25 C.F.R. § 211 *et seq*. 25 C.F.R. §§ 211, 212 and 225 have been the subject of recent Department of the Interior action. New Final Rules, promulgated in August of 1987 to be effective on October 23, 1987 (52 Fed. Reg. 31916, August 24, 1987), were withdrawn and reproposed with a 60-day comment period ending December 21, 1987. As of January 15, 1988, there had been no subsequent action.

individual Indians pursuant to Federal statute or treaty providing for the distribution of tribal property in severalty or pursuant to the General Allotment Act of 1887.[61] In the 1868 Treaty, the Utes agreed to the allotment of Reservation lands. This new form of Indian ownership was designed to assimilate Indian tribal members into the American system by transferring parcels of land to individual Indians.[62]

An allotted parcel of land may be owned by the United States in trust for an individual Indian (trust allotment) or owned by the Indian subject to a restriction on alienation in favor of the United States (restricted fee allotment).[63] Either type of allotment may be owned beneficially by more than one Indian or as cotenant by devise, descent, or conveyance. Because all such allotments came from tribal lands, such allotments conveyed the entire fee interest in the land, including the minerals. Lands can lose their allotted status and hence the governmental trust supervision and restriction on alienation in a number of ways, including patenting, competency determinations, sales to non-trust status, death and inheritance by non-Indians, mortgage foreclosures, condemnations, leasing, and easements.[64] If the restrictions on alienation are removed, the land and minerals under the lands may be conveyed as if the land were fee land.

a. Coal-bed Gas Development on Allotted Lands. Allotted Indian lands (which include the entire interest in the land, surface, and minerals) may be leased for the development of oil and gas[65] and other minerals[66] pursuant to the Indian Leasing Act of 1909[67] or the Indian Mineral Development Act of 1982.[68] The definition of gas in such leases is broad enough to include coal-bed gas, but does not specifically so state. As with tribal lands, these rules and regulations do not specifically address the rights of the oil and gas lessee and (or) the coal lessee to develop the coal-bed gas. The rules and regulations allow for the issuance of oil and gas leases and leases for other minerals under allotted lands, but the priority or conflict between lessees is not addressed.

The legal risks associated with developing coal-bed gas on Indian allotted lands are similar to those on Indian tribal lands: the risk that the specified leasing procedures were not correctly followed; the risk that a coal lessee might challenge the oil and gas lessee's right to develop the coal-bed gas; and the risk that the allottee, through BLM, BIA or tribal action, may move to protect the coal resource, affecting the oil and gas lessee's access to the coal.

If lands have passed out of allotted status, Federal rules and regulations are no longer applicable and the lands may be transferred in the same manner as fee lands. The instruments transferring title to the lands should be carefully examined to determine the actual property rights transferred and, if it can be discerned, the intent of the parties regarding the right to develop coal-bed gas. The issue of the right to develop the coal-bed gas under fee lands is discussed in Section VI of this paper.

3. Ownership of the Severed Coal Rights Under the Southern Ute Reservation.

The existence of severed Tribal coal on the Reservation presents a unique situation for the coal-bed gas developer. In 1938, the Federal government granted the previously reserved coal rights under approximately 500,000 acres to the Southern Ute Tribe.[69] Although part of the remaining mineral interest and (or) the surface has been purchased by the Southern Ute Tribe, much of the ownership remains split, with the Southern Utes owning the coal and non-Indian third parties owning the surface and remaining minerals, including oil and gas.

The right to develop coal-bed gas in this situation depends upon the precise mineral rights originally reserved by the Federal government. The majority of Federal reservations involved coal under patented lands reserved pursuant to the Act of March 3, 1909 or the Act of June 22, 1910, and oil and gas reserved pursuant to the Act of July 17, 1914.[70] The Opinion of the Solicitor referred to above[71] examined the language of the reservations in the first two Acts and concluded that Congress intended to reserve only the coal underlying the patented lands and not the coal-bed gas or other minerals associated with the coal.[72] The Solicitor also examined the reservation of oil and gas pursuant to the Act of July 7, 1914, and concluded that Congress intended to reserve oil and gas, which included the coal-bed gas. These reserved mineral rights were then transferred to the Tribe in 1938.[73] If the Solicitor's Opinion is relied upon for the three analyzed Acts, the grant

[61] 24 Stat. 388 (1887) codified as amended at 25 U.S.C. §§ 331-355 (1986). The General Allotment Act is commonly referred to as the Dawes Act; it established the legal framework for most present day allotment law. Before 1887, allotment of Indian lands required the consent of the Tribe, and provisions to that effect were included in many treaties, including the Ute Treaty of 1868. After 1887, allotment of Reservation lands was mandatory.

[62] As a companion policy to allotment, the Federal government sought the cession of Tribal Lands after each Tribal member had received an allotment. These surplus lands were sold to or homesteaded by non-Indians. By 1920, it was generally recognized that the allotment policy had failed, and in 1934 Congress passed the Indian Reorganization Act (IRA), 48 Stat. 984, ch. 576, codified at 25 U.S.C. § 461 *et seq.*, repudiating the policy of allotment of tribal lands. Application of the IRA depended on tribal consent and many tribes rejected it. Despite these rejections, the Interior Department has not initiated any new allotment programs since 1934, and has routinely extended the trust or restricted period on allotments to non-IRA tribes. The current extension is until January 1, 1989, for all trust and restricted patents which would have expired prior to 1989. 48 Fed. Reg. 34,026 (1983).

[63] Cohen at 615-16.

[64] Moore at § 26.01[3].

[65] 25 C.F.R. § 212.4 (1987).

[66] 25 C.F.R. § 212.6 (1987).

[67] 25 U.S.C. § 396.

[68] 25 U.S.C. § 2101 *et seq.* Section 2102(b) specifically allows an individual Indian owning a beneficial or restricted interest in a mineral resource to include such resource in tribal minerals agreements.

[69] *See* notes 38 through 44 and accompanying text.

[70] Minerals were reserved pursuant to many laws, some of which are the Acts of March 3, 1909 (30 U.S.C. § 81), June 22, 1910 (30 U.S.C. §§ 83-85), July 17, 1914 (30 U.S.C. §§ 121-123), February 27, 1917 (30 U.S.C. §§ 86-89) and December 29, 1916 (43 U.S.C. §§ 291-301). The precise nature of each reservation must be examined to ascertain the nature of the mineral reservation.

[71] *See* notes 20 through 27, *supra*, and accompanying text.

[72] 88 I.D. at 541.

of coal from the Federal government to the Tribe did not contain the coal-bed gas. Accordingly, the oil and gas owner or lessee of the tract, not the Tribe, has the right to develop the coal-bed gas.

a. Coal-bed Gas Development From Severed Indian Coal. The two specific legal risks associated with the development of coal-bed gas in this situation are fairly clear. First, there is the theoretical risk that the Tribe may dispute the Solicitor's analysis and claim the rights to develop the coal-bed gas by virtue of their ownership of the coal. For the purposes of this article, we assume that a court analyzing this mineral reservation will follow the Solicitor's position, based on the deference given to agency determinations and the fact that the Solicitor's interpretation does not favor the United States. This analysis might be different if the Tribe had reserved the minerals, based on the general rule of construction liberally construing statutes and conveyances in favor of Indians. However, this rule is not applicable in this situation because the United States, not the Indians, reserved the minerals. Moreover, to take the tack that the coal owner has the rights to develop the coal-bed gas would be a complete reversal of prior leasing policy and, accordingly, would cause tremendous problems.

The second legal risk is that the Tribe will move to protect its coal resource from damage, or require the oil and gas lessee to compensate the Tribe for any such damage. This risk is discussed above in the context of coal-bed gas development on Tribal lands, and appears manageable at present.

V. COAL-BED GAS ACTIVITIES ON STATE LANDS.

Turning now to the situation where the coal-bed gas developer has targeted a coal-bed gas prospect involving state lands, the discussion focuses on Colorado and New Mexico.[74] In the San Juan Basin, state lands include primarily the school sections (16 and 36).

A. Introduction.

Very few states have faced the issue of whether coal-bed gas is included in a state coal leasehold estate or, on the other hand, whether the resource is covered by a state oil and gas lease. To the extent that any trend can be determined, it appears that most states in the West recognize that coal-bed gas is covered by a state oil and gas lease, subject to the state's right to protect its coal resources and, possibly, subject also to the coal lessee's right to ventilate the gas during mining. Research indicates that Colorado has specifically addressed this issue and has taken this approach. Assuming that other states follow Colorado's lead, the legal risks in developing coal-bed gas from state lands will not involve the ownership question, but rather the question of the state oil and gas lessee's ability to have access to the gas found in the state's coal deposit. This risk will surface in situations where the state has leased, or where there is a potential that it will lease, the coal.

However, even with respect to the access issue, there are good legal guideposts to follow. Commentators on this subject indicate that the Western States are recognizing that the swing of Western jurisprudence, consistent with the nation's energy needs and public policy regarding the development of energy resources, is slowly moving toward the adoption of multiple-use concepts and away from the traditional first in time, first in right theory of access to different mineral deposits.[75] However, with respect to state-owned minerals, the current status of this balancing act is varied, depending on the particular state involved.

B. Colorado.

Colorado is one of the few Western States that has faced the issue of coal-bed gas development under state lands head-on. It is the authors' understanding that this is because the state sought to avoid disputes when some relatively aggressive oil and gas development occurred in the early 1980s on the western slope of Colorado, which development was right in the middle of some longtime producing coal fields. The problem was addressed by amending the terms of the state coal mining and oil and gas leases. Paragraph 22 of the current coal mining lease issued by the State Board of Land Commissioners[76] provides as follows:

> COAL MINE GAS—Methane Gas or other volatile gases produced, saved, and/or sold by the coal mining Lessee from minable coal measures and from roofs and floors of minable coal measures shall be the property of that Lessee provided that the gas is removed as a mining safety procedure prior to mining and that a royalty be paid to Lessor per the terms set forth in Lessor's then current oil and gas leases. Gas that is uneconomical to produce may be vented or flared provided that the venting or flaring complies with all Federal, State and Local requirements. Methane gas or other volatile gases produced by the oil and gas Lessees from the minable coal measures and from roofs and floors of minable coal measures prior to mining shall be the property of the oil and gas Lessee under the terms of the oil and gas lease, so long as the terms in this paragraph are consistent within the law.

Similarly, the current Colorado oil and gas lease issued by the State Board of Land Commissioners[77] contains a provision as follows:

> Coal Mine Gas—Coal Mine Gas is included herein, however, volatile coal mine gas may

[73] Mineral reservations pursuant to other Acts of Congress should be analyzed in the same manner as did the Solicitor. The authors have not undertaken an analysis of those other Acts because the majority of the Federal reservations involved the three Acts analyzed in the Solicitor's Opinion.

[74] For an excellent survey of multiple mineral development issues as they pertain to twelve states in the Rocky Mountain region, *see* Lear at 154-98.

[75] *Id.*

[76] State of Colorado, State Board of Land Commissioners, Coal Mining Lease, form SLB 237-(8/72) (rev. 3/77, 3/81, 2/85). Credit in this regard should go to Tom Bretz, State Mineral Director, and his staff for their foresight in trying to avoid disputes before they happened.

[77] State of Colorado Board of Land Commissioners, Oil and Gas Lease, form 193-5 (Rev. 4/84).

> be produced, saved and/or sold by a coal mining Lessee from minable coal measures and from roofs and floors of minable coal measures and the gas shall be the property of that Lessee, provided that the gas is removed only as a mining safety procedure prior to or during mining and that a royalty be paid to Lessor per the terms set forth in Lessor's then current oil and gas leases. Gas that is uneconomical to produce may be vented or flared, provided that such venting or flaring complies with all Federal and State requirements. Gases produced by the oil and gas Lessee from the minable coal measures and from roofs and floors of minable coal measures prior to, during or after mining shall be the property of the oil and gas Lessee under the terms of the lease. Oil and gas operations shall not render coal seams unminable.

In drafting the above-referenced lease provisions, the State of Colorado attempted to settle the question of ownership of coal-bed gas and define certain of the rights of a coal lessee and an oil and gas lessee with respect to access to and right to extract the gas prior to, during, and after coal mining operations. This has eliminated one of the common legal risks associated with coal-bed gas development—that the coal lessee will challenge the oil and gas lessee's rights to develop the coal-bed gas. Thus, one of the two remaining risks associated with coal-bed gas development concerns the questions of access and (or) protection of the coal. The lease provisions set forth above do not define the various rights and obligations of the parties in sufficient detail to enable an oil and gas lessee to precisely quantify the legal risks associated with developing the coal-bed gas where a prior coal lease has been issued. For example, the lease terms do not settle the question of whether hydrofracturing of the coal seam renders coal unminable. These issues would have to be faced on a case-by-case basis and may, in a contentious situation, be finally resolved by order of the Board of Land Commissioners.

The remaining legal risk associated with developing coal-bed gas on State of Colorado lands concerns a possible conflict between the lease provisions and Colorado statutes. The State Land Board's lease provisions respecting the development of coal-bed gas from minable coal seams must be read in conjunction with Article 61 of Title 34 of the Colorado Revised Statutes, which requires that any oil and gas operator must conduct his operations and employ casing and cementing programs so as to protect workable coal beds. This requirement is echoed by the Colorado Oil and Gas Conservation Commission in its Rule 209.[78] The statutory requirement that the oil and gas lessee protect workable coal seams includes a provision that "coal seams or worked out coal seams shall be sufficiently protected by casing so that the gas or oil shall not come into contact with the coal seams." It appears there may at least be an argument that any completion of a well in the coal seam is inconsistent with this statutory requirement that the coal deposit be cased-through. Presumably, however, the State Land Board, as the leasing agency for the sovereign, has the authority to waive compliance with the statutory protection scheme if, in its judgment as trustee of the State's resources, it determines that the highest and best use of its mineral resources is to allow development of the coal-bed gas prior to mining the coal. This risk, therefore, must be evaluated on a case-by-case basis, depending on the minability of the coal.

[78] Rules and Regulations of the Oil and Gas Conservation Commission of the State of Colorado, effective October 1, 1987.

C. New Mexico.

Although New Mexico has not addressed the issues of ownership and access to coal-bed gas in as comprehensive a manner as Colorado, the regulatory scheme in New Mexico appears to be one which embraces the notion of accommodation. Accordingly, the development of coal-bed gas on State lands in New Mexico would probably proceed along the same lines as those in Colorado, with the caveat that more negotiations might be involved.

The provisions of the State of New Mexico Oil and Gas Lease[79] and the statutory provisions addressing the leasing of State-owned coal[80] do not specifically address the issue of which lessee has the right to develop coal-bed gas. It has been the policy of the New Mexico State Land Office, however, to allow development of coal-bed gas pursuant to the terms of the State of New Mexico Oil and Gas Lease. Thus, the legal risks associated with development of coal-bed gas on lands owned by the State of New Mexico are similar to those faced by a coal-bed gas developer on fee lands.

VI. COAL-BED GAS ACTIVITIES ON FEE LANDS.

This section focuses on the particular legal risks involved where the coal-bed gas developer has targeted a coal-bed gas prospect involving fee lands.

A. Introduction.

A question common to almost all jurisdictions, including the Southern Ute Indian Reservation, concerns the ownership of coal-bed gas under fee lands: Is the substance owned by the coal owner or the oil and gas owner?[81] This question may arise in a situation where the coal and the oil and gas have been severed. However, even in the case where the coal and oil and gas are owned by the same person or entity, the question may be important where the mineral owner has leased the coal to one party and the oil and gas to another. Because of the lack of judicial precedent in Colorado, New Mexico, and elsewhere, the answer is not readily apparent.

Any judicial examination of the issue of ownership of coal-bed gas will necessarily turn on the resolution of two issues: First, is coal-bed gas a gas or is it an integral and intrinsic part of the coal? The answer turns on the legal definition of coal-bed gas. Eventually, in every jurisdiction where coal-bed gas is found, that definition will have to be enunciated by judicial decree and (or) legislation before coal-bed gas development

[79] N.M. Stat. § 19-10-4.1, .2 and .3 (1985).

[80] N.M. Stat. § 19-9-1 *et seq.* (1985).

[81] *See also* McGinley at 370. Mr. McGinley also considers the possibility that the coal-bed resource may belong to the surface owner.

can proceed entirely risk-free. However, such judicial or legislative edicts are relatively unlikely in many states, at least in the near future. Until then, the coal-bed gas developer can consider the following legal analysis.

The second issue that may be the focus of any litigation relating to the ownership of coal-bed gas is the intent of the parties to the particular grants, reservations, or conveyances of the coal rights on the one hand, and the gas rights on the other. The precise issue here is whether the parties to a particular mineral conveyance (be it a coal grant or an oil and gas lease) intended coverage of coal-bed gas. Resolution of this issue will be on a case-by-case basis.[82] Again, analysis of existing precedent can serve to illuminate the true level of the risk.

Specifically, recent decisions of the Pennsylvania Supreme Court and a Federal district court in Alabama, both discussed below, present an opportunity to view the way in which courts may approach and decide these questions of defining coal-bed gas and construing the intent of the parties to a particular mineral conveyance.

B. **The U.S. Steel Decision.**

In a case of first impression, the Pennsylvania Supreme Court rendered what is considered to be the leading decision regarding ownership of methane found in coal seams. The trial court and an intermediate appellate court had ruled in favor of the gas rights owner and against the owner of the coal vein.[83] In its holding, the Pennsylvania Supreme Court reversed the lower court findings.[84]

In the *U.S. Steel* decision, the Pennsylvania Supreme Court held that coal-bed gas belonged to the owner of the coal vein, which had previously been severed from the other mineral strata. Although the decision was based on construction of a mineral severance deed with unique language, the court appears to adopt a broad rule that ownership of gas, in general, follows ownership of the stratum in which the gas is found.

The decision is not binding on courts in other states. Its significance in Colorado, New Mexico, and other states in the Rocky Mountain region will lie in the persuasiveness of its reasoning that gas is not to be treated as a mineral in and of itself but rather as merely a captive of the rock in which it is found. The authors think it is probable that a Colorado court would reject this reasoning, since it is based on precedent which courts in Western States have not followed. Further, it is a decision which, by its terms, deals with severance by strata, not by mineral, a situation not often encountered in the West.

More significantly, where there has been no severance of the coal from the mineral estate at the time of issuance of any fee oil and gas lease owned by the coal-bed gas developer, the *U.S. Steel* decision may have no applicability. This is because the issue in such cases is whether the gas covered by the coal-bed gas developer's oil and gas lease includes coal-bed gas. Resolution of that question will turn on interpretation of the language of the particular lease involved, not on construction of previous conveyances having language broad enough to have arguably conveyed the coal-bed gas. Even in those cases where there has been a severance of coal, in the West such a severance will almost invariably be a severance by patent, *i.e.*, a reservation of coal by the United States. The *U.S. Steel* decision may not affect a court's or the Federal government's interpretation of the language and legislative history of the appropriate Federal statutes which effected the severance.

The fact remains, however, that the *U.S. Steel* decision does constitute an adverse precedent for the coal-bed gas developer, as it raises the specter of competing claims to the coal-bed gas by a coal owner or lessee.

1. **Analysis of the U.S. Steel Decision.**

The 1980 trial court decision in the *U.S. Steel* litigation represented the first judicial analysis of the coal-bed gas ownership problem where gas and coal rights are separately owned.[85] The plaintiff, United States Steel Corporation (U.S. Steel), owned a tract of Pittsburgh vein coal by virtue of a grant contained in a 1920 coal severance deed to its predecessor. The defendants, owners of other rights in the tract of land, leased all oil and gas rights to defendant Mary Cunningham, in 1976. U.S. Steel operated a coal mine near the defendant's tract. In 1978, the defendant drilled into the plaintiff's coal seam to extract the entrapped coal-bed gas. In response, U.S. Steel sought an injunction to prevent both the surface rights owners and the oil and gas lessees from drilling into and removing gas from the coal seam. In addition, U.S. Steel sought to quiet its title to the ownership of the gas located within the coal seam.

U.S. Steel argued that the 1920 conveyance of the coal seam included title to the gas in the seam because the parties were aware that the gas existed when the transaction was made, and that the right to ventilate the coal-bed gas, expressly granted in the deed, necessarily implied such a degree of control as to be tantamount to ownership. Coal-bed gas, the plaintiff argued, traditionally has not been viewed as a natural gas. U.S. Steel also contended that use of artificial stimulation methods to remove coal-bed gas could make it more hazardous and expensive to later mine the coal.

Defendant Cunningham argued that the right to extract all gas underlying the tract she had leased included the gas contained in coal seams, that coal-bed gas is a natural gas, and that its removal, with or without hydrofracturing, does not harm the structural integrity of the coal seam. She also argued that the plaintiff's failure to object to her application for State drilling permits estopped U.S. Steel from subsequently judicially attacking the defendant's drilling operations.

In its holding, the trial court addressed two central issues: (1) the legal definition of coal-bed gas (*i.e.*, is it a natural gas or a part of the coal?) and (2) the intent of the parties with regard to the specific mineral conveyances and reservations

[82] For discussions of the general rules relating to the construction of legal instruments and the principles governing the interpretation of mineral conveyances, see Craig and Meyers at 776-91; McGinley at 387-92; Olson at 392-97.

[83] *See* note 5, *supra*, regarding the trial court decision; to review the intermediate appellate court opinion, *see* U.S. Steel Corp. v. Hoge, et al., 450 A.2d 162 (Pa. Super. Ct. 1982).

[84] 468 A.2d 1380 (1983).

[85] Specific page citations to the trial court rulings in the *U.S. Steel* case are omitted.

involved. The trial court held that coal-bed gas is a natural gas by scientific definition and should be considered such by legal definition as well. It followed, then, that a grant of the coal in which the gas was situated could not possibly be a grant of the gas, unless the specific conveyance revealed that the grantor intended to include the gas in the grant. The court then held that the severance deed to U.S. Steel's predecessor did not reveal any intent by the grantor to include the gas in the property transferred to the grantee.

U.S. Steel appealed the lower court decision, but the Superior Court of Pennsylvania affirmed.[86] The appellate court held that: (1) U.S. Steel's fee simple interest in the coal did not embrace everything in the geological stratum comprising the coal vein; (2) at the time of execution of the severance deed, although coal seams were known to contain dangerous gases, there was no common understanding that the coal-bed gas was intimately connected with the coal itself; (3) the grantor's reservation of the right to drill through the coal to recover gas included the reservation of the right to recover the coal-bed gas; (4) the grant of the right to ventilate the coal-bed gas created no property right in the coal-bed gas; and (5) the equitable considerations and policy arguments made by U.S. Steel were insufficient to mandate a result different from the one reached under the applicable legal principles.[87]

U.S. Steel then appealed to the Supreme Court of Pennsylvania. In the *U.S. Steel* decision, by a 3-to-2 vote, the Pennsylvania Supreme Court reversed the holding of the lower appellate court and remanded the case to the trial court, with orders to quiet the title to the coal-bed gas in the owner of the coal, U.S. Steel.[88]

Citing early Pennsylvania precedent, the majority opinion of the Pennsylvania Supreme Court first determined that, under Pennsylvania law,

> gas necessarily belongs to the owner in fee, so long as it remains part of the property; ownership in it will be lost only upon grant or upon the gas leaving the property through migration.[89]

Since the gas was part of the fee, the court reasoned, a conveyance of that part of the fee in which the gas was situated necessarily included the gas as well:

> When a landowner conveys a portion of his property, in this instance coal, to another, it cannot thereafter be said that the property conveyed remains as part of the former's land, since title to the severed property rests solely in the grantee. In accordance with the foregoing principles governing gas ownership, therefore, *such gas as is present in coal must necessarily belong to the owner of the coal*, so long as it remains within his property and subject to his exclusive dominion and control. The landowner, of course, has title to the property surrounding the coal, and owns such of the coal-bed gas as migrates into the surrounding property.[90]

As the above language indicates, the court recognized that the owner of the coal could allow others certain rights respecting the gas.

The remainder of the opinion analyzes the provisions of the severance deed to determine the effect of the reservation, by the grantor, of the right to drill through the granted coal stratum in order to produce oil and gas. On this issue, described by the court as a question of the intent of the parties, the court first held that, although the unrestricted term "gas" was used in the reservation clause of the severance deed, when viewed in light of the conditions existing at the time of its execution, the court deemed it inconceivable that the parties intended a reservation of ownership of the right to extract a "valueless waste product with the attendant potential for damages resulting from its dangerous nature."[91]

Second, the court held that the fact that the grantor of the severance deed reserved the right to drill through the severed coal seam for oil and gas implied that the parties intended the gas to which the grantor retained the fee simple title to be "that which was generally known to be commercially exploitable."[92]

In sum, the Pennsylvania Supreme Court's majority decision in *U.S. Steel* appears to be founded on the broad rule that ownership of gas, in general, follows ownership of the stratum in which the gas is found. Although there is nothing in the opinion stating that a grantor of a coal severance deed could not maneuver around that rule by granting a fee interest in the coal vein and expressly reserving to himself the gas found therein, the court found that such was not the specific intent of the parties nor the effect of the language of the severance deed in question.

Two justices of the Pennsylvania Supreme Court dissented from the majority. Significantly, however, the dissenting justices also embraced the general rule adopted by the majority that "subterranean gas is owned by whoever has title to the property in which the gas is resting."[93] The dissent disagreed only with the majority's interpretation of the parties' intent with respect to the reservation of the right to drill through the coal to produce gas, believing that the effect of this language was to reserve the gas present in the coal seam.[94]

2. Analysis of the Impact of the U.S. Steel Decision on Western Coal-bed Gas Operations.

In analyzing the possible impact of the *U.S. Steel* decision on coal-bed gas operations in the Rocky Mountain region, it should first be said that the decision, although it represents statewide judicial authority in Pennsylvania, does not

[86] 450 A.2d 162 (Pa. Super. Ct. 1982).

[87] *Id.* at 166-73.

[88] 468 A.2d at 1380-85.

[89] 468 A.2d at 1383, *citing* Hamilton v. Foster, 272 Pa. 95, 102, 116 A.50, 52-53 (1922), and Westmoreland & Cambria Natural Gas Co. v. DeWitt, 130 Pa. 235, 249, 18 A. 724, 725 (1889).

[90] 468 A.2d at 1383 (emphasis in the original).

[91] *Id.* at 1384-85.

[92] *Id.* at 1385.

[93] *Id.* at 1388.

[94] *Id.* at 1389.

represent binding judicial precedent for Colorado, New Mexico, or other Western States. Its significance in this part of the country will depend entirely on its persuasiveness.

Second, the authors' experience indicates that most of the fee oil and gas leases in western areas of coal-bed gas operations cover lands in which ownership of the coal has not been severed from the oil and gas. In such cases, unless there is a coal lease which predates the coal-bed gas developer's oil and gas lease and which arguably includes the right to extract coal-bed gas, the only risk is that the oil and gas lessor may assert that he did not intend to include coal-bed gas in the oil and gas lease. This risk predated the *U.S. Steel* decision; thus, that decision should have little, if any, relevance to the resolution of this issue, the question being one of intent of the parties to the lease as reflected by the language of the lease.

Third, in those situations where the coal has been severed prior to the grant of the coal-bed gas developer's oil and gas lease, the *U.S. Steel* decision may well be rejected by western courts as based on the holdings of several very old Pennsylvania cases which may be at odds with the development of mineral law in the West. This point can be explained further.

The Pennsylvania decision is based on the general rule in that state that subterranean gas is owned by whoever has title to the property in which the gas is resting.[95] In other words, Pennsylvania has apparently adopted the rule that ownership of gas follows the ownership of the stratum in which it is found. Professors Williams and Meyers' treatise on oil and gas law states that certain cases in other jurisdictions hint at a similar result.[96] However, the authors go on to say that "other opinions in these states [Kentucky, Illinois and Kansas] suggest, however, the adoption of one of the other [ownership] theories."[97] Colorado, New Mexico, and the majority of the other coal and oil and gas producing jurisdictions in the West have not followed the ownership of strata theory; rather, such jurisdictions have adopted the ownership in place theory, which holds that oil and gas are owned by the surface owner and are capable of being sold or leased in place.[98]

The Pennsylvania rule as adopted by the court in *U.S. Steel* appears to be based on an early case which sets forth the governing principle of gas ownership in Pennsylvania as follows:

> Water and oil, and still more strongly gas, may be classed by themselves, if the analogy be not too fanciful, as minerals *ferae naturae*. In common with animals, and unlike other minerals, they have a power and a tendency to escape without the volition of the owner. . . . They belong to the owner of the land and are part of it, so long as they are on or in it, and are subject to his control; but when they escape and go to other land, or come under another's control, the title to the former owner is gone.[99]

A close look at that early decision suggests that the court may have had in mind simply the vertical severance of lands and the minerals underlying them. That is, the *Westmoreland and Cambria Natural Gas Co. v. DeWitt* decision supports the producibility of oil and gas by the party having the right to enter upon and drill into the lands under which the oil and gas are situated, similar to the theory of ownership in place adopted by Colorado and most of the other oil and gas producing jurisdictions in the West. The *Westmoreland* decision clearly states: "If an adjoining, or even a distant, owner drills his own land, and taps your gas, so that it comes into his well and under his control, it is no longer yours, it is his."[100] Thus, the view can be supported that the *Westmoreland* decision does not necessarily require recognition of the principle of the horizontal severance of oil and gas or of the rule that ownership of a subsurface mineral stratum includes ownership of the oil and gas captured within it.

It seems likely that courts in the Rocky Mountain states would not follow the rule adopted by the Pennsylvania Supreme Court that gas belongs to the owner of the stratum in which it is found. Rather, western courts would continue to treat the severance of coal or oil or gas or any other specifically named mineral resource as creating a separate corporeal estate in fee simple in that resource. Under this rule, a showing that coal-bed gas is a gas and that as such it is mineralogically different from coal should result in a holding that the owner of the coal does not own the gas trapped therein purely by virtue of his ownership of the coal.

Fourth, it may also prove significant that *U.S. Steel* turns on the majority's interpretation of the intent of the parties to a severance deed involving the carving out of one specifically defined vein of coal. That situation is not likely to be found in a severance deed in the Western States; rather, the usual severance deed in the Rocky Mountain region severs the coal by grant or reservation from the surface and (or) other minerals. This, of course, would allow the owner or lessee of the gas in the affected lands to focus the debate on the properties of the respective minerals rather than on the constituents of stratum ownership.

Fifth, it is doubtful that coal-bed gas litigation in the Rocky Mountain area would involve the complication of the coal owner being granted the express right to ventilate the gas found in the coal seams. This seems particularly true in light of the relative depth and unminability of many of the Colorado and New Mexico coal seams which are now the target of coal-bed gas developers.

Sixth, western courts might well reject the application of the specific intent test employed by the court in *U.S. Steel* and use a general intent test instead. Some commentators have criticized the use of the specific intent test to determine how parties to a conveyance intended to dispose of substances unknown (or substances bearing little commercial value) at the time of the conveyance.[101] Under the application of a test of general intent, construction of a severance deed would focus

[95] *Id.* at 1383, *citing* Kier v. Peterson, 41 Pa. 357 (1862) (owner of subterranean salt entitled to oil commingled with it).

[96] *See* 1 Williams & Meyers, Oil and Gas Law § 203.4 (1983 ed.).

[97] *Id.* (brackets added).

[98] *Id.* at § 203.3.

[99] 468 A.2d at 1303, quoting Westmoreland & Cambria Natural Gas Co. v. Dewitt, *supra*, note 89.

[100] *Id.*

on the proper definition of gas and coal, *i.e.*, on whether the parties to the instrument used language which naturally and inevitably encompassed the mineral in question, and not on what the parties to the instrument thought they were doing. Application of a general intent test would likely impel a western court to find that a gas grant (or reservation) encompasses coal-bed gas, subject perhaps to the gas grantee's duty to protect the structural integrity of the coal.

One specter that the Pennsylvania decision raises, of course, is the possibility that the coal-bed gas developer's oil and gas lessor, if he owns all of the minerals under a particular leased tract of fee lands, would challenge the developer's right to extract coal-bed gas. He could use the *U.S. Steel* decision as the cornerstone of his argument that gas found in the coal seam was not intended to be covered by the lease. Where no severance of coal from other minerals has occurred, and the developer owns an oil and gas lease covering gas, the issue in any such dispute will not, as in *U.S. Steel*, be who owns the gas found in the seam but, rather, whether the lessor intended to convey the rights to such gas. Absent the existence of a prior coal lease which arguably conveys rights to extract gas found in the coal deposit, the *U.S. Steel* decision does not alter the probability that a western court would find that the lessor of a standard-form oil and gas lease generally intends to convey leasehold ownership of all substances produced in a gaseous state, regardless of the stratum from which the gas is produced.

In this regard, several western courts have applied the general intent test to find that a lease of gas includes all kinds of natural gases, whether hydrocarbon or nonhydrocarbon in nature and whether or not commercially exploitable at the time of lease execution.[102] For example, in the *Northern Natural* case, the United States Court of Appeals for the Tenth Circuit held that helium was covered by the oil and gas leases at issue, stating:

> Nothing in the leases shows an intent to convey only combustible gas. If that is what the lessors meant, they should have said so. The evidence of the landowners is that they knew nothing of helium. If that is true, they could not have had any intent with regard to it. Reliance on specific intent is not compatible with denial of any intent. We believe that the search should be for general intent rather than for a supposed, but unexpressed, specific intent. General intent should be discovered not by defining and redefining the terms used but by considering the purposes of the grant in terms of enjoyment of the rights created. . . .
>
> We will not import into the grants of oil and gas a subjective intent to convey only those components of the gas which comport to a subjective notion of the commercial end uses at the time of lease execution.[103]

Under this reasoning, a holding that the coal-bed gas developer's oil and gas lease covers coal-bed gas would not be defeated by a showing that the lessor did not know about either the developer's intent to extract the gas found in coal beds or the commercial exploitability of coal-bed gas.

The prediction that western courts would not follow the *U.S. Steel* precedent should be accompanied by a word of caution. The *U.S. Steel* decision impacts only the issue of ownership of gas found in coal deposits.[104] The coal-bed gas developer must remember that there are numerous issues relating to access to and one's right to extract the gas, and these may still come into play even if it is certain that the lessor of an oil and gas lease owns the gas in the coal seam. This is true because of the basic axiom which is applicable where the coal and oil and gas rights are separately owned (either by severance or by two separate leases from the same owner covering the different minerals). That axiom, stated simply, is that although the owner of the oil and gas may be found to have an implied right of ingress and egress or a way of necessity and therefore may be allowed to drill into the coal seam to extract the gas, the coal owner has available remedies to protect against damage to his coal. General remedies to the coal owner include an injunction to protect against undue and unreasonable interference and damages where liability is found for necessary injury to the coal.[105] This means that, even if the legal risks relative to ownership of the coal-bed gas are acceptable to the developer, he should remember that, under certain circumstances, he may still have to face considerable legal risks relative to the extraction of the gas. Mitigation of these risks is discussed in detail below in Part VIII.

Finally, a word should be said about the possible impact of the *U.S. Steel* decision on the Federal situation, discussed in Part III of this article. We think there is no significant risk that the *U.S. Steel* decision will cause the Federal government to reconsider its view that its patent reservations of coal did not include coal-bed gas, because the principles of State law which governed the Pennsylvania Supreme Court's decision are not generally applicable to Federal statutory construction issues governing reservations in Federal patents.

C. Other Cases.

Although *U.S. Steel* remains the leading case on the issue of

[101] *See* Comment, *New Values Under Old Oil and Gas Leases: Helium, Who Owns It?* 61 Mich. L. Rev. 1168 (1964); *see also* Craig and Meyers, *supra* note 2.

[102] *See, e.g.*, Northern Natural Gas Co. v. Grounds, 441 F.2d 704 (10th Cir. 1971), *affirming* 292 F. Supp. 619 (D. Kan. 1968); Navajo Tribe of Indians v. U.S., 364 F.2d 320, 176 Ct. Cl. 502 (1966). It is also the policy of the State of Colorado and the BLM to permit helium to be developed pursuant to the terms of an oil and gas lease. In addition, the Regional Solicitor in Denver, Colorado, has concluded that carbon dioxide is included in a reservation of oil and gas in a land patent under the Act of July 17, 1914 (30 U.S.C. § 121). Reservation of Carbon Dioxide Gas in Land Patent, Memorandum to the Colorado State Director, BLM, from Regional Solicitor, Denver, Colorado (July 12, 1979).

[103] 441 F.2d at 714-15.

[104] The lower court decisions which were reversed, because they awarded the coal-bed gas to the gas owner, concerned themselves with not only the ownership question, but also access issues, and fashioned a holding that protected the integrity of the coal. When the Pennsylvania Supreme Court awarded the gas to the coal owner, the access issues became irrelevant.

[105] For a discussion of cases involving these issues, *see* Lear at 198-209.

coal-bed gas ownership, this issue has been before a Federal district court in Alabama.

The dispute in *Rayburn, et. al. v. USX Corporation*[106] involved a deed from the plaintiffs' predecessor in interest to USX Corporation (USX) conveying the minerals in approximately 2,874 acres of land in Tuscaloosa County, Alabama, reserving the oil and gas. The plaintiffs claimed the coal-bed gas fell within the definition of gas excepted from the deed to USX. Conversely, USX claimed that coal-bed gas is inherent to the coal bed and, as such, it too passed with the mineral rights. Despite urgings by the parties to decide the broader issue of coal-bed gas ownership as a matter of law, the court specifically confined its decision to the language of the deed in question.

To interpret the deed, the court first attempted to determine the intent of the parties from the circumstances surrounding the conveyance. The court examined all information available to the parties on the date the deed was executed (1960) and found the record bare of any indicia of common knowledge of the extraction of coal-bed gas for commercial purposes, or for future possibilities. The court then examined the language of the deed in question, which clearly stated that the grantors reserving the oil and gas rights were "subject to the requirement that all coal seams located in said land penetrated in such exploration or drilling operations shall be encased or grouted off. . . ."[107] The court found this language determinative; it clearly expressed the intention that the coal-bed gas not be available to any oil and gas well drilled by the grantors or their lessees. Relying on this language, coupled with the fact that coal-bed gas was not considered to be commercially recoverable at the time of the reservation, the court concluded that it was the intention of the parties that coal-bed gas not be included in the reservation. This case has been appealed to the Eleventh Circuit Court of Appeals.

Unlike the *U.S. Steel* decision, the *Rayburn* court confined its decision to the language of the instruments in question and specifically declined to answer the broader question of whether coal-bed gas is or is not included in the term gas, and is or is not severed with the mineral coal. This type of decision, based solely on a deed executed before the commercial potential of coal-bed gas was generally known, sheds little light on the issue of coal-bed gas ownership in general, but may be indicative of how courts handle such disputes.

VII. LEGISLATIVE TRENDS AND THE JURISDICTION OF STATE OIL AND GAS REGULATORY AGENCIES.

One of the side issues relating to coal-bed gas is whether State oil and gas regulatory agencies have authority to regulate the spacing, drilling, and production of coal-bed gas wells.[108] State legislatures have been slow to act in expressly defining coal-bed gas or methane as a natural gas. Nonetheless, many state statutes governing oil and gas conservation and production seem to include methane or coal-bed gas when gas is defined.[109] For example, gas is defined in Colorado as "all natural gas and all hydrocarbons not defined in this section as oil."[110] In New Mexico, natural gas, as used in the State Oil and Gas Leasing forms, is defined as covering carbon dioxide and helium gas as well as gas of the hydrocarbon kind.[111]

In a few states, regulation of coal-bed gas has already been undertaken by oil and gas regulatory agencies. By Memorandum dated November 24, 1980, Alabama's State Oil and Gas Board has done so. In early 1984, the State Oil and Gas Board of Alabama approved rules and regulations governing the permitting, drilling, and production of coal-bed gas.[112] In West Virginia, the State Oil and Gas Conservation Commission has assumed control of methane well permitting.[113] There, completion requirements for wells producing from coal are similar to those from other sources; a standard oil and gas lease is interpreted to give the operator "the same rights to produce gas from black rock as from brown rock."[114] In many instances, statutory provisions authorizing administrative bodies to establish drilling units and prescribe the location of wells seem broad enough to allow the agency to promulgate statewide rules governing production of gas from coal seams. This seems true in Colorado and New Mexico, where parties have complied with spacing orders from the state conservation agencies without the issue of jurisdiction being raised.

A. Fee and State-owned Lands in Colorado and New Mexico.

In Colorado, the Oil and Gas Conservation Commission (COGCC) has not specifically promulgated any rules dealing with coal-bed gas development. However, as is discussed above, the COGCC's Rules and Regulations do speak to the issue of oil and gas wells penetrating coal seams and the protection of workable coal seams. Further, the COGCC has assumed regulation of the spacing of coal-bed gas wells. On May 21, 1984, a hearing was held before the COGCC regarding the application of Resource Enterprises, Inc. (REI) for an order exempting REI from the spacing requirements of two orders of the Commission, insofar as those orders applied to the coal-bed gas wells which REI planned to drill on certain lands in Mesa County, Colorado. REI's application also sought an exemption from the well location rules of the Commission. REI's drilling program was in connection with the deep seam coal-bed gas research and development program instituted and supported by the Gas Research Institute (GRI).

[106] Rayburn, et. al. v. USX Corporation, Civ. Act. No. 85-G-2661-W (N.D. Ala. 1987). This case has been appealed to the 11th Circuit (Case No. 87-75-05), and as of mid-January 1988, no action had been taken.

[107] *Id.* at 2.

[108] Farnell at 20-23 and 26-28.

[109] *See, e.g.,* Code of Virginia, § 55-154.1 (1977); Okla. Stat. tit. 52 § 231 (1971).

[110] C.R.S. § 34-60-103(5). Oil is defined in C.R.S. § 34-60-103(6) as "crude petroleum oil and any other hydrocarbons, regardless of gravities which are produced at the well in liquid form by ordinary production methods, and which are not the result of condensation of gas before or after it leaves the reservoir."

[111] N.M. Stat. Ann. § 19-10-2 (1985).

[112] State of Alabama, Oil and Gas Board Rules and Regulations, § 400-4, *et seq.* (January 3, 1984 revision); for a more detailed discussion of these rules, see SOMED Research Report at 48, n. 62 and Appendix 4-B.

[113] Memorandum of West Virginia Oil and Gas Conservation Commission to Alabama State Oil and Gas Board, dated October 22, 1980, cited in Farnell at 21.

[114] *Id.*

At the conclusion of the hearing, the REI/GRI application was granted, although the Commission disclaimed that their actions constituted an opinion on REI's legal authority to proceed with its drilling plans under standard-form oil and gas leases. In sum, experiences of companies in New Mexico and Colorado should serve as examples of ways in which state oil and gas conservation agencies may influence the progress of coal-bed gas development projects.

Although REI's application covered lands in Mesa County, Colorado, at the present time other negotiations are underway concerning spacing in the San Juan Basin. Currently, in the Ignacio Blanco field, the COGCC has spaced the wells drilled into the Fruitland Formation on 320-acre units, with an infill well located on the undrilled quarter section.[115]

New Mexico's Oil and Gas Act,[116] like its counterpart in Colorado, does not specifically address the development of coal-bed gas. The New Mexico Oil Conservation Commission (NMOCC) has designated spacing for at least two gas pools in the San Juan Basin—the Cedar Hill Basal Fruitland Coal gas pool and the Undesignated Fruitland Coal gas pool. The NMOCC has also assumed jurisdiction over coal-bed gas development on fee, Federal, State and Indian lands within New Mexico, and has followed the Federal position that coal-bed gas should be developed pursuant to the terms of the applicable oil and gas lease. It is our understanding that, in connection with its coal-bed gas drilling activities in the San Juan Basin, Amoco Production Company has been actively seeking and obtaining appropriate field-wide spacing orders from the NMOCC.[117]

The experimental status of methane production fosters the caution shown by New Mexico and Colorado. The legislatures of both states have yet to deal specifically with the issue of coal-bed gas ownership, development and production. The fact that states such as Alabama and West Virginia are attempting to deal with coal-bed gas development questions gives us hope that Colorado and New Mexico may soon focus on the issue. As coal-bed gas development increases, the pressure to resolve outstanding issues will increase, particularly from coal operators who seek to prevent the permitting of wells which might interfere with or damage coal mining operations.

B. **Federal Lands.**

To the extent that any Federal lands are targeted by the coal-bed gas developer, the applicable state oil and gas conservation agency may have limited jurisdiction. This is an interesting issue, as one could foresee an aggressive coal-bed gas developer trying to convince the BLM that coal-bed gas development should take place on spacing units smaller than those designated by the state oil and gas conservation commission. In Colorado, the statute authorizing the COGCC applies to all lands within the State, except "as to lands of the United States or lands which are subject to its supervision, this article shall apply only to the extent necessary to permit the commission to protect the correlative rights of each owner and producer within a pool . . ." or if the officer of the United States having jurisdiction approves the order of the commission which purports to affect such lands.[118]

The New Mexico Oil Conservation Commission has jurisdiction over all lands located in New Mexico, including public domain lands owned by the Federal government held merely in a proprietary capacity to prevent waste of oil and gas, so far as such regulations do not conflict with congressional enactments.[119] If application of any State rule or regulation is in direct conflict with a Federal statute or regulation, the Federal statute would prevail. If there is no conflict, the State statute would prevail.

This principle was upheld in the recent case of *California Coastal Commission v. Granite Rock Co.*,[120] where the United States Supreme Court upheld the authority of the California Coastal Commission to require that a mining operator conducting mineral exploration operations on Federal land obtain a State permit. This holding appears to permit states to regulate the environmental impacts, but not land-use planning restrictions, of mineral development on public lands.

C. **Indian Lands.**

The jurisdiction of state oil and gas conservation agencies over Indian lands is questionable.[121] This is an interesting issue, as one could foresee an aggressive coal-bed gas developer trying to convince an Indian tribe that, say, the Fruitland coal formation should be drilled out at 80 acres rather than the 320-acre units required by the COGCC. This is a complex jurisdictional issue and, therefore, outside the scope of this paper. Nevertheless, it remains a definite legal issue which a coal-bed gas developer must face when developing coal-bed gas on Indian land. At present, the common approach to this issue, compliance with all possible state, Federal and tribal rules and regulations, may mitigate the risk, but provides no real answer.

VIII. CONCLUSION: MITIGATING THE RISKS.

Assuming that coal-bed gas development in the San Juan Basin continues to attract government and industry interest as an energy resource, the complex legal issues surrounding ownership of and access to coal-bed gas will become of increasing concern. Currently, although those issues are far from being resolved, the risks from a legal standpoint are ascertainable

[115] *See* In the Matter of the Promulgation and Establishment of Field Rules to Govern Operations in the Ignacio Blanco Field, La Plata and Archuleta Counties, Colorado, Colorado Oil and Gas Conservation Commission Order Nos. 112-6 and 112-46.

[116] N.M. Stat. § 70-1-1 *et seq.*

[117] Memorandum from Utah Dept. of Natural Resources to Alabama State Oil and Gas Board, dated September 24, 1980, cited in Farnell at 21.

[118] C.R.S. § 34-60-120.

[119] Opinion of the Attorney General 75 (1935-6).

[120] California Coastal Commission v. Granite Rock Co., 107 S. Ct. 1419, 94 L. Ed. 577 (1987).

[121] *See* Moore at 35. "The regulation of oil and gas drilling and spacing on tribal and allotted lands is presently outside the purview and authority of State oil and gas conservation laws unless consented to by the Indian owners." *See generally*, Montana v. United States, 450 U.S. 544 (1981) ("State conservation laws are applicable to fee lands within Indian Reservations and apparently tribes have acquiesced in State commission jurisdiction over fee lands for a number of years".) and Confederated Bands of Yakima Indians v. Whiteside, et al., 828 F.2d 529 (9th Cir. 1987).

and, in most cases, manageable.

In the area of fee coal and oil and gas, the ownership question has been decided judicially only by the Pennsylvania courts in favor of the owner of the coal seam. In the West, the issue has not been litigated. Even if a dispute as to ownership were to be settled judicially in favor of the owner of the gas rights, many issues would undoubtedly remain relating to access to and methods of extracting the coal-bed gas, and the corresponding rights of the coal owner.

As to Federal minerals, the Solicitor of the Department of the Interior has determined that coal-bed gas is leasable under the oil and gas provisions of the Mineral Leasing Act, but his interpretation has not been tested in the courts. Moreover, for access to the coal-bed gas, the BLM requires the coal-bed gas developer to jump through some fairly significant procedural hoops before the coal-bed gas can be removed.

As to Indian-owned minerals, current BIA, BLM and Southern Ute Indian Tribal policy gives the allotted or Tribal oil and gas lessee the right to develop the coal-bed gas. This policy has not been litigated or challenged to date.

As to State-owned minerals, Colorado has declared that its oil and gas leases cover coal-bed gas, but the coal lessee has the right to the gas during mining and, furthermore, the oil and gas lessee may not render the coal unminable. Although not enunciated as a formal policy, coal-bed gas development of State-owned minerals in New Mexico is done pursuant to an oil and gas lease rather than a coal lease.

Thus, with respect to Federal, Indian, fee and State domains of coal and oil and gas in Colorado and New Mexico, while it is the view of these authors that ownership of coal-bed gas may be and now is (at least in practice) awarded to the holder of the oil and gas rights, a significant amount of law remains to be settled. Assuming continued development of the resource, the coal-bed gas developer should keep abreast of new developments by paying close attention to the courts, state legislatures, and conservation agencies, tribal councils, the Department of the Interior, and Congress.

Taking all this into account, the coal-bed gas developer should proceed by identifying and mitigating the special legal and business risks associated with exploration and production of coal-bed gas by anticipating disputes and resolving them in advance of exploration and production. This is discussed in more detail below.

A. Tract Identification: The Key to Determining the Risks.

The key to anticipating disputes regarding coal and coal-bed gas rights lies with the early identification of possible conflicting claimants to those rights. To identify the various owners of and conflicts involving the target prospect, the coal-bed gas developer's land personnel should go beyond the usual search of the county real estate records and (or) the Federal, Indian and state oil and gas leasing records. Additional steps should be taken to determine not only who to lease from, but also who may, under certain circumstances, have conflicting claims to the coal deposit. In areas where the Federal and (or) state governments or Indian tribes own the coal deposit, a detailed surface inspection to identify the presence of any coal mining activity, as well as a search of the respective coal leasing records, should be made.

If no special coal interests can be identified, such as fee simple conveyances to known coal operators, conveyances or reservations of coal rights which severed the coal and oil and gas rights, coal leases or surface disturbances indicating coal mining activity, no further title search is necessary. On the other hand, if some evidence of a special interest in the coal is found, the landman's work should be broadened to determine the intention of the coal owner or lessee. Such work might include, depending on the ownership of the subject lands, examination of the state mining and reclamation agency records, the Indian tribal records, the county mining records, the county and U.S. Forest Service water use records or the county zoning records, as appropriate. In addition, the coal owner or lessee can be approached directly about its intentions to develop the coal. Where the coal is determined to be minable, as a basic minimum it must be determined whether (1) coal mining operations are currently underway or planned; (2) sufficient time remains for coal-bed gas recovery before the reservoir is to be mined; and (3) the coal operator currently has an economic coal-bed gas recovery operation planned or in effect.

The importance of this tract identification process can be explained in greater detail by examining the more common land situations in the San Juan Basin which the coal-bed gas developer may encounter.

B. Low-risk Situations: Non-segregated Ownership.

The simplest and lowest-risk route from tract identification through recovery of coal-bed gas occurs where the coal-bed gas developer leases a tract (Federal, Indian, state or fee) having all the oil, gas, and coal rights vested in the lessor(s), and none of the mineral owners are associated with development of coal.

This situation may be the result of either an unminable coal seam or a mineral owner uninterested in coal development. Even in this low-risk situation, however, one cannot unequivocally conclude that the granting clause of a typical oil and gas lease or minerals agreement includes coverage of coal-bed gas. While it may be an acceptable business risk to proceed with coal-bed gas development under a standard-form oil and gas lease, inclusion of certain language in the oil and gas lease may reduce the risks to near zero, depending on certain factors. If it is determined that the mineral owner is uninterested in coal development, or takes a position which is consistent with that of the coal-bed gas developer and supports production of coal-bed gas under the lease, any such risk may be almost totally eliminated by the inclusion of lease language to that effect. Such language should specifically grant the right to explore for and produce coal-bed gas and specifically subordinate any subsequent grant of coal rights to the rights of the oil and gas lessee. In most instances, inclusion of these two special provisions should not materially affect or alter the progress or expense of lease negotiations.

Coal seams below 3,000 feet or that dip steeply may not be economic to develop. In these situations, the coal owner or lessee may even encourage coal-bed gas development. Certain problems, however, may be present in the situation where the coal-bed gas developer targets a deep, unminable coal seam and only wants to drill through the shallower, minable coal zones to reach his objective horizon. The access question presents the legal risks here, with the ultimate risk being that the non-oil and gas minerals owner will not execute a lease or will exact too heavy a price for limited access. Risks can be mitigated in this situation if the coal-bed gas developer finds a mineral owner willing to negotiate a reasonable oil and gas lease. The oil and gas lease in this situation should include language addressing (1) the timing of and the operations the coal rights owner is permitted to undertake; (2) which coal operations are prohibited during the life of the oil and gas lease; (3) the exact location of the wells to optimize coal-bed gas production and conform with the design of the coal mine; and (4) the extent of and limits on liability for coal seam damage.[122]

In the San Juan Basin, the unified ownership situation will be most commonly found on Southern Ute Tribal lands, due to the lack of coal leases on such lands and hence the lack of coal development activity. Unified ownership may also be found on tracts where the coal resource would not be economic to mine, due to the low grade of the coal, the great depth of the coal, or the location of the coal seam.

The notable exception to the above generalizations occurs where a coal operator or railroad owns all the coal rights under the target prospect. In this case, the would be coal-bed gas developer may confront risks associated with the unsettled legal questions and issues surrounding access to and extraction of coal-bed gas. In this situation, the coal-bed gas developer will face an uphill struggle when attempting to lease and drill for coal-bed gas. Negotiations with the mineral owner may be very difficult; as a general rule, most coal companies or coal developers (1) view coal-bed gas development as a needless and possibly costly headache; (2) will go to great lengths to avoid having wellbores drilled into or through their target coal seam; and (3) feel (probably with some justification) that standard ventilation technology is sufficient to alleviate all but the most severe methane problems. The benefits of possible royalty income and degasification of the coal seam before mining offer the coal interest owner little incentive to cooperate.

C. Higher-risk Situations: Segregated Ownership.

The coal-bed gas developer enters into an arena of higher risk where the tract identification process reveals that the ownership of or rights to develop the coal have been severed from the ownership of or rights to develop the oil and gas. If certain factors are present, the risk may be minimal. Such factors might include a target coal seam which is unminable by current standards or the absence of any intention on the part of the coal owner to develop the coal.

On the Southern Ute Reservation, this situation will be most commonly encountered when the coal is owned by the Tribe and the oil and gas is owned by a non-Indian third party. In the San Juan Basin, it may be an acceptable business risk to proceed with coal-bed gas drilling activities on the premise that a coal interest owner receives no interest in the coal-bed gas by virtue of a conveyance or reservation of coal only. The issue of access to and protection of the coal resource, however, should be discussed by the Tribe and the coal-bed gas developer. The tract identification process setting forth the exact nature of the conveyance or reservation, as well as the nature of the coal resource, should indicate the likelihood of any dispute.

D. Highest-risk Situation: The Presence of Coal Development.

In the event that the tract identification process indicates that coal operations are pending or underway or that the coal or coal rights are owned by a person or entity intending or likely to mine the coal, the coal-bed gas developer faces a potentially difficult situation. Certain measures, however, can be taken to mitigate the likelihood of a dispute which would interrupt the planned coal-bed gas operations.

Where a dispute is likely because of the nature and scope of existing or anticipated coal mining operations, the coal-bed gas developer should attempt negotiations with the coal owner to provide for the orderly and equitable development of both resources. Such negotiations should address, at a minimum, the timing and priority of respective operations, permitted activities and locations, and liability and damages. Bearing this in mind, once a lease has been obtained from the oil and gas owner, the coal-bed gas developer should attempt to secure a subordination agreement from any prior segregated coal interest owner. Ideally, such a subordination should provide for a disclaimer of ownership of the coal-bed gas, a prohibition of mining operations during the life of the lease, and payment of any damages associated with coal left in place. It should also recite a specific grant of the right to stimulate any wells drilled on the leased lands.

It is entirely possible, if not probable, that a coal interest owner may elect not to sign a subordination agreement (perhaps arguing that the *U.S. Steel* precedent should be followed in the particular jurisdiction involved). That risk is inherent in targeting a coal-bed gas prospect containing segregated ownership of coal and oil and gas where coal development is possible. Unless the nonsubordinating parties own horizons which may be isolated from the wellbore without destroying the economic viability of the prospect, the coal-bed gas developer may face such a risk of litigation that it will not make sense to proceed without careful consideration of the situation, including the advice of counsel.

In the San Juan Basin, the coal-bed gas developer will not encounter any coal mining activity on lands owned by the Tribe. This does not mean, however, that the Tribe will not insist on protection for its coal resources. There is a possibility of encountering such situations where the coal is owned in fee; coal mining operations do exist in some fee areas targeted by coal-bed gas developers.

E. Access to and Protection of Coal Beds.

It is clear from the preceding pages that certain legal risks may be anticipated and certain procedural hurdles will have to be cleared by the coal-bed gas developer in the early stages of planning a coal-bed gas prospect. These legal risks and procedural hurdles relate to the concept of multiple-use or multiple mineral development, and pertain to the ways in which the two resources, coal and coal-bed gas, should be developed. Determining who owns and has the right to develop the coal-bed gas is not the end of the story: questions of access will move to the fore. How to develop both the coal

[122] For a discussion of possible lease provisions, see SOMED Research Report at 23-43.

and coal-bed gas has been dealt with by the Federal government where the coal is federally owned. The development situation where an Indian tribe, the state or an individual owns the coal is less well-defined.

Multiple mineral development issues, where gas or oil is one of the minerals and coal is the other, typically relate to the issue of access through the coal to reach the gas or oil. The wellbore is the bane of the coal operator and usually the focus of his opposition. These issues have typically been resolved in favor of such access; provided, however, that the oil and gas developer compensates the coal owner for damage to his resource. In the area of coal-bed gas, however, more than access through is involved. In the coal-bed gas situation we are talking about access into the coal and the extraction techniques employed, which may or may not render the coal unminable or more expensive to mine.

As is discussed below in greater detail, the Federal government seems to afford great protection to its coal resource. The Indian and state protections are not as clearly delineated.

1. Federal Coal. Where the coal is federally owned and there is a Federal coal lease outstanding, the coal-bed gas developer can expect significant problems in gaining unfettered access to the resource.[123] These problems will be greater if the coal lease was issued prior to the Federal oil and gas lease under which the coal-bed gas developer desires to drill. However, even where there is no Federal coal lease in effect covering the target coal seam, our experience indicates that the BLM may take specific actions to protect the Federal coal resource from possible damage resulting from drilling and completion operations, notwithstanding that the target coal seam is below the currently accepted minable depth standard of 3,000 feet. Such protective actions are normally accomplished through restrictive stipulations attached to a Permit to Drill as a condition of approval. Those stipulations can take the form of requiring submission of well logs to the BLM District Mining Supervisor, approval prior to hydrofracture stimulation of the well, and casing and cementing the wellbore through any coal seams encountered. In any event, case-through stipulations and restrictions against hydrofracturing will, in all likelihood, have to be removed before the coal-bed gas developer commences actual drilling operations. If this cannot be accomplished by negotiation and (or) submission of fracture design information, an administrative appeal may be necessary to protect the coal-bed gas developer's rights as lessee.

Where the coal is federally owned but the oil and gas is owned in fee, and the coal-bed gas developer contemplates drilling a well into a Federal coal seam as lessee under a fee oil and gas lease, a search of the BLM records should be made to determine whether a Federal coal lease has been issued. The authors have cautioned clients (in Drilling Title Opinions, for example) to notify the BLM of their drilling plans even if no coal lease has been issued. In an abundance of caution, it has also been suggested that they obtain BLM approval of extraction techniques which arguably might damage the coal, particularly where the coal seam is above 3,000 feet (although neither of the authors, in practice, has been involved with any western coal-bed gas projects where the target seam is so shallow). A more conservative approach would be to attempt to obtain a subordination agreement from the BLM. Such an agreement should, ideally, provide for the BLM's disclaimer of any ownership interest in the coal-bed gas. The agreement might also preclude the commencement of mining operations during the life of the oil and gas lease, provide for the payment of any damages associated with well stimulation, and recite a specific grant of the right to stimulate any wells drilled on the leased lands. There is a significant likelihood of delays if such an avenue is taken, because no codified procedures exist which might guide the BLM in negotiating such an agreement.

Where the coal-bed gas developer's search of the BLM coal leasing records reveals that a coal lease has been issued, prior to drilling under a fee oil and gas lease, the BLM will likely urge that any and all access and extraction issues be resolved by negotiation between the oil and gas lessee and the Federal coal lessee.

2. Indian Coal. Where the coal is owned by the Southern Ute Indian Tribe, the problems a coal-bed gas developer may face are not as clearly defined. There are no regulatory provisions either authorizing or prohibiting multiple mineral leasing on either allotted or Tribal lands, but any Indian mineral leases are made expressly subject to such restrictions or stipulations of the Secretary of the Interior as may be necessary for the protection of natural resources. The regulations applicable to oil and gas operations on Indian lands require the operations to be conducted in a "good and workmanlike manner";[124] however,the regulations do not specifically address multiple mineral development, in particular, the development of coal and coal-bed gas.

The BIA does have authority to create and enforce stipulations pertaining to multiple mineral development conflicts on Indian lands, but no standard stipulations have been found. In a situation where all the minerals are Indian-owned, it would be prudent to negotiate leases or minerals agreements with language addressing multiple mineral development. In situations where the Indian tribe or allottee owns only the coal, and coal development is likely, prudence would dictate that the fee oil and gas lessee approach the Indian tribe, BIA, and BLM and inform them of its proposed plan of development. If it is likely that there will be no coal development on the tract, it may be an acceptable business risk to ignore multiple mineral development issues and proceed with development. If negotiations between the parties take place, they should focus on the multiple mineral development issues of access and extraction as outlined above.

3. State-owned Coal. The New Mexico statutes do not address protection of a coal seam from damage that may be caused by an oil and gas well. If the targeted prospect involves coal owned by the State of Colorado and that coal seam is deemed workable, pursuant to statute, the oil and gas developer is required to protect the coal seam.[125] This statute seems to address the traditional oil and gas well situation rather than the coal-bed gas situation.

[123] *See* Lear at 89-95.

[124] 25 C.F.R. § 211.19 (1987).

[125] C.R.S. §§ 34-61-101 through 108.

If such a situation is encountered, the oil and gas lessee may be forced to negotiate with the State Land Board, the Colorado Oil and Gas Conservation Commission, and the chief inspector of coal mines to address the multiple mineral development questions of access and extraction. The deeper and more unminable the coal seam, the more unlikely any significant restriction will be imposed upon the oil and gas lessee.

F. **Conclusion.**

Although many legal questions remain to be settled here in the West, the law has developed sufficiently to properly analyze the issues and foresee with relative accuracy the degree of risk and even the outcome of disputes between owners of oil and gas rights and owners of coal rights regarding the ownership of the coal-bed gas, or the right to drill into the coal seam and extract such gas. Where the Western coal-bed gas developer has targeted a prospect, it must determine such risks and the likelihood of such disputes and, where they exist in sufficient degree, negotiate solutions to them in advance of coal-bed gas exploration and production. Aside from certain procedural differences, the same strategy, combining vigilance and, where the likelihood of a dispute arises, advance negotiation, applies equally to mineral resources under Federal, Indian, state and privately owned lands.

Southern Ute Fair and Rodeo during the early 1900s. The Fair is a social gathering for all the Ute bands with similar activities to those of other fairs and rodeos. Events include men's and women's horse racing, "Shinny" a women's game similar to Lacrosse, card games and others, all with betting. Additional activities include art, craft and produce competitions. Indian dances were held following the Fair and Rodeo events. Photo courtesy of the Colorado Historical Society.

The UTE BEAR DANCE, oldest of the Ute Dances and still celebrated each May, tells a story of a man's dream. The Ute man dreamed that if he traveled to a specific location on the mountain during the spring he would see a bear just when bears are awakening from their winter hibernation, (Bears represent strength and power to the Utes).

The Ute traveled to the spot he dreamed of and there was a bear, having recently emerged from his winter den, shuffling back and forth. The bear taught the Ute this dance and the appropriate songs. The bear then instructed the Ute to return to his people and teach them this dance, the Bear Dance. Every spring since, when bears emerge from their sleep, each of the seven bands of Utes dance the Bear Dance. The dance usually lasts for three days and three nights.

On the final day of the dance, if a couple falls, the Ute Medicine man will pray over the dancers and bless them. A feast is held following this three day ceremony.

The participants dress in their finest outfits for the dance. As shown in the photo, (date unknown), two lines are formed, men on the north women on the south. The women select the men they want to dance with by waving their shawl toward them or pointing toward them. After selecting partners, the women line up in the center of the dance area and join hands. When the song begins, they dance two steps forward and three steps backwards. As the men move forward, the women retreat, and vice-versa. Photo courtesy of the Colorado Historical Society.

State Regulations Affecting Fruitland Formation Coal-Bed Methane Gas Development in the San Juan Basin, New Mexico and Colorado

Katy Templeton-Buell[1]
and
Ernie Busch[2]

ABSTRACT

Rules and Regulations covering gas produced from Fruitland Formation coal beds in the San Juan Basin are reviewed in detail as they apply in December 1987. The Fruitland Coal-Bed Methane Committee was formed to advise the States of Colorado and New Mexico on the unique aspects of coal-bed methane production and to help them make informed decisions concerning future regulations. The Committee's goals and progress as well as some decision deadlines are reviewed.

INTRODUCTION

In November 1986, the New Mexico Oil Conservation Division (NMOCD) called a meeting of concerned parties to discuss Fruitland Formation coal-bed methane production. The States of Colorado and New Mexico realized at the time that coal-bed methane is a unique resource, possibly requiring some new and innovative regulations. A committee was formed to study the issues and make recommendations to the States for new rules and regulations to cover Fruitland coal-bed methane production.

PRESENT REGULATIONS

New Mexico and Colorado presently have regulations covering drilling and production from the Fruitland coal. Each State has different regulations and their work with the committee has shown that a comprehensive review of the current regulatory policies in both States is in order.

The Colorado Oil and Gas Conservation Commission (COGCC) addressed spacing issues in the Ignacio Blanco field in 1952-54 under Order 3 and later under Order 112. At that time, the Dakota-Morrison pool was spaced on 640 acre units and the Mesaverde pool was spaced on 320 acre units. The Dakota-Morrison was subsequently downspaced to 320 acre units and both pools were allowed an optional infill well. The Fruitland-Pictured Cliffs pool was defined by log picks from the Stanolind-Ute Indian B-1 well located in the NE¼ sec. 18, T. 33 N., R. 7 W., NMPM. Spacing was set on 320 acre units based on legal subdivisions of U.S. Land Surveys. Wells were to be located either in the NW¼ or the SE¼ of the section and were to be located no less than 990 feet from the boundaries of the quarter section. An infill well was later granted in the offsetting quarter section. It is significant to note that the definition of the Fruitland Formation and the Pictured Cliffs Sandstone as a pool allows for blanket down-hole commingling and prevents separate wells to the Fruitland and Pictured Cliffs on 160 acre quarter sections.

Some recent wells have been drilled outside the spaced area and fall under the General Rules. In these cases, any well drilled to a source of supply greater than 2,500 feet in depth shall be located not less than 600 feet from the nearest lease line and 1,200 feet from any other producible or drilling oil or gas well in the same source of supply. Wells drilled to a source of supply less than 2,500 feet must be "located not less 200 feet from any lease line and not less than 300 feet from any other producible oil or gas well or drilling well in said source of supply"[1].

Casing requirements were also established in the 1950s. A minimum of two strings of casing are to be set in each hole. Surface casing must be at least 200 feet below the surface or to a depth to confine all fresh waters present. Cement is to be set by the pump-and-plug method to the surface and allowed to set for 24 hours. The Director has waived the 200 foot minimum surface casing requirement when the second string of pipe is set and cemented to the surface.

Water disposal has become a significant issue with the large quantities of water produced from many coal-bed completions. Several alternatives exist for disposing of produced waters. On-site pits are generally the most economical method for water disposal. On private and State land an on-site pit must be permitted through the COGCC if the pit receives more than five barrels a day. Liners are required if the total dissolved solids of the water is greater than 5,000 ppm or if there is less than 50 feet of impermeable material between the bottom of the pit and the top of the nearest potential aquifer. Pits on Indian and Federal land are permitted through the Bureau of Land Management. Under Bureau of Land Management guidelines, only pits receiving less than 5 barrels per day are allowed to be unlined. Commercial pits are permitted through the County Health Departments.

Discharge permits can sometimes be obtained through the State Health Department and are worth investigating with the

[1] Katy Templeton-Buell
Colorado Oil and Gas Conservation Commission
Loma, Colorado 81524
[2] Ernie Busch
New Mexico Oil Conservation Division
Aztec, New Mexico 87410

[1] Colorado Oil & Gas Conservation Commission Rules and Regulations, October, 1987.

Table 1. Disposal wells for Fruitland coal-bed water

Well name	Location	Formation	Comments
Salvador A-1	sec. 28, T. 33 N., R. 7 W.	Entrada	872,116 Bbls injected through 10-87
San Juan 1-12Y	sec. 26, T. 30 N., R. 6 W	Entrada/Morrison	675,020 Bbls injected through 12-87
McCaw #1	sec. 20, T. 34 N., R. 8 W.	Mesa Verde	1,314,428 Bbls injected through 10-87
S.E. Cha Cha #37	sec. 15, T. 28 N., R. 13 W.	Gallup	124,578 Bbls injected through 11-87
USG Lease #17	sec. 19, T. 29 N., R. 16 W.	Dakota	17,985,103 Bbls injected through 12-87
San Juan 30-6	sec. 31, T. 30 N., R. 6 W.	Ojo Alamo	22,742 Bbls injected through 12-87 To be plugged and abandoned
Rea 1-32	sec. 32, T. 34 N., R. 8 W.	Mesa Verde	148,900 Bbls injected through 10-87
S. Ute 15-9	sec. 15, T. 32 N., R. 9 W.	Pictured Cliffs	Completing 1-88
McGrath #4	sec. 34, T. 30 N., R. 12 W.	Mesa Verde	140,165 Bbls injected through 12-87
Echols-Ute 1-12U	sec. 12U, T. 34 N., R. 7 W.	Pictured Cliffs	Permitted conversion, No injection
Sun-Tyner Lunt #1	sec. 18U, T. 34 N., R. 8 W.	Mesa Verde	Proposed conversion
Ohnemus #1	sec. 4, T. 33 N., R. 7 W.	Mesa Verde	Proposed conversion
Los Pinos #1	sec. 31, T. 35 N., R. 6 W	Mesa Verde	Proposed new well
Simon Land & Cattle #1	sec. 15U, T. 34 N., R. 7 W.	Mesa Verde	Proposed new well
Bonds 5-1 WDW	sec. 1, T. 32 N., R. 10 W.	Ojo Alamo	Proposed new well
Disposal #1	sec. 3, T. 29 N., R. 11 W.	Mesa Verde	Proposed new well

relatively fresh water produced from coal. Detailed water analyses are required relatively frequently for the discharge permit and the permitting process has been fairly lengthy. Discharge permits have been obtained for coal-bed wells in the Piceance and Raton basins of Colorado.

Water is also being disposed of in disposal wells. Disposal wells on private, State, and Federal lands are permitted and regulated by the COGCC under the Underground Injection Control (UIC) program. There are presently several disposal wells operating in the Ignacio Blanco field in Colorado using the Mesaverde Group and Entrada Sandstone (table 1). Several other disposal wells have been permitted to date including one projected to the Animas Formation. Injection wells south of the the northern boundary of the Southern Ute Indian Reservation, whether on fee or Indian land, will be permitted and regulated by the Environmental Protection Agency in conjunction with the Southern Ute Indian Tribe and the COGCC.

The NMOCD has regulations that are significantly different from those of the COGCC. As of December 1987, 21 Fruitland pools had been defined. All but the Cedar Hill pool are spaced on 160 acres with Cedar Hill spaced on 320 acres. Six additional pools (Ojo, WAW, South Los Pinos, South Gallegos, Twin Mounds, and Harper Hill) are presently commingled Fruitland-Pictured Cliffs pools spaced on 160 acres. In those pools that are not commingled, the top of the Pictured Cliffs Sandstone is presently defined as the top of the stratigraphically highest marine sandstone bed.

Spacing of wildcat wells or wells located more than one mile from the outer boundary of a designated pool "shall be located on a designated drilling tract consisting of 160 surface contiguous acres, more or less, substantially in the form of a square which is a quarter section, being a legal subdivision of the U.S. Public Land surveys, and shall be located not closer than 790 feet to any outer boundary of the tract nor closer than 130 feet to any quarter-quarter section or subdivision inner boundary."[2]

None of the Fruitland coal-bed methane production is prorated. An initial seven day shut-in pressure build up is required on all new wells in accordance with the *Gas Well Testing Manual for Northwest New Mexico* (1987). Nonprorated wells are required to have a minimum seven day biennial shut-in test preceded by seven days of uninterrupted production. Hardship wells and wells incapable of producing more than 250 MCF per month are exempt from biennial testing. The NMOCD is presently reviewing shut-in requirements for coal-bed methane wells due to possible damage that might be incurred during prolonged shut-in periods.

Water disposal in pits on fee and State lands fall into two categories in New Mexico. A vulnerable area has been defined as 100 vertical feet from the Animas, San Juan, and La Plata rivers. The NMOCD is also reviewing whether other drainages should be included in the vulnerable area. In the vulnerable area, all pits receiving more than five bbls per day must be lined. If the total dissolved solids exceed 10,000 ppm the pit must be lined regardless of water volume. Pits outside the vulnerable area are not required to be permitted but must be registered with the NMOCD (with the exception of central and commercial pits). The NMOCD Environmental Bureau handles the permitting of both central and commercial facilities.

Several injection wells have been drilled to dispose of produced Fruitland water into the Entrada and Ojo Alamo formations (table 1). Attempts to dispose of water in the Ojo Alamo Sandstone have thus far been found to be unsuccessful. Until recently the UIC program in New Mexico considered a two-mile radius of investigation when permitting a disposal well. Because of the intense oil and gas development of both the Dakota and Mesaverde pools, disposal in these formations was essentially ruled out. The UIC program now considers each application on an individual basis with respect to radius of influence.

The NMOCD is presently considering requiring coal and water samples from all wells being drilled. The intent is to collect information from all coal seams encountered while drilling to assist in the evaluation of coal-bed wells. The U. S. Geological Survey has shown interest in helping collect and evaluate samples submitted from drilling operations. This information would be used to better define the production

[2] State of New Mexico, Energy and Minerals Department, Oil Conservation Division Rules and Regulations, March 1987.

characteristics of all coal-bed methane wells and provide better information on gas-production potential from coals other than the Fruitland.

POTENTIAL REGULATIONS

The Fruitland Coal-Bed Committee was formed in early 1987 to identify problems unique to coal-bed methane production and, it is hoped, to find some solutions to these problems in the early stages of development. The NMOCD and COGCC met with interested operators producing gas from Fruitland coal. The Southern Ute Indian Tribe and Bureau of Land Management have also been participating in the meetings. Some of the committee's goals are to define vertical and horizontal limits, identify alternative water disposal methods, define Fruitland coals and their extent as they apply to pools and boundaries, determine well spacing, address multi-mineral management, address mineral ownership of coal-bed methane, study potential uses of CO_2 coproduction, and study commingled Fruitland and Pictured Cliffs production.

The Committee has run into many stumbling blocks in its attempts to address these issues. The available data base is quite limited, primarily because too few wells had been drilled and completed in the Fruitland coal by early 1987. In 1986, the State of Colorado permitted nine Fruitland coal-bed methane wells of which two were Pictured Cliffs commingled wells and two were never drilled. New Mexico permitted only 3 Fruitland wells during 1986. Both Colorado and New Mexico saw substantially increasing interest in Fruitland coal-bed gas in 1987, with approximately 52 and 26 permits being filed, respectively. Projections for 1988 are in the range of 300 permits for the basin. Another problem is the sensitive nature of proprietary information. Most of the information concerning the 52 wells permitted during 1987 in Colorado is still confidential. We anticipate that as the play develops, confidentiality will become less of an issue. Pipeline limitations, lack of continuous production, and reporting problems further hamper progress.

In an attempt to identify and differentiate the Fruitland coal-bed gas production, some research has been done on gas typing using information available from pipeline companies. Coal gas typically has a lower Btu content, very low ethane content, and some CO_2. Water quantity and quality are also fingerprints for coal-gas production. It is hoped that some of this information will help to identify anomalous wells and answer some questions about the integrity of producing zones and the potential for out-of-zone frac jobs.

The Committee has determined that the Fruitland pool needs to be separated from the Fruitland-Pictured Cliffs pool in Colorado. Commingling is beneficial in some, but not all situations and should probably be allowed on a well-by-well or pool basis. In some cases, interfingering of coal seams and marine sandstones makes the geologic definition of the top of the Pictured Cliffs Sandstone an impractical boundary from a regulatory point of view. The Committee will make a recommendation to the COGCC concerning new definitions for the Fruitland and Pictured Cliffs pools shortly.

In November 1987, Frank Chavez, District Supervisor, NMOCD and Bill Smith, Director, COGCC directed the Committee to seriously consider the spacing issue. The Committee was advised that recommendations are due on April 1, 1988. Plans are to study traditional as well as unconventional spacing alternatives. It is hoped that the Committee's recommendations will be flexible enough to encourage some of the new and innovative technologies being developed in the San Juan Basin while promoting prudent development of oil and gas and protecting correlative rights.

Thus far, many issues have been discussed affecting the development of Fruitland coal gas. Within the next year some preliminary decisions concerning pool limits and spacing requirements will be made. In addition, more should be known about water production and disposal including emphasis on the economic feasibility of different disposal alternatives and the effects of downtime on production rates.

CONCLUSIONS

Current regulations covering Fruitland coal-gas production may not be adequate to promote production of this sizable resource while still protecting correlative rights. The Fruitland Coal-Bed Methane Committee was formed in early 1987 to address many of the questions, problems, and possibilities raised by the intense interest in developing this resource. To date, many issues have been discussed and few answers found but by the time this paper is in print, some significant decisions in regards to spacing and pooling should have been reached.

View looking south over Durango at Smelter Mountain, which has at its base, the last standing remnant of the San Juan Smelter, the smokestack. The outcropping ledge located slightly below the ridge top is the Mesaverde Point Lookout sandstone. A small displacement is observable in the gully on the left (east) side of the photo. Down side of fault is left. The uranium mill tailing piles are the mounds in right center up against the base of Smelter Mountain. Removal of the tailings began in early fall of 1987. Photo courtesy of William E. Hanson.

An Overview of Geothermal Studies in the San Juan Basin, New Mexico and Colorado

Gerry Clarkson[1] and Marshall Reiter[2]

ABSTRACT

The geothermal history of the San Juan Basin-San Juan volcanic field region has been investigated by several authors using heat flow and hydrocarbon maturation data. The regional patterns of these data suggest a significant influence, from San Juan volcanic field heat sources, on both the present and past subsurface temperatures within the basin. An examination of heat sources, such as heat conduction from the cooling of the San Juan batholith, heat conduction from deep seated thermal sources beneath the San Juan volcanic field, and increased burial depths (for maturation of the Fruitland Formation coals) shows that these sources cannot completely explain the observed heat flow and coal maturation patterns. Heat advection by ground-water flow may have played an important role in the geothermal history of the San Juan region. The present observed heat flow in the San Juan volcanic field appears to be influenced by ground-water movement, and even deep heat-flow measurements in the San Juan Basin may also be influenced by water flow. Heat advection by ground-water flow may also have contributed a significant amount heat for hydrocarbon maturation in the San Juan Basin. Heat advection into the basin by water flow is reasonable in terms of the probable ground-water flow direction since Late Cretaceous time, the amount of heat available for redistribution by ground-water flow, and the potential for alteration of the thermal regime of sedimentary basins by ground-water flow. Detailed modeling of temperature distributions and ground-water flow over time will be needed to analyze the role of heat advection in the thermal history of the San Juan Basin.

INTRODUCTION

An appreciation of the geothermal history of the San Juan Basin and the surrounding region is of importance not only for understanding the general geologic history of the region but also for understanding the generation and maturation of hydrocarbons in the basin (fig. 1). Geothermal studies concerning the San Juan Basin have addressed both the present day heat flow and the thermal history using levels of hydrocarbon maturation. The thermal history of a region is reflected in the level of hydrocarbon maturation since the maturation level is exponentially dependent on temperature and linearly dependent on time. Although the thermal history of the San Juan Basin has been studied using maturation levels of hydrocarbons of different ages, most of the studies have considered the level of maturation of the Fruitland Formation coals of Late Cretaceous age.

The purpose of this paper is to characterize some of the general results of previous studies of the geothermal history in the San Juan Basin and to discuss implications of these results. In particular, the influence of events to the north, in the region of the San Juan volcanic field, on the basin's thermal history will be considered. Prior to reviewing the geothermal studies of the basin, a brief geologic history of the region as it relates to geothermal studies will be presented. A short summary of the role of temperature and time in the maturation of hydrocarbons will also be given.

[1] Gerry Clarkson
Department of Geology
Western Michigan University
Kalamazoo, Michigan 49008

[2] Marshall Reiter
New Mexico Bureau of Mines and Mineral Resources
Socorro, New Mexico 87801

GEOLOGIC BACKGROUND

The San Juan Basin is a structural and sedimentary basin located in the southeastern part of the Colorado Plateau. The Basin contains up to 4.3 km of Phanerozoic rocks. In terms of hydrocarbon occurrence, the most important rocks are Late Cretaceous with some commercial production of oil and gas from Pennsylvanian and Jurassic rocks (Rice, 1983). The Jurassic rocks were deposited in various desert environments with local alluvial and fluvial deposition continuing into early Cretaceous time (Stone and others, 1983). This period was also one of local nondeposition and erosion (Stone and others, 1983). A major change in depositional style occurred in the Late Cretaceous when the shoreline of a large, shallow sea covering a large part of North America fluctuated across the San Juan region. These Late Cretaceous rocks form a sequence of intertonguing marine and nonmarine deposits. The Dakota Sandstone was deposited during the first invasion of this sea and is overlain by the marine siltstones and shales of the Mancos Shale (Rice, 1983). The Mancos Shale is overlain by the Mesaverde Group, which contains the Point Lookout Sandstone (regressive shoreline), the nonmarine Menefee Formation and the Cliff House Sandstone (transgressive shoreline) which is in turn overlain by the marine Lewis Shale. In Late Cretaceous time, during the final regression of the sea, the marine Pictured Cliffs Sandstone and the nonmarine Fruitland

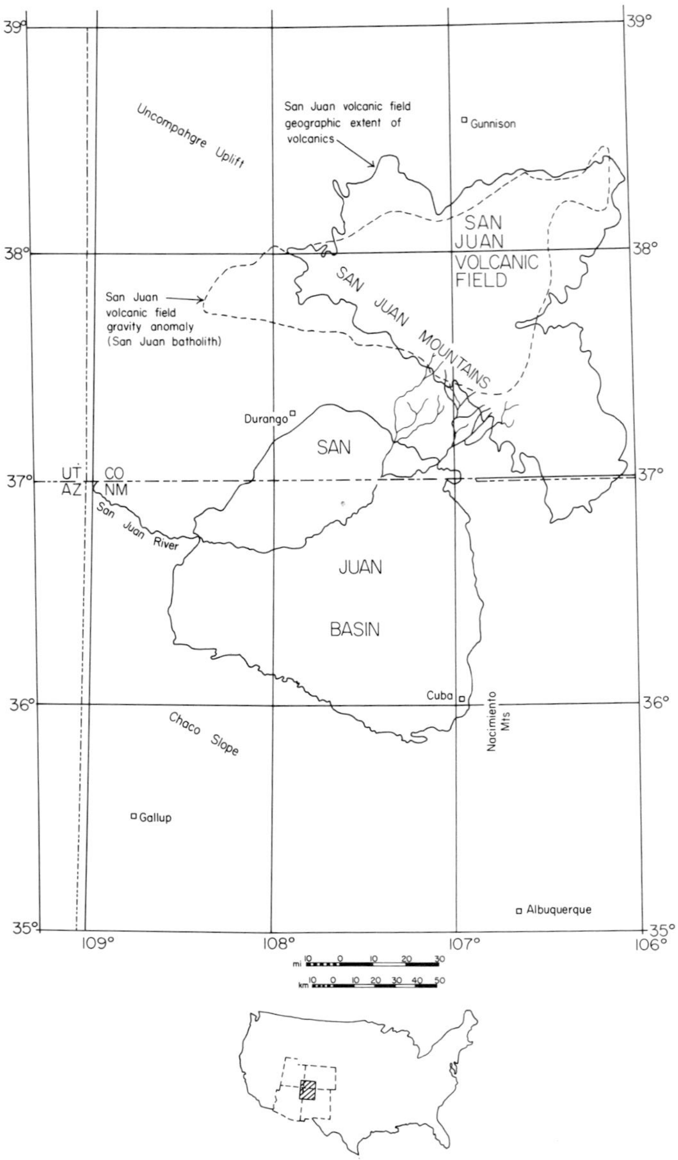

Figure 1. Map showing the location and some of the principal geologic features of the San Juan region. Extent of the San Juan volcanic field is after Steven (1975) and the San Juan batholith gravity anomaly is after Plouff and Pakiser (1972).

Formation and Kirtland Shale were deposited in the basin (Molenaar, 1983). Following deposition of the Kirtland Shale, erosion removed as much as 0.6 km of Late Cretaceous sediments from the basin (Fassett and Hinds, 1971). In the Paleocene and Eocene, stream channel deposits were laid down. Since the deposition of the San Jose Formation in the Eocene, the basin has been a region of nondeposition and erosion with perhaps up to 0.3 km of the San Jose Formation being eroded in the central part of the Basin (Stone and others, 1983). The northern San Juan Basin was probably covered by some thickness of volcanic debris from the Oligocene San Juan Mountains volcanism (Steven, 1975; Bond, 1984). Any volcanic cover in the area has subsequently been removed by erosion.

The San Juan Basin began to subside during the Laramide (Late Cretaceous-early Tertiary), the subsidence accelerated in the early Tertiary and ended by the late Paleocene or early Eocene (Tweto, 1975; Stone and others, 1983). The maximum structural relief in the basin is about 3 km along the San Juan uplift (Kelly, 1950). The elevation in the basin ranges from 1.6 to 2.4 km and the average elevation in the Colorado Plateau is 2 km (Fassett and Hinds, 1971; Bird, 1979). The uplift in the Colorado Plateau has occurred since the deposition of Late Cretaceous sediments; however there is little agreement as to the timing of the uplift (see for example, King, 1959; Eardley, 1962; P.E. Damon, in Bird, 1979; Lucchita, 1979; C.E. Dutton, in Bodell and Chapman, 1982).

The San Juan Mountains experienced magmatic and tectonic activity in the Laramide, Oligocene, and Miocene-Pliocene (Steven, 1975; Tweto, 1975; Lipman and others, 1978). Just prior to the Laramide orogeny the region was a flat, slowly subsiding marine environment. The Laramide orogeny began about 72 Ma and continued into the Eocene with associated igneous activity occurring 70 to 63 Ma (Tweto, 1975). During the Oligocene, between 35 and 26 Ma, 60,000 km^3 of volcanic material was erupted (Lipman and others, 1978). The present San Juan volcanic field is the remnant of this material. It has been suggested that the Oligocene volcanism was associated with the rise and emplacement of a large crustal batholith and that the batholithic magmas were ultimately mantle derived and perhaps associated with subduction processes (Steven, 1975; Lipman and others, 1978). In the Miocene and Pliocene, the San Juan Mountains were within an extensional environment characterized by bimodal basalt-rhyolite volcanism (Lipman and others, 1978). This later volcanism appears to be genetically unrelated to the Oligocene volcanism (Lipman and others, 1978).

TIME AND TEMPERATURE IN HYDROCARBON MATURATION

Hydrocarbon maturation is exponentially dependent on temperature and linearly dependent on time (Phillipi, 1965; Bostick, 1973; Lopatin and Bostick, 1974; Waples, 1980). Thus the observed level of hydrocarbon maturation in an area may be used to constrain various models of thermal histories for the region. Conversely, a knowledge of the thermal history of a region may be used to predict the level of hydrocarbon maturation which is to be expected in that region. Models of the thermal history of a number of areas have been generated using various methods of integrating the effect of time and temperature on hydrocarbon maturation (see for example, Hood and others, 1975; Buntebarth, 1978; Bond, 1984; Cercone, 1984). One particularly useful technique for relating thermal history and hydrocarbon maturation is the time-temperature index (TTI) (Waples, 1980). If the maturation process is taken to be a continuously varying function of time, the TTI value can be expressed as:

$$\mathrm{TTI} = \int 2^{T(t)-105} dt$$

where temperature (T) is in °C and time (t) is in millions of years. Waples (1980) gives a relation between TTI values and vitrinite reflectance, thus the effects of time and temperature can be related to a physically observable measure of maturation. In studying the thermal history of an area and the related hydrocarbon maturation, various parameters such as the thermal gradient and burial history of the rocks must be considered. Thermal gradients can be influenced by a number of factors including mantle heat flow, radiogenic heat production, changes in surface temperature related to climate or elevation changes, and so on.

GEOTHERMAL STUDIES

Heat flow

Several studies have reported measured heat-flow values in the San Juan region. These include Sass and others (1971), Decker and Birch (1974), Reiter and others (1975), Edwards and others (1978), Decker and Bucher (1979), Galloway (1980) and Reiter and Mansure (1983). Some of the measured heat-flow values are shown in figure 2. (For a more complete compilation see Reiter and Clarkson, 1983c.) Two major concerns regarding the heat-flow data will be addressed; one is the potential for vertical water movement to affect the heat flow data, the second is the regional pattern in the heat-flow data.

The effect of water movement on observed thermal gradients has long been recognized (see for example, Van Orstrand, 1934; Stallman, 1963; Bredehoeft and Papadopulos, 1965). Vertical water movement can cause the measured conductive heat flow in an area to vary considerably. Lachenbruch and Sass (1977) show that conductive heat flux in the near surface may be reduced to zero by downward groundwater flow at the rate of 1 foot/year through a layer a few hundred feet thick. Alternatively, the upward flow of ground water may create very high apparent heat-flow values. Of course the influence of water movement is not always dramatic and subtle effects due to ground-water flow may be difficult to discern in many cases.

In studying the heat flow in the San Juan Basin, Reiter and Mansure (1983) presented a set of deep heat-flow measurements where thermal gradients were measured to greater than 750 meters depth (fig. 2). The use of deep measurements allowed them to avoid much of the disturbance to the heat flow from near surface ground-water movement. In examining this data set they concluded that the deep heat-flow measurements gave a better representation of the regional heat flow in the basin, although it is possible that deep ground-water flow is still affecting the results (Reiter and Mansure, 1983). The deep heat-flow values show a relatively smooth trend of increasing heat flow from south to north across the Basin. Reiter and Mansure (1983) note that this trend of increasing heat flow is statistically valid.

The heat flow in the San Juan volcanic field appears to be high; the average value for all locations is 127 milliwatts per meter^{-2} (mWm^{-2}), although there is a considerable range of values, 88 to 255 mWm^{-2}. None of the values in the San Juan volcanic field are deep measurements and the variability in the data suggests that ground-water movement may significantly influence measured values. The high average heat-flow value in the San Juan volcanic field is consistent with the heat-flow trend across the San Juan Basin.

The high heat flow in the San Juan volcanic field and the trend of increasing heat flow across the San Juan Basin towards the San Juan volcanic field suggest that thermal sources associated with the volcanism have a significant influence on the regional heat-flow pattern. That an area such as the San Juan volcanic field, which has been the site of repeated and extensive volcanism since Late Cretaceous time, should have a high heat flow and exercise an influence on the

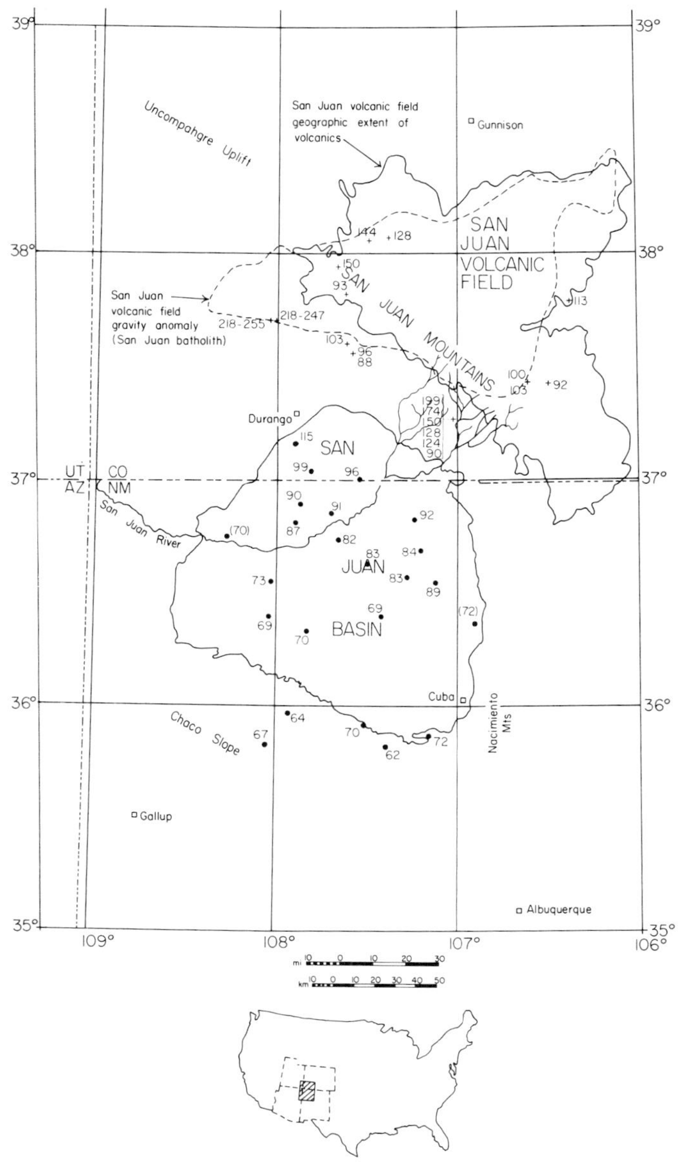

Figure 2. Heat-flow values in milliwatts per meter^{-2} (mWm^{-2}) and locations of data sites. Heat-flow values in the San Juan Basin are from deep measurements (thermal gradients measured to greater than 750 meters) only. Values are from Decker and Birch (1974), Decker and Bucher (1979), Edwards and others (1978), Galloway (1980), Reiter and Mansure (1983), Reiter and others (1975), and Sass and others (1971).

thermal regime of surrounding areas is intuitively reasonable. However, the nature of this influence and the source of anomalous heat in the area merits some examination.

One source of anomalous heat in the region is the cooling of the San Juan batholith. Although the batholith is extensive and was emplaced relatively recently (Oligocene time), heat released by the cooling of the batholith is not sufficient to produce the present observed heat flow in the San Juan volcanic field (Reiter and Clarkson, 1983b). Consequently there must be other sources of heat as well. Two potential additional heat sources are suggested by Reiter and Clarkson (1983b).

One possible heat source is an increased concentration of radiogenic heat-producing elements associated with magmatic activity. A second possible heat source is a deep seated thermal anomaly beneath the region. Increased radiogenic heat production for the region is difficult to justify, however the presence of a deep seated thermal anomaly beneath the San Juan region is supported by the large lateral dimension of the heat-flow trend across the San Juan Basin. Studies of this heat-flow trend show that a large (radius on the order of at least 60 km) plume-like thermal anomaly originating from at least 100 km depth and rising to about 35 km depth would be compatible with the observed heat-flow trend (Reiter and Clarkson, 1983b; Clarkson and Reiter, 1987). Modeling of the thermal regime also suggests that this deep thermal anomaly must be quasi-steady state to produce the observed heat-flow pattern. Near steady-state conditions would imply replenishment of heat since the major (Oligocene) volcanism in the region.

Hydrocarbon maturation

The pattern of hydrocarbon maturation in the San Juan Basin shows a regional trend which is similar to the pattern of heat flow. Maturation within the Basin has been studied largely from the standpoint of coal maturation. The Late Cretaceous Fruitland Formation coals have been particularly well studied, with numerous maturation values reported in Fassett and Hinds (1971) and Rice (1983). A few studies have considered the maturation of liquid hydrocarbons, for example, Reiter and Clarkson (1983a) and Rice (1983).

The maturation pattern of the Fruitland coals is indicated in figure 3. Again, a regional trend of increasing maturation to the north in the basin is noted. A similar trend of increasing maturation for samples from the Dakota Sandstone was also noted by Rice (1983). Although maturation trends from oil and gas generation and preservation are more problematic than those for coal (because migration of the oil and gas may occur), the trends in oil and gas maturation are similar to those observed for the coal. In addition to the expected increase in maturation in older formations due to greater burial depths, there is also a trend of increasing maturation from south to north (cooler to warmer areas) in the basin (Reiter and Clarkson, 1983a; Rice, 1983). In general, oil generation and preservation is noted in the peripheral areas of the western, southern, and southeastern basin while gas typically occurs in the central and northern basin where subsurface temperatures are hotter than in other areas.

The observed trends in hydrocarbon maturation have been attributed to two sources. One source is an increasing thickness of overburden in the northern San Juan Basin. The San Juan Basin is not a symmetric structure, the deepest part of the basin being located along a northwest-trending axis in the northeastern half of the Basin. The regional variation of burial depth does not entirely explain the maturation trend however, since the greatest maturation values are not found in the deepest parts of the basin. Clarkson and Reiter (1987) also show that reasonable burial histories for the Fruitland coals will not lead to the observed maturation pattern. Additional increases in hydrocarbon maturation have been attributed to factors associated with the San Juan volcanic field (Reiter and Clarkson, 1983a; Rice, 1983; Bond, 1984). Both heating from the San Juan volcanic field and increased burial depth beneath

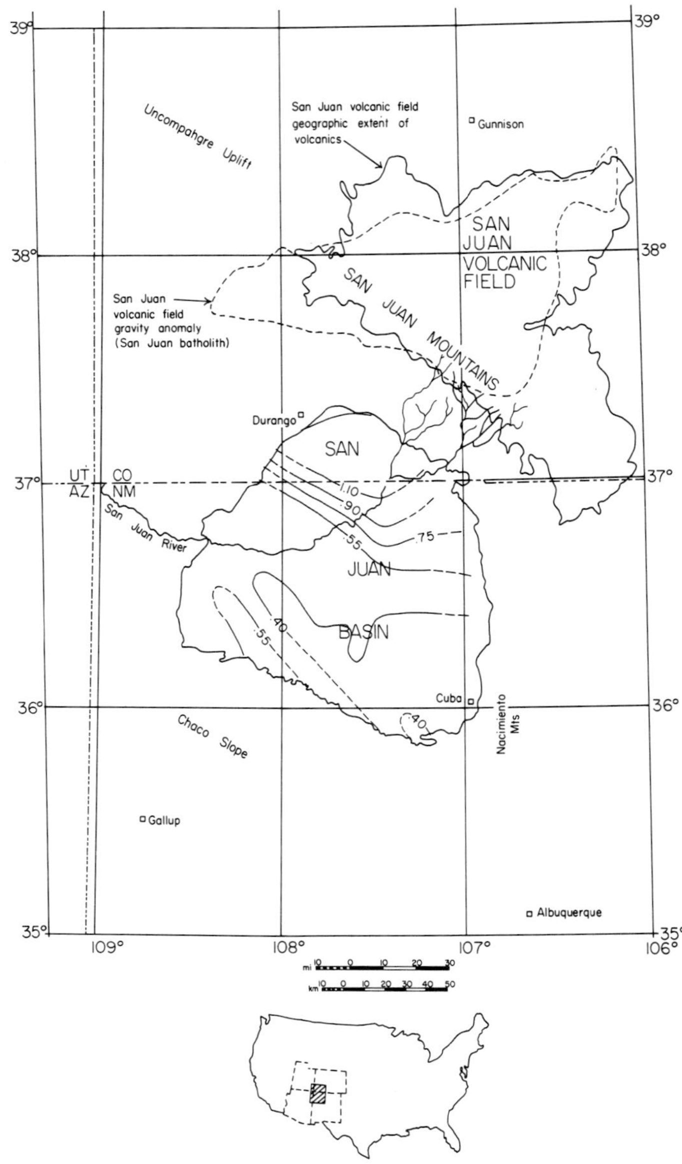

Figure 3. Coal maturation contours, in percent vitrinite reflectance, for Fruitland Formation coals (modified from Fassett and Hinds, 1971; Rice, 1983).

volcaniclastic debris from the volcanic field could contribute to elevated temperatures and higher maturation levels within the basin. As with the heat-flow pattern, an influence on the maturation pattern from sources related to the San Juan volcanic field seems reasonable but the exact nature of this influence needs further examination.

The effect of increased burial depths due to the presence of volcaniclastic debris is difficult to establish because any material which may have been present in the San Juan Basin after the Oligocene volcanism has been removed by subsequent erosion. The maximum thickness and time duration of such cover for a given location in the basin are important questions relating to coal maturation. Steven (1975) indicates that the likely extent of outflow from the San Juan volcanic field would have covered the northern flank of the basin. If this is accurate then such additional burial would contribute to the elevated maturation levels in the northern basin. Bond

(1984), using a variety of geologic data including stratigraphic information and shale bulk densities coupled with a thermal history model, calculated a thickness of just under 500 meters for Oligocene volcanic material at one location in the northern basin. Bond (1984) suggests that this value may be representative for the northern basin. Although such burial may contribute significantly to the maturation level in the northern basin, it does not appear likely that a significant thickness of Oligocene volcanic debris extended into the central or perhaps even the north-central San Juan Basin.

Since the depth of burial cannot totally explain the observed hydrocarbon maturation levels within the San Juan Basin, an additional thermal source is required. A reasonable source may be heat associated with the emplacement and cooling of the San Juan batholith and associated thermal events. However, modeling of thermal sources in the San Juan volcanic field by Clarkson and Reiter (1987) demonstrate that conductive heat transport from the area of the San Juan volcanic field, even including heat from a deep, steady-state source (the model discussed above in relation to the heat-flow data) is insufficient to explain levels of maturation in the San Juan Basin, in particular for the central San Juan Basin.

DISCUSSION

Although the patterns of heat flow and hydrocarbon maturation in the San Juan Basin indicate the significant influence of a heat source in the San Juan volcanic field, the exact nature of this influence is unclear. Even though the present day heat flow can be modeled as discussed above, questions exist concerning the perturbation to the heat flow by ground-water movement, even for the deep heat-flow data in the basin. The level of hydrocarbon maturation in the central and northern basin raises questions concerning the thermal history of the basin as well. In summarizing the thermal history of the basin, Clarkson and Reiter (1987) note that the present day heat flow is not sufficient to totally account for the observed pattern in maturation levels, even when allowing for greater burial depths prior to Eocene erosion. In addition, an analysis of the temperatures produced by thermal sources associated with the San Juan volcanic field indicate that conductive heat transport does not transfer enough heat into the basin to adequately explain the maturation pattern. Some other source of heat is therefore required to complete the thermal history of the basin. Magmatic activity within the basin itself could be suggested as an additional source of heat, however there is no indication of any major igneous activity within the basin. The requirement for an additional source of heat in the San Juan Basin has led to the proposal that heat advection due to ground-water flow has played a role in the thermal history of the San Juan Basin (Clarkson and Reiter, 1987).

A detailed analysis of ground-water flow and subsurface temperature distributions through time in the San Juan region will be required to definitively answer how large a role heat advection by ground-water flow has played in the maturation process. Such an analysis is beyond the scope of this paper, however some basic questions concerning the role of heat advection can be addressed in a first order fashion to determine the significance of such a mechanism. One basic question is the availability of a sufficient amount of heat to produce the observed maturation levels if the heat were redistributed by ground-water flow. A second question is whether heat advection by water flow is in general a reasonable mechanism for heat transport on a regional scale.

The temperatures which are required to produce the observed maturation level for the Fruitland coals in the northern San Juan Basin range from 110° to 240° C for heating times of from 70 m.y. to 10,000 years. Temperatures for Fruitland coals in the central basin would of course be somewhat lower. These temperatures represent an approximate increase over present day temperatures in the range of 50° to 150° C. If some additional heating due to increased burial depths of the Fruitland coals is also considered, the increase in temperature due to heat advection would be proportionally lower.

To address the question of the amount of available heat from San Juan thermal sources, first consider the amount of heat associated with the cooling of the San Juan batholith. The amount of heat available from this source is related to the volume of the batholith, the temperature difference between the intruded material and the surrounding rock, and the density and specific heat of the magma. The San Juan batholith has an approximate areal extent of 10^4 km^2 and extends from at least 7 km to 19 km depth (fig. 1; Plouff and Pakiser, 1972), giving a volume on the order of 10^5 km^3. Prior to heating, the rocks in the San Juan region were probably in the temperature range of 200° to 400° C at 7 to 19 km depth. If magma was initially intruded at temperatures of about 800° to 1,000° C, an average temperature difference on the order of 500° C seems reasonable. Using a density of 2.5 g/cm^3, a specific heat of $1.0 \cdot 10^3$ Joules per kilogram-degree Kelvin (J/kgK) and assuming only 25 percent of the volume of the batholithic material initially had the 500° C temperature difference, one can calculate a total anomalous quantity of heat on the order of 10^{22} to 10^{23} J. If this extra heat were evenly distributed within the upper 10 km of the crust over an area within a 150 km radius of the center of the batholith, it would be enough heat to raise the temperature throughout the crustal volume by about 100° C (assuming adiabatic conditions). Although this heat would most likely not be evenly distributed over the volume, it is only a part of the total heat available from the San Juan volcanic field source. As noted above, the present day elevated heat flow in the region requires more heat than would be produced by just the cooling of the batholith. Reiter and Clarkson (1983b) note that the amount of heat required by the present day heat flow is at least twice that produced by just the cooling of the batholith. Although the above calculations are only first order estimates, it is reasonable to assume that there was sufficient heat from the San Juan thermal sources to significantly contribute to the observed levels of maturation in the basin if the heat were redistributed properly.

Given that enough heat is available from the San Juan thermal source to greatly influence hydrocarbon maturation in the basin, the question of advection of this heat by ground-water flow may be considered. It should first be noted that the general ground-water flow pattern in the northern San Juan region is from the San Juan Mountains towards the north-central San Juan Basin (Stone and others, 1983). Considering the probable topography of the region since Late Cretaceous time, this general flow pattern would appear to be reasonable over much of the Cenozoic history of the basin.

Several recent studies have demonstrated the potential for ground-water flow to significantly perturb the regional heat flow in sedimentary basins (see for example, Chapman and others, 1984; Majorowicz and others, 1985, 1986). Numerical modeling by Smith and Chapman (1983) shows that for a basin 40 km wide and 5 km deep, with a water table relief of 500 meters across the basin and with a permeability of $2.0 \cdot 10^{-16}$ m^2 for the material in the basin, the water flow will alter an initial uniform regional heat flow of 60 mWm^{-2} by about ± 30 percent in the recharge and discharge areas. The measured surface heat flow would depart from the regional heat flow by more than 10 percent over about 40 percent of the basin. These studies demonstrate that ground-water flow may have a significant regional influence on the thermal regimes of sedimentary basins.

A significant difference exists between the San Juan Basin and the regions in the above studies in that a major thermal source is adjacent to the basin. Modeling which allows for a heat source and its interaction with ground-water flow will be necessary to evaluate the potential for the San Juan volcanic field heat source to influence the basin. It does appear reasonable, however, that heat advection by ground-water flow could have played a role in the maturation of hydrocarbons in the San Juan Basin.

SUMMARY AND CONCLUSIONS

The observed patterns of heat flow and hydrocarbon maturation in the San Juan region suggest the influence of a thermal source associated with the San Juan volcanic field. Part of the hydrocarbon maturation pattern can be explained by increased burial depths due to thicker overburden which was subsequently removed by post-Oligocene erosion. In the northern basin, burial beneath Oligocene volcaniclastic material from the San Juan volcanic field may have contributed significantly to the maturation level. The heat-flow pattern and a significant part of the observed maturation pattern cannot be explained by depth of burial within much of the basin. Increases in subsurface temperature within the basin due to heating from the San Juan volcanic field is considered to be significant. However heat conduction from the area of the volcanic field probably had a limited regional effect on the thermal history of the San Juan Basin. Therefore the role of heat advection by ground-water flow into the basin was probably significant (and may still be important). Such a role for heat advection appears reasonable in terms of the amount of heat available and the ability of ground-water flow to significantly influence the thermal regime of sedimentary basins. Detailed modeling of temperature distributions and ground-water flow over time will be needed to analyze the role of heat advection in the thermal history of the San Juan Basin.

A greater understanding of the role of burial depths and ground-water flow in the thermal history of rock formations within the San Juan Basin will be required to better understand the geothermal history of the basin. Such an understanding of the thermal history of the basin will be important in studying the pattern of hydrocarbon occurrences within the basin.

REFERENCES CITED

Bird, P., 1979, Continental delamination and the Colorado Plateau: Journal of Geophysical Research, v. 84, p. 7561-7571.

Bodell, J.M., and Chapman, D.S., 1982, Heat flow in the north-central Colorado Plateau: Journal of Geophysical Research, v. 84, p. 2869-2884.

Bond, W.A., 1984, Application of Lopatin's method to determine burial history, evolution of the geothermal gradient, and timing of hydrocarbon generation in Cretaceous source rocks in the San Juan Basin, northwestern New Mexico and southwestern Colorado, *in* Woodward, J., Meisner, F.F., and Clayton, J.L., eds., Hydrocarbon source rocks of the Greater Rocky Mountain region: Rocky Mountain Association of Geologists, p. 433-447.

Bostick, N.H., 1973, Time as a factor in thermal metamorphism of phytoclasts (coaly particles), Congres International de Stratigraphie et de Geologie Carbonifere, 7th, Krefield, 1971, Compte Rendu, v. 2, p.183-193.

Bredehoeft, J.D., and Papadopulos, I.S., 1965, Rates of vertical groundwater movement estimated from the Earth's thermal profile: Water Resources Research, v. 1, p. 325-328.

Buntebarth, G., 1978, The degree of metamorphism of organic matter in sedimentary rocks as a paleo-geothermometer, applied to the upper Rhine Graben: Pure and Applied Geophysics, v. 117, p. 83-91.

Cercone, K.R., 1984, Thermal history of Michigan Basin: American Association of Petroleum Geologists Bulletin, v. 68, p. 130-136.

Chapman, D.S., Keho, T.H., Bauer, M.S., and Picard, M.D., 1984, Heat flow in the Uinta Basin determined from bottom hole temperature (BHT) data: Geophysics, v. 49, p. 453-466.

Clarkson, G., and Reiter, M., 1987, The thermal regime of the San Juan Basin since Late Cretaceous times and its relationship to San Juan Mountains thermal sources: Journal of Volcanology and Geothermal Research, v. 31, p. 217-237.

Decker, E.R., and Birch, F., 1974, Basic heat-flow data from Colorado, Minnesota, New Mexico, and Texas, *in* Sass, J.H. and Munroe, R.J., compilers, United States Geological Survey, Open-File Report 74-9, p. 5.1-5.59.

Decker, E.R., and Bucher, G.J., 1979, Thermal gradients and heat-flow data in Colorado and Wyoming: A preliminary report: Los Alamos Scientific Laboratory, LA-7993-MS, 9 p.

Eardley, A.J., 1962, Structural geology of North America: Harper and Row, New York, 743 p.

Edwards, C.L., Reiter, M., Shearer, C., and Young, W., 1978, Terrestrial heat flow and crustal radioactivity in northeastern New Mexico and southeastern Colorado: Geological Society of America Bulletin, v. 898, p. 1341-1350.

Fassett, J.E., and Hinds, J.S., 1971, Geology and fuel resources of the Fruitland Formation and Kirtland Shale of the San Juan Basin, New Mexico and Colorado: United States Geological Survey Professional Paper 676, 76 p.

Galloway, M.J., 1980, Hydrogeologic and geothermal investigation of Pagosa Springs, Colorado: Colorado Geological Survey Special Publication 10, 95 p.

Hood, A., Gutjahr, C.C.M., and Heacock, R.L., 1975, Organic metamorphism and the generation of petroleum: American Association of Petroleum Geologists Bulletin, v. 59, p. 986-996.

Kelly, V.C., 1950, Regional structure of the San Juan Basin: New Mexico Geological Society First Field Conference Guidebook, p. 101-108.

King, P.B., 1959, The evolution of North America: Princeton University Press, Princeton, N. J., 189 p.

Lachenbruch, A.H., and Sass, J.H., 1977, Heat flow in the United States and the thermal regime of the crust, *in* Heacock, J.G., ed., The nature and properties of the earth's crust: American Geophysical Union Monograph 20, p. 626-678.

Lipman, P.W., Doe, B.R., Hedge, C.E., and Steven, T.A., 1978, Petrologic evolution of the San Juan volcanic field, southwestern Colorado; Pb and Sr isotope evidence: Geological Society of America Bulletin, v. 89, p. 59-82.

Lopatin, H.V., and Bostick, N.H., 1974, The geologic factors in coal catagenesis: Illinois State Survey Reprint 1974Q, p. 1-16.

Lucchita, I., 1979, Late Cenozoic uplift of the southwestern Colorado Plateau and adjacent lower Colorado River region: Tectonophysics, v. 61, p. 63-95.

Majorowicz, J.A., Jones, F.W., Lam, H.L., and Jessop, A.M., 1985, Regional variations of heat flow differences with depth in Alberta, Canada: Geophysical Journal of the Royal Astronomical Society, v. 81, p. 479-487.

Majorowicz, J.A., Jones, F.W., and Jessop, A.M., 1986, Geothermics of the Williston Basin in Canada in relation to hydrodynamics and hydrocarbon occurrence: Geophysics, v. 51, p. 767-779.

Molenaar, C.M., 1983, Major depositional cycles and regional correlations of Upper Cretaceous rocks, southern Colorado Plateau and adjacent areas, *in* Reynolds, M.W., and Dolly, E.D., eds., Mesozoic paleogeography of west-central United States: Rocky Mountain Section, Society of Economic Paleontologists and Mineralogists, p. 201-224.

Phillipi, G.T., 1965, On the depth, time, and mechanisms of petroleum generation: Geochimica et Cosmochemica Acta, v. 29, p. 1021-1049.

Plouff, D., and Pakiser, L.C., 1972, Gravity study of the San Juan Mountains, Colorado: United States Geological Survey Professional Paper 800-B, p. B183-B190.

Reiter, M., and Clarkson, G., 1983a, Geothermal studies in the San Juan Basin and the Four Corners area of the Colorado Plateau, II, Steady state models of the thermal source of the San Juan volcanic field: Tectonophysics, v. 91, p. 253-269.

Reiter, M., and Clarkson, G., 1983b, Relationships between heat flow, paleotemperatures, coalification, and petroleum maturation in the San Juan Basin, northwest New Mexico and southwest Colorado: Geothermics, v. 12, p. 323-339.

Reiter, M., and Clarkson, G., 1983c, A note on terrestrial heat flow in the Colorado Plateau: Geophysical Research Letters, v. 10, p. 929-932.

Reiter, M., and Mansure, A.J., 1983, Geothermal studies in the San Juan Basin and the Four Corners area of the Colorado Plateau, I, Terrestrial heat-flow measurements: Tectonophysics, v. 91, p. 233-251.

Reiter, M., Edwards, C.L., Hartman, H., and Weidman, C., 1975, Terrestrial heat flow along the Rio Grande rift, New Mexico and southern Colorado: Geological Society of America Bulletin, v. 86, p. 811-818.

Rice, D.D., 1983, Relation of natural gas composition to thermal maturity and source rock type in San Juan Basin, northwestern New Mexico and southwestern Colorado: American Association of Petroleum Geologists Bulletin, v. 67, p. 1199-1218.

Sass, J.H., Lachenbruch, A.H., Munroe, R.J., Greene, G.W., and Moses, T.H., Jr., 1971, Heat flow in the western United States: Journal of Geophysical Research, v. 76, p. 6376-6413.

Smith, L., and Chapman, D.S., 1983, On the thermal effects of groundwater flow: 1. Regional scale systems: Journal of Geophysical Research, v. 88, p. 593-608.

Stallman, R.W., 1963, Computation of ground-water velocity from temperature data, *in* Bentall, R., compiler, Methods of collecting and interpreting ground-water data: United States Geological Survey Water Supply Paper 1544H, p. H36-H46.

Steven, T.A., 1975, Mid-Tertiary volcanic field in the southern Rocky Mountains, *in* Curtis, B.F., ed., Cenozoic history of the southern Rocky Mountains: Geological Society of America Memoir 144, p. 74-94.

Stone, W.J., Lyford, F.P., Frenzel, P.F., Mizell, N.H., and Padgett, E., 1983, Hydrogeology and water resources of the San Juan Basin, New Mexico: New Mexico Bureau of Mines and Mineral Resources Hydrogeologic Report 6, 70 p.

Tweto, O., 1975, Laramide (Late Cretaceous-Early Tertiary) orogeny in the southern Rocky Mountains, *in* Curtis, B.F., ed., Cenozoic history of the southern Rocky Mountains: Geological Society of America Memoir 144, p. 1-44.

Van Orstrand, C.E., 1934, Temperature gradients, *in* Problems of Petroleum Geology: American Association of Petroleum Geologists, p. 989-1021.

Waples, D.W., 1980, Time and temperature in petroleum formation: Application of Lopatin's method to petroleum exploration: American Association of Petroleum Geologists Bulletin, v. 64, p. 916-926.

Buckskin Charley with members of his band. The man in the center is wearing a Benjamin Harrison peace medal. Designed in 1890, it was the last one to be specifically produced for the Indians. Photo courtesy of the Colorado Historical Society and the Smithsonian Institution.

Preliminary Interpretation of Short-Wavelength Magnetic and Gravity Anomalies on the Southern Ute Indian Reservation, Southwestern Colorado

Michael W. Webring[1]

Magnetic anomalies with low amplitudes and spectral energy concentrated at short wavelengths indicative of shallow sources exist in the southwest part of the Southern Ute Indian Reservation. These anomalies are of interest because hydrocarbon microseepage is thought to cause the emplacement of magnetic minerals and alter the amount and composition of rock cements (Donovan and others, 1979). The basic hypothesis is that all hydrocarbon traps leak and that the escaped hydrocarbons and dissolved gases can be detected at the surface. Mechanisms have been proposed whereby the deposition of magnetic minerals (magnetite and pyrrhotite) is enhanced by escaping hydrocarbons in horizontal zones near the water table causing magnetic anomalies with specific characteristics. Oil and gas are produced from structural and stratigraphic traps in this area and a study of the anomalies caused by shallow magnetic and gravity sources was carried out to see if detection of hydrocarbon microseepage was feasible in this locality. If so, previously unrecognized areas of petroleum potential might thereby be delineated.

Aeromagnetic data were used to select a target area on the Southern Ute Reservation approximately 25 kilometers wide for detailed ground work. The aeromagnetic survey was flown at 125-meter ground clearance and 800-meter line spacing and was digitally processed to enhance anomalies caused by shallow sources while suppressing 2-dimensional anomalies such as those caused by hogbacks, arroyos, and pipelines. The remaining anomalies can then be attributed to thin tabular bodies with limited horizontal extent. Ensembles of these thin and shallow sources create anomalies which persist from flight line to flight line and appear as irregular cloudlike areas on contour maps. Based on the results of this analysis, an area west of Redmesa, Colorado, roughly centered on T. 33 N., R. 13 W., was chosen for further study. This area is just north of the gas-productive Barker Dome and west of the Hogback Monocline and contains a band of northeast-aligned cloud anomalies.

Ground-level magnetic data were obtained using a truck-mounted proton precession magnetometer driven along section roads in the vicinity of the most promising anomaly centered on Wheeler Draw, 10 kilometers west of Redmesa, Colorado. The maximum depth to magnetic sources in this area was determined by modeling a continuous magnetic source (an equivalent source layer) at various depths and noting the degree of fit between the observed residual magnetic anomaly and that calculated from the model. Based on this analysis magnetic basement was determined to be 3 kilometers deep, which agrees with a nearby well. The remaining magnetic anomalies were determined to be caused by sources shallower than 1 kilometer (± 200 meters). There do not seem to be any significant magnetic sources in the sedimentary section between depths of 1 and 3 kilometers. Based on this analysis, 200 meters is the minimum depth for sources with normal remanent magnetization. If reversed remanent magnetization is considered then anomalies with shorter wavelength can be modeled and the minimum depth decreases to 90 meters. However, the anomalies fit by the 90-meter source have short enough wavelengths to be caused by surficial magnetic placers or by cultural sources.

These depth estimates were derived using a continuous layer consisting of individual 2-dimensional bodies. These bodies are smaller than the resolving limit determined by depth of burial and therefore do not provide information on the detailed distribution of magnetic material. For instance, gradational distribution of magnetic material (grading from a central more concentrated area outward to a diffuse halo) in a shallow host rock can mimic the magnetic field caused by a deep sharp-edged source. In the hydrocarbon-microseepage hypothesis, diffusion of fluids through pore spaces would result in such gradational distribution and the depths given above, based on disjunct non-gradational models, would be too deep. The 200-meter-deep source layer has some gradational transitions whereas the 90-meter-deep source layer has almost none, so these two limits probably bracket the minimum shallow-magnetic-source location. In summary, there are magnetic sources at shallow depths within the sedimentary section (less than 200 meters deep) that could be, but probably are not, due to surficial sources.

A gravity survey was also conducted to test for changes in density caused by emplacement or removal of rock cements. Three north-trending and two connecting east-trending profiles with surveyed elevation control were set up with station spacing varying from 200 to 500 meters. Accuracy among stations in a single profile is estimated to be ±.05 mGal with the terrain correction the main contribution to error. Other stations were located at spot elevations to give an overall station density of 1.6 kilometer (1 mile). Regional gravity anomalies associated with the basement, 3 kilometers deep, are consistent with regional magnetic anomalies. Residual gravity maps generated by highpass filters with cutoffs in the 2 to 5 kilometer range contain very low amplitude (.2 to .3 mGal) anomalies in the vicinity of the Wheeler Draw magnetic anomaly that are not directly related to topography or subsurface structure. The shape and connectivity of these local gravity anomalies varies depending on the type of filter and cutoff wavelength because of the lack of stations between profiles; but when several maps with different filtering parameters are compared, common features emerge.

[1] U.S. Geological Survey
Box 25046, MS 964, DFC
Lakewood, CO 80225

A generalized residual gravity map in the Wheeler Draw area compared to the short-wavelength aeromagnetic map contains some weak correlations in the form of magnetic halos near gravity highs. When the gravity is compared to the soil helium map (Cunningham, this volume) a weak inverse correlation can be observed. Short-wavelength aeromagnetic data for the whole 25 kilometer target area show a northeast trend of anomalies that correlates well with helium anomalies. Aside from these general observations, it is not clear what the sources of the magnetic and gravity anomalies are. The shallow depth to magnetic sources, less than 1 kilometer, would provide good drilling targets for core samples.

REFERENCES CITED

Cunningham, K. I., 1988, Preliminary interpretation of soil-gas data and relationships to other near-surface hydrocarbon microseepage indicators, Four Corners Platform-San Juan Basin transitional areas, southwest Colorado, *in* Fassett, J. E., ed., Geology and coal-bed methane resources of the northern San Juan Basin, Colorado and New Mexico: Rocky Mountain Association of Geologists Guidebook

Donovan, T. J., Forgey, R. L., Roberts, A.A., 1979, Aeromagnetic detection of diagentic magnetite over oil fields: American Association of Petroleum Geologists Bulletin, v. 63, p. 245-248.

Distribution of Exposed Limonitic Rocks and Soils From Landsat Multispectral Scanner Data on the Southern Ute Indian Reservation, Southwestern Colorado

Daniel H. Knepper, Jr.[1]

INTRODUCTION

A Landsat multispectral scanner (MSS) scene is acquired in four spectral bands and covers an area approximately 160 km on a side. Digital image-processing techniques can be applied to Landsat MSS data to extract spectral information that allows discrimination between different types of materials. Rowan and others (1974) demonstrated that ferric iron oxide and hydroxide minerals (limonite) have unique spectral reflectance characteristics that can be identified using MSS data. Consequently, Landsat MSS data can be used to map the distribution of limonite on regional and local scales.

Oxidation of iron-bearing minerals to limonite has been associated with the leakage of hydrocarbon products to the surface in areas of petroleum accumulations. Areas of hydrocarbon microseepage have also been associated with shallow-source magnetic and soil-gas anomalies. However, the relationships between these phenomena and their geological and geochemical controls are poorly understood. The potential applications of hydrocarbon microseepage phenomena to petroleum exploration and resource assessment is a strong incentive to better understand these phenomena. As part of a U. S. Geological Survey (U.S.G.S.) project to investigate the relationships between various microseepage phenomena, Landsat MSS data on the Southern Ute Indian Reservation in southwestern Colorado were used to map the distribution of limonitic materials at the surface.

BASIC IMAGE PROCESSING

Landsat MSS data covering the Southern Ute Indian Reservation (scene ID 20637-16584) were digitally processed at the U.S.G.S., Branch of Geophysics Remote Sensing Laboratory in Denver, Colorado. Prior to analysis, the Landsat MSS data were digitally registered to a Universal Transverse Mercator (UTM) grid to simplify the subsequent location of limonite concentrations. The registered MSS digital data numbers (DN) were also converted to reflectance values using the calibration factors determined by Robinove (1982). The use of calibrated MSS data allows the direct comparison of spectral characteristics of different materials between Landsat scenes acquired by different satellites and at different times. Consequently, experience in mapping limonite in many other areas could be applied to identifying exposed limonitic rocks and soils on the Southern Ute Indian Reservation.

LIMONITE MAPPING

Ferric oxide and hydroxide minerals have their most intense absorption in the ultraviolet portion of the spectrum, which is responsible for a rapid falloff of reflected energy from the red to the blue part of the spectrum for these minerals (Hunt and Ashley, 1979). The steep slope of the spectral reflectance curves of the common ferric oxide and hydroxide minerals (hematite, goethite, jarosite) in the visible spectrum is characteristic of these minerals and can be used to identify their presence in digital Landsat MSS data (fig. 1). The ratio of calibrated Landsat MSS band 4 and band 5 (4/5) provides a measure of the slope of the reflectance curves in the visible region for each pixel. Ratios less than 0.8 to 0.9 have been found during other on-going studies to be characteristic of limonitic rocks and soils.

The calibrated Landsat MSS 4/5 ratio was prepared for the Southern Ute Indian Reservation and viewed on a video display. Limonite distribution was viewed using various upper 4/5 ratio cutoffs between 0.8 and 0.9, and an upper cutoff of 0.8 was selected as best displaying limonite concentrations while eliminating numerous isolated limonite pixels. The limonite distribution map of the Southern Ute Indian Reservation (fig. 2), therefore represents those limonitic pixels having only the strongest limonite response (steepest slopes on the reflectance curve) on the MSS data.

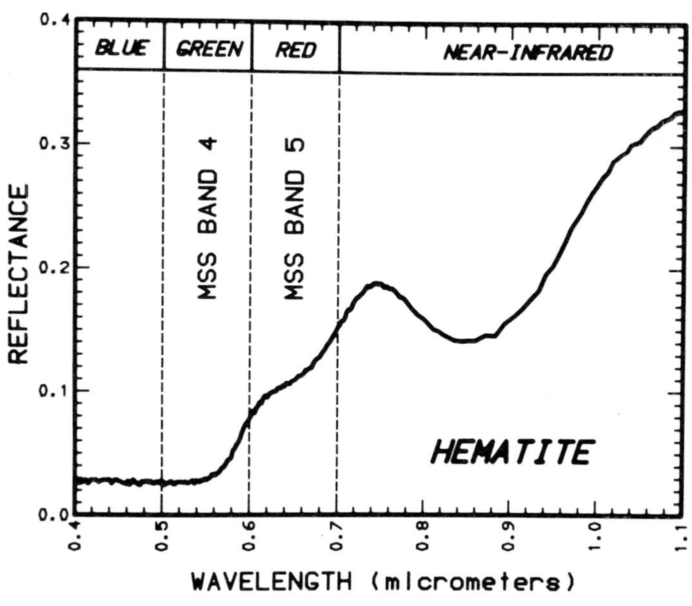

Figure 1. Spectral reflectance curve of hematite showing the steep slope in the visible part of the spectrum characteristic of ferric oxides and hydroxides. The Landsat band-4/band-5 ratio provides a measure of the steepness of the slope and a means of identifying limonitic rocks and soils on regional and local scales.

[1] U. S. Geological Survey
P.O. Box 25046, MS 964, DFC
Lakewood, Colorado 80225

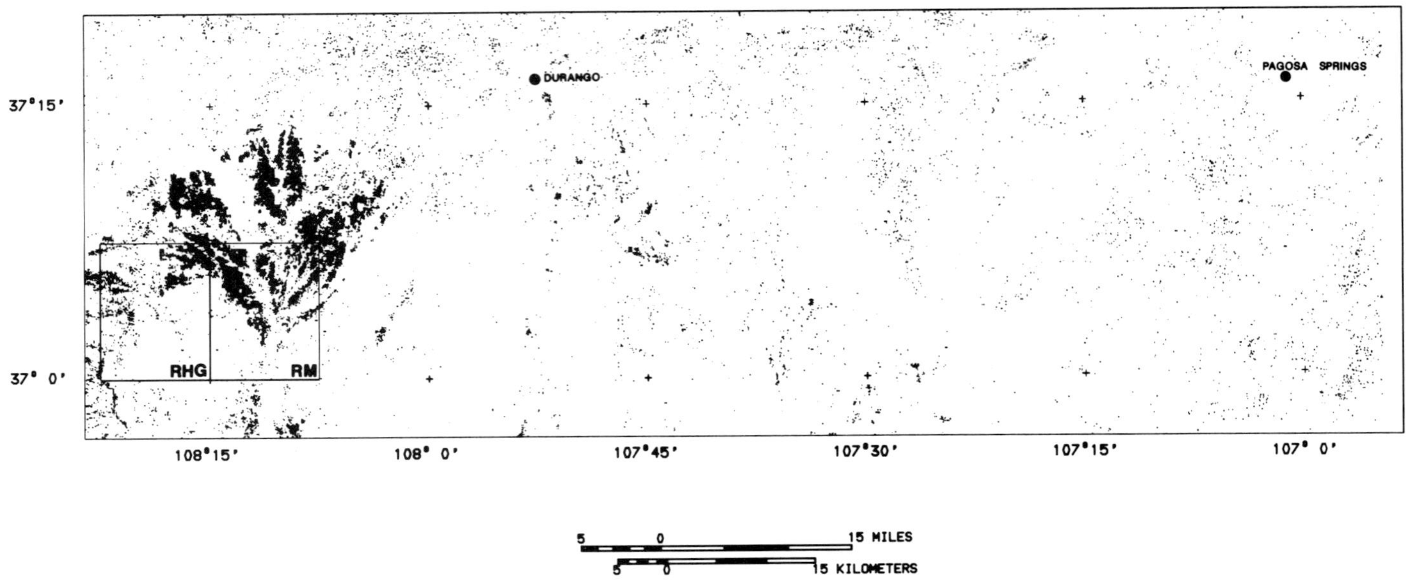

Figure 2. Limonite map of the Southern Ute Indian Reservation, southwestern Colorado, derived from the analysis of a Landsat MSS band-4/band-5 ratio image. Limonite is black; RM, Redmesa quadrangle; RHG, Red Horse Gulch quadrangle.

Limonitic rocks and soils are indicated in many parts of the Southern Ute Indian Reservation; however, the major limonite concentrations are in the western part (fig. 2). Based on aeromagnetics and gravity (Webring, this volume) and soil gas anomalies (Cunningham, this volume), the area including these prominent limonite anomalies was independently selected for more detailed study. Current work is primarily focused in the Redmesa and Redhorse Gulch quadrangles outlined in fig. 2. The spatial correlation of limonite distribution with other suggestions of hydrocarbon microseepage indicates a possible relationship that should be further studied. If the relationship is positive, then the application of digital satellite imagery for mapping limonite distribution will be a powerful tool for the exploration and assessment of gas and oil resources. The application of Landsat Thematic Mapper data, with more spectral bands than MSS data and more than twice the spatial resolution, would allow more detailed mapping of limonite distribution.

REFERENCES CITED

Cunningham, K. I., 1988, Preliminary interpretation of soil-gas data and relationships to other near-surface hydrocarbon microseepage indicators, Four Corners Platform-San Juan Basin transitional areas, southwest Colorado, *in* Fassett, J. E., ed., Geology and coal-bed methane resources of the northern San Juan Basin, Colorado and New Mexico: Rocky Mountain Association of Geologists Guidebook

Hunt, G. R., and Ashley, R. P., 1979, Spectra of altered rocks in the visible and near infrared: Economic Geology, v. 74, p. 1613-1629.

Robinove, C. J., 1982), Computation with physical values from Landsat digital data: Photogrammetric Engineering and Remote Sensing, v. 48, no. 5, p.781-784.

Rowan, L. C., Wetlaufer, P. H., Goetz, A. F. H., Billingsley, F. C., and Stewart, J. H., 1974, Discrimination of rock types and detection of hydrothermally altered areas in south central Nevada by use of computer-enhanced ERTS images: U. S. Geological Survey Professional Paper 883, 35p.

Webring, M. W., 1988, Preliminary interpretation of high wavenumber magnetic and gravity anomalies on the Southern Ute Indian Reservation, southwestern Colorado, *in* Fassett, J. E., ed., Geology and coal-bed methane resources of the northern San Juan Basin, Colorado and New Mexico: Rocky Mountain Association of Geologists Guidebook

A Preliminary Interpretation of Carbon and Oxygen Isotopic Data from Surface Rocks, Southern Ute Indian Reservation, Southwest Colorado

Mitchell E. Henry[1]

ABSTRACT

Surface samples of calcium-carbonate cemented sandstone collected over the Southern Ute Indian Reservation in southwest Colorado, suggest the presence of surface geochemical anomalies related to hydrocarbon microseepage. Carbon isotopic values, reported in standard δ notation relative to the Peedee Belemnite (PDB) standard, range from 3.2/mil to -16.47/mil and average -2.2/mil. Oxygen isotopic values, also reported in standard δ notation relative to the PDB standard, range from -0.89/mil to -16.27/mil and average -6.9/mil. These values, which are positively correlated to one another, lie between values of known Cretaceous marine and freshwater limestones; they may simply represent mixtures of variable proportions of these two components.

When these data are displayed as contour maps, a major pattern of north-trending isotopic lows and a minor pattern of northeast-trending lows are produced.

Published accounts of similar studies over known oil fields show isotopic anomalies that are generally greater than those found in this study. Isotopic values reported from the Southern Ute Indian Reservation may represent a poorly developed geochemical anomaly related to hydrocarbon microseepage.

INTRODUCTION

Purpose Of Study

This study was begun as part of a larger project designed to evaluate the hydrocarbon potential of the Southern Ute Indian Reservation in southwestern Colorado. The San Juan Basin lies to the southeast and the Paradox basin lies to the northwest of the area (fig. 1). Both of these basins contain commercial quantities of hydrocarbons that may also have migrated into the study area.

An isotopic study was included in the larger project because this type of analysis has been shown, in other areas, to be related to hydrocarbon microseepage and therefore compliments the helium and hydrocarbon soil-gas studies that are also being conducted for this project. This project was a geochemical study of carbon and oxygen stable isotopic ratios measured from calcareous material present in surface rocks of the study area.

Geologic Setting

Surface rocks in the area are primarily Upper Cretaceous units of the Mesaverde Group (Tweto, 1979). The Mesaverde Group is a regressive-transgressive sequence and is represented by a lower formation, the Point Lookout Sandstone (regressive marine), a middle unit, the Menefee Formation (continental), and an upper unit, the Cliff House Sandstone (transgressive marine) (Baars and Ellingson, 1984). Sandstone samples collected during this study were from these three formations.

[1] U. S. Geological Survey
P.O. Box 25046, M.S. 939, DFC
Lakewood, Colorado 80225

Previous Work

Donovan (1974); Donovan and others (1974); and Donovan and Dalziel (1977) have shown that anomalous ratios of stable carbon and oxygen isotopes exist in calcareous cements found within sandstones overlying known hydrocarbon deposits at several locations.

At the Cement oil field in Oklahoma, these anomalies are present in calcareous cements that are enriched in the light isotope of carbon (^{12}C) and that may be enriched or depleted in the light isotope of oxygen (^{16}O) as compared to laboratory standards. The ^{12}C enrichment is attributed to the presence of hydrocarbon-derived carbon which escapes from a reservoir, is oxidized, and forms ^{12}C-rich carbon dioxide which is then incorporated in the calcareous cement (Donovan, 1974). The variability of the oxygen isotopic data is attributed to the relative importance of two isotopic fractionation mechanisms; membrane effects of clay layers and gas induced evaporation of subsurface waters (Donovan, 1974).

Carbon and oxygen isotopic anomalies in surface rocks have been reported from the Davenport oil field, Oklahoma. (Donovan and others, 1974). Calcareous cements from surface rocks at the Davenport oil field are less enriched in ^{12}C and ^{16}O as compared to cements from the Cement oil field (Donovan and others, 1974). Rapid hydrocarbon loss in the form of gas is suggested as the mechanism causing the more subtle carbon isotopic anomaly and the heavier oxygen isotopic signatures at the Davenport oil field (Donovan and others, 1974).

Donovan and Dalziel (1977) also reported on a study of the Bisti oil field, New Mexico. This field is a stratigraphically controlled accumulation in the central part of the San Juan Basin (Sabins, 1963). The field produces oil and some gas and

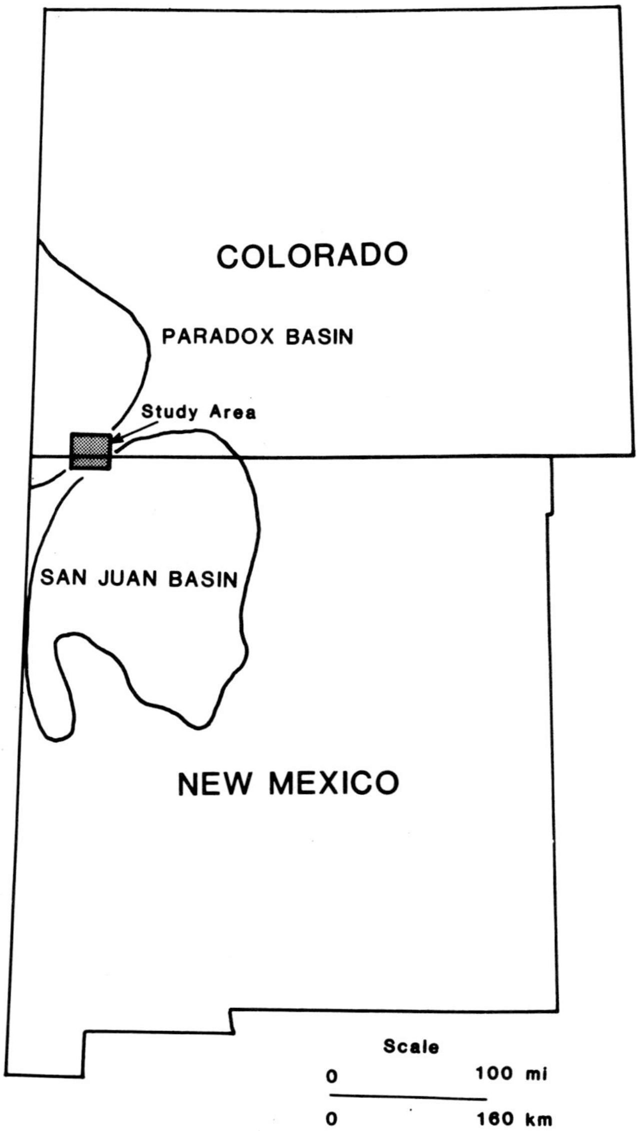

Figure 1. Location of study area relative to the San Juan and Paradox basins.

has an estimated ultimate recovery of 45 million barrels of oil (Sabins, 1978). Carbon isotopic data from this field range from $\delta(PDB)$ = −18.9/mil to +0.1/mil and average −10.1/mil. Oxygen isotopic data from the Bisti field range from $\delta(SMOW)$ = +10.9/mil to +26.7/mil and average +19.6/mil. Thus, the isotopic signature of the Bisti oil field is a moderately low $\delta^{13}C$ and $\delta^{18}O$ value, similar to anomalous areas discussed later within the present study area. The relatively light oxygen isotopic values reported from the Bisti field are interpreted to be the result of the predominance of a micropore filtration fractionation mechanism (Donovan and Dalziel, 1977).

A procedure somewhat similar to the method used by Donovan (1974) and Donovan and others (1974) was more recently used as part of a program designed to evaluate a variety of remote and direct detection methods as applied to petroleum exploration (Lang and others, 1984a; 1984b). In the two studies discussed below, soil samples rather than rock samples were collected for analysis.

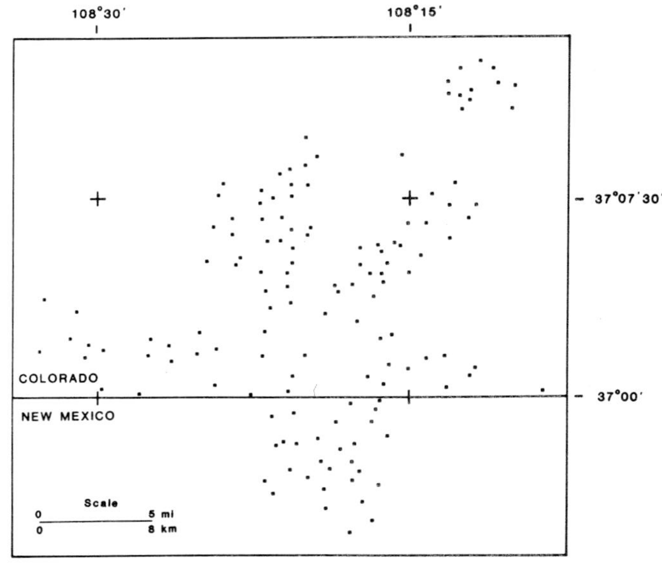

Figure 2. Map showing locations of sites where rock samples were collected for this study.

Thirty-eight soil samples from the Patrick Draw, Wyoming, petroleum test site were collected and analyzed to determine their carbon and oxygen isotopic composition. The isotopic values reported for these samples do not indicate a hydrocarbon related source of the carbon. They appear to be isotopically closely related to freshwater limestones. No relationship is noted between the amount of methane in absorbed soil gas and carbon isotope values. The conclusion is that isotopic data from soil carbonates from the Patrick Draw test site imply a caliche or detrital freshwater limestone origin for the carbonates (Lang and others, 1984a).

A second soil study was conducted at the Coyanosa, Texas, petroleum test site (Lang and others,1984b). Samples collected from 138 locations at this site were analyzed for carbon and oxygen isotopic and elemental composition. Carbon isotopic values are in the range of those known to exist in freshwater limestones. Statistical analysis of all the geochemical data indicates the absence of surface anomalies attributable to hydrocarbon microseepage.

METHODS

A suite of 162 samples was collected from rock outcrops in the study area (fig. 2). A preliminary map, supplied by CoGeomap Inc.[2] showing the locations of near-surface magnetic anomalies was used as a guide in selecting sample locations because of the possible relationship between those anomalies and hydrocarbon microseepage (Donovan and others, 1979). Samples were also collected in areas where magnetic anomalies were not detected to serve as background samples. The attempt to collect samples in a square grid pattern with an approximate spacing of 1 mile between samples was hampered by the absence of outcrop in many areas, poor access, and time limitations.

Samples were crushed, pulverized, and sieved. The material

[2] The use of trade names or products is for descriptive purposes only and does not constitute an endorsement by the U. S. Geological Survey.

which passed a 100 mesh sieve was analyzed by standard mass-spectrometric techniques to determine their carbon and oxygen isotopic composition. The values are given in δ notation:

$$\delta = \frac{R\ (\text{sample}) - R\ (\text{standard})}{R\ (\text{standard})} \times 1000$$

where R = the ratio of the heavy isotope to the light isotope. Values are reported relative to the PDB Chicago standard. Carbon isotopic ratios are still normally given relative to the PDB standard, even though the standard no longer exists. Secondary standards which are well defined relative to PDB are used as working standards and allow the measured ratios to be calculated in terms of the primary standards (PDB or Vienna Standard Mean Ocean Water (V-SMOW)). Analyses reported were performed using National Bureau of Standards-19 (NBS-19) (marble) as the working standard. Isotopic values of the standard determined before, during, and at the end of the sample analyses were within ± 0.05/mil for $\delta^{13}C$ and ± 0.05/mil for $\delta^{18}O$ of the accepted value of the standard. Replicates of 20 samples were measured to determine the reproducibility of the samples. Variability between two or more analyses of the same sample was found to range from 0.02/mil to 0.35/mil for $\delta^{13}C$ and from 0.02/mil to 0.64/mil for $\delta^{18}O$. The assumption that the isotopic data are reliable to ± 0.5 $\delta^{13}C$ unit and ± 1 $\delta^{18}O$ unit is justified based on replicate sample analyses and is adequate for the purposes of this study.

Petrographic analyses of the rocks have not been completed. Observations of the effect of dilute hydrochloric acid on the rocks suggest that calcium carbonate is uniformly distributed within the samples. Fossils, fossil fragments, or other calcareous clasts that would alter the isotopic signature of diagenetic cements were not observed with a low-power reflected light microscope.

RESULTS AND DISCUSSION

Results of the isotopic analyses are shown in appendix 1. Values of $\delta^{13}C(PDB)$ range from +3.2/mil to −16.47/mil and values of $\delta^{18}O(PDB)$ range from −0.89/mil to −16.27/mil. These carbon isotopic values are not exceptionally light; they occur generally in an intermediate position between known Cretaceous marine and Cretaceous freshwater carbonates (appendix 2). The oxygen isotopic values also range from those similar to known Cretaceous marine carbonates to those similar to known Cretaceous freshwater carbonates (appendix 2).

The isotopic data from this study (appendix 1) are plotted with data from appendix 2, (fig. 3). Figure 3 suggests that samples from this study may simply represent a mixture of variable proportions of isotopically normal marine Cretaceous carbonate and isotopically normal freshwater Cretaceous carbonate and do not require the addition of light-hydrocarbon-derived carbon to account for the values measured.

It should be pointed out that during the early work done by Donovan and his colleagues (Donovan, 1974, Donovan and others, 1974, Donovan and Dalziel, 1977), an emphasis was placed on the establishment of the relationship between hydrocarbon leakage and carbon and oxygen isotopic anomalies. This early work required that the majority of the investi-

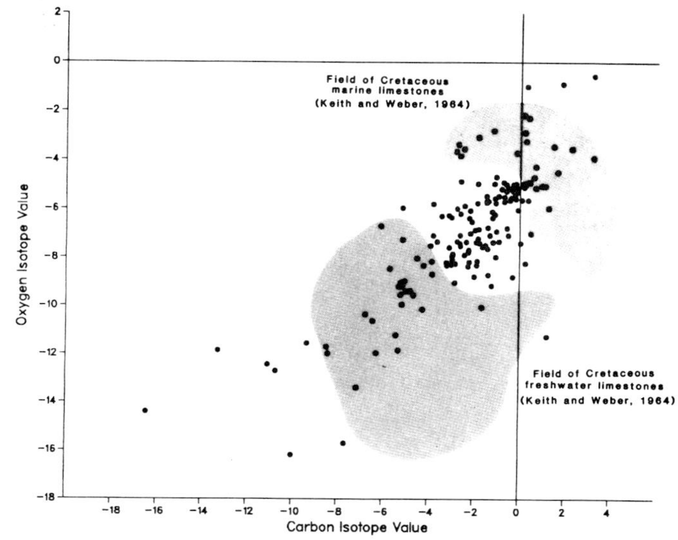

Figure 3. Plot of carbon isotopic values versus oxygen isotopic values for carbonate-cemented sandstones in the western part of the Southern Ute Indian Reservation. The data for this plot are shown in appendix 1 and appendix 2.

gations be carried out over and near known accumulations of hydrocarbons. These studies were usually fairly detailed and somewhat restricted to the main producing area with some samples taken away from the main producing area or to serve as background samples. For the isotopic anomalies discussed by Donovan and Dalziel (1977) the average $\delta^{13}C$ value ranged from −8.3/mil to −23.2/mil (table 1). These low field averages may be related to the relative proximity of background samples to the hydrocarbon accumulation, or in the case of the Oklahoma anomalies, it may represent an anomaly superimposed on an already low background level for the area in general. Although carbonate cements with $\delta^{13}C$ values of less than −20/mil are exceptionally light, values as high as −5/mil to −10/mil may have significance in an area where background or average values are about −2/mil.

Following this line of reasoning, the carbon and oxygen isotopic data were computer contoured using a commercial software package named Interactive Surface Modeling (ISM) sold by Dynamics Graphics, Inc.[2] The procedure used was to load the scattered data as an x, y, z data set, calculate a uniform grid, and then contour the resultant calculated grid data.

The non-uniform distribution of samples has been mentioned previously. This distribution may not fit rigid rules for automatic (computer) contouring of digital data (Sharpe, 1987). The use of computer contouring does not imply that the resulting map represents, in detail, an actual surface as does a topographic map. These data are presented as a contour map only to show general trends in the data.

The carbon isotopic data are contoured on figure 4. There is a pattern of isotopic lows that trends nearly north-south in the western part of the study area. This pattern does not follow any known structural element. There is a suggestion of other, less pronounced patterns of lows, which trend to the north-east.

In the lower center of figure 4, the area with the most negative (in absolute value) closed carbon isotopic contours is

Table 1. Carbon and oxygen isotopic data from known hydrocarbon accumulations. (Modified from Donovan and Datziel 1977.)

	$\delta^{13}C$(PDB)		$\delta^{18}O$(SMOW)		Ratio of most negative $\delta^{13}C$ value to mean $\delta^{13}C$ value
	Range	Mean	Range	Mean	
Oklahoma					
Cement	−39.2 to −5.5	−22.2	22.5 to 36.5	27.4	1.8
Davenport	−11.3 to −3.3	−9.1	27.5 to 48.8	38.5	1.2
Doyle	−26.3 to −3.1	−10.1	26.9 to 36.9	32.5	2.6
Fox-Graham	−34.2 to −2.0	−13.7	21.6 to 37.0	30.2	2.7
Velma	−37.6 to −10.9	−23.2	21.6 to 37.9	29.0	1.6
New Mexico					
Bisti	−18.9 to 0.1	−10.1	10.9 to 26.7	19.6	1.9
Texas					
Jameson reef	−38.0 to −0.4	−20.5	22.9 to 28.0	26.0	1.8
Garza	−32.0 to −6.2	−9.2	22.6 to 36.0	25.3	3.5
California					
Kettleman Hills North Dome	−22.1 to −4.0	−12.6	21.9 to 30.0	25.2	1.8
Colorado					
Gunbarrel Hill	−18.0 to −3.0	−8.3	12.2 to 24.1	16.0	2.1

found. This is significant because this area is nearly coincident with the productive area of the Barker Creek gas field, and thus represents an internal calibration or standard for this entire study in terms of how a known hydrocarbon accumulation is expressed at the surface by carbon and oxygen isotopic values.

The Barker Creek accumulation represents the most likely known setting (given sufficient time) for the development of a strong carbon isotopic anomaly in this study area. Factors supporting this statement are: 1) An anticlinal trap with the possibility of fractures and (or) faults at the apex of the flexure, 2) a wet gas accumulation providing ease of leakage as compared to liquid hydrocarbons, and 3) shallow depth providing shorter migration path. If this is the best known opportunity for the development of a surface carbon isotopic anomaly in the study area, and it is represented by $\delta^{13}C$ values which are only about −10/mil, then $\delta^{13}C$ values as negative as those reported by Donovan and Dalziel (1977) will probably not be found in this study area.

A comparison of the ratio of the mean isotopic value of all samples from a particular oil field and the most anomalous isotopic values of samples from the same field is an alternative way of evaluating the data. The data presented by Donovan and Dalziel (1977) for 10 oil fields (see table 1, this report) show a wide range in absolute values of $\delta^{13}C$ from +0.1/mil for the Bisti field, New Mexico, to −38/mil for the Jameson Reef field, Texas; however, the ratio of the extreme low values to the mean value for these same 10 fields is fairly constant with a low of 1.2 (Davenport field, Oklahoma), a high of 3.47 (Garza field, Texas), and an average of 2.

This relationship between the extreme and the mean isotopic values of a particular area lends some support for applying the term anomaly to closed areas bounded by contours of $\delta^{13}C$ values of less than −5/mil (the mean value of $\delta^{13}C$ for all samples in this study was −2.2/mil).

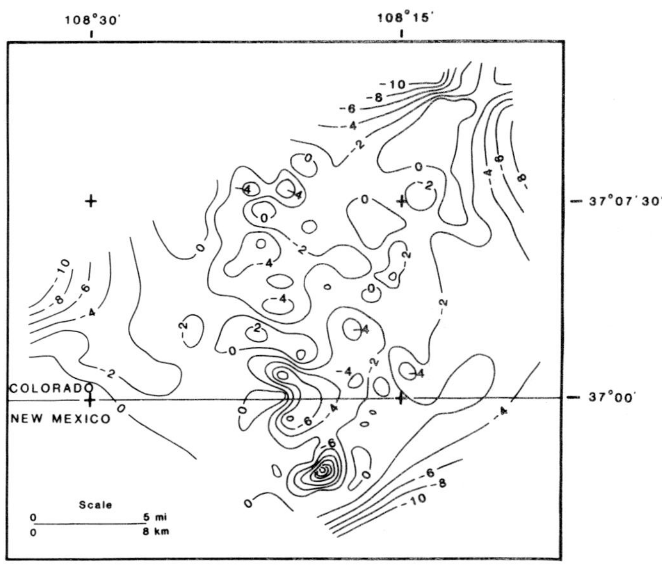

Figure 4. Contour map showing distribution of carbon isotopic data from surface sandstones, Southern Ute Indian Reservation, southwestern Colorado.

If one assumes that the patterns in the carbon isotopic data are related to hydrocarbon microseepage, then a reasonable explanation for the lack of exceptionally light carbon isotopic values here as compared to those in Oklahoma may be related to: 1) The length of time over which the leakage has occurred, 2) the efficiency of the leakage process, 3) the efficiency of the processes by which the hydrocarbons are oxidized and incorporated into the calcium carbonate cement, 4) the hydrocarbon type, or, 5) a combination of the above. Donovan and others (1974) proposed that $\delta^{13}C$ values from rocks overlying the Davenport oil field, Oklahoma, were not as negative as

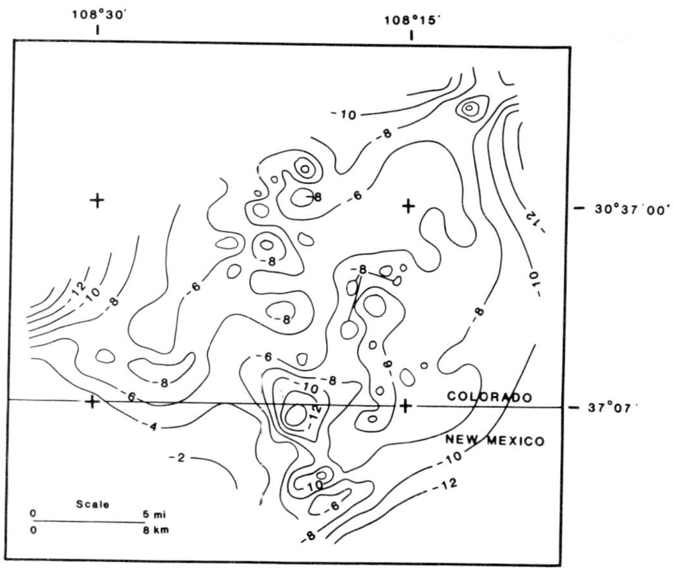

Figure 5. Contour map showing distribution of oxygen isotopic data, from surface sandstones, Southern Ute Indian Reservation, southwestern Colorado.

values from other fields in Oklahoma because the accumulation at Davenport was gas and a rapid leakage rate may not allow sufficient time for complete oxidation of the hydrocarbon.

Oxygen isotopic data were computer drawn to create a contour map (fig. 5) by the method described for the carbon isotopic contour map. Oxygen isotopic data are highly correlated with the carbon isotopic data (see fig. 3), hence, the similarity between the two maps. A slightly arcuate (concave to the east) pattern of low oxygen isotopic values which trends north is present in figure 5 in nearly the same position as the carbon isotopic pattern from figure 4. The suggestion of minor northeast-trending oxygen isotope lows, similar to those discussed previously in figure 4, can be seen in figure 5. The oxygen isotopic data are more difficult to interpret because of the two fractionation mechanisms which may operate singly or together and may operate independently from the hydrocarbon microseepage mechanism (Donovan and Dalziel, 1977). The correlation between the carbon and oxygen isotopic data, however, suggests a relationship between the two and the mechanism causing the range of values reported is driving the system in the same direction; lower carbon isotope values generally occur with lower oxygen isotope values.

SUMMARY AND CONCLUSIONS

Carbon and oxygen isotope values from carbonate-cemented sandstones of the Southern Ute Indian Reservation do not unequivocally support the thesis of input of hydrocarbon-derived light carbon or the operation of an oxygen fractionation mechanism to yield the isotopic signatures found in these sandstone cements. The values for $\delta^{13}C$ range from +3.2/mil to −16.47/mil and for $\delta^{18}O$ range from −0.89/mil to −16.27/mil and fit neatly between the cluster of known Cretaceous marine carbonates and the cluster of known Cretaceous freshwater carbonates. These data may represent a simple mixture of varying proportions of the two end members.

However, when these values are contoured, the contour maps exhibit major north-trending patterns of isotopic lows that emerge from both the carbon and oxygen data sets. A secondary northeast-trending pattern (a major structural trend in the area shown by Matheny, 1978a; 1978b) is also present in both isotopic data sets. The lack of spatial coincidence between the isotopic lows shown here and known structural elements is not a severe problem for the interpretation because stratigraphic as well as structural traps are known to display surface isotopic anomalies (Donovan and Dalziel, 1977).

The magnitude of the anomalies reported here is less than that for most of the anomalies reported in the literature, however, if the most negative values are compared to background values, these anomalies are in the same general range as those from the literature. The values representing anomalous areas defined by the ratio of the most anomalous values to the average background values will often be of only local importance, but, that may be sufficient to make the method a valuable part of an exploration package. In areas like southern Oklahoma, where large oil-filled structures are comparatively common, large scale faults are common, oil seeps at the surface are or were common, and reservoirs, source rocks, and presumably hydrocarbon accumulations have been present for hundreds of millions of years, it does not seem unusual that surface geochemical indicators would be better developed there. Similarly, in an area such as the northwestern part of the San Juan Basin, an area likely to have subtle stratigraphic traps, significant quantities of gas, and a shorter geologic history, surface geochemical anomalies may not be as pronounced.

If the patterns of low carbon and oxygen isotope values in calcareous sandstone cements described in this paper are, in fact, related to hydrocarbon microseepage, the oil field outside this study area which represents the best known analog for comparison, is the Bisti oil field (Donovan and Dalziel, 1977), also located in the San Juan Basin. Further studies planned in this area should help determine the nature and paragenesis of the calcareous material.

***APPENDICES

Appendix 1. Carbon and oxygen isotopic data from carbonate-cemented sandstones from the Southern Ute Indian Reservation. Values are given in standard notation relative to the PDB standard.

Sample	$\delta^{13}C$	$\delta^{18}O$	Sample	$\delta^{13}C$	$\delta^{18}O$
MH-1-87	−1.76	−7.40	MH-21-87	−0.93	−5.47
MH-2-87	−2.21	−8.18	MH-22-87	+0.14	−4.96
MH-3-87	+0.50	−5.11	MH-23-87	−0.31	−5.26
MH-4-87	−1.19	−6.36	MH-24-87	−0.08	−6.03
MH-5-87	+1.26	−5.96	MH-25-87	N.D.	N.D.
MH-6-87	−3.81	−5.82	MH-26B-87	−1.02	−6.91
MH-7-87	−2.21	−6.37	MH-27-87	+0.48	−7.01
MH-8-87	+0.69	−5.41	MH-28-87	+0.04	−7.40
MH-9-87	−3.44	−6.29	MH-29-87	−7.14	−13.40
MH-10-87	N.D.	N.D.	MH-30-87	−1.16	−7.36
MH-11-87	+0.96	−5.05	MH-31-87	+3.20	−0.56
MH-12-87	−3.13	−6.26	MH-32-87	−2.76	−5.98
MH-13-87	−2.42	−5.87	MH-33-87	−1.39	−6.76
MH-14-87	−3.64	−7.34	MH-34-87	−5.17	−9.95

MH-15-87	−1.79	−6.83
MH-15A-87	+0.26	−8.23
MH-16-87	−3.21	−9.32
MH-17-87	−1.06	−5.75
MH-18-87	−1.37	−8.71
MH-19-87	−0.15	−3.71
MH-20-87	−2.42	−7.69
MH-42-87	−0.44	−5.64
MH-43-87	+0.34	−4.90
MH-44-87	+0.43	−4.91
MH-45-87	−5.71	−8.46
MH-46-87	+0.14	−2.16
MH-47-87	+0.36	−5.64
MH-48-87	−5.43	−11.22
MH-49-87	−0.68	−5.50
MH-50-87	−3.87	−8.16
MH-51-87	+2.78	N.D.
MH-52-87	−1.17	−2.80
MH-53-87	−4.67	−9.56
MH-54-87	−1.45	−5.68
MH-55-87	−3.11	−7.58
MH-56-87	−1.84	−4.96
MH-57-87	−1.86	−5.76
MH-58-87	−5.17	−9.16
MH-59-87	−3.81	−7.09
MH-60A-87	+0.27	−1.00
MH-60B-87D	−2.59	−4.87
MH-61A-87	−4.27	−10.15
MH-61B-87	−6.45	−10.63
MH-62-87	−0.60	−4.98
MH-86-87	−2.72	−3.36
MH-87-87	−1.90	−7.36
MH-88-87	N.D.	N.D.
MH-89-87	−4.81	−9.38
MH-90-87	−2.98	−7.94
MH-91-87	−5.16	−5.96
MH-92-87	−0.80	−5.18
MH-93-87	−5.22	−9.03
MH-94-87	N.D.	N.D.
MH-95-87	−2.10	−6.52
MH-96-87	−1.82	−7.21
MH-97-87	−2.23	−7.68
MH-98-87	−13.27	−11.86
MH-99-87	−0.29	−4.99
MH-99C-87	−6.77	−10.35
MH-100-87	−1.04	−7.97
MH-101-87	−2.17	−6.06
MH-102-87	+0.20	−2.87
MH-103-87	−1.50	−7.48
MH-104-87	−2.95	−7.85
MH-105D-87	+1.46	−3.44
MH-106-87	−4.51	−8.03
MH-107-87	−2.76	−7.33
MH-131-87	−16.44	−14.41
MH-132-87	−2.97	−8.09
MH-133-87	−2.46	−3.55
MH-134-87	−0.68	−5.70
MH-135-87	−2.85	−9.03
MH-136-87	−11.08	−12.44
MH-137-87	N.D.	N.D.
MH-138-87	−10.73	−12.71
MH-35-87	−3.23	−8.20
MH-36-87	−3.01	−6.34
MH-37-87	−8.47	−11.71
MH-38-87	N.D.	N.D.
MH-39-87	−6.11	−6.71
MH-40-87	−1.98	−8.78
MH-41-87	N.D.	N.D.
MH-63-87	+0.18	−5.11
MH-64-87	−2.81	−3.65
MH-65-87	−5.06	−8.97
MH-66-87	−2.62	−3.83
MH-67-87	−3.92	−7.51
MH-68-87	−0.28	−5.18
MH-69-87	+0.08	−5.63
MH-70-87	−2.34	−7.35
MH-71-87	−0.10	−5.30
MH-72-87	−1.86	−8.14
MH-73-87	−2.63	−7.41
MH-74-87	N.D.	N.D.
MH-75-87D	+1.82	−0.89
MH-76-87	−2.61	−7.76
MH-77-87	+0.25	−3.23
MH-78-87	+1.11	−5.06
MH-79-87	+1.64	−4.49
MH-80-87	−5.32	−11.86
MH-81-87	−0.90	−7.33
MH-82-87	−1.40	−7.63
MH-83-87	N.D.	N.D.
MH-84-87	−3.85	−8.69
MH-85-87D	−1.09	−5.56
MH-108-87	N.D.	N.D.
MH-109-87	−0.50	−7.08
MH-110-87	−5.50	−7.26
MH-111-87	−1.64	−10.06
MH-112-87	−7.66	−15.71
MH-113-87	−1.63	−7.55
MH-114-87	−0.29	−8.97
MH-115-87	−2.25	−7.78
MH-116-87	−8.42	−11.99
MH-117-87	−1.45	−5.73
MH-118-87	+1.22	−11.27
MH-119-87	−9.36	−11.56
MH-120-87	+2.26	−3.54
MH-121-87	−6.29	−11.97
MH-122-87	−0.19	−5.37
MH-123-87	−0.19	−5.58
MH-124-87	−1.12	−4.96
MH-125-87	−0.18	−4.96
MH-126-87	+0.19	−4.98
MH-127-87	+0.61	−4.70
MH-128-87	−1.83	−3.07
MH-129-87	−0.59	−5.26
MH-130-87	−1.21	−6.10
MH-151-87	−10.01	−16.19
MH-152-87	−5.00	−9.40
MH-153-87	−5.23	−9.54
MH-154-87	−1.09	−6.22
MH-155-87	−1.95	−6.86
MH-156-87	−1.22	−9.16
MH-157-87	+0.36	−2.28
MH-158-87	−0.51	−5.14
MH-139-87	−0.72	−4.90
MH-140-87	−0.12	−5.15
MH-141-87	−1.38	−5.60
MH-142-87	−2.95	−7.89
MH-143-87	−1.88	−8.31
MH-144-87	−2.62	−7.08
MH-145-87	−2.26	−8.36
MH-146-87	−1.06	−4.86
MH-147-87	−3.10	−8.29
MH-148-87	−2.84	−8.25
MH-149-87	+0.68	−4.27
MH-150-87	−0.38	−5.53
MH-159-87	−5.31	−9.20
MH-160-87	−0.56	−6.78
MH-161-87	−2.16	−7.00
MH-162-87	−1.45	−5.47
MH-163D-87	+3.22	−3.91
MH-164-87	−4.23	−8.33
MH-165-87	−2.60	−6.19

Appendix 2. Carbon and oxygen isotopic data from selected Cretaceous marine and non-marine limestones. Values are given in standard δ notation relative to the PDB standard.*

Cretaceous Limestones			
Marine		Freshwater	
$\delta^{13}C$	$\delta^{18}O$	$\delta^{13}C$	$\delta^{18}O$
−13.75	−4.52	−0.81	−14.08
+0.40	−4.81	+1.26	−9.49
+0.40	−4.81	−4.50	−14.10
+1.42	−5.00	−4.28	−9.38
−2.54	−2.34	−4.30	−10.72
−2.12	−9.74	−6.22	−15.72
+1.58	−7.39	−4.55	−12.60
+3.85	−5.60	−3.93	−10.50
+0.29	−3.82	−0.91	−9.83
+3.82	−8.71	−2.70	−10.75
−2.79	−3.93	−3.34	−11.48
+2.34	−5.76	+0.28	−11.31
+0.66	+5.44	−7.43	−13.02
+0.75	−4.48	−1.40	−15.10
+1.64	−5.14	−0.42	−12.15
+1.56	−7.16	−3.73	−16.16
+2.58	−3.08	−6.19	−10.93
−0.74	−1.64	−7.71	−8.80
+0.84	−2.79	−4.48	−8.32
−1.64	−3.24	−4.02	−8.38
+1.97	−4.48	−12.72	−10.97
+1.96	−4.38	−2.64	−15.74
+1.72	−3.74	−4.72	−6.84
+2.35	−3.45	−6.02	−7.34
+1.68	−3.08	−5.14	−6.64
+0.20	−3.10	−5.26	−3.64
+1.96	−2.42	−7.20	−3.76
+1.62	−3.26	−6.34	−9.16
+1.70	−3.34	−8.26	−10.00
+2.12	−4.34	−1.27	−4.10
+0.69	−3.35	−4.29	−8.68
+0.28	−6.58		
+1.58	−3.82		
−0.06	−4.42		
+2.84	−3.45		
−9.32	−5.29		
−1.10	−3.83		
+0.32	−4.54		

* Data from Keith and Weber, 1964

REFERENCES CITED

Baars, D. L., and Ellingson, J. A., 1984, Geology of the western San Juan Mountains, *in* Brew, D. C., ed., Field Trip Guide Book, 37th Annual Meeting, Rocky Mountain Section, Geological Society of America: Four Corners Geological Society, p. 1-32.

Donovan, T. J., 1974, Petroleum microseepage at Cement, Oklahoma—evidence and mechanism: American Association of Petroleum Geologists Bulletin, v. 58, p. 429-446.

Donovan, T. J., Friedman, I., and Gleason, J. D., 1974, Recognition of petroleum-bearing traps by unusual isotopic compositions of carbonate-cemented surface rocks: Geology, v. 2, p. 351-354.

Donovan, T. J., and Dalziel, M. C., 1977, Late diagenetic indicators of buried oil and gas: U.S. Geological Survey Open-File Report 77-817, 44 p.

Donovan, T. J., Forgey, R. A., and Roberts, A. A., 1979, Aeromagnetic detection of diagenetic magnetite over oil fields: American Association of Petroelum Geologists Bulletin, v.63, no.2, p. 245-248.

Keith, M. S., and Weber, J. N., 1964, Carbon and oxygen isotopic composition of selected limestones ans fossils: Geochemica et Cosmochemica Acta, v. 28, p. 1787-1816.

Lang, H. R., Alderman, W. H., and Sabins, F. F. Jr., 1984a, Patrick Draw, Wyoming, petroleum test case report, *in* Abrams, M. J., Conel, J. E., and Lang, H. R., The Joint NASA/Geosat Test Case Project: American Association of Petroleum Geologists, v II, part 2, sec. 11.

Lang, H. R., Nicolais, S. M., and Hopkins, R., H., 1984b, Coyanosa, Texas, petroleum test site report, *in* Abrams, M. J., Conel, J. E., and Lang, H. R., The Joint NASA/Geosat Test Case Project: American Association of Petroleum Geologists, v. II, part 2, sec. 13.

Matheny, M. L., 1978a, Barker Creek Dakota, *in* Fassett, J. E., ed., Oil and gas fields of the Four Corners area: Four Corners Geological Society, p. 198-200.

Matheny, M. L., 1978b, Barker Creek Paradox, *in* Fassett, J. E., ed., Oil and gas fields of the Four Corners area: Four Corners Geological Society, p. 201-203.

Sabins, F. F., 1963, Anatomy of a stratigraphic trap, Bisti Field, New Mexico: American Association of Petroleum Geologists Bulletin, v. 47, p. 193-228.

Sharp, W. E., 1987, Two basic rules for valid contouring: Geobyte, v. 2, no. 4, p. 11-15.

Tweto, O. T., 1979, Geologic map of Colorado: U.S. Geological Survey.

Ute dwelling. Photo courtesy of the Colorado Historical Society.

Preliminary Interpretation of Soil-Gas Data and Relationships to other Hydrocarbon Microseepage Indicators, Four Corners Platform-San Juan Basin Transitional Area, Southwest Colorado and Northwest New Mexico

Kimberley I. Cunningham[1]

ABSTRACT

A soil-gas survey was conducted over a 320-square-mile (930 km^2) area on the Ute Indian Reservations in southwest Colorado and northwest New Mexico. The study area contains a group of northeast-trending shallow magnetic anomalies that may be related to hydrocarbon-induced production of magnetic minerals in the oxidizing near-surface environment. Helium anomalies in soil-gas correlate with this magnetic trend and also occur over the Red Mesa and Cinder Butte oil and gas fields. These associations are contrasted with deeper Pennsylvanian gas fields, such as Alkali Gulch and Barker Dome which display offset helium anomalies that marginally correlate with field position. The helium-anomaly trends also correlate with lineament-fracture patterns and gravity-derived structural trends on the Four Corners platform suggesting some structural control over the near-surface distribution of helium. Support for some of these helium anomalies being related to hydrocarbon microseepage is obtained from C^{13} and O^{18} isotopic compositions of carbonate cements in outcropping rocks, areas of excess free CO_2 and N_2 in soil-gas, near-surface gravity and magnetic data, regional gravity interpretation, and LANDSAT imaging data enhanced to show areas of surficial limonitic mineralization.

INTRODUCTION: MICROSEEPAGE MECHANICS AND EFFECTS ON SUPERCUMBENT RESERVOIR ROCKS

Hydrocarbon microseepage is the vertical movement of minute amounts of hydrocarbons (C_1-C_6) and associated compounds (H, He, CO_2, N_2, H_2S, etc.) from a buried petroleum accumulation to the surface. Macroseeps, or liquid petroleum following high permeability pathways to the surface, have been used extensively as direct indicators of subsurface deposits (Barton, 1927; Coomber, 1938; DeGolyer, 1940; and Link, 1952). Microseeping hydrocarbons follow microfracture pathways via various migration mechanisms to the surface where chemical reactions with groundwater leaves characteristic signatures in surface rocks and soils. Proposed mechanisms for migration include effusive mass transport of hydrocarbons through high-permeability channels, hydrodynamic or chemical potential drive of low-weight hydrocarbons dissolved in formation waters, and diffusion (Donovan and Dalziel, 1977; Reitsema and others, 1981; Stahl and others, 1981; Leythaeuser and others, 1983; Faber and Stahl, 1984). All, none, or combinations of these mechanisms may be operational in any given area. The production of CO_2 from the near-surface oxidation of microseeping hydrocarbons results in the biologically-mediated production of isotopically distinctive pore-filling carbonate cements in the overlying rocks (Donovan, 1974; Donovan and others, 1974). The responsible microseepage mechanism is indicated by the relative enrichment or depletion of O^{18} in C^{13}-depleted carbonate cements (Donovan and Dalziel, 1977).

Microseeping hydrocarbons indirectly create a locally reducing environment in host rocks causing the reduction, dissolution, and partial removal of iron and manganese from minerals and the transformation of some mineral phases by chemical reduction (Donovan and Dalziel, 1977). The effects of these reduction-initiated processes can be observed in the field as: 1) systematic variations in iron and manganese concentrations in carbonate cements from outcropping rocks (Cheney, 1964; Donovan, 1974; Donovan and others, 1974; Donovan and Roberts, 1980), 2) the vertical redistribution of iron and manganese in rocks, soils, and plants with significant bleaching of iron-rich rocks (Donovan, 1974; Dalziel and Donovan, 1980), and 3) high-wavenumber magnetic anomalies arising from the production of epigenetic magnetite and related sulfur-bearing magnetic minerals (Donovan and others, 1979; McCabe and others, 1983; Donovan and others, 1984; Ellwood and others, 1986; Elmore and others, 1986; McCabe, 1986; Reynolds and others, 1986).

These surface phenomena are indirect or fossil indicators of microseepage; they represent continuous or episodic processes through time with the amount of material emplaced dependant on microseepage intensity, duration, and mechanism. It is not

[1] Kimberley I. Cunningham
U.S. Geological Survey
P.O. Box 25046, MS 939, DFC
Lakewood, Colorado 80225

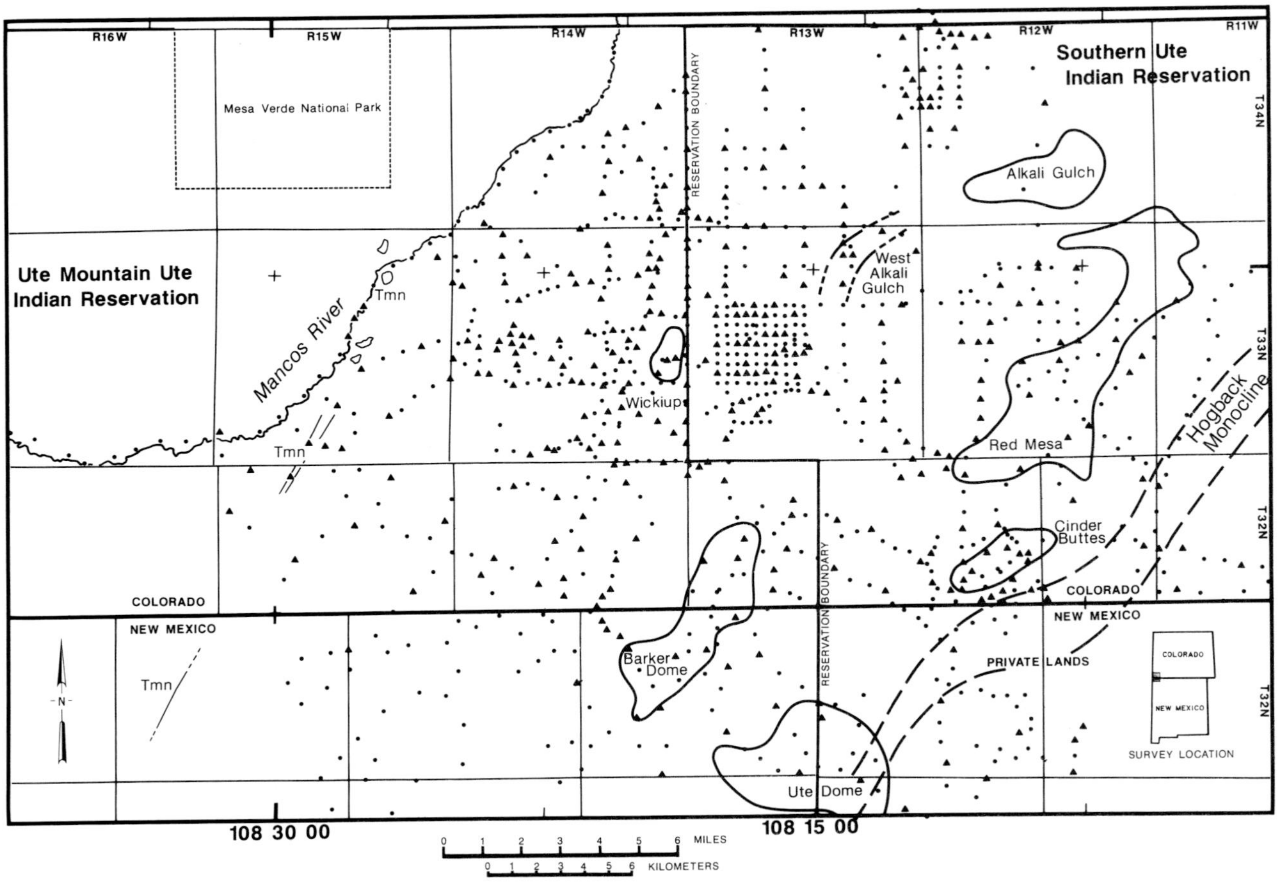

Figure 1. Map showing distribution of soil-gas samples analyzed for helium (solid circles) and helium plus gas chromatographic analyses (solid triangles), Southern Ute Indian Reservation, southwest Colorado and northwest New Mexico. Heavy dashed line in western survey area represents surface trace of Tertiary minette intrusives. Heavy dashed line in eastern survey area represents surface trace of Hogback monocline.

possible to determine from these data if the responsible reservoir may still contain hydrocarbons. Near-surface soil gas concentrations reflect reservoir composition (Laubmeyer, 1933; Kartsev and others, 1959; Sokolov, 1971; Stahl and others, 1981; Jones and Drozd, 1983; Faber and Stahl, 1984) and provide a direct or modern expression of gas microseepage controlled by regional structure and porosity reduction in the column. Helium and hydrogen are used as direct indicators because they commonly occur in most petroleum reservoirs with helium having nominal concentrations greater than 0.01 percent or 100 ppm by volume (Dyck, 1976; Moore, 1976, 1982). Anomalous helium and light-hydrocarbon concentrations in soil-gas have been associated with petroleum accumulations and substantiate modern leakage patterns (Laubmeyer, 1933; Kartsev and others, 1959; Debnam, 1969; Horvitz, 1969; McCrossan and others, 1971; Ball and Snowdon, 1973; Palacas and Roberts, 1980; Pogorski and Quirt, 1981; Roberts, 1981; Jones and Drozd, 1983; Roberts and Roen, 1985; Cunningham and others, 1987; Cunningham and Roberts, 1987; Cunningham, 1988). Fault and fracture systems have been delineated by employing similar techniques (Bulashevich and Bashorin, 1973; Ovchinnikov and others, 1972; Plyusnin and others, 1972; Eremeev and others, 1973; Dikun and others, 1976; Jones and Drozd, 1983; Roberts and Roen, 1985; McCarthy and others, 1986) and provide valuable structural information.

These relationships between soil-gas constituents and reservoir gas composition are poorly understood due to the large number of lithologic, hydrologic, and tectonic variables controlling migration from the reservoir to the surface. However, numerous studies have demonstrated firm relationships between many of these direct and indirect indicators and the locations of known hydrocarbon occurrences in diverse geologic settings.

The study area (fig. 1) is centered on a group of shallow magnetic anomalies delineated from reprocessed aeromagnetic data (figs. 2-5). These anomalies trend northeast along the Montezuma-La Plata County line and may be related to a microseepage indicator (number 3 above) since no correlation with topography, culture, intrusives, clinker, magnetic placers, or other related geologic phenomena can be demonstrated. Five magnetic anomalies are present on the Southern Ute Reservation and other anomalies occur on the Ute Mountain Ute Reservation. The Ute Mountain Ute Reservation is not discussed in this report.

EXPERIMENTAL METHOD

Approximately 950 soil-gas samples were collected during the period of May 28-June 13, 1987 in an area roughly centered over the common reservation boundary between the Ute Mountain Ute and the Southern Ute Indian Reservations. The survey area covers approximately 360 square miles (930 km^2) extending south past the Barker Dome gas field, west to the Mancos River, north of the Alkali Gulch gas field, and east to the Red Mesa oil and gas field (fig. 1). Soil-gas sampling density was approximately one sample per 0.25 square mile (0.65 km^2) while within a magnetic anomaly; this density was decreased to one sample per 0.5-0.75 square mile (1.3-1.95 km^2) outside of these areas. Soil-gas samples were obtained by pounding a hollow steel probe 2.5 feet (.75 m) into the ground. The probe is sealed by securing an airtight septum to the top. An initial 10 cc sample is withdrawn and discarded to purge the dead-air volume of the probe. A second 10 cc sample is collected, sealed by soft silicone tubing, and returned to the mobile lab where mass spectrometric and gas chromatographic analyses are performed within 4-6 hrs. Sample collection procedures are identical for permanent probes used to monitor the diurnal variation of helium.

ANALYTICAL TECHNIQUES

Soil-gas samples for this microseepage study were collected and analyzed using procedures developed by the U. S. Geological Survey (Reimer, 1976). Over 950 soil-gas samples were acquired and analyzed for helium (fig. 1). Of these, 360 samples were analyzed for CO_2, O_2 (as O_2 + Ar), N_2, and light (C_1-C_3) hydrocarbons by gas chromatography (fig. 1). Analytical precision is 10 ppb v/v for helium, 10 ppm v/v for CO_2, 1,000 ppm v/v for O_2 and N_2, and 1 ppm v/v for C_1-C_3 hydrocarbons. Helium is reported in parts per billion above or below the concentration of helium in the atmosphere (5,220 ppb), CO_2 above or below 0.0335 percent, oxygen (reported as the sum of oxygen and argon) above or below 21.881 percent ppm, and N_2 above or below 78.084 percent. Fewer than 14 percent of the 360 soil-gas samples contained measurable hydrocarbons at our detection limits.

HELIUM

The raw helium data was corrected for diurnal variations using procedures after Reimer and Roberts (1985). Permanent probes are placed within the survey area and sampled from 3 to 6 times per day. Measured helium concentrations are plotted against time of day and a regression line calculated. Decline rates for helium in soil-gas ranged from 14-43 ppb/hr with an average decline rate of 20.7 ppb/hr. Diurnal correction was not attempted for other gas species due to an inadequate daily sample population. After correction, a series of regional trend (nth order polynomial) surfaces are calculated and residual values obtained by subtracting the regional from the diurnally corrected raw data. Comparison of the diurnally-corrected and regionally-corrected anomaly maps show minimal change in anomaly position. A map of 2nd order residual helium values contoured above the 0 ppb regional isopleth (fig. 2) shows subsurface anomalies which display northeast and northwest trends in areas of good sample control. Anomaly trends having values greater than 1 standard deviation above the background population mean (approximately 25 ppb) are considered significant. Anomalous areas on the Southern Ute Reservation include the Wheeler Draw, Red Mesa, and west of upstream San Juan Arroyo areas (near magnetic anomaly C).

STABLE CARBON AND OXYGEN ISOTOPES

Over 150 samples of carbonate cements from the outcropping Upper Cretaceous Cliff House Sandstone of the Mesaverde Group were analyzed for stable carbon and oxygen isotopes. Values range from -16.47 to 3.20 (relative to PDB standard) for C^{13}, and from -16.27 to -0.89 for O^{18} (SMOW). These samples have C^{13}/O^{18} ratios similar to those at the Bisti oil field in the central San Juan basin (Donovan and Dalziel, 1977). Southern Ute sandstone-cement populations display C^{13}/O^{18} ratios that may be representative of intermediate marine- and freshwater-derived carbon and oxygen values (Henry, this volume). According to Donovan (1974) and Donovan and Dalziel (1977) these C^{13}/O^{18} ratios and degree of O^{18}/O^{16} fractionation provide evidence for hydrodynamic or chemical potential drive of dissolved gases through semipermeable shale membranes. Microseeping hydrocarbons following high-permeability pathways above Pennsylvanian reservoirs will ultimately first reach the Mancos Shale and then the Lewis Shale. The Lewis Shale is absent in the study area west of the Hogback due to erosion leaving the Mancos Shale as the principal O^{18}/O^{16} fractionating membrane. Anomaly maps of C^{13} and O^{18} abundance show that the more depleted the C^{13}, the lower the O^{18} (Henry, this volume). Both species display north and northeast trends. While these trends correspond with the trends of other microseepage indicators, sample density is inadequate to provide a more rigorous correlation.

SHALLOW MAGNETIC ANOMALIES: COMPARISON WITH HELIUM AND ISOTOPE TRENDS

The high-wavenumber magnetic features in the Southern Ute study area (fig. 2) range from 1 to 4 square miles (2.5—10.4 km^2) and occur as five discrete anomalies: three in T. 33 N., R. 13 W. (anomalies A,B, and C), and two in T. 34 N., Rs. 12 and 13 W. (anomalies D and E). Anomalies A and C are coincident with the locations of the Dietrich Aztec Ute No. 1 and the West Alkali Gulch No. 1 gas wells, respectively. These anomalies are in a relatively magnetically quiet area where influences from near-surface clinker, culture, and heavy-mineral placers is minimized or absent. Shallow magnetic anomalies were not detected over other local producing oil and gas fields. The absence of significant magnetic signal may be attributable to: 1) Lack of sufficient iron in the column available for dissolution and transport, 2) removal or alteration of the diagenetic products by oxidative weathering, 3) removal of the host rocks by erosion, or 4) inability to extract a mappable signal due to processing limitations. Truck magnetometer traverses over the large anomaly in Wheeler Draw (anomaly B, fig. 2) provided a more sensitive look at the shallow source(s). Analysis of the magnetic power spectrum shows a shallow anomaly source at a depth of approximately 295 feet (90 m) with a second significant zone at approximately 2,500-3,000 feet (760-915 m) (Webring, this volume). These depths roughly correspond with the Menefee Formation-Cliff House Sandstone and Dakota Sandstone-Mancos Shale contacts and may represent geochemically favorable zones for development of epigenetic minerals. The northeast alignment of the anomaly trend corresponds with the orientation of other geomorphic and geologic elements west of (and including) the Mancos River (that is, trends of canyons, the Hogback monocline, Tertiary minette intrusives, and segments of the Mancos River) and suggests structural control (Donovan and others, 1986). Anomalous C^{13} and O^{18}

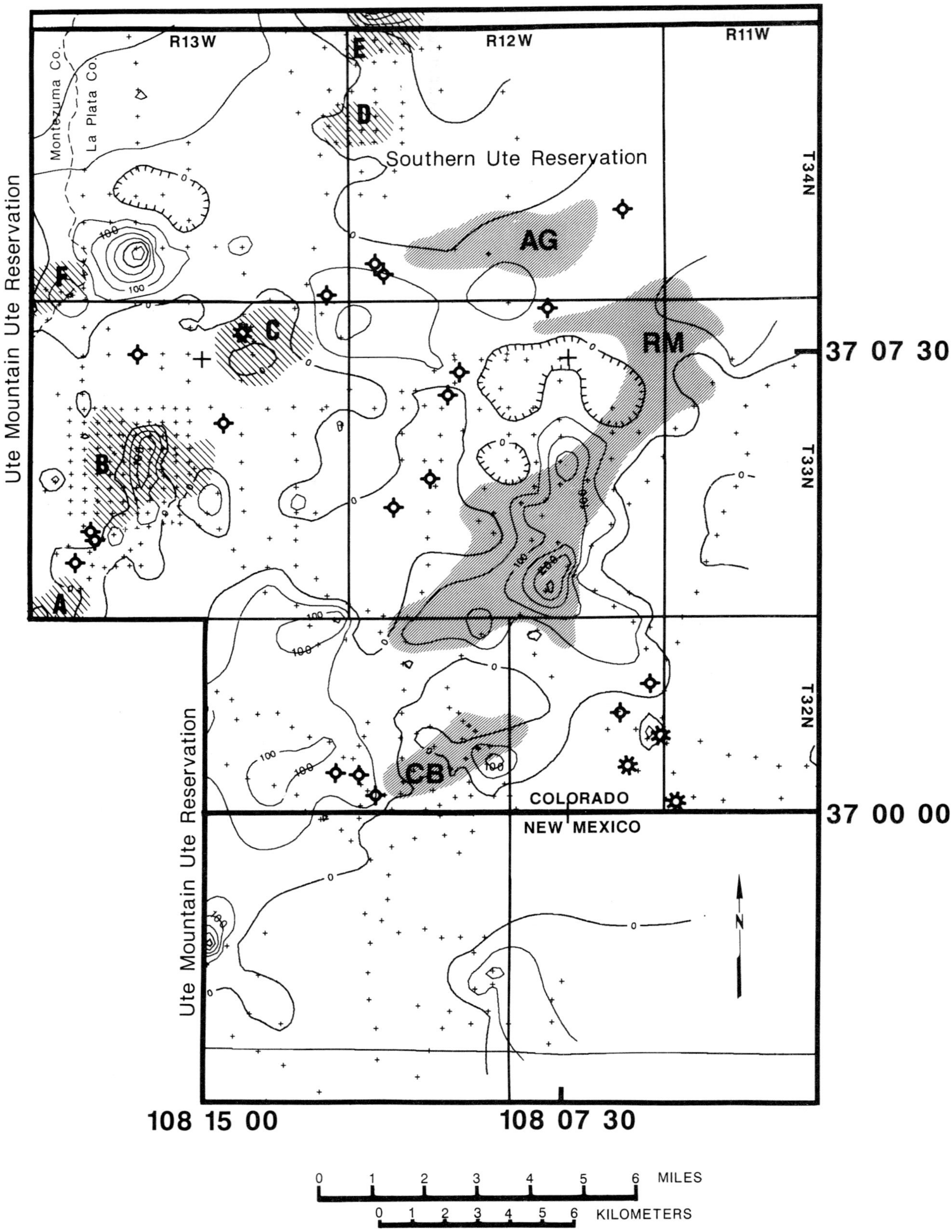

Figure 2. Contour map of helium concentrations in soil-gas, Southern Ute Indian Reservation and vicinity, southwest Colorado and northwest New Mexico. Helium contour interval = 50 ppb. Hachures indicate area of closed low. Anomalous shallow magnetic areas A-E indicated by coarse-line pattern. Local oil and gas fields indicated by fine-line pattern: AG = Alkali Gulch, BD = Barker dome Paradox, CB = Cinder Buttes, RM = Red Mesa, and W = Wickiup (abandoned).

values are coincident with this magnetic trend, and extend northeast 18 miles (29 km) to the north of the Alkali Gulch gas field. High correlation with helium is also observed and anomalous values occur discontinuously in a line from Barker Arroyo through Wheeler Draw to San Juan Arroyo, diffusing into an area of relatively high values just north of the Alkali Gulch gas field. The coincident orientations of these three microseepage indicators suggest the presence of a primary migration avenue which has provided access to reducing fluids moving into the near-surface over geologic time.

HELIUM ANOMALY TRENDS AND FRACTURE AND JOINT ORIENTATIONS

Kelly and Clinton (1960) documented several dominant fracture and joint group orientations in the study area which include a major set at N70°W and a minor set oriented at N30°-40°E. Three subordinate fracture systems are N20°-45°W, N30°-35°E, and N70°E. Knepper (1982) utilized LANDSAT imagery to resolve five main lineament trends in this area: N10°W-N16°E, N35°-72°E, N33°-59°W, N74°-83°W, and N89°-90°W. These trends are best characterized by a line of Tertiary minette intrusives, southeast of the Mancos River (fig. 1), and other similar or complimentary geomorphic elements such as the orientation of the Mancos River (oriented N30°E), adjoining canyons to the southeast of Mesa Verde National Park (MVNP), and the Hogback monocline from the Colorado border northeast to Durango. This structural grain may have part of its origin in recurrent basement faulting patterns documented in the Paradox basin where the direction of flexural strikes average N60°W (Szabo and Wengerd, 1975; Baars and Ellingson, 1984). The orientation of the high-helium trends with these aerially- and ground-derived trends supports structural control over the distribution of helium into the near-surface and establishes a link between areas of high permeability, high helium, and mapped lineament and fracture orientations.

HELIUM ANOMALY TRENDS AND GRAVITY

The approximate northeast and northwest patterns over the survey area are observed on contour maps of regional gravity data using a 2.45 gm/cc average stack density and a 3.4 km (2.1 mile) equivalent source (Webring, 1988). Several features directly coincide with positions of deep (greater than 2 km) Paleozoic domes and anticlinal flexures. Upper Cretaceous features such as the Mesa Verde basin, Red Horse syncline, and Barker Creek anticline are not evident suggesting that most of the map topography is basement-related and does not represent features higher in the stack where sediments exhibit lower density contrasts. Comparison of regional gravity features and helium indicates helium anomalies occurring in structurally favorable positions on the flanks, noses, or culminations of some structural features such as in the Red Mesa and Cinder Butte areas or along inferred block boundaries. The belt of anomalous helium and shallow magnetic sources lies on a northeast-trending, high-gradient structural feature that may represent a boundary between two blocks. The absence of appreciable helium directly above the Alkali Gulch field as opposed to the upper Cretaceous Red Mesa and Cinder Buttes fields may be due to porosity reduction in the sediments over the reservoirs. Older fields which have leaked longer should reflect greater emplacement of diagenetic products that forces modern gas migration to the outside of the chimney (Roberts, 1981). This relationship is not evident at Alkali Gulch where the helium sample density is low. In the Wheeler Draw area (magnetic anomaly B), additional gravity stations were acquired at approximately .1 mile spacing with laser-theodolite controlled elevations. These data indicate a slight, shallow-source density contrast with surrounding rocks (Upper Cretaceous Cliff House Sandstone of the Mesaverde Group) of approximately 0.5 gm/cc (Webring, this volume). This is hardly definitive, but lends additional support for reduction-initiated microseepage processes being operational in the survey area and affecting subsequent migration of gases to the surface.

MICROSEEPAGE ANOMALIES AND PROXIMITY TO SEISMIC STRUCTURES

Preliminary analysis of seismic data in the Red Mesa area indicates a persistent northeast-trending structural high in T. 33 N., R. 12 W. which is probably a northerly extension of the Pennsylvanian Barker Dome anticlinal trend. This culmination first appears in the Mississippian Leadville Limestone and propagates upsection through all Pennsylvanian Paradox members of the Hermosa Formation. The high is reduced to a minor saddle in the Jurassic Entrada Sandstone and is absent in all post-Jurassic formations (Taylor, 1988). Faulting cannot be resolved from the data. This structural trend lies approximately 2 miles southeast of the magnetic, helium, and carbon and oxygen isotopic anomaly trend. Upper Cretaceous beds on the monocline typically have dips from 30°-50° to the southeast and may reflect the attitude of deeper faults in the section. If assumptions concerning the structurally-controlled origination of the microseepage anomalies are valid, the displacement of the anomalies to the northwest could be explained by basement-involved normal faults intercepting leakage pathways above Pennsylvanian reservoirs. A 45° normal fault occurring in beds immediately above the Pennsylvanian reservoirs in Barker dome would intersect the surface in the general area of Barker Arroyo and could explain the anomaly displacement.

COMPARISON OF HELIUM, ATMOSPHERIC, AND LIGHT HYDROCARBON DATA

Without carbon isotope determinations of soil-gas hydrocarbons and carbon dioxide, or a larger analytical database of C_1-C_3 values from which ratios may be obtained (Jones and Drozd, 1983), it is not possible to determine the thermogenic vs. biogenic origin of these compounds. Light hydrocarbons were measured at several sample localities in the Cinder Buttes and Wheeler Draw areas with values as high as 50 ppm for methane and 10-20 ppm for C^2 and C^3 hydrocarbons. It is stressed that 86 percent of the samples analyzed had hydrocarbon concentrations less than our detection limit of 1 ppm. Significant variations may occur in the ppb range (Jones and Drozd, 1983), or alternatively, soil-gas hydrocarbons are not present in the survey area possibly due to effective near-surface biological utilization. Contour maps of soil-gas CO_2, O_2, and N_2 values show significant anomaly trends with predictable relationships for the more abundant O_2 and N_2; that is, high nitrogen areas are low oxygen areas and vice versa. Carbon dioxide anomalies (fig. 3) occur directly above the Cinder Buttes gas field and over part of the Red Mesa oil and

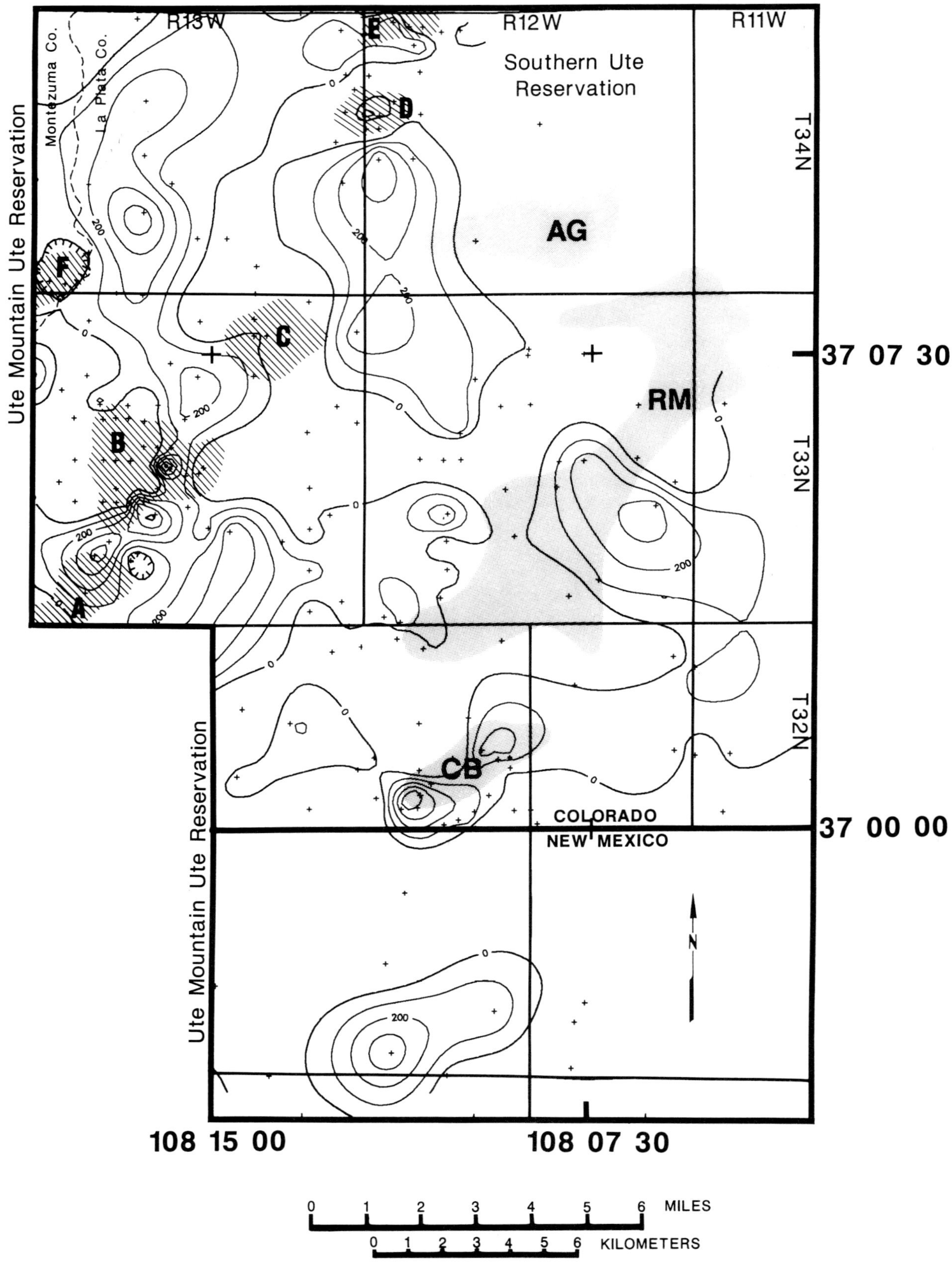

Figure 3. Contour map of carbon dioxide concentrations in soil gas, Southern Ute Indian Reservation and vicinity, southwest Colorado and northwest New Mexico. Contour interval = 100 ppm. Hachures indicate area of closed low. Areas A-E indicate shallow magnetic anomalies. Oil and gas fields indicated by fine-line pattern: AG = Alkali Gulch, BD = Barker dome Paradox, CB = Cinder Buttes, RM = Red Mesa, and W = Wickiup (abandoned).

gas field. Extremely high CO_2 values are coincident with the northeast-trending helium and high-wavenumber magnetic anomalies in T. 33 N., R. 13 W. and continue northward into T. 34 N., R. 13 W. The isopleth map of nitrogen concentrations in soil-gas (fig. 4) is similar to the one for CO_2 except that nitrogen maps the production limits of the Red Mesa field better. An approximate inverse relationship appears to exist between these two gases and the distribution of the shallow magnetic anomalies. These unique associations, particularly in T. 33 N., R. 13 W., suggest that some fraction of the nitrogen and carbon dioxide may be following the same high-permeability pathways as helium and reflect a common migration route. An isopleth map of carbon dioxide/oxygen ratios (fig. 5) is similar to the CO_2 map and confirms that carbon dioxide highs are at the expense of oxygen. Areas of high CO_2 and low O_2 may be related to subsurface oxidation of sulfide bodies (Lovell and others, 1980; May and Schmidt, 1982; Lovell and Hale, 1983; McCarthy and Reimer, 1986).

ASSOCIATIONS OF SURFICIAL LIMONITE ANOMALIES AND OTHER MICROSEEPAGE INDICATORS

LANDSAT multi-spectral scanner (MSS) data provided information in the study area about the distribution of ferric iron oxides and hydroxides in soils (Knepper, this volume). The distribution of these alteration products may be controlled in part by microseepage and thus may reflect either modern or paleo-soil zones where oxidation of ferric iron compounds may have occurred or is occurring. Anomaly distribution to the west of the Hogback monocline is extremely sparse with most of the San Juan Basin area reflecting small random occurrences. This is contrasted with the area immediately west of the Hogback monocline where hydrated ferric oxides are a major component of the soil spectrum. Comparison of these anomaly locations with other microseepage indicators is inconclusive; anomaly intersections appear to randomly correlate with helium, isotope, or magnetic trends and generally appear to reflect the distribution of iron-rich soils on mesa tops. These soils may be eolian in origin and reflect compositions of parent rocks from localities southwest of the study area. If these soils have been transported, any pre-existing signatures have been covered and detection by remote sensing is precluded.

SUMMARY

Preliminary interpretations of shallow magnetic, helium, and stable carbon and oxygen isotope data qualified with known structural, seismic, imaging, and fracture-lineament orientation information are consistent with a model in which reducing fluids with dissolved hydrocarbons have moved vertically through microfractures or high-permeability migration routes (faults and (or) fracture zones) to the surface. The microseepage mechanism is probably dominated by solutional transport of dissolved hydrocarbon gases through semi-permeable shale beds. A northeast-trending group of shallow magnetic anomalies of unknown origin coincides with helium and stable isotope anomaly positions and supports the existence of a high-permeability migration route trending northeast across the western part of the Southern Ute Reservation. The structurally-controlled formation of these anomalies is further supported by LANDSAT lineament strike-frequency analysis and fracture orientations which display coincident trends with the helium and magnetic anomaly orientations. While soil-gas hydrocarbon analyses were inconclusive, soil-gas carbon dioxide and nitrogen display anomaly patterns similar to helium and may reflect common migration pathways. Simultaneous comparison of these data sets suggests an exploration fairway running northeast across the western part of the Reservation which may be related to Pennsylvanian structural trends and hydrocarbon objectives.

REFERENCES CITED

Baars, D.L., and Ellingson, J.A., 1984, Geology of the western San Juan Mountains, *in* 37th Annual Meeting Field Trip Guide Book, Rocky Mountain Section, Geological Society of America: Four Corners Geological Society, Durango, Colorado, p. 1-42.

Ball, N.L., and Snowdon, L.R., 1973, A preliminary evaluation of the applicability of the helium survey technique to prospecting for petroleum: Geological Survey of Canada Paper 73-1B, p. 199-202.

Barton, D.C., 1927, The indications of the oilfield in the mid-continent and Gulf Coastal Plain of the United States: Journal of the Institute of Petroleum Technology, v. 13, p. 333-339.

Bulashevich, Yu. P., and Bashorin V.N., 1973, On the detection of faults along the Sverdlovsk DSS profile from high concentrations of helium in underground water: Physics of the Solid Earth, Academy of Sciences, USSR-Izvestiya, March, p. 185-189.

Cheney, E.S., 1964, Stable isotopic geology of the Gas Hills Uranium District, Wyoming: Yale University, Ph. D. thesis, 342 p.

Coomber, S.E., 1938, Surface indications of oil, *in* The science of petroleum: Oxford University Press, London, v. 1, p. 291-293.

Cunningham, K.I., 1988, Helium and C_1-C_3 hydrocarbon concentrations in permafrost ice, Marsh Creek area, Arctic National Wildlife Refuge, Alaska: U.S. Geological Survey Open-File Report 88-001, 43 p.

Cunningham, K.I., and Roberts, A.A., 1987, Helium concentrations in soil gas over known petroleum accumulations, WaKeeney area, Trego Co., Kansas: U.S. Geological Survey Open-File Report 87-254.

Cunningham, K.I., Roberts, A.A., and Donovan, T.J., 1987, Horizontal-gradient magnetic and helium surveys, *in* Bird, K.B., and Magoon, L.B., eds., Petroleum geology of the northern part of the Arctic National Wildlife Refuge, northeastern Alaska: U.S. Geological Survey Bulletin 1778, ch. 14, p. 209-218.

Dalziel, M.C., and Donovan, T.J., 1980, Biogeochemical evidence for subsurface hydrocarbon occurrence, Recluse oil field, Wyoming—preliminary results: U.S. Geological Survey Circular 837, 11 p.

Debnam, A.H., 1969, Geochemical prospecting for petroleum and natural gas in Canada: Geological Survey of Canada Bulletin v. 177, p. 1-26.

DeGolyer, E., 1940, Future position of petroleum geology in the oil industry: American Association of Petroleum Geologists Bulletin, v. 24, p. 1389-1399.

Dikun, A.V., Korobeynik, V.M., and Yanitskii, I.N., 1976, Some features of development of helium surveying: Interna-

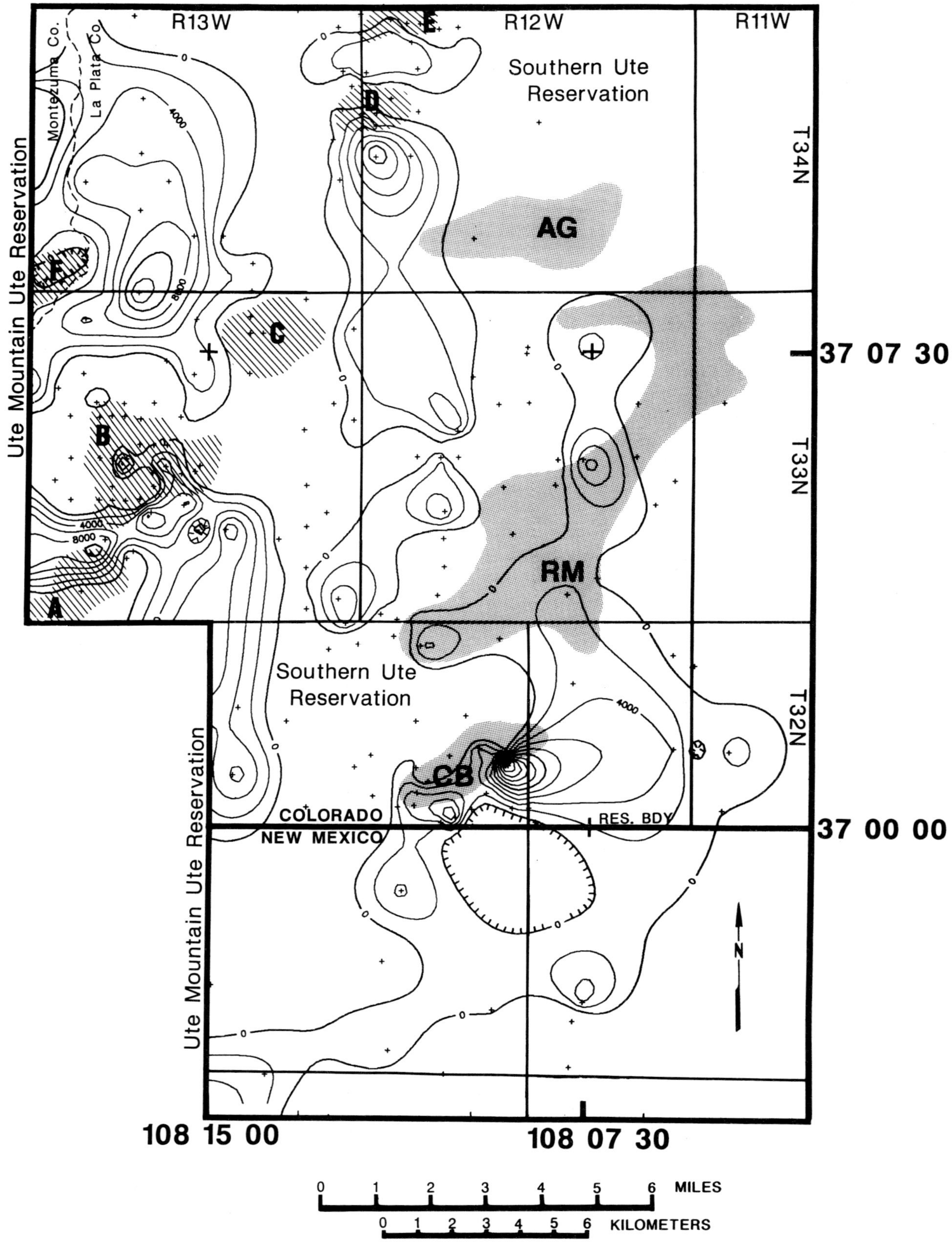

Figure 4. Contour map of nitrogen concentrations in soil gas, Southern Ute Indian Reservation and vicinity, southwest Colorado and northwest New Mexico. Contour interval = 2,000 ppm. Hachures indicate area of closed low. Areas A-G indicate shallow magnetic anomalies. Oil and gas fields indicated by fine-line pattern: AG = Alkali Gulch, BD = Barker dome Paradox, CB = Cinder Buttes, RM = Red Mesa, and W = Wickiup (abandoned).

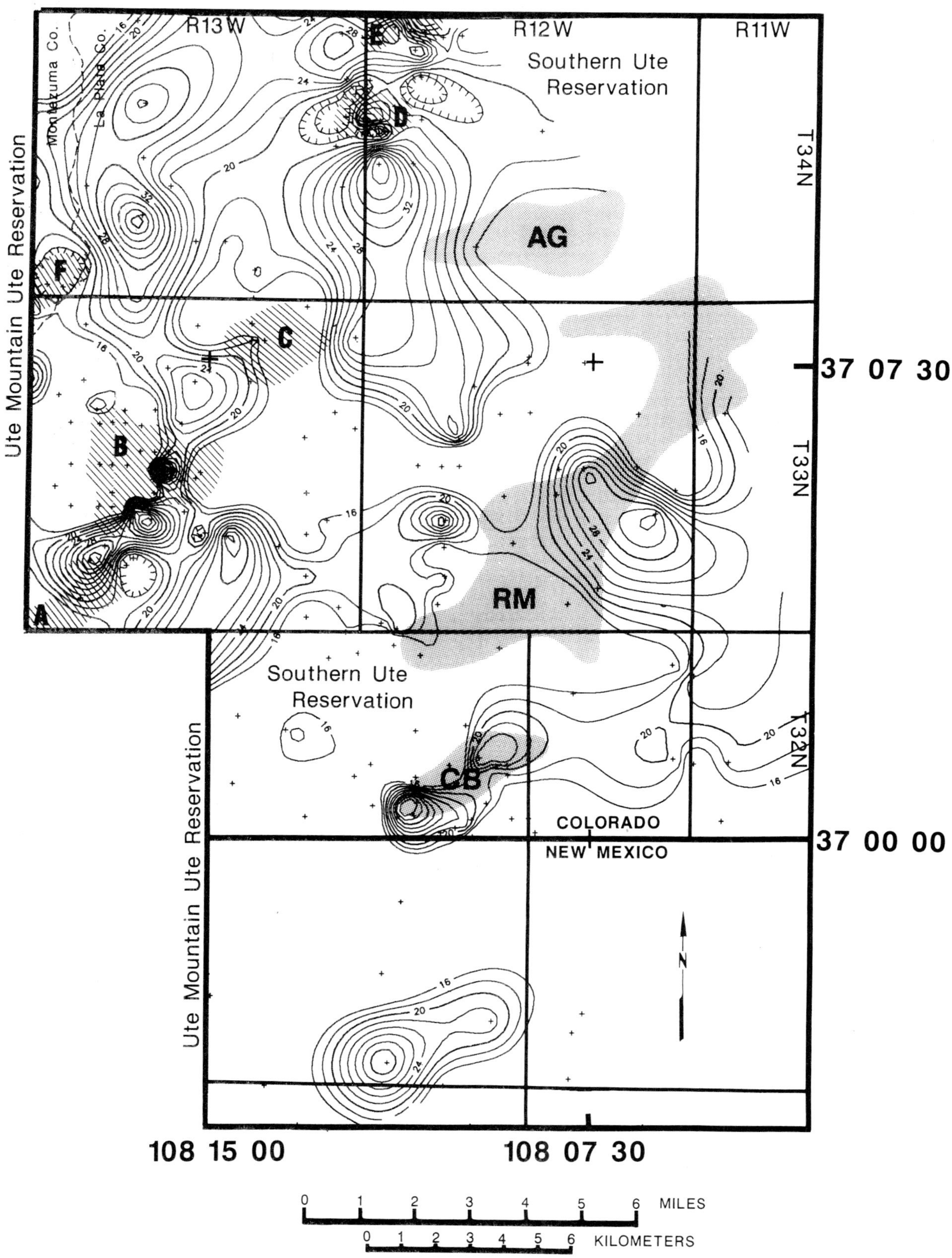

Figure 5. Contour map of carbon dioxide to oxygen ratios above the approximate atmospheric ratio of 0.0016 , Southern Ute Indian Reservation and vicinity, southwest Colorado and northwest New Mexico. All ratio values multiplied by $1x10^4$. Contour interval = 2 ratio units. Hachures indicate areas with less than 16 ratio units. Areas A-G indicate shallow magnetic anomalies. Oil and gas fields indicated by fine-line pattern: AG = Alkali Gulch, BD = Barker dome Paradox, CB = Cinder Buttes, RM = Red Mesa, and W = Wickiup (abandoned).

tional Geology Review, v. 18, no. 1, p. 98-100.

Donovan, T.J., 1974, Petroleum microseepage at Cement, Oklahoma—evidence and mechanism: American Association of Petroleum Geologists Bulletin, v. 58, p. 429-446.

Donovan, T.J., Friedman, I., and Gleason, J.D., 1974, Recognition of petroleum-bearing traps by unusual isotopic compositions of carbonate-cemented surface rocks: Geology, v. 2, p. 351-354.

Donovan, T.J., and Dalziel, M.C., 1977, Late-diagenetic indicators of buried oil and gas: U.S. Geological Survey Open-File Report 77-817, 44 p.

Donovan, T.J., Forgey, R.L., and Roberts, A.A., 1979, Aeromagnetic detection of diagenetic magnetite over oil fields: American Association of Petroleum Geologists Bulletin, v. 63, p. 245-248.

Donovan, T.J., and Roberts, A.A., 1980, Stable isotope anomalies in surface rocks and helium anomalies in soil-gas over oil fields—causes and correlations [abs.]: American Chemical Society Annual Meeting, Houston, Texas, March 23-28, 1980, Abstracts of papers (GEOC), unpaged.

Donovan, T.J., Hendricks, J.D., Roberts, A.A., and Eliason, P.T., 1984, Low-altitude aeromagnetic reconnaissance for petroleum in the Arctic National Wildlife Refuge, Alaska: Geophysics, v.49, no. 8, p. 1338-1353.

Donovan, T.J., O'Brien, D.P., Bryan, J.G., and Cunningham, K.I., 1986, Near-surface magnetic indicators of buried hydrocarbons—aeromagnetic detection and separation of spurious signals: Association of Petroleum Geochemical Explorationists, v. 2, no. 1, p. 1-20.

Dyck, W., 1976, The use of helium in mineral exploration: Journal of Geochemical Exploration, v. 5, p. 3-10.

Ellwood, B.B., Balsam, W., Burkart, B., Long, G.J., and Buhl, M.L., 1986, Anomalous magnetic properties in rocks containing the mineral siderite—paleomagnetic implications: Journal of Geophysical Research, v. 91.

Elmore, R.D., Cochran, K.A., and Nick, K.E., 1986, Authegenic magnetite and hydrocarbon migration—testing the hypothesis in the lower Ordovician Arbuckle Group, Oklahoma: EOS, v. 67, no. 16, p. 265.

Eremeev, A.N., Sokolov, V.A., Solovov, A.P., Yanitskii, I.N., 1973, Application of helium surveying to structural mapping and ore deposit forecasting: Fourth International Geochemical Exploration Symposium, London, Proceedings, p. 183-192.

Faber, E., and Stahl, W., 1984, Geochemical surface exploration for hydrocarbons in North Sea: American Association of Petroleum Geologists Bulletin, v. 68, no. 3, p. 363-386.

Henry, M., 1988, Preliminary interpretation of stable carbon and oxygen isotopes in carbonate cements, Southern Ute Reservation, southwest Colorado, *in* Fassett, J.E., ed., Geology and Coal-Bed Methane resources of the northern San Juan Basin, Colorado and New Mexico: Rocky Mountain Association of Geologists Guidebook.

Horvitz, L., 1969, Hydrocarbon geochemical prospecting after 30 years, *in* Unconventional methods in exploration for petroleum and natural gas: Southern Methodist University, Institute for the Study of Earth and Man, SMU Press, p. 205-218.

Jones, V.T, and Drozd, R.J., 1983, Predictions of oil and gas potential by near-surface geochemistry: American Association of Petroleum Geologists Bulletin, v. 67, no. 6, p. 932-952.

Kartsev, A.A., Tabarsaranskii, Z.A., Subbota, M.I., and Mogilevskii, G.A., 1959, Geochemical methods of prospecting and exploration for petroleum and natural gas: Berkeley-Los Angeles, University of California Press, 347 p.

Kelley, V.C., and Clinton, N.J., 1960, Fracture systems and tectonic elements of the Colorado Plateau: University of New Mexico Publications in Geology, University of New Mexico Press, No. 6, 104 p.

Knepper, D.H., 1982, Lineaments derived from analysis of linear features mapped from LANDSAT images of the Four Corners of the southwest U.S.: U.S. Geological Survey Open File Report 82-849, 79 p.

Knepper, D.H., 1988, Distribution of exposed limonitic rocks and soils from LANDSAT MSS data of the Southern Ute Indian Reservation, southwestern Colorado, *in* Fassett, J.E., ed., Geology and Coal-Bed Methane resources of the northern San Juan Basin Colorado and New Mexico: Rocky Mountain Association of Geologists Guidebook.

Laubmeyer, G., 1933, A new geophysical prospecting method: Zeitschrift fur Petroleum, v. 29, no. 18, p. 1-4.

Leythaeuser, D., Schaeffer, R.G., and Pooch, H., 1983, Diffusion of light hydrocarbons in subsurface sedimentary rocks: American Association of Petroleum Geologists Bulletin, v. 67, no. 6, p. 889-895.

Link, W.K., 1952, Significance of oil and gas seeps in world oil exploration: American Association of Petroleum Geologists Bulletin, v. 36, p. 1505-1541.

Lovell, J.S., Hale, M., and Webb, J.S., 1980, Soil air disequilibria as a guide to concealed mineralization at Keel, Erie, Prospecting in areas of glaciated terrain 1979: Transactions of the Institute of Mining and Metallurgy Section B, p. 45-50.

Lovell, J.S., and Hale, M., 1983, Application of soil-air carbon dioxide and oxygen measurements to mineral exploration: Transactions of the Institute of Mining and Metallurgy, Section B, v. 92, p. 28-32.

May, E.R., and Schmidt, P.G., 1982, The discovery, geology, and mineralogy of the Crandon Precambrian massive sulfide deposit, Wisconsin in Precambrian massive sulfide deposits, H.S. Robinson Memorial volume: Special Papers of the Geological Association of Canada, v. 25, p. 446-480.

McCabe, C., Van der Voo, R., Placor, D.R., Scotese, C.R., and Freeman, R., 1983, Diagenetic magnetite carries ancient yet secondary remanence in some Paleozoic sedimentary carbonates: Geology, v.11, p. 221-223.

McCabe, C., 1986, Do hydrocarbons cause secondary magnetizations?: EOS, v. 67, no. 16, p. 265.

McCarthy, J.H., and Reimer, G.M., 1986, Advances in soil gas geochemical exploration for natural resources—some current examples and practices: Journal of Geophysical Research, v. 91, no. B12, p. 12327-12338.

McCarthy, J.H., Cunningham, K.I., Roberts, A.A., and Dietrich, J.A., 1986, Soil-gas studies around hydrogen-rich natural gas wells in northern Kansas: U.S. Geological Survey Open-File Report 86-461, 21 p., 4 pls.

McCrossan, R.G., Ball, N.L., and Snowdon, L.R., 1971, An evaluation of surface geochemical prospecting for petroleum, Olds-Caroline area, Alberta: Geological Survey of Canada Professional Paper 71-31, p. 4-9.

Moore, B.J., 1976, Analyses of natural gases: U.S. Bureau of

Mines computer printout, 76 p.
Moore, B.J., 1982, Analyses of natural gases, 1917-80: U.S. Bureau of Mines Information Circular 8870, 1055 p.
Ovchinnikov, L.N., Sokolov, V.A., Friedman, A.I., and Yanitskii, I.N., 1972, Gaseous geochemical methods in structural mapping and prospecting for ore deposits: Fourth International Geochemical Exploration Symposium, London, Proceedings, p. 177-182.
Palacas, J.G., and Roberts, A.A., 1980, Helium anomaly in surficial deposits of south Florida—possible indicator of deep subsurface petroleum or shallow uranium-associated phosphate deposits: U.S. Geological Survey Open-File Report 80-91, 14 p.
Pogorski, L.A., and Quirt, G.S., 1981, Helium emanometry in exploring for hydrocarbons, part I, *in* Gottlieb, B.M., ed., Unconventional methods in exploration for petroleum and natural gas: Southern Methodist University Press, p. 124-135.
Plyusnin G.S., Volkova, N.V., Godumskii, G.P., Smirnov, V.N., Papolitov, E.I., and Brandt, S.B., 1972, Trends in the distribution of argon and helium in contact regions and fault zones: Geokhimiya, no. 10, p. 1189-1196.
Reimer, G.M., 1976, Design and assembly of a portable helium detector for evaluation as a uranium exploration instrument: U. S. Geological Survey Open-File Report 76-398, 18 p.
Reimer, G.M., and Roberts, A.A., 1985, Correcting soil-gas helium concentrations for diurnal variations: Association of Petroleum Geochemical Explorationists Bulletin, v.1, no. 1, p. 1-17.
Reitsema, R.H., Kaltenback, A.J., and Lindberg, F.A., 1981, Source and migration of light hydrocarbons indicated by carbon isotopic ratios: American Association of Petroleum Geologists Bulletin, v. 65, no. 9, p. 1536-1542.
Reynolds, R.L., Fishman, N.S., Hudson, M.R., Karachewski, J.A., and Goldhaber, M.B., 1986, Magnetic minerals and hydrocarbon seepage—possibilities for magnetic detection of oil fields [abs.]: U.S. Geological Survey Circular 974, p. 58-59.
Roberts, A.A., 1981, Helium emanometry in exploring for hydrocarbons, part II, *in* Gottlieb, B.M., ed., Unconventional methods in exploration for petroleum and natural gas: Southern Methodist University Press, p. 136-149.
Roberts, A.A., and Roen, J.B., 1985, Near-surface helium anomalies associated with faults and gas accumulations in western Pennsylvania: U.S. Geological Survey Open-File Report 85-546, 23 p.
Sokolov, V.A., Geodekyan, A.A., Grigoryev, G.G., Krems, A. Ya., Stroganov, V.A., Zorkin L.M., Zeidelson, M.I., and Yesenev, B.P., 1971, The new methods of gas surveys, gas investigations of wells, and some practical results: CIM Special Volume 11, p. 538-544.
Stahl, W., Faber, E., Carey, B.D., and Kirksey, D.L., 1981, Nearsurface evidence of migration of natural gas from deep reservoir and source rocks: American Association of Petroleum Geologists Bulletin, v. 65, no. 9, p. 1543-1550.
Szabo E., and Wengerd, S.A., 1975, Stratigraphy and tectogenesis of the Paradox basin, *in* Fassett, J.E., ed., Canyonlands Country: Four Corners Geological Society Guidebook, p. 193-210.
Taylor, D., 1988, Preliminary interpretation of regional seismic data, Southern Ute Reservation, southwest Colorado: U.S. Geological Survey Administrative Report (in press).
Webring, M., 1988, Preliminary interpretation of regional magnetic and gravity data, Southern Ute Reservation, southwest Colorado, *in* Fassett, J.E., ed., Geology and Coal Bed Methane resources of the northern San Juan Basin, Colorado and New Mexico: Rocky Mountain Association of Geologists Guidebook.

Photo of Electra Lake, taken in 1917, located between Durango, Colorado and Purgatory Ski Area. The lake, actually a reservoir, received its name from the very steeply cascading outlet which was harnessed to generate electricity. The electrical generating station is located on the edge of the Animas River at Tacoma. The effective head is 790 feet with 5200 cubic feet of available water per minute for a generating capacity of approximately 50,000 volts. The generating station can be viewed from the Durango to Silverton train as it passes by. Photo courtesy of the Center for Southwest Studies.

Utes at the Denver National Mining Exposition of 1882. Photo courtesy of the Colorado Historical Society.

First Day Road Log from Durango, Colorado around Northwest Rim of San Juan Basin Via Cedar Hill, Aztec, and La Plata, New Mexico and Soda Springs, Marvel, and Breen, Colorado

James E. Fassett[1]

TUESDAY, JUNE 7, 1988

ASSEMBLY POINT: Red Lion Inn parking lot, Durango, Colorado

DEPARTURE TIME: 8:00 a.m.

DISTANCE: 93.5 miles

NUMBER OF STOPS: 5

INTRODUCTION

Today's trip will begin in the Upper Cretaceous Mancos Shale and will go up section through the Mesaverde Group, Lewis Shale, Pictured Cliffs Sandstone, Fruitland Formation, Kirtland Shale, and McDermott Member of the Animas Formation. At the top of the McDermott we cross a major unconformity marking the Cretaceous-Tertiary boundary into the Paleocene upper part of the Animas Formation. As we go south across the central San Juan Basin we will move laterally from the Animas into the time-equivalent Nacimiento Formation and up section into the basal part of the Eocene San Jose Formation near the Colorado-New Mexico State line. Traveling further south we will drop down section into the Nacimiento and as we go northwest from Aztec to La Plata, New Mexico and north into Colorado we will generally drop down section again until we finally come back down into the Mancos Shale as we complete our circle trip into Durango from the west.

Rob Zech (USGS), Brad Boyce (geologist, Southern Ute Indian Tribe), Clarence Harr (consulting geologist), and Mary Gillam (consulting geologist) assisted in the preparation of the road log. Fred Samson provided information about the coal on the La Plata mine coal lease.

We will be driving across the Southern Ute Indian Reservation on parts of today's field trip. For those of you using this road log on your own, you should obtain permission from the Tribe at their Tribal headquarters in Ignacio, Colorado before doing geologic work on the Reservation, especially in those areas off of Federal, State, or County highways.

0.0 **TURN RIGHT ONTO U.S. 550** on leaving the Red Lion Inn parking lot.

0.2

0.2 Traffic light at intersection of U.S. 550 and 160; go south on U.S. 550.

0.1

0.3 Uranium mill tailings pile on right. Beginning in the 1880s, the mill processed metallic ores from base-metal mines in the Durango area. The mill closed in the mid 1930s but was reopened during World War II under contract to the Manhattan Project. Uranium-vanadium ore processed here was mostly from the Colorado Plateau (Moore and Gillam, 1984). The toe of the pile appears to rest on the late Holocene floodplain. After several years of study, the department of Energy is now having the tailings trucked about 3 miles to a disposal site in Bodo Canyon, on the opposite side of Smelter Mountain. (We will pass by Bodo Canyon at milepoint 85.7 of today's road log.)

0.5

0.8 On right across Animas River we can see exposures of Upper Cretaceous Mancos Shale, Point Lookout Sandstone, and Menefee Formation. Note coal dump from abandoned mine in the lower part of the Menefee. Note the fault in the gully which is down to the left (south) with about 50 feet displacement.

0.1

0.9 **TURN LEFT on access road to old U.S. 550 (Colorado Rt. 3).**

0.2

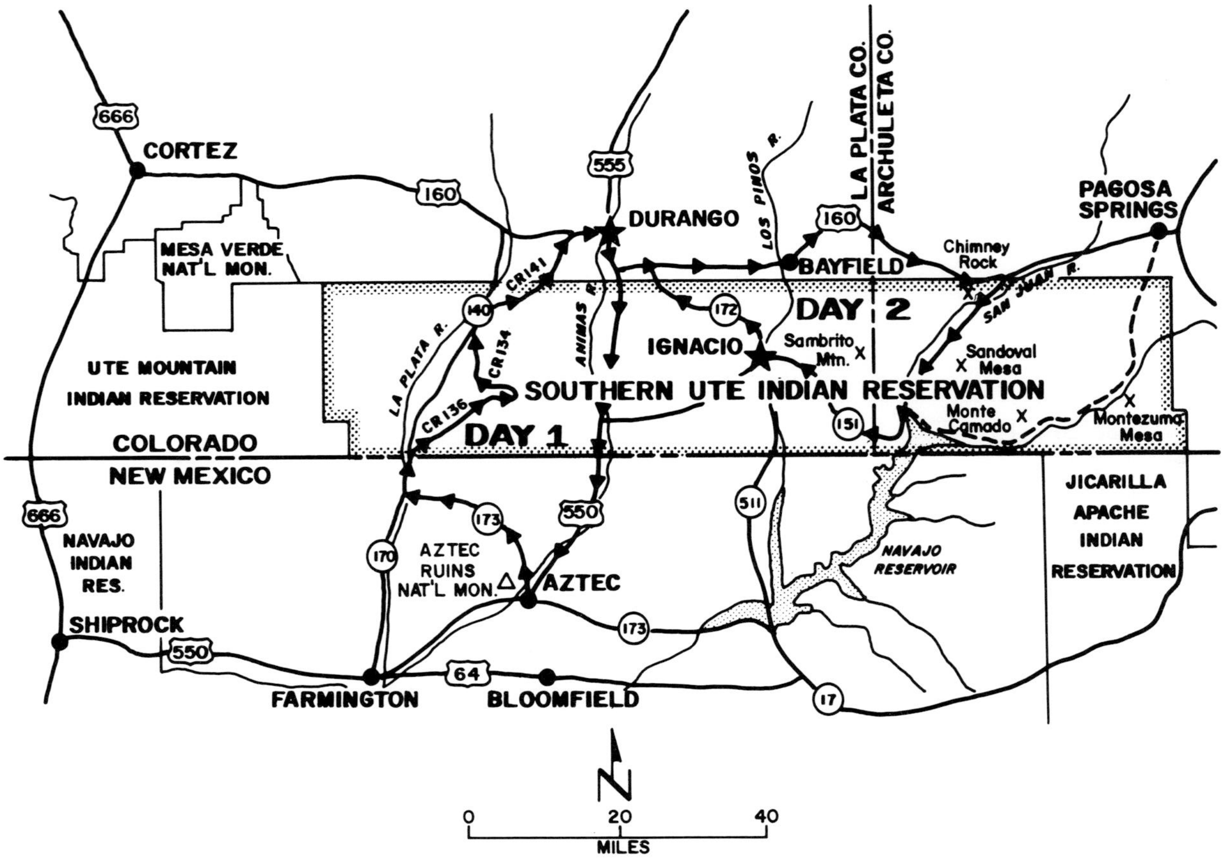

Road map showing the field-trip routes around the northern part of the San Juan Basin in southwestern Colorado and northwestern New Mexico.

1.1 **TURN RIGHT on Colo. Rt. 3.**
0.1

1.2 Point Lookout Sandstone in roadcut on left.
0.6

1.8 Abandoned coal-mine dump in Menefee Formation on right above steeply south-dipping Point Lookout Sandstone across the Animas River.
0.4

2.2 We are now opposite the contact between the Menefee Formation and the overlying Cliff House Sandstone. The Cliff House is not well exposed here.
0.2

2.4 Lewis Shale is poorly exposed in valley to right between Cliff House Sandstone and Pictured Cliffs Sandstone.
0.3

2.7 Animas River gravels on Lewis Shale in roadcut on left.
0.1

2.8 In roadcut on left we pass into the transition zone between the offshore-marine Lewis Shale and the shoreface-marine Pictured Cliffs Sandstone.
0.2

3.0 **STOP 1: Pull off on right and disembark.** We will walk across the road (watch out for cars) and then walk southward (stratigraphically up section) through the lower, middle, and upper shoreface facies of the Pictured Cliffs Sandstone. This is a spectacular section through a typical regressive marine sandstone sequence.
0.25

3.25 Top of Pictured Cliffs Sandstone—base of Fruitland Formation. The view to the west across the Animas River includes, from right to left (north to south) the Upper Cretaceous Mancos Shale, Point Lookout Sandstone, Menefee Formation, Cliff House Sandstone (difficult to see), Lewis Shale, Pictured Cliffs Sandstone, Fruitland Formation, Kirtland Shale (lower shale member and the Farmington Sandstone Member are visible), and the McDermott Member of the Animas Formation; overlying the McDermott is the Tertiary upper member of the Animas Formation. The unconformable contact between the McDermott and the upper Animas marks the Cretaceous-Tertiary boundary in this part of the San Juan Basin. Figure 3.25 is a photograph of this section.

We will examine the thick basal-Fruitland coal zone exposed here and discuss the relationship between the fracture patterns in the Pictured Cliffs and the cleat in the Fruitland coal beds. It is thought that the fractures in the Fruitland coal beds (cleat) may largely control the production trends of coal-bed methane in Fruitland coals in the subsurface of the San Juan Basin. The density of the cleats in these coals is also thought to be a significant factor controlling the rate at which gas is desorbed from the coal beds and moves to the well bore in a Fruitland coal-bed methane well.

Following discussion at this stop return to busses and continue field trip.
0.25

Mile 1.2—Aerial view looking northeast across the Animas River valley (visible in left foreground) south of Durango, Colorado. We are looking at the northwest flank of the Hogback monocline which marks the north rim of the San Juan Basin. The Upper Cretaceous rocks visible here comprise, from left to right, the Dakota Sandstone (Kd), Mancos Shale (Km), Mesaverde Group (Kmv), Lewis Shale (Kl), Pictured Cliffs Sandstone (Kpc), Fruitland Formation (Kf), Kirtland Shale (Kk) including the Farmington Sandstone Member (Kkf), and the McDermott Member of the Animas Formation (Kam). The snow-capped San Juan Mountains are visible on the skyline. Photo courtesy of Bruce Conrad.

3.5 On right across Animas River, Upper Cretaceous Farmington Sandstone Member of Kirtland Shale, overlain by upper shale member of Kirtland Shale, overlain by McDermott Member of the Animas Formation (purple beds), overlain, unconformably by Tertiary upper part of Animas Formation. McDermott was originally named McDermott Formation by Reeside (1934). The McDermott was redesignated the lower member of the Animas Formation by Barnes and others (1954). This redesignation was unfortunate because it placed the Cretaceous-Tertiary boundary, represented here by an unconformity of around 6 million years, within the Animas (Fassett 1987, Fassett and others 1987, Newman 1987).

0.8

4.3 Base of upper member of Animas Formation visible on right.

0.4

4.7 Animas River terrace gravels and exposures of upper member of Animas Formation in roadcuts on left. Note volcaniclastic nature of Animas rocks.

0.3

5.0 **TURN RIGHT (south) on U.S. 550.** Proceed uphill through Animas Formation and Animas River gravels in roadcuts on left.

0.4

5.4 Note the La Plata Mountains on the skyline to right-rear.

0.2

5.6 Terrace gravels visible in roadcuts on left cap the lower bench of Florida Mesa. Nearly the entire route from the Red Lion Inn to mile 37.2 (turnoff from U.S. 550 onto N.M. 173 in Aztec) lies on glacial outwash terraces of the Animas River.

Mile 2.8-3.25a—Good exposure of the Pictured Cliffs Sandstone forming the white ledges which here dip to the southeast. View is to the northeast looking across the south-flowing Animas River; Colorado Route 3 (old US 550) and location of STOP 1 of the Day 1 field trip is barely visible above tops of cottonwood trees in river bottom. The lower part of the Fruitland Formation, which directly overlies the Pictured Cliffs, is less resistant to erosion and has been removed here leaving a strike-valley in which the dip slope of the top of the Pictured Cliffs is well exposed. (See the detailed measured section at Mile 2.8-3.25b for a detailed description of this section.) Photo courtesy of Bruce Conrad.

0.1

5.7 Entering Southern Ute Indian Reservation. Road to left is County Road 220. We are now on top of Florida Mesa.

1.3

7.0 On the right, note the prominent purple-colored McDermott rocks cropping out in the distance; to the north, the Pictured Cliffs Sandstone outcrop forms the hogback marking the northern structural rim of the San Juan Basin.

1.5

8.5 On the right, the Pictured Cliffs Sandstone hogback curves around to the southwest on the horizon.

0.7

9.2 Well on left with pump jack is the Amoco Arthur Mason No. 1; this well was completed in the Fruitland at 2,800 feet in July 1986. This well produces coal-bed methane at a rate of 50-100 MCFPD and 100 BWPD from coal beds in the Fruitland Formation. The sight of a pump jack on gas wells is unusual in most areas that produce natural gas from conventional sandstone reservoirs. Gas normally produces at high pressures from those reservoirs directly into a pipeline (or in some instances through a separator into the pipeline). For Fruitland coal-bed methane wells however, production of the gas requires the pumping of some amount of water out of the coal beds in order to lower the pressure in the coal bed. Once the pressure is lowered sufficiently, the coal gas begins to desorb from the coal and flow to the wellbore through the coal cleats.

Another unusual feature of coal-bed methane wells is that the rate of gas production usually increases through

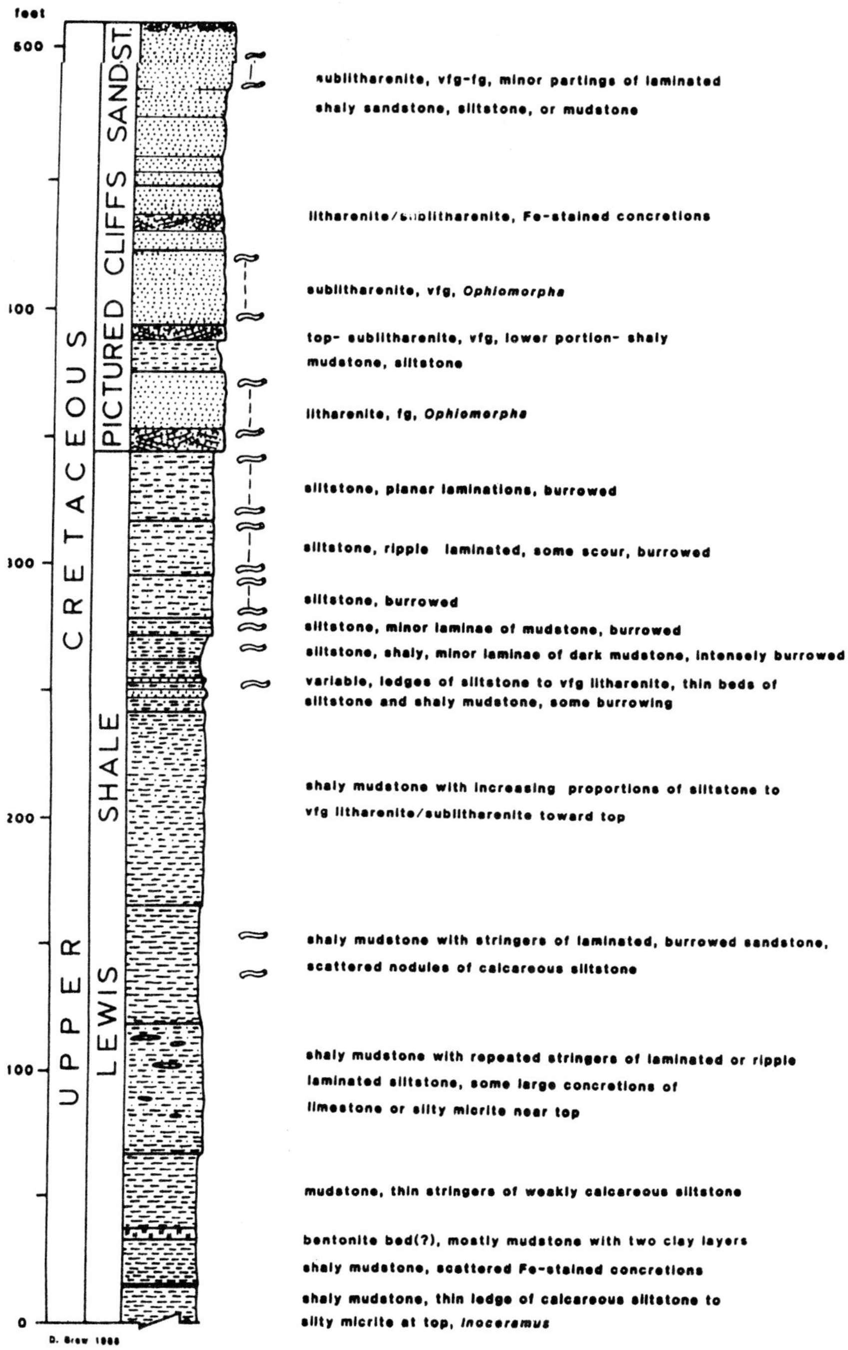

Mile 2.8-3.25b—Measured section of the Upper Cretaceous Pictured Cliffs Sandstone and uppermost part of the Lewis Shale at STOP 1 of the Day 1 field trip. This section was measured by Doug Brew, Fort Lewis College, along Colorado Route 3 (old US 550) south of Durango, Colorado.

time as more and more water is pumped from the coal, and so rather than having a decline curve, as is the case with a typical sandstone-reservoir gas well, coal-bed methane wells typically have an incline curve. Thus, the trademark for this field trip and symposium (note the cover on this guidebook) is a pump jack and gas production graph showing the rate of gas production increasing through time.

1.0

10.2 On left is gravel pit in upper terrace bench of Florida Mesa. Overlying sandy tributary-stream alluvium contains reworked Lava Creek B ash from the Yellowstone caldera, dated at roughly 610 ka by potassium-argon and fission- track methods (Izett, 1981). Rare outcrops of the ash in the Animas and La Plata valleys provide an important datum for terrace correlation, age estimation, and incision-rate calculations. Inferred Animas River incision rates for the last 630 ka are about 1.0 ft/1,000 years near Durango and 0.5 ft/1,000 years near Aztec and Farmington, New Mexico.

1.0

11.2 On right are exposures of variegated shales of the

Mile 16.6—Faulted channel sandstone in the Nacimiento Formation in roadcut on east side of US 550 just below and south of the top of Bondad Hill. Massive sandstone bed is down on the right about 18 feet. Note the smaller fault, down to the left a few feet, which angles to the right into the aforementioned, nearly vertical fault plane. Faults such as these may have created enhanced coal-cleat porosity and permeability at depth in the Fruitland Formation coal beds. Photo courtesy of Bruce Conrad.

Nacimiento Formation. The Nacimiento is laterally equivalent to the Animas Formation to the north but contains no macroscopic volcaniclastic material. See Fassett (1985) for a discussion of the history of deposition of these rocks.

1.6

12.8 On left, Northwest Pipeline Corp. Bondad 33-9 No. 6 well produces gas from the Mesaverde Group at a depth of about 5,200 feet. This well has produced 2.2 BCF and it is estimated to have an additional 2.2 BCF of gas remaining. This well site is a dual location and also contains a Fruitland coal-bed methane well, the Amoco Clovis No. 1 completed in 1987; top of Pictured Cliffs Sandstone at 2,856 feet; initial production 200 MCFPD, 400 BWPD. Eureka Grange hall on right.

1.1

13.9 On left is the basal contact of the Eocene San Jose Formation; the contact is marked by a transition upward from the predominantly shaly beds of the Nacimiento Formation into the overlying predominantly sandy beds of the San Jose. This contact appears to be gradational here in the structurally flatter part of the basin, whereas it is an unconformity around the rim of the basin (Fassett 1985, Fassett and Hinds, 1971).

2.2

16.1 Top of Bondad Hill.

0.4

16.5 On left, road cut in upper part of the Nacimiento Formation; note channel sandstone. Florida Mesa terrace gravels rest on the eroded Nacimiento Formation.

0.1

16.6 On left, fault in roadcut. Displacement of channel sandstone bed is down to the south (right) about 18 feet.

0.1

16.7 On left fluvial channel sandstone in the Nacimiento Formation in roadcut. Terraces of the last glaciation are

Mile 18.4—Fault in the Nacimiento Formation on the east side of US 550 about one-half mile south of the Animas River. The fault is down about 15 feet, to the north, and offsets fluvial-channel sandstones. Photo courtesy of Bruce Conrad.

well exposed at bottom of hill (with original braided bar-and-swale topography) and across river (partly covered by alluvial fans). The last glaciation is not well dated in the San Juan Mountains, but occurred roughly 35 ka to 16 ka.

0.4

17.1 On left La Plata County Road 310 leads to Ignacio, Colorado, about 12 miles to the east. Ignacio is the headquarters of the Southern Ute Indian Reservation. We will pass through Ignacio tomorrow on Day 2 of this field trip.

0.8

17.9 Cross Animas River.

0.3

18.2 Carbon River Energy Bonds No. 1-1 well on right. This well is completed in Fruitland Formation coal beds at a depth of less than 2,900 feet. This well had an initial potential of 400 MCFPD and it is still being tested.

0.2

18.4 On left is a fault in the Nacimiento Formation which has a displacement of about 15 feet, up to the south (right). To the right of the highway a gulley has eroded along the fault plane; note that none of the several small faults in this area displace Pleistocene terraces. The lowermost San Jose Formation crops out on right.

0.2

18.6 To left is a fault or sandstone channel edges in the Nacimiento Formation.

0.7

19.3 Base of the San Jose Formation in roadcut on left(?). Note constriction in valley width and light pinkish-brown sands on right along next curve (see below).

0.5

19.8 Twin Crossings restaurant on right. Slightly lower terrace to left exhibits hundreds of boulders 3-20 feet in maximum diameter, some of which are visible at edge of trees and (farther south) beneath a large barn. Boulders appear to form a flood bar 100-300 feet wide and at least 0.7 miles long that aligns with valley constriction upstream. The boulders' long axes are roughly

Mile 25.4a—Aerial view looking east at Amoco's Cedar Hill field which produces Fruitland Formation coal-bed gas from depths of around 2,000 feet. The highway visible at the top of the photograph is US 550. The Schneider B No. 1S well is in the left-central part of the photograph and the Cahn No. 1 well is in the upper right, just west of US 550. The Animas River is visible at the top-center, east of highway US 550. Photo courtesy of Bruce Conrad.

transverse to flow and some appear imbricated; also, boulders tend to become smaller and more rounded downstream. Pinkish sands north of restaurant may represent eddy deposits of the same flood. Preliminary calculations suggest a flow depth of 30-60 feet and a discharge of hundreds of thousands of cubic feet per second, compared to about 25,000 cubic feet per second for the largest recorded flood on the Animas River in 1911.

0.7

20.5 Dick and Jane's Place on left.

1.1

21.6 State line; leave "Colorful Colorado" and enter "Land of Enchantment". Outcrops of San Jose Formation visible to right and left.

1.4

23.0 New Mexico rest stop and weigh station on right.

0.6

23.6 Contact between San Jose Formation and underlying Nacimiento Formation on left. Slight constriction in valley floor ahead was once considered for a damsite by the Bureau of Reclamation. Top of Mt. Nebo, highest point ahead on left, is capped by three terraces of the Animas River. Their height of 1,000 to 1,070 feet above the present channel suggests an age of about 1.7 Ma, based on estimated incision rates.

0.3

23.9 Contact between San Jose Formation and underlying Nacimiento Formation on right.

0.6

24.5 Entering Amoco Cedar Hill field. This field produces coal-bed methane from Fruitland Formation coal beds at depths averaging around 2,000 feet.

0.9

25.4 **TURN RIGHT on San Juan County Road 2300.** Proceed 0.3 mi. to STOP 2 at Schneider Gas Com. "B" No. 1S

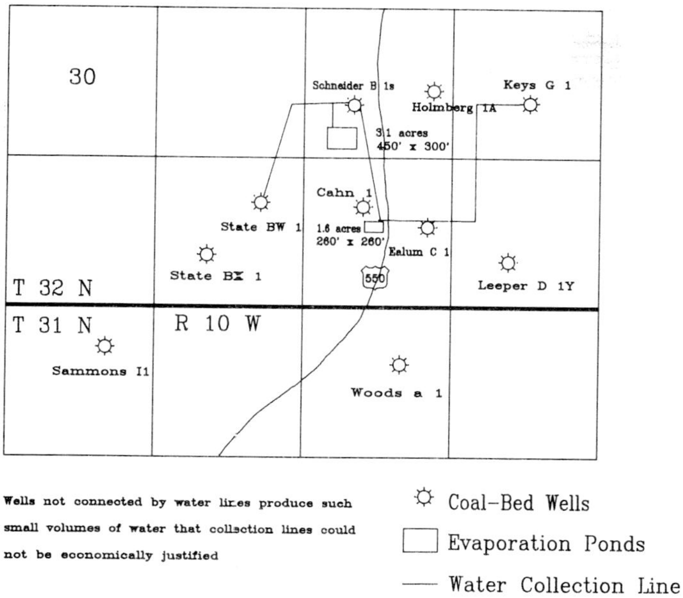

Mile 25.4b—Map of Cedar Hill field showing well locations and highway US 550. Note the locations of the Schneider B No. 1S and the Cahn No. 1 wells to help orient the photograph of part of this field (Mile 25.4a).

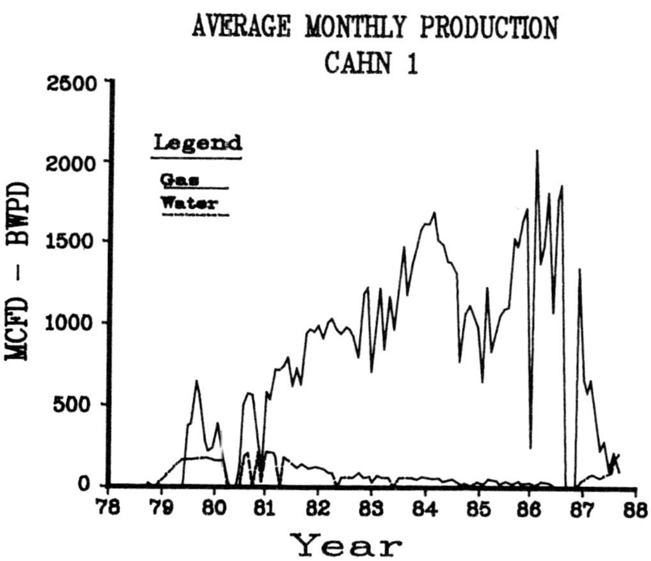

Mile 25.4c—Production graph for the Cahn No. 1 well in the Cedar Hill field.

Well. The discovery well for this field, the Amoco Cahn No. 1 was completed May 1979. Production for this well peaked in 1986 at nearly 2 million CFGPD. See papers by Hale and Decker (in this volume) for more information on this field. Disembark for a discussion of the production history of the Cedar Hill field. After discussion return to U.S. 550, turn right (south), and resume field trip.

0.6

26.0 Enter Cedar Hill and leave Cedar Hill field. Cedar Hill is the home town of the present Governor of New Mexico, Gary Carruthers. Carruthers also served as Assistant-Secretary of Interior under Secretary James Watt.

0.6

26.6 Cross Animas River. Between here and Aztec, oldest preserved terraces of the Animas River are occasionally visible, capping the highest mesas along the drainage divide 4 to 6 miles east of U.S. 550. The height of these terraces; 1,200 to 1,400 feet above the present channel of the Animas River suggests an age of about 2.0 Ma (late Pliocene).

1.0

27.6 Nacimiento Formation on right contains variegated shales.

1.3

28.9 Contact of San Jose Formation on the Nacimiento Formation on left; the source area for this part of the San Jose was to the north, thus, it generally becomes less sandy southward across the basin (Fassett, 1985). The San Jose has been subdivided into five complexly interbedded members in the southern part of the basin (Baltz, 1967). These members have not been carried into this part of the basin.

1.7

30.6 Animas River terraces are well developed to right and left.

1.3

31.9 Nacimiento Formation outcrops on right.

3.1

35.0 Enter Aztec, New Mexico. Reduce speed and carefully observe speed limits.

1.3

36.3 Junction of N.M 44 on left (to Cuba and Albuquerque, New Mexico); continue ahead on U.S. 550.

0.5

36.8 Cross Animas River.

0.1

36.9 **TURN RIGHT at traffic light.** Proceed .7 miles to Aztec Ruins National Monument parking lot (New Mexico Energy and Minerals office, .1 mi. on left). STOP 3 for lunch and tour of the ruins. Return to U.S. 550, turn right, and resume field trip.

0.3

37.2 **TURN RIGHT at traffic light on N.M. 173.** We will proceed on this road to La Plata, New Mexico.

0.8

38.0 Estes Arroyo on right. Animas River terraces to right and left; outcrops are Nacimiento Formation.

0.5

38.5 Numerous gas wells along this road produce from the Dakota Sandstone, Mesaverde Group, and Pictured Cliffs Sandstone. We are passing through the heart of the San Juan gas field which has produced (through 1983) more than 14 trillion cubic feet of gas making it the second largest gas field in the conterminous United States. Minor amounts of gas have also been produced from the Farmington Sandstone Member of the Kirtland Shale south and east of here; these channel sandstones have been found to be overpressured in places and have occasionally blown out and burned down drilling rigs. See Fassett (1978, 1983) for detailed information on all producing oil and gas fields in the San Juan Basin and see Dugan and Williams (this volume) for an entertaining account of the history of gas production in the

Mile 36.9—Aerial view of Aztec Ruins National Monument; lunch stop (STOP 3) of Day 1 field trip. Note Animas River in the background. Photo courtesy of Bruce Conrad.

basin.

2.1

40.6 Nacimiento Formation contains purple beds in this area which may represent finer-grained equivalents of Animas Formation volcaniclastic material present north of here. The same southward-flowing streams that deposited the coarser-grained sediment of the Animas to the north were depositing the finer-grained sediment that now constitutes the Nacimiento in this area.

3.3

43.9 Barker Dome on skyline ahead. The dip-slope of this structural feature is capped by the Cliff House Sandstone, the uppermost formation of the Mesaverde Group. Barker Dome is outside the structural central San Juan Basin on the Four Corners Platform; it is on the Ute Mountain Ute Indian Reservation and produces gas from the Paradox Member of Middle Pennsylvanian age at depths from 7,500 to nearly 9,000 feet and from the Upper Cretaceous Dakota Sandstone at depths of 2,000 to 2,500 feet.

1.0

44.9 Cross Farmington Glade Arroyo. The Farmington Glade is a strikingly linear, northeast-trending feature. The origin of this drainage and especially its linearity continues to be a puzzle to friends of the recent. The rocks exposed here are Nacimiento Formation.

0.5

45.4 Cattleguard; panoramic view ahead across the La Plata River valley consists of the La Plata Mountains to the right (north) and Barker Dome ahead.

0.2

45.6 Dip-slope of the Hogback Monocline formed by Pictured Cliffs Sandstone visible ahead to right.

1.4

47.0 From left to right ahead: Pinyon Mesa, an outlier capped by the Paleocene Ojo Alamo Sandstone; Shiprock, a volcanic plug of late Oligocene age; Barker Dome; and to the northeast, spoil piles of the La Plata coal mine are visible. On the skyline the lacolithic Lukachukai Mountains are visible to the left (south) and

the Carrizo Mountains to the right (north).
0.9

47.9 La Plata river terrace ahead on left.
1.9

49.8 Cross McDermott Arroyo. The type area of the McDermott Formation of Reeside (1924) (now called the McDermott Member of the Animas Formation, see discussion at mile 3.5) is located just across the Colorado State line, just a few miles north of here.
1.2

51.0 Cross the La Plata River.
0.3

51.3 **TURN RIGHT at intersection of N.M. 173 and N.M. 170 (the La Plata Highway).**
0.9

52.2 La Plata River terraces are well developed in this area. Many of these terraces contain prehistoric Anasazi Indian ruins. The La Plata valley was heavily populated by the Anasazi until around 1300 at which time they mysteriously vanished (Hewett 1977).
1.5

53.7 **TURN RIGHT on the La Plata Mine road.** The bridge over N.M. 170 is the La Plata coal mine haulage road. This road goes from the La Plata mine to the San Juan Generating Station, 22 miles to the southwest. Proceed 1.8 mi. to the La Plata Mine for STOP 4. At this stop, we will tour the La Plata coal mine and examine Fruitland Formation coal beds in detail. Discussions will include the geology of the coal, coal fracturing, coal quality, coal mining methods, and coal reserves. The road between N.M. 170 and the La Plata mine is on channel sandstones of the Farmington Sandstone Member of the Kirtland Shale. Following the mine tour, return to N.M. 170, turn right (north), and resume field trip.

Four coal seams are mined at the La Plata mine. They are numbered from the bottom up, as follows; Seam 1 is 32 feet thick and is located close to the top of the underlying Pictured Cliffs Sandstone; Seam 2 is 50 feet above seam 1, is 13 feet thick, and has the highest ash content of the four seams; Seam 3 is 50 feet above seam 2 and is 6 feet thick; and Seam 4 is 15 feet above seam 3 and is 5 feet thick. The dip of the beds in the mine area is 10 degrees to the southeast. The coal lease covers 2,310 acres and contains 83 million tons of recoverable coal. Four generally north-trending faults offset the coal beds on the eastern side of the lease. The heating value of the coal is 8,600 to 10,700 Btu with the average being 9,600 Btu. Expected mine life is to the year 2017. The total Fruitland coal thickness of 57 feet is somewhat greater than published outcrop and subsurface measurements (Fassett and Hinds, 1971) have shown in this area.
0.4

54.1 Cross the La Plata River. Farmington Sandstone Member of Kirtland Shale crops out on right. The Farmington Sandstone and upper shale member of the Kirtland Shale reach their maximum thickness of more than 1,700 feet thick in this part of the basin, the lower shale member of the Kirtland is 250 feet thick, and the underlying Fruitland Formation is about 400 feet thick.

On the east rim of the basin, the Kirtland, Fruitland, and underlying Pictured Cliffs Sandstone are absent. There, the Tertiary Ojo Alamo Sandstone rests on the Lewis Shale. An erosional episode, beginning in latest Cretaceous and possibly continuing into earliest Tertiary time occurred as the result of uplift and tilting of the basin area, up on the east, resulted in beveling of the Kirtland-Fruitland rocks across the basin. The hiatus between Cretaceous and Tertiary rocks in the San Juan Basin may represent from 6 to 7 million years. (See Fassett and Hinds (1971) for a detailed discussion of these rocks.)

Here, and to the north in Colorado, the volcaniclastic McDermott Member of the Animas Formation overlies the Kirtland; the McDermott is the youngest Cretaceous rock unit in the San Juan Basin.
0.7

54.8 Base of Farmington Sandstone crops out on right. Contact of Farmington with underlying lower shale member of Kirtland Shale is gradational.
0.1

54.9 View of lower Fruitland Formation coal in small rincon ahead to right. Red clinker on hills to right is common where Fruitland crops out around the basin. The clinker formed when outcropping Fruitland coal beds spontaneously ignited and burned and oxidized the iron-rich shales overlying the coals. The clinker is mined in places for road metal and for landscaping material.
0.2

55.1 Lower part of Fruitland on right is quite sandy in this area. Note the Fruitland coal-mine dump on right; coal bed was close to the top of the Pictured Cliffs Sandstone.
0.1

55.2 Top of the Pictured Cliffs Sandstone on right. The Pictured Cliffs is unusually thick and massive at this locality.
0.4

55.6 Contact of base of Pictured Cliffs and underlying Lewis Shale.
0.6

56.2 La Plata River terraces on left and right.
0.2

56.4 Cross New Mexico-Colorado State line. We will not return to New Mexico again on this field trip; hasta la vista. The La Plata Highway changes here from N.M. 170 to Colo. 140. We also enter the Southern Ute Indian Reservation when we cross the State line.
1.5

57.9 **TURN RIGHT on La Plata County Road 136.** We will be on the Southern Ute Indian Reservation for the next several miles. **TO THOSE OF YOU USING THIS ROAD LOG ON YOUR OWN: YOU MUST OBTAIN CROSSING PERMITS FROM THE SOUTHERN UTE INDIAN TRIBE, NATURAL RESOURCES DIVISION, BOX 737, IGNACIO, COLORADO 81137 (OR CALL (303) 563-4525) PRIOR TO DOING GEOLOGIC INVESTIGATIONS ON TRIBAL PROPERTIES.** Rocks along this road are the Lewis Shale.
1.7

59.6 Cinder Buttes on right. These hills are outliers of Fruit-

Mile 53.7a—Aerial view of the La Plata coal mine looking west near the outcrop of the Fruitland Formation; STOP 4 of Day 1 field trip. The Fruitland can be seen to be dipping to the left (south), into the basin. Two coal benches have been uncovered here; an upper bench on the left and a lower bench in the center foreground. Photo courtesy of Bruce Conrad.

Mile 53.7b—Aerial view of two "coal trains" carrying coal from the La Plata coal mine to the New Mexico Public Service Company San Juan Generating Station, 22 miles to the southwest; capacity of each train is 200 tons, 100 tons per trailer. This road was built solely as a coal-haulage road. Photo courtesy of Bruce Conrad.

land on the Pictured Cliffs Sandstone; they are capped by red clinker formed when near-surface lower Fruitland Formation coal beds burned and oxidized the overlying shales. Northwest-trending faults (not visible from here) cut the Pictured Cliffs and Fruitland at right angles to the strike of the Pictured Cliffs at Cinder Buttes.

0.5

60.1 La Plata Mountains are visible to the north. Bridge Timber Mountain is straight ahead; there, the San Jose Formation rests unconformably on Fruitland Formation rocks. Bridge Timber Mountain is capped by the oldest preserved terrace gravel of the La Plata River, the Bridge Timber Gravel. This unit is approximately 1,000 feet above the modern channel at this location. This height suggests a Pliocene age for the Bridge Timber Gravels greater than 2 Ma.

0.4

60.5 We are passing through the southern part of the Red Mesa oil field. This field produces oil and gas from a 20-mile long, northeast-trending, anticlinal structural complex. This structure is located northwest of the monoclinal rim of the central structural San Juan Basin on the Four Corners Platform. The Red Mesa field produces gas from the Paradox Member Middle (Pennsylvanian) at 10,600 feet; oil and gas from the Upper Cretaceous Dakota Sandstone at 3,400 feet; oil from fractured shale in the El Vado Member of the Mancos Shale (formerly called "Gallup" production) at 2,600 feet; and gas from the Mesaverde Group at 1,200 feet. The Dakota Sandstone has been the most prolific producer in the field. See Lauth (1983) for more detail on the Red Mesa field.

The spelling of Red Mesa or Redmesa is confusing in this area. The name of the town is Redmesa; the geographic feature is called Red Mesa. The U.S.G.S. 7½ minute topographic quadrangle is named for the town, and is thus Redmesa. The geologic map of this area by Barnes and others (1954) is titled the "Geology and fuel resources of the Red Mesa area". The oil field is Red Mesa. Had enough?

2.3

62.8 **TURN RIGHT on secondary road at compressor station and cross cattle guard.** (Intersection of La Plata County roads 140, 134, and 136.)

0.1

62.9 Pyramid Oil Co. Ferguson no. 2 well.

0.2

63.1 Enter Southern Ute Indian Reservation (again); bear left. We are leaving public access roads here; permission to use this road must be obtained from the Southern Ute Tribe as explained above. Road to right goes to Southern Ute A no. 1 well. This well was completed in the late 1950's and produces about 50 BOPD from the Dakota Sandstone at around 3,500 feet.

0.2

63.3 Pictured Cliffs Sandstone caps ridge ahead.

1.1

64.4 Lewis Shale in roadcut on left; good view of Pictured Cliffs Sandstone capping ridge ahead with Lewis Shale visible below. Iron Springs Gulch on left.

0.6

65.0 Cross Iron Springs Gulch.

0.1

65.1 Pictured Cliffs Sandstone crops out on left; the Pictured Cliffs is clearly much thinner here than at our first stop this morning just south of Durango (mile 3.0) and at the outcrop on the La Plata Highway (mile 55.2). The thickness of the Pictured Cliffs is quite variable throughout the basin ranging from 0 to 400 feet. This thickness variation is the result of the interplay of the rate of basin subsidence and the rate of sediment influx to the shoreline during the time the Pictured Cliffs was deposited. The Pictured Cliffs represents deposition at the shoreface during the final regression of the sea from this part of the Western Interior. At times when the sediment influx rate equaled the rate of subsidence of the seaway, the Pictured Cliffs built up vertically into massive, thick, sandstone piles; when sediment influx was greater, the seaway was more rapidly filled in, the Pictured Cliffs shoreline regressed rapidly to the northeast, and the Pictured Cliffs was relatively thin. In the southeast part and along the east side of the basin (in New Mexico) the Pictured Cliffs is absent because of non-deposition in some places and due to erosion in others. See Fassett and Hinds (1971), and Fassett (1986, 1987, and this volume) for discussions of the basin-wide geometry and mode of deposition of the Pictured Cliffs Sandstone and its relation to the location of thick or thin Fruitland Formation coal beds.

0.1

65.2 Top of the Pictured Cliffs Sandstone; Soda Springs Gulch on left follows a strike valley in the Fruitland Formation. Note the basal Fruitland coal bed on top of the Pictured Cliffs on right.

0.2

65.4 Cross Soda Springs valley. **PULL OFF ON LEFT FOR STOP 5,** our final stop for today. At this stop we will examine in detail the Fruitland coal exposed here and some of the other fluvial features of the lower Fruitland as well. After discussion, turn around and return to County road 134.

3.2

68.6 **TURN LEFT on La Plata County road 134.** La Plata Mountains on skyline to north.

0.4

69.0 Drop down into Long Hollow Canyon; caution sharp curves and steep descent. Sleeping Ute Mountain is visible to the west across the Four Corners platform.

0.2

69.2 Roadcuts are in Lewis Shale; note terrace gravels.

0.2

69.4 Pyramid Oil Co. well on right produces oil from the Dakota Sandstone.

0.2

69.6 Cross Long Hollow Creek. This unusual flowing stream is fed by irrigation-ditch water. The intersection to the right here is with a driveway leading to a ranch.

0.4

70.0 Third Canyon on right; Lewis Shale outcrops.

0.1

70.1 Cannibal Canyon on left; roadcuts in Lewis Shale on

Mile 65.2—Aerial view looking northeast toward the San Juan Mountains. Iron Springs Gulch is the white, treeless drainage in the left-center of the photograph; this area is on the northwest rim of the San Juan Basin just north of the New Mexico-Colorado State line. The white cliffs to the right of Iron Springs Gulch are held up by the Pictured Cliffs Sandstone; the outcrop of the Pictured Cliffs can be traced to the northeast to where it forms the white band seen in the lower part of Bridge Timber Mountain. STOP 5, of the Day 1 field trip is just off of this photograph to the lower right. Photo courtesy of Bruce Conrad.

left. Road climbs onto Red Mesa, a broad terrace of the La Plata River that correlates with the upper bench of Florida Mesa in the Animas River valley.

0.3

70.4 La Plata County Sanitary Land Fill on left.

0.8

71.2 Road turns left—oil tanks on right. The majority of the oil is trucked to the Permian Corporation Refinery east of Farmington, New Mexico. The La Plata Mountains are straight ahead.

Bridge Timber Mountain is to the right; the Pictured Cliffs Sandstone forms the lowest white cliff. On the right side of the mountain, the Eocene San Jose Formation rests unconformably on lower Kirtland rocks. This locality provides important information regarding the structural evolution of the San Juan Basin; the presence of this angular unconformity here at the base of the San Jose tells us that the monoclinal northwest rim of the basin was forming through Paleocene time because the Paleocene Nacimiento Formation, present both north and south of here, is missing on Bridge Timber Mountain. See Fassett (1985) for more detail on the structural evolution of the San Juan Basin.

0.7

71.9 Good view of Bridge Timber Mountain. Meridian Oil Co. yard on left.

0.4

72.3 Pavement begins.

0.4

72.7 Marvel, Colorado; intersection of county roads 100, 133, and 134; continue straight ahead on La Plata 134.

1.1

73.8 Red soils in this area are loess deposits composed primarily of silt and very fine sand. Sources probably

Mile 65.4—Fruitland Formation coal bed exposed near Soda Springs just north of the New Mexico-Colorado State line on the Southern Ute Indian Reservation. The base of the coal bed is just below the white-colored coal layer at the lower left. Near the top of the photo is a slump feature thought to have formed as the result of the underground burning of the upper part of this coal bed. Note the hammer for scale in the center of the photograph. Photo courtesy of William Hanson.

Mile 74.5—Aerial view looking north at the La Plata Mountains. The snow-covered flat area in the central part of the photograph is the Perrins Peak syncline. The rocks forming this feature are the cliff-forming Point Lookout Sandstone, the lower formation of the Mesaverde Group. The lower part of the coal-bearing Menefee Formation overlies the Point Lookout here. The Point Lookout is underlain by the marine, Mancos Shale. Photo courtesy of Bruce Conrad.

include rock flour blown off the active braidplain of the La Plata River during several post-Red Mesa glaciations. Loess is thought to form as the result of glacial scouring and then to be transported by wind. The loess deposits of the La Plata valley and southwest Colorado have supported dry-land pinto-bean farming for many years; Dove Creek, Colorado, northwest of here calls itself "the pinto bean capitol of the world". The Red Mesa area would be irrigated under the proposed Animas-La Plata Irrigation Project.

0.6

74.4 **TURN RIGHT AT intersection with La Plata County road 140.** Sleeping Ute Mountain ahead before turn. The Sleeping Ute is a lacolithic mountain complex which formed during early to mid-Tertiary time. Ute Mountain is on the Ute Mountain Ute Indian Reservation.

0.1

74.5 La Plata Mountains are visible ahead as we drive north. (The following material is from Molenaar and Werts, 1968 p. 19-20.) The La Platas are an excellent example of a lacolithic mountain group; they were carved from a domal uplift of sedimentary rocks intruded by numerous stocks, dikes, and sills. The igneous rocks are all of Late Cretaceous or Tertiary age and are all intrusive. They vary widely in composition and form, but two general types are easily distinguished; porphyritic and nonporphyritic. The porphyritic rocks, most of which are intermediate between diorite and monzonite in composition, are more abundant than the nonporphyritic rocks and occur as more or less contemporaneous stocks, sills, and dikes. The nonporphyritic rocks, in general somewhat younger than the porphyries, form irregular stocks and associated dikes. They consist of syenite, monzonite, and diorite.

The La Plata mining district is best known for its

veins and replacement deposits of gold-bearing and silver- bearing telluride ores, from which the greater part of the production has come. In addition to these it contains a surprising variety of types of deposits within a small area. These types include deposits of disseminated platinum-bearing chalcopyrite, gold-bearing contact ore, veins of mixed base-metal sulfides containing silver or native gold, chalcocite veins, and veins of ruby silver ore. The gold-bearing placers have not been very productive.

The first mining took place in the La Platas in 1873 and between then and 1900 many small mines were opened. Between 1900 and 1937 several larger deposits were uncovered. By the end of 1937 nearly $6,000,000 worth of ore had been mined. Gold has always been the most valuable product of most of the mines, but over 2,000,000 ounces of silver and several hundred thousand pounds of lead and copper have also been recovered. For many years all of the mines were closed down, but with recent increases in the prices of gold and silver some activity is again commencing.

0.2

74.7 Red Mesa field on right.

1.2

75.9 Pinto-bean fields to right and left; Bridge Timber Mountain to right.

0.5

76.4 Alkali Gulch field. This field produces gas from the Barker Creek zone of the Pennsylvanian Paradox Member. Production is from 9,500 feet and consists mostly of sweet gas; however gas from one zone contains 12 percent hydrogen sulfide from which sulfur is recovered. See Squires (1983) for more detailed information on the Alkali Gulch field.

2.8

79.2 Breen, Colorado

0.4

79.6 **TURN RIGHT on La Plata County road 141, the Wildcat Canyon road.**

1.0

80.6 Loess in roadcuts.

0.8

81.4 Note Pictured Cliffs Sandstone outcrop on Bridge Timber Mountain.

1.6

83.0 Intersection with La Plata 136, Long Hollow road on right, continue straight ahead. The Pictured Cliffs Sandstone thickens in a stratigraphic rise on Bridge Timber Mountain ahead. Fruitland Formation coals are usually thicker southwest (shoreward) of these rises. Just beyond the intersection, Lava Creek B ash is exposed in edge of Red Mesa terrace on left, but is not visible from road. This is the only known occurrence of this ash in the La Plata valley and suggests an incision rate of about .2 foot per 1,000 years; a much slower rate than in the Animas valley.

0.4

83.4 Head of Long Hollow Creek on right. This area was considered as a disposal site for the Durango uranium tailings (mile 0.3 of road log) and may have been more suitable than the one selected; however, it was rejected primarily because the tailings would have been trucked along a heavily used road.

0.5

83.9 Perrins Peak Syncline ahead on skyline. This feature contains basal Menefee Formation rocks in the saddle resting on Point Lookout Sandstone. Perrins City, located in this saddle, was an old coal-mining town; coal was mined from lower Menefee coal beds. The Pictured Cliffs Sandstone outcrop swings eastward on the skyline ahead to the right. Needle Mountains on skyline are part of the San Juan Mountains.

0.6

84.5 Lewis Shale in roadcut on left.

0.6

85.1 Bodo Creek and Park on right.

0.6

85.7 Road to right—La Plata 211—leads to Bodo Park. If constructed, Ridges Basin Reservoir of the proposed AnimasLa Plata Project would fill most of this basin with water pumped uphill from the Animas River, substantially decreasing its natural discharge. Durango uranium tailings are being trucked to a site in upper Bodo Canyon, across a low divide from the proposed reservoir.

0.2

85.9 Roadcut on left is Cliff House Sandstone; this transgressive-marine shoreface sandstone consists of thin-bedded silty sandstone and shale interbeds. Throughout a large part of the basin there is little or no shoreface sandstone representing the Cliff House transgression; in those areas, the continental Menefee Formation rests directly on the marine Lewis Shale. In some areas where the Cliff House consists of thick massive sandstones, at Mesa Verde Park, for example, transgression of the shoreline probably paused, temporarily, and the geographically-stable shoreline, built up vertically, as the basin subsided, into a thick massive sand pile. In some instances regressive sandstone beds extend seaward (northeast) in the Lewis Shale for a few to tens of miles from thick, massive, Cliff House Sandstone bodies. One of these is the "Chacra Sandstone"; a regressive sandstone bed of the La Ventana Tongue of the Cliff House Sandstone. The "Chacra" is an important gas producer in the San Juan Basin. See Fassett (1977) for a detailed depiction and discussion of the Cliff House Sandstone, basinwide and Fassett (1983) for a discussion of the stratigraphy of the oil and gas producing rocks of the San Juan Basin.

1.0

86.9 La Plata 125 on left; continue ahead on La Plata 141. Covered Cliff House Sandstone on left.

0.3

87.2 Well on right was drilled in 1960 to test the Dakota Sandstone.

0.5

87.7 Cliff House Sandstone on left.

0.4

88.1 Road on left is La Plata 142; continue on La Plata 141. Roadcuts are in Cliff House Sandstone.

0.5

88.6 Good exposure in basal Cliff House Sandstone reveals

thin-bedded sandstones interbedded with marine shales.
0.1

88.7 Roadcut on left in basal Cliff House Sandstone.
0.2

88.9 Roadcut on left in basal Cliff House Sandstone.
0.1

89.0 On right, contact of Cliff House Sandstone and overlying Menefee Formation. Note carbonaceous shales in the Menefee.
0.3

89.3 On left, coal mine in basal Menefee Formation is for sale. The road ahead descends through Menefee coals, carbonaceous shales, and channel sandstones.
0.3

89.6 Top of Point Lookout Sandstone; the Point Lookout and the overlying Menefee intertongue in this area.
0.2

89.8 Top of massive Point Lookout Sandstone.
0.2

90.0 Basal transition zone of the Point Lookout Sandstone into the underlying Mancos Shale. The Point Lookout, like the younger Pictured Cliffs Sandstone, is a regressive-marine shoreface sandstone which extends across the entire central San Juan Basin. The basal coal-bearing part of the Menefee Formation is similar to the coal-bearing Fruitland Formation in its mode of deposition, but the Menefee contains far less coal than does the Fruitland. For a comparison of the Fruitland and Menefee coals, see Fassett (1986).
0.9

90.9 Twin Buttes to left on skyline are capped by the Point Lookout Sandstone. Perrin's Peak is on skyline to right with white radio towers on top.
0.2

91.1 **TURN RIGHT at intersection with U.S. 160 (Colo. 789)**. Proceed back to Durango. Roadcuts between here and Durango are all in the Mancos Shale.
2.0

93.1 Cross Animas River; uranium-tailings pile on right.
0.1

93.2 **TURN LEFT (north) at traffic light onto U.S. 550.** The Red Lion Inn and other riverfront buildings are on a late Holocene floodplain that was raised with artificial fill. Historic maps and photos show channel changes in this area during the largest recorded flood in 1911. Other facilities within the 100-year floodplain include the new sewage treatment plant, new Chamber of Commerce building, and the State trout hatchery.
0.1

93.3 **TURN LEFT at traffic light into Red Lion Inn parking lot.**
0.2

93.5 Front door of Red Lion Inn. End of field trip for day 1; disembark and enjoy the town of Durango this evening. See you here tomorrow at 8:00 a.m. for field trip around the northeast part of the San Juan Basin.

REFERENCES CITED

Baltz, E. H., Jr., 1967, Stratigraphy and regional tectonic implications of part of Upper Cretaceous and Tertiary rocks, east-central San Juan basin, New Mexico: U.S. Geological Survey Professional Paper 552, 101 p.

Barnes, Harley, Baltz, E.H., Jr., and Hayes, P.T., 1954, Geology and fuel resources of the Red Mesa area, La Plata and Montezuma Counties, Colorado: U.S. Geological Survey Oil and Gas Investigations Map OM-149.

Fassett, J.E., 1977, Geology of the Point Lookout, Cliff House, and Pictured Cliffs Sandstones of the San Juan Basin, New Mexico and Colorado, *in* San Juan Basin III: New Mexico Geological Society 28th Field Conference Guidebook, p. 193-197.

Fassett, J.E., 1978, ed., Oil and gas fields of the Four Corners area (volumes I, II): Four Corners Geological Society, 727 p.

Fassett, J.E., 1983a, ed., Oil and gas fields of the Four Corners area (volume III): Four Corners Geological Society, p. 729-1143.

Fassett, J.E., 1983b, Stratigraphy and oil and gas production of northwest New Mexico updated through 1983, *in* Fassett, J.E., ed., Oil and gas fields of the Four Corners Area: Four Corners Geological Society, p. 849-863.

Fassett, J. E., 1985, Early Tertiary paleogeography and paleotectonics of the San Juan Basin area, New Mexico and Colorado, *in* Flores, R. M., and Kaplan, S. S., eds., Cenozoic paleogeography of west-central United States; Rocky Mountain Section of Society of Exploration Paleontologists and Mineralogists, p. 317-334.

Fassett, J.E., 1986, The non-transferability of a Cretaceous coal model in the San Juan Basin of New Mexico and Colorado, *in* Lyons, P.C. and Rice, C.I., eds., Paleoenvironmental and tectonic controls in coal-forming basins in the United States: Geological Society of America Special Paper 210, p. 155-171.

Fassett, J. E., 1987, The ages of the continental, Upper Cretaceous, Fruitland Formation and Kirtland Shale based on a projection of ammonite zones from the Lewis Shale, San Juan Basin New Mexico and Colorado, *in* Fassett, J.E., ed., The Cretaceous-Tertiary boundary in the San Juan and Raton Basins, New Mexico and Colorado; Geological Society of America Special Paper 209, p. 5-16.

Fassett, J. E., and Hinds, J. S., 1971, Geology and fuel resources of the Fruitland Formation and Kirtland Shale of the San Juan Basin, New Mexico and Colorado: U.S. Geological Survey Professional Paper 676, 76 p.

Fassett, J. E., Lucas, S. G., and O'Neill, F. M., 1987, Dinosaurs, pollen and spores, and the age of the Ojo Alamo Sandstone, San Juan Basin, New Mexico, *in* Fassett, J.E., ed., The CretaceousTertiary boundary in the San Juan and Raton Basins, New Mexico and Colorado; Geological Society of America Special Paper 209, p. 17- 34.

Hewett, N.S., 1977, The prehistory of the San Juan Basin, *in* San Juan Basin III: New Mexico Geological Society, 28th Field Conference Guidebook, p. 65-82.

Izett, G. A., 1981, Volcanic ash beds: Recorders of upper Cenozoic silicic pyroclastic volcanism in the western United States: Journal of Geophysical Research, V. 86, no. B 11, p. 10,200-10,222.

Lauth, R.L., 1983, Red Mesa oil and gas field, *in* Fassett, J.E., ed., Oil and gas fields of the Four Corners Area, Volume III: Four Corners Geological Society, p. 899-900.

Molenaar, C. M., and Werts, L., 1968, First day road log from Farmington, New Mexico to Cortez, Colorado, via

Four Corners Power Plant, La Plata Canyon, La Plata mining area, and Mancos, Colorado, in San Juan-San Miguel-La Plata Region: New Mexico Geological Society 19th Field Conference Guidebook, p. 11-23.

Moore, D. W. and Gillam, M. L., 1984, Road log—Quaternary deposits and soils in the Durango area, southwestern Colorado, *in* Brew, Douglas C., ed., Field Trip Guidebook, 37th Annual Meeting, Rocky Mountain Section, Geological Society of America, Durango, Colorado: Four Corners Geological Society, p. 183-209.

Newman, K. R., 1987, Biostratigraphic correlation of Cretaceous- Tertiary boundary rocks, Colorado to San Juan Basin, New Mexico, *in* Fassett, J.E., ed., The Cretaceous-Tertiary boundary in the San Juan and Raton Basins, New Mexico and Colorado: Geological Society of America Special Paper 209, p. 151-164.

Reeside, J.B., Jr., 1924, Upper Cretaceous and Tertiary formations of the western part of the San Juan Basin of Colorado and New Mexico: U. S. Geological Survey Professional Paper 134, p. 1-70.

Squires, B.W., 1983, Alkali Gulch gas field, *in* Fassett, J.E., ed, Oil and gas fields of the Four Corners Area, Volume III: Four Corners Geological Society, p. 103-104.

Second Day Road Log From Durango, Colorado around Northeast Rim of San Juan Basin Via Bayfield, Chimney Rock, Arboles, Allison, and Ignacio, Colorado and back to Durango

James E. Fassett[1]

WEDNESDAY, JUNE 8, 1988

ASSEMBLY POINT: Red Lion Inn parking lot, Durango, Colorado

DEPARTURE TIME: 8:00 a.m.

DISTANCE: 103.6 miles

NUMBER OF STOPS: 3

INTRODUCTION

Today's trip will begin in the Upper Cretaceous Mancos Shale and will proceed up section through the Upper Cretaceous Mesaverde Group, Lewis Shale, Pictured Cliffs Sandstone, Fruitland Formation, Kirtland Shale, and McDermott Member of the Animas Formation. At the top of the McDermott, we cross a major unconformity marking the Cretaceous-Tertiary boundary into the Paleocene upper part of the Animas Formation. As we go eastward we will be traversing Animas Formation rocks for several miles until east of Yellowjackett Pass, where we drop down through the section from the Animas, into the Upper Cretaceous Kirtland Shale (the McDermott Member of the Animas is not present this far east), Fruitland Formation, Pictured Cliffs Sandstone, and Lewis Shale. East of Chimney rock we turn south, and climb up section through the Fruitland and Kirtland and will again cross the Cretaceous-Tertiary boundary into the Paleocene part of the Animas Formation and on up into the Eocene San Jose Formation. As we travel west from Arboles through Ignacio, Colorado and back toward Durango, we will be in the San Jose Formation most of the way until we drop back down section into the Mancos Shale as we approach Durango from the southeast.

Brad Boyce (geologist, Southern Ute Tribe), Clarence Harr (consulting petroleum geologist), Rob Zech (U. S. G. S.), and Mary L. Gillam (consulting geologist) assisted in the preparation of the road log.

We will be driving across the Southern Ute Indian Reservation on parts of today's field trip. For those of you using this road log on your own, you should obtain permission from the Tribe at their Tribal headquarters in Ignacio, Colorado before doing geologic work on the Reservation, especially in those areas off of Federal, State, or County highways. A Crossing Permit is required to traverse Tribal Lands, this permit can be obtained from the Director of Natural Resources, Box 737, Southern Ute Indian Tribe, Ignacio, Colorado 81137, or call (303) 563-4525.

0.0 **TURN RIGHT onto U.S. 550** as you leave the Red Lion Inn parking lot.
1.1

1.1 Cross Animas River; Point Lookout Sandstone is exposed on the right and left.
0.3

1.4 Traffic light, continue straight ahead. Abandoned Menefee Formation coal mine on right.
0.3

1.7 The Cliff House Sandstone is present in the poor exposures on the right. The Cliff House is thin-bedded and silty in this area, consequently it has little-to-no topographic expression here. (See discussion of Cliff House Sandstone at mile 85.9; Day 1 of this field trip, for more information about Cliff House Sandstone deposition throughout the basin.)
0.3

2.0 Lewis Shale is present in the valley to the right.
0.2

2.2 Traffic light; continue straight ahead.

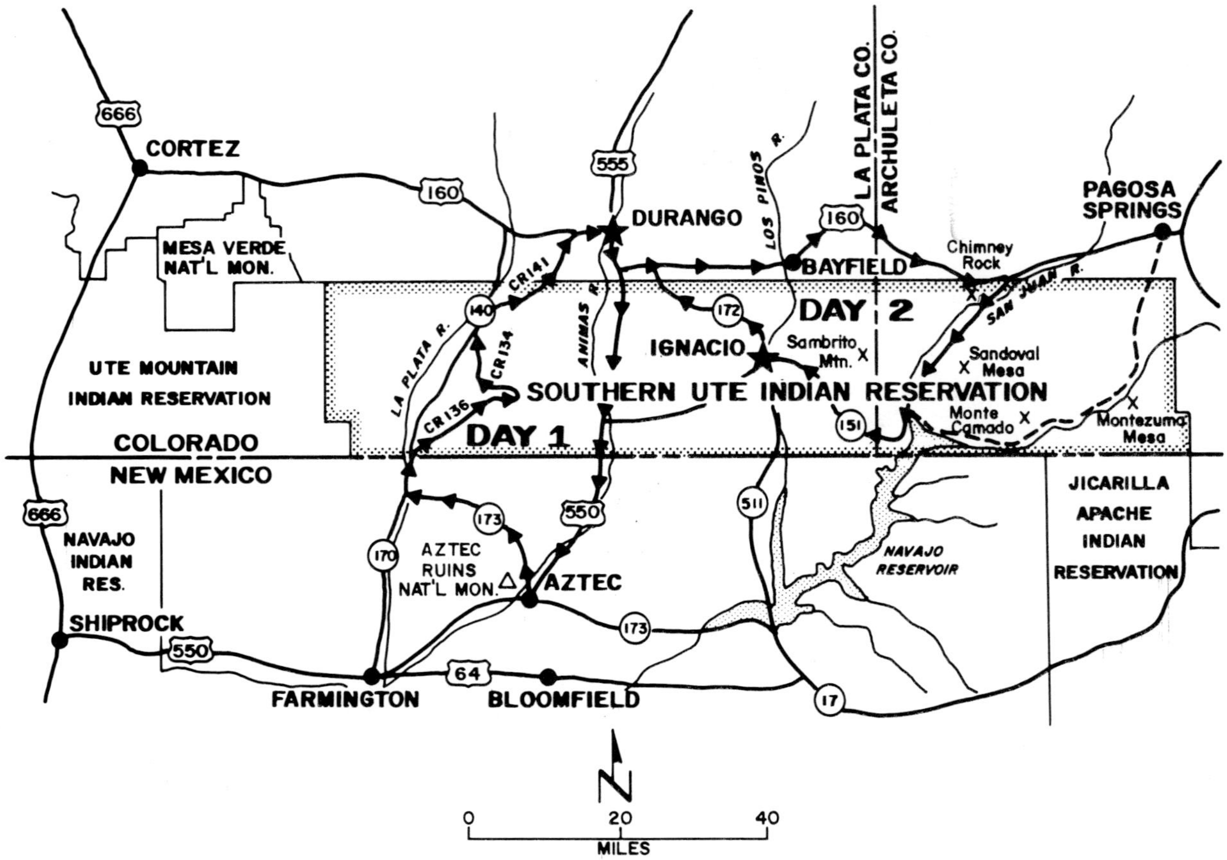

Road map showing the field-trip routes around the northern part of the San Juan Basin in southwestern Colorado and northwestern New Mexico.

0.6

2.8 Cross Animas River; Upper Cretaceous Pictured Cliffs Sandstone, Fruitland Formation, Kirtland Shale, McDermott Sandstone Member of Animas Formation, and Tertiary upper member of Animas Formation are visible on the right, ahead; coal beds in the lower part of the Fruitland are visible in the cut above the firing range. An earthflow of Fruitland rock rubble is visible on top of the Pictured Cliffs. According to Moore and Gillam (1984) this landslide began in 1932 and continued intermittently for several years. The slide received much publicity and was named "Moving Mountain" because of the minor earth tremors that accompanied its movement and for smoke from coal beds set afire by the movement. For years there was fear that this slide would dam the Animas River and create "Lake Durango"; obviously, it hasn't happened yet.

0.9

3.7 McDermott Member of the Animas Formation (purple beds) is exposed on the right. The purple beds seem to grade laterally into more normal, buff-colored fluvial sandstones. The reason for this color change has not yet been determined; is it the result of fluvial cut- and-fill activity and deposition of the buff-colored sandstones after the purple beds were deposited or could it be the result of cut-and-fill associated with deposition of the purple volcaniclastic material, after the buff-colored sandstones were deposited? Or is the answer, none of the above? What do you think? The Tertiary upper member of the Animas Formation overlies the McDermott here with apparent angular unconformity.

0.8

4.5 Animas Formation exposed in roadcut on left.

0.1

4.6 Intersection of U.S. 550 and 160; continue ahead on U.S. 160. U.S. 550, to the right (south) goes to Aztec, New Mexico and from there west to Farmington, New Mexico. (We traveled this road to Aztec on Day 1 of this field trip.) Roadcuts ahead are in the upper member of the Animas Formation.

0.2

4.8 Animas Formation overlain by Florida Mesa terrace gravel in roadcut on left. For more information on the Quaternary gravels and soils in the Durango area, see papers by Gillam and others (1984) and Moore and Gillam (1984).

0.1

4.9 Gravel pit on left in Florida Mesa terrace gravel.

0.3

5.2 In the right-rear of the embayment to the left (the last before the valley widens), Lava Creek B ash (dated at 610 ka) is visible as a white band roughly 30-50 feet below the top of the gullied area. This ash is within locally-derived fan alluvium overlying the Animas River terrace gravel.

0.4

Mile 3.7—McDermott Member of the Animas Formation (Upper Cretaceous) exposed west of the Animas River south of Durango, Colorado. View is to the west. The McDermott is dipping to the south here and is underlain by the Kirtland Shale and overlain, unconformably, by the Tertiary (Paleocene) upper member of the Animas Formation. Western Gas Supply pipe yard is in the foreground. Photo courtesy of Bruce Conrad.

5.6 Pictured Cliffs Sandstone dip slope is visible to left.
0.3
5.9 Tower on horizon to left is on the Dakota Sandstone dip slope.
1.3
7.2 On the left, the purple-colored McDermott Member is visible; the McDermott has not been mapped very far east of here. East of the termination of the McDermott, the Animas Formation is mapped as a single, undivided unit. See Plate I, Geologic map of the San Juan Basin, in Fassett and Hinds (1971), for more detail on the geology of this area.
0.4
7.6 On right, Colo. 172 leads to Ignacio, Colorado to the south and east. Continue ahead on U.S. 160. Ignacio is the headquarters of the Southern Ute Indian Tribe. Highway 172 also provides access to the Durango airport.
1.4
9.0 Sambrito Mountains on skyline ahead; these mountains are composed mostly of Paleocene part of Animas Formation capped by the Eocene San Jose Formation.
0.7
9.7 Roadcut is in Animas overlain by Florida terrace gravel.
0.4
10.1 Cross Florida River.
1.4
11.5 Animas Formation in roadcuts.
0.9
12.4 The well on the left is the Federal-Texaco 1-11; it has produced gas from the Dakota Sandstone. This gas well was completed in 1980 by Natomas North American Oil Co. at a depth of 7,370 feet. Production was marginal and the well was recently purchased by Amoco to be recompleted as a Fruitland Formation coal-bed methane well. Mesa Mountain is visible in the distance on the right (south); the swaths of cleared vegetation on this mountain are not ski runs, they are gas pipelines. The

Animas is exposed in roadcuts.
1.7

14.1 Cross Dry Creek; Animas in roadcuts. The wells ahead are Fruitland coal-bed methane wells.
1.3

15.4 Pig farm on right!
0.6

16.0 Animas in roadcuts; Dry Creek on right.
0.6

16.6 Enter Gem Village. (The following historical account was provided by Fred Kopeland, Gem Village resident.)

In the early 1940s, Frank Morris bought 160 acres and subdivided lots to form a small village; he named it Gem Village. Mr. Morris was from Wisconsin and was a "rock hound". He chose this area because of its location close to gems, minerals, and rocks of interest. Silverton was close by, a source of minerals; Wolf Creek pass was also close by where geodes had been found. Also, the area was on the old Navajo Trail (Highway 160 was called the Navajo Trail until recently).

Mr. Morris and most of the other people who bought land from him had a great interest in gems, minerals, and rocks. They formed the Navajo Trail Gem and Mineral Club and built a clubhouse, which is the log cabin building that you can see from the highway. The club was active until about two years ago. Many of the members live in Gem Village. Mr. Armer Gerhardt, the oldest living member, is 101 and still lives in Gem Village. The blue house next to the Clubhouse Building was built and lived in by Mr. Morris.

Lottie Shipley owned and ran Shipley's Mineral House from about 1948 until just a few years ago. Shipley's Mineral House was started in her front room and grew until it was a nationally recognized business. Rock enthusiasts from all over the country and foreign countries made a special effort to come to Shipley's Mineral House. It was featured in many rock hound publications.

There are still Indian artifacts to be found around Gem Village. In the Pine River area some of the hardest rocks known are found; they are a conglomerate of granite. Mr. Kopeland related that two rock crushing businesses have gone bankrupt trying to crush rock from this area.

There are no longer any rock shops in Gem Village and the long-term residents are saddened by this fact; times have changed and rock collecting is no longer a favorite hobby. However, Gem Village is alive and well as it is directly aligned with Bayfield and Durango, which assures its livelihood. Gem Village remains a nice, quiet, scenic place to live.
1.1

17.7 Intersection on right of Business Route 160 to beautiful downtown Bayfield, stay to left. Fruitland Formation coal-bed methane well on left. This well produces from a depth of 2,155 feet.
0.7

18.4 Pass under irrigation pipe crossing road.
0.2

18.6 Cross Los Pinos River.
0.6

19.2 Intersect La Plata County Road 5 (leads to downtown Bayfield to right). To the left, road leads to Vallecito Reservoir and trail heads into the San Juan Mountains Wilderness Area via Vallecito Creek and Los Pinos River. Excellent trout fishing and spectacular mountain scenery up these drainages for those willing to hike in a ways. Continue ahead on U.S. 160. From Bayfield on to the east, U.S. 160 follows the Beaver Creek drainage which is controlled by the northern structural rim of the San Juan Basin.
1.7

20.9 Enter San Juan National Forest; hills ahead are dip slopes of the Dakota Sandstone, Mesaverde Group, and Pictured Cliffs Sandstone, which form the northern rim of the San Juan Basin. The Mesaverde Group is mapped undivided in the northeastern part of the basin because in this area, the middle unit of the Mesaverde, the Menefee Formation, has thinned to the extent that it is difficult to separate the Point Lookout Sandstone (lower unit) from the Cliff House Sandstone (upper unit). As discussed in the road log for Day 1 (mile 85.9), in large areas of the basin, there is no shoreface sandstone representing the Cliff House transgression, thus it is difficult in this part of the basin, where the Mesaverde Group is thinning, to differentiate shoreface sandstone tongues of the Point Lookout in the very thin Menefee Formation from possible shoreface sandstone tongues of the Cliff House. Further east, in the vicinity of Pagosa Springs, the Mesaverde Group has totally wedged out; there, there is no Mesaverde between the Lewis Shale and the underlying Mancos Shale. This eastern wedge-edge of the Mesaverde Group marks the place where the Point Lookout shoreline ceased its northeastward regression and the Cliff House shoreline (or Lewis shoreline, where there is no Cliff House) began its southwestward transgression.
0.7

21.6 Beaver Creek on right.
0.5

22.1 The roadcuts ahead for the next 10 miles will be in the Animas Formation. The Animas reaches a thickness of 2,700 feet in the northern San Juan Basin as reported in Reeside (1924). The Animas was deposited on an erosion surface that has truncated underlying Upper Cretaceous rocks; this erosion event also apparently removed the lowermost-Tertiary Ojo Alamo Sandstone throughout the northern San Juan Basin. The Animas is composed of volcaniclastic debris that was produced by widespread eruptions in the San Juan volcanic field to the northeast. These eruptions apparently continued throughout most of Paleocene time. The Animas was deposited in the rapidly-downwarping northern part of the San Juan Basin; this downwarping was probably the result of flank subsidence south of the San Juan volcanic field. The structural evolution of the San Juan Basin and the relations of the Animas with underlying rocks in this area is discussed in Fassett and Hinds (1971) and Fassett (1985).
0.5

22.6 Fruitland coal-bed methane well on right is the Payne C

Mile 25.6—Aerial view looking southeast along the northern rim of the San Juan Basin. Highway US 160 is visible to the right. The ridges to the left of the highway are held up by the Fruitland Formation and the Pictured Cliffs Sandstone; the valley to the left of the Pictured Cliffs outcrop is in the Lewis Shale; and the more massive ridge to the left is the Mesaverde Group. The broad valley to the left of the Mesaverde is the Mancos Shale underlain by the more gently-dipping, tree-covered slopes of the Dakota Sandstone. Photo courtesy of Bruce Conrad.

No. 1, completed in May 1987. The top of the Pictured Cliffs Sandstone (base of Fruitland Formation) is at 1,436 feet. The total depth of the well is 1,750 feet. The initial production from Fruitland coal beds was 400 MCFPD and 450 BWD. This large well pad area includes space for a Mesaverde Group water injection well for disposal of Fruitland coal-bed water.

0.5

23.1 Fruitland coal-bed methane well on right is the Parry Land Gas Unit B No. 1. It was completed in May 1987; the top of the Pictured Cliffs is at 1,482 feet and total depth is 1,640 feet. Initial production was 50 MCFPD and 1,000 BWD.

2.5

25.6 On the left, the Pictured Cliffs forms the dip slope on the monocline in the foreground. The dip slope on the higher ridge behind the Pictured Cliffs is the Mesaverde Group.

2.1

27.7 Sambrito Mountains on the right; Animas Formation capped by the San Jose Formation. Pargin Mountain, the highest point in the Sambritos, has an elevation of 8,936 feet.

0.2

27.9 Enter Archuleta County.

2.7

30.6 Cross Yellowjacket Pass; we now enter the Squaw Creek drainage. Squaw Creek drains into Yellowjacket Creek, which drains into the Piedra River. The Piedra is a tributary of the San Juan River.

0.7

31.3 Steeply-dipping Animas Formation in roadcut on left. We will drop rapidly down section ahead.

0.8

Mile 35.3—Aerial view looking east with the San Juan Mountains, east of Pagosa Springs, Colorado on the skyline. Highway US 160 is visible in center-foreground; the highway is in a strike valley of the Lewis Shale. The Pictured Cliffs Sandstone is visible to the right and the Mesaverde Group forms the steep dipslope to the left of the highway. The south-flowing Piedra River emerges from its beautiful mountain canyons at the gap through the Mesaverde Group in the central part of the photograph. The curve of the dipping beds of the Mesaverde Group in the near foreground represents an anticlinal feature that extends into the San Juan Basin to the south. Photo courtesy of Bruce Conrad.

32.1 Contact of Animas Formation and underlying Kirtland Shale in roadcut on left. The fault exposed in this roadcut is down to the left (west) about 75 feet.

0.1

32.2 We are dropping down section rapidly through the Farmington Sandstone Member of the Kirtland Shale.

0.3

32.5 Carbonaceous shale on left marks the top of the Fruitland Formation and the base of the overlying Kirtland Shale.

0.4

32.9 Coal beds in the lower Fruitland are exposed on the left.

0.1

33.0 Red clinker on left formed as Fruitland coals burned, oxidizing adjacent rocks.

0.1

33.1 Fruitland coal beds on left.

0.1

33.2 Pictured Cliffs Sandstone on left.

0.1

33.3 Fruitland Formation coal bed rests directly on Pictured Cliffs Sandstone on right. This occurrence of coal on top of shoreface sandstones of the Pictured Cliffs is not uncommon in the basin and would seem to indicate little or no tidal sea-level rise and fall at the Pictured Cliffs shoreline; otherwise fresh-water swamps could not have existed right up to the edge of the shore.

0.2

33.5 Pictured Cliffs Sandstone contact with a tongue of the Lewis Shale on left.

0.4

Mile 37.6—Highway US 160 below the towering dipslope of the Mesaverde Group; highway is in a Lewis Shale strike valley. View is to the west. Photo courtesy of Bruce Conrad.

33.9 Tongue of Pictured Cliffs Sandstone in the Lewis Shale on the left; Lewis Shale is exposed in roadcuts ahead.
0.3

34.2 Mesaverde Group dip slope ahead.
0.6

34.8 Cross Yellowjacket Creek.
0.5

35.3 Flatirons of Lewis Shale on the Mesaverde dip slope on the left.
0.8

36.1 Piedra River cuts through the Mesaverde Group ahead on left forming a notch in this part of the basin's northern rim.
1.1

37.2 Pictured Cliffs Sandstone caps hills on right.
0.4

37.6 Road to left is Piedra road (sign says Piedra Post Office). This road passes through the hogback forming the northern rim of the San Juan Basin. Good exposures of the Mesaverde Group, lower part of the Mancos Shale, Dakota Sandstone, and Jurassic and older rocks are present on the Piedra road. The Piedra-River notch has cut into Precambrian rocks further upstream.
0.2

37.8 Pictured Cliffs Sandstone caps ridge on skyline to right on top of Lewis Shale.
0.2

38.0 Cross Piedra River. Road to left leads to trailhead for backpack trail to upper reaches of the Piedra and tributaries; Weminuche Creek and Williams Creek. An easy trail and excellent trout fishing.
1.0

39.0 Chimney Rock ahead on right. Chimney Rock consists of an outlier of Pictured Cliffs Sandstone resting on the Lewis Shale. Chimney Rock is one of the most prominent landmarks of the northern San Juan Basin. Probably one of the first historical records of this feature is in the diary of Fathers Dominguez and Escalante who passed south of the rock on August 4, 1776. The following quotation from their diary is from a paper by

Dane (1960): "We traveled about a quarter of a league to the northwest and went through the pass where begins another lake which we called Laguna de Olivares. It must be about a quarter of a league long and two hundred varas wide, more or less. Although its water has not a very pleasant taste it is fit to drink. From the lake and little pass we continued north half a league, then turned northeast, leaving the road which goes to the Piedra Parada (a place known to those of us who have traveled through here)." Piedra Parada, which means standing rock, is Chimney Rock; the lake referred to as Laguna de Olivares is Horse Lake.

Dane reports that more than 80 years later, the Macomb expedition passed by Chimney Rock in 1859: "From the camp of the 12th on the San Juan River at the site of modern Pagosa Springs the party moved westward, probably along a trail followed by the Indians long before the first Spanish travelers through this area, and after one intermediate camp, stopped on the Piedra River near Chimney Rock, the striking obelisk of Pictured Cliffs sandstone, which was sketched and described by Newberry, 1876, (p. 76-78)." Newberry was Professor J. S. Newberry, the geologist on the Macomb expedition.

1.1

40.1 Devil Creek on right.

1.3

41.4 Chimney Rock Store and Campground on right; Lewis Shale in roadcuts on left.

0.3

41.7 Cross Devil Creek.

1.0

42.7 Mesaverde Group forms dip slope ahead on left; Lewis Shale in roadcuts ahead.

1.0

43.7 **TURN RIGHT (south) onto Colo. 151.** Chimney Rock on skyline on right. Note the color change of the Lewis Shale about ¾ of the way down the Lewis slope from the base of the Pictured Cliffs. The lighter-colored shales generally contain more silt-size clastic material and less organic material than the overlying dark shales and could record a time of severe storms on the shelf of the Lewis seaway which spread a layer of coarser clastic debris far out to sea. These clastic-rich layers sometimes show up on geophysical logs and can often be traced laterally to the southwest to where they intersect the time-equivalent Pictured Cliffs shoreface sandstone beds; these layers have been called "phantom horizons" by well-log analysts. Phantom horizons are usually subtle and require experience to detect and correlate, but they are useful marker beds for the careful geophysical log analyst because they are time horizons.

0.4

44.1 Entrance to Lake Capote on left. This lake is on Southern Ute Indian land and has picnic and campground facilities and excellent fishing. The lake and campground are open to the public; an entrance fee is required.

0.2

44.3 Ridge line on horizon to left is Pictured Cliffs Sandstone; Lewis Shale in roadcuts. Stollsteimer Creek on left.

1.8

46.1 Note two light-colored zones (phantom-horizon beds) in the lower part of the Lewis Shale on right.

0.6

46.7 Outcrop of massive Pictured Cliffs Sandstone ahead.

0.2

46.9 Road on right leads to the Chimney Rock Archeological Site. The gate across this road is locked and permission to enter the site must be obtained from the National Forest Service office in Pagosa Springs, Colorado. We will drive up this road to the archaeological site at the base of Chimney Rock for our lunch stop later.

0.2

47.1 Cross Stollsteimer Creek.

0.6

47.7 Massive Pictured Cliffs Sandstone on right.

0.2

47.9 Fruitland coal beds on top of Pictured Cliffs Sandstone ahead.

0.5

48.4 Highwall of abandoned Chimney Rock coal mine ahead on right.

0.4

48.8 Contact of Pictured Cliffs and Fruitland.

0.2

49.0 Reclaimed waste dump from Chimney Rock mine on right.

0.4

49.4 **ENTRANCE TO CHIMNEY ROCK MINE, TURN RIGHT INTO MINE FOR STOP 1.** We will drive into the abandoned coal mine to examine the Fruitland coal beds and their relationship to the underlying Pictured Cliffs Sandstone. We will also examine some well exposed, fluvial channel sandstones in the Fruitland exposed in the highwall of the mine.

The Chimney Rock coal mine was opened in November 1976 as the Martinez strip mine which provided coal for the Four Corners area power plants and local consumers. The mine subsequently was purchased by Chimney Rock Coal Company and then by Kaiser Steel of Colorado Springs, Colorado. The mine comprises both private and Federal lands. The coal was all on Federal land. The mine closed when it reached the lease boundaries, and due to archaeological constraints, no further leasing was allowed. Cumulative production from the mine was around 1.3 million tons of coal.

Three seams of coal were mined, labeled from top down, the A, B, and C seams. The lower Fruitland section containing these seams is as follows, from the bottom up: Top of Pictured Cliffs Sandstone; 10 feet shale; coal seam C (6 feet thick); 22 feet shale, thin-bedded limestone, and sandstone (this interval contains marine mollusks); coal seam B (12 feet thick); 10 feet shale and thin-bedded sandstone; coal seam A (5 feet thick). Coal seams A and B contain the best coal; these coals have a heating value of approximately 13,200 Btu and are good quality coking coals. Seam C coals have a heating value of 11,600 Btu and were washed to improve their quality prior to sale. The 22-foot-thick interbed between coal

Mile 46.9—Aerial view of the Chimney Rock Archaeological Site (STOP 2). The ruins are on a sandstone ledge of the Pictured Cliffs Sandstone; the underlying softer beds of the Lewis Shale are visible below the ruins. Photo courtesy of Bruce Conrad.

seams B and C is interesting because it is, at least in part, marine. This unit must represent a minor transgression of the shoreline to the southwest following deposition of seam A, and the thin sandstone bed at the top of this unit is a thin tongue of Pictured Cliffs Sandstone representing the final regression of the sea from this area. This marine unit may be the youngest and stratigraphically highest marine rock unit in the San Juan Basin.

At the conclusion of our discussion at this stop, we will board the busses to drive to the Chimney Rock Archaeological Site for **STOP 2;** a lunch with a view and discussion of the structure of the north rim of the San Juan Basin. After lunch, the field trip will resume at the intersection of Colo. 151 and the road into the Chimney Rock mine.

0.1

49.5 Cross Stollsteimer Creek

0.2

49.7 Gate on right. The Farmington Sandstone Member of the Kirtland Shale is in contact with the lower shale member of the Kirtland in cliff to right. The Farmington consists of interbedded fluvial sandstones and finer-grained overbank deposits; the basal contact of the Farmington Member with the lower shale member is gradational and stratigraphically erratic.

0.2

49.9 Approximate location of contact of Fruitland Formation and overlying lower shale member of Kirtland Shale. Because this contact is marked by the stratigraphically highest occurrence of coal or carbonaceous shale in the Fruitland, it is sometimes a difficult contact to locate or map.

0.3

50.2 Contact of Paleocene Animas Formation on Upper Cretaceous Kirtland Shale on hill to right. This contact is an unconformity which truncates the underlying Kirtland Shale and then the Fruitland Formation from west to east across the basin. The Fruitland and Kirtland are nearly 500 feet thick here, whereas south of Durango,

Mile 49.4—Highwall of the Chimney Rock coal mine (STOP 1). The rocks exposed here are the Fruitland Formation and possibly the lower Kirtland Shale at the top of the highwall. Fruitland coal beds are visible in the lower part of the highwall interbedded with fluvial channel sandstone beds. Note the beautiful point-bar sandstone bed in the left-center of the highwall. Accretion-bedding surfaces are clearly visible in this bed. This highwall will be destroyed when the Chimney Rock Mine is reclaimed later this year. Photo courtesy of Bruce Conrad.

these rocks are around 1,500 feet thick. In the vicinity of the Colorado-New Mexico border on the west rim of the basin, the Kirtland-Fruitland interval is over 2,000 feet thick. On the east edge of the basin (southeast of here), the Kirtland-Fruitland interval is 0 feet thick. At least two episodes of erosion truncated the Kirtland and Fruitland: the first was during Late Cretaceous-very early Tertiary time, prior to deposition of the Ojo Alamo Sandstone; the second was during later early-to-mid-Paleocene time, prior to deposition of the Animas Formation. The Ojo Alamo Sandstone is missing in the northern part of the basin, probably having been removed by the pre-Animas erosion episode which was associated with the initial uplift of the northern rim of the basin in early-to-mid-Paleocene time (Fassett and Hinds 1971, Fassett 1985).

0.1

50.3 Cross Stollsteimer Creek.

0.4

50.7 Roadcuts in the Animas Formation here and ahead for several miles.

0.6

51.3 Gravel pit in terrace gravels on ridge to right.

0.2

51.5 Road to right leads to gravel pit.

0.4

51.9 Sambrito Mountains ahead to right on skyline; San Jose Formation on the Animas Formation.

0.2

52.1 Stollsteimer Creek joins the Piedra River on the right.

0.4

52.5 Animas Formation in hills on right is dipping south, into the basin. Piedra River is on the right.

2.3

54.8 Fossett Gulch road to right, continue straight ahead. The Animas Formation in hills to right is flat-lying.

0.6

55.4 Well on right is Sun Oil Co. Mildred A. Wright No. 1 completed in 1974. This well is a Dakota Sandstone and Mancos Shale gas well which has been shut in since its completion due to lack of a pipeline.

0.6

56.0 Piedra River parallels road on right. All of the lands along the river bottom are fee land resulting from

homesteading between 1879-1938. The lands away from the river on both sides are Southern Ute Indian Reservation. See the paper by Boyce and Burch (this volume) for a detailed account of the history of the Southern Ute Reservation.
2.1

58.1 Approximate contact between the Animas Formation and overlying San Jose Formation. This contact appears to be conformable in this part of the basin; away from the basin's structural rim. Contrast this with the angular unconformity at the base of the San Jose seen on Bridge Timber Mountain on Day 1 of this field trip (mile 71.2 of Day 1 road log) where the San Jose rests on the lower part of the Kirtland Shale.
0.2

58.3 Archuleta County road 500 on left, continue straight ahead. This road parallels the San Juan River and leads to Pagosa Junction (Gato), Juanita, and ultimately to Pagosa Springs.
0.7

59.0 Cross Piedra River. The trestle on left was built after Navajo Dam was built on the San Juan River (to the south, in New Mexico) because the waters of the lake would drown the old railroad bridge, two miles south of here near the confluence of the Piedra and San Juan rivers. Three years after the new trestle was completed, the Denver and Rio Grande Western Railroad abandoned their narrow-gauge line here.
0.6

59.6 San Jose Formation in roadcut on right capped by terrace gravels.
0.3

59.9 Navajo Lake on left; roadcuts in terrace gravels and variegated shales of the San Jose Formation in roadcuts ahead. Navajo Lake, the upper end of which we see here, was created by the construction of Navajo Dam, just below the confluence of the San Juan and Pine rivers, about 20 miles southwest of here in New Mexico. The dam was built in the 1960's to provide water for the Navajo Indian Irrigation Project. Navajo Lake water flows to the irrigation project area through a series of canals, tunnels, and siphons. The project was originally designed to irrigate more than 100,000 acres in an area south of the San Juan River roughly between Bloomfield and Farmington, New Mexico. The first block of 10,000 acres was placed under irrigation in May 1977 and additional blocks have been developed since that time.

Navajo Lake has come to be one of the major recreation areas of northwestern New Mexico with opportunities for boating and fishing with a lake shore that is almost all still totally natural and undeveloped. The only established marinas are at the southern end of the lake near the dam and here at the northern end at Arboles. The quality fishing waters on the San Juan River below Navajo Dam provide some of the best trout fishing in the United States.
1.0

60.9 Entering Arboles, Colorado. The well on the left is in the Ignacio-Blanco Fruitland gas field. This Amoco well produces gas from Fruitland Formation coal beds.

Between here and Durango, we will pass numerous Fruitland coal-gas wells with their distinctive pump jacks. The history of the coal-gas development in this area is described in detail in the papers by Harr and Boyce and Burch (this volume).
0.6

61.5 County road 982 to Navajo Lake on left; continue straight ahead.
0.2

61.7 Lakeside Cocina on the left.
2.3

64.0 Piedra Peaks on skyline on right.
0.6

64.6 Roadcut in San Jose Formation.
0.4

65.0 La Plata County line.
0.4

65.4 Allison, Colorado.
1.0

66.4 Roadcut in San Jose Formation; we will be driving through the San Jose for many more miles.
1.2

67.6 La Plata Mountains ahead on skyline.
0.5

68.1 County road 328 on left goes to Tiffany, Colorado. The name Tiffany has become immortalized because fossil mammals were found in Animas Formation rocks north of here and the faunal zone containing these fossils was called the Tiffanian, after the town of Tiffany. Sloan (1987, p.176-178) discusses the Tiffanian in detail and the following is from his report: "Wood and others (1941) defined the Tiffanian age based on the Tiffany local fauna of La Plata and Archuleta Counties, Colorado, described by Granger (1917), Matthew and Granger (1921), and Simpson (1935). Most of the specimens came from the Mason Pocket, a carnivore den(?) of small animal remains; the others came from eight surface localities of large animal remains scattered over a distance of 80 km and a vertical interval of at least 300 m. The specimens occur in the middle of the Animas Formation of the northern San Juan Basin." The Tiffanian, according to Sloan, is the fifth highest of six mammal zones making up the Paleocene.
1.1

69.2 Fruitland Formation coal-bed methane well on right is the Taichurt 32-2 drilled by the William Perlman Co. in 1983. The well was bought by Amoco in 1986. It produces 100 MCFPD of gas and 20 BWD.
0.5

69.7 Big roadcut in the San Jose Formation.
0.5

70.2 Fruitland coal-bed methane well on right. This well is the Hott 29-2, drilled by Perlman in 1982 and acquired by Amoco in 1986. It produces 100 MCFPD and 40 BWD.
0.6

70.8 Cross Devil Creek.
0.7

71.5 Fruitland coal-bed methane well on left.
1.4

72.9 County road 334 on right, continue straight ahead.

Mile 60.9—Aerial view of an Amoco coal-bed methane well near the shore of Navajo Reservoir (upper left of photograph). Road is Colorado 151. Photo courtesy of Bruce Conrad.

Shellhammer Ridge is ahead on left. This ridge separates the Spring Creek drainage from the Los Pinos River drainage.
0.5
73.4 River terraces on left.
0.2
73.6 Cross Spring Creek.
0.9
74.5 County road 324; continue straight ahead.
1.2
75.7 Gravel pit on road to left (Gosney and Sons) mines gravels from terrace gravels capping the top of Shellhammer Ridge.
0.4
76.1 Amoco Production field office on right.
1.3
77.4 County road 521 on right; continue straight ahead.
0.4
77.8 Cross Los Pinos River.
0.1
77.9 Enter Ignacio.
0.1
80.0 **TURN RIGHT at intersection with Colorado 172.**
0.3
80.3 Southern Ute buffalo herd on right.
0.3
80.6 Sky Ute lodge and Convention center on right.
0.1
80.7 Southern Ute Tribal Headquarters on right.
0.1
80.8 Southern Ute Tribal Health Center on right. River terraces ahead on left.
0.2
81.0 Shellhammer Ridge on right, Los Pinos River valley below ridge.
0.7
81.7 Top of gravel-capped terrace.
1.2
82.9 Terrace gravel on Animas Formation in roadcuts on right.

Mile 89.1a— Aerial view of Amoco's McCaw water-injection facility (STOP 3). View is to the north with Colorado 172 at the bottom of the photograph. Photo courtesy of Bruce Conrad.

1.1

84.0 Hills to left on skyline are Mesa Mountains.

0.9

84.9 Cross abandoned Denver and Rio Grande Western narrow-gauge railroad grade.

1.0

85.9 Alfalfa fields are on private land which was homesteaded before the present Southern Ute Indian Reservation was defined. These fields yield up to 5 cuttings per year.

1.6

87.5 Oxford, Colorado.

1.6

89.1 **TURN OFF ON RIGHT for STOP 3 at the McCaw No. 1 Amoco water-injection well.** We will disembark here to discuss the disposal of water from Fruitland coal-bed methane wells. (The following information is summarized from the paper by Zimpfer and others in this volume; see that report for a comprehensive discussion of water disposal in the northern San Juan Basin.)

The McCaw water injection facility is one of two operating water-injection facilities owned by Amoco and designed to dispose of water produced from Fruitland Formation coal-bed methane wells. This system utilizes low-pressure polyethylene or fiberglass pipes to collect brines from adjacent gas-production wells. Brines are collected in holding tanks at the injection site and filtered prior to injection.

The McCaw system receives water from 27 coal-bed methane wells. The daily water production from each gas well ranges from less than 10 barrels to greater than 400 barrels. The McCaw well, injecting into the locally-gas-depleted Point Lookout Sandstone of the Mesaverde Group, has, as of January 1, 1988, disposed of approximately 1.5 million barrels of water. Maximum daily injection capacity is 9,600 barrels of water, the limit of the surface equipment.

0.6

89.7 Fruitland Formation coal-bed methane well on left is the

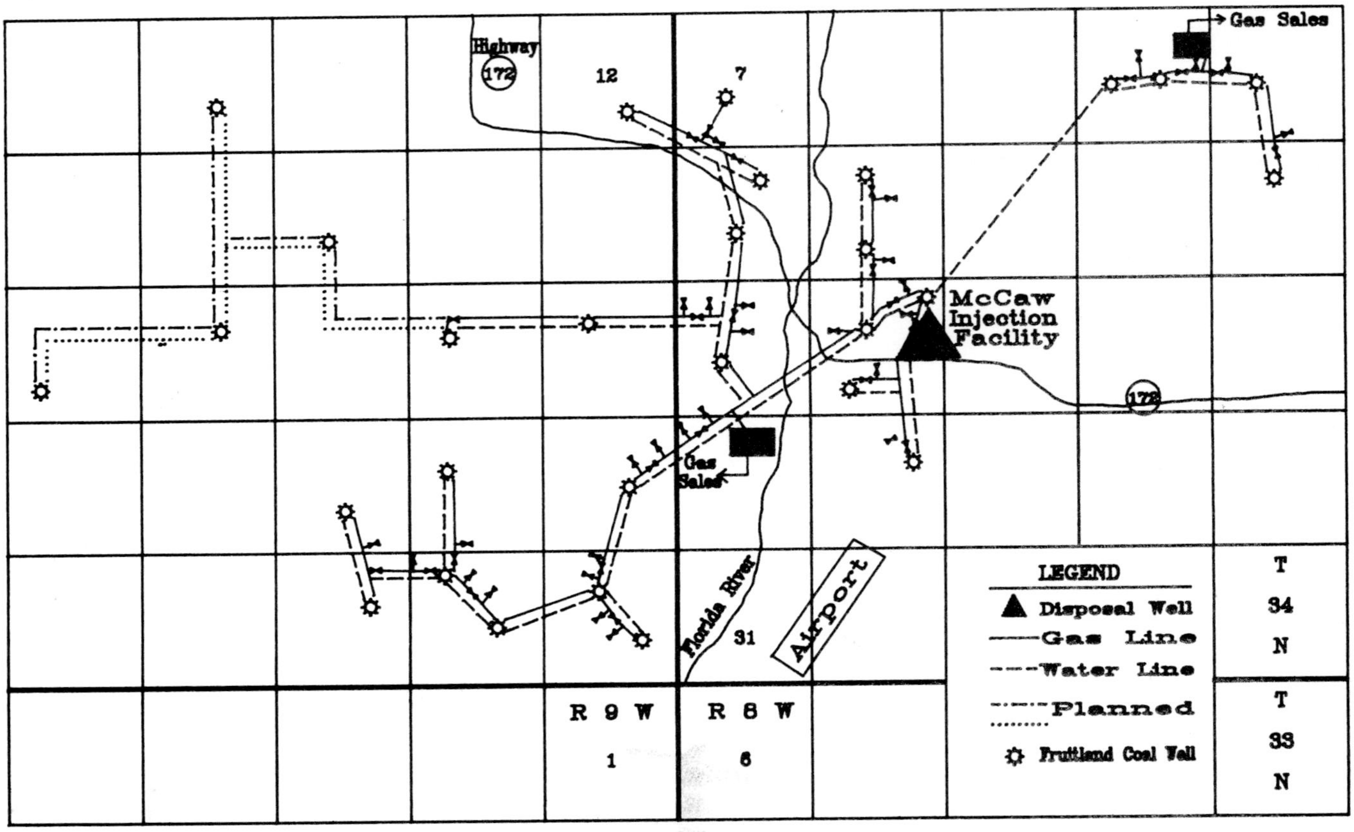

Mile 89.1b—Map of the McCaw water-injection facility.

Amoco McCaw 20-1 No. 2.
0.2

89.9 Road to Durango-La Plata County Airport to left; continue straight ahead.
0.5

90.4 Cross Florida River.
0.3

90.7 Several Fruitland Formation coal-bed methane wells can be seen up the Florida River valley.
0.4

91.1 Spring on left produces good water; the spring water comes from a perched water table in terrace gravels.
1.4

92.5 Pictured Cliffs Sandstone flatirons on dip slope on right; Dakota Sandstone forms dip slope on horizon. The Mesaverde Group is not visible here.
1.4

93.9 Pictured Cliffs flatirons on left. Dakota dip slope dead ahead.
0.7

94.6 Cross northern boundary of Southern Ute Indian Reservation.
0.1

94.7 Loma Linda, Colorado.
0.9

95.6 **TURN LEFT at intersection of Colorado 160.** From here we will proceed west to Durango. We will repeat the first part of today's field trip from here to Durango, but in reverse.
1.4

97.0 Basin Mountain on skyline ahead to left.
1.2

98.2 Descend hill through roadcuts of Animas Formation and terrace gravels. Note abandoned railroad grade on right near bottom of hill.
0.6

99.8 Intersection with U.S. 150 on left to Aztec, New Mexico; continue straight ahead.
1.5

100.3 Animas Formation (upper member), McDermott Member of Animas, Kirtland Shale, Fruitland Formation, Pictured Cliffs Sandstone, and Lewis Shale outcrop on left.
0.2

100.5 Fruitland coal beds on top of Pictured Cliffs Sandstone.
0.2

100.7 Cross Animas River.
0.5

101.2 Traffic light; Durango Mall to right.
0.8

102.0 Traffic light; Point Lookout Sandstone overlain by Menefee Formation coal beds in river cut on right.
0.2

102.2 Possible tongue of Point Lookout Sandstone in the Menefee Formation on right?
0.1

102.3 Cross Animas River.
0.4

102.7 Fault in Point Lookout Sandstone and underlying Mancos Shale to left. The left (east) side is down about 50 to 60 feet. The Point Lookout caps Perrins Peak on

skyline ahead on left.
0.6

103.3 Traffic light; intersection of U.S. 160 to left (west) goes to Cortez, Colorado. Continue on U.S. 550.
0.2

103.5 **TURN LEFT at traffic light into Red Lion Inn parking lot.**
0.1

103.6 Red Lion Inn; end Day 2 field trip.

REFERENCES CITED

Boyce, B.C., and Burch, L.C., 1988, The Southern Utes: An economically and socially successful Indian Nation building upon its history and challenging the future, *in* Fassett, J.E., ed., Geology and coal-bed methane resources of the northern San Juan Basin, Colorado and New Mexico: Rocky Mountain Association of Geologists Guidebook.

Fassett, J.E., 1985, Early Tertiary paleogeography and paleotectonics of the San Juan Basin area, New Mexico and Colorado, *in* Flores, R.M., and Kaplan, S.S., eds., Cenozoic paleogeography of west-central United States: Rocky Mountain Section of Society of Exploration Paleontologists and Mineralogists, p. 317-334.

Fassett, J.E., and Hinds, J.S., 1971, Geology and fuel resources of the Fruitland Formation and Kirtland Shale: U.S. Geological Survey Professional Paper 676, 76 p.

Gillam, M.L., Moore, D.W., and Scott, G.R., 1984, Quaternary deposits and soils in the Durango area, southwestern Colorado, *in* Brew, D.C., ed., Field trip guidebook for 37th Annual Meeting, Rocky Mountain Section, Geological Society of America: Four Corners Geological Society, p. 149-182.

Granger, W., 1917, Notes on Paleocene and lower Eocene mammal horizons of northern New Mexico and southern Colorado: American Museum of Natural History Bulletin, v. 34, p. 821-830.

Harr, C.L., 1988, The Ignacio Blanco gas field, northern San Juan Basin, Colorado, *in* Fassett, J.E., ed., Geology and coal-bed methane resources of the northern San Juan Basin, Colorado and New Mexico: Rocky Mountain Association of Geologists Guidebook.

Matthew, W.D., and Granger, W., 1921, New genera of Paleocene mammals: American Museum of Natural History Novitates 13, p. 2-7.

Moore, D.W., and Gillam, M.L., 1984, Road log—Quaternary deposits and soils in the Durango area, southwestern Colorado, *in* Brew, D.C., ed., Field trip guidebook for 37th Annual Meeting, Rocky Mountain Section, Geological Society of America: Four Corners Geological Society, p. 149-182.

Reeside, J.B., Jr., 1924, Upper Cretaceous and Tertiary formations of the western part of the San Juan Basin of Colorado and New Mexico: U. S. Geological Survey Professional Paper 134, p. 1-70.

Simpson, G.G., 1935, The Tiffany fauna, Upper Paleocene; Parts I, II, III: American Museum of Natural History novitates, no. 795, p. 1-19; no 816, p. 1-30; no 817, p. 1-28.

Sloan, R.E., 1987, Paleocene and latest Cretaceous mammal ages, biozones, magnetozones, rates of sedimentation, and evolution, in The Cretaceous-Tertiary boundary in the San Juan and Raton basins, New Mexico and Colorado: Geological Society of America Special Paper 209, p. 165-200.

Wood, H.E., and 6 others, 1941, Nomenclature and correlation of the North American continental Tertiary: Geological Society of America Bulletin, v. 52, p. 1-48.

Zimpfer, G.I., Harmon, E.I., and Boyce, B.C., Disposal of water produced from oil and gas wells in the northern San Juan Basin, Colorado, *in* Fassett, J.E., ed., Geology and coal-bed methane resources of the northern San Juan Basin, Colorado and New Mexico: Rocky Mountain Association of Geologists Guidebook.